Lecture Notes in Computer Science 15620

The series Lecture Notes in Computer Science (LNCS), including its subseries Lecture Notes in Artificial Intelligence (LNAI) and Lecture Notes in Bioinformatics (LNBI), has established itself as a medium for the publication of new developments in computer science and information technology research, teaching, and education.

LNCS enjoys close cooperation with the computer science R & D community, the series counts many renowned academics among its volume editors and paper authors, and collaborates with prestigious societies. Its mission is to serve this international community by providing an invaluable service, mainly focused on the publication of conference and workshop proceedings and postproceedings. LNCS commenced publication in 1973.

Nicole Megow · Amitabh Basu
Editors

Integer Programming and Combinatorial Optimization

26th International Conference, IPCO 2025
Baltimore, MD, USA, June 11–13, 2025
Proceedings

Springer

Editors
Nicole Megow
University of Bremen
Bremen, Germany

Amitabh Basu
Johns Hopkins University
Baltimore, MD, USA

ISSN 0302-9743 ISSN 1611-3349 (electronic)
Lecture Notes in Computer Science
ISBN 978-3-031-93111-6 ISBN 978-3-031-93112-3 (eBook)
https://doi.org/10.1007/978-3-031-93112-3

This Springer imprint is published by the registered company Springer Nature Switzerland AG
The registered company address is: Gewerbestrasse 11, 6330 Cham, Switzerland

Preface

This volume contains the 33 extended abstracts presented at IPCO 2025, the 26th Conference on Integer Programming and Combinatorial Optimization, held on June 11–13, 2025 at Johns Hopkins University in Baltimore.

The IPCO conference is under the auspices of the Mathematical Optimization Society. It brings together researchers and practitioners working on various aspects of integer programming and combinatorial optimization. It has become an important forum for presenting recent developments in theory, computation, and applications in these areas. Traditionally, IPCO consists of three days of non-parallel sessions, with no invited talks. More information on IPCO and its history can be found at www.mathopt.org/?nav=ipco.

This year, we received 109 submissions. In a double-blind review process, each submission was reviewed by at least three Program Committee members, often with the help of external reviewers. The PC members met in Aussois in January 2025 and, after thorough discussions, selected 33 papers, with an acceptance ratio of 30%. Many excellent submissions could not be accepted.

As usual, this volume is edited jointly by the PC chair (Nicole Megow) and the main local organizer (Amitabh Basu, also a PC member). We expect the full versions of the accepted extended abstracts that appear in this volume to be published in refereed journals. A special issue of Mathematical Programming B containing such full papers is in process.

The Program Committee decided that the Best Paper Award for IPCO 2025 be given to Lars Rohwedder and Karol WĘgrzycki for their paper "Space-Efficient Algorithm for Integer Programming with Few Constraints". Congratulations!

We would like to thank:

- All authors who submitted extended abstracts to IPCO; we are happy how active all areas of integer programming and combinatorial optimization are;
- The members of the Program Committee, who graciously gave their time and energy;
- The external reviewers, whose expertise was instrumental in guiding our decisions;
- The organizers of the *Aussois* Workshop: Volker Kaibel, Gautier Stauffer, and Paolo Ventura, for inviting us to hold the PC meeting in Aussois;
- EasyChair for making the entire reviewing process so pleasant and easy;
- Springer for their efficient cooperation in producing this volume and for sponsoring the Best Paper Award;
- The sponsors:
  - Air Force Office of Scientific Research
  - Department of Applied Mathematics & Statistics, Johns Hopkins University
  - Office of Naval Research
  - Department of Computer Science, Johns Hopkins University
  - Gurobi Optimization
  - MOSEK
  - The Optimization Firm;

- The members of the Local Organizing Committee at Johns Hopkins University who made this conference possible.
- The speakers of the summer school preceding IPCO: Daniel Dadush, Aida Khajavirad, and Thomas Rothvoss;
- The Mathematical Optimization Society and in particular the members of its IPCO Steering Committee: Karen Aardal, Jochen Könemann, and Giacomo Zambelli, for their help and advice.

March 2025

Nicole Megow
Amitabh Basu

Organization

Program Committee

Amitabh Basu	Johns Hopkins University, USA
Deeparnab Chakrabarty	Dartmouth College, USA
Friedrich Eisenbrand	École Polytechnique Fédérale de Lausanne, Switzerland
Vineet Goyal	Columbia University, USA
Robert Hildebrand	Virginia Tech, USA
Christopher Hojny	Eindhoven University of Technology, The Netherlands
Ivana Ljubic	ESSEC Business School, France
Nicole Megow (chair)	University of Bremen, Germany
Benjamin Moseley	Carnegie Mellon University, USA
Giacomo Nannicini	University of Southern California, USA
Laura Sanità	Bocconi University, Italy
Chaitanya Swamy	University of Waterloo, Canada
Laura Vargas Koch	University of Bonn, Germany
José Verschae	Pontifical Catholic University of Chile, Chile
Angelika Wiegele	University of Klagenfurt, Austria
Rico Zenklusen	ETH Zurich, Switzerland

Organizing Committee

Amitabh Basu (Chair)	Johns Hopkins University, USA
Tinglong Dai	Johns Hopkins University, USA
Michael Dinitz	Johns Hopkins University, USA
Kimia Ghobadi	Johns Hopkins University, USA

Additional Reviewers

Abrahamsen, Mikkel
Ahipasaoglu, Selin
Aleman Espinosa, David
Ales, Zacharie
Aliev, Iskander
Alon, Tal
Aprile, Manuel
Arslan, Ayse Nur
Awasthi, Pranjal
Balkanski, Eric

Bamas, Etienne
Barbato, Michele
Barvinok, Alexander
Basu Roy, Aniket
Beck, Yasmine
Bercea, Ioana
Bhattiprolu, Vijay
Black, Alexander
Blasiok, Jaroslaw
Blikstad, Joakim
Bolusani, Suresh
Bordewich, Magnus
Brosch, Daniel
Buchbinder, Niv
Buchheim, Christoph
Buhai, Rares-Darius
Carvalho, Margarida
Cen, Ruoxu
Chalermsook, Parinya
Chandrasekaran, Karthekeyan
Chekuri, Chandra
Chen, Rui
Cheng, Hongyu
Dahlmeier, Niklas
Das Gupta, Shuvomoy
de Roux, Daniel
Del Pia, Alberto
Delle Donne, Diego
Deo-Campo Vuong, Ramiro
Disser, Yann
Eberle, Franziska
Eickhoff, Katharina
Faenza, Yuri
Fattahi, Salar
Feige, Uriel
Feldman, Moran
Feng, Yiding
Foussoul, Ayoub
Friggstad, Zachary
Gaar, Elisabeth
Gersing, Timo
Ghuge, Rohan
Gleixner, Ambros
Gomez, Andres
Gong, Evelyn Xiao-Yue
Goujaud, Baptiste
Grappe, Roland
Gribanov, Dmitriy
Grinberg, Vadim
Guenin, Bertrand
Gupta, Anupam
Haddadan, Arash
Han, Shaoning
Har-Peled, Sariel
Hathcock, Daniel
Hatzel, Meike
He, Chengyue
Hershkowitz, D. Ellis
Hommelsheim, Felix
Hrga, Timotej
Huang, Neng
Huiberts, Sophie
Hurkens, Cor
Inamdar, Tanmay
Jabal Ameli, Afrouz
Jain, Rhea
Kaibel, Volker
Kakimura, Naonori
Kamiyama, Naoyuki
Kazachkov, Aleksandr
Kelk, Steven
Kerger, Phillip
Kerivin, Herve
Kesselheim, Thomas
Khajavirad, Aida
Khan, Arindam
Király, Csaba
Király, Tamás
Kiss, Peter
Klein, Kim-Manuel
Klein, Nathan
Kobayashi, Yusuke
Kober, Stefan
Koenemann, Jochen
Korhonen, Tuukka
Kronqvist, Jan
Kuhlmann, Stefan
Kumar, Nikhil
Lacroix, Mathieu
Laddha, Aditi

Laekhanukit, Bundit
Lambert, Amélie
Lassota, Alexandra
Lee, Hyunwoo
Letchford, Adam
Li, Yongchun
Liers, Frauke
Linderoth, Jeff
Liu, Siyue
Lodi, Andrea
Louis, Anand
Luedtke, James
MacRury, Calum
Maggiori, Andreas
Merkert, Maximilian
Mnich, Matthias
Moreno, Eduardo
Muir, Christopher
Muñoz, Gonzalo
Mömke, Tobias
Nadara, Wojciech
Nagarajan, Viswanath
Nakajima, Tamio-Vesa
Newman, Alantha
Nikolov, Aleksandar
Nutov, Zeev
Nägele, Martin
Nöbel, Christian
Oertel, Timm
Oliveira, Fabricio
Oliveira, Fernando
Oliveira Reis, Victor
Onn, Shmuel
Ordyniak, Sebastian
Paat, Joseph
Pabst, Philipp
Pashkovich, Kanstantsin
Patel, Neel
Peis, Britta
Pferschy, Ulrich
Polak, Adam
Poremba, Joseph
Potechin, Aaron
Raach, Stephen
Rieken, Niklas
Roghani, Mohammad
Rohwedder, Lars
Rossmanith, Peter
Rujeerapaiboon, Napat
Salvagnin, Domenico
Sankaranarayanan, Sriram
Santiago, Richard
Saranurak, Thatchaphol
Schewe, Lars
Schwarcz, Tamas
Schwartz, Roy
Shafiee, Soroosh
Shen, Haoming
Singla, Sahil
Sitters, Rene
Skarlatos, Antonis
Skutella, Martin
Sokołowski, Marek
Sorensen, Michael M
Soto, José A.
Spieksma, Frits
Sudoso, Antonio M.
Szigeti, Zoltan
Tanigawa, Shin-Ichi
Tanneau, Mathieu
Tiegel, Stefan
Toth, Csaba
Traub, Vera
Traversi, Emiliano
Tröbst, Thorben
Tu, Ta-Wei
Turner, Mark
Tóthmérész, Lilla
Umboh, Seeun William
van Doornmalen, Jasper
van Rhijn, Jesse
Vercesi, Eleonora
Verdugo, Victor
Villa, Tullio
Vogl, Lukas
Vondrak, Jan
Wahlström, Magnus
Walter, Matthias
Wang, Jiaqi
Wang, Yining

Wei, Jiaye
Weltge, Stefan
Weninger, Dieter
Wietheger, Simon
Xiao, Yang
Xu, Pan
Ye, Chun
Yokoi, Yu
Yuditsky, Yelena
Zhou, Hang
Zhou, Rudy
Zhou, Samson
Zielinski, Pawel

Contents

Strongly Connected Orientations and Integer Lattices

Ahmad Abdi[1(✉)], Gérard Cornuéjols[2], Siyue Liu[2], and Olha Silina[2]

[1] Department of Mathematics, LSE, London, UK
a.abdi1@lse.ac.uk

[2] Tepper School of Business, Carnegie Mellon University, Pittsburgh, USA
{gc0v,siyueliu,osilina}@andrew.cmu.edu

Abstract. Let $D = (V, A)$ be a digraph whose underlying graph is 2-edge-connected, and let P be the polytope whose vertices are the incidence vectors of arc sets whose reversal makes D strongly connected. We study the lattice theoretic properties of the integer points contained in a proper face F of P not contained in $\{x : x_a = i\}$ for any $a \in A, i \in \{0,1\}$. We prove under a mild necessary condition that $F \cap \{0,1\}^A$ contains an *integral basis* B, i.e., B is linearly independent, and any integral vector in the linear hull of F is an integral linear combination of B. This result is surprising as the integer points in F do not necessarily form a Hilbert basis. In proving the result, we develop a theory similar to Matching Theory for degree-constrained dijoins in bipartite digraphs. Our result has consequences for head-disjoint strong orientations in hypergraphs, and also to a famous conjecture by Woodall that the minimum size of a dicut of D, say τ, is equal to the maximum number of disjoint dijoins. We prove a relaxation of this conjecture, by finding for any prime number $p \geq 2$, a p-adic packing of dijoins of value τ and of support size at most $2|A|$. We also prove that the all-ones vector belongs to the lattice generated by $F \cap \{0,1\}^A$, where F is the face of P satisfying $x(\delta^+(U)) = 1$ for every minimum dicut $\delta^+(U)$.

1 Introduction

Let $D = (V, A)$ be a digraph whose underlying undirected graph is 2-edge-connected. A *strengthening set* is an arc subset J such that the digraph obtained from D after reversing the arcs in J is strongly connected. Observe that $J \subseteq A$ is a strengthening set if, and only if, its indicator vector x satisfies the following *generalized set covering inequalities*:

$$\sum_{a \in \delta^+(U)} x_a + \sum_{b \in \delta^-(U)} (1 - x_b) \geq 1 \quad \forall U \subset V, U \neq \emptyset . \qquad \text{(CUT)}$$

In words, (CUT) asks that after reversing the arcs of J in D, every nonempty proper vertex subset U has at least one incoming arc. Observe that (CUT) can be

N. Megow and A. Basu (Eds.): IPCO 2025, LNCS 15620, pp. 1–14, 2025.
https://doi.org/10.1007/978-3-031-93112-3_1

rewritten as $x(\delta^+(U)) - x(\delta^-(U)) \geq 1 - |\delta^-(U)|$; as the right-hand sides correspond to a crossing supermodular function, the system above may be viewed as a *supermodular flow system*. Let

$$\mathrm{SCR}(D) := [0,1]^A \cap \{x : x \text{ satisfies } (CUT)\}.$$

It is well-known that SCR(D) is a nonempty integral polytope, and so its vertices are precisely the indicator vectors of the strengthening sets of D ([9], see [17], §60.1). This polytope and its variants have played an important role in graph orientations, combinatorial and matroid optimization; see ([17], Chapters 60–61) and ([10], Chap. 16).

In this paper, we study the lattice theoretic properties of the integer points in SCR(D). Given a rational linear subspace $S \subseteq \mathbb{R}^A$, an *integral basis for* S is a subset $B \subseteq S \cap \mathbb{Z}^A$ of linearly independent vectors such that every vector in $S \cap \mathbb{Z}^A$ is an integral linear combination of B.

Theorem 1. *Let $D = (V, A)$ be a digraph whose underlying undirected graph is 2-edge-connected. Let $\mathcal{F}$ be a nonempty family over ground set V such that $\emptyset, V \notin \mathcal{F}$, and the following face of* SCR(D) *is nonempty:*

$$F := \mathrm{SCR}(D) \cap \{x \in \mathbb{R}^A : x(\delta^+(U)) - x(\delta^-(U)) = 1 - |\delta^-(U)|, \ \forall U \in \mathcal{F}\}.$$

Suppose $\gcd\{1 - |\delta^-(U)| : U \in \mathcal{F}\} = 1$. *Then $F \cap \{0,1\}^A$ contains an integral basis for* $\mathrm{lin}(F)$.

Above, $\mathrm{lin}(F)$ refers to the linear hull of F. It can be readily checked that the GCD condition is necessary for $F \cap \{0,1\}^A$ to contain an integral basis for $\mathrm{lin}(F)$. Theorem 1 is a consequence of a more general theorem about the lattice generated by the integer points in any face of SCR(D) where the GCD condition is replaced by '$1 - |\delta^-(U)| \neq 0$ for some $U \in \mathcal{F}$'. This theorem is stated in Sect. 3.

Theorem 1 is best possible in two different ways. First, the result does not extend to faces F involving both $\mathbf{0} \leq x \leq \mathbf{1}$ and (CUT) inequalities. Secondly, for the face F from Theorem 1, the integer points in F do not necessarily form a *Hilbert basis*, so the result cannot be strengthened in this direction either. These points are explained further in the full version of this extended abstract [2].

1.1 Three Applications of the Main Theorem

Woodall's Conjecture. Let $D = (V, A)$ be a digraph whose underlying undirected graph is connected. A *dicut* is the set of arcs leaving a nonempty proper vertex subset with no incoming arc, i.e., it is of the form $\delta^+(U)$ where $U \subset V, U \neq \emptyset$ and $\delta^-(U) = \emptyset$. A *dijoin* is an arc subset whose contraction makes the digraph strongly connected. Subsequently, every strengthening set is a dijoin. It can be readily checked that J is a dijoin if, and only if, J intersects every dicut at least once.

A famous conjecture by Douglas Woodall states that the maximum number of disjoint dijoins is equal to the minimum size of a dicut [19]. This conjecture has a convenient reformulation that appears in an unpublished note by Lex Schrijver.

Conjecture 2 ([16]). Let $\tau \geq 2$ be an integer, and let $D = (V, A)$ be a digraph, where every dicut has size at least τ. Then A can be partitioned into τ strengthening sets.

Note the difference between the original formulation of Woodall's conjecture and Conjecture 2. While the former is concerned with *packing* dijoins, the latter seeks a *partition* into strengthening sets. This subtle difference comes from the key distinction that while every superset of a dijoin is also a dijoin, a superset of a strengthening set may not necessarily remain a strengthening set.

As a consequence of Theorem 1, we obtain the following relaxation of this conjecture. For a subset $J \subseteq A$, $\mathbf{1}_J \in \{0,1\}^A$ denotes the indicator vector of J.

Theorem 3. *Let $\tau \geq 2$ be an integer, and let $D = (V, A)$ be a digraph where the minimum size of a dicut is τ. Then there exists an assignment $\lambda_J \in \mathbb{Z}$ to every strengthening set J that intersects every minimum dicut exactly once, such that $\sum_J \lambda_J \mathbf{1}_J = \mathbf{1}$, $\mathbf{1}^\top \lambda = \tau$, and $\{\mathbf{1}_J : \lambda_J \neq 0\}$ is an integral basis for its linear hull.*

Observe that Conjecture 2 states that one can replace $\lambda_J \in \mathbb{Z}$ by $\lambda_J \in \mathbb{Z}_{\geq 0}$ in this theorem. This result does not extend to the capacitated setting [2].

p-Adic Programming. Given a prime number $p \geq 2$, a rational number is *(finitely) p-adic* if it is of the form r/p^k for some integer r and nonnegative integer k, and a vector is *p-adic* if each entry is a p-adic rational number. The 2-adic, or *dyadic*, rationals are important for numerical computations because they have a finite binary representation, and therefore can be represented exactly on a computer in floating-point arithmetic. Recent research has characterized when a linear program admits an optimal solution that is p-adic, and furthermore, it has provided a polynomial algorithm for solving a linear program whose domain is restricted to the set of p-adic vectors [1].

Theorem 1 implies, for any prime number $p \geq 2$, the existence of a sparse p-adic optimal solution to a linear program related to packing dijoins. To elaborate, let $D = (V, A)$ be a digraph whose underlying undirected graph is connected. Denote by M the matrix whose columns are labeled by A, and whose rows are the indicator vectors of the dijoins of D. Consider the following pair of dual linear programs,

$$(P) \quad \min\left\{\mathbf{1}^\top x : Mx \geq \mathbf{1}, x \geq \mathbf{0}\right\} \qquad\qquad (D) \quad \max\left\{\mathbf{1}^\top y : M^\top y \leq \mathbf{1}, y \geq \mathbf{0}\right\}$$

where $\mathbf{1}, \mathbf{0}$ denote the all-ones and all-zeros vectors of appropriate dimensions, respectively. A seminal theorem is that the primal linear program (P) models exactly the *minimum dicut problem*, i.e., (P) admits an integral optimal solution ([14], see [5], §1.3.4). Woodall's conjecture equivalently states that the dual linear program (D), in turn, computes the maximum number of pairwise disjoint dijoins, that is, (D) admits an integral optimal solution [19]. The main result of this paper implies some number-theoretic evidence for this conjecture, as it has the following consequence. A † means the proof appears in [2].

Theorem 4 (†). *For any prime number $p \geq 2$, (D) admits a p-adic optimal solution with at most $2|A|$ nonzero entries.*

Observe that Carathéodory's theorem guarantees an optimal solution to (D) with at most $|A|$ nonzero entries. Theorem 4 guarantees a p-adic optimal solution to (D), all the while losing only a factor 2 in the guarantee for the number of nonzero entries.

This theorem does not extend to the capacitated setting. More specifically, if the objective function of (P) is replaced by $c^\top x$ for a nonnegative integral vector c, then (D) may not necessarily have a p-adic optimal solution, for any prime number $p \neq 2$ [2]. Interestingly, it has very recently been shown that (D) always admits a dyadic optimal solution in the capacitated setting [12]; the techniques do not seem to yield a guarantee on the number of nonzero entries of a solution.

Hypergraph Orientations. Let $H = (V, \mathcal{E})$ be a hypergraph. An *orientation of H* consists in designating to each hyperedge $E \in \mathcal{E}$ a node inside as the *head* of E, i.e., it is a mapping $O : \mathcal{E} \to V$ such that $O(E) \in E$ for each $E \in \mathcal{E}$. The orientation is *strongly connected* if for each $X \subset V, X \neq \emptyset$, there exists a hyperedge whose designated head is inside X, and has at least one node outside X.

Two orientations of H are *head-disjoint* if no hyperedge has the same head in both orientations. It is well-known that a graph, which is simply a 2-uniform hypergraph, has 2 head-disjoint strongly connected orientations if, and only if, the graph is 2-edge-connected. The following unpublished conjecture by Bérczi and Chandrasekaran attempts to extend one direction of this to general τ-uniform hypergraphs.

Given two subsets $X, E \subseteq V$, we say that X *separates* E if $E \cap X \neq \emptyset$ and $E \not\subseteq X$. For $X \subseteq V$, denote by $d_H(X)$ the sum of $|X \cap E|$ ranging over all hyperedges $E \in \mathcal{E}$ separated by X.

Conjecture 5 Let $H = (V, \mathcal{E})$ be a τ-uniform hypergraph such that $d_H(X) \geq \tau$ for all $X \subset V, X \neq \emptyset$. Then H has τ pairwise head-disjoint strongly connected orientations.

For $\tau = 3$, a weaker form of this conjecture appears explicitly in ([10], Conjecture 9.4.15). We prove the following relaxation of Conjecture 5; the conjecture states that one can replace $\lambda_O \in \mathbb{Z}$ by $\lambda_O \in \mathbb{Z}_{\geq 0}$ below.

Theorem 6 (†). *Let $\tau \geq 2$ be an integer, and let $H = (V, \mathcal{E})$ be a τ-uniform hypergraph such that $d_H(X) \geq \tau$ for all $X \subset V, X \neq \emptyset$. Then there exists an assignment $\lambda_O \in \mathbb{Z}$ to every strongly connected orientation $O : \mathcal{E} \to V$ such that*

$$\sum (\lambda_O : \textit{strongly connected orientation } O, O(E) = v) = 1 \qquad \forall E \in \mathcal{E}, \forall v \in E,$$

and $|\{O : \lambda_O \neq 0\}| \leq (\tau - 1)|\mathcal{E}| + 1$.

1.2 The Dijoin Polyhedron and Digrafts

Theorem 1 is a consequence of a lattice theoretic result about the dijoin polyhedron of bipartite digraphs. To this end, for a digraph $D = (V, A)$, let

$$\mathrm{DIJ}(D) := \left\{x \in \mathbb{R}^A : x(\delta^+(U)) \geq 1, \forall \text{ dicut } \delta^+(U); x \geq \mathbf{0}\right\}.$$

It is known that $\mathrm{DIJ}(D)$ is an integral polyhedron, and its vertices are precisely the indicator vectors of the (inclusionwise) minimal dijoins of D ([14], see [5], §1.3.4).

A digraph is *bipartite* if every node is a source or a sink. Recent research has demonstrated the importance of bipartite digraphs in studying Woodall's conjecture, by reducing the conjecture to nearly regular bipartite digraphs [3]. We follow this approach by studying faces of $\mathrm{DIJ}(D)$ of a bipartite digraph D.

Definition 7 (digraft). A *digraft* is a pair $(D{=}(V, A), \mathcal{F})$ where D is a bipartite digraph, the underlying undirected graph of D is 2-edge-connected, and $\mathcal{F}$ is a family over ground set V such that (a) $\emptyset, V \notin \mathcal{F}$, (b) if $U \in \mathcal{F}$ then $\delta^-(U) = \emptyset$, (c) $V \setminus v \in \mathcal{F}$ for every sink v of D, and (d) the following face of $\mathrm{DIJ}(D)$ is nonempty:

$$F(D, \mathcal{F}) := \mathrm{DIJ}(D) \cap \left\{x \in \mathbb{R}^A : x(\delta^+(U)) = 1, \forall U \in \mathcal{F}\right\} \subseteq [0, 1]^A.$$

The choice of the 'digraft' terminology mirrors that of a 'graft', an object that shows up in the context of the *minimum T-join problem*, and is loosely related to the *minimum dijoin problem* (see [5], §1.3.5). We prove the following theorem about digrafts.

Theorem 8. *Let $(D = (V, A), \mathcal{F})$ be a digraft. Then $F(D, \mathcal{F}) \cap \{0, 1\}^A$ contains an integral basis for* $\mathrm{lin}(F(D, \mathcal{F}))$.

Theorem 1 and Theorem 8 are in fact equivalent. To de-mystify the connection between these results, let $(D = (V, A), \mathcal{F})$ be a digraft. As the underlying undirected graph of D is 2-edge-connected, every minimal dijoin is a strengthening set (see [17], Theorem 55.1). Furthermore, every strengthening set that has exactly one arc incident with every sink, is also a minimal dijoin. Subsequently, $F(D, \mathcal{F}) = \mathrm{SCR}(D) \cap \left\{x \in \mathbb{R}^A : x(\delta^+(U)) - x(\delta^-(U)) = 1 - |\delta^-(U)|, \forall U \in \mathcal{F}\right\}$. Furthermore, $\gcd\{1 - |\delta^-(U)| : U \in \mathcal{F}\} = 1$. Thus, Theorem 8 follows from Theorem 1. The converse implication is shown in Sect. 3, after we prove Theorem 8 directly.

We find Theorem 8 more convenient to work with than Theorem 1. At the highest level, one explanation for this is that every point in $F(D, \mathcal{F}) \cap \{0, 1\}^A$ is the indicator vector of an arc subset J that has degree one at every sink of D, and degree at least one at every source of D; so that J may be viewed as a perfect b_J-matching in a bipartite graph, for some degree vector b_J. Fixing the degree of J at each sink to one has advantages: first, the cardinality of J becomes invariant and equal to the number of sinks of D; secondly, in this case, J is a minimal dijoin if and only if it is a strengthening set, an equivalence which we utilized above.

2 Proof Overview of Theorem 8

Let us provide an overview of the proof of Theorem 8. To this end, let $(D = (V, A), \mathcal{F})$ be a digraft. Our goal is to find an integral basis in $F(D,\mathcal{F}) \cap \{0,1\}^A$ for $\mathrm{lin}(F(D,\mathcal{F}))$.

Dicut Contractions. A dicut $\delta^+(U)$ is *tight* for $(D,\mathcal{F})$ if $F(D,\mathcal{F}) \subseteq \{x : x(\delta^+(U)) = 1\}$. A tight dicut $\delta^+(U)$ is *non-trivial* if $1 < |U| < |V| - 1$, otherwise it is *trivial.* The proof proceeds by first decomposing the digraft into 'basic' pieces along non-trivial tight dicuts. Once we find integral bases for the basic pieces, then by composing the bases together in a natural manner, we obtain an integral basis in $F(D,\mathcal{F}) \cap \{0,1\}^A$. Though this part of the proof is fairly routine for the most part, there are some subtle points about when and how a digraft can be decomposed; let us give a following flash summary of these subtleties. A dicut $\delta^+(U)$ is *contractible* for $(D,\mathcal{F})$ if $1<|U|<|V|-1$ and $F(D,\mathcal{F}) \cap \{x : x(\delta^+(U))=1\} \neq \emptyset$. The *closure of* $\mathcal{F}$ for $(D,\mathcal{F})$ is the family of subsets $U \subset V$, $U \neq \emptyset$ such that $\delta^-(U)=\emptyset$ and $F(D,\mathcal{F}) \subseteq \{x : x(\delta^+(U)) = 1\}$.

Definition 9. **($(U, V \setminus U)$-contractions).** Suppose $\delta^+(U)$ is a contractible dicut of $(D,\mathcal{F})$. Let $\overline{\mathcal{F}}$ be the closure of $\mathcal{F} \cup \{U\}$ for the digraft $(D, \mathcal{F} \cup \{U\})$. Let $U_1 := U$ and $U_2 := V \setminus U$. Let $D_i = (V_i, A_i)$ be the bipartite digraph obtained from D after shrinking U_i to a single node u_i; so $V_i = \{u_i\} \cup U_{3-i}$. Let

$$\mathcal{F}_i := \{W : W \cap U_i = \emptyset, W \in \overline{\mathcal{F}}\} \cup \{(W \setminus U_i) \cup \{u_i\} : U_i \subseteq W, W \in \overline{\mathcal{F}}\}.$$

We refer to $(D_i, \mathcal{F}_i), i = 1, 2$ as the $(U, V \setminus U)$*-contractions of* $(D,\mathcal{F})$.

Thus, a digraft can be decomposed along not only non-trivial tight dicuts but any contractible dicut; this will be particularly useful in Phase 3 of the proof. More specifically, a $(U, V \setminus U)$-contraction decomposes the digraft $(D,\mathcal{F})$, as well as the $0,1$ points in $F(D,\mathcal{F}) \cap \{x : x(\delta^+(U)) = 1\}$, into digrafts $(D_i, \mathcal{F}_i)$, and $0,1$ points in $F(D_i, \mathcal{F}_i)$, respectively. Furthermore, thanks to our careful choice of $\mathcal{F}_i, i = 1, 2$, we can ensure that under some mild conditions, two $0,1$ points in $F(D_i, \mathcal{F}_i), i = 1, 2$ can be composed to give a $0,1$ point in $F(D,\mathcal{F}) \cap \{x : x(\delta^+(U)) = 1\}$. This composition is crucial in the following summary of a more technical ingredient that appears in the full version of this article [2].

Lemma 10. **(†).** *Let* $(D = (V, A), \mathcal{F})$ *be a digraft, and suppose* $\delta^+(U)$ *is a contractible dicut. Let* $(D_i = (V_i, A_i), \mathcal{F}_i), i = 1, 2$ *be the* $(U, V \setminus U)$*-contractions of* $(D,\mathcal{F})$*. Suppose* $B_i \subseteq F(D_i, \mathcal{F}_i) \cap \{0,1\}^{A_i}$ *is an integral basis for* $\mathrm{lin}(F(D_i, \mathcal{F}_i))$*,* $i=1,2$*. Then there exists an integral basis* $B \subseteq F(D,\mathcal{F}) \cap \{0,1\}^A \cap \{x : x(\delta^+(U))=1\}$ *for* $\mathrm{lin}(B)$ *where* $|B| = |B_1| + |B_2| - |\delta^+(U)|$*. Moreover, if* $\delta^+(U)$ *is a tight dicut, then* $\mathrm{lin}(B) = \mathrm{lin}(F(D,\mathcal{F}))$*.*

The proof of Theorem 8 is inductive and divided into three phases. Before explaining this, note that without loss of generality we may assume that whenever $\delta^+(U)$ is tight for $(D,\mathcal{F})$, then $U \in \mathcal{F}$; otherwise we add U to $\mathcal{F}$ without changing $F(D,\mathcal{F})$. (Maximality of $\mathcal{F}$ ensures we obtain a basic digraft in Phase 1.)

Phase 1. In this phase, we look for a non-trivial tight dicut. If there is one, say $\delta^+(U)$, then let $D_i = (V_i, A_i), i = 1,2$ be the $(U, V \setminus U)$-contractions of $(D, \mathcal{F})$. By induction, $F(D_i, \mathcal{F}_i) \cap \{0,1\}^{A_i}$ contains an integral basis B_i for $\text{lin}(F(D_i, \mathcal{F}_i))$, for $i \in \{1,2\}$. Then by Lemma 10, there exists $B \subseteq F(D, \mathcal{F}) \cap \{0,1\}^A$ that is an integral basis for $\text{lin}(B) = \text{lin}(F(D, \mathcal{F}))$, thus completing the induction step, and proving Theorem 8 in this phase.

Otherwise, $(D, \mathcal{F})$ has no non-trivial tight dicut, thus yielding the following.

Definition 11. (basic digraft). A digraft $(D, \mathcal{F})$ is *basic* if every tight dicut $\delta^+(U)$ is trivial and $U \in \mathcal{F}$.

The challenge for the next two phases is to find an integral basis for a basic digraft. Here comes a key idea of the proof: to study the facet-defining inequalities of $F(D, \mathcal{F})$.

Definition 12. (basic robust digraft). A basic digraft $(D{=}(V, A), \mathcal{F})$ is *robust* if every facet-defining inequality for $F(D, \mathcal{F})$ is equivalent to $x_a \geq 0, a \in A$, or $x(\delta^+(u)) \geq 1$ for some source u of D.

Phase 2 (Base Case). In this phase, which is the base case of our induction, we assume that the basic digraft is robust. In this case, we prove that $F(D, \mathcal{F})$ is a very special polyhedron. To elaborate, for a polyhedron $P \subseteq \mathbb{R}^n$ and $k \geq 0$, define kP as the set of all points of the form $\sum_{p \in P} \lambda_p p$ where $\lambda \in \mathbb{R}^P_{\geq 0}$ and $\mathbf{1}^\top \lambda = k$. P has the *integer decomposition property* if for every integer $k \geq 1$, every integral point in kP can be written as the sum of k integral points in P. The inequality description of $F(D, \mathcal{F})$, along with a classic result of de Werra [18] on balanced edge-colourings of bipartite graphs, allows for the following theorem.

Theorem 13. (†). *Let $(D = (V, A), \mathcal{F})$ be a basic robust digraft. Then $F(D, \mathcal{F})$ has the integer decomposition property, and* $\text{aff}(F(D, \mathcal{F})) = \{x : Mx = \mathbf{1}\}$ *for some $M \in \mathbb{Z}^{m \times n}$ with $m \geq 1$.*

Theorem 8 for basic robust digrafts now follows from the following general-purpose result about polyhedra with the integer decomposition property. The theorem below is obtained by first proving that $P \cap \mathbb{Z}^n$ forms an *integral generating set for a cone*, better known as a *Hilbert basis*, and then using a result of Gerards and Sebő [11] about such sets to finish the proof.

Theorem 14. (†). *Let $P \subseteq \mathbb{R}^n$ be a pointed polyhedron with the integer decomposition property, where* $\text{aff}(P) = \{x : Ax = b\}$ *for $A \in \mathbb{Z}^{m \times n}, b \in \mathbb{Z}^m$ such that $m \geq 1$, $b \neq \mathbf{0}$, and* $\gcd\{b_i : i \in [m]\} = 1$. *Then $P \cap \mathbb{Z}^n$ contains an integral basis for* $\text{lin}(P)$.

Phase 3. Thus comes the most challenging yet insightful part of our proof, where the basic digraft $(D = (V, A), \mathcal{F})$ is not robust, so there is a facet-defining inequality $x(\delta^+(U)) \geq 1$ that is not equivalent to $x(\delta^+(u)) \geq 1$ for any source u. We decompose $(D, \mathcal{F})$ along the contractible dicut $\delta^+(U)$ into two smaller digrafts. By the induction hypothesis and Lemma 10, we then compose the two integral bases of the $(U, V \setminus U)$-contractions to obtain a linearly independent set $B' \subseteq F(D, \mathcal{F}) \cap \{0,1\}^A$. However, there are two key challenges to turn B' into an integral basis B for $\mathrm{lin}(F(D, \mathcal{F}))$. First, B' is at least one vector away from forming a linear basis for $\mathrm{lin}(F(D, \mathcal{F}))$, and a priori, we do not know the number of extra vectors we would need to add. Secondly, a linear basis is a long way from an integral one, so we need to extend B' very carefully. We have two lemmas that address these issues.

Phase 3, Issue 1. The first issue stems from the fact that the two $(U, V \setminus U)$-contractions are not necessarily basic digrafts. This can be fatal as we could lose our guarantee on the size of $|B \setminus B'|$. However, we will be able to prove that both of these pieces share a key property with basic digrafts, thus allowing us to guarantee that $|B \setminus B'| = 1$. To describe this property, we need a few notions. A node $u \in V$ is *tight* for the digraft if $F(D, \mathcal{F}) \subseteq \{x : x(\delta(u)) = 1\}$; the node is *active* for the digraft if it is not tight. Note that all sinks of a digraft are tight.

Definition 15. (affine critical digraft). Let $(D = (V, A), \mathcal{F})$ be a digraft, and let V^t be the set of tight nodes. The digraft is *affine critical* if $\mathrm{aff}(F(D, \mathcal{F})) = \{x : x(\delta(v)) = 1, \forall v \in V^t\}$.

In particular, a basic digraft is affine critical (the converse does not necessarily hold). The following important lemma addresses the first issue.

Lemma 16 (Affine Critical Lemma †). *Let $(D = (V, A), \mathcal{F})$ be a basic digraft that is not robust. Let $x(\delta^+(U)) \geq 1$ be a facet-defining dicut inequality for $F(D, \mathcal{F})$ not equivalent to $x(\delta^+(u)) \geq 1$ for any active source u. Then $(D, \mathcal{F})$ and its $(U, V \setminus U)$-contractions $(D_i, \mathcal{F}_i), i = 1, 2$ are affine critical digrafts each of which contains at least one active source. Furthermore, for $i \in \{1, 2\}$, every active source for $(D, \mathcal{F})$ in U_{3-i} is an active source for $(D_i, \mathcal{F}_i)$, and vice versa.*

This lemma is a byproduct of a careful analysis of the dimension of $F := F(D, \mathcal{F})$. Let V^t be the set of tight nodes. Define $\kappa^t(D, \mathcal{F})$ to be 1 if $V = V^t$, and 0 otherwise. Denote by $\dim(F)$ the dimension of the affine hull $\mathrm{aff}(F)$, which satisfies the inequality $|A| - |V^t| + \kappa^t(D, \mathcal{F}) \geq \dim(F)$.

Definition 17 (slack). The *slack* of $(D = (V, A), \mathcal{F})$ is

$$s(D, \mathcal{F}) := |A| - |V^t| + \kappa^t(D, \mathcal{F}) - \dim(F(D, \mathcal{F})) \geq 0.$$

The slack is a characteristic quantity associated with a digraft, which has an intuitive interpretation. The affine hull of $F(D, \mathcal{F})$ is described by two types of constraints: $x(\delta(v)) = 1, \ \forall v \in V^t$, and $x(\delta^+(U)) = 1$ for a non-trivial tight dicut $\delta^+(U)$. The slack computes the additional contribution of non-trivial tight dicuts—in terms of rank increase—in defining the affine hull. In particular, a digraft is affine critical if and only if it has slack zero. Furthermore, if $s := s(D, \mathcal{F}) \geq 1$, then there exists a cross-free family of s non-trivial tight dicuts, which can be used to decompose the digraft into $s + 1$ pieces partitioning the active sources of $(D, \mathcal{F})$ into $s + 1$ nonempty parts, such that each piece of the form $(D', \mathcal{F}')$ satisfies $s(D', \mathcal{F}') = \kappa^t(D', \mathcal{F}') = 0$.

Phase 3, Issue 2. The Affine Critical Lemma guarantees that to turn B' into a linear basis for $\mathrm{lin}(F(D,\mathcal{F}))$, we just need to add one more vector b from $F(D,\mathcal{F}) \cap \{0,1\}^A$, which must inevitably satisfy $b(\delta^+(U)) > 1$. However, we need not just a linear basis but an integral one. As it turns out, integrality of the basis can be guaranteed if $b(\delta^+(U)) = 2$, whose existence is implied by the following lemma.

Lemma 18 (Jump-Free Lemma †). *Let $(D=(V,A),\mathcal{F})$ be a digraft, let $\delta^+(U)$ be a dicut, and let $x, y \in F(D,\mathcal{F}) \cap \{0,1\}^A$ where $\lambda_1 := x(\delta^+(U)) < y(\delta^+(U)) =: \lambda_2$. Then for any integer $\lambda \in (\lambda_1, \lambda_2)$, there exists $z \in F(D,\mathcal{F}) \cap \{0,1\}^A$ such that $z(\delta^+(U)) = \lambda$.*

To prove this lemma, we map $F(D,\mathcal{F}) \subseteq \mathbb{R}^A$ to a base polyhedron in $\mathbb{R}^V$ via a suitable affine function f, which also maps $F(D,\mathcal{F}) \cap \{0,1\}^A$ to an M-convex set in $\mathbb{Z}^V$. The exchange axiom for M-convex set then allows us to construct a 'jump-free' sequence from $f(x)$ to $f(y)$ in $\mathbb{Z}^V$. From there, the theory of perfect b-matchings in bipartite graphs allows us to construct a (not necessarily unique) inverse to f, thus obtaining a jump-free sequence from x to y in $F(D,\mathcal{F}) \cap \{0,1\}^A$, and proving the Jump-Free Lemma.

Now that we have dealt with the two issues, we are ready to prove Theorem 8 for basic non-robust digrafts. Recall that the basic digraft $(D=(V,A),\mathcal{F})$ is not robust, and $x(\delta^+(U)) \geq 1$ is a facet-defining inequality that is not equivalent to $x(\delta^+(u)) \geq 1$ for any active source u. Let $(D_i=(V_i,A_i),\mathcal{F}_i), i=1,2$ be the $(U, V \setminus U)$-contractions of $(D,\mathcal{F})$. By the induction hypothesis, $F(D_i,\mathcal{F}_i) \cap \{0,1\}^{A_i}$ contains an integral basis B_i for $\mathrm{lin}(F(D_i,\mathcal{F}_i)), i = 1,2$. Then by Lemma 10, there exists $B' \subseteq F(D,\mathcal{F}) \cap \{0,1\}^A$ forming an integral basis for $\mathrm{lin}(B')$.

By the Jump-Free Lemma, there exists $b \in F(D,\mathcal{F}) \cap \{0,1\}^A$ such that $b(\delta^+(U)) = 2$. We claim that $B := B' \cup \{b\} \subseteq F(D,\mathcal{F}) \cap \{0,1\}^A$ is an integral basis for $\mathrm{lin}(F(D,\mathcal{F}))$, thus finishing the proof.

First, we prove that B is a linear basis for $\mathrm{lin}(F(D,\mathcal{F}))$. Linear independence can be checked through a routine argument. To show that B is a linear basis, we count the linear dimension of $F := F(D,\mathcal{F})$, which is $d := 1 + \dim(F)$. We claim that $d = |B|$. To this end, let $F_i := F(D_i,\mathcal{F}_i), i = 1,2$ and $d_i := 1 + \dim(F_i), i = 1,2$.

By the Affine Critical Lemma, $(D,\mathcal{F}), (D_i,\mathcal{F}_i), i = 1,2$ are affine critical digrafts each of which contains an active source. Thus, $\kappa^t(D,\mathcal{F}) = \kappa^t(D_1,\mathcal{F}_1) = \kappa^t(D_2,\mathcal{F}_2) = 0$, and $s(D,\mathcal{F}) = s(D_1,\mathcal{F}_1) = s(D_2,\mathcal{F}_2) = 0$. Subsequently, by the slack formula in Definition 17, $d = 1 + |A| - |V^t|$, $d_1 = 1 + |A_1| - |V_1^t|$, and $d_2 = 1 + |A_2| - |V_2^t|$, where $V^t, V_i^t, i = 1,2$ denote the sets of tight nodes of $(D,\mathcal{F}), (D_i,\mathcal{F}_i), i = 1,2$, respectively. By the Affine Critical Lemma, $V_1^t \cup V_2^t = V^t \cup \{u_1, u_2\}$, implying in turn the first equality below:

$$d = d_1 + d_2 - |\delta^+(U)| + 1 = |B_1| + |B_2| - |\delta^+(U)| + 1 = |B'| + 1 = |B|.$$

The third equality follows from Lemma 10.

It remains to prove that B is an *integral* basis. To this end, pick an integral vector f in $\mathrm{lin}(F(D,\mathcal{F}))$. We now know that f can be expressed as a linear combination of the vectors in B; let $\lambda_z \in \mathbb{R}$ be the coefficient of $z \in B$. Given that f is integral, $f(\delta^+(U))$ is an integer, which can be calculated alternatively as follows, where we have used $z(\delta^+(U))=1$ for all $z\in B'$, guaranteed by Lemma 10.

$$f(\delta^+(U)) = \sum_{z\in B} \lambda_z z(\delta^+(U)) = 2\lambda_b + \sum_{z\in B'} \lambda_z = \lambda_b + \mathbf{1}^\top\lambda.$$

As $B \subseteq F(D,\mathcal{F})$, we have $z(\delta(v)) = 1, \forall z \in B$ for any fixed tight node v. Subsequently, $\mathbf{1}^\top\lambda = \sum_{z\in B}\lambda_z = f(\delta(v))$ is an integer, implying in turn that $\lambda_b = f(\delta^+(U)) - \mathbf{1}^\top\lambda$ is an integer. Now let $f' := f - \lambda_b b \in \mathbb{Z}^A$. Evidently, $f' \in \mathrm{lin}(B') \cap \mathbb{Z}^A$, so given that B' is an integral basis for its linear hull, f' is an integer linear combination of the vectors in B', implying in turn that λ is an integral vector. Thus, $B \subseteq F(D,\mathcal{F}) \cap \{0,1\}^A$ is an integral basis for $\mathrm{lin}(F(D,\mathcal{F}))$, thereby completing the induction step, hence proving Theorem 8. □

3 Proofs of Theorem 1 and Theorem 3

In this section, we prove a general theorem about proper faces of SCR(D) not contained in proper faces of the unit hypercube. We then prove Theorem 1 and one of the three applications, namely Theorem 3. We start off with a useful mapping of the strengthening sets of a digraph to dijoins of another digraph. The mapping is routine and has appeared before in the literature, e.g., [6,16].

Theorem 19 (†). *Let $D=(V,A)$ be a digraph whose underlying undirected graph is 2-edge-connected. Let $\mathcal{F}$ be a family over ground set V such that $\emptyset, V \notin \mathcal{F}$, and the following face of* SCR(D) *is nonempty:*

$$F := \mathrm{SCR}(D) \cap \left\{x \in \mathbb{R}^A : x(\delta^+(U)) - x(\delta^-(U)) = 1 - |\delta^-(U)|,\ \forall U \in \mathcal{F}\right\}.$$

Then there exists a digraft $(D',\mathcal{F}')$ such that the mapping $x \mapsto \binom{x}{\mathbf{1}-x}$ defines a bijection between the face F of SCR(D) *and the face $F(D',\mathcal{F}')$ of* DIJ(D').

Primer on Integer Lattices. Let us recall some basic concepts of the theory of integer lattices; for a reference textbook we recommend ([15], Chap. 1). A subset $L \subseteq \mathbb{R}^A$ is a *lattice* if it is the set of integer linear combinations of finitely many vectors. Alternatively, L is a lattice if it forms a subgroup of $\mathbb{R}^A$ under addition that is *discrete*, that is, there exists an $\varepsilon > 0$ such that every pair of distinct vectors in L are at distance $\geq\varepsilon$. Given a finite subset $G \subset \mathbb{R}^A$, the *lattice generated by* G, denoted $\mathrm{lat}(G)$, is the set of all integer linear combinations of the vectors in G. A *lattice basis for* L is a set B of linearly independent vectors that generates the lattice, i.e., $L = \mathrm{lat}(B)$. A nontrivial fact is that a lattice basis always exists.

Suppose now L is an *integer lattice*, that is, L is a lattice and $L \subseteq \mathbb{Z}^A$. Let $\overline{L} := \mathrm{lin}(L) \cap \mathbb{Z}^A$ which is another integer lattice that contains L. Note that $\overline{L}$ is the 'densest' integer lattice in $\mathrm{lin}(L)$. It is known that $\overline{L}$ can be partitioned into a finite number of lattices, each of which is an integral shift of L, i.e., of the form $L + w := \{v + w : v \in L\}$ for some $w \in \mathrm{lin}(L) \cap \mathbb{Z}^A$. We refer to the number of parts in this partition as the *index of* L and denote it by $\text{ind}(L) \in \mathbb{Z}_{\geq 1}$. Thus, the smaller the index of L, the denser the lattice is. Of particular interest is the case when L is densest possible. Observe that L has index 1 if, and only if, L contains an integral basis for $\text{lin}(L)$.

Theorem 20. *Let $D = (V, A)$ be a digraph whose underlying undirected graph is 2-edge-connected. Let $\mathcal{F}$ be a family over ground set V such that $\emptyset, V \notin \mathcal{F}$, $1 - |\delta^-(U)| \neq 0$ for some $U \in \mathcal{F}$, and the following face of* SCR(D) *is nonempty:*

$$F := \text{SCR}(D) \cap \{x \in \mathbb{R}^A : x(\delta^+(U)) - x(\delta^-(U)) = 1 - |\delta^-(U)|, \forall U \in \mathcal{F}\}.$$

Then the following statements hold:

1. *The lattice generated by $F \cap \{0,1\}^A$ has a lattice basis contained in $F \cap \{0,1\}^A$.*
2. *Let $g := \gcd\{1 - |\delta^-(U)| : U \in \mathcal{F}\}$. Then $gx \in \text{lat}\left(F \cap \{0,1\}^A\right)$ for all $x \in \text{lin}(F) \cap \mathbb{Z}^A$.*

Proof. Let L be the lattice generated by $F \cap \{0,1\}^A$, $\overline{L} := \text{lin}(F) \cap \mathbb{Z}^A$, and $g := \gcd\{1 - |\delta^-(U)| : U \in \mathcal{F}\}$. By Theorem 19, there exists a digraft $(D' = (V', A'), \mathcal{F}')$ such that for $F' := F(D', \mathcal{F}')$, the mapping $f : F \to F'$ defined as $f(x) = \binom{x}{1-x}$ is a bijection. Let L' be the lattice generated by $F(D', \mathcal{F}') \cap \{0,1\}^{A'}$. By Theorem 8, there is an integral basis $B' \subseteq F' \cap \{0,1\}^{A'}$ for $\text{lin}(F')$.

Claim. *Let w be an integral vector in $\text{lin}(F)$, expressed as $w = \sum_{x \in F} \lambda_x x$. Then $\mathbf{1}^\top \lambda$ is $\frac{1}{g}$-integral. Furthermore, if $w = \mathbf{0}$, then $\mathbf{1}^\top \lambda = 0$.*

Proof of Claim. Let $\tau := \sum_{x \in F} \lambda_x$. Note that

$$w(\delta^+(U)) - w(\delta^-(U)) = \sum_{x \in F} \lambda_x (1 - |\delta^-(U)|) = \tau(1 - |\delta^-(U)|) \quad \forall U \in \mathcal{F}.$$

As w is integral, we have $\tau(1 - |\delta^-(U)|) \in \mathbb{Z}$ for all $U \in \mathcal{F}$, and so since $\gcd\{1 - |\delta^-(U)| : U \in \mathcal{F}\} = g$, it follows that τ is $\frac{1}{g}$-integral. Furthermore, if $w = \mathbf{0}$, then as $1 - |\delta^-(U)| \neq 0$ for some $U \in \mathcal{F}$, we have $0 = w(\delta^+(U)) - w(\delta^-(U)) = \tau(1 - |\delta^-(U)|)$, implying in turn that $\tau = 0$.

Let B be the pre-image of B' under f. Observe that $B \subseteq F \cap \{0,1\}^A$. We shall prove that (a) B is linearly independent, (b) B is a lattice basis for L, and (c) $gw \in L$ for all $w \in \text{lin}(F) \cap \mathbb{Z}^A$.

(a) Suppose $\sum_{b \in B} \lambda_b b = \mathbf{0}$. It follows from the claim above that $\mathbf{1}^\top \lambda = 0$, so $\sum_{b \in B} \lambda_b f(b) = \mathbf{0}$. The linear independence of $B' = \{f(b) : b \in B\}$, along with the bijectivity of f, implies that $\lambda = \mathbf{0}$.

(b) By (a), it suffices to show that $\text{lat}(B) = L$. Clearly, $\text{lat}(B) \subseteq L$. For the reverse inclusion, let $w \in L$. Then $w = \sum_{x \in F \cap \{0,1\}^A} \lambda_x x$ for some integers $\lambda_x, x \in F \cap \{0,1\}^A$. Let $w' := \sum_{x \in F \cap \{0,1\}^A} \lambda_x f(x)$. As λ is integral, $w' \in L'$, so $w' = \sum_{b \in B} \alpha_b f(b)$ for some integers $\alpha_b, b \in B$. Restricting to the coordinates in A, we obtain that $w = \sum_{b \in B} \alpha_b b \in \text{lat}(B)$.

(c) Let $w \in \text{lin}(F) \cap \mathbb{Z}^A$. Write $w = \sum_{x \in F \cap \{0,1\}^A} \lambda_x x$, and let $\tau := \mathbf{1}^\top \lambda$ which is $\frac{1}{g}$-integral by the claim above. Let $w' := \sum_{x \in F \cap \{0,1\}^A} \lambda_x f(x)$, which is $\frac{1}{g}$-integral as $\tau \in \frac{1}{g}\mathbb{Z}$. Subsequently, gw' is an integral vector in $\text{lin}(F')$, so $gw' = \sum_{b \in B} \alpha_b f(b)$ for some integers $\alpha_b, b \in B$, as B' is an integral basis for $\text{lin}(F')$. Restricting to the coordinates in A, we obtain that $gw = \sum_{b \in B} \alpha_b b \in L$, as promised.

Observe that (b) proves part (1), and (c) proves part (2) of the theorem. □

Let us point out a subtle detail about part (1) of Theorem 20. A set G of generators may not necessarily contain a lattice basis for $\text{lat}(G)$. For instance, $\text{lat}(\{2,3\}) = \mathbb{Z}$, yet $\{2,3\}$ does not contain a lattice basis for $\mathbb{Z}$. Thus, the claim that $F \cap \{0,1\}^A$ contains a lattice basis is non-trivial.

Let's look at part (2) of Theorem 20. This part equivalently states that, for $L := \text{lat}\left(F \cap \{0,1\}^A\right)$ and $\overline{L} := \text{lin}(F) \cap \mathbb{Z}^A$, the quotient group $\overline{L}/L$ is an abelian group where the order of every element divides g. Subsequently, every elementary divisor of $\overline{L}/L$ divides g. This implies in turn that $\text{ind}(L)$ is the product of some divisors of g. Furthermore, if g is a prime number, then $\overline{L}/L$ is an elementary p-primary group. For more on concepts relating to group theory, we refer the interested reader to [7], more specifically, Chap. 5, Theorem 5.

We are ready to prove the main theorem.

Proof of Theorem 1. Let $D = (V, A)$ be a digraph whose underlying undirected graph is 2-edge-connected. Let $\mathcal{F}$ be a nonempty family over ground set V such that $\emptyset, V \notin \mathcal{F}$, and the following face of $\text{SCR}(D)$ is nonempty:

$$F := \text{SCR}(D) \cap \left\{x \in \mathbb{R}^A : x(\delta^+(U)) - x(\delta^-(U)) = 1 - |\delta^-(U)|,\ \forall U \in \mathcal{F}\right\}.$$

Suppose $\gcd\{1 - |\delta^-(U)| : U \in \mathcal{F}\} = 1$. It then follows from Theorem 20 part (1) that the lattice L generated by $F \cap \{0,1\}^A$ has a lattice basis $B \subseteq F \cap \{0,1\}^A$. Furthermore, by part (2), $L = \text{lin}(F) \cap \mathbb{Z}^A$, so $\text{ind}(L) = 1$, implying in turn that B is an integral basis for $\text{lin}(F)$ contained in $F \cap \{0,1\}^A$. □

We are also ready to prove one of the applications of the main theorem.

Proof of Theorem 3. Let $\tau \geq 2$ be an integer, and let $D = (V, A)$ be a digraph where the minimum size of a dicut is τ. We will prove that there exists an assignment $\lambda_J \in \mathbb{Z}$ to every strengthening set J intersecting every minimum dicut exactly once, such that $\sum_J \lambda_J \mathbf{1}_J = \mathbf{1}$, $\mathbf{1}^\top \lambda = \tau$, and $\{\mathbf{1}_J : \lambda_J \neq 0\}$ will be an integral basis for its linear hull.

Let $\mathcal{F}$ be the family of sets $U \subset V, U \neq \emptyset$ such that $\delta^-(U) = \emptyset$ and $|\delta^+(U)| = \tau$. Let $F := \text{SCR}(D) \cap \{x : x(\delta^+(U)) - x(\delta^-(U)) = 1 - |\delta^-(U)|, \forall U \in \mathcal{F}\}$. Observe that $F \cap \{0,1\}^A$ corresponds to the strengthening sets of D that intersect every minimum dicut exactly once.

Since every dicut of D (if any) has size at least τ, it follows that $|\delta^+(U)| + (\tau - 1)|\delta^-(U)| \geq \tau$ for all $U \subset V, U \neq \emptyset$, implying in turn that $x^\star := \frac{1}{\tau}\mathbf{1} \in F$.

Since $\mathcal{F} \neq \emptyset$, then $\gcd\{1 - |\delta^-(U)| : U \in \mathcal{F}\} = 1$, so we may apply Theorem 1 to conclude that $F \cap \{0,1\}^A$ contains an integral basis B for $\text{lin}(F)$. This implies that $\mathbf{1} = \tau x^\star \in \tau F$ is an integral linear combination of the vectors in B, say $\sum_{b \in B} \lambda_b \cdot b$. Furthermore, $\mathbf{1}^\top \lambda = \tau$, because $\mathbf{1}^\top \lambda = \sum_{b \in B} \lambda_b \cdot b(\delta^+(U)) = \mathbf{1}(\delta^+(U)) = \tau$ for any given $U \in \mathcal{F}$. Given that B is an integral basis for $\text{lin}(F)$, it follows that $\{b : \lambda_b \neq 0\}$ is also an integral basis for its linear hull, so we are done. □

Acknowledgement. Our work is inspired by the works of Edmonds, Lovász, and Pulleyblank [8], Lovász [13], and de Carvalho, Lucchesi, and Murty [4] on the matching lattice of a matching-covered graph. We would like to thank Kristóf Bérczi, Karthekeyan Chandrasekaran, Bertrand Guenin, and Levent Tunçel for fruitful discussions about various aspects of this work. This work was supported in part by EPSRC grant EP/X030989/1 and ONR grant N00014-22-1-2528.

Data Availability Statement. No data are associated with this article. Data sharing is not applicable to this article.

References

1. Abdi, A., Cornuéjols, G., Guenin, B., Tunçel, L.: Dyadic linear programming and extensions. Math. Program. (2024). https://doi.org/10.1007/s10107-024-02146-4
2. Abdi, A., Cornuéjols, G., Liu, S., Silina, O.: Strongly connected orientations and integer lattices (2024). https://arxiv.org/abs/2410.13665
3. Abdi, A., Cornuéjols, G., Zlatin, M.: On packing dijoins in digraphs and weighted digraphs. SIAM J. Discret. Math. **37**(4), 2417–2461 (2023)
4. de Carvalho, M.H., Lucchesi, C.L., Murty, U.: Optimal ear decompositions of matching covered graphs and bases for the matching lattice. J. Combin. Theory Ser. B **85**(1), 59–93 (2002). https://doi.org/10.1006/jctb.2001.2090
5. Cornuéjols, G.: Combinatorial Optimization: Packing and Covering. Society for Industrial and Applied Mathematics (2001)
6. Cornuéjols, G., Liu, S., Ravi, R.: Approximately packing dijoins via nowhere-zero flows. In: Integer Programming and Combinatorial Optimization: 25th International Conference, IPCO 2024, Wroclaw, Poland, 3–5 July 2024, Proceedings, pp. 71–84. Springer, Heidelberg (2024)
7. Dummit, D.S., Foote, R.M.: Abstract Algebra, 3rd edn. Wiley (2004)
8. Edmonds, J., Pulleyblank, W.R., Lovász, L.: Brick decompositions and the matching rank of graphs. Combinatorica **2**(3), 247–274 (1982). https://doi.org/10.1007/BF02579233
9. Edmonds, J., Giles, R.: A min-max relation for submodular functions on graphs. In: Studies in Integer Programming (Proceedings of Workshop, Bonn, 1975). Annals of Discrete Mathematics, vol. 1, pp. 185–204 (1977)
10. Frank, A.: Connections in Combinatorial Optimization. Oxford Lecture Series in Mathematics and Its Applications. Oxford University Press (2011)
11. Gerards, A., Sebő, A.: Total dual integrality implies local strong unimodularity. Math. Program. **38**(1), 69–73 (1987). https://doi.org/10.1007/BF02591852
12. Guenin, B., Hwang, S.: Dyadic packing of dijoins (2024). Submitted
13. Lovász, L.: Matching structure and the matching lattice. J. Combin. Theory Ser. B **43**(2), 187–222 (1987). https://doi.org/10.1016/0095-8956(87)90021-9
14. Lucchesi, C.L., Younger, D.H.: A minimax theorem for directed graphs. J. London Math. Soc. (2) **17**(3), 369–374 (1978). https://doi.org/10.1112/jlms/s2-17.3.369

15. Martinet, J.: Perfect Lattices in Euclidean Spaces. Springer, Heidelberg (2003)
16. Schrijver, A.: Observations on Woodall's conjecture. https://homepages.cwi.nl/~lex/files/woodall.pdf
17. Schrijver, A.: Combinatorial Optimization. Polyhedra and Efficiency. Springer, Heidelberg (2003)
18. de Werra, D.: Equitable colorations of graphs. R.A.I.R.O. **5**(R-3), 3–8 (1971)
19. Woodall, D.: Menger and König systems. In: Alavi, Y., Lick, D. (eds.) Theory and Applications of Graphs. Lecture Notes in Mathematics, vol. 642. Springer, Heidelberg (1978)

Prize-Collecting Forest with Submodular Penalties: Improved Approximation

Ali Ahmadi, Iman Gholami, MohammadTaghi Hajiaghayi, Peyman Jabbarzade(✉), and Mohammad Mahdavi

University of Maryland, College Park, MD 20742, USA
{ahmadia,igholami,hajiagha,peymanj,mahdavi}@umd.edu

Abstract. *Constrained forest problems* form a class of graph problems where specific connectivity requirements for certain cuts within the graph must be satisfied by selecting the minimum-cost set of edges. The prize-collecting version of these problems introduces flexibility by allowing penalties to be paid to ignore some connectivity requirements.

Goemans and Williamson [8] introduced a general technique and developed a 2-approximation algorithm for constrained forest problems. Further, Sharma, Swamy, and Williamson [16] extended this work by developing a 2.54-approximation algorithm for the prize-collecting version of these problems. Motivated by the generality of their framework, which includes problems such as Steiner trees, Steiner forests, and their variants, we pursued further exploration.

We present a significant improvement by achieving a 2-approximation algorithm for this general model, matching the approximation factor of the constrained forest problems. Notably, the best-known approximation factor for a specific case, the Steiner forest, has remained at 2 since it was established in 1991 by Agrawal, Klein, and Ravi [1].

Keywords: Steiner · Approximation algorithm · Prize-collecting · Submodularity

1 Introduction

In the context of connectivity problems, many algorithms focus on satisfying specific demands while minimizing costs. These problems typically involve weighted graphs, where we select a subset of edges to fulfill these demands. The Minimum Spanning Tree (MST) is a well-known example where efficient algorithms can find the optimal solution. However, numerous generalizations of MST are NP-hard, meaning finding guaranteed optimal solutions becomes computationally intractable. Therefore, for these problems, researchers have shifted their focus to developing effective approximate solutions. This work addresses constrained forest problems with submodular penalty functions for unmet demands, achieving a 2-approximation guarantee.

We formally define the problem on a weighted graph $G = (V, E)$ with a weight function $c : E \to \mathbb{R}_{\geq 0}$ assigning a cost to each edge. Additionally, we have a penalty function $\pi : 2^{2^V} \to \mathbb{R}_{\geq 0}$ defined on families of subsets of vertices. The objective is to

N. Megow and A. Basu (Eds.): IPCO 2025, LNCS 15620, pp. 15–28, 2025.
https://doi.org/10.1007/978-3-031-93112-3_2

find a subset of edges $F \subseteq E$ and define a family of unsatisfied sets $\mathcal{S} \subseteq 2^V$ such that $\mathcal{S}$ contains all sets that do not cut F (i.e., no edge in F has exactly one endpoint in those sets). We aim to minimize the total cost, which is the sum of the cost of the selected edges, $c(F)$, plus the penalty for unsatisfied demands, $\pi(\mathcal{S})$. For the rest of the paper, we refer to this problem as the *prize-collecting forest problem (PCF).*

The penalty function captures the cost associated with leaving certain connectivity requirements unmet. Since its domain is exponential in size, storing penalties for all inputs is impossible. Thus, we assume the availability of an oracle that, given a family of subsets of vertices, returns its penalty. This assumption is reasonable for all submodular problems, as they have an exponential domain size, and unlike additive functions, the value of any subset of elements cannot be computed as a function of the values for individual elements. The penalty function also satisfies the following properties:

- Empty set property: $\pi(\emptyset) = 0$.
- Submodularity: $\forall \mathcal{S}_1, \mathcal{S}_2 \subseteq 2^V$: $\pi(\mathcal{S}_1) + \pi(\mathcal{S}_2) \geq \pi(\mathcal{S}_1 \cup \mathcal{S}_2) + \pi(\mathcal{S}_1 \cap \mathcal{S}_2)$.
- Monotonicity: $\forall \mathcal{S}_1 \subseteq \mathcal{S}_2 \subseteq 2^V$: $\pi(\mathcal{S}_1) \leq \pi(\mathcal{S}_2)$.
- Union property: $\forall S_1, S_2 \subseteq V$: $\pi(\{S_1, S_2\}) = \pi(\{S_1, S_2, S_1 \cup S_2\})$.
- Complementarity: $\forall S \subseteq V$: $\pi(\{S\}) = \pi(\{S, S^c\})$.

The empty set property establishes the baseline: no penalty is incurred if all connectivity requirements are satisfied. Submodularity indicates that having a larger set of unsatisfied sets reduces the marginal cost of not satisfying additional sets. Monotonicity ensures that the penalty increases as the number of unsatisfied sets grows.

The last two properties, union and complementarity, relate to how the penalty treats unsatisfied sets. The union property states that the penalty for violating the connectivity constraint on the union of two unsatisfied sets remains unchanged compared to the penalties of the individual sets. This is because the union inherits the lack of connectivity from its subsets. Complementarity asserts that the penalty for a set is the same as the penalty for its complement. This holds true because both the set and its complement cut the same edges of the graph, and failing to connect either results in the same connectivity situation. These properties together imply that for any $\mathcal{S} \subseteq 2^V$, we have $\pi(\mathcal{S}) = \pi(\textsc{closure}(\mathcal{S}))$, where $\textsc{closure}(\mathcal{S})$ is the smallest superset of $\mathcal{S}$ that is closed under union and complement, and consequently, intersection and set difference operations.

While numerous studies focus on specific cases of our model, some of which are discussed later, the current state-of-the-art for this general problem is a 2.54-approximation algorithm by Sharma, Swamy, and Williamson [16], which employs LP rounding. They also developed a 3-approximation algorithm based on a primal-dual approach, which our method frequently uses as a black-box subroutine. Through a combinatorial analysis, we demonstrate that our approach achieves a better approximation factor by establishing tighter bounds on both the optimal solution and our algorithm's output.

Several prior studies in connectivity research address similar problems. Notably, Agrawal, Klein, and Ravi [1] explored a generalization of Steiner forests, focusing on connecting vertex pairs with r paths. Building on this work, Goemans and Williamson [8] introduced a powerful technique that yielded constant-factor approximation algorithms for various special cases of our problem, inspiring numerous subsequent studies including ours. Their model leverages vertex subsets to define connectivity demands. More precisely, they define a function $f : 2^V \rightarrow \{0, 1\}$ such that for any subset $S \subseteq V$, $f(S) = 1$ if there is a requirement to connect vertices within S to those outside, and $f(S) = 0$ otherwise. We extend their model by assigning penalties to each subset of vertices, allowing these penalties to be zero if connecting the set's vertices to the rest of the graph is not required (i.e., $f(S) = 0$). Additionally, beyond prize-collecting, there are other generalizations where the function f can exceed 1 (see [12, 17]). This means that a minimum number of edges is mandated for each cut.

Expanding upon the work of Goemans and Williamson [8], Sharma, Swamy, and Williamson [16] are not the only ones who have considered the prize-collecting version of the generalized forest problem. Other works [5, 10, 11, 14] explore variations of the prize-collecting version, differing in the definition of the penalty function and the connectivity constraint. For example, [14] study a generalization of [16] with a K connectivity requirement. They use the algorithm from [16] as a black box and provide a $2.54 \log(K)$ approximation algorithm. However, since the primal-dual properties of [16] are not trivially observed in our algorithm, a key question is whether it is possible to improve the result of [14] and achieve a $2 \log(K)$ approximation algorithm for their problem.

One of the well-known special cases of our model is the Steiner tree problem, where the goal is to connect a designated set of vertices. This problem has an obvious 2-approximate solution. Many works have addressed Steiner tree [13, 15, 18], leading to a $(\ln 4 + \epsilon)$-approximation algorithm [7]. Similarly, for the prize-collecting Steiner tree problem, where a penalty must be paid for each unconnected vertex, studies began with a 3-approximate solution by [6], continued with a 2-approximation algorithm by [8], and then [4] broke the 2-approximation barrier. Recently, [3] achieved a 1.79-approximation algorithm. These advancements in approximating special cases of our model, all achieving factors below 2, naturally raise the question: can we develop an algorithm with an approximation factor strictly less than 2 for our general problem?

The Steiner forest problem, a more general model than Steiner trees, deals with a set of demands, each specifying a set of vertices that must be connected. Notably, the 2-approximation algorithm for Steiner forests of [8] has remained unimproved for over three decades. This suggests that our proposed algorithm, which tackles a generalization of Steiner forests, might be challenging to improve significantly in the future, potentially indicating the tightness of our approximation factor.

It is worth noting that the prize-collecting version of the Steiner forest (PCSF), another special case of our model, had a 2.54-approximation algorithm [9] until a recent work by Ahmadi, Gholami, Hajiaghayi, Jabbarzade, and Mahdavi [2] achieved a 2-approximation for this problem. This development opens doors to potential improvements in our model's approximation factor as well. Our algorithm leverages their general recursive approach, a powerful tool for designing algorithms in connectivity problems (e.g., [3]). Additionally, we demonstrate the applicability of their approach to a more general model.

Compared to PCSF problem, designing an algorithm for our model introduces significantly greater complexity. One major source of this complexity is the penalty structure. In PCSF, penalties are defined for n^2 elements, while in our model, penalties are defined for 2^n elements, with the penalty function applicable to any subset of them. This leads to 2^{2^n} potential combinations, which considerably increases the computational complexity. Attempting to access or compute all penalties would result in exponential runtime, which is infeasible.

Additionally, the use of submodular penalty functions further complicates the analysis. With additive penalty functions, if both our solution and the optimal solution pay penalties for overlapping sets of elements, we can simplify the analysis by ignoring those elements, as both solutions are charged the same fixed penalty. However, with submodular functions, the penalty for a set of elements is not independent; it depends on the presence of other elements in the penalty calculation. This dependency implies that the penalty for a given set of elements may vary between our solution and the optimal solution, based on which other elements are penalized. Consequently, a more refined analysis is required to account for these interactions accurately.

1.1 Overview of Contribution

In this paper, we present a 2-approximation algorithm for *prize-collecting forest problem (PCF)*, formally proving the following theorem:

Theorem 1. *There exists a deterministic polynomial-time 2-approximation algorithm for the* prize-collecting forest problem *with a submodular penalty function defined on the family of subsets of vertices for unsatisfied connectivity requirements.*

We use the 3-approximation algorithm from [16] as a black box, as explained in our full version. Our algorithm, described in Sect. 2, is inspired by the iterative approach of the 2-approximation algorithm for prize-collecting Steiner forests [2]. However, our model introduces additional complexities due to the submodular penalty function and exponential connectivity requirements, leading to a more challenging analysis compared to the special case addressed in [2]. The analysis of our algorithm is presented in Sect. 3, where Theorem 2 demonstrates the approximation guarantee, and Theorem 3 establishes the polynomial-time complexity of our algorithm.

Sharma, Swamy and Williamson (SSW) Algorithm. We briefly explain the 3-approximation algorithm for the prize-collecting forest problem developed by Sharma, Swamy, and Williamson [16]. We refer to this algorithm as SSW and will call it multiple times in our algorithm.

The algorithm operates on a graph where edges are assumed to be curves with lengths equal to their costs. It starts with an empty forest, and some of its connected components are chosen as active sets. At each moment, active sets gradually color portions of uncolored edges that have exactly one endpoint within them.

There are two events that trigger changes in the algorithm:

- An edge becomes fully colored and is added to the forest, merging its endpoints' connected components. The resulting new component becomes an active set.
- A family of subsets of vertices becomes tight, meaning the total duration of their activation equals the penalty for violating connectivity requirements for the entire family. The algorithm deactivates these sets (now called tight sets) as paying the penalty becomes more cost-effective than further coloring edges with those sets.

Once there are no more active sets left, the algorithm enters a pruning phase. Here, it examines tight sets. While tight sets stop coloring their adjacent edges, other active sets may continue to color these edges and eventually fully color them, leading to their inclusion in the forest. If a tight set cuts exactly one edge in the forest, that edge is removed. Since the penalty for the tight set will be paid anyway, there is no need to pay the additional cost of including the edge. However, if a tight set cuts multiple edges, they are left untouched. This is because vertices outside the tight set might rely on those edges, along with a path within the tight set, to connect to each other.

Finally, the algorithm returns the remaining forest (the fully colored edges that were not removed during pruning) and pays the penalty associated with all tight sets. The key idea is that all sets not cutting any edges in the forest belong to the closure of the tight sets. This ensures that paying the penalty of tight sets along with the obtained forest is sufficient to satisfy all connectivity requirements.

Our Iterative Algorithm. We propose an iterative algorithm inspired by [2] that calls SSW on a series of newly defined penalty functions. We begin by calling SSW and consider its resulting forest as the initial solution. We also assume SSW provides the family of tight sets it identified (denoted by $\mathcal{D}_{\text{SSW}}$). For the rest of our algorithm, we pay the penalty of $\mathcal{D}_{\text{SSW}}$, and since its penalty is already paid, we do not need to satisfy the connectivity of sets in $\textsc{closure}(\mathcal{D}_{\text{SSW}})$.

A key component of our approach is a newly defined penalty function, denoted by $\pi_R : 2^{2^V} \to \mathbb{R}_{\geq 0}$. The value, $\pi_R(\mathcal{S})$, represents the marginal cost of adding a new family $\mathcal{S}$ to the existing paid sets from SSW, $\mathcal{D}_{\text{SSW}}$. Formally, $\pi_R(\mathcal{S}) = \pi(\mathcal{S} \cup \mathcal{D}_{\text{SSW}}) - \pi(\mathcal{D}_{\text{SSW}})$. Since π_R inherits the desired properties of a penalty function, we can recursively call our algorithm using π_R as the new penalty function. This recursive call involves running SSW with π_R and then creating another penalty function based on the new tight sets identified by that SSW call.

The recursion continues until a call to SSW results in a forest with zero penalty for tight sets. This indicates that the newly created penalty function in this step is identical to the previous one, and further recursion would not yield improvement. Once this termination condition is met, we stop the recursion and return the best forest encountered throughout the iterative process.

Analysis. First, note that to prove the approximation guarantee of the SSW algorithm, it can be shown that:

- The optimal solution is at least equal to the total coloring duration of sets since every set colors at least one edge of the optimal solution at each moment, or the optimal solution pays penalty proportional to that amount of coloring.
- The forest SSW returns costs at most twice the total coloring duration since, on average, every set colors at most two edges of the forest.
- The penalty SSW pays is at most equal to the total coloring duration as SSW only pays for tight sets whose coloring duration is equal to their penalty.

This information is sufficient to conclude that SSW returns a 3-approximate solution.

For the analysis of our algorithm, we leverage its recursive nature by employing induction on the recursive steps. In the base case where $\pi(\mathcal{D}_{\text{SSW}}) = 0$, we demonstrate that the SSW algorithm returns a 2-approximate solution, which follows trivially from the earlier analysis. For our induction hypothesis, we assume that the recursive call provides a 2-approximate solution for the instance with the penalty function π_R. In the induction step, we need to demonstrate that the best solution, chosen between the one returned by the recursive call and the initial solution provided by SSW, is a 2-approximate solution for the initial instance with the penalty function π.

To prove the induction step, we focus on $\mathcal{D}_{\text{SSW}}$. Since we pay the penalty of $\mathcal{D}_{\text{SSW}}$, it's crucial to ensure that paying the penalty for each set in $\mathcal{D}_{\text{SSW}}$ leads to a good solution, whether from the recursive call or the initial SSW algorithm. We partition sets in $\mathcal{D}_{\text{SSW}}$ into $\mathcal{D}_{paid}$ and $\mathcal{D}_{unpaid}$ based on whether the optimal solution pays their penalty or connects them, respectively. Sets in $\mathcal{D}_{unpaid}$ are further divided into $\mathcal{D}_{single}$ and $\mathcal{D}_{multiple}$ based on whether they cut exactly one edge or multiple edges of the optimal solution, respectively. Now, let's focus on each class separately.

- For sets in $\mathcal{D}_{paid}$, since the optimal solution pays their penalty, it's logical for our algorithm to do the same.
- Regarding sets in $\mathcal{D}_{multiple}$, since they cut more than one edge of the optimal solution, a better lower-bound for the optimal solution can be found in the analysis of the SSW, resulting in an improved approximation from SSW's output.
- As for sets in $\mathcal{D}_{single}$, which only cut one edge of the optimal solution, we demonstrate that removing these edges from the optimal solution does not increase its penalty in the recursive instance where the penalty of $\mathcal{D}_{\text{SSW}}$ has already been paid. Thus, a solution for the recursive instance can be achieved by removing these edges from the initial instance's optimal solution, resulting in a significantly reduced cost due to the removal of edges. By leveraging the induction hypothesis that the recursive call provides a 2-approximate solution for its instance, we conclude that the recursive call yields a good solution for the initial instance.

1.2 Preliminaries

We introduce some notation and elementary lemmas used in the subsequent sections. Proofs of them, along with some additional lemmas, are presented in our full version.

Penalty Function Properties.

Definition 1. *For a family $\mathcal{S} \subseteq 2^V$, we use* CLOSURE($\mathcal{S}$) *to denote the smallest subset of 2^V that contains $\mathcal{S}$ and is closed under union and complement operations. That is, for any sets $S_1, S_2 \in$* CLOSURE($\mathcal{S}$)*, $S_1 \cup S_2$ and $S_1{}^c$ are also in* CLOSURE($\mathcal{S}$)*. It is worth noting that* CLOSURE($\mathcal{S}$) *is closed under intersection and set difference operations as well, since these operations can be written as a combination of union and complement operations.*

Unless otherwise specified, when a penalty function π is used hereinafter, we assume the function satisfies the required properties: empty set property, submodularity, monotonicity, union property, and complementarity.

Lemma 1. *For any family $\mathcal{S}$, sets $S_1, S_2 \in \mathcal{S}$, and family $\mathcal{S}'$ with $\pi(\mathcal{S}') = 0$ we have*

$$\pi(\mathcal{S}) = \pi(\mathcal{S} \cup \{S_1 \cup S_2\}) = \pi(\mathcal{S} \cup \mathcal{S}') = \pi(\mathcal{S} \cup \{S_1^C\}).$$

Consequently, we have $\pi(\mathcal{S}) = \pi(\text{CLOSURE}(\mathcal{S}))$.

Definition 2. *We use the following notation to denote the marginal cost of the penalty function π in relation to a given family $\mathcal{S}_2$:*

$$\pi(\mathcal{S}_1 \mid \mathcal{S}_2) := \pi(\mathcal{S}_1 \cup \mathcal{S}_2) - \pi(\mathcal{S}_2).$$

We can show that this marginal cost function retains the required properties of the original cost function π.

Lemma 2. *If π satisfies the required properties (empty set property, submodularity, monotonicity, union property, and complementarity), then for any fixed family $\mathcal{S}$, the function $\pi(\,\cdot \mid \mathcal{S})$ will also satisfy these properties.*

Graph Notations. For a set $S \subseteq V$, we use $\delta(S)$ to denote the set of edges cut by S, i.e., edges with exactly one endpoint in S. We also say that an edge $e \in E$ and set $S \subseteq V$ are adjacent if and only if $e \in \delta(S)$. For a forest F, we use $CC(F)$ to denote the connected components of F. For a subgraph H of graph G, we use $c(H)$ to denote the total weight $\sum_{e \in H} c(e)$ of this subgraph.

Definition 3. *For a forest F, we use*

$$\mathcal{P}(F) := \{S \subseteq V \mid |\delta(S) \cap F| = 0\}$$

to denote the family for which the penalty π needs to be paid.

Lemma 3. *For any forest F, we have*

$$\mathcal{P}(F) = \text{CLOSURE}(CC(F)).$$

For an instance $I = (G = (V, E, c), \pi)$ of the PCF problem and a solution SOL using forest F_{SOL}, we use $cost_I(\text{SOL}) = c(F_{\text{SOL}}) + \pi(\mathcal{P}(F_{\text{SOL}}))$ to denote the cost of this solution.

2 2-Approximation Algorithm for Submodular PCF

In this section, we present RecursivePCF, a 2-approximation algorithm for our prize-collecting forest problem with submodular penalties, utilizing a recursive approach. This algorithm utilizes the 3-approximation algorithm described in our full version as a key component. Algorithm 1 illustrates the steps of obtaining a 2-approximation solution.

Recall that the 3-approximation algorithm explained in the previous section is denoted as SSW. For a given instance $I = (G, \pi)$, we first call SSW (Line 2). It is worth noting that we use $\mathcal{D}_{\text{SSW}}$ to denote the union of tight families in the 3-approximation algorithm i.e., for instance I, $\mathcal{D}_{\text{SSW}}$ is the final $\mathcal{D}_m$ returned by SSW (explained in our full version). If $\pi(\mathcal{D}_{\text{SSW}}) = 0$, the total penalty that SSW pays is 0 which means we have a 2-approximate solution (based on Lemma 6). Therefore, we can immediately return the output of SSW as the solution (Lines 4 and 5).

On the other hand, since SSW suggests paying the penalty for the family $\mathcal{D}_{\text{SSW}}$, we intuitively consider the penalty of this family to be already paid and solve the problem to satisfy other demands. Since the penalty is a function of families, it is not trivial to simply subtract the penalty of these sets. However, the *marginal cost* provides a natural way to account for the cost of a family, allowing the introduction of a new penalty function. Consequently, we define a new instance $R = (G, \pi_R)$ where π_R is as follows. For all $\mathcal{S} \subseteq 2^V$:

$$\pi_R(\mathcal{S}) = \pi(\mathcal{S} \mid \mathcal{D}_{\text{SSW}}).$$

Based on Lemma 2, the new penalty function has the required properties for being a penalty function for our problem. Lines 6 and 7 of the algorithm outline the construction of π_R and the instance R.

Next, we recursively find the 2-approximate solution for instance R (Line 8). Assume the solution for this instance is F_{REC}. This solution is also valid for the input instance since the graph remains the same in instance R. The final solution is the one with the lower cost between the SSW solution and the solution returned by the recursive call (Lines 10 to 13). In the next section, we provide the analysis of the described algorithm.

Algorithm 1 PCF 2-approximation algorithm

Input: An undirected graph $G = (V, E, c)$ with edge costs $c : E \rightarrow \mathbb{R}_{\geq 0}$ and penalties $\pi : 2^{2^V} \rightarrow \mathbb{R}_{\geq 0}$.

Output: A forest F with the minimum cost.

1: **procedure** RecursivePCF($I = (G, \pi)$)
2: $(F_{\text{SSW}}, \mathcal{D}_{\text{SSW}}) \leftarrow \text{SSW}(I)$
3: $cost(\text{SSW}) \leftarrow c(F_{\text{SSW}}) + \pi(\mathcal{P}(F_{\text{SSW}}))$
4: **if** $\pi(\mathcal{D}_{\text{SSW}}) = 0$ **then**
5: **return** F_{SSW}
6: Define $\pi_R : 2^{2^V} \rightarrow \mathbb{R}_{\geq 0}$ as $\pi_R(\mathcal{S}) = \pi(\mathcal{S} \mid \mathcal{D}_{\text{SSW}})$
7: Let instance R of the problem be the graph G with penalty function π_R
8: $F_{\text{REC}} \leftarrow$ RecursivePCF(R)
9: $cost(\text{REC}) \leftarrow c(F_{\text{REC}}) + \pi(\mathcal{P}(F_{\text{REC}}))$
10: **if** $cost(\text{SSW}) \leq cost(\text{REC})$ **then**
11: **return** F_{SSW}
12: **else**
13: **return** F_{REC}

3 Analysis of the 2-Approximation Algorithm

In this section, we analyze the approximation ratio achieved by Algorithm 1. Proofs of these lemmas appear in our full version.

To analyze our algorithm, we compare the cost of the solution returned by this algorithm on an instance $I = (G, \pi)$ to the cost of an optimal solution OPT. In our analysis, we focus on the algorithm's variables and output in a single RecursivePCF call where I is the input and use induction to analyze the recursive call. Additionally, we use the variables y_S from SSW (explained in our full version), executed during the call to the SSW procedure in Line 2, to bound the cost of the different solutions we obtain. The variable y_S represents the coloring duration of a set $S \subseteq V$ in SSW. We use SSW and REC to refer to the solutions obtained using the call to the 3-approximation algorithm and the recursive call, respectively.

We begin by defining some notations and classifying sets in $\mathcal{D}_{\text{SSW}}$. Let F_{OPT} denote the forest used in the optimal solution. Similarly, let $\mathcal{D}_{\text{OPT}}$ denote the family for which the optimal solution pays the penalty, i.e., $\mathcal{D}_{\text{OPT}} = \mathcal{P}(F_{\text{OPT}})$.

Definition 4. *Recall that for instance I, $\mathcal{D}_{SSW}$ is the family returned by* SSW *in the output (Line 2 of Algorithm 1). We partition $\mathcal{D}_{SSW}$ into two disjoint families as follows:*

$$\mathcal{D}_{paid} = \{S \in \mathcal{D}_{SSW} \mid \pi(\{S\} \mid \mathcal{D}_{OPT}) = 0\}$$
$$\mathcal{D}_{unpaid} = \{S \in \mathcal{D}_{SSW} \mid \pi(\{S\} \mid \mathcal{D}_{OPT}) > 0\}$$

In the following lemma, we show that all sets in $\mathcal{D}_{unpaid}$ cut at least one edge of the forest in the optimal solution.

Lemma 4. *For any set $S \in \mathcal{D}_{unpaid}$, there is at least one edge e in F_{OPT} such that $e \in \delta(S)$.*

Now, since all sets in $\mathcal{D}_{unpaid}$ cut at least one edge of F_{OPT}, we can divide this family into two families as follows:

Definition 5. *Let us partition sets in family $\mathcal{D}_{unpaid}$ into two disjoint families $\mathcal{D}_{single}$ and $\mathcal{D}_{multiple}$ based on the number of edges of the forest F_{OPT} they cut. Formally,*

$$\mathcal{D}_{single} = \{S \in \mathcal{D}_{unpaid} \mid |\delta(S) \cap F_{OPT}| = 1\}$$
$$\mathcal{D}_{multiple} = \{S \in \mathcal{D}_{unpaid} \mid |\delta(S) \cap F_{OPT}| > 1\}.$$

In addition, we define y_{single} and $y_{multiple}$ based on the y_S values for sets in $\mathcal{D}_{single}$ and $\mathcal{D}_{multiple}$:

$$y_{single} = \sum_{S \in \mathcal{D}_{single}} y_S \qquad\qquad y_{multiple} = \sum_{S \in \mathcal{D}_{multiple}} y_S$$

Up to this point, our approach involves deciding to pay the penalty of $\mathcal{D}_{\text{SSW}}$, which we then divide into sets based on whether paying their penalty increases the penalty of the family that the optimal solution pays. We have shown that sets increasing the optimal solution's penalty must cut at least one edge of the optimal solution. Thus, we classify these sets based on whether they cut exactly one edge or more than one edge. Figure 1 demonstrates the classification of sets in $\mathcal{D}_{\text{SSW}}$.

Next, we use this classification to establish a lower bound for the cost of the optimal solution and upper bounds for SSW and REC solutions. We will employ induction to complete the proof of the approximation guarantee. We begin by providing a lower bound for the cost of the optimal solution.

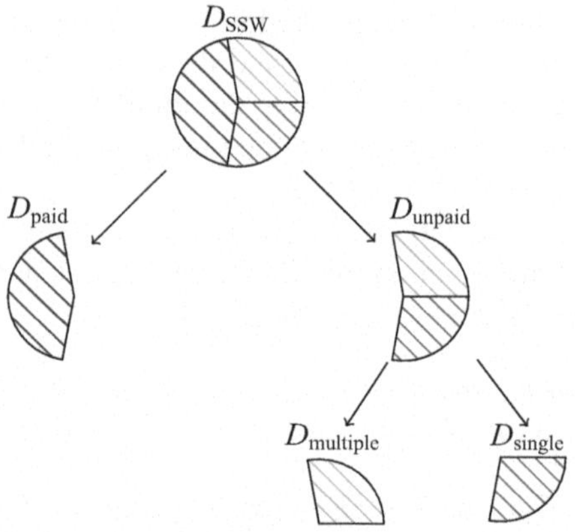

Fig. 1. Tight sets returned by SSW, denoted by $\mathcal{D}_{\text{SSW}}$, are partitioned into $\mathcal{D}_{paid}$ and $\mathcal{D}_{unpaid}$, with $\mathcal{D}_{unpaid}$ further divided into $\mathcal{D}_{single}$ and $\mathcal{D}_{multiple}$.

Lemma 5. *For instance I, the cost of the optimal solution can be bounded as follows:*

$$cost(OPT) \geq \sum_{S \subseteq V} y_S + y_{multiple}.$$

Next, we give upper bounds for the costs of solutions SSW and REC which are used in our algorithm. We begin with an upper bound for SSW which uses the lower bound obtained for OPT.

Lemma 6. *For instance I, we have the following bound for the cost of the solution SSW:*

$$cost(SSW) \leq 2cost(OPT) - 2y_{multiple} + \pi(\mathcal{D}_{SSW}).$$

To analyze the cost of the solution REC, we first introduce the solution SOL_R below to establish an upper bound on the cost of the optimal solution for instance R. Assuming by induction that our algorithm achieves a 2-approximate solution for instance R, we then use this upper bound to bound the cost of REC on the original instance I.

Definition 6. *Consider the forest F_{OPT} selected by the optimal solution OPT. For any set $S \in \mathcal{D}_{single}$, remove the unique edge in F_{OPT} cut by S. Let F_R be the resulting forest and $\mathcal{D}_R = \mathcal{P}(F_R)$ denote the family for which the penalty needs to be paid for forest F_R. We define SOL_R as the solution represented by the forest F_R, which pays the penalty for family $\mathcal{D}_R$.*

We analyze the cost of SOL_R in the next two lemmas. First, we introduce a lower bound for the cost of the removed edges to analyze how the cost of the forest F_R compares to F_{OPT}. Then, we show that removing these edges from F_{OPT} does not affect the penalty that needs to be paid when considering the penalty function for instance R. Note that the inequality in Lemma 8 is in fact an equality, as $\mathcal{D}_{\text{OPT}} \subseteq \text{CLOSURE}(\mathcal{D}_R)$. However, proving the inequality is sufficient for our analysis.

Lemma 7. *The weight of the forest F_R is at most $c(F_{OPT}) - y_{single}$.*

Lemma 8. *For instance R, the penalty paid by SOL_R is no more than the penalty paid by OPT:*

$$\pi_R(\mathcal{D}_R) \leq \pi_R(\mathcal{D}_{OPT}).$$

Now, we use the solution SOL_R to bound the optimal solution of instance R, which costs less than SOL_R due to its optimality. Note that we use $cost_R$ to denote the cost of a solution for instance R.

Lemma 9. *Let OPT_R be an optimal solution for instance R. Then, we have*

$$cost_R(OPT_R) \leq cost(OPT) - (\pi(\mathcal{D}_{OPT}) - \pi(\mathcal{D}_{OPT} \mid \mathcal{D}_{SSW})) - y_{single}.$$

We use the following lemma to bound the cost of the solution REC on instance I based on its cost on instance R.

Lemma 10. *For any solution SOL,*

$$cost(SOL) \leq cost_R(SOL) + \pi(\mathcal{D}_{SSW}).$$

Next, we relate y_{single} and $y_{multiple}$ so we can give an upper bound for REC in terms of $y_{multiple}$, which is consistent with our upper bound for SSW.

Lemma 11. *We have*

$$y_{single} + y_{multiple} \geq \pi(\mathcal{D}_{SSW} \mid \mathcal{D}_{OPT}).$$

Finally, we use induction on the recursive steps of the algorithm to prove that RECURSIVEPCF returns a 2-approximate solution. The base case of our induction is when RECURSIVEPCF does not recursively call itself. In the induction step, we show that if the recursive call returns a 2-approximate solution for the new instance R, then the minimum solution between SSW and REC is a 2-approximate solution for instance I. The following lemma is crucial for the induction step as it helps bound the cost of REC using the induction hypothesis. After establishing this lemma, we proceed to prove the approximation guarantee in Theorem 2.

Lemma 12. *If a solution SOL is a 2-approximate solution for instance R, we have*

$$cost(SOL) \leq 2cost(OPT) + 2y_{multiple} - \pi(\mathcal{D}_{SSW})$$

for instance I.

Finally, we combine the upper bounds for the two solutions SSW and REC to establish the approximation ratio of the solution returned by Algorithm 1. We note that we assume the algorithm terminates in the proof of Theorem 2, before formally proving this in Theorem 3.

Theorem 2. *The solution SOL returned by the algorithm* RECURSIVEPCF *has a cost of at most $2cost(OPT)$.*

Proof. We prove this by induction on the recursive tree of the algorithm. In the base case, which is the last call to the RecursivePCF the algorithm returns the solution SSW without making a recursive call. In this case, $\pi(\mathcal{D}_{\text{SSW}}) = 0$, so the algorithm does not pay any penalties. Then, by Lemma 6, we have

$$cost(\text{SOL}) = cost(\text{SSW}) \leq 2cost(\text{OPT}) - 2y_{multiple} + \pi(\mathcal{D}_{\text{SSW}}) \leq 2cost(\text{OPT}).$$

Now, consider the case where we also obtain a recursive solution REC. In this case, by Lemma 6, we have the following bound for the non-recursive solution SSW

$$cost(\text{SSW}) \leq 2cost(\text{OPT}) - 2y_{multiple} + \pi(\mathcal{D}_{\text{SSW}}).$$

By induction, the recursive solution REC is a 2-approximation for instance R. So, by Lemma 12 we have

$$cost(\text{REC}) \leq 2cost(\text{OPT}) + 2y_{multiple} - \pi(\mathcal{D}_{\text{SSW}}).$$

Combining the two bounds, we can see that

$$\begin{aligned} cost(\text{SOL}) &= \min(cost(\text{SSW}), cost(\text{REC})) \\ &\leq \frac{cost(\text{SSW}) + cost(\text{REC})}{2} \\ &\leq \frac{2cost(\text{OPT}) - 2y_{multiple} + \pi(\mathcal{D}_{\text{SSW}}) + 2cost(\text{OPT}) + 2y_{multiple} - \pi(\mathcal{D}_{\text{SSW}})}{2} \\ &\leq 2cost(\text{OPT}). \end{aligned}$$

Lastly, we show that our algorithm operates in polynomial time.

Theorem 3. *Algorithm 1 always terminates and runs in polynomial time.*

Acknowledgments. The work is partially supported by DARPA QuICC, ONR MURI 2024 award on Algorithms, Learning, and Game Theory, Army-Research Laboratory (ARL) grant W911NF2410052, NSF AF:Small grants 2218678, 2114269, 2347322.

Disclosure of Interests. The authors have no competing interests to declare that are relevant to the content of this article.

References

1. Agrawal, A., Klein, P.N., Ravi, R.: When trees collide: an approximation algorithm for the generalized Steiner problem on networks. SIAM J. Comput. **24**(3), 440–456 (1995). https://doi.org/10.1137/S0097539792236237
2. Ahmadi, A., Gholami, I., Hajiaghayi, M., Jabbarzade, P., Mahdavi, M.: 2-approximation for prize-collecting Steiner forest. In: Woodruff, D.P. (ed.) Proceedings of the 2024 ACM-SIAM Symposium on Discrete Algorithms, SODA 2024, Alexandria, VA, USA, 7–10 January 2024, pp. 669–693. SIAM (2024). https://doi.org/10.1137/1.9781611977912.25

3. Ahmadi, A., Gholami, I., Hajiaghayi, M., Jabbarzade, P., Mahdavi, M.: Prize-collecting Steiner tree: a 1.79 approximation. In: Mohar, B., Shinkar, I., O'Donnell, R. (eds.) Proceedings of the 56th Annual ACM Symposium on Theory of Computing, STOC 2024, Vancouver, BC, Canada, 24–28 June 2024, pp. 1641–1652. ACM (2024). https://doi.org/10.1145/3618260.3649789
4. Archer, A., Bateni, M., Hajiaghayi, M., Karloff, H.J.: Improved approximation algorithms for prize-collecting Steiner tree and TSP. SIAM J. Comput. **40**(2), 309–332 (2011). https://doi.org/10.1137/090771429
5. Bateni, M., Chekuri, C., Ene, A., Hajiaghayi, M.T., Korula, N., Marx, D.: Prize-collecting Steiner problems on planar graphs. In: Randall, D. (ed.) Proceedings of the Twenty-Second Annual ACM-SIAM Symposium on Discrete Algorithms, SODA 2011, San Francisco, California, USA, 23–25 January 2011, pp. 1028–1049. SIAM (2011). https://doi.org/10.1137/1.9781611973082.79
6. Bienstock, D., Goemans, M.X., Simchi-Levi, D., Williamson, D.P.: A note on the prize collecting traveling salesman problem. Math. Program. **59**, 413–420 (1993). https://doi.org/10.1007/BF01581256
7. Byrka, J., Grandoni, F., Rothvoß, T., Sanità, L.: An improved LP-based approximation for Steiner tree. In: Schulman, L.J. (ed.) Proceedings of the 42nd ACM Symposium on Theory of Computing, STOC 2010, Cambridge, Massachusetts, USA, 5–8 June 2010, pp. 583–592. ACM (2010). https://doi.org/10.1145/1806689.1806769
8. Goemans, M.X., Williamson, D.P.: A general approximation technique for constrained forest problems. SIAM J. Comput. **24**(2), 296–317 (1995). https://doi.org/10.1137/S0097539793242618
9. Hajiaghayi, M.T., Jain, K.: The prize-collecting generalized Steiner tree problem via a new approach of primal-dual schema. In: Proceedings of the Seventeenth Annual ACM-SIAM Symposium on Discrete Algorithms, SODA 2006, Miami, Florida, USA, 22–26 January 2006, pp. 631–640. ACM Press (2006). http://dl.acm.org/citation.cfm?id=1109557.1109626
10. Hajiaghayi, M.T., Khandekar, R., Kortsarz, G., Nutov, Z.: Prize-collecting Steiner network problems. In: Eisenbrand, F., Shepherd, F.B. (eds.) IPCO 2010. LNCS, vol. 6080, pp. 71–84. Springer, Heidelberg (2010). https://doi.org/10.1007/978-3-642-13036-6_6
11. Hayrapetyan, A., Swamy, C., Tardos, É.: Network design for information networks. In: Proceedings of the Sixteenth Annual ACM-SIAM Symposium on Discrete Algorithms, SODA 2005, Vancouver, British Columbia, Canada, 23–25 January 2005, pp. 933–942. SIAM (2005). http://dl.acm.org/citation.cfm?id=1070432.1070567
12. Jain, K.: A factor 2 approximation algorithm for the generalized Steiner network problem. Comb. **21**(1), 39–60 (2001). https://doi.org/10.1007/S004930170004
13. Karpinski, M., Zelikovsky, A.: New approximation algorithms for the Steiner tree problems. J. Comb. Optim. **1**(1), 47–65 (1997). https://doi.org/10.1023/A:1009758919736
14. Nagarajan, C., Sharma, Y., Williamson, D.P.: Approximation algorithms for prize-collecting network design problems with general connectivity requirements. In: Bampis, E., Skutella, M. (eds.) WAOA 2008. LNCS, vol. 5426, pp. 174–187. Springer, Heidelberg (2009). https://doi.org/10.1007/978-3-540-93980-1_14
15. Robins, G., Zelikovsky, A.: Tighter bounds for graph Steiner tree approximation. SIAM J. Discret. Math. **19**(1), 122–134 (2005). https://doi.org/10.1137/S0895480101393155
16. Sharma, Y., Swamy, C., Williamson, D.P.: Approximation algorithms for prize collecting forest problems with submodular penalty functions. In: Bansal, N., Pruhs, K., Stein, C. (eds.) Proceedings of the Eighteenth Annual ACM-SIAM Symposium on Discrete Algorithms, SODA 2007, New Orleans, Louisiana, USA, 7–9 January 2007, pp. 1275–1284. SIAM (2007). http://dl.acm.org/citation.cfm?id=1283383.1283520

17. Williamson, D.P., Goemans, M.X., Mihail, M., Vazirani, V.V.: A primal-dual approximation algorithm for generalized Steiner network problems. In: Kosaraju, S.R., Johnson, D.S., Aggarwal, A. (eds.) Proceedings of the Twenty-Fifth Annual ACM Symposium on Theory of Computing, 16–18 May 1993, San Diego, CA, USA, pp. 708–717. ACM (1993). https://doi.org/10.1145/167088.167268
18. Zelikovsky, A.: An 11/6-approximation algorithm for the network Steiner problem. Algorithmica **9**(5), 463–470 (1993). https://doi.org/10.1007/BF01187035

On the Approximability of Unsplittable Flow on a Path with Time Windows

Alexander Armbruster[1], Fabrizio Grandoni[2], Edin Husić[2], Antoine Tinguely[2(✉)], and Andreas Wiese[1]

[1] Technical University of Munich, Munich, Germany
{alexander.armbruster,andreas.wiese}@tum.de
[2] USI-SUPSI, IDSIA, Lugano, Switzerland
{fabrizio,antoine.tinguely}@idsia.ch

Abstract. In the Time-Windows Unsplittable Flow on a Path problem (twUFP) we are given a resource whose available amount changes over a given time interval (modeled as the edge-capacities of a given path G) and a collection of tasks. Each task is characterized by a demand (of the considered resource), a profit, an integral processing time, and a time window. Our goal is to compute a maximum profit subset of tasks and schedule them non-preemptively within their respective time windows, such that the total demand of the tasks using each edge e is at most the capacity of e.

We prove that twUFP is APX-hard which contrasts the setting of the problem without time windows, i.e., Unsplittable Flow on a Path (UFP), for which a PTAS was recently discovered [Grandoni, Mömke, Wiese, STOC 2022]. Then, we present a quasi-polynomial-time $2 + \varepsilon$ approximation for twUFP under resource augmentation. Our approximation ratio improves to $1+\varepsilon$ if all tasks' time windows are identical. Our APX-hardness holds also for this special case and, hence, rules out such a PTAS (and even a QPTAS, unless $\mathsf{NP} \subseteq \mathrm{DTIME}(n^{\mathrm{poly}(\log n)})$) *without* resource augmentation.

Keywords: Approximation algorithm · APX-hardness · Quasi-polynomial algorithm · Scheduling · Unsplittable Flow on a Path

1 Introduction

In the well-studied Unsplittable Flow on a Path problem (UFP), we are given a path graph $G = (V, E)$ with m edges, a capacity $u(e) \in \mathbb{N}$ for each edge $e \in E$, and a collection T of n tasks. Each task $i \in T$ is characterized by a *weight* (or *profit*) $w(i) \in \mathbb{N}$, a *demand* $d(i) \in \mathbb{N}$, and a subpath $P(i) \subseteq E$ of G.[1] A feasible solution consists of a subset of tasks $S \subseteq T$ such that $\sum_{i \in S: e \in P(i)} d(i) \leq u(e)$ for each $e \in E$. In other words, the total demand of the tasks in S whose subpath

[1] Given a subpath P of G, we will sometimes use P also to denote the corresponding set of edges $E(P)$. The meaning will be clear from the context.

N. Megow and A. Basu (Eds.): IPCO 2025, LNCS 15620, pp. 29–42, 2025.
https://doi.org/10.1007/978-3-031-93112-3_3

contains e does not exceed the capacity of e. Our goal is to compute a feasible solution OPT of maximum profit $w(\mathsf{OPT}) := \sum_{i \in \mathsf{OPT}} w(i)$. One can naturally interpret G as a time interval subdivided into time slots (the edges). At each time slot, a given amount of a considered resource (e.g., energy) is available. Each task i corresponds to a job that we can execute (or not) in a fixed time interval: if i is executed, it consumes a fixed amount of the considered resource during its entire execution and generates a profit of $w(i)$. UFP is strongly NP-hard [8,11], and a lot of attention was devoted to the design of approximation algorithms for it [1–3,6,8,14,16–18], culminating in a recent PTAS for the problem [15].

In UFP, we have no flexibility for the time interval during which each selected task i is executed. In practice, it makes sense to consider scenarios where i has a given length (or processing time) and a *time window* during which it needs to be executed. In this setting, for each selected task i we need to specify a starting time for i such that i is processed completely within its time window. This leads to the Time-Windows UFP problem (TWUFP). Here, we are given the same input as in UFP, with the difference that for each task i, instead of a subpath $P(i)$ we are given a *length* (or *processing time*) $p(i) \in \{1, \ldots, m\}$, and a subpath $\mathrm{tw}(i) \subseteq E$ with at least $p(i)$ edges (the *time window* of i). A *scheduling* of i is a subpath $P(i) \subseteq \mathrm{tw}(i)$ containing precisely $p(i)$ edges. A feasible solution for the given instance is a pair $(S, P(\cdot))$ such that for each $i \in S$ the path $P(i)$ is a feasible scheduling of i and $\sum_{i \in S: e \in P(i)} d(i) \leq u(e)$ for each $e \in E$. Our goal is to maximize $w(S)$, like in UFP. Observe that UFP is the special case of TWUFP where for each task $i \in T$ the time window $\mathrm{tw}(i)$ contains exactly $p(i)$ edges; hence, TWUFP is strongly NP-hard. The best known approximation algorithm for it is a $O(\log n / \log\log n)$-approximation [13], improving on a prior result in [9] (both results hold for a more general problem, BAGUFP, which is defined later). However, no result in the literature excludes the existence of a much better approximation ratio for TWUFP, including possibly a PTAS. In this paper, we make progress on a better understanding of the approximability of TWUFP.

1.1 Our Results and Techniques

Our first main result is that TWUFP does *not* admit a PTAS, which in particular implies that it is strictly harder than UFP (unless $\mathsf{P} = \mathsf{NP}$). We show that this already holds when the time window of each task spans all the edges of G, i.e., if $\mathrm{tw}(i) = E$ for each $i \in T$; we denote this special case of the problem by Spanning UFP (SPANUFP). Our result even holds in the cardinality setting, i.e., when all tasks have unit weight, and for polynomially bounded input data.

Theorem 1. SPANUFP *(thus also* TWUFP*) is* APX*-hard and does not admit a (polynomial-time)* $\frac{2755}{2754}$*-approximation algorithm (unless* $\mathsf{P} = \mathsf{NP}$*), even in the cardinality case and if demands and the number of edges is polynomially bounded in* n.

The proof of Theorem 1 follows from a reduction from the Maximum 3-Dimensional Matching (3-DM) problem and the bound in [10]. In our reduction, we model 3-DM instances as instances of SPANUFP in which the capacity profile models bins, reminiscent of the 2-Dimensional Vector Bin Packing (2-VBP) problem, and adapt the construction in [23] (which refutes the existence of an asymptotic PTAS for 2-VBP).

Our second main contribution is a constant factor approximation algorithm for TWUFP with resource augmentation which runs in quasi-polynomial time. More specifically, for any constant $\varepsilon > 0$, we compute a $(2+\varepsilon)$-approximate solution in time $2^{O_\varepsilon(\text{poly}(\log(nm)))}$ which violates the edge capacities at most by a factor $1 + O(\varepsilon)$. As usual in the setting of resource augmentation, the approximation ratio is computed with respect to an optimal solution which is *not* allowed to violate the edge capacities.

Theorem 2. *For any* $\varepsilon > 0$, *there is a* $(2+\varepsilon)$*-approximation for* TWUFP *under* $(1+O(\varepsilon))$*-resource augmentation with running time* $O\big((mn)^{O(\log^2 n \log^5 m/\varepsilon^4)} \cdot \log U\big)$, *where* $U := \max_{e\in E} u(e)$ *is the maximum capacity of the instance.*

For the special case of SPANUFP, the approximation ratio of our algorithm improves to $1+\varepsilon$, i.e., we achieve a quasi-PTAS (QPTAS) under resource augmentation for SPANUFP. In contrast (unless $\mathsf{NP} \subseteq \mathrm{DTIME}(n^{\text{poly}(\log n)})$), a QPTAS for SPANUFP is impossible *without* resource augmentation since SPANUFP is APX-hard.

Theorem 3. *For any* $\varepsilon > 0$ *there is a QPTAS for* SPANUFP *under* $(1+O(\varepsilon))$*-resource augmentation.*

We leave as an interesting open problem to find a polynomial-time (or even quasi-polynomial-time) constant factor approximation for TWUFP *without* resource augmentation.

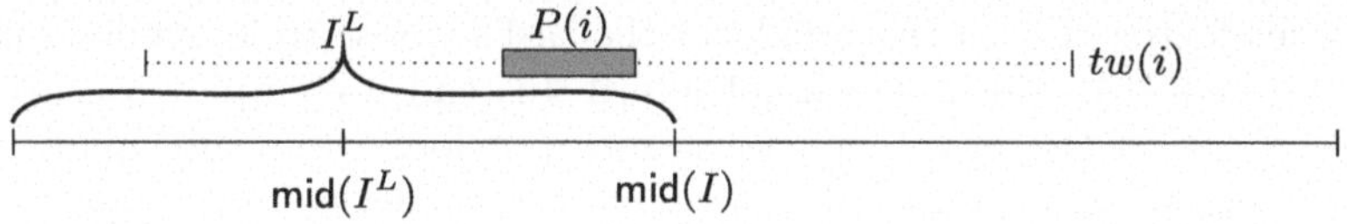

Fig. 1. The interval I and its subdivision together with a task i.

We describe now the key ideas in our $(2+\epsilon)$-approximation for TWUFP. The previous QPTASes for UFP [2,6] are based on the following approach (written in slightly different terminology here). We interpret the path E as an interval I, i.e., with one subinterval of unit length for each edge $e \in E$. We consider its midpoint which we denote by $\mathsf{mid}(I)$ and all input tasks i whose path $P(i)$ contains $\mathsf{mid}(I)$. These tasks are partitioned into polylogarithmically many groups such that within each group, all tasks have roughly the same weight and demand (up to a factor of $1+\epsilon$). For each group, we guess an estimate for the amount of capacity they use in the optimal solution on each edge in E. Given this, we compute the most profitable set of tasks for which these edge capacities are sufficient. Then, the remaining problem splits into two *independent* subproblems: one for all input tasks whose paths lie on the left of $\mathsf{mid}(I)$ and one for all input tasks whose paths lie on the right of $\mathsf{mid}(I)$. Therefore, we can easily recurse on these two subproblems and solve them independently.

Unfortunately, this approach does not extend to TWUFP. A natural adaption for TWUFP would be to consider all input tasks i for which the path $P(i)$ *in the optimal solution* contains the midpoint of I which we denote by $\mathsf{mid}(I)$. However, then the remaining problem does *not* split nicely into two independent subproblems: there can be a task i whose time window $\mathrm{tw}(i)$ contains $\mathsf{mid}(I)$ and which we could schedule entirely on the left of $\mathsf{mid}(I)$ or entirely on the right of $\mathsf{mid}(I)$ (see Fig. 1). Thus, i appears in the input of the left *and* the right subproblem, even though we are allowed to select i only once. Hence, these subproblems are no longer independent. Even more, this issue for i can arise again in each level of the recursion, which yields many interconnected subproblems.

Therefore, we use a different approach to obtain a $(2+\epsilon)$-approximation for TWUFP. We consider all input tasks i whose *time window* contains $\mathsf{mid}(I)$. By losing a factor of 2 on the weight of these tasks in the optimal solution, we can sacrifice either all of them that are scheduled on the left of $\mathsf{mid}(I)$ or all of them that are scheduled on the right of $\mathsf{mid}(I)$. We guess which case applies and we assume it in the following to be the latter case w.l.o.g. For the tasks i for which $P(i)$ contains $\mathsf{mid}(I)$ in the optimal solution we guess an estimate for the amount of capacity they use on the edges, similar to the QPTASs for UFP. Then we recurse on the subproblems corresponding to the left and the right of $\mathsf{mid}(I)$. For the right subproblem, we do not allow to select tasks whose time window uses $\mathsf{mid}(I)$. Therefore, our two subproblems are now independent! Let us consider the left subproblem, denote its corresponding interval by I^L and its midpoint by $\mathsf{mid}(I^L)$. If we continued natively in the same fashion, we would lose another factor of 2 on the weight of the tasks whose time windows contain $\mathsf{mid}(I)$ and $\mathsf{mid}(I^L)$, which we cannot afford. Instead, we employ a crucial new idea. We partition the mentioned tasks into polylogarithmically many groups such that within each group, all tasks have the same demand (using resource augmentation) and roughly the same weight (up to a factor of $1+\epsilon$). Since the time window of each such task i contains $\mathsf{mid}(I)$ and $\mathsf{mid}(I^L)$, we can schedule it freely within the interval $[\mathsf{mid}(I^L), \mathsf{mid}(I)]$. Therefore, for each group, we consider its tasks that are scheduled entirely during $[\mathsf{mid}(I^L), \mathsf{mid}(I)]$ in the optimal solution and round their processing times via linear grouping [21,22] to $O(1/\epsilon)$ different values. For each resulting combination of demand, weight, and rounded processing time, we guess for the right subproblem of I^L (i.e., the "left-right subproblem") how many tasks it schedules and give it these tasks as part of the input. To the left subproblem of I^L (i.e., the "left-left subproblem") we give the additional constraint that it needs to leave the corresponding number of tasks from each group unassigned. An important aspect is that the processing times of these tasks are *not* rounded in the left-left subproblem, but they *are* rounded in the left-right subproblem. This is a crucial difference of our approach in comparison to other rounding methods from the literature. Thanks to this technique, we avoid to lose another factor of 2 on the mentioned tasks. We apply it recursively for $O(\log m)$ levels and obtain an approximation ratio of $2 + \epsilon$ overall.

1.2 Related Work

The BAGUFP problem is a generalization of UFP where we are given the same input as in UFP plus a partition of the tasks into subsets, the *bags*, and we are allowed to select at most one task per bag. This also generalizes TWUFP by creating bags that model each possible way to schedule a task within its time window. It was already known that BAGUFP is APX-hard [20] (which is now also implied by our result). The current best approximation ratio for BAGUFP is $O(\log n/\log\log n)$ [13] (and 17 under the no-bottleneck assumption [9]). A constant factor approximation algorithm for BAGUFP (hence for TWUFP) is known for the cardinality case of the problem [13].

For the special case of TWUFP when all the demands and capacities are 1 (also known as the non-preemptive throughput maximization problem), the best-known approximation factor for this problem is 2 [4,5,7]. Notice that we give a $(2 + \varepsilon)$-approximation for a much more general problem, however, in quasi-polynomial time and using resource augmentation. In the cardinality case of the problem, there is an algorithm with an approximation ratio of $e/(e-1) + \varepsilon$ [12] (also for bags instead of time windows), which was later slightly improved in [19].

2 A $(2 + \varepsilon)$-Approximation for TWUFP

In this section, we present a $(2 + \varepsilon)$-approximation algorithm with a quasi-polynomial running time for TWUFP under $(1 + \varepsilon)$-resource augmentation, for any constant $\varepsilon > 0$. W.l.o.g. assume $1/\varepsilon \in \mathbb{N}$. Formally, let $(\mathsf{OPT}, P^*(\cdot))$ denote an optimal solution for the given instance. We compute a solution $(S, P(\cdot))$ consisting of a set of tasks S and for each task $i \in S$ a subpath $P(i) \subseteq \mathrm{tw}(i)$ of length $p(i)$ such that $w(\mathsf{OPT}) \leq (2 + \varepsilon)w(S)$ and $\sum_{i \in S: e \in P(i)} d(i) \leq (1 + \varepsilon)u(e)$ for every $e \in E$.

First, we use resource augmentation in order to round the edge capacities and task demands. We also round the tasks' weights.

Lemma 1. *Let $\alpha \geq 1$. Assume that there is an α-approximation algorithm running in time $T(n, \varepsilon)$ for the special case of* TWUFP *in which*

- *for each $e \in E$ we have that $u(e) \in [1, (n/\varepsilon)^{1/\varepsilon})$*
- *for each task $i \in T$ we have that*
  - *$d(i) \in [1, (n/\varepsilon)^{1/\varepsilon})$ and $d(i)$ is a power of $1 + \varepsilon$, and*
  - *$w(i) \in [1, n/\varepsilon]$ and $w(i)$ is a power of $1 + \varepsilon$.*

Then there is a $(1 + 5\varepsilon)\alpha$-approximation algorithm for TWUFP *under $(1 + 4\varepsilon)$-resource augmentation with a running time of $O(\mathrm{poly}(n)^{\frac{\log U}{\log n}} T(n, \varepsilon))$, where $U = \max_{e \in E} u(e)$.*

In the following, we will present a $(2+\varepsilon)$-approximation algorithm for the special case of TWUFP defined in Lemma 1; this yields a $(2+O(\varepsilon))$-approximation for arbitrary instances of TWUFP under $(1+O(\varepsilon))$-resource augmentation.

We first define a hierarchical decomposition of E. Assume w.l.o.g. that $m = |E|$ is a power of 2. We define a (laminar) family of subpaths $\mathcal{I}$ which intuitively form a tree. We will refer them also as *intervals*. There is one interval $I_r \in \mathcal{I}$ such that $I_r = E$ which intuitively forms the root of the tree. For each $I \in \mathcal{I}$, we will ensure that $|I|$ is a power of 2. Consider an interval $I \in \mathcal{I}$ with $|I| \geq 2$ and let $\mathsf{mid}(I)$ denote its middle vertex. Given I, we define recursively that there is an interval $I^L \in \mathcal{I}$ induced by the edges of I to the left of $\mathsf{mid}(I)$, and an interval $I^R \in \mathcal{I}$ induced by the edges of I to the right of $\mathsf{mid}(I)$. We say that I^L and I^R are the two *children* of I. We apply this definition recursively. As a result, for each edge $e \in E$ there is an interval $\{e\} \in \mathcal{I}$; such intervals form the leaves of our tree. We say that an interval $I \in \mathcal{I}$ is of *level* ℓ for some $\ell \in \mathbb{N}_0$ if $|I| = 2^{\log m - \ell}$, e.g., the interval I_r is of level 0. Note that we have $1 + \log m$ levels (for which there is an interval in $\mathcal{I}$).

We group the tasks into a polylogarithmic number of groups. For each task $i \in T$ we define that it is of *level* $\ell(i)$ if $\mathrm{tw}(i)$ is contained in an interval $I \in \mathcal{I}$ of level $\ell(i)$ but not in an interval $I' \in \mathcal{I}$ of level $\ell(i)+1$. We define $T_\ell := \{i \in T : \ell(i) = \ell\}$ for each integer ℓ. Also, for each combination of values w, d such that w represents a weight and d a demand, we define the set $T_{w,d} = \{i \in T : w(i) = w \wedge d(i) = d\}$. Note that there are only a polylogarithmic number of combinations for these pairs w, d for which $T_{w,d} \neq \emptyset$ by Lemma 1.

Recall that $(\mathsf{OPT}, P^*(\cdot))$ denotes an optimal solution. Consider a task $i \in \mathsf{OPT}$ of some level $\ell(i)$. Let I be the (unique) interval of level $\ell(i)$ that contains $\mathrm{tw}(i)$ and let I^L and I^R denote the children of I. Observe that $P^*(i)$ must satisfy one of the following three conditions:

1. $P^*(i)$ is contained in I^L, or
2. $\mathsf{mid}(I)$ is in the interior of $P^*(i)$, i.e., $\mathsf{mid}(I)$ is a vertex of $P^*(i)$ but it is neither the leftmost nor the rightmost vertex of $P^*(i)$, or
3. $P^*(i)$ is contained in I^R.

Let $\mathsf{OPT}^{(L)}, \mathsf{OPT}^{(M)}$, and $\mathsf{OPT}^{(R)}$ denote the tasks in OPT for which case 1, 2, and 3 applies, respectively. In particular, $w(\mathsf{OPT}) = w(\mathsf{OPT}^{(M)}) + w(\mathsf{OPT}^{(L)}) + w(\mathsf{OPT}^{(R)})$ and thus $w(\mathsf{OPT}^{(L)}) \leq \frac{1}{2}w(\mathsf{OPT})$ or $w(\mathsf{OPT}^{(R)}) \leq \frac{1}{2}w(\mathsf{OPT})$. The first step of our algorithm is to guess which of these cases applies. Assume w.l.o.g. that $w(\mathsf{OPT}^{(R)}) \leq \frac{1}{2}w(\mathsf{OPT})$. In the following, we will essentially aim to select only tasks from $\mathsf{OPT}^{(L)} \cup \mathsf{OPT}^{(M)}$. In this step we lose a factor of (up to) 2 on the obtained profit.

Formally, we say that a solution $(S, P(\cdot))$ is *left constrained* if the following holds for each task $i \in S$: For I being the unique interval of level $\ell(i)$ containing $\mathrm{tw}(i)$, we require that $P(i)$ is not contained in the right child of I, i.e., in the subinterval of I containing all edges of I on the right of $\mathsf{mid}(I)$. We will later define artificial tasks (not contained in T) for which this is not required. Our algorithm will return a left constrained solution by construction and we will compare its profit with the left constrained solution $\mathsf{OPT}^{(L)} \cup \mathsf{OPT}^{(M)}$.

Our algorithm is recursive. We start by describing the root subproblem in the next subsection; it is simpler than a generic subproblem, however it is convenient to introduce certain notions that will be needed also in the general case. In the subsequent subsection, we will describe a generic subproblem.

2.1 Root Subproblem

The root subproblem corresponds to the interval $I_r = E$ (of level 0). Let $(\overline{\mathsf{OPT}}_r, P^*(\cdot))$ be an optimal left constrained solution. Let $\overline{\mathsf{OPT}}_r^{(M)}$ be all the tasks $i \in \overline{\mathsf{OPT}}_r$ such that $\mathsf{mid}(I_r)$ is in the interior of $P^*(i)$. Consider a group $T_{w,d}$. We want to guess a set of *boxes* that delimit space that we reserve for tasks in $T_{w,d} \cap \overline{\mathsf{OPT}}_r^{(M)}$. Formally, a box b is characterized by a path $\mathrm{pb}(b) \subseteq E$, a height $h(b) > 0$, a demand $d(b) > 0$, and a weight $w(b) > 0$. We will consider only boxes b for which $w(b)$ is some task weight, $d(b)$ is some task demand, and $h(b)$ is some integer multiple of $d(b)$ upper bounded by $n \cdot d(b)$. In the following, we will assign certain sets of tasks to some boxes such that, intuitively, these tasks are stacked on top of each other inside the box and each of them has a weight of $w(b)$, a demand of $d(b)$, and their total demand is at most $h(b)$, i.e., they are at most $h(b)/d(b)$ many. Furthermore, it is possible to schedule each such task within the path of the box.

More explicitly, for any set of tasks S, using the notation $d(S) := \sum_{i\in S} d(i)$ and $w(S) := \sum_{i\in S} w(i)$, we say that S *fits inside a box* b if

- $d(S) \leq h(b)$,
- $|\mathrm{tw}(i) \cap \mathrm{pb}(b)| \geq p(i)$ for each $i \in S$, and
- $w(i) = w(b)$ and $d(i) = d(b)$ for each $i \in S$.

We use the next lemma to show that there is a set of constantly many boxes $B_{w,d}$ in which almost all tasks in $T_{w,d} \cap \overline{\mathsf{OPT}}_r^{(M)}$ fit, i.e., up to a factor of $1+\varepsilon$ in the profit. Also, the capacity used by these boxes is at most the capacity used by the tasks in $T_{w,d} \cap \overline{\mathsf{OPT}}_r^{(M)}$ on each edge. Formally, we apply the following lemma to the set $T_{w,d} \cap \overline{\mathsf{OPT}}_r^{(M)}$ and obtain a set of at most $1/\varepsilon^2$ boxes that we denote by $B_{w,d}$. We prove this lemma with a similar linear grouping scheme as in [21] (see also [22]) to show that we can approximate the demand profile of such a set of tasks by a simpler profile with a constant number of steps only.

Lemma 2. *Let T' be a set of tasks, all having identical weight w and identical demand d, and a schedule $P(i) \subseteq E$ for every $i \in T'$ such that there is an edge f that is contained in each path $P(i)$. Then, there exists a set $S \subseteq T'$, a set of boxes B with $|B| \leq 1/\varepsilon^2$, and a partition of S into sets $\{S_b\}_{b\in B}$ such that:*

(1) $\sum_{b\in B: e\in \mathrm{pb}(b)} h(b) \leq \sum_{i\in T': e\in P(i)} d(i)$ for each $e \in E$,
(2) $w(S) \geq (1 - O(\varepsilon))w(T')$,
(3) the tasks in S_b fit into b for each $b \in B$.

We guess the at most $1/\varepsilon^2$ boxes $B_{w,d}$. For each box $b \in B_{w,d}$, let $S_b \subseteq T_{w,d} \cap \overline{\mathsf{OPT}}_r^{(M)}$ be the subset of $T_{w,d} \cap \overline{\mathsf{OPT}}_r^{(M)}$ that is assigned to b when applying Lemma 2 to $T_{w,d} \cap \overline{\mathsf{OPT}}_r^{(M)}$; we guess $|S_b| =: n_b$. We do this procedure for the set $T_{w,d}$ for each combination of w and d. Let B_0 denote the set of all boxes that we guessed this way.

Let I_r^L and I_r^R be the two children of I_r. We recurse on a *left subproblem* corresponding to I_r^L and on a *right subproblem* corresponding to I_r^R. In the right subproblem, we are given as input the set consisting of each task $i \in T$ such that $\text{tw}(i) \subseteq I_r^R$. Note that such a task must have level 1 or larger, so the tasks of level 0 are not considered in the subproblem corresponding to I_r^R. The capacity of each edge $e \in I_r^R$ is defined as $u'(e) := u(e) - \sum_{b \in B_0 : e \in \text{pb}(b)} h(b)$, that is, the boxes in B_0 use a certain amount of the capacities of the edges that we cannot use. The goal is to compute a feasible solution to this smaller instance, i.e., a set of tasks Q^R with a path $P(i) \subseteq \text{tw}(i) \subseteq I_r^R$ for each task $i \in Q^R$ that respects the edge capacities $u'(e)$. The objective is to maximize $w(Q^R)$.

In the left subproblem, we are given as input the set of all tasks $i \in T$ such that $\text{tw}(i) \subseteq I_r^L$ or $\ell(i) = 0$; let T^L denote this set of tasks. The capacity of each edge $e \in I_r^L$ is defined as $u'(e) := u(e) - \sum_{b \in B_0 : e \in \text{pb}(b)} h(b)$ analogously to the right subproblem. Also, we are given the set of boxes B_0 and a value n_b for each box b. The goal is to select a subset $Q \subseteq T^L$, a partition of Q into a set Q^L and a set Q_b for each box $b \in B_0$, and a path $P(i)$ for each task $i \in Q$ such that

- the set Q^L with the path $P(i) \subseteq \text{tw}(i) \cap I_r^L$ for each task $i \in Q^L$ forms a feasible solution for the given edge capacities $u'(e)$, i.e., for each edge $e \in I_r^L$ we have $\sum_{i \in Q^L : e \in P(i)} d(i) \leq u'(e)$, and
- for each box $b \in B_0$ we have that $|Q_b| = n_b$, Q_b fits into b, and $P(i) \subseteq \text{tw}(i) \cap \text{pb}(b)$ for each $i \in Q_b$.

Observe that the given boxes B_0 do not interact at all with the edges I_r^L and their given capacities. However, the subpath $\text{pb}(b)$ for a box $b \in B_0$ is important since a task i fits into a box b only if $|\text{tw}(i) \cap \text{pb}(b)| \geq p(i)$.

Given a solution to the left and the right subproblems, we combine them to a solution to the overall problem in the obvious way: the solution is given by the tasks $Q = Q^L \cup Q^R \cup (\cup_{b \in B_0} Q_b)$ with the respective paths $P(i)$. Each of our quasi-polynomially many guesses yields a candidate solution; among all feasible candidate solutions, we output the solution with maximum profit $w(Q)$.

2.2 Arbitrary Subproblems

In the following, we describe our routine for arbitrary subproblems. The reader may think of the subproblem for the left child of the root I_r, i.e., the interval I_r^L (which already incorporates all the challenges of a generic subproblem). We assume that the input consists of

- an interval $I \in \mathcal{I}$ of level ℓ with a capacity $u'(e)$ for each edge $e \in I$,
- a set of tasks T', and

- a set of boxes B and a value n_b for each box $b \in B$.

We remark that, for a box $b \in B$, not necessarily $\text{pb}(B) \subseteq I$. Furthermore, given a task $i \in T'$, it might happen that $\text{tw}(i) \not\subseteq I$; in this case, we will ensure that $\text{tw}(i)$ contains the rightmost edge of I by construction. We remark that it might happen that $i \notin T$, namely i is not one of the input task. This happens because during the process, we introduce some *artificial tasks* whose role will become clear later. Like the regular tasks, each artificial task i has a weight $w(i)$, a demand $d(i)$, a length $p(i)$, and a time window $\text{tw}(i)$.

Let I^L and I^R denote the two children of I and let $\mathsf{mid}(I)$ denote the middle vertex of I. The goal is to compute a set of tasks $Q \subseteq T'$, a partition of Q into sets Q_I and a set Q_b for each box $b \in B$, and a path $P(i)$ for each task $i \in Q$ such that

- the tasks in Q_I with their paths $P(i) \subseteq \text{tw}(i) \cap I$ obey the edge capacities of I, i.e., for each edge $e \in I$ we have $\sum_{i \in Q_I : e \in P(i)} d(i) \leq u'(e)$,
- for each box $b \in B$ the tasks in Q_b fit into B, $|Q_b| = n_b$, and $P(i) \subseteq \text{tw}(i) \cap \text{pb}(b)$ for each $i \in Q_b$.

The objective is to maximize $w(Q_I)$, i.e., the weight of the tasks in Q_I. Note that the tasks assigned to the boxes B do not yield any profit in this subproblem. Intuitively, we accounted the profit of such tasks already in subproblems of higher levels in the recursion.

Let $(\overline{\mathsf{OPT}}, P^*(\cdot))$ denote an optimal left constrained solution to the given subproblem. We denote by $\overline{\mathsf{OPT}}_I$ and $\{\overline{\mathsf{OPT}}_b\}_{b \in B}$ the corresponding partition of $\overline{\mathsf{OPT}}$ (i.e., $\overline{\mathsf{OPT}}_I$ consists of the tasks in $\overline{\mathsf{OPT}}$ scheduled on I and $\overline{\mathsf{OPT}}_b$ contains the task in $\overline{\mathsf{OPT}}$ that are placed inside b). The base cases of our recursion are there subproblems where $|I| = 1$. In this case, we can solve the subproblem exactly in quasi-polynomial time. We guess for each combination of a weight w and a demand d how many tasks with this weight and demand are contained in $\overline{\mathsf{OPT}}_I$. Then, the remaining problem can be reduced easily to an instance of b-matching.

Lemma 3. *Any subproblem with $|I| = 1$ can be solved exactly in $n^{O(WD)}$ time, where W and D denote the number of distinct weights and demands, respectively.*

Assume now that $|I| > 1$. For each task $i \in \overline{\mathsf{OPT}}_I$ there are three possibilities:

1. $P^*(i)$ is contained in I^L, or
2. $\mathsf{mid}(I)$ is in the interior of $P^*(i)$, i.e., $\mathsf{mid}(I)$ is a vertex of $P^*(i)$ but it is neither the leftmost nor the rightmost vertex of $P^*(i)$, or
3. $P^*(i)$ is contained in I^R.

Let $\overline{\mathsf{OPT}}^{(L)}, \overline{\mathsf{OPT}}^{(M)}$, and $\overline{\mathsf{OPT}}^{(R)}$ denote the tasks in $\overline{\mathsf{OPT}}_I$ for which case 1, 2 and 3 applies, respectively. Note that the partition of $\overline{\mathsf{OPT}}$ into $\overline{\mathsf{OPT}}^{(L)}, \overline{\mathsf{OPT}}^{(M)}$, and $\overline{\mathsf{OPT}}^{(R)}$ is different from the partition of OPT into $\mathsf{OPT}^{(L)}, \mathsf{OPT}^{(M)}$, and $\mathsf{OPT}^{(R)}$, since the former is with respect to $\mathsf{mid}(I)$ for each task $i \in \overline{\mathsf{OPT}}$ of *any* level. Recall also that $\overline{\mathsf{OPT}}$ is left constraint and, hence, assuming that I is an interval of level ℓ, for no task i of level ℓ, its scheduled path $P^*(I)$ in the optimal solution is contained in $\overline{\mathsf{OPT}}^{(R)}$. More precisely, for each task $i \in \overline{\mathsf{OPT}}^{(R)}$, it must be the case that either $\mathrm{tw}(i) \subseteq I^R$ (so i is of level $\ell+1$ or deeper), $I^R \subseteq \mathrm{tw}(i)$ (so i is of level $\ell-1$ or higher) or the leftmost edge of $\mathrm{tw}(i)$ is in I^R but not all edges of $\mathrm{tw}(i)$ (so i is of level $\ell-1$ or higher).

First, we guess boxes for the tasks in $\overline{\mathsf{OPT}}^{(M)}$. Formally, for each combination of a weight w and demand d, we apply Lemma 2 to the set $\{i \in \overline{\mathsf{OPT}}^{(M)} : w(i) = w \wedge d(i) = d\}$. Let $\bar{B}_{w,d}$ denote the resulting set of boxes and for each box $b \in \bar{B}_{w,d}$ let $\overline{\mathsf{OPT}}_b^{(M)}$ be the resulting set of tasks. We guess the boxes in $\bar{B}_{w,d}$ and $|\overline{\mathsf{OPT}}_b^{(M)}| =: n_b$ for each box $b \in \bar{B}_{w,d}$. Let $\bar{B}$ denote the union of all sets of boxes $\bar{B}_{w,d}$ that we guessed in this way.

Let $\overleftarrow{T}' \subseteq T'$ denote all tasks in $i \in T'$ such that $I^R \subseteq \mathrm{tw}(i)$ and $I^L \cap \mathrm{tw}(i) \neq \emptyset$, i.e., task whose time windows stick into I from the right and intersect I^L. For the subproblem for I_r^L, the reader may imagine that $\overleftarrow{T}'$ are all tasks in T_0 whose time windows start in the left child of I_r^L. For a task $i \in \overleftarrow{T}' \cap \overline{\mathsf{OPT}}_I$ it is possible that $i \in \overline{\mathsf{OPT}}^{(L)}$ or that $i \in \overline{\mathsf{OPT}}^{(R)}$. Therefore, at first glance it would make sense to pass i to our subproblem for I^L and to our subproblem for I^R. However, we must avoid that our subproblem for I^L and our subproblem for I^R both select i. On the other hand, we do not want to sacrifice a factor of 2 by, e.g., omitting $\overline{\mathsf{OPT}}^{(L)}$ or $\overline{\mathsf{OPT}}^{(R)}$. For example, if $\overleftarrow{T}' = T_0$ in the subproblem for I_r^L, then this would lose a second factor of 2 on the profit of the tasks in T_0, while we want to bound our approximation ratio by $2 + O(\varepsilon)$.

We execute the following steps instead:

I. Perform a linear grouping step in which we round up the task lengths so that there are at most polylogarithmically many distinct lengths in $\overleftarrow{T}' \cap \overline{\mathsf{OPT}}^{(R)}$;
II. For each combination of a demand, weight and (rounded) length, guess how many such tasks are contained in $\overleftarrow{T}' \cap \overline{\mathsf{OPT}}^{(R)}$;
III. Give a suitable number of these rounded tasks as *artificial input tasks* to our right subproblem for I^R;
IV. Enforce our left subproblem for I^L to leave sufficiently many tasks from $\overleftarrow{T}'$ unassigned such that we can schedule these tasks later in the places in which the solution to the right subproblem schedules the artificial tasks (we model this requirement for I^L via suitable additional box constraints).

Formally, for each combination of a weight w and demand d we apply the following lemma to the set $\overleftarrow{T}'_{w,d} = \{i \in \overleftarrow{T}' \cap \overline{\mathsf{OPT}}^{(R)} : w(i) = w \wedge d(i) = d\}$.

Lemma 4. *Let T' be a set of tasks, all having identical weight w and identical demand d, and a schedule $P'(i) \subseteq I \subseteq \mathrm{tw}(i)$ for every $i \in T'$. Then, there exists a set of* artificial tasks *$\tilde{T}$ (not contained in T), a set of boxes $\tilde{B}$ and a value $n_b \in \mathbb{N}$ for each box $b \in \tilde{B}$ with $\sum_{b \in \tilde{B}} n_b = |\tilde{T}|$ such that*

(1) $(1-\varepsilon)|T'| \leq |\tilde{T}| \leq |T'|$, *and the tasks in* $\tilde{T}$ *have at most* $1/\varepsilon$ *distinct lengths,*
(2) $\mathrm{tw}(i) = I$, $d(i) = d$, *and* $w(i) = w$ *for each task* $i \in \tilde{T}$,
(3) there is a solution $(\tilde{T}, \tilde{P}(\cdot))$ *for* $\tilde{T}$ *such that* $\sum_{i \in \tilde{T}: e \in \tilde{P}(i)} d(i) \leq \sum_{i \in T': e \in P'(i)} d(i)$ *for each edge* $e \in I$,
(4) there is a collection $\{Q'_b\}_{b \in \tilde{B}}$ *of pairwise disjoint subsets of* T' *such that for each* $b \in \tilde{B}$ *the set* Q'_b *fits into* b *and* $|Q'_b| = n_b$,
(5) given a combination of a solution $(\tilde{S}, \tilde{P}(\cdot))$ *for a subset* $\tilde{S} \subseteq \tilde{T}$ *and any collection of pairwise disjoint sets of tasks* $\{Q_b\}_{b \in \tilde{B}}$, *in which for each box* $b \in \tilde{B}$, *the set* Q_b *fits into* b, $|Q_b| = n_b$, *and each task* $i \in Q_b$ *has a time window containing the leftmost edge of* I*; then there is an injective function* $f : \tilde{S} \to \bigcup_{b \in \tilde{B}} Q_b$ *such that* $f(i)$ *can be scheduled instead of* i, *i.e.,*$|\tilde{P}(i) \cap \mathrm{tw}(f(i))| \geq p(f(i))$, *for each* $i \in \tilde{S}$.

We sketch now how we will apply Lemma 4. Since we are unable to guess each set $\overleftarrow{T}'_{w,d}$ at this stage, we rather guess the corresponding set of artificial tasks $\tilde{T}_{w,d}$ according to Lemma 4. Notice that this is doable in polynomial time (per pair (w, d)) since it is sufficient to guess the distinct lengths $p_1, ..., p_{1/\varepsilon}$ and the respective number of tasks $\tilde{n}_1, ..., \tilde{n}_{1/\varepsilon}$. We will pass on the tasks $\tilde{T}_{w,d}$ to our right subproblem as placeholders: intuitively, they will reserve some capacity on the right subproblem which will eventually be occupied by actual tasks (potentially the tasks $\overleftarrow{T}'_{w,d}$) computed in the left subproblem: this replacement exploits Property 5 of the lemma. Also, we guess the boxes $\tilde{B}_{w,d}$ and the number $n_b \in \mathbb{N}$ for each $b \in \tilde{B}_{w,d}$ corresponding to $\tilde{T}_{w,d}$, which also takes polynomial time. These boxes will be passed on to the left subproblem: intuitively, the tasks placed in these boxes in the left subproblem will replace the artificial tasks in $\tilde{T}_{w,d}$.

Let $\tilde{T}$ denote the union of all sets $\tilde{T}_{w,d}$ guessed in this way and let $\tilde{B}$ denote the union of all the sets of boxes $\tilde{B}_{w,d}$ guessed in this way. We will call the tasks in $\tilde{T}$ *artificial tasks.*

We will recurse on a left and a right subproblem, corresponding to I^L and I^R, respectively. For each box $b \in B$, we intuitively guess how many input tasks from the left and the right subproblem are assigned to b in $\overline{\mathsf{OPT}}$. Formally, we guess values $n_b^L, n_b^R \in \mathbb{N}_0$ where n_b^L is the number of tasks $i \in \overline{\mathsf{OPT}}_b$ for which $\mathrm{tw}(i) \cap I^L \neq \emptyset$ and n_b^R is the number of tasks $i \in \overline{\mathsf{OPT}}_b$ for which $\mathrm{tw}(i) \cap I^L = \emptyset$ (in particular, $n_b^L + n_b^R = n_b$). The input for the left subproblem consists of

- the interval I^L, where each edge $e \in I^L$ has capacity $u'(e) - \sum_{b \in \bar{B}: e \in \mathrm{pb}(b)} h(b)$,
- the task set $\{i \in T' : \mathrm{tw}(i) \cap I^L \neq \emptyset\}$,
- the boxes $B \cup \bar{B} \cup \tilde{B}$, the value n_b for each box $b \in \bar{B} \cup \tilde{B}$, and the value n_b^L for each box $b \in B$.

The input for the right subproblem consists of

- the interval I^R, where each edge $e \in I^R$ has capacity $u'(e) - \sum_{b \in \bar{B}: e \in \mathrm{pb}(b)} h(b)$,
- the task set $\{i \in T' : \mathrm{tw}(i) \cap I^L = \emptyset\} \cup \tilde{T}$,
- the boxes B and the value n_b^R for each box $b \in B$.

Suppose that recursively we computed a solution to the left and the right subproblem. We combine them to a solution to the (given) subproblem corresponding to I as follows. Let $Q_{I^L}, \{Q_b\}_{b \in B \cup \bar{B} \cup \tilde{B}}$ denote the computed tasks in the left subproblem, and let $Q'_{I^R}, \{Q'_b\}_{b \in B}$ denote the computed tasks in the right subproblem. For each task i in these computed sets of tasks, the returned solutions include a computed path $P(i)$. We compute the injective function f due to the last property of Lemma 4 by solving a b-matching instance. If such a function f does not exist, we simply discard the considered combination of guesses.

We want to return a solution that consists of the tasks in $Q_{I^L} \cup \left(\bigcup_{b \in \bar{B}} Q_b\right) \cup (Q_{I^R} \setminus \tilde{T})$ and additionally a set of tasks $\tilde{Q}$ that we use to replace the artificial tasks in $\tilde{T}$ in the solution Q_{I^R}. For each task $i \in Q_{I^L} \cup \left(\bigcup_{b \in \bar{B}} Q_b\right) \cup (Q_{I^R} \setminus \tilde{T})$ we keep the path $P(i)$ that we obtained from the solution of the left or right subproblem, respectively. We want to replace the tasks in $Q_{I^R} \cap \tilde{T}$ and, to this end, we compute the set of tasks $\tilde{Q} := f(Q_{I^R} \cap \tilde{T}) \subseteq \bigcup_{b \in \tilde{B}} Q_b$ and define paths for them as follows. For each task $i \in Q_{I^R} \cap \tilde{T}$ we schedule the task $f(i) \in \tilde{Q}$ within the edges of $P(i)$, i.e., we compute a path $P(f(i))$ for $f(i)$ such that $P(f(i)) \subseteq P(i)$. This is possible since $|P(i) \cap \mathrm{tw}(f(i))| \geq p(f(i))$ due to the last property of Lemma 4. Notice that this replacement does not violate the edge capacities since we omit $i \in \tilde{T}$ from our solution and $d(i) = d(f(i))$.

This yields a candidate solution consisting of the tasks $Q_I := Q_{I^L} \cup \left(\bigcup_{b \in \bar{B}} Q_b\right) \cup (Q_{I^R} \setminus \tilde{T}) \cup \tilde{Q}$ and for each box $b \in B$ the tasks in $Q_b \cup Q'_b$ and their respective computed paths $P(i)$. Each combination of our quasi-polynomial number of guesses yields a candidate solution. We reject a candidate if the resulting solution is infeasible; among the remaining candidate solution, we return the solution that maximizes $w(Q_I)$. Recall here that $\tilde{Q}$ are real tasks that we used to replace the artificial tasks $\tilde{T}$; we count the profit of the tasks in $\tilde{Q}$ but not the profit of $\tilde{T}$. Also, we do not consider the profit of the tasks in B.

2.3 Analysis

We argue that our algorithms runs in quasi-polynomial time and that it returns a (feasible) $(2+\varepsilon)$-approximate solution. The bound on the running time holds easily since in each recursive call we make quasi-polynomially many guesses and the recursion depth is $O(\log m)$, which yields the total quasi-polynomial running time bound. Recall that we reject any infeasible candidate solution, so our returned solution is assuredly feasible.

Lemma 5. *Let W be the number of different profits, and let D be the number of different demands of the tasks in our* TWUFP *instance. Our algorithm runs in time $O((mn)^{O(WD\log^2 m)/\varepsilon^2})$ and its returned solution is always feasible.*

We now bound our approximation ratio. Assuming now that we guess all boxes b, their corresponding values n_b, n_b^L, n_b^R and the artificial tasks correctly at every stage of the algorithm (i.e., that match Lemma 2, Lemma 4, and $\overline{\mathsf{OPT}}_r$), at each level we lose at most an $O(\varepsilon)$-fraction of the profit when we replace the artificial tasks by real tasks and place tasks in the boxes (see Lemma 4). Since we have $O(\log m)$ levels, this yields a profit of at least $(1 + O(\varepsilon \log m))^{-1} \cdot w(\overline{\mathsf{OPT}}_r) \geq (2+O(\varepsilon \log m))^{-1} w(\mathsf{OPT})$. We rescale ε by a factor $\log m$ to obtain our approximation ratio of $2 + O(\varepsilon)$.

Theorem 2. *For any $\varepsilon > 0$, there is a $(2+\varepsilon)$-approximation for* TWUFP *under $(1 + O(\varepsilon))$-resource augmentation with running time $O\big((mn)^{O(\log^2 n \log^5 m/\varepsilon^4)} \cdot \log U\big)$, where $U := \max_{e \in E} u(e)$ is the maximum capacity of the instance.*

We adopt our algorithm to a $(1 + \varepsilon)$-approximation for SPANUFP under resource augmentation: we pad E with dummy edges on the right such that all feasible solutions become left constrained; thus, we avoid losing the factor of 2.

Theorem 3. *For any $\varepsilon > 0$ there is a QPTAS for* SPANUFP *under $(1+O(\varepsilon))$-resource augmentation.*

Acknowledgments. The authors thank the anonymous reviewers for their helpful comments. Fabrizio Grandoni, Edin Husić and Antoine Tinguely were supported by the Swiss National Science Foundation (SNSF) Grant 200021 200731/1.

References

1. Anagnostopoulos, A., Grandoni, F., Leonardi, S., Wiese, A.: A mazing 2+ε approximation for unsplittable flow on a path. In: SODA, pp. 26–41 (2014)
2. Bansal, N., Chakrabarti, A., Epstein, A., Schieber, B.: A quasi-PTAS for unsplittable flow on line graphs. In: STOC, pp. 721–729 (2006)
3. Bansal, N., Friggstad, Z., Khandekar, R., Salavatipour, R.: A logarithmic approximation for unsplittable flow on line graphs. In: SODA, pp. 702–709 (2009)
4. Bar-Noy, A., Guha, S., Naor, J., Schieber, B.: Approximating the throughput of multiple machines in real-time scheduling. SIAM J. Comput. **31**(2), 331–352 (2001). https://doi.org/10.1137/S0097539799354138
5. Bar-Noy, A., Bar-Yehuda, R., Freund, A., Naor, J., Schieber, B.: A unified approach to approximating resource allocation and scheduling. J. ACM (JACM) **48**(5), 1069–1090 (2001)
6. Batra, J., Garg, N., Kumar, A., Mömke, T., Wiese, A.: New approximation schemes for unsplittable flow on a path. In: SODA, pp. 47–58 (2015). https://doi.org/10.1137/1.9781611973730.5
7. Berman, P., DasGupta, B.: Multi-phase algorithms for throughput maximization for real-time scheduling. J. Comb. Optim. **4**(3), 307–323 (2000). https://doi.org/10.1023/A:1009822211065
8. Bonsma, P., Schulz, J., Wiese, A.: A constant-factor approximation algorithm for unsplittable flow on paths. SIAM J. Comput. **43**, 767–799 (2014)
9. Chakaravarthy, V.T., Choudhury, A.R., Gupta, S., Roy, S., Sabharwal, Y.: Improved algorithms for resource allocation under varying capacity. In: ESA, pp. 222–234 (2014)
10. Chlebík, M., Chlebíková, J.: Complexity of approximating bounded variants of optimization problems. Theor. Comput. Sci. **354**(3), 320–338 (2006). https://doi.org/10.1016/j.tcs.2005.11.029

11. Chrobak, M., Woeginger, G., Makino, K., Xu, H.: Caching is hard, even in the fault model. In: ESA, pp. 195–206 (2010)
12. Chuzhoy, J., Ostrovsky, R., Rabani, Y.: Approximation algorithms for the job interval selection problem and related scheduling problems. Math. Oper. Res. **31**(4), 730–738 (2006). https://doi.org/10.1287/MOOR.1060.0218
13. Grandoni, F., Ingala, S., Uniyal, S.: Improved approximation algorithms for unsplittable flow on a path with time windows. In: WAOA, pp. 13–24 (2015)
14. Grandoni, F., Mömke, T., Wiese, A.: Faster $(1+\epsilon)$-approximation for unsplittable flow on a path via resource augmentation and back. In: ESA, vol. 204, pp. 49:1–49:15 (2021)
15. Grandoni, F., Mömke, T., Wiese, A.: A PTAS for unsplittable flow on a path. In: STOC, pp. 289–302. ACM (2022)
16. Grandoni, F., Mömke, T., Wiese, A.: Unsplittable flow on a path: The game! In: SODA, pp. 906–926. SIAM (2022)
17. Grandoni, F., Mömke, T., Wiese, A., Zhou, H.: To augment or not to augment: solving unsplittable flow on a path by creating slack. In: SODA, pp. 2411–2422 (2017)
18. Grandoni, F., Mömke, T., Wiese, A., Zhou, H.: A $(5/3 + \epsilon)$-approximation for unsplittable flow on a path: placing small tasks into boxes. In: STOC, pp. 607–619 (2018)
19. Im, S., Li, S., Moseley, B.: Breaking 1–1/e barrier for nonpreemptive throughput maximization. SIAM J. Discret. Math. **34**(3), 1649–1669 (2020). https://doi.org/10.1137/17M1148438
20. Spieksma, F.: On the approximability of an interval scheduling problem. J. Sched. **2**, 215–227 (1999)
21. de la Vega, W.F., Lueker, G.S.: Bin packing can be solved within $1 + \varepsilon$ in linear time. Combinatorica **1**, 349–355 (1981)
22. Williamson, D.P., Shmoys, D.B.: The Design of Approximation Algorithms. Cambridge University Press (2011)
23. Woeginger, G.J.: There is no asymptotic PTAS for two-dimensional vector packing. Inf. Process. Lett. **64**(6), 293–297 (1997). https://doi.org/10.1016/S0020-0190(97)00179-8

A Constant-Factor Approximation for Pairwise Fair k-Center Clustering

Sayan Bandyapadhyay[1](✉), Tianzhi Chen[1](✉), Zachary Friggstad[2](✉), and Mahya Jamshidian[2](✉)

[1] Portland State University, Portland, OR, USA
{sayanb,chench}@pdx.edu
[2] University of Alberta, Edmonton, AB, Canada
{zacharyf,mjamshidian}@ualberta.ca

Abstract. We study k-center clustering with an additional fairness constraint: for any cluster Γ and any two groups a, b, the number of points of group a in Γ is at most t times the number of points of group b in Γ. The problem is known to admit $O(1)$-approximation in some special cases including (i) when the number of groups $c = 2$ and t is an integer, and (ii) when $t = 1$ and c is arbitrary. We obtain the first $O(1)$-approximation for this problem with arbitrary c and integer t. Our algorithm is based on a novel tree-based LP rounding scheme which also works in a more general setting with outliers. In contrast, we show an integrality gap of $\Omega(k)$ when t is not an integer, even when $c = 2$.

Keywords: k-center · fair clustering · LP rounding

1 Introduction

In this work, we study the following generalization of k-center.

> **Pairwise Fair k-center.** We are given a set of points C, called *clients*, partitioned into c disjoint *color classes* (or *groups*) $C_1, \ldots, C_c$, a set of points F, called *facilities*, a metric $d : C \cup F \rightarrow \mathbb{R}^+ \cup \{0\}$, a fairness parameter $t \geq 1$, and an integer $k \geq 1$. For any subset $S \subseteq F$, an assignment $\sigma : C \rightarrow S$ is called *fair* if for any two colors $1 \leq a, b \leq c$ and any $i \in S$, we have $|\sigma^{-1}(i) \cap C_a| \leq t \cdot |\sigma^{-1}(i) \cap C_b|$, i.e., the number of points of color a assigned to i is at most t times the number of points of color b assigned to i. The cost of σ is $\max_{j \in C} d(j, \sigma(j))$, i.e., the maximum distance any client travels. The goal is to find a subset S of F with at most k facilities (or *centers*) and a fair assignment $\sigma : C \rightarrow S$ whose cost is minimized over all such subsets of F and assignments of the clients.

The work supported by an NSERC Discovery Grant and the NSF grant 2311397.

N. Megow and A. Basu (Eds.): IPCO 2025, LNCS 15620, pp. 43–57, 2025.
https://doi.org/10.1007/978-3-031-93112-3_4

We assume $|C_a| \leq t \cdot |C_b|$ for any two colors a, b (i.e. the input itself is fair) or else there is no feasible solution. The partition $\cup_{i \in S}\{\sigma^{-1}(i)\}$ of C w.r.t. a fair σ is called a *fair clustering* of the clients. PAIRWISE FAIR k-CENTER was introduced by Chierichetti et al. [15] who considered the problem with two colors. The problem is NP-hard to approximate within a factor less than 2, as it generalizes (vanilla) k-center clustering [18]. Chierichetti et al. designed a polynomial-time 4-approximation for this problem with two colors (red and blue) and integer t, based on a *fairlet decomposition* method. Roughly, a fairlet is a *t-fair* set of points, i.e., the ratio of red and blue is at most t and at least $1/t$. In fact, it is possible to partition C into fairlets using a minimum cost flow algorithm such that the points in each fairlet are in a 2·OPT radius ball, where OPT is the optimal cost. Then each such fairlet can be treated as a point and a clustering of them can be computed using a 2-approximation algorithm for k-center [18], which leads to a fair clustering of C of cost at most 4·OPT.

Unfortunately, it is not clear how to extend the algorithm of Chierichetti et al. [15] to more than two colors. In particular, it is not clear how to efficiently compute a fairlet decomposition for three colors where the points in each fairlet is in a sufficiently small radius ball. Such a decomposition can be computed easily in two special cases: (i) $t = 1$: here again it is possible to compute the fairlets using network flow [8]; (ii) there is a fixed color a such that a contributes the minimum number of points to each optimal cluster. This version was considered by Bercea et al. [6] who obtained a 4-approximation for any arbitrary c colors using a similar method. We note that in PAIRWISE FAIR k-CENTER, it is not guaranteed that such a color a exists. Hence, the approximability status of this problem remained open. We also point out that, PAIRWISE FAIR k -MEDIAN has also been studied for integers t. The best-known approximation factor is $O(t)$ when $c = 2$ [15] and a function of t, c, and k for arbitrary number of colors c [3].

Our Results and Contributions. We study PAIRWISE FAIR k-CENTER with the focus of designing approximation algorithms whose output clustering fully satisfies the fairness constraints. We prove the following theorem.

Theorem 1. *There is a polynomial-time 33-approximation algorithm for PAIRWISE FAIR k-CENTER when t is an integer.*

Our algorithm is based on rounding a solution to a natural LP relaxation and has two major steps. The first step is a modification of standard filtering used for k-center rounding. This filtering process produces a set of up to k balls with radius at most 5·OPT and a fractional fair assignment of points to these balls. The second step aims to obtain a fair integral assignment. To do this, we design a new tree-based rounding technique, which is one of our main contributions.

We consider a suitably defined rooted tree over the centers of the balls. As with other tree-based rounding schemes for k-median variants, [7,17,24], nodes are processed in a bottom-up manner. Once processed, each node is assigned a set of points satisfying the constraints of the problem. Excess points are passed up to the parent. In all such schemes, it has been shown that a point is never passed up twice, making it to be assigned within one hop of its original center.

In our approach, we devise a *fractional fairlet decomposition* scheme: For each node i, we will take its currently-assigned fractional flow plus some unassigned fractional flow from its children; We calculate a t-fair integer vector and assign exactly this much from each color to i. A small (t-fair and fractional) amount may remain unassigned, that amount is propagated up the tree to i's parent. Moreover, we prove that no flow that is initially assigned to i is sent more than three hops up the tree. This algorithm is described in Sect. 3.

We also study a version with outliers. Namely, we are given some parameter $\gamma \geq 0$ and the goal is to pick a set of clients $C' \subseteq C$ with $|C'| \geq |C| - \gamma$ and then to find a PAIRWISE FAIR k-CENTER solution for C' while minimizing the maximum distance traveled by any $j \in C'$. Our earlier algorithm can be extended to the more general setting with outliers (Sect. 4).

Theorem 2. *There is a polynomial-time* 63*-approximation algorithm for* PAIRWISE FAIR k-CENTER *with outliers when t is an integer.*

The approximation factor is slightly larger, but is still a constant. This case is more challenging, as here points can be only partly assigned to the facilities by the LP. We note that this might be a more practical model where the input itself does not have to be t-fair (i.e. some colors are overrepresented in the input).

We also consider some negative results. In particular, for non-integer values t it could be that the LP we use has an integrality gap of $\Omega(k)$ (Sect. 5). Further, our LP for the outlier setting is easily generalized to the variant of the problem where we have a separate outlier bound for each color class. We show the integrality gap is not bounded by any function of k and n.

Related Work. Fair clustering has been studied with all popular objectives [14]. Due to space constraints, below we focus on results for the k-center objective. Besides fairlet decomposition, a tree based LP rounding technique has been used for k-center with a different fairness model, called *exact fairness* [7]. In the exact fairness model, one seeks a clustering where for any fair cluster T and color a, $|C_a \cap T|/|T| = |C_a|/|C|$. Notably, such techniques were already popular for another constrained version, called capacitated k-center [17,24]. Bercea et al. [7] obtained a 5-approximation for exactly fair k-center improving an earlier guarantee of 14 by Rösner and Schmidt [27]. In independent work, Bercea et al. [7] and Bera et al. [5] introduce another fair clustering model. Given balance parameters $l_a, u_a \in [0, 1]$ for $1 \leq a \leq c$, a clustering is called fair if, in each cluster T, the number of points of color a is at least $l_a \cdot |T|$ and at most $u_a \cdot |T|$, for all $1 \leq a \leq c$. Bercea et al. gave a polynomial-time 3-approximation for this model, albeit with a bi-criteria kind of guarantee where in each cluster T, the number of points of color a is at least $l_a \cdot |T| - 1$ and at most $u_a \cdot |T| + 1$, for all $1 \leq a \leq c$. The approach of Bera et al. also leads to the same bi-criteria kind of guarantee. Ahmadian et al. [1] considered a model with a fairness parameter α that denotes the maximum fraction of points in each cluster from a single color class. They also obtained a bi-criteria approximation. Many other notions of fairness in several different settings have also been studied, e.g., fair center [16,20,23,25] and individual fairness [9,26,28].

LP based techniques are fairly popular in solving clustering problems including variants of k-center. Recently, Chakrabarty et al. [10] and Harris et al. [19] obtained a 2-approximation algorithm based on LP rounding for k-center with outliers. They also extend their techniques to more general problems including a non-uniform version of k-center [10] and a lottery model for k-center [19]. k-center has been studied with colorful outliers as well where certain number of points of each color must be clustered [2,4,22]. Bandyapadhyay et al. [4] designed a pseudo-approximation with $k+c-1$ centers exploiting the sparsity of a carefully chosen LP. Subsequently, Jia et al. [22] extended this to a true 3-approximation by designing an algorithm for well-separated instances. Anegg et al. [2] independently designed a 4-approximation based on a round-or-cut method. In round-or-cut, if it is not possible to round the current fractional solution, we can find a linear constraint that cuts (or separates) this solution from an integral optimal solution. See also [11–13,21] for other applications of the technique for variants of k-center.

2 A Polytope for Pairwise Fair k-center

For a particular "guess" D for the optimum solution, we consider the following polytope. Here, we have a variable y_i for each $i \in F$ indicating we select i to be opened and a variable $x_{j,i}$ for each $j \in C, i \in F$ indicating we assign j to i.

$$\begin{aligned}
\textstyle\sum_{i\in F} x_{j,i} &= 1 && \forall\, j \in C \\
x_{j,i} &\le y_i && \forall\, j \in C, i \in F \\
\textstyle\sum_i y_i &\le k && \\
\textstyle\sum_{j\in C_a} x_{j,i} &\le t \cdot \textstyle\sum_{j \in C_b} x_{j,i} && \forall\, 1 \le a,b \le c, i \in F \\
x_{j,i} &= 0 && \forall\, j \in C, i \in F \text{ s.t. } d(j,i) > D \\
x, y &\ge 0 &&
\end{aligned} \qquad (\mathcal{P}(D))$$

To see how this models fair k-center (we drop the term pairwise for simplicity), consider an integer solution. The first constraint ensures every client is assigned to a center, the second ensures a center is opened if a client is assigned to it, the third ensures at most k centers are opened (it could be that $y_i > 1$ in a feasible integer solution but feasibility would not be changed if y_i was then reduced to 1), and the fourth ensures the assignment of clients to each center is fair. The fifth constraint ensures no client travels distance more than D. Conversely, the natural $\{0,1\}$-variable assignment corresponding to any feasible fair k-center solution that has each client travel distance at most D satisfies all constraints.

As is typical with k-center approximations based on similar polytopes, we find the smallest value D such that the corresponding polytope is not empty and then round the resulting fractional point to an integer point while maintaining feasibility in all constraints except, perhaps, assigning clients to centers that are distance $O(D)$ away. Since the polytope is feasible for any $D \ge \text{OPT}$, this would yield an $O(1)$-approximation. To find the minimum value of D, there are at most $|C| \cdot |F|$ distinct distances to consider namely those of the form $d(j,i)$ for $j \in C, i \in F$. A binary search for the smallest such value of D results in only $O(\log |C| + \log |F|)$ different polytope feasibility queries to be solved.

3 The Algorithm for Integer t

In this section, first we present a polynomial-time algorithm that converts a point $(x, y) \in \mathcal{P}(D)$ into an integral fair solution with cost at most $35 \cdot D$. Later, we argue how to improve the factor from 35 to 33. We assume $t \geq 2$, as the $t = 1$ case is already solved.

Throughout, a vector $v \in \mathbb{R}^{\ell}_{\geq 0}$ for some $\ell \geq 1$ is said to be *t-fair* if $v_i \leq t \cdot v_{i'}$ for any $1 \leq i, i' \leq \ell$. We emphasize that an all-0 vector is t-fair.

Claim 1. For t-fair vectors $v^1, v^2, \ldots, v^m \in \mathbb{R}^c_{\geq 0}$, the sum $v = \sum_{i=1}^m v^i$ is t-fair.

Proof. For any $1 \leq a, b \leq c$ we have $v_a = \sum_{i=1}^m v^i_a \leq t \cdot \sum_{i=1}^m v^i_b = t \cdot v_b$.

3.1 Filtering

The first step is to open a set S of at most k centers and find a fractional fair assignment $x'_{j,i}$ supporting a maximum distance of $O(D)$. One detail that differs from typical filtering steps is that we also need to ensure each $i \in S$ receives at least one unit of assignment in x' from each color, a property that is required for the second step of the algorithm.

Let $(x, y) \in \mathcal{P}(D)$ and assume that y is minimal, i.e. $y_i = \max_{j \in C} x_{j,i}$. In particular, $y_i > 0$ means some clients are at least partially assigned to i. Algorithm 1 describes how to construct S and x'.

Algorithm 1. Filtering Step

Input: A feasible solution $(x, y) \in \mathcal{P}(D)$ with $y_i = \max_{j \in C} x_{j,i}$
Output: A set $S \subseteq F$ and a fair fractional assignment $x'_{j,i}, j \in C, i \in S$

1: $S \leftarrow \emptyset$.
2: **while** There is some $i \in F - S$ with $y_i > 0$ and $d(i, S) > 4 \cdot D$ **do**
3: $\quad S \leftarrow S \cup \{i\}$
4: $x'_{j,i} \leftarrow 0$ for each $j \in C, i \in S$
5: **for** each $j \in C$ and $i' \in F$ **do**
6: $\quad$ let $i \in S$ satisfy $d(i, i') = d(i', S)$ $\qquad \triangleright$ breaking ties arbitrarily
7: $\quad x'_{j,i} \leftarrow x'_{j,i} + x_{j,i'}$
8: **return** S, x'

First, we observe x' provides a fair assignment of clients to centers in S.

Claim 2. Let S, x' be returned from Algorithm 1. For each $j \in C$ we have $\sum_{i \in S} x'_{j,i} = 1$ and for each $i \in S$ and any two colors a, b we have $\sum_{j \in C_a} x'_{j,i} \leq t \cdot \sum_{j \in C_b} x'_{j,i}$.

Proof. Clearly $\sum_{i \in S} x'_{j,i} = 1$ for each $j \in C$ (i.e. all clients are fractionally assigned to an extent of 1) since x' was constructed by redirecting the fractional assignment x. It is also fair by Claim 1, i.e. if we define $v_{i,a}(x) = \sum_{j \in C_a} x_{j,i}$ to be the amount of assignment from color a to center i under x then for each $i \in F$, $\langle v_{i,1}(x), v_{i,2}(x), \ldots, v_{i,c}(x)\rangle$ is a t-fair vector by the constraints of $(\mathcal{P}(D))$ and $v_{i,a}(x')$ is the sum of $v_{i',a}(x)$ values for the various centers i' that had i as their closest center in S. That is, $\langle v_{i,1}(x'), v_{i,2}(x'), \ldots, v_{i,c}(x')\rangle$ is also t-fair.

Other important properties of the algorithm are summarized below.

Claim 3. Let S, x' be returned from Algorithm 1. The following hold (i) $\sum_{j \in C_a} x'_{j,i} \geq 1$ for each $i \in S$ and color $1 \leq a \leq c$, (ii) $|S| \leq k$, and (iii) $x'_{j,i} > 0$ only if $d(j,i) \leq 5 \cdot D$.

Proof. Since $y_i > 0$ for any $i \in S$, then minimality of y means there is some $j \in C$ with $x_{j,i} > 0$ as well. The fairness constraints then ensure there is a client j_a of each color C_a with $x_{j_a,i} > 0$ which also must satisfy $d(j_a, i) \leq D$ since x can only support distances at most D. All of j_a's assignment is sent to $i' \in F$ within distance D of j_a, so it must be within distance at most $2 \cdot D$ of i, meaning all of j_a's assignment is redirected to i, i.e. $x'_{j_a,i} = 1$. This proves point (i).

For each $i \in S$, let j again be any client with $d(j,i) \leq D$. Since $x'_{j,i'} > 0$ only for $i' \in B_F(i, 2 \cdot D) = \{i'' \in F : d(i,i'') \leq 2 \cdot D\}$, then $\sum_{i' \in B_F(i,2 \cdot D)} y_{i'} \geq \sum_{i' \in F} x_{j,i'} = 1$. The balls $B_F(i, 2 \cdot D)$ are disjoint for the various centers in S as $d(i,i') > 4 \cdot D$ for any distinct $i, i' \in S$ and each captures at a total y-value of at least 1, so $|S| \leq k$ proving (ii), as $\sum_{i \in F} y_i \leq k$.

Finally, consider any $j \in C$ and $i' \in F$ with $x_{j,i'} > 0$ and say $x_{j,i'}$ was moved to $i \in S$ with $i \neq i'$ in the second loop. It must be $d(i,i') \leq 4 \cdot D$ (otherwise $i' \notin S$ yet $y_{i'} \geq x_{j,i'} > 0$ so the first loop would not have terminated), so $d(j,i) \leq d(j,i') + d(i',i) \leq 5 \cdot D$ proving (iii).

3.2 Tree Rounding

Let S and x' be what is returned from Algorithm 1. Consider the graph $G = (S, E)$ where $E = \{(i,i') : d(i,i') \leq 10 \cdot D\}$. For any $S' \subseteq S$, let $C_{S'} = \{j \in C : x'_{j,i} > 0 \text{ for some } i \in S'\}$ be the clients sending at least some flow to S'. We claim $C_{S'}$ partitions C as S' ranges over the components of G. That is, if $x_{j,i} > 0$ and $x_{j,i'} > 0$ for different $i, i' \in S$, then $d(i,i') \leq d(i,j) + d(j,i') \leq 5 \cdot D + 5 \cdot D = 10 \cdot D$, so $(i,i') \in E$ already and i, i' are in the same component.

Also note that since the fractional assignments to each $i \in S$ is fair then in any connected component $S' \subseteq S$ of G and any two colors a, b we have $|C_a \cap C_{S'}| \leq t \cdot |C_b \cap C_{S'}|$ (cf. Claim 1). So we focus on one component $S' \subseteq S$ of G and the clients $C_{S'}$, and show how to find a feasible fair $|S'|$-center assignment with cost at most $35 \cdot D$. Doing so for all components S' finishes the algorithm. If $|S'| = 1$ then all clients in $C_{S'}$ are already fully sending their entire flow to the unique center in S' so there is nothing more to do.

So from now on suppose $|S'| \geq 2$. We will show how to redistribute the fractional assignment x' to nodes in S' to get a fair integer assignment while ensuring any portion of flow is moved along at most three edges in G from its original destination under x'. Since each unit of flow initially travels at most $5 \cdot D$ via x' and then hops across at most three edges of G, each having distance at most $10 \cdot D$, then no unit of flow travels more than $35 \cdot D$ in this final assignment.

Consider any spanning tree T' of the component S' in G and root it at an arbitrary root node $r \in S'$. Intuitively, for each node $i \in S'$ we will take its currently-assigned fractional flow plus some unassigned fractional flow from its children. We calculate a t-fair integer vector and assign exactly this much from each color to i. A small (t-fair and fractional) amount may remain unassigned, that amount is propagated up the tree to i's parent.

3.3 Processing Fairlets

Next, we define a concept that helps us track movement of flow across tree nodes.

Definition 1. *A* **fairlet** *is a vector $f \in [0,1]^{|C|}$ such that the total count of each color is a t-fair vector. Formally, if we define $f(b) = \sum_{j \in C_b} f_j$ then the vector $\langle f(1), f(2), \ldots, f(c) \rangle$ is t-fair. An* **integral fairlet** *is a fairlet that further satisfies $f(b) \in \mathbb{Z}$ for each color b.*

One can view a fairlet as a partial assignment of clients that do not have any particular destination (yet) but the overall assignments to the colors is fair. We emphasize that saying an f is integral simply means the total flow from each color is integral, it does not necessarily mean each entry f_j is an integer. It is straightforward to verify the sum of two fairlets is again a fairlet using Claim 1.

Our algorithm will process nodes in a leaf-to-root fashion. When processing a node, a fairlet f will be formed by summing a fairlet obtained from its current assignment under x' and fairlets pushed up to i from all of its children. From f, an integer fairlet f'^i will be formed, this will indicate the amount $f'^i(b)$ of each color b that should be assigned to i at the end of the algorithm. The remaining portion will be another fairlet f' with bounded entries that will be pushed up to the parent of i. Thus, f is decomposed into f'^i and f' where f' may be fractional.

The formation of f'^i and f' are described in Algorithm 2. Intuitively, the goal is to draw out as large of an integer fairlet as possible from f while ensuring the remaining fairlet that is pushed up to the parent of i has at least 1 total assignment from each color. Ultimately this property helps us ensure no fractional assignment travels too far up the tree from its original destination $x'_{j,i}$ before it is left at some $i' \in S$ in the integer fairlet.

The term big push is used because f' will be "pushed up" the tree from i to its parent, and a color b processed as a big push could have $f'(b)$ being close to $2 \cdot t$ where as in the small push case we would only have $f'(b) < 3$. This, and further relevant properties of Algorithm 2 are summarized as follows.

Lemma 1. *During the algorithm, $n_i \geq 0$ and $n_i \leq M_b \leq t \cdot n_i$. Also, $M_b \leq f(b)$ for each color b (i.e. there is enough of color b to move to f'^i). The vectors f'^i and f' constructed from this procedure are both fairlets. Furthermore, f'^i is*

Algorithm 2. Fractional Fairlet Decomposition

Input: A node $i \in S$ and a fairlet f with $f(b) \geq 1$ for each color b and a node $i \in S$.
Output: An integer fairlet f'^i and another fairlet f' with $f'^i + f' = f$.

Let f'^i, f' initially be the all-0 vectors in $[0,1]^{|C|}$.
Let a be any color with minimum value $f(a)$ in the given fairlet.
$n_i \leftarrow \lfloor f(a) \rfloor - 1$
for each color $1 \leq b \leq c$ **do**
  if $a = b$ **then**
    $M_b \leftarrow n_i$
  else if $f(b) \geq t \cdot n_i + (f(a) - n_i)$ **then**
    $M_b \leftarrow t \cdot n_i$ ▷ Call this a **big push** for color b.
  else
    $M_b \leftarrow \lfloor f(b) - (f(a) - n_i) \rfloor$ ▷ Call this a **small push** for color b.
  for each $j \in C_b$ in non-increasing order of distance $d(j, i)$ **do**
    $f_j'^i \leftarrow \min\{M_b - f'^i(b), f_j\}$
$f' \leftarrow f - f'^i$ ▷ The leftover parts of f that are not contributing to f'^i.
return f'^i, f'

an integer fairlet, $f'(a) \in [1, 2)$ where a is the color with minimum value $f(a)$, $f'(b) \in [1, 2 \cdot t)$ if b is processed as a big push, and $f'(b) \in [1, 3)$ if b is processed as a small push.

Proof. Since $f(a) \geq 1$ (by assumption on the input fairlet), $n_i \geq 0$ by construction. It is straightforward to verify $M_b \leq f(b)$ for each color, so it is possible to have $f(b)$ units of assignment of color b move to $f'^i(b)$. Next we claim $n_i \leq M_b \leq t \cdot n_i$, so in fact f'^i is an integer fairlet. This is immediate for $b = a$ or when b is processed as a big push. When b is processed as a small push, we have $f(b) < t \cdot n_i + f(a) - n_i$, so $f(b) - (f(a) - n_i) \leq t \cdot n_i$, thus $M_b \leq t \cdot n_i$. On the other hand, $f(b) \geq f(a)$ by our choice of a, so $f(b) - (f(a) - n_i)) \geq n_i$. Thus, $M_b = \lfloor f(b) - (f(a) - n_i) \rfloor \geq n_i$ as well since n_i is an integer.

We now establish that f' is a fairlet satisfying the claimed color bounds. For color a itself, note $f(a) - 1 < \lfloor f(a) \rfloor \leq f(a)$ which immediately shows $f'(a) = f(a) - n_i = f(a) - (\lfloor f(a) \rfloor - 1) \in [1, 2)$. For $b \neq a$ processed as a big push, since $f(b) \geq t \cdot n_i + f'(a)$ then $f'(b) = f(b) - t \cdot n_i \geq f'(a) \geq 1$. We also claim $f'(b) \leq t \cdot f'(a)$ which would also show $f'(b) < 2 \cdot t$. Suppose otherwise, i.e. that $f'(b) > t \cdot f'(a)$. We would then have $f(b) = f'^i(b) + f'(b) > t \cdot n_i + t \cdot f'(a) = t \cdot (n_i + f'(a)) = t \cdot f(a)$, which contradicts that f itself is a fairlet. So it must be that $f'(b) \leq t \cdot f'(a)$.

Finally, consider $b \neq a$ is processed as a small push. We claim $f'(b) \in [f'(a), f'(a) + 1)$, which suffices since $f'(a) + 1 < 2 + 1 = 3$. But $f'(b) \in [f'(a), f'(a) + 1)$ simply follows because $f'(b) = f(b) - M_b = f(b) - \lfloor f(b) - f'(a) \rfloor$.

The last thing to mention is that in the course of verifying the claimed color bounds for f', we have also shown $f'(b) \in [f'(a), t \cdot f'(a))$ for $b = a$ and $b \neq a$ processed as a big push. For $b \neq a$ processed as a small push, we saw $f'(b) \in [f'(a), f'(a) + 1)$ which suffices since $f'(a) + 1 \leq 2 \cdot f'(a) \leq t \cdot f'(a)$.

3.4 Processing a Tree

Process the nodes of S' from the deepest layer (bottom) to the shallowest layer (top). For each node i in this ordering, and execute the following instructions depending on the role i in the spanning tree. For $i \in S$, let $\hat{f}^i$ be the **initial fairlet** for i defined by $\hat{f}^i_j = x'_{j,i}$.

- **Case: i is a leaf node:** Use the initial fairlet $\hat{f}^i$ as input to the procedure in Sect. 3.3. Let f'^i and f' be the resulting fairlets. Leave f'^i at i and push f' up to the parent of i.
- **Case: i is an internal, non-root node:** Suppose i has m children and let $f_1, f_2, \ldots, f_m$ be the fairlets received from them. Let $f = \hat{f}^i + \sum_{\ell=1}^{m} f_\ell$. Process f according to the procedure in Sect. 3.3. Keep the fairlet f'^i at i and send f' upward to the parent of i.
- **Case: i is a root node:** Form f in the same way as in the previous case, using the assignment to i under x' plus the fairlets received from children. We note f is already an integer fairlet because $f'^{i'}$ is an integer fairlet for each $i' \in S' - \{i\}$ (which have been processed already) and the total assignment of all clients into S' is integer, so the portion that is not yet left at an integer fairlet, namely f, must itself be an integer fairlet.

This procedure defines an integer fairlet for each $i \in S'$. Define a final fractional assignment x'' by $x''_{j,i} = f'^i_j$ for each $j \in C_{S'}$ and each $i \in S'$. This defines a feasible solution for $(\mathcal{P}(D))$ except using a larger distance. We will show below this distance is at most $35 \cdot D$.

The algorithm concludes by using integrality of the b-matching polytope. That is, consider the bipartite graph where we have a copy of $C_{S'}$ on the left side and nodes $\{(i,a) : i \in S', 1 \leq a \leq c\}$ on the right side. Connect $j \in C_{S'}$ to a node (i,a) if $j \in C_a$ and $d(j,i) \leq 35 \cdot D$. The vector $x''_{i,j}$ gives a feasible fractional perfect matching where each $j \in C_{S'}$ has degree requirement 1 and each (i,a) on the right side has degree requirement $f'^i(a)$. By integrality of this polytope, there is an integer perfect b-matching in this graph. This describes an integer t-fair solution where each client is assigned at a distance of at most $35 \cdot D$.

3.5 Cost Analysis

We prove for any $i \in S'$ and $j \in C_{S'}$ that $f'^i_j > 0$ only if $d(j,i) \leq 35 \cdot D$ by observing further structure about the fairlets that are passed up the tree. For a color b, a node i, a fairlet f, and a radius r, we let $f(i,b,\leq r) = \sum_{j \in C_b : d(i,j) \leq r} f_j$ be the total amount color b contributes to the fairlet from clients within distance r of i. Similarly, $f(i,b,> r) = \sum_{j \in C_b : d(i,j) > r} f_j$, which equals $f(b) - f(i,b,\leq r)$.

Intuitively what we will show is that any initial assignment $x'_{j,i}$ will be redirected up at most three edges of the tree. Since the initial assignment travels distance at most $5 \cdot D$ by Claim 3 and since each tree edge has distance at most $10 \cdot D$, we get our final bound of $35 \cdot D$ for the final assignment. To show this, we will essentially show that only a limited amount of each color is passed up the tree in the fairlets f' formed at each node, and that this amount diminishes as the distance it has traveled increases. Formally, we prove the following which is also illustrated in Fig. 1.

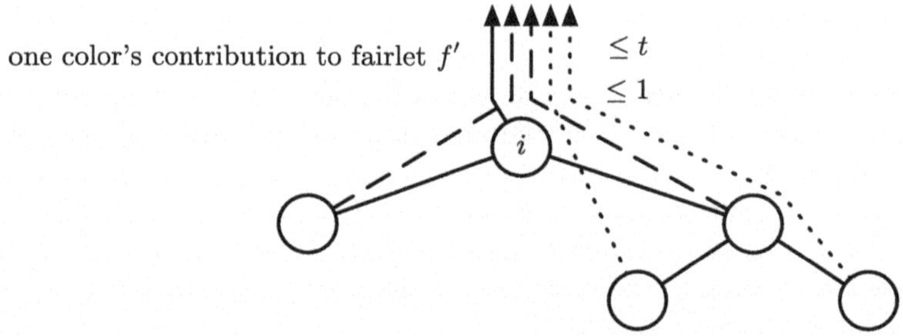

Fig. 1. A depiction of a color's contribution to fairlet f' sent up from center i. Intuitively, Lemma 2 proves at most t flow from any color comes from flow initially sent (via x') to children or grandchildren and at most 1 from grandchildren.

Lemma 2. *For any non-root node i let f' be the fairlet it passes up the tree. For any color b, $f'(i, b, \leq 5 \cdot D) \geq 1$ and we also have the upper bounds $f'(i, b, > 5 \cdot D) \leq t, f'(i, b, > 15 \cdot D) \leq 1$, and $f'(i, b, > 25 \cdot D) = 0$.*

Proof. This is trivially true for the fairlets f' sent up from a leaf node i since $f_j'^i > 0$ only if $x'_{j,i} > 0$ which can only happen if $d(j, i) \leq 5 \cdot D$ and also since $\sum_{j \in C_a} x'_{j,i} \geq 1$ by Claim 3.

Now let i be an internal node that is not the root and let $i_1, i_2, \ldots, i_m$ be its children. By induction, the result holds for all fairlets $f_1, f_2, \ldots, f_m$ received from the children. Since the edge connecting the child to i has distance at most $10 \cdot D$, any client contributing to one of the fairlets f_ℓ is at most $10 \cdot D$ farther from i than from i_ℓ.

Recall notation from the procedure in Sect. 3.3 when applied to $f = \hat{f^i} + \sum_\ell f_\ell$, i.e. let a be the color with minimum value $f(a)$ and let n_i be the amount of its total assignment left in f'^i in Algorithm 2. Observe $n_i \geq m$ because each f_i contributes at least 1 to color a as does $\hat{f^i}$. So $n_i = \lfloor f(a) \rfloor - 1 \geq m + 1 - 1 = m$.

In all cases we verify below, it is important to remember that in Algorithm 2, while forming the integer fairlet f'^i, flow was assigned from the clients in non-increasing order of distance from i. So, the fractional fairlet $f' = f - f'^i$ produced by the algorithm consists of the flow from the input fairlet f coming from the clients nearest to i.

Verifying $f'(i, b, \leq 5 \cdot D) \geq 1$

By Lemma 1 and the construction of $f' = f - f'^i$, $f'(b) \geq 1$ and it must come from the $f'(b)$ fractional assignments in f from color b clients that are closest to i. We also have $\hat{f^i}(i, b, \leq 5 \cdot D) \geq 1$ as x' sends i at least 1 unit of flow from each color within distance at most $5 \cdot D$. So $f(i, b, \leq 5 \cdot D) \geq 1$ meaning f' gets at least 1 total assignment from clients $j \in C_b$ with $d(j, i) \leq 5 \cdot D$.

Verifying $f'(i, b, > 25 \cdot D) = 0$
For $b = a$, $f'(a) < 2$ (Lemma 1) but $\hat{f}^i(i, a, \leq 5 \cdot D) \geq 1$ and $f_\ell(i, a, \leq 15 \cdot D) \geq f_\ell(i_\ell, a, \leq 5 \cdot D) \geq 1$ for any $1 \leq \ell \leq m$, so $f(i, a, \leq 15 \cdot D) \geq m + 1 \geq 2$ meaning all flow of color a sent to f' comes from clients within distance $15 \cdot D$ of i.

Now consider $b \neq a$. Recall $n_i \geq m$. By induction, each $f_\ell(i_\ell, b, > 15 \cdot D) \leq t$ so $f_\ell(i, b, > 25 \cdot D) \leq t$. Also by construction and Claim 3, $\hat{f}^i(i, b, > 25 \cdot D) = 0$. If b was processed as a big push, then exactly $t \cdot n_i$ units of its flow from f are sent to f'^i. Since $n_i \geq m$, then all of the at most $t \cdot m$ units of flow of distance $> 25 \cdot D$ to i are left in f'^i, so $f'(i, b, 25 \cdot D) = 0$ in this case.

Finally suppose b was processed as a small push, so $f'(b) < 3$ by Lemma 1. If $m \geq 2$ we are done since the nearest 1 unit in each of $\hat{f}, f_1$, and f_2 comes from a client with distance at most $15 \cdot D$ from i. Otherwise, suppose $m = 1$. In this case, $f_1(i, b, > 25 \cdot D) \leq f_1(i_1, b, > 15 \cdot D) \leq 1$. Also, $n_i \geq m = 1$ means $f'^i(b) \geq 1$ (see Lemma 1 again). So all the at most 1 unit of flow in f coming from clients with distance $> 25 \cdot D$ of i will be left in f'^i.

Verifying $f'(i, b, > 15 \cdot D) \leq 1$
The previous case shows in fact $f(i, b, > 15 \cdot D) = 0$ if $b = a$ or b was processed as a big push. So suppose b was processed as a small push. Again, recall $\hat{f}^i(i, b, \leq 5 \cdot D) \geq 1$ and $f_1(i_1, b, \leq 5 \cdot D) \geq 1$ so at least two of the < 3 units of flow of color b in f' come from distance at most $15 \cdot D$.

Verifying $f'(i, b, > 5 \cdot D) \leq t$
This is trivial for $b = a$ since $f'(a) < 2 \leq t$. For $b \neq a$, if b was processed as a small push then, again, $\hat{f}^i(i, b, \leq 5 \cdot D) \geq 1$ and $f'(b) < 3$ means at most $2 \leq t$ units of flow come from further away.

Finally, if b is a big push then assume, for the sake of contradiction, that $f'(i, b, > 5 \cdot D) > t$. Since $\hat{f}^i(i, b, > 5 \cdot D) = 0$ (by Claim 3) and since f'^i is formed using assignments from the clients farthest from i, we then have all of $f'^i(b)$ comes from the child fairlets $f_1, \ldots, f_m$ as does the assignment to f' contributing to $f'(i, b, > 5 \cdot D)$. That is, $\sum_{\ell=1}^m f_\ell(b) > t \cdot n_i + t$.

On the other hand, $\sum_{\ell=1}^m f_\ell(a) = f(a) - \hat{f}^i(a) < n_i + 2 - \hat{f}^i(a) \leq n_i + 1$ where the first bound uses the definition of n_i from Algorithm 2 and the second bound is by Claim 3.

Thus, we see $\sum_{\ell=1}^m f_\ell(b) > t \cdot n_i + t > t \cdot \sum_{\ell=1}^m f_\ell(a)$ which contradicts that $\sum_{\ell=1}^m f_\ell$ is a fairlet. So we must have $f'(i, b, > 5 \cdot D) \leq t$.

Corollary 1. *$d(j, i) \leq 35 \cdot D$ for any $i \in S'$ and $j \in C$ with $f'^i_j > 0$.*

Proof. Using the same notation as before for i, $\hat{f}^i_j > 0$ only for $j \in C$ with $d(j, i) \leq 5 \cdot D$ by Claim 3. For each child's fairlet f_ℓ, we have $f_\ell(i_\ell, b, > 25 \cdot D) = 0$ for each color b, meaning any $j \in C$ contributing to f_ℓ is within distance at most $25D$ from i_ℓ. Since $d(i, i_\ell) \leq 10 \cdot D$, $d(j, i) \leq 35 \cdot D$ as required.

3.6 Slightly Improving the Ratio

We sketch the details on how to improve the ratio to 33. When S is formed, we moved assignments from F to their nearest center in S to form x'. For each $i \in S$, it in fact receives at least 1 from each color from facilities i' within distance $2 \cdot D$ and any client j partially assigned to such an i' has $d(j, i') \leq 3 \cdot D$. So the total x' assigned to i naturally partitions into two fair allocations: that which was reassigned from facilities with distance $\leq 2 \cdot D$ from i and the others (which are still within $4 \cdot D$ of i). The important part here is that the fair allocation that represented flow redirected from facilities with distance $\leq 2 \cdot D$ is indeed a fairlet, i.e. it assigns at least one from each color.

One can then slightly modify the properties in Lemma 2 to ultimately ensure any flow of distance > 30 that is transferred to f^i in fact was initially redirected to its center in S from a facility with distance $\leq 2 \cdot D$ from that facility. So the distance of any client's assignment to i is in fact at most $33 \cdot D$.

4 The Algorithm with Outliers

In this case, we can discard up to γ points and fairly allocate the rest to at most k centers. We use the same rounding framework. The filtering step is more involved, as the points might not be assigned up to an extent of 1 to the centers by the LP. Here we describe the difference. Consider any center i. Let N_i be the centers (including i) whose D radius balls intersect with the D radius ball at i. If $\sum_{i' \in N_i} y_{i'} \geq 1$, then we can select a ball of radius $3D$ at i and reassign the x-values of the centers in N_i to i. Moreover, if we can partition the balls in the LP solution in this way into neighborhoods around central balls, we can show that we select at most k balls and cover as many points as covered by the LP. In fact, without outliers, picking a subset of *well-separated* centers pairwise at least $4D$ distance apart is sufficient. With outliers, for a center i in such a well-separated subset, $\sum_{i' \in N_i} y_{i'}$ is not-necessarily at least 1. Thus, if we select each such center, we might end up selecting far more than k. To deal with this, we show that it is possible to modify the solution to one that has an additional structure: for any facility i with $\sum_{i' \in N_i} y_{i'} < 1$, we can modify the solution so that $x_{j,i} \in \{0, y_i\}$ for each $j \in C$. An immediate consequence is that the total amount of points (or *load*) assigned to i is an integer multiple of y_i. Thus, for two such well-separated centers, it is possible to transfer y-value from the one with the smaller load to the other, without decreasing the total coverage. We can stop when no such transfer is possible, at which moment one of the two statements is true: (i) there is an i with $\sum_{i' \in N_i} y_{i'} \geq 1$, and (ii) all centers that have non-zero loads are $O(D)$ away from a chosen center. This suffices to obtain the desired set of centers and a fractional fair assignment of at least $|C| - \gamma$ amount of points.

The tree rounding step is exactly the same, except we cannot argue that at the end the root contains integral amount of points for all colours. However, we show that it is possible to modify the load of the root for all colours to integers, without decreasing the total coverage. Hence, again we obtain the desired integral assignment applying integrality of flow argument.

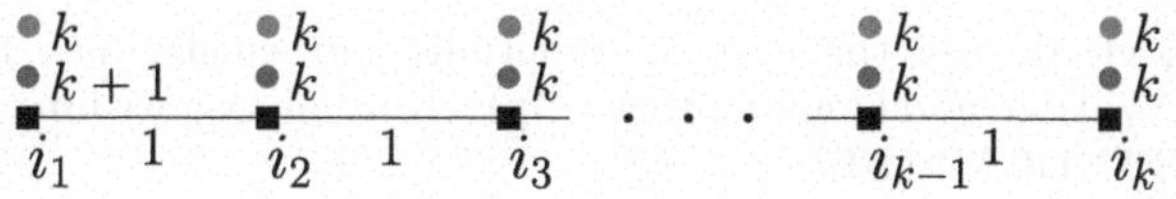

Fig. 2. Each location except i_1 has k red and k blue. i_1 has k red and $k+1$ blue. (Color figure online)

5 A Bad Gap Example for Non-Integer t

We give a bad integrality gap example for ($\mathcal{P}(D)$) even when $c = 2$. We have locations $i_1, i_2, \ldots, i_k$ as shown in Fig. 2. $t = 1 + 1/k^2$. It can be shown that any fair clustering has exactly one non-empty cluster, and thus the optimum cost is $\Omega(k)$. However, there is an LP solution with cost 1: For each location i_ℓ, $1 \le \ell \le m$ we set $y_{i_\ell} = 1$ (i.e. it is fully opened). For $\ell < k$, exactly one blue client collocated with i_ℓ has $x_{j,i_\ell} = \frac{\ell}{k}$ and $x_{j,i_{\ell+1}} = \frac{k-\ell}{k}$. All other clients collocated with i_ℓ have $x_{j,i_\ell} = 1$. Additionally, all clients j collocated with the last center i_k have $x_{j,i_k} = 1$. Every other assignment variable is 0. We also show this is a worst-case instance, i.e. the integrality gap is always $O(k)$ for any t and any number of colors which, essentially, follows by proving the (unweighted) support graph of any feasible point has diameter $O(k)$.

References

1. Ahmadian, S., Epasto, A., Kumar, R., Mahdian, M.: Clustering without over-representation. In: Proceedings of the 25th ACM SIGKDD International Conference on Knowledge Discovery & Data Mining, pp. 267–275 (2019)
2. Anegg, G., Angelidakis, H., Kurpisz, A., Zenklusen, R.: A technique for obtaining true approximations for k-center with covering constraints. In: International Conference on Integer Programming and Combinatorial Optimization, pp. 52–65. Springer (2020)
3. Bandyapadhyay, S., Chlamtáč, E., Makarychev, Y., Vakilian, A.: A polynomial-time approximation for pairwise fair k-median clustering. arXiv preprint arXiv:2405.10378 (2024)
4. Bandyapadhyay, S., Inamdar, T., Pai, S., Varadarajan, K.: A constant approximation for colorful k-center. In: 27th Annual European Symposium on Algorithms (ESA 2019). Schloss Dagstuhl-Leibniz-Zentrum fuer Informatik (2019)
5. Bera, S., Chakrabarty, D., Flores, N., Negahbani, M.: Fair algorithms for clustering. In: Advances in Neural Information Processing Systems, pp. 4954–4965 (2019)
6. Bercea, I.O., et al.: On the cost of essentially fair clusterings. arXiv preprint arXiv:1811.10319 (2018)

7. Bercea, I.O., et al.: On the cost of essentially fair clusterings. In: Approximation, Randomization, and Combinatorial Optimization. Algorithms and Techniques (APPROX/RANDOM) (2019)
8. Böhm, M., Fazzone, A., Leonardi, S., Schwiegelshohn, C.: Fair clustering with multiple colors. arXiv preprint arXiv:2002.07892 (2020)
9. Brubach, B., Chakrabarti, D., Dickerson, J., Khuller, S., Srinivasan, A., Tsepenekas, L.: A pairwise fair and community-preserving approach to k-center clustering. In: International Conference on Machine Learning (ICML), pp. 1178–1189 (2020)
10. Chakrabarty, D., Goyal, P., Krishnaswamy, R.: The non-uniform k-center problem. ACM Trans. Algorithms (TALG) **16**(4), 1–19 (2020)
11. Chakrabarty, D., Krishnaswamy, R., Kumar, A.: The heterogeneous capacitated k-center problem. In: International Conference on Integer Programming and Combinatorial Optimization, pp. 123–135. Springer (2017)
12. Chakrabarty, D., Negahbani, M.: Generalized center problems with outliers. ACM Trans. Algorithms (TALG) **15**(3), 1–14 (2019)
13. Chakrabarty, D., Negahbani, M.: Robust k-center with two types of radii. Math. Program. **197**(2), 991–1007 (2023)
14. Chhabra, A., Masalkovaitė, K., Mohapatra, P.: An overview of fairness in clustering. IEEE Access (2021)
15. Chierichetti, F., Kumar, R., Lattanzi, S., Vassilvitskii, S.: Fair clustering through fairlets. In: Advances in Neural Information Processing Systems, pp. 5029–5037 (2017)
16. Chiplunkar, A., Kale, S., Ramamoorthy, S.N.: How to solve fair k-center in massive data models. In: Proceedings of the International Conference on Machine Learning (ICML), pp. 1877–1886 (2020)
17. Cygan, M., Hajiaghayi, M.T., Khuller, S.: LP rounding for k-centers with non-uniform hard capacities. In: 53rd Annual IEEE Symposium on Foundations of Computer Science, FOCS 2012, New Brunswick, NJ, USA, 20–23 October 2012, pp. 273–282. IEEE Computer Society (2012)
18. Gonzalez, T.F.: Clustering to minimize the maximum intercluster distance. Theor. Comput. Sci. **38**, 293–306 (1985)
19. Harris, D.G., Pensyl, T., Srinivasan, A., Trinh, K.: A lottery model for center-type problems with outliers. ACM Trans. Algorithms (TALG) **15**(3), 1–25 (2019)
20. Hotegni, S.S., Mahabadi, S., Vakilian, A.: Approximation algorithms for fair range clustering. In: International Conference on Machine Learning, pp. 13270–13284. PMLR (2023)
21. Jia, X., Rohwedder, L., Sheth, K., Svensson, O.: Towards non-uniform k-center with constant types of radii. In: Symposium on Simplicity in Algorithms (SOSA), pp. 228–237. SIAM (2022)
22. Jia, X., Sheth, K., Svensson, O.: Fair colorful k-center clustering. Math. Program. 1–22 (2021)
23. Jones, M., Nguyen, H., Nguyen, T.: Fair k-centers via maximum matching. In: Proceedings of the International Conference on Machine Learning (ICML), pp. 4940–4949 (2020)
24. Khuller, S., Sussmann, Y.J.: The capacitated K-center problem. SIAM J. Discret. Math. **13**(3), 403–418 (2000)
25. Kleindessner, M., Awasthi, P., Morgenstern, J.: Fair k-center clustering for data summarization. In: 36th International Conference on Machine Learning, ICML 2019, pp. 5984–6003. International Machine Learning Society (IMLS) (2019)

26. Negahbani, M., Chakrabarty, D.: Better algorithms for individually fair k-clustering. In: Advances in Neural Information Processing Systems (NeurIPS), vol. 34, pp. 13340–13351 (2021)
27. Rösner, C., Schmidt, M.: Privacy preserving clustering with constraints. In: International Colloquium on Automata, Languages, and Programming (ICALP) (2018)
28. Vakilian, A., Yalçıner, M.: Improved approximation algorithms for individually fair clustering. In: International Conference on Artificial Intelligence and Statistics (AISTATS), pp. 8758–8779. PMLR (2022)

A Global Analysis of the Primal-Dual Method for Edge Augmentation Problems

Ishan Bansal(✉)

Amazon, Bellevue, WA, USA
ib332@cornell.edu

Abstract. We study an algorithmic problem in network design called $\mathcal{F}$-augmentation where a family of cuts $\mathcal{F}$ is given and the goal is to find a minimum-cost edge set that covers each cut in $\mathcal{F}$.

When $\mathcal{F}$ is a so-called uncrossable family, Williamson et al. (Combinatorica 1995) provided a 2-approximation algorithm based on the primal-dual method. Extending their results to families that are non-uncrossable has remained a challenging question.

In this paper, we introduce the notion of the crossing density of a set family and present a new approach to analyzing primal-dual algorithms based on this notion.

We focus on pliable families, a strict generalization of uncrossable families, and provide the first $O(1)$-approximation algorithm for $\mathcal{F}$-augmentation of pliable families. Pliable families were introduced by Bansal et al. (Algorithmica 2024).

We also improve on the results in Bansal et al. (Algorithmica 2024) by providing a 6-approximation algorithm for the $\mathcal{F}$-augmentation problem when $\mathcal{F}$ is a family of small cuts. Our improvement implies improved approximation algorithms for the Capacitated Network Design problem. Finally, we study the $(p, 3)$-flexible graph connectivity problem. By carefully analyzing the structure of feasible solutions and using the techniques developed in this paper, we provide the first $O(1)$-approximation algorithm for this problem, exhibiting a 12-approximation algorithm.

Keywords: Network Design · Primal-Dual Method · Approximation Algorithms · Edge Augmentation

1 Introduction

A branch of network design involves finding the optimal cost of maintaining and designing networks to meet certain connectivity requirements. A well known and widely used algorithmic idea to design approximation algorithms for such network design problems is that of edge augmentation - instead of finding a network that immediately meets all the given connectivity requirements, we start with an infeasible solution and increase the connectivity of this solution one step at a time. This methodology of gradually increasing connectivity has been

N. Megow and A. Basu (Eds.): IPCO 2025, LNCS 15620, pp. 58–71, 2025.
https://doi.org/10.1007/978-3-031-93112-3_5

used for decades starting from classical k-edge/node connectivity and survivable network design problems [9,13,17] to more recent non-uniform fault tolerant models like flexible graph connectivity and bulk-robust network design problems [1,4,5,8,15,16]. Furthermore, for some of these problems like flexible graph connectivity, this augmentation methodology is currently the only known technique to obtain good approximation algorithms. A formal description of this augmentation framework is given by the **$\mathcal{F}$-augmentation** problem[1].

Problem 1 ($\mathcal{F}$-augmentation).
Input: Graph $G = (V, E)$ with edge costs $\{c_e\}_{e \in E}$ and a family of cuts $\mathcal{F} \subseteq 2^V$.
Output: Subset of the edges $F \subseteq E$ with minimum $c(F) = \sum_{e \in F} c_e$ such that for every $S \in \mathcal{F}$, the set of edges $F \cap \delta(S)$[2] is non-empty.

The $\mathcal{F}$-augmentation problem captures the hitting set problem [4]. Unfortunately, this implies that unless $P = NP$, it is impossible to find an approximation algorithm for the problem with approximation factor better than $O(\log |V|)$. Thus, one needs to exploit underlying structures of the family of cuts $\mathcal{F}$ to be able to design good approximation algorithms. In a seminal paper, Williamson, Goemans, Mihail and Vazirani (WGMV) [17] considered a versatile property for a family of cuts $\mathcal{F}$ that in particular captures the family of minimum cuts of any graph G. They called such families of cuts **uncrossable**.

Definition 1 (Uncrossable Family). *A family of sets $\mathcal{F}$ is called uncrossable if $A, B \in \mathcal{F}$ implies that either (i) $A \cup B \in \mathcal{F}$ and $A \cap B \in \mathcal{F}$ or (ii) $A \setminus B \in \mathcal{F}$ and $B \setminus A \in \mathcal{F}$.*

Building on the works of [1,10], they provided a primal-dual 2-approximation algorithm for the $\mathcal{F}$-augmentation problem when $\mathcal{F}$ is an uncrossable family. We will refer to their algorithm as the *WGMV primal-dual algorithm* and a brief description of the algorithm is provided in Sect. 2. Very recently, Bansal, Cheriyan, Grout and Ibrahimpur (BCGI) [4] considered a natural generalization of uncrossable families that they called **pliable** families. For any two cuts $A, B \subseteq V$, the four cuts $A \cup B, A \cap B, A \setminus B, B \setminus A$ are referred to as the *corner cuts.* While uncrossability requires that for $A, B \in \mathcal{F}$, one of two particular pairs of corner cuts also lie within $\mathcal{F}$, pliability relaxes this condition to allow for any pair of corner cuts to lie within $\mathcal{F}$.

Definition 2 (Pliable Family). *A family of sets $\mathcal{F}$ is called pliable if $A, B \in \mathcal{F}$ implies that at least two of the four cuts $A \cup B, A \cap B, A \setminus B, B \setminus A$ also lie in $\mathcal{F}$.*

Pliable families generalize uncrossable families in the most natural way but also manage to capture the family of α-min cuts. A cut is an α-min cut if its size is at most α times the size of a minimum cut. We can see that α-min cuts are pliable using the submodularity (1) and posimodularity (2) of the cut function. Indeed if A and B are α-min cuts, then $|\delta(A)| \leq \alpha\lambda$ and $|\delta(B)| \leq \alpha\lambda$ where λ is

[1] $\mathcal{F}$-augmentation has also been referred to as the 0–1 f-connectivity problem in the literature.

[2] $\delta(S)$ refers to the set of edges with exactly one end point in the set S.

the size of a minimum cut. Then, Eq. (1) tells us that at least one of $|\delta(A \cup B)|$ or $|\delta(A \cap B)|$ is at most $\alpha\lambda$ and similarly equation (2) tells us that at least one of $|\delta(A \setminus B)|$ or $|\delta(B \setminus A)|$ is at most $\alpha\lambda$. Hence at least two of the four corner cuts are also α-min cuts.

$$|\delta(A)| + |\delta(B)| \geq |\delta(A \cup B)| + |\delta(A \cap B)| \tag{1}$$

$$|\delta(A)| + |\delta(B)| \geq |\delta(A \setminus B)| + |\delta(B \setminus A)| \tag{2}$$

The main contribution of BCGI was to extend the WGMV primal-dual algorithm to work for a sub-family of pliable families that also captures the family of α-min cuts. They called such families γ-pliable families and these were characterized by a local property of cuts (a technical definition is provided in the next section). They showed that the WGMV primal-dual algorithm is a 16-approximation algorithm for the $\mathcal{F}$-augmentation problem when $\mathcal{F}$ is a γ-pliable family. Subsequently, Nutov [14] considered a further sub-family of γ-pliable families called semi-uncrossable families and showed that the WGMV primal-dual algorithm is a 2-approximation algorithm in this case.

Prior to our work however, nothing was known about the performance of the WGMV primal-dual algorithm for general pliable families. We bridge this gap by introducing a novel concept known as the crossing density of a set family that captures a notion of how often sets within the family cross[3]. We show that the WGMV primal-dual algorithm for the $\mathcal{F}$-augmentation problem when $\mathcal{F}$ is a general pliable family provides an approximation ratio parametrized by the crossing density of the family. In addition, we show that attempts to generalize pliable families further by requiring only one of the four corner cuts to lie within the family are futile, as the $\mathcal{F}$-augmentation problem then captures the hitting set problem and has an $\Omega(\log |V|)$-hardness of approximation (See Fig. 1). We also apply our techniques to specific network design problems to obtain current best approximation ratios. In particular, we study the small cut augmentation problem, the capacitated k-edge connectivity problem and the flexible graph connectivity problem.

1.1 Our Contribution

Crossing Density. Our first major contribution is the identification of a right parameter to study when analyzing the WGMV primal-dual algorithm for non-uncrossable families. The main obstacle with non-uncrossable families is that inclusion-wise minimal sets of the family $\mathcal{F}$ (also referred to as *active sets*) start to cross other sets within the family. BCGI attempted to control for such 'problematic' crossings by considering a sub-family of pliable families characterized by a set property they called **property γ**. For any set of edges $D \subseteq E$, define $\mathcal{F}_D := \{S \in \mathcal{F} : \delta(S) \cap D = \emptyset\}$ i.e. the family of sets in $\mathcal{F}$ that have not yet been augmented by edges in D. Define $\mathcal{C}_D$ to be the inclusion-wise minimal sets in $\mathcal{F}_D$.

[3] Recall that two sets A and B cross if all four sets $A \setminus B, B \setminus A, A \cap B, \overline{A \cup B}$ are non-empty.

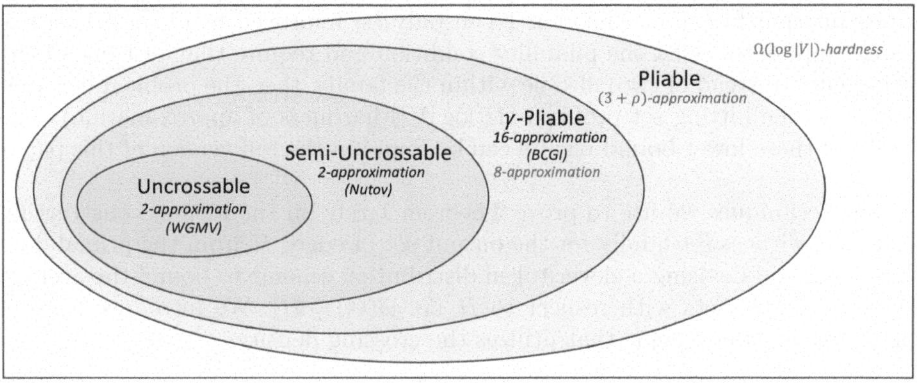

Fig. 1. Performance of the WGMV primal-dual algorithm for the $\mathcal{F}$-augmentation problem. Red text indicates work in this paper.

Definition 3 (γ-Pliable Family). *A set family $\mathcal{F}$ is γ-pliable if it is pliable and also satisfies*
Property γ*: For any edge set $D \subseteq E$, sets $S_1 \subseteq S_2 \in \mathcal{F}_D$, and $C \in \mathcal{C}_D$ that crosses both S_1 and S_2, the set $S_2 \setminus (S_1 \cup C)$ is either empty or lies in $\mathcal{F}$.*

Property γ is a local property looking at 'problematic' crossings arising from triplets of sets. Instead, we explicitly and globally account for all such crossings by defining the crossing density of a set family.

Definition 4 (Crossing Density). *The crossing density of a set family $\mathcal{F}$ is the smallest number ρ such that for any edge set $D \subseteq E$ and a laminar subfamily $\mathcal{F}_D^{lam}$ of $\mathcal{F}_D$, we have*

$$\textit{number of crossing pairs } (S, C) \textit{ where } S \in \mathcal{F}_D^{lam} \textit{ and } C \in \mathcal{C}_D \leq \rho|\mathcal{C}_D|$$

The definition of crossing density above compares the number of crossings between active sets and other sets within the family with the total number of active sets. The smaller the density, the more 'manageable' a set family is. Note that the crossing density of uncrossable families is zero since active sets do not cross other sets. The reason we restrict our attention to only laminar sub-families $\mathcal{F}_D^{lam}$ of $\mathcal{F}_D$ is due to the analysis technique of the primal-dual algorithm we utilize that relies on the construction of laminar sub-families of $\mathcal{F}_D$. If we consider all sets $S \in \mathcal{F}_D$, then the crossing density could be made artificially larger. We prove the following.

Theorem 1. *The WGMV primal-dual algorithm is a $(3 + \rho)$-approximation algorithm for the $\mathcal{F}$-augmentation problem when $\mathcal{F}$ is a pliable family with crossing density ρ.*

We also show that there exists a family of instances of the $\mathcal{F}$-augmentation problem on pliable families where the crossing density is $O(|V|)$ and the WGMV primal-dual algorithm provides an approximation ratio of $\Omega(\sqrt{|V|})$. Hence, the approximation factor in Theorem 1 can only be improved to $O(\sqrt{\rho})$. Furthermore, suppose we relax the pliability condition and require that just one of the corner cuts (instead of two) also lie within the family, then the problem becomes as hard as the hitting set problem ($\Omega(\log|V|)$-hardness of approximation). The details of these lower bound results can be found in the full version of this paper [2].

The techniques we use to prove Theorem 1 rely on the idea of constructing a laminar witness set family for the output set of edges H from the primal-dual algorithm, and devising a clever token distribution scheme to bound the average degree of active sets with respect to H i.e. $|\delta(C) \cap H|$. We introduce a novel token distribution scheme that utilizes the crossing density.

Small Cut Augmentation

Our second major contribution is in the context of α-min cuts. BCGI introduced the **Small Cut Augmentation** problem to capture the problem of augmenting α-min cuts.

Problem 2 (Small Cut Augmentation).
Input: Graph $G = (V, E \cup E_0)$ with edge costs $\{c_e\}_{e \in E}$, edge capacities $\{u_e\}_{e \in E_0}$ and a threshold $k > 0$.
Output: Subset of the edges $F \subseteq E$ with minimum $c(F) = \sum_{e \in F} c_e$ such that for every non-empty $S \subsetneq V$ with capacity $\sum_{e \in \delta(S) \cap E_0} u_e < k$, the set of edges $F \cap \delta(S)$ is non-empty.

The family $\mathcal{F}$ of cuts S with capacity $\sum_{e \in \delta(S) \cap E_0} u_e < k$ are referred to as *small cuts* and it is easy to see that changing the threshold k captures the family of α-min cuts for any $\alpha > 0$. BCGI showed that the family of small cuts is a γ-pliable family and so were able to provide a 16-approximation algorithm[4]. To improve on these approximation factors, we identify an additional property that small cuts satisfy.

Definition 5 (Sparse Crossing Property). *A set family $\mathcal{F}$ has the sparse crossing property if for any subset of the edges $D \subseteq E$, a set $S \in \mathcal{F}_D$ crosses at most one set in $\mathcal{C}_D$.*

We show that the WGMV algorithm for augmenting a γ-pliable family with the sparse crossing property is a 6-approximation algorithm leading to the following theorem.

Theorem 2. *There exists a 6-approximation algorithm for the Small Cut Augmentation Problem.*

BCGI also showed how an algorithm for the small cut augmentation problem can be used to solve the capacitated k-edge connected spanning subgraph problem. Here one is given a graph Graph $G = (V, E)$ with edge costs $\{c_e\}_{e \in E}$, edge capacities $\{u_e\}_{e \in E}$ and the goal is to find a cheapest subgraph so that the min cut in the graph has capacity at least k. Using Theorem 2, we can obtain a current best $6\lceil k/u_{min} \rceil$-approximation algorithm.

[4] Nutov [16] improved this to a 10-approximation algorithm recently.

Flexible Graph Connectivity (FGC)
Our final major contribution is in the study of the FGC problem introduced by Adjiashvili et al. [1], and generalized by Boyd et al. [5].

Problem 3 ((p,q)-Flexible Graph Connectivity).
Input: Graph $G = (V,E)$ with edge costs $\{c_e\}_{e\in E}$, and a partition of the edge set into safe edges $\mathcal{S}$ and unsafe edges $\mathcal{U}$.
Output: Subset of the edges $F \subseteq E$ with minimum $c(F) = \sum_{e\in F} c_e$ such that (V,F) is p-edge connected even after the removal of any q unsafe edges from F.

We provide the first constant factor approximation for the $(p,3)$-FGC problem improving on the previous best $O(p)$ factor [8]. Our techniques here are different depending on the parity of p. When p is even, we start with a 6-approximate solution to the $(p,2)$-FGC problem [4] and attempt to augment the family of cuts that are still violated. We show that the family of violated cuts forms a pliable family with the sparse crossing property and leverage the techniques developed in the proof of Theorem 2. When p is odd, we start with a $7+\epsilon$-approximate solution to the $(p,2)$-FGC problem [15]. Unfortunately the resulting family of violated cuts is not pliable anymore. Instead, we split these cuts into two uncrossable subfamilies based on how they cross each other and then apply the results of WGMV.

Theorem 3. *$(p,3)$-FGC admits an 12-approximation algorithm when p is even and an $11+\epsilon$-approximation algorithm when p is odd.*

Improvements for γ-Pliable Families
As a byproduct of our analysis, we also improve the results by BCGI [4] and Nutov [16] on the $\mathcal{F}$-augmentation problem when $\mathcal{F}$ is a γ-pliable family. BCGI had shown a 16-factor approximation and Nutov tightened their analysis to obtain a 10-approximation. We prove the following theorem in the full version of this paper [2].

Theorem 4. *The WGMV primal-dual algorithm is an 8-approximation algorithm for the $\mathcal{F}$-augmentation problem when $\mathcal{F}$ is a γ-pliable family (Table 1).*

Table 1. Summary of results in this paper. ρ is the crossing density of a set family and u_{min} is the minimum capacity of an edge

Problem	previous	this paper
$\mathcal{F}$-augmentation of pliable families	–	$3+\rho$
$\mathcal{F}$-augmentation of γ-pliable families	10 [16]	8
Small Cut Augmentation	10 [16]	6
Capacitated k-ECSS	$10 \cdot \lceil k/u_{min} \rceil$ [16]	$6 \cdot \lceil k/u_{min} \rceil$
$(p,3)$-FGC	$O(p)$ [8]	12

1.2 Related Work

The $\mathcal{F}$-augmentation problem was introduced by WGMV and they provided a primal dual 2-approximation algorithm for $\mathcal{F}$-augmentation when $\mathcal{F}$ is uncrossable. Bansal et al. [4] studied the $\mathcal{F}$-augmentation problem beyond uncrossable families and introduced pliable families. They provided a 16-approximation algorithm for $\mathcal{F}$-augmentation of γ-pliable families. Their analysis was later refined by Nutov [16] who showed a 10-approximation factor. Nutov [14] studied *semi-uncrossable* and showed that the primal-dual algorithm is a 2-approximation algorithm here. Jain [12] introduced a different approach to solving augmentation and connectivity problems known as iterative rounding. He provided a 2-approximation algorithm for a variety of network design problems including the survivable network design problem.

The capacitated k-edge connected spanning subgraph problem was introduced by Goemans et al. [9] who showed that there is a $2k$-approximation algorithm for the problem. Carr et al. [6] introduced the very versatile knapsack cover inequalities to strengthen the LP relaxation of the problem and exhibited an LP rounding scheme to provide an approximation factor characterized by the size of the largest *bond* in the graph. Chakrabarty et al. [7] used these knapsack cover inequalities to provide an $O(\log|V|)$-approximation algorithm based on randomized rounding. Boyd et al. [5] exhibited a k-approximation and Bansal et al. [4] provided a $16 \cdot \lceil k/u_{min} \rceil$ approximation. This was later improved by Nutov [16] to $10 \cdot \lceil k/u_{min} \rceil$.

The small cut augmentation problem was introduced by Bansal et al. [4] who provided a 16-approximation algorithm for the problem. This was improved to a factor of 10 by Nutov [16]. Nutov [15] used a repeated augmentation framework to provide a $(\alpha - \lambda_0)$-approximation algorithm where α is the threshold defining the small cuts and λ_0 is the size of the mincut in the graph.

The flexible graph connectivity problem was introduced by Adjiashvili et al. [1] and generalized by Boyd et al. [5] to the (p, q)-FGC model. Adjiashvili et al. [1] presented the first constant factor approximation for $(1, 1)$-FGC. This was improved to a factor of 2 by Boyd et al. [5] who also provided a $(q + 1)$-approximation for $(1, q)$-FGC, a 4-approximation for $(p, 1)$-FGC and an $O(q \log n)$-approximation for the general (p, q)-FGC problem. Chekuri and Jain [8] presented a $(2q + 2)$-approximation algorithm for $(2, q)$-FGC and also an $O(p)$-approximation for $(p, 2)$-FGC, $(p, 3)$-FGC and $(2p, 4)$-FGC. Bansal et al. [4] provided the first constant factor approximation for $(p, 2)$-FGC exhibiting a 6-approximation when p is even and a 20-approximation when p is odd. This was later improved by Nutov [15] to a $(7 + \epsilon)$-approximation when p is odd. Hyatt-Denesik et al. [11] study an unweighted version of $(1, q)$-FGC and exhibit a $1 + O(1/\sqrt{q})$-approximation algorithm in this case. Bansal et al. [3] study a Steiner version of $(1, 1)$-FGC from a parameterized algorithms lens.

2 Preliminaries

We begin by re-explaining the WGMV [17] primal-dual algorithm at a high level. The WGMV algorithm is based on the following linear program relaxation for the $\mathcal{F}$-augmentation problem and its dual.

Primal LP

$$\begin{aligned} \min \quad & \sum_{e \in E} c_e x_e \\ \text{subject to:} \quad & \sum_{e \in \delta(S)} x_e \geq 1 \quad \forall S \in \mathcal{F} \\ & 0 \leq x_e \leq 1 \quad \forall e \in E \end{aligned}$$

Dual LP

$$\begin{aligned} \max \quad & \sum_{S \in \mathcal{F}} y_S \\ \text{subject to:} \quad & \sum_{S \in \mathcal{F} : e \in \delta(S)} y_S \leq c_e \quad \forall e \in E \\ & y_S \geq 0 \quad \forall S \in \mathcal{F} \end{aligned}$$

The algorithm starts with an infeasible primal solution $F = \emptyset$, which corresponds to $x = \chi^F = \mathbf{0}$, and a feasible dual solution $y = \mathbf{0}$. At any time, we refer to the sets in $\mathcal{F}_F = \{S \in \mathcal{F} : \delta(S) \cap F = \emptyset\}$ as *violated sets* and its inclusion-wise minimal elements as *active sets*. We will denote the family of active sets using $\mathcal{C}$. The algorithm has two stages. In the first stage, the algorithm iteratively increases the dual values of all active sets uniformly till an edge $e \in E \setminus F$ becomes tight, and adds edge e to the set F. The first stage ends when $\mathcal{F}_F = \emptyset$. In the second stage, the algorithm removes redundant edges in reverse order of additions. This final set of edges F is returned by the algorithm.

Theorem 5 ([17]). *Let F be the set of edges returned by the primal-dual algorithm. If in every iteration of the first stage of the primal-dual algorithm,*

$$\sum_{C \in \mathcal{C}} |\delta(C) \cap F| \leq \alpha \, |\mathcal{C}|$$

where $\mathcal{C}$ is the set of active sets in that iteration, then the algorithm is an α approximation-algorithm for the $\mathcal{F}$-augmentation problem.

They showed that $\alpha = 2$ is valid when $\mathcal{F}$ is uncrossable and Bansal et al. [4] showed that $\alpha = 16$ is valid when $\mathcal{F}$ is γ-pliable. A note on oracle access required by the WGMV algorithm and future works on the primal-dual algorithm: For any subset of the edges $D \subseteq E$, the algorithm needs oracle access to the inclusion-wise minimal sets from the set family $\mathcal{F}_D$. We will assume throughout this paper that such an oracle exists.

2.1 Witness Sets and Their Red-Green Coloring

We mention some lemmas from [4] that will be useful to us. We use the shorthand $\delta_F(C)$ to mean $\delta(C) \cap F$.

Lemma 1 ([4]). *Suppose $\mathcal{F}$ is a pliable family. For any edge $e \in \cup_{C \in \mathcal{C}} \delta_F(C)$, there exists a witness set $S_e \in \mathcal{F}$ such that (i) S_e is violated in the current iteration and (ii) $\delta_F(S_e) = \{e\}$.*

Furthermore, there exists a laminar family of witness sets for edges in $\cup_{C\in\mathcal{C}}\delta_F(C)$.

Let $\mathcal{L}$ be the laminar family of the witness sets from Lemma 1 together with the node set V. This laminar family can be represented by a rooted directed out-tree $\mathcal{T}$ with a node v_S for every set $S \in \mathcal{L}$. The root is the node v_V and there is an edge from v_S to v_T if and only if S is the smallest set in $\mathcal{L}$ that properly contains T. A node v_S in $\mathcal{T}$ is called active if there exists an active set $C \in \mathcal{C}$ such that S is the smallest set in $\mathcal{L}$ that contains C. We color all active nodes using **red**, and color all non-red nodes with out degree at least 2 using **green**. Every other node is colored black (uncolored). Let n_R, n_G and n_B be the number of red, green and black nodes respectively. Let n_L be the number of leaves in $\mathcal{T}$.

Lemma 2 ([4]). *The following statements are true.*

(i) The sets in $\mathcal{C}$ *are pairwise disjoint.*
(ii) Every leaf node of $\mathcal{T}$ *is colored red.*
(iii) $n_G \le n_L \le n_R \le |\mathcal{C}|$.

Lemma 3. *Let* $C \in \mathcal{C}$ *and* $S \in \mathcal{L}\setminus\{V\}$. *If* $S\cap C$ *and* $C\setminus S$ *are non-empty then* S *crosses* C.

Proof. If $S\setminus C$ is empty, then $S \subsetneq C$ contradicting the minimality of C. If $\overline{S\cup C}$ is empty, then one of $S\cap C$ or $C\setminus S$ must also be violated due to pliable property. But then again, a proper subset of C is violated which is a contradiction. □

3 Crossing Density

In this section, we prove Theorem 1. We start by distributing some tokens to $\mathcal{T}$ and make some observations and claims.

> *Token Distribution Phase I*
> For every active set $C \in \mathcal{C}$ and $e \in \delta_F(C)$, assign a token to the node v_{S_e} where S_e is the witness set such that $\delta_F(S_e) = \{e\}$.

Observe that the number of tokens distributed is exactly equal to $\sum_{C\in\mathcal{C}} \delta_F(C)$. Furthermore, the node v_{S_e} has tokens if there are active sets C with $e \in \delta(C)$. There can be at most two such active sets each containing an end point of edge e exclusively (since active sets are disjoint by Lemma 2). Thus, each node in $\mathcal{T}$ has at most two tokens. Additionally, each non-root node v_S in $\mathcal{T}$ has at least one token since S is a witness set and so $\delta_F(S) = \{e\}$ for some edge $e \in \cup_{C\in\mathcal{C}}\delta_F(C)$. Note that the root node v_V does not have tokens.

Claim 1. *Let* $v_{S_1}, v_{S_2}, \ldots, v_{S_k}$ *be a directed path in* $\mathcal{T}$ *where none of* $v_{S_1}, v_{S_2}, \ldots, v_{S_{k-1}}$ *is red. Let* $e_k = (u, v)$ *be the edge such that* $\delta_F(S_k) = \{e_k\}$ *with* $u \in S_k$. *Then,*

(i) If there exists $C \in \mathcal{C}$ *with* $v \in C, u \notin C$, *then* C *crosses all the sets* $S_1, S_2, \ldots, S_{k-1}$.

(ii) If there exists $C \in \mathcal{C}$ with $u \in C, v \notin C$ and v_{S_k} is not red, then C crosses all the sets $S_1, S_2, \ldots, S_k$.

Proof. (i) The active set C cannot be contained in S_k since $v \in C \setminus S_k$. Also C is not contained in any of $S_1, \ldots, S_{k-1}$, as otherwise one of these sets must be the smallest witness set containing C and the corresponding node v_S will be colored red. Furthermore, $v \in S_i$ for every $i \leq k-1$ since otherwise $e_k \in \delta(S_i)$ contradicting property (ii) of witness sets in Lemma 1. $C \setminus S_i$ are non-empty and so C crosses S_i (by Lemma 3).

(ii) The active set C cannot be contained in S_k since otherwise there would have to exist a witness set S such that $C \subseteq S \subsetneq S_k$. But then $e_k \in \delta(S)$ contradicting property (ii) of witness sets in Lemma 1. As before, this implies that C is not contained in any of $S_1, \ldots, S_{k-1}$. But now for all $i \leq k$, $C \cap S_i$ is non-empty as $u \in C \cap S_i$ and $C \setminus S_i$ is non-empty. Hence C crosses all the sets $S_1, \ldots, S_k$. □

3.1 Proof of Theorem 1

We first decompose the tree $\mathcal{T}$ into vertex-disjoint paths as follows. Initially, our set of paths $\mathcal{P}$ is empty. While there exists a red node v_{S_k} that is not included in some path in $\mathcal{P}$, consider the unique path $v_{S_1}, v_{S_2}, \ldots, v_{S_k}$ in $\mathcal{T}$ such that none of the nodes $v_{S_1}, v_{S_2}, \ldots, v_{S_{k-1}}$ is colored red, and the parent of v_{S_1} either belongs to a path in $\mathcal{P}$ or is colored red or is v_V. Add this path to $\mathcal{P}$ and repeat. Note that at termination, every non-root node in $\mathcal{T}$ is part of a unique path in $\mathcal{P}$ since all leaves of $\mathcal{T}$ are colored red (by Lemma 2). Now consider a path $v_{S_1}, v_{S_2}, \ldots, v_{S_k}$ in $\mathcal{P}$. Here, the node v_{S_k} is colored red. We will analyze the number of crossings between active sets and witness sets $S_1, \ldots, S_k$ and compare it with the number of tokens on this path.

Claim 2. *Let $v_{S_1}, v_{S_2}, \ldots, v_{S_k}$ be a path in $\mathcal{P}$. Let t be the number of tokens on this path. Then, the number of crossing pairs (C, S_i) where $C \in \mathcal{C}$ and $1 \leq i \leq k$ is at least $t-3$.*

Proof. Case 1: Suppose each node $v_{S_1}, \ldots, v_{S_{k-1}}$ has one token. Let e_{k-1} be the edge such that $\delta_F(S_{k-1}) = \{e_{k-1}\}$. Edge e_{k-1} is incident to some active set C, and this active set must cross $S_1, \ldots, S_{k-2}$ by Claim 1. Since v_{S_k} has at most two tokens, the number of tokens on this path is at most $k+1$ and the number of crossing pairs is at least $k-2$.

Case 2: Let i be the largest index such that v_{S_i} has two tokens and $i \leq k-1$ so that v_{S_i} is not colored red. Let e_i be the edge such that $\delta_F(S_i) = \{e_i\}$. Then, there are two active sets C_1, C_2 incident to each endpoint of e_i respectively and by Claim 1, C_1 crosses all the sets $S_1, \ldots, S_i$ and C_2 crosses all the sets $S_1, \ldots, S_{i-1}$.

Case 2.1: Suppose v_{S_k} has two tokens, then by Claim 1, there exists an active set C that crosses all the sets $S_1, \ldots, S_{k-1}$. Note that C could be the same as C_1 or C_2. But now, the number of tokens on the path is at most $2i+k-i+1 = k+i+1$ and the number of crossing pairs is at least $i-1+k-1 = k+i-2$.

Case 2.2: Suppose v_{S_k} has one token. Let e_{k-1} be the edge such that $\delta_F(S_{k-1}) = \{e_{k-1}\}$. As earlier, by Claim 1, there exists an active set that crosses $S_1, \ldots, S_{k-2}$. But now, the number of tokens on the path is at most $2i+k-i = k+i$ and the number of crossing pairs is at least $i-1+k-2 = k+i-3$. □

For any path in $\mathcal{P}$, Claim 2 shows that the number of tokens on this path is at most 3 plus the number of crossing pairs (C, S) such that $C \in \mathcal{C}$ and v_S is on this path. Since each path contains exactly one red node, and paths in $\mathcal{P}$ are vertex disjoint, this implies that the total number of tokens is bounded by $3n_R + a$ where a is the number of crossing pairs (S, C) where $S \in \mathcal{L}$ and $C \in \mathcal{C}$. This is upper bounded by $3|\mathcal{C}| + \rho|\mathcal{C}|$ and so $\sum_{C \in \mathcal{C}} |\delta_F(C)| \leq (3+\rho)|\mathcal{C}|$. Hence, Theorem 5 implies that we have proven Theorem 1.

4 Small Cut Augmentation

In this section, we prove theorem 2. Recall the definition of sparse crossing properties mentioned in the introduction.

Lemma 4. *The family of small cuts satisfies the sparse crossing property.*

Proof. Let $\mathcal{F}$ be a small cut family. Let $D \subseteq E$ and $S \in \mathcal{F}_D$ and suppose that S crosses two minimal sets of $\mathcal{F}_D$ say C_1 and C_2. Define $u(T_1, T_2)$ for sets $T_1, T_2 \subseteq V$ to be $\sum_{e \in E_0(T_1, T_2)} u_e$ where $E_0(T_1, T_2)$ is the set of edges in E_0 with exactly one end point in T_1 and exactly the other end point in T_2. Also define $u(T) = u(T, \overline{T})$. The following equation is easy to verify:

$$u(C_1 \setminus S) + u(S \cap C_1) = u(C_1) + 2u(S \cap C_1, C_1 \setminus S)$$

Since C_1 is minimal in $\mathcal{F}_D$, we can conclude that $S \cap C_1$ and $C_1 \setminus S$ are not in $\mathcal{F}_D$. Hence, $u(S \cap C_1)$ and $u(C_1 \setminus S)$ are both at least the threshold k. Furthermore $u(C_1) < k$ since $C_1 \in \mathcal{F}$. Hence, $u(S \cap C_1, C_1 \setminus S) \geq k/2$. Similarly $u(S \cap C_2, C_2 \setminus S) \geq k/2$. Now since C_1 and C_2 are disjoint (By Lemma 2), we can conclude that $u(S) \geq k$. But this contradicts the fact that $S \in \mathcal{F}$. Hence S cannot cross two minimal sets C_1 and C_2. □

Using this sparse crossing property, one can refine the analysis in BCGI [4] that showed a 16-approximation algorithm to a 6-approximation algorithm. The proof is technical and lengthy and is deferred to the full version of this paper [2].

5 Flexible Graph Connectivity

In this section, we prove Theorem 3. The cases when p is even and odd are handled separately. We begin by introducing some notation that will be useful in the discussions to follow.

5.1 Notation and Terminology

We will refer to the set of unsafe edges as $\mathcal{U}$. A cut of a graph is simply $\delta(A)$ for some subset of the vertices $A \subseteq V$, and when the context is clear we might refer to the cut as A instead of $\delta(A)$. For $A, B \subseteq V$ with $A \cap B = \emptyset$, we define $\mathbf{E(A, B)}$ to be the set of edges with one end point in A and the other in B.

Let $\mathcal{F}$ be a family of cuts of a graph G. We define an equivalence relation between vertices of the graph G as follows: $\mathbf{u} \equiv \mathbf{v}$ if and only if for all $C \in \mathcal{F}$, either $u, v \in C$ or $u, v \notin C$. The residual graph of G with respect to $\mathcal{F}$ is obtained by contracting each equivalence class of vertices into a single vertex. We will call this graph $\mathbf{G}_{\mathcal{F}} = (\mathbf{V}_{\mathcal{F}}, \mathbf{E}_{\mathcal{F}})$. We will call a vertex (edge) in $G_{\mathcal{F}}$ a **residual vertex (edge)**. Note that any residual vertex (edge) is a subset of vertices (edges) in the original graph G. Hence the cardinality of a residual edge makes sense and so does the cardinality of the unsafe edges in a residual edge. We will call a residual edge **thin** if it contains at most one unsafe edge.

For two sets $A, B \subseteq V$, we refer to the cuts $\delta(A \setminus B), \delta(A \cap B), \delta(B \setminus A), \delta(A \cup B)$ as **corner cuts**. In the graph $G_{\{A,B\}}$, we refer to the residual edges $(A \cap B, \overline{A \cup B})$ and $(A \setminus B, B \setminus A)$ as **diagonal edges**. We refer to the residual edges $(A \cap B, A \setminus B)$ and $(A \cap B, B \setminus A)$ as **internal edges** and the residual edges $(\overline{A \cup B}, A \setminus B)$ and $(\overline{A \cup B}, B \setminus A)$ as **external edges**. We say that a set $A \subseteq V$ **amenably crosses** a set $B \subseteq V$ if (i) A crosses B and (ii) either the residual internal edge $(A \cap B, B \setminus A)$ or the residual external edge $(A \setminus B, \overline{A \cup B})$ does not contain any unsafe edges.

We say that a graph $G = (V, E)$ is $\mathbf{(p, q)}$**-flex-connected** if for any subset of the unsafe edges $F \subseteq \mathcal{U}$ with $|F| \leq q$, the graph $G = (V, E \setminus F)$ is p-edge connected. For a graph G, we call a cut $\mathbf{(p, q)}$**-violated** if it contains exactly $p + q - 1$ edges, at least q of which are unsafe.

5.2 Approximation Algorithms

We employ the augmentation framework for flexible graph connectivity utilized by [4,5,8,15]. We start with a solution $H = (V, F)$ that is $(p, 2)$-flex-connected using the 6-approximation algorithm in [4] when p is even and the $7 + \epsilon$-approximation algorithm in [15] when p is odd. At this point, each cut either has p safe edges or $p + 2$ edges in total. Hence, the only cuts which require more edges to make H $(p, 3)$-flex-connected are cuts with fewer than p safe edges and exactly $p + 2$ edges in total. These are precisely the $(p, 3)$-violated cuts and we will refer to these as simply violated cuts henceforth. Furthermore, adding just a single edge (safe or unsafe) to these violated cuts will ensure that they have $p + 3$ edges in total and so H would become $(p, 3)$-flex-connected. Thus, we only need to solve the $\mathcal{F}$-augmentation problem where $\mathcal{F}$ is the family of $(p, 3)$-violated cuts of a $(p, 2)$-flex-connected graph H. We prove the following theorems. The proofs are quite technical and can be found in the full version of this paper [2].

Theorem 6. *Let $p \geq 2$ be even and let $G = (V, E)$ be a $(p, 2)$-flex-connected graph. Then the family of violated cuts forms a γ-pliable family with the sparse crossing property.*

Theorem 7. *Let $p \geq 3$ be odd let $G = (V, E)$ be a $(p, 2)$-flex-connected graph. Let $\mathcal{F}$ be the family of violated cuts and let $\mathcal{F}_1 \subset \mathcal{F}$ be the family violated cuts that either (i) do not cross any other violated set or (ii) cross at least one other violated set amenably. Then, $\mathcal{F}_1$ and $\mathcal{F} \setminus \mathcal{F}_1$ are both uncrossable families.*

Theorem 6 allows us to augment the family of violated cuts incurring an approximation factor of 6 via techniques developed for Theorem 2. Thus, we obtain an 12-approximation algorithm for $(p, 3)$-FGC when p is even. Theorem 7 allows us to augment each of the families $\mathcal{F}_1$ and $\mathcal{F} \setminus \mathcal{F}_1$ incurring an approximation factor of 2 each. Thus, we obtain an $11 + \epsilon$-approximation algorithm for $(p, 3)$-FGC when p is odd.

References

1. Adjiashvili, D., Hommelsheim, F., Mühlenthaler, M.: Flexible graph connectivity. Math. Program. **192**, 409–441 (2022). https://doi.org/10.1007/s10107-021-01664-9
2. Bansal, I.: A global analysis of the primal-dual method for pliable families. **2308**, 15714 (2024)
3. Bansal, I., Cheriyan, J., Grout, L., Ibrahimpur, S.: Algorithms for 2-connected network design and flexible steiner trees with a constant number of terminals. In: Megow, N., Smith, A.D. (eds.) Approximation, Randomization, and Combinatorial Optimization. Algorithms and Techniques, APPROX/RANDOM 2023, 11–13 September 2023, Atlanta, Georgia, USA. LIPIcs, vol. 275, pp. 14:1–14:14. Schloss Dagstuhl - Leibniz-Zentrum für Informatik (2023). https://doi.org/10.4230/LIPICS.APPROX/RANDOM.2023.14
4. Bansal, I., Cheriyan, J., Grout, L., Ibrahimpur, S.: Improved approximation algorithms by generalizing the primal-dual method beyond uncrossable functions. Algorithmica **86**(8), 2575–2604 (2024). https://doi.org/10.1007/S00453-024-01235-2
5. Boyd, S.C., Cheriyan, J., Haddadan, A., Ibrahimpur, S.: Approximation algorithms for flexible graph connectivity. Math. Program. (2023). https://doi.org/10.1007/s10107-023-01961-5
6. Carr, R.D., Fleischer, L.K., Leung, V.J., Phillips, C.A.: Strengthening integrality gaps for capacitated network design and covering problems. In: Proceedings of the Eleventh Annual ACM-SIAM Symposium on Discrete Algorithms, pp. 106–115. Society for Industrial and Applied Mathematics, USA (2000)
7. Chakrabarty, D., Chekuri, C., Khanna, S., Korula, N.: Approximability of capacitated network design. Algorithmica **72**(2), 493–514 (2014). https://doi.org/10.1007/s00453-013-9862-4
8. Chekuri, C., Jain, R.: Approximation algorithms for network design in non-uniform fault models. In: Proceedings of the 50th International Colloquium on Automata, Languages, and Programming, vol. 261, Article 36, pp. 1–20 (2023). https://doi.org/10.4230/LIPIcs.ICALP.2023.36

9. Goemans, M.X., Goldberg, A.V., Plotkin, S.A., Shmoys, D.B., Tardos, É., Williamson, D.P.: Improved approximation algorithms for network design problems. In: Proceedings of the 5th Symposium on Discrete Algorithms, pp. 223–232 (1994)
10. Goemans, M.X., Williamson, D.P.: A general approximation technique for constrained forest problems. SIAM J. Comput. **24**(2), 296–317 (1995). https://doi.org/10.1137/S0097539793242618
11. Hyatt-Denesik, D., Ameli, A.J., Sanità, L.: Improved approximations for flexible network design. In: Chan, T.M., Fischer, J., Iacono, J., Herman, G. (eds.) 32nd Annual European Symposium on Algorithms, ESA 2024, 2–4 September 2024, Royal Holloway, London, United Kingdom. LIPIcs, vol. 308, pp. 74:1–74:14. Schloss Dagstuhl - Leibniz-Zentrum für Informatik (2024). https://doi.org/10.4230/LIPICS.ESA.2024.74
12. Jain, K.: A factor 2 approximation algorithm for the generalized Steiner network problem. Combinatorica **21**(1), 39–60 (2001). https://doi.org/10.1007/s004930170004
13. Jain, K., Măndoiu, I., Vazirani, V.V., Williamson, D.P.: A primal-dual schema based approximation algorithm for the element connectivity problem. J. Algorithms **45**(1), 1–15 (2002)
14. Nutov, Z.: Extending the primal-dual 2-approximation algorithm beyond uncrossable set families (2023). https://arxiv.org/abs/2307.08270
15. Nutov, Z.: Improved approximation algorithms for some capacitated k edge connectivity problems. CoRR abs/2307.01650 (2023). https://doi.org/10.48550/ARXIV.2307.01650
16. Nutov, Z.: Improved approximation ratio for covering pliable set families. CoRR (2024). https://doi.org/10.48550/arXiv.2404.00683, https://arxiv.org/abs/2404.00683
17. Williamson, D.P., Goemans, M.X., Mihail, M., Vazirani, V.V.: A primal-dual approximation algorithm for generalized Steiner network problems. Combinatorica **15**(3), 435–454 (1995). https://doi.org/10.1007/BF01299747

Monotone Submodular Multiway Partition

Richard Bi[1], Karthekeyan Chandrasekaran[1], and Soham Joshi[2](✉)

[1] Grainger College of Engineering, University of Illinois, Urbana-Champaign, USA
rbi3@illinois.edu, karthe@illinois.edu
[2] IIT Bombay, Mumbai, India
sohamjoshi@cse.iitb.ac.in

Abstract. In submodular multiway partition (Sub-MP), the input is a non-negative submodular function $f : 2^V \to \mathbb{R}_{\geq 0}$ given by an evaluation oracle along with k terminals $t_1, t_2, \ldots, t_k \in V$. The goal is to find a partition $V_1, V_2, \ldots, V_k$ of V with $t_i \in V_i$ for every $i \in [k]$ in order to minimize $\sum_{i=1}^{k} f(V_i)$. In this work, we focus on Sub-MP when the input function is monotone (termed Mono-Sub-MP). Mono-Sub-MP formulates partitioning problems over several interesting structures—e.g., matrices, matroids, graphs, and hypergraphs. Mono-Sub-MP is NP-hard since the graph multiway cut problem can be cast as a special case. We investigate the approximability of Mono-Sub-MP: we show that it admits a 4/3-approximation and does not admit a $(10/9 - \epsilon)$-approximation for every constant $\epsilon > 0$. Next, we study a special case of Mono-Sub-MP where the monotone submodular function of interest is the coverage function of an input graph, termed Graph-Cov-MP. Graph-Cov-MP is equivalent to the classic multiway cut problem for the purposes of exact optimization. We show that Graph-Cov-MP admits a 1.125-approximation and does not admit a $(1.00074 - \epsilon)$-approximation for every constant $\epsilon > 0$ assuming the Unique Games Conjecture. These results separate Graph-Cov-MP from graph multiway cut in terms of approximability.

Keywords: Approximation Algorithms · Multiway Partition · Monotone Submodular Functions

1 Introduction

A set function $f : 2^V \to \mathbb{R}$ is *submodular* if $f(A) + f(B) \geq f(A \cap B) + f(A \cup B)$ for every $A, B \subseteq V$. We focus on the submodular multiway partition problem denoted Sub-MP: The input here is a non-negative submodular function $f : 2^V \to \mathbb{R}_{\geq 0}$ given by an evaluation oracle[1] and a collection $T = \{t_1, t_2, \ldots, t_k\} \subseteq$

R. Bi and K. Chandrasekaran—Supported in part by NSF grant CCF-2402667.
S. Joshi—Work done during REU while visiting UIUC. Supported in part by NSF grant CCF-2402667.

[1] The evaluation oracle for a function $f : 2^V \to \mathbb{R}$ takes a set $S \subseteq V$ as input and returns $f(S)$.

N. Megow and A. Basu (Eds.): IPCO 2025, LNCS 15620, pp. 72–85, 2025.
https://doi.org/10.1007/978-3-031-93112-3_6

V of terminals. The goal is to find a partition $V_1, V_2, \ldots, V_k$ of V with $t_i \in V_i$ for every $i \in [k]$ in order to minimize $\sum_{i=1}^{k} f(V_i)$, i.e.,

$$\min \left\{ \sum_{i=1}^{k} f(V_i) : V_1, V_2, \ldots, V_k \text{ is a partition of } V \text{ such that } t_i \in V_i \ \forall\, i \in [k] \right\}.$$

Throughout this work, we assume that the input function f is non-negative, i.e., $f(A) \geq 0$ for every $A \subseteq V$ and normalized, i.e., $f(\emptyset) = 0$. SUB-MP was introduced by Zhao, Nagamochi, and Ibaraki [17] as a unified generalization of several partitioning problems—e.g., edge multiway cut in graphs/hypergraphs and node multiway cut in graphs. We recall that a set function $f : 2^V \to \mathbb{R}$ is *symmetric* if $f(A) = f(V \setminus A)$ for every $A \subseteq V$ and is *monotone* if $f(A) \leq f(B)$ for every $A \subseteq B \subseteq V$. In SUB-MP, if the input function is symmetric, then we call the problem as symmetric submodular multiway partition, denoted SYM-SUB-MP and if the input function is monotone, then we call the problem as monotone submodular multiway partition, denoted MONO-SUB-MP. The focus of this work is on MONO-SUB-MP.

Motivations and Connections. We motivate MONO-SUB-MP via three important special cases: (1) In Matrix Multiway Partition, denoted MATRIX-MP, the input is a matrix $M \in \mathbb{R}^{n \times m}$ along with k specified rows $r_1, r_2, \ldots, r_k \in [n]$ and the goal is to partition the n rows into $V_1, V_2, \ldots, V_k$ with $r_i \in V_i$ for every $i \in [k]$ in order to minimize the sum of the dimensions of the subspaces spanned by the parts, i.e., minimize $\sum_{i=1}^{k} \text{dimension}(V_i)$. This is a special case of MONO-SUB-MP since the dimension function is non-negative, normalized, monotone, and submodular. MATRIX-MP is a fundamental linear algebra problem and is likely to have applications in data analysis. (2) In Matroid Multiway Partition, denoted MATROID-MP, the input is the rank function $r : 2^V \to \mathbb{Z}_{\geq 0}$ of a matroid (given by an evaluation oracle) along with a collection $t_1, \ldots, t_k \in V$ of elements. The goal is to partition the ground set V into k parts $V_1, V_2, \ldots, V_k$ with $t_i \in V_i$ for every $i \in [k]$ in order to minimize $\sum_{i=1}^{k} r(V_i)$. MATROID-MP is a special case of MONO-SUB-MP since the matroid rank function is non-negative, normalized, monotone, and submodular. We note that MATRIX-MP is a special case of MATROID-MP. (3) In Coverage Multiway Partition, denoted COV-MP, the input is a hypergraph $H = (V, E)$ along with k terminals $t_1, t_2, \ldots, t_k \in V$ and the goal is to partition the vertex set into k parts $V_1, V_2, \ldots, V_k$ in order to minimize the sum of the coverage values of the parts, i.e., minimize $\sum_{i=1}^{k} b(V_i)$, where $b(X)$ is the number of hyperedges in H intersecting the vertices in X. COV-MP is a special case of MONO-SUB-MP since the coverage function $b : 2^V \to \mathbb{Z}_{\geq 0}$ is non-negative, normalized, monotone, and submodular. We emphasize that all three special cases mentioned above are themselves fairly general and numerous other partitioning problems can be cast as special cases of these problems as well. Thus, our motivation behind studying MONO-SUB-MP is to understand whether we can get good approximations for all these problems simultaneously.

In COV-MP, if the input hypergraph is simply a graph, then we denote the resulting problem as GRAPH-COV-MP. GRAPH-COV-MP is an important special case of Mono-Sub-MP, so we will study this special case as well in this work. Graph-Cov-MP is closely related to the well-studied graph multiway cut problem (denoted Graph-Multiway-Cut). The input in both problems is a graph with edge weights $G = (V, E, w : E \to \mathbb{R}_+)$ and a collection of k terminals $t_1, \ldots, t_k \in V$. The goal in both problems is to find a partition $V_1, \ldots, V_k$ of the vertex set with $t_i \in V_i$ for every $i \in [k]$ in order to minimize an objective function. The objective function in Graph-Cov-MP is $\sum_{i=1}^{k} b(S_i)$, where $b : 2^V \to \mathbb{R}_{\geq 0}$ is the graph coverage function defined as $b(S) := \sum_{e \in E: e \cap S \neq \emptyset} w_e$ for all $S \subseteq V$. The objective function in Graph-Multiway-Cut is $(1/2) \sum_{i=1}^{k} d(S_i)$, where $d : 2^V \to \mathbb{R}_{\geq 0}$ is the graph cut function defined as $d(S) := \sum_{e \in E: e \cap S \neq \emptyset, e \setminus S \neq \emptyset} w_e$ for all $S \subseteq V$. We recall that the graph coverage function is monotone submodular while the graph cut function is symmetric submodular. Moreover, the objective in Graph-Cov-MP is a translation of the objective in Graph-Multiway-Cut by an additive $w(E) := \sum_{e \in E} w_e$. This connection immediately implies that Graph-Cov-MP is NP-hard (since Graph-Multiway-Cut is known to be NP-hard). However, there is no approximation preserving reduction between Graph-Multiway-Cut and Graph-Cov-MP (in either direction). This situation raises the following question: to what extent do the approximation and inapproximability techniques for Graph-Multiway-Cut extend to Graph-Cov-MP?

Approximability. Sub-MP, Sym-Sub-MP, and Mono-Sub-MP are all NP-hard since the classic graph multiway cut problem can be cast as a special case of all these problems. Zhao, Nagamochi, and Ibaraki [17] introduced a greedy splitting approach and showed that it implies a $(k-1)$-approximation for Sub-MP and a $(2-2/k)$-approximation for both Sym-Sub-MP and Mono-Sub-MP. Chekuri and Ene [6] introduced the Lovász extension based convex programming relaxation for Sub-MP. They designed rounding algorithms for this convex program that achieved a 2-approximation for Sub-MP and a 3/2-approximation for Sym-Sub-MP—we note that their rounding algorithms are different for Sub-MP and Sym-Sub-MP. In subsequent work, Ene, Vondrák, and Wu [10] improved the rounding algorithm for Sub-MP to achieve a $(2-2/k)$-approximation. On the hardness side, Ene, Vondrák, and Wu [10] showed that for every constant $\epsilon > 0$, (1) a $(2-2/k-\epsilon)$-approximation for Sub-MP requires exponentially many function evaluation queries and (2) a $(1.268-\epsilon)$-approximation for Sym-Sub-MP requires exponentially many function evaluation queries.

We now discuss the complexity and approximability of Mono-Sub-MP. Graph-Cov-MP is NP-hard and consequently, Mono-Sub-MP is also NP-hard. As mentioned in the previous paragraph, two independent prior approaches for Mono-Sub-MP achieved a $(2-2/k)$-approximation: (i) the greedy splitting technique [17] and (ii) rounding an optimum solution to the Lovász extension based convex program for Sub-MP [6]. We observe that there is a considerably simpler and faster (linear-time) algorithm for Mono-Sub-MP that achieves a $(2-1/k)$-approximation: order the terminal set as $t_1, t_2, \ldots, t_k$ where $f(\{t_1\}) \leq f(\{t_2\}) \leq \ldots \leq f(\{t_k\})$ and return the partition $V_1 := \{t_1\}$,

$V_2 := \{t_2\}, \ldots, V_{k-1} := \{t_{k-1}\}, V_k := (V \setminus T) \cup \{t_k\}$ (see full version for the approximation factor analysis of this algorithm). Given this simple and linear-time 2-approximation and the plethora of applications that can be cast as MONO-SUB-MP, we focus on understanding the approximability of MONO-SUB-MP in this work. Another technical motivation to study MONO-SUB-MP arises from the convex program that is used to approximate SUB-MP. If the input function is monotone in addition to being submodular, then can we design a rounding algorithm for the convex program with better approximation guarantee (i.e., better than what is possible for submodular or symmetric submodular functions)?

GRAPH-COV-MP is a special case of MONO-SUB-MP and is closely related to the well-studied GRAPH-MULTIWAY-CUT. GRAPH-MULTIWAY-CUT is a fundamental graph partitioning problem and has been studied extensively in the literature. Its rich structure enables an approximation preserving reduction to a geometric problem, namely simplex partitioning [5,7,12]. In spite of this reduction to a geometric problem, pinning down the approximability of GRAPH-MULTIWAY-CUT has remained a tantalizing open question with several works devoted to narrowing the gap between the upper and lower bounds [1,2,4,11,14,15]. For GRAPH-MULTIWAY-CUT, the current largest inapproximation factor is 1.20016 and the current smallest approximation factor is 1.2965. Although GRAPH-COV-MP is equivalent to GRAPH-MULTIWAY-CUT for the purposes of exact optimization, we will later see that they are not equivalent in terms of approximability.

1.1 Results and Techniques

As mentioned previously, there are three known 2-approximations for MONO-SUB-MP—the greedy splitting approach, the convex program approach, and the simple linear-time greedy approach described above. Our first result improves the approximability of MONO-SUB-MP.

Theorem 1. *MONO-SUB-MP admits a* $4/3$*-approximation.*

This result is based on rounding an optimum solution to the Lovász extension based convex program for SUB-MP introduced by Chekuri and Ene [6]. We use a threshold rounding algorithm similar to Chekuri and Ene. However, the interval for the threshold that we use is different from theirs. For SUB-MP, Chekuri and Ene pick a uniform random threshold from the interval $[1/2, 1]$ in order to ensure that sets obtained by rounding do not overlap and hence, give a partition. For SYM-SUB-MP, they pick a uniform random threshold from the interval $[0, 1]$ which leads to overlapping sets, but these sets can be uncrossed to obtain a partition without increasing the objective value owing to the posimodularity property of symmetric submodular functions. For MONO-SUB-MP, we pick a uniform random threshold from the interval $[1/4, 1]$. This again leads to overlapping sets, but we uncross these sets arbitrarily without increasing the objective by exploiting the monotonicity of the function. We emphasize that our approximation factor analysis differs considerably from that of Chekuri and Ene as well as the subsequent work of Ene, Vondrák, and Wu [10] (although we use some of their ingredients). Our analysis differs since our threshold interval is different and we need to exploit monotonicity to bound the approximation factor. Moreover, the approximation factor that we achieve for MONO-SUB-MP is 4/3 which is smaller than the known approximation factor for SUB-MP and SYM-SUB-MP. We show that the approximation factor analysis of our algorithm is tight via an example. Hence, improving on the approximability requires new ideas.

Our next result shows a lower bound on the approximability assuming that the algorithm makes polynomial number of function evaluation queries.

Theorem 2. *For every constant $\epsilon > 0$, every algorithm for MONO-SUB-MP that achieves an approximation factor of $(\frac{10}{9} - \epsilon)$ requires $2^{\Omega(n)}$ function evaluation queries, where n is the size of the ground set.*

This lower bound result on the inapproximability of MONO-SUB-MP is based on the *symmetry gap* machinery introduced by Vondrák [16]. Vondrák introduced the symmetry gap machinery to show inapproximability results in the evaluation oracle input model (i.e., assuming that the algorithm makes only polynomial number of function evaluation queries) for constrained submodular *maximization* problems. Subsequently, Ene, Vondrák, and Wu [10] adapted this machinery to show inapproximability results in the evaluation oracle model for SUB-MP and SYM-SUB-MP, which are *minimization* problems related to submodular objectives. Informally, the symmetry gap of a function relative to an optimization problem is the largest ratio between the objective values of an optimum symmetric solution (under an appropriately chosen notion of symmetry) and an optimum solution to the problem. We adapt Vondrák's [16] and Ene, Vondrák, and Wu's [10] machinery in a straightforward way to conclude that the symmetry gap of MONO-SUB-MP is a lower bound on its inapproximability factor. Given this status, the technical challenge in using the symmetry gap machinery for inapproximability involves constructing instances of the problem which have large symmetry gap. This is our key contribution: we construct an instance of MONO-SUB-MP that has symmetry gap at least 10/9. We note that our analysis of the symmetry gap of our instance is tight. Thus, improving the inapproximability factor beyond 10/9 via the symmetry gap machinery requires fundamentally new types of instances. We mention that the symmetry gap machinery allows query complexity based inapproximability results to be converted to computation based inapproximability results assuming NP $\neq$ RP [9].

Next, we turn to GRAPH-COV-MP which is a special case of MONO-SUB-MP. Its close relationship to GRAPH-MULTIWAY-CUT inspires our results. We note that NP-hardness of GRAPH-MULTIWAY-CUT implies the NP-hardness of GRAPH-COV-MP. It is easy to show that if GRAPH-MULTIWAY-CUT admits an α-approximation, then GRAPH-COV-MP admits a $(1 + \alpha)/2$-approximation (see full version for a proof). Thus, the current best 1.2965-approximation for GRAPH-MULTIWAY-CUT implies a 1.14825-approximation for GRAPH-COV-MP. We improve the approximability of GRAPH-COV-MP beyond this trivial guarantee.

Theorem 3. *Graph-Cov-MP admits a* 1.125*-approximation.*

Our algorithm for Graph-Cov-MP is based on rounding an optimum solution to a linear programming relaxation. The linear programming relaxation associated with Graph-Cov-MP has a different objective compared to that of Graph-Multiway-Cut, but the constraints are the same. Buchbinder, Naor, and Schwartz [4] designed the exponential clocks rounding scheme for the LP-relaxation of Graph-Multiway-Cut and showed that it achieves an approximation factor of 1.32388. The analysis was subsequently improved by Sharma and Vondrák [15] to show that it achieves a 1.309017-approximation for Graph-Multiway-Cut. Our main contribution is analyzing the approximation factor of the same algorithm for the objective of Graph-Cov-MP and bounding it by 1.125.

Next, we turn to the hardness of Graph-Cov-MP. It is known that Graph-Multiway-Cut is APX-hard via reduction from the max-cut problem [8]. We adapt the same reduction to conclude the following result.

Theorem 4. *For every constant* $\epsilon > 0$*, Graph-Cov-MP does not admit a* $(1.00036-\epsilon)$*-approximation assuming* $P \neq NP$ *and does not admit a* $(1.00074-\epsilon)$*-approximation assuming the unique games conjecture.*

Our lower and upper bounds for Graph-Cov-MP shows a separation of this problem from Graph-Multiway-Cut in terms of approximability although they are equivalent in terms of exact optimization. We conclude the section with a table of results showing the current best upper and lower bounds on the approximability of submodular partitioning problems that are of interest to this work (see Table 1). Due to space limitations, we present our proof of Theorem 1 here and refer the reader to the full version of the paper [3] for proofs of the rest of the theorems and further discussion.

Table 1. Upper bound and lower bound on the approximability of submodular partitioning problems. We have included known results for Sub-MP, Sym-Sub-MP, and Graph-Multiway-Cut for the sake of comparison. The lower bounds for Sub-MP, Sym-Sub-MP, and Mono-Sub-MP are assuming polynomial number of function evaluations. The lower bounds for Graph-Cov-MP and Graph-Multiway-Cut are assuming the unique games conjecture.

Problem	Lower Bound	Upper Bound
Sub-MP	2 [10]	2 [6]
Sym-Sub-MP	1.268 [10]	1.5 [6]
Mono-Sub-MP	1.1111 (Theorem 2)	1.33333 (Theorem 1)
Graph-Cov-MP	1.00074 (Theorem 4)	1.125 (Theorem 3)
Graph-Multiway-Cut	1.20016 [2,14]	1.2965 [15]

2 Approximation Algorithm for Mono-Sub-MP

In this section, we describe our algorithm for Mono-Sub-MP and analyze its approximation factor. Let $f : 2^V \to \mathbb{R}_{\geq 0}$ be the input monotone submodular function with terminals $t_1, \ldots, t_k \in V$. For $\mathbf{x} \in [0,1]^V$ and $\theta \in [0,1]$, we define $\mathbf{x}^\theta \in \{0,1\}^V$ as

$$(\mathbf{x}^\theta)_v := \begin{cases} 1 & \text{if } \mathbf{x}_v \geq \theta, \\ 0 & \text{otherwise.} \end{cases}$$

We also define $f\left(\mathbf{x}^\theta\right) := f\left(\{v \in V : (\mathbf{x}^\theta)_v = 1\}\right)$. We define the Lovász extension $\hat{f} : [0,1]^V \to \mathbb{R}$ of f as follows: For all $\mathbf{x} \in [0,1]^V$, $\hat{f}(\mathbf{x}) := \mathbb{E}_{\theta \in [0,1]}\left[f(\mathbf{x}^\theta)\right] = \int_0^1 f\left(\mathbf{x}^\theta\right) d\theta$. We introduce decision variable vector $\mathbf{x}_i = \{x(v,i)\}_{v \in V}$ for each $i \in [k]$ and consider the convex program below:

$$\text{minimize} \sum_{i=1}^{k} \hat{f}\left(\mathbf{x}_i\right) \qquad \text{(Sub-MP-Rel)}$$

$$\begin{aligned} \sum_{i=1}^{k} x(v,i) &= 1 \quad \forall v \in V \\ x(t_i,i) &= 1 \quad \forall i \in [k] \\ x(v,i) &\geq 0 \quad \forall v \in V, i \in [k]. \end{aligned}$$

The convex program is identical to the convex programming relaxation of Sub-MP introduced by Chekuri and Ene and it can be solved in polynomial time [6]. We present a rounding algorithm for Mono-Sub-MP based on this relaxation. Our rounding algorithm is similar to the θ-rounding algorithms of Chekuri and Ene for Sub-MP and Sym-Sub-MP [6], but the interval from which we choose θ is different. Our choice for the interval of θ is designed to exploit monotonicity in order to achieve a better approximation factor. Below, we present our rounding algorithm for Mono-Sub-MP:

Algorithm 1. Mono-Sub-MP Rounding

Let $\mathbf{x}$ be an optimal solution to Sub-MP-Rel
Choose $\theta \in [\frac{1}{4}, 1]$ uniformly at random
for $i \in [k]$ **do**
  $A(i,\theta) \leftarrow \{v \in V : x(v,i) \geq \theta\}$
$U(\theta) \leftarrow V - \bigcup_{i=1}^{k} A(i,\theta)$
for $v \in V$ **do**
  $I_v \leftarrow \{i \in [k] : v \in A(i,\theta)\}$
  if $|I_v| > 1$ **then**
    delete v from all but one of the sets $\{A(i,\theta)\}_{i \in I_v}$
return $(A_1, \ldots, A_{k-1}, A_k) = \left(A(1,\theta), \ldots, A(k-1,\theta), A(k,\theta) \cup U(\theta)\right)$

We observe that the rounding algorithm guarantees that each $v \in V$ is assigned to exactly one of the returned sets. Additionally, for each $i \in [k]$, we have that $t_i \in A(i, \theta)$ since $x(t_i, i) = 1$. Hence, the algorithm returns a feasible multiway partition of V. Moreover, the algorithm can be implemented to run in polynomial time. The following theorem bounds the approximation factor of Algorithm 1 relative to the optimum value of Sub-MP-Rel.

Theorem 5. *Let* OPT_{frac} *be the optimum value of Sub-MP-Rel and* $A_1, \dots, A_k$ *be the solution returned by Algorithm 1. Then,*

$$\mathbb{E}\left[\sum_{i=1}^{k} f(A_i)\right] \leq \frac{4}{3}\text{OPT}_{frac}.$$

The rest of this section is devoted to proving Theorem 5. Our approximation factor analysis uses certain ingredients of Chekuri and Ene [6], but differs from their analysis to exploit monotonicity and the different threshold range. We will point out the differences as we proceed with the analysis.

We start with a lemma about the truncated mass of a discrete probability distribution. A vector $\mathbf{p} \in \mathbb{R}^k_{\geq 0}$ with $\sum_{i \in [k]} p_i = 1$ can be viewed as a probability distribution over the set $\{1, 2, \dots, k\}$ with p_i being the probability of picking item i. For $\delta \in [0, 1]$, the quantity $\sum_{i=1}^{k} \min\{\delta, p_i\}$ is the truncated probability mass. We show that if $\delta \geq 1/4$, then the truncated probability mass is at least $1 - \max_{i \in [k]} p_i$. The choice of $\delta \in [1/4, 1]$ in the rounding algorithm emanates from this lemma. This lemma does not appear in Chekuri and Ene [6] and is our contribution.

Lemma 1. *Let* $\mathbf{p} \in \mathbb{R}^k_{\geq 0}$ *such that* $\sum_{i=1}^{k} p_i = 1$ *and* $\delta \geq 1/4$*. Then,*

$$\sum_{i=1}^{k} \min\{\delta, p_i\} \geq 1 - \max_{i \in [k]} p_i.$$

Proof. The LHS can be rewritten as follows:

$$\sum_{i=1}^{k} \min\{\delta, p_i\} = \sum_{i=1}^{k} p_i - \sum_{i \in [k]: p_i > \delta} (p_i - \delta) = 1 - \sum_{i \in [k]: p_i > \delta} (p_i - \delta).$$

Let $L := \{i \in [k] : p_i > \delta\}$ and let $j \in [k]$ such that $p_j = \max_{i \in [k]} p_i$. It suffices to show that $p_j \geq \sum_{i \in L}(p_i - \delta)$. For the sake of contradiction, suppose that $p_j < \sum_{i \in L}(p_i - \delta)$. We case based on the size of $L - \{j\}$ and arrive at a contradiction in all cases.

Case 1. Suppose $L - \{j\} = \emptyset$. If $j \in L$, then $L = \{j\}$, and hence, $p_j < \sum_{i \in L}(p_i - \delta) = p_j - \delta$, a contradiction. If $j \notin L$, then $L = \emptyset$, and hence, $0 \leq p_j < \sum_{i \in L}(p_i - \delta) = 0$, a contradiction.

Case 2. Suppose $|L - \{j\}| = 1$. Let $L - \{j\} = \{\ell\}$. Therefore, $L = \{\ell, j\}$. Consequently, we have that $p_j < \sum_{i \in L}(p_i - \delta) = p_j + p_\ell - 2\delta$. Hence, we have that $p_\ell > 2\delta \geq 1/2$ since $\delta \geq 1/4$. Also, by definition of j, we have that $p_j \geq p_\ell$. Hence, we have that $p_j \geq p_\ell > 1/2$ and therefore, $\sum_{i=1}^{k} p_i \geq p_j + p_k > 1$, a contradiction.

Case 3. Suppose $|L - \{j\}| \geq 2$. Then, by definition of j, we have that $j \in L$ and hence, $|L| \geq 3$ and $p_j > \delta$. Thus, we have that $\delta < p_j < \sum_{i\in L}(p_i - \delta) \leq \left(\sum_{i\in L} p_i\right) - 3\delta$, where the last inequality is because $|L| \geq 3$. From these inequalities, we have that $\sum_{i=1}^{k} p_i \geq \sum_{i\in L} p_i > 4\delta \geq 1$, a contradiction. □

Notation. We now set up the notation (similar to that of Chekuri and Ene [6]) to analyze the approximation factor of our algorithm. Let $\mathbf{x}$ be an optimum solution to SUB-MP-REL. We assume that $V = \{v_1, \ldots, v_n\}$. For each $v_j \in V$, we define $\alpha_j = \max_{i=1}^{k} x(v_j, i)$. We assume that the elements $v_1, v_2, \ldots, v_n$ of the ground set V are ordered such that $0 \leq \alpha_1 \leq \alpha_2 \leq \ldots \leq \alpha_n \leq 1$. For convenience, we define $\alpha_0 := 0$ and $\alpha_{n+1} := 1$. Let $\theta \in [0, 1]$. For each $i \in [k]$, we define $A(i, \theta) = \{v \in V : x(v, i) \geq \theta\}$ as in the algorithm. Additionally, for each $j \in [n]$, we let $V_j := \{v_1, \ldots, v_j\}$, and for convenience, we define $V_0 := \emptyset$. Finally, we let $A_j(i, \theta) := A(i, \theta) \cap V_j$.

With the above notation, we show the following technical lemma based on monotonicity and submodularity of f and by using Lemma 1. Since we rely on monotonicity and Lemma 1, the following lemma does not follow from the results in Chekuri and Ene [6].

Lemma 2. *For every $\delta \in [\frac{1}{4}, 1]$ and $j \in [n]$, we have that*

$$\sum_{i=1}^{k} \int_0^{\delta} \Big(f\left(A_j(i,\theta)\right) - f\left(A_{j-1}(i,\theta)\right) \Big)\, d\theta \geq (1 - \alpha_j)\left(f(V_j) - f(V_{j-1})\right).$$

Proof. Let $\delta \in [1/4, 1]$ and $j \in [n]$. We show the following claim using submodularity of f—see full version for the proof [3].

Claim 1. *For every $i \in [k]$, we have*

$$\int_0^{\delta} \Big(f\left(A_j(i,\theta)\right) - f\left(A_{j-1}(i,\theta)\right) \Big)\, d\theta \geq \min\{\delta, x(v_j, i)\} \Big(f\left(V_j\right) - f\left(V_{j-1}\right) \Big).$$

Proof. Now, we use the above claim to lower bound the LHS of the lemma as follows:

$$\begin{aligned}
&\sum_{i=1}^{k} \int_0^{\delta} \Big(f\left(A_j(i,\theta)\right) - f\left(A_{j-1}(i,\theta)\right) \Big)\, d\theta \\
&\quad\geq \sum_{i=1}^{k} \min\{\delta, x(v_j, i)\} \Big(f\left(V_j\right) - f\left(V_{j-1}\right) \Big) \qquad \text{(by Claim 1)} \\
&\quad= \Big(f\left(V_j\right) - f\left(V_{j-1}\right) \Big) \left(\sum_{i=1}^{k} \min\{\delta, x(v_j, i)\} \right). \qquad (1)
\end{aligned}$$

We know that $x(v_j, i) \geq 0$ for every $i \in [k]$ and $\sum_{i=1}^{k} x(v_j, i) = 1$ since x is a feasible solution to the convex program. Hence, using Lemma 1, we get

$$\sum_{i=1}^{k} \min\{\delta, x(v_j, i)\} \geq 1 - \max_{i \in [k]} x(v_j, i) = 1 - \alpha_j.$$

Further, by monotonicity of f, we have that $f(V_j) - f(V_{j-1}) \geq 0$. Hence,

$$\Big(f\left(V_j\right) - f\left(V_{j-1}\right)\Big) \left(\sum_{i=1}^{k} \min\{\delta, x(v_j, i)\}\right) \geq (1 - \alpha_j) \Big(f\left(V_j\right) - f\left(V_{j-1}\right)\Big). \tag{2}$$

The lemma holds by inequalities (1) and (2). □

We need the following Proposition from Chekuri and Ene [6].

Proposition 1. [6] *Let f be an arbitrary non-negative submodular function, $r \in [0, 1]$, and let h be the largest value of $j \in [n]$ such that $\alpha_j \leq r$. We have*

$$\int_0^r f\left(U(\theta)\right) d\theta = \sum_{j=1}^{h} \alpha_j \left(f(V_{j-1}) - f(V_j)\right) + r f(V_h).$$

Lemma 2 and Proposition 1 lead to the following corollary.

Corollary 1. *For every $\delta \in [\frac{1}{4}, 1]$,*

$$\sum_{j=1}^{n} \left(\sum_{i=1}^{k} \int_0^{\delta} \left(f(A_j(i, \theta)) - f(A_{j-1}(i, \theta))\right) d\theta\right) \geq \int_0^1 f(U(\theta)) d\theta - f(\emptyset).$$

Proof. We have that

$$\begin{aligned}
&\sum_{j=1}^{n} \left(\sum_{i=1}^{k} \int_0^{\delta} \Big(f\left(A_j(i, \theta)\right) - f\left(A_{j-1}(i, \theta)\right)\Big) d\theta\right) \\
&\quad \geq \sum_{j=1}^{n} (1 - \alpha_j) \left(f(V_j) - f(V_{j-1})\right) \quad \text{(by Lemma 2)} \\
&\quad = \sum_{j=1}^{n} \left(f(V_j) - f(V_{j-1})\right) + \sum_{j=1}^{n} \alpha_j \left(f(V_{j-1}) - f(V_j)\right) \\
&\quad = f(V_n) - f(V_0) + \sum_{j=1}^{n} \alpha_j \left(f(V_{j-1}) - f(V_j)\right) \\
&\quad = \int_0^1 f\left(U(\theta)\right) d\theta - f(V_0) = \int_0^1 f\left(U(\theta)\right) d\theta - f(\emptyset).
\end{aligned}$$

The last but one equation above is by using $r = 1$ and $h = n$ in Proposition 1 and the last equation is by using $V_0 = \emptyset$. □

We have the following theorem as a consequence of Corollary 1. We observe that this theorem upper bounds the expected function value of the set of unassigned elements.

Theorem 6. *For all* $\delta \in [\frac{1}{4}, 1]$,

$$\sum_{i=1}^{k} \int_{0}^{\delta} f(A(i,\theta))d\theta \geq (k\delta - 1)f(\emptyset) + \int_{0}^{1} f(U(\theta))d\theta.$$

Proof. We have that $\sum_{i=1}^{k} \int_{0}^{\delta} f(A(i,\theta))d\theta$

$$
\begin{aligned}
&= \sum_{i=1}^{k} \int_{0}^{\delta} f(A_n(i,\theta))d\theta \\
&= \sum_{i=1}^{k} \int_{0}^{\delta} f(A_0(i,\theta))d\theta + \sum_{j=1}^{n} \left(\sum_{i=1}^{k} \int_{0}^{\delta} f(A_j(i,\theta)) - \sum_{i=1}^{k} \int_{0}^{\delta} f(A_{j-1}(i,\theta)) \right) \\
&= \sum_{i=1}^{k} \int_{0}^{\delta} f(A_0(i,\theta))d\theta + \sum_{j=1}^{n} \left(\sum_{i=1}^{k} \int_{0}^{\delta} \left(f(A_j(i,\theta)) - f(A_{j-1}(i,\theta)) \right) d\theta \right) \\
&\geq \sum_{i=1}^{k} \int_{0}^{\delta} f(A_0(i,\theta))d\theta + \int_{0}^{1} f(U(\theta))d\theta - f(\emptyset) \qquad \text{(by Corollary 1)} \\
&= (k\delta - 1)f(\emptyset) + \int_{0}^{1} f(U(\theta))d\theta. \qquad \text{(since } A_0(i,\theta) = \emptyset \text{ for every } i \in [k] \text{)}
\end{aligned}
$$

□

We now restate and prove Theorem 5.

Theorem 5. *Let* OPT_{frac} *be the optimum value of Sub-MP-Rel and* $A_1, \ldots, A_k$ *be the solution returned by Algorithm 1. Then,*

$$\mathbb{E}\left[\sum_{i=1}^{k} f(A_i)\right] \leq \frac{4}{3}\mathrm{OPT}_{frac}.$$

Proof. By Theorem 6, $f(\emptyset) = 0$, and f is non-negative, we have

$$\sum_{i=1}^{k} \int_{0}^{\frac{1}{4}} f(A(i,\theta))d\theta \geq \left(\frac{1}{4}k - 1\right) f(\emptyset) + \int_{0}^{1} f(U(\theta))d\theta \geq \int_{\frac{1}{4}}^{1} f(U(\theta))d\theta. \quad (3)$$

By definition, $\text{OPT}_{frac} = \sum_{i=1}^{k} \hat{f}(\mathbf{x}_i) = \int_0^1 \sum_{i=1}^{k} f(A(i,\theta))d\theta$. Hence, $\mathbb{E}_{\theta\in[\frac{1}{4},1]}\left[\sum_{i=1}^{k} f(A_i)\right]$

$$\leq \mathbb{E}_{\theta\in[\frac{1}{4},1]}\left[\sum_{i=1}^{k-1} f(A(i,\theta)) + f(A(k,\theta)\cup U(\theta))\right] \qquad \text{(sinc } f \text{ is monotone)}$$

$$\leq \mathbb{E}_{\theta\in[\frac{1}{4},1]}\left[\sum_{i=1}^{k} f(A(i,\theta)) + f(U(\theta))\right]$$

(since f is non-negative and submodular)

$$= \frac{4}{3}\int_{\frac{1}{4}}^{1}\left(\sum_{i=1}^{k} f(A(i,\theta)) + f(U(\theta))\right)d\theta$$

$$= \frac{4}{3}\left(\text{OPT}_{frac} - \sum_{i=1}^{k}\int_0^{\frac{1}{4}} f(A(i,\theta))d\theta + \int_{\frac{1}{4}}^{1} f(U(\theta))d\theta\right)$$

$$\leq \frac{4}{3}\text{OPT}_{frac}. \qquad \text{(by inequality (3))}$$

□

3 Conclusion

In this work, we improved on the approximation factor and the inapproximability of Mono-Sub-MP. Further, our results for Graph-Cov-MP showed a separation of the problem from Graph-Multiway-Cut in terms of approximability. Our results raise several interesting directions for future research. Firstly, can we obtain approximation guarantees for Matroid-MP, Matrix-MP, and Cov-MP that are better than the guarantee of 4/3 for Mono-Sub-MP? Secondly, can we extend the known results for Graph-Multiway-Cut to Graph-Cov-MP? We mention two fundamental results for Graph-Multiway-Cut that do not seem to extend to Graph-Cov-MP. Călinescu, Karloff, and Rabani [5] formulated a linear programming relaxation for Graph-Multiway-Cut that is referred to in the literature as the CKR-relaxation. A similar LP-relaxation can also be formulated for Graph-Cov-MP (it has the same constraints but different objective). Firstly, the integrality gap of the CKR-relaxation for $k = 3$ terminals has been determined exactly [7,13]. However, we are unable to determine the integrality gap of the LP-relaxation for Graph-Cov-MP for $k = 3$ terminals. This is because it is unclear if the worst-case integrality gap instance will be *axis-aligned* (as is the case for the CKR-relaxation). Secondly, the inapproximability factor of Graph-Multiway-Cut is known to coincide with the integrality gap of the CKR-relaxation [14]. We are unable to show a similar connection between the inapproximability factor of Graph-Cov-MP and the integrality gap of its LP-relaxation. This is because we do not have a metric associated with edge costs in the coverage setting. The issue underlying our inability to extend both results seem related. Extending either of these results from Graph-Multiway-Cut to Graph-Cov-MP would reveal further similarities/differences between the two problems.

Acknowledgments. All authors were supported in part by the NSF grant CCF-2402667.

Disclosure of Interests. The authors have no competing interests to declare that are relevant to the content of this article.

References

1. Angelidakis, H., Makarychev, Y., Manurangsi, P.: An improved integrality gap for the Călinescu-Karloff-Rabani relaxation for multiway cut. In: Integer Programming and Combinatorial Optimization, pp. 39–50. IPCO (2017)
2. Bérczi, K., Chandrasekaran, K., Király, T., Madan, V.: Improving the integrality gap for multiway cut. Math. Program. **183**, 171–193 (2020)
3. Bi, R., Chandrasekaran, K., Joshi, S.: Monotone submodular multiway partition (2024). https://arxiv.org/abs/2411.05255
4. Buchbinder, N., Naor, J., Schwartz, R.: Simplex partitioning via exponential clocks and the multiway cut problem. In: Proceedings of the Forty-Fifth Annual ACM Symposium on Theory of Computing, pp. 535–544. STOC (2013)
5. Călinescu, G., Karloff, H., Rabani, Y.: An improved approximation algorithm for multiway cut. J. Comput. Syst. Sci. **60**(3), 564–574 (2000)
6. Chekuri, C., Ene, A.: Approximation algorithms for submodular multiway partition. In: Proceedings of the 2011 IEEE 52nd Annual Symposium on Foundations of Computer Science, pp. 807–816. FOCS (2011)
7. Cheung, K., Cunningham, W., Tang, L.: Optimal 3-terminal cuts and linear programming. Math. Program. **106**(1), 1–23 (2006)
8. Dahlhaus, E., Johnson, D.S., Papadimitriou, C.H., Seymour, P.D., Yannakakis, M.: The complexity of multiterminal cuts. SIAM J. Comput. **23**(4), 864–894 (1994)
9. Dobzinski, S., Vondrak, J.: From query complexity to computational complexity. In: Proceedings of the Forty-Fourth Annual ACM Symposium on Theory of Computing, pp. 1107–1116. STOC (2012)
10. Ene, A., Vondrák, J., Wu, Y.: Local distribution and the symmetry gap: approximability of multiway partitioning problems. In: Proceedings of the 2013 Annual ACM-SIAM Symposium on Discrete Algorithms, pp. 306–325. SODA (2013)
11. Freund, A., Karloff, H.: A lower bound of $8/(7+1/(k-1))$ on the integrality ratio of the Călinescu-Karloff-Rabani relaxation for multiway cut. Inf. Process. Lett. **75**(1), 43–50 (2000)
12. Karger, D., Klein, P., Stein, C., Thorup, M., Young, N.: Rounding algorithms for a geometric embedding of minimum multiway cut. Math. Oper. Res. **29**(3), 436–461 (2004)
13. Karger, D.R., Klein, P., Stein, C., Thorup, M., Young, N.E.: Rounding algorithms for a geometric embedding of minimum multiway cut. In: Proceedings of the Thirty-First Annual ACM Symposium on Theory of Computing, pp. 668–678. STOC (1999)

14. Manokaran, R., Naor, J., Raghavendra, P., Schwartz, R.: SDP gaps and UGC hardness for multiway cut, 0-extension, and metric labeling. In: Proceedings of the Fortieth Annual ACM Symposium on Theory of Computing, pp. 11–20. STOC (2008)
15. Sharma, A., Vondrák, J.: Multiway cut, pairwise realizable distributions, and descending thresholds. In: Proceedings of the Forty-Sixth Annual ACM Symposium on Theory of Computing, pp. 724–733. STOC (2014)
16. Vondrák, J.: Symmetry and approximability of submodular maximization problems. SIAM J. Comput. **42**(1), 265–304 (2013)
17. Zhao, L., Nagamochi, H., Ibaraki, T.: Greedy splitting algorithms for approximating multiway partition problems. Math. Program. **102**(1), 167–183 (2005)

Exponential Lower Bounds for Many Pivot Rules for the Simplex Method

Alexander E. Black(✉)

ETH Zürich, 8092 Zürich, Switzerland
alexander.black@math.ethz.ch

Abstract. The existence of a pivot rule for the simplex method that guarantees a polynomial run-time is a longstanding, fundamental open problem in the theory of linear programming. The most popular pivot rule for theoretical analysis is the shadow pivot rule, which solves a linear program by projecting the feasible region onto a polygon. It has been shown to perform in expected strongly polynomial time on uniformly random instances and in smoothed analysis. In practice, the pivot rule of choice is the steepest edge rule, which normalizes the set of improving neighbors and then chooses a maximally improving normalized neighbor. Exponential lower bounds are known for both rules in worst-case analysis. However, for the shadow simplex method, all exponential examples were only proven for one choice of projection, and for the steepest edge rule, the lower bounds were only proven for the Euclidean norm. In this work, we construct linear programs for which any choice of projection for shadow rule variants will lead to an exponential run-time and exponential examples for any choice of norm for a steepest edge variant.

Keywords: Simplex Method · Parametric Optimization

1 Introduction

The simplex method is a foundational 75 year old algorithm for linear optimization still used in practice today, yet it is open whether one can even find even a weakly polynomial worst-case guarantee for its run-time. In part, this is due to the many possible variants of the simplex method. Namely, geometrically, the simplex method solves a bounded, feasible linear program $\max(\mathbf{c}^\intercal \mathbf{x})$ subject to $A\mathbf{x} \leq \mathbf{b}$ by generating an initial vertex of the polytope defined by $\{\mathbf{x} : A\mathbf{x} \leq \mathbf{b}\}$ and then walking from vertex to vertex along edges such that at each step the linear objective function increases. Regardless of how the path is chosen, so long as the linear objective function increases at each step, the path will eventually reach an optimum. Such a path is called a **monotone path** and a way of choosing a monotone path is called a **pivot rule**. The length of the monotone path together with the complexity of making a single step determines the run-time of the simplex method.

N. Megow and A. Basu (Eds.): IPCO 2025, LNCS 15620, pp. 86–99, 2025.
https://doi.org/10.1007/978-3-031-93112-3_7

There is a long history of finding superpolynomial lower bounds for the performance of individual pivot rules for the simplex method starting with the 1972 work of Klee and Minty [30] with many follow up works [1,2,18,19,21,22,25,26, 28,33,34,40,41] covering the standard pivot rules used in theory and practice [38]. However, as noted in Shamir's 1987 survey [35] an "interesting question is whether one can construct a single family of examples which are "bad" for many Simplex variants. The examples constructed so far were 'tailored' to behave "badly" when one specific variant is used, but they often are extremely easy to solve by other variants." The goal of this work is to partially address Shamir's concern that has remained as relevant yet unaddressed more than 30 years later. In particular, if one has a constant number of pivot rules, one can easily construct a uniform example for each of them by combining linear programs that are exponential for each of them by taking products of their underlying polytopes. Though doing this may break the structure of the problem, so it is still interesting to do this for structured examples as in [19]. The nontrivial case of Shamir's question for general linear programs is for infinite families of pivot rules. Here we construct exponential examples for two continuous, infinite families of pivot rules motivated by well known pivot rules in the literature.

Shadow pivot rules choose a monotone path by projecting the polytope onto some two-dimensional polygon called the **shadow** implicitly and then lifting a path on the polygon back to a path on the polytope. The strongest positive results for the simplex method rely on variants of the shadow simplex method. The first major breakthrough in this direction is due to Borgwardt in [10,11] in the 1970's and 1980's in which he showed the shadow simplex method on a random linear program taken from any spherically symmetric probability distribution takes an expected $\Theta(n^2 m^{\frac{1}{n-1}})$ steps for a linear program defined by m facets in n variables. Spielman and Teng introduced the smoothed analysis of algorithms in [36] with the key tool being the shadow simplex method in their analysis and all subsequent analyses [15,17,27,29,39]. In this case, each entry of A and $\mathbf{b}$ for the linear program $\max(\mathbf{c}^\intercal \mathbf{x})$ such that $A\mathbf{x} \leq \mathbf{b}$ is perturbed by a Gaussian with standard deviation σ, and the best known upper bounds on the expected number of steps in this framework is $O(\sigma^{-3/2} n^{13/4} \log^{7/4}(m))$ from [27]. Thus, in those two models of random linear programs, the simplex method performs in polynomial time in expectation. There has furthermore been successful work on using random choices of shadows for a fixed linear program. Bounds in that case are in terms of m and n alongside a measure of the size of the data given by Δ, the maximum absolute value of a subdeterminant of the constraint matrix, or other similar parameters [12,14] and tend to be of the form $\text{poly}(m, n, \Delta)$. Deterministic versions of the shadow simplex method take an optimal, linear number of non-degenerate steps on 0/1-polytopes [7] with more general bounds shown for lattice polytopes [6]. Given all of these positive results, we ask here whether there is a version of the shadow simplex method guaranteed to solve any linear program in polynomial time.

For 50 years, it has been known that the number of steps the shadow simplex method could take is exponential with many examples [3,25,33,40]. Superpolynomial lower bounds for the shadow simplex method appear further in the parametric optimization literature and hold even for combinatorial linear programs such as shortest paths, flows, and matchings [13,23,32].

However, the path chosen by the shadow pivot rule depends on the choice of shadow. One thing these pathological examples all have in common is that they only show a lower bound for one bad choice of shadow, but one could potentially choose a different shadow. For example, there exist 0/1-polytopes with superpolynomial sized shadows [13,31,32], but in [7] they showed that one could always avoid that bad case by choosing a different shadow. Here we construct explicit linear programs for which the shadow simplex method takes exponentially many steps regardless of the choice of shadow.

Theorem 1. *For each $n \in \mathbb{N}$, there exists an n-dimensional simple polytope P with $3n$ facets of polynomial encoding length, a vertex $\mathbf{v}$ of P, and $\mathbf{c} \in \mathbb{R}^n$ such that any shadow path on P starting at $\mathbf{v}$ to optimize $\mathbf{c}^\intercal\mathbf{x}$ is of length at least 2^n. Hence, there is no weakly polynomial time shadow simplex algorithm for linear programming regardless of how the shadow is chosen.*

The steepest edge pivot rule given by maximizing $\frac{\mathbf{c}^\intercal(\mathbf{u}-\mathbf{v})}{\|\mathbf{u}-\mathbf{v}\|_2}$ is the pivot rule of choice in practice [5,20]. Goldfarb and Sit constructed an exponential example [26] in 1979 via a modification of Klee and Minty's construction method. However, the choice of 2-norm here is restrictive. We generalize their result to an infinite class of pivot rules.

Theorem 2. *For any $n \in \mathbb{N}$ and any continuous function $\eta : \mathbb{R}^n \to \mathbb{R}_{\geq 0}$ such that $\eta(\mathbf{x}) = 0$ if and only if $\mathbf{x} = 0$ and $\eta(\lambda\mathbf{x}) = |\lambda|\eta(\mathbf{x})$ for all $\lambda \in \mathbb{R}$ and $\mathbf{x} \in \mathbb{R}^n$, there exists an n-dimensional combinatorial cube P_n and $\mathbf{c} \in \mathbb{R}^n$ for which the pivot rule given by*

$$\mathbf{v}^{i+1} = \underset{\mathbf{u} \in N_\mathbf{c}(\mathbf{v}^i)}{\operatorname{argmax}} \frac{\mathbf{c}^\intercal(\mathbf{u}-\mathbf{v}^i)}{\eta(\mathbf{u}-\mathbf{v}^i)},$$

where $N_\mathbf{c}(\mathbf{v}^i)$ is the set of $\mathbf{c}$-improving neighbors of $\mathbf{v}^i$, follows a path through all vertices of C_n to maximize $\mathbf{c}^\intercal\mathbf{x}$. In particular, such a lower bound holds whenever η is a norm.

To our knowledge, lower bounds were not known for any norm other than the 2-norm including the 1-norm used in [7,16] and p-steepest edge pivot rules studied in [37]. That exponential lower bounds can be proven for every pivot rule in this class is surprising, since, by the argument in the proof of Theorem 1.3 of [8], finding a generic linear program where every positively homogeneous, continuous normalization function such that $\eta(\mathbf{x}) = 0$ if and only if $\mathbf{x} = 0$ yields an exponential length path is equivalent to the polynomial monotone Hirsch conjecture. Hence, providing a single example that exhibits exponential behavior for any function is this class is very hard or impossible. However, we can construct such an example for a large subfamily in pursuit of Shamir's question. Namely, we call a norm η **regular** if $\eta(e_i) = 1$ for all $i \in [n]$. Then there is a family of examples that yields exponential run-time for all regular norms simultaneously:

Theorem 3. *For each $n \in \mathbb{N}$, there exists an n-dimensional combinatorial cube and $\mathbf{c} \in \mathbb{R}^n$ such that every pivot rule of the form:*

$$\mathbf{v}^{i+1} = \operatorname*{argmax}_{\mathbf{u} \in N_{\mathbf{c}}(\mathbf{v}^i)} \frac{\mathbf{c}^\intercal(\mathbf{u} - \mathbf{v}^i)}{\eta(\mathbf{u} - \mathbf{v}^i)},$$

where $N_{\mathbf{c}}(\mathbf{v}^i)$ is the set of $\mathbf{c}$-improving neighbors of $\mathbf{v}^i$ and η is an arbitrary regular norm, follows a path through all vertices of the combinatorial cube to maximize $\mathbf{c}^\intercal\mathbf{x}$. In particular, the p-steepest edge simplex method will follow an exponential length path for all choices of p.

This is a mild assumption on our norm in that any norm satisfies it up to a precomposing with a positive diagonal matrix. Our strategy to prove all of our main theorems is to take a linear program with a single exponential length shadow path from [33] and modify it.

2 Background

We rely on the simplex method at the level of [4] and then discuss pivot rules. The simplex method is generally defined for standard form linear programs $\max(\mathbf{c}^\intercal\mathbf{x})$ subject to $A\mathbf{x} = \mathbf{b}, \mathbf{x} \geq 0$ for $\mathbf{x} \in \mathbb{R}^m$ and A an $n \times m$ matrix. A basic feasible solution $\mathbf{v} \in \mathbb{R}^m$ is any vector satisfying $A\mathbf{v} = \mathbf{b}$, $\mathbf{v} \geq 0$, and $|\text{supp}(\mathbf{v})| \leq n$. For our purposes here, we may assume without loss of generality that the linear program is non-degenerate meaning that $|\text{supp}(\mathbf{v})| = n$ for all basic feasible solutions $\mathbf{v}$. Then the vertices of the polytope $P = \{\mathbf{x} : A\mathbf{x} = \mathbf{b}, \mathbf{x} \geq \mathbf{0}\}$ correspond to basic feasible solutions, and two basic feasible solutions $\mathbf{u}$ and $\mathbf{v}$ differ by an edge if $|\text{supp}(\mathbf{u}) \cap \text{supp}(\mathbf{v})| = n - 1$.

We also assume the simplex method is given a starting basic feasible solution $\mathbf{v}$ with support B and A_i denote the ith column of A. We call B the corresponding **basis**. Finding an initial basis is referred to as phase 1 of the simplex method. For our analysis, we assume the simplex method could start from any arbitrary initial basis. Hence, to prove lower bounds on the simplex method's performance, it suffices to find one initial basis from which the number of steps taken is exponential. A **pivot** is a change from the current basic feasible solution to another. Since we assumed the linear program is nondegenerate, pivoting corresponds to walking along an edge of the polytope. Each possible pivot is determined uniquely by a new choice of variable to add to the basis. Namely, let A_B be the $n \times n$ matrix given by restricting A to the columns indexed by B. Then, from the standard theory of the simplex method assuming nondegeneracy, for each $i \in [m] \setminus B$, there exists a unique $\lambda_i > 0$ such that $\mathbf{v} + \lambda_i(e_i - A_B^{-1}A_i)$ is a basic feasible solution with support given by i and all but one element of B. Hence, any choice of $i \in [m] \setminus B$ gives an edge, and all edges arise in that way.

The simplex method solves the linear program by pivoting from basis to basis such that for each new i added to a basis, $\mathbf{c}^\intercal(e_i - A_B^{-1}A_i) > 0$. If no such i exists, the current basis is optimal. A **pivot rule** is a way of choosing i. For example, Dantzig's pivot rule says to choose i maximizing $\mathbf{c}^\intercal(e_i - A_B^{-1}A_i)$, the greatest improvement rule chooses the maximizer of $\mathbf{c}^\intercal\lambda_i(e_i - A_B^{-1}A_i)$, and the steepest edge pivot rule chooses the maximizer of $\frac{\mathbf{c}^\intercal(e_i - A_B^{-1}A_i)}{\|e_i - A_B^{-1}A_i\|_2}$.

Here we consider a generalized steepest edge rule, which chooses i maximizing $\frac{\mathbf{c}^\intercal(e_i - A_B^{-1}A_i)}{\eta(e_i - A_B^{-1}A_i)}$ for some choice of function $\eta : \mathbb{R}^n \to \mathbb{R}$ is any function satisfying for all $\mathbf{x} \in \mathbb{R}^n$ and $\lambda \in \mathbb{R}$, $\eta(\mathbf{x}) \geq 0$ with equality if and only if $\mathbf{x} = \mathbf{0}$, positive homogeneity ($\eta(\lambda\mathbf{x}) = |\lambda|\eta(\mathbf{x})$ for all $\lambda \in \mathbb{R}$), Note that η is a norm if it also satisfies the triangle inequality ($\eta(\mathbf{x} + \mathbf{y}) \leq \eta(\mathbf{x}) + \eta(\mathbf{y})$).

Shadow pivot rules require more explanation. Namely, there are many choices of a shadow rule. Each is given by a choice of $\mathbf{w}_j > 0$ for each $j \in [m] \setminus B_0$ for the initial basis B_0 to yield a vector $\mathbf{w} = \sum_{j\in[m]\setminus B_0} \mathbf{w}_j e_j$. Then each subsequent basis after a basis B is chosen by maximizing $\frac{\mathbf{c}^\intercal(e_i - A_B^{-1}A_i)}{\mathbf{w}^\intercal(e_i - A_B^{-1}A_i)}$. Note that $\mathbf{w}$ is fixed in the first step and does not change later. Thus, computing the shadow pivot rule at each step may be done by the simplex method nearly as efficiently as computing Dantzig's original pivot rule. In particular, unlike greatest improvement, it does not require computing λ_i.

It remains to explain the connection to shadows. Consider the map $\pi : \mathbb{R}^m \to \mathbb{R}^2$ given by $\pi(\mathbf{x}) = (\mathbf{w}^\intercal\mathbf{x}, \mathbf{c}^\intercal\mathbf{x})$. Then $\pi(P)$ is a polygon. Note that we call the optimum of a linear program for maximizing or minimizing $\mathbf{c}^\intercal\mathbf{x}$, a $\mathbf{c}$-maximizer or $\mathbf{c}$-minimizer respectively. We may assume without loss of generality that the $\mathbf{c}$-maximizer is unique. Furthermore, by construction, the initial vertex $\mathbf{v}$ is a $\mathbf{w}$-minimizer, since the support of $\mathbf{w}$ is $[m] \setminus B_0$. Hence, under the projection $\pi(\mathbf{v})$ minimizes the first coordinate. Similarly $\pi(\mathbf{v}^*)$ maximizes the second coordinate, where $\mathbf{v}^*$ is the $\mathbf{c}$-maximizer. Then the shadow pivot rule maximizes $\frac{\mathbf{c}^\intercal(\mathbf{u}-\mathbf{v})}{\mathbf{w}^\intercal(\mathbf{u}-\mathbf{v})}$ over all $\mathbf{c}$-improving neighbors of $\mathbf{v}$. The quantity $\frac{\mathbf{c}^\intercal(\mathbf{u}-\mathbf{v})}{\mathbf{w}^\intercal(\mathbf{u}-\mathbf{v})}$ is precisely the slope of the edge in the projection π. An edge of maximal slope must stay on the boundary of the polygon. Hence, the pivot rule will follow the path on the boundary of the polygon from the point minimizing the first coordinate to the point maximizing the second. As a result, the total number of vertices of the shadow yields an upper bound on the length of the path.

Finally, a key tool in our proofs will be a third interpretation of shadow pivot rules on a dual view of polytopes. In fact, Gass and Saaty originally introduced the shadow simplex method in 1955 in this framework [24] as parametric linear programming. Namely, one can form a path from the starting vertex $\mathbf{u}$ to the optimum $\mathbf{v}$ by taking the path consisting of every vertex that maximizes $\lambda\mathbf{w} + (1 - \lambda)\mathbf{c}$ for some $\lambda \in [0, 1]$, where $\mathbf{w}$ and $\mathbf{c}$ are uniquely maximized at $\mathbf{u}$ and $\mathbf{v}$ respectively. This path is precisely the path followed by the shadow simplex method for $\pi(\mathbf{x}) = (-\mathbf{w}^\intercal\mathbf{x}, \mathbf{c}^\intercal\mathbf{x})$ and is best understood in terms of the normal fan of the polytope. Namely, given a polytope P, for each vertex $\mathbf{v}$, there is a cone given by $C_\mathbf{v} = \{\mathbf{z} \in \mathbb{R}^m : \mathbf{z}^\intercal\mathbf{v} \geq \mathbf{z}^\intercal\mathbf{x} \text{ for all } \mathbf{x} \in P\}$. That is $C_\mathbf{v}$ is the cone of all vectors maximized at $\mathbf{v}$. The extreme rays of $C_\mathbf{v}$ are precisely the facet defining inequalities tight at $\mathbf{v}$. The set of all such cones is called the **normal fan** of P. Then, by definition, the shadow path for $\mathbf{w}$ consists of all vertices whose normal cones intersect the line segment from $\mathbf{w}$ to $\mathbf{c}$. See Fig. 1 for an illustration of shadow paths from this perspective. Fundamentally for this work, the choice of $\mathbf{w}$ is not unique, and one may choose any $\mathbf{w}$ in the interior of $C_{\mathbf{u}}$ to choose a shadow path.

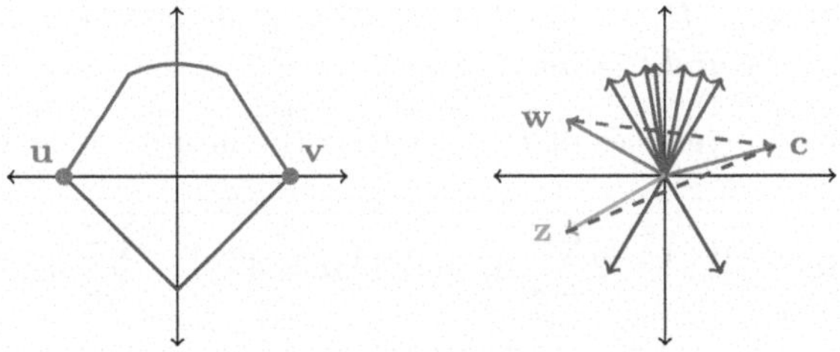

Fig. 1. Pictured on the left is a polygon. The path on the top of the polygon corresponds to the shadow path chosen by interpolating between $\mathbf{w}$ and $\mathbf{c}$, while the lower path corresponds to the shadow path chosen by interpolating between $\mathbf{z}$ and $\mathbf{c}$. As one can see from this picture, different choices of auxiliary vectors may yield different shadow paths.

3 Exponential Sized Shadows

The main idea of the proof is as follows. We start with a polytope with an exponentially large shadow from [33]. Then, by the parametric definition of the shadow pivot rule, there exist $\mathbf{w}, \mathbf{c} \in \mathbb{R}^n$ such that $\mathbf{u}$ and $\mathbf{v}$ are a unique $\mathbf{w}$-maximizer and a unique $\mathbf{c}$-maximizer respectively. Using a polyhedral construction, we can construct a new polytope P' with the same normal fan as P except the normal cones $C_{\mathbf{u}}$ is subdivided. The resulting polytope then has a vertex $\mathbf{a}$ such that $\mathbf{w} \in C_{\mathbf{a}} \subset C_{\mathbf{u}}$. We construct $C_{\mathbf{a}}$ to be so thin that any shadow path from $\mathbf{a}$ to maximize $\mathbf{c}$ must be as long as the shadow path $\mathbf{w}$ to $\mathbf{c}$ in P. In [9,12,14], each result relies on an assumption that all normal cones are sufficiently wide to ensure the existence of small shadows. Similarly, to force all shadows to be large, we only require one particularly thin cone.

For our proof, we modify the following linear program appearing in a dual form in [33]:

$$\max\left(\sum_{i=1}^{n}\mathbf{x}_i\right) \quad \text{s.t. } 0 \leq \mathbf{x}_k \leq 4^{k-1} - 2\sum_{i=1}^{k-1}\mathbf{x}_i \text{ for all } k \in [n].$$

We call the feasible region of this linear program P_n. The following lemma may be easily verified.

Lemma 1. *For all $n \in \mathbb{N}$, P_n is a combinatorial cube with the defining inequalities for each pair of disjoint facets being given by the bounds on each $\mathbf{x}_k$.*

In [33], Murty showed in a dual form that the shadow path interpolating between the pair of vectors $(-2, -4, \ldots, -2^n)$ and $(1, 1, \ldots, 1)$ passed through every vertex of P_n. Our goal is to generalize this result. To start, it will help to better understand the orientation induced by the vector $\mathbf{1} = \sum_{i=1}^n e_i$.

Lemma 2. *Let* $\mathbf{u}, \mathbf{v} \in P_n$*. Then* $\mathbf{1}^\intercal \mathbf{u} < \mathbf{1}^\intercal \mathbf{v}$ *if and only if* $\mathbf{u}_k < \mathbf{v}_\ell$*, where* $k = \max(\{i : \mathbf{u}_i > 0\})$ *and* $\ell = \max(\{j : \mathbf{v}_j > 0\})$*.*

Proof. Let $\mathbf{u}$ and $\mathbf{v}$ be vertices of P_n. Let k be the largest nonzero coordinate of $\mathbf{u}$. Then

$$\mathbf{u}_k = 4^{k-1} - 2\sum_{i=1}^{k-1} \mathbf{u}_i \text{ and } \mathbf{1}^\intercal \mathbf{u} = 4^{k-1} - \sum_{i=1}^{k-1} \mathbf{u}_i.$$

Let ℓ be the largest nonzero coordinate of $\mathbf{v}$. Suppose that $k < \ell$. Then

$$\mathbf{u}_k = 4^{k-1} - 2\sum_{i=1}^{k-1} \mathbf{u}_i \leq 4^{k-1} < 4^{\ell-1} - \sum_{i=1}^{\ell-1} 4^{i-1} \leq 4^{\ell-1} - 2\sum_{i=1}^{\ell-1} \mathbf{v}_i = \mathbf{v}_\ell.$$

By similar reasoning, $\mathbf{1}^\intercal \mathbf{u} < \mathbf{1}^\intercal \mathbf{v}$.

Suppose that $k = \ell$. Then $\mathbf{1}^\intercal \mathbf{u} < \mathbf{1}^\intercal \mathbf{v}$ if and only if $\sum_{i=1}^{k-1} \mathbf{u}_i > \sum_{i=1}^{k-1} \mathbf{v}_i$, which is true if and only if $\mathbf{u}_k < \mathbf{v}_k$. Hence, $\mathbf{1}^\intercal \mathbf{u} < \mathbf{1}^\intercal \mathbf{v}$ if and only if $\mathbf{u}_k < \mathbf{v}_\ell$. □

In order to construct our exponential example, we need a cone of vectors that all yield an exponential run-time as we find in the following lemma.

Lemma 3. *Consider the linear program* $\max(\mathbf{1}^\intercal \mathbf{x})$ *such that* $\mathbf{x} \in P_n$*, where* $\mathbf{1} = \sum_{i=1}^n e_i$*. Let* $\mathbf{r} = (r_1, \ldots, r_n)$ *be any nonzero vector in* $\mathbb{R}^n$ *such that* $2r_i \leq r_{i+1}$ *for all* $i \in [n-1]$ *and* $r_1 > 0$*. Then* $\mathbf{r}$ *is minimized uniquely at* $\mathbf{0}$ *and the shadow simplex path interpolating between* $-\mathbf{r}$ *and* $\mathbf{1}$ *contains every vertex of* P_n*.*

Proof. Clearly for P_1 the shadow simplex path traces every vertex. Furthermore, the slope of the edge from 0 to 1 is $\frac{1}{r_1} > \frac{1}{r_2}$, and the first edge goes from 0 to 1.

Proceed by induction, and suppose that the result holds for P_n, the slope of the final edge in P_n is greater than $\frac{1}{r_{n+1}}$, and the first edge in the path on P_n goes from $(0, 0, \ldots, 0)$ to $(1, 0, \ldots, 0)$. By Lemma 1 P_{n+1} has two facets, P_{n+1}^0 and P_{n+1}^φ, that are affine isomorphic to P_n with 0 in the final coordinate and to P_n with the last coordinate given by applying the function $\varphi : \mathbf{R}^n \to \mathbb{R}$ given by $\varphi(\mathbf{x}) = 4^n - \sum_{i=1}^n \mathbf{x}_i$ respectively. Then given a vertex $\mathbf{v}$ of P_{n+1}^0 its neighbors are precisely its corresponding neighbors in P_n and $\mathbf{v} + \varphi(\mathbf{v})e_{n+1}$. Then the slope from $\mathbf{v}$ to $\mathbf{v} + \varphi(\mathbf{v})e_{n+1}$ is precisely

$$\frac{\mathbf{1}^\intercal(\mathbf{v} + \varphi(\mathbf{v})e_{n+1} - \mathbf{v})}{\mathbf{r}^\intercal(\mathbf{v} + \varphi(\mathbf{v})e_{n+1} - \mathbf{v})} = \frac{\varphi(\mathbf{v})}{r_{n+1}\varphi(\mathbf{v})} = \frac{1}{r_{n+1}}.$$

By our inductive hypothesis, $\frac{1}{r_{n+1}}$ is less than the slope of the final edge of the shadow simplex path in P_n. In particular, since the slopes decrease along shadow simplex paths, $\frac{1}{r_{n+1}}$ is less than the slope of the maximum slope neighbor of $\mathbf{v}$ in P_{n+1}^0. Hence, if $\mathbf{v}$ is not the optimum in P_{n+1}^0, $\mathbf{v}$ will always choose the shadow simplex neighbor in P^0_{n+1}. It follows that the path starting at $\mathbf{0}$ in P^0_{n+1} will be precisely the shadow simplex path from P_n, which passes through every vertex by our inductive hypothesis.

By Lemma 2, the maximizer of $\mathbf{1}$ on P_n is the same as the maximizer for e_n. Hence, the $\mathbf{1}$-optimum of P^0_{n+1} is $(0, 0, \ldots, 0, 4^{n-1}, 0)$, since the maximum value of the $(n-1)$st coordinate is 4^{n-1}. The next step in the path is the vertex $(0, \ldots, 0, 4^{n-1}, 4^n - 4^{n-1})$ of P^{φ}_{n+1}. The remaining vertices have the same total ordering given by e_{n+1} and $\mathbf{1}$ by Lemma 2. Note that for all $\mathbf{v}$ in P^{φ}_{n+1}, $\mathbf{v}_{n+1} = 4^n - 2\sum_{i=1}^n \mathbf{v}_i$. Hence, the same ordering on the vertices on P^{φ}_{n+1} is also induced by the function $\psi(\mathbf{v}) = -\sum_{i=1}^n \mathbf{v}_i$. It follows that the orientation of the graph of P^{φ}_{n+1} is the reverse of the orientation of the graph on P^0_{n+1}.

Now consider a vertex $\mathbf{u}$ of P^{φ}_{n+1} and an improving neighbor $\mathbf{v}$ of P^{φ}_{n+1}. Let $\mathbf{u}^0$ and $\mathbf{v}^0$ denote the corresponding vectors in P_n given by deleting the $(n+1)$st coordinates of $\mathbf{u}$ and $\mathbf{v}$ respectively. Then the slope from $\mathbf{u}^0$ to $\mathbf{v}^0$ is given by

$$s(\mathbf{u}^0, \mathbf{v}^0) = \frac{\sum_{i=1}^n \mathbf{v}_i - \mathbf{u}_i}{\sum_{i=1}^n \mathbf{r}_i(\mathbf{v}_i - \mathbf{u}_i)}.$$

It follows that the slope from $\mathbf{u}$ to $\mathbf{v}$ is:

$$\begin{aligned}
\frac{\mathbf{1}^\intercal(\mathbf{v}-\mathbf{u})}{\mathbf{r}^\intercal(\mathbf{v}-\mathbf{u})} &= \frac{\mathbf{v}_{n+1} - \mathbf{u}_{n+1} + \sum_{i=1}^n \mathbf{v}_i - \mathbf{u}_i}{r_{n+1}(\mathbf{v}_{n+1} - \mathbf{u}_{n+1}) + \sum_{i=1}^n \mathbf{r}_i(\mathbf{v}_i - \mathbf{u}_i)} \\
&= \frac{-\sum_{i=1}^n \mathbf{v}_i - \mathbf{u}_i}{-2r_{n+1}(\sum_{i=1}^n \mathbf{v}_i - \mathbf{u}_i) + \sum_{i=1}^n \mathbf{r}_i(\mathbf{v}_i - \mathbf{u}_i)} \\
&= \frac{1}{2r_{n+1} - \frac{\sum_{i=1}^n \mathbf{r}_i(\mathbf{v}_i - \mathbf{u}_i)}{\sum_{i=1}^n \mathbf{v}_i - \mathbf{u}_i}} = \frac{1}{2r_{n+1} - \frac{1}{s(\mathbf{u}^0,\mathbf{v}^0)}}.
\end{aligned}$$

Note that the function $f : \mathbb{R} \setminus \{0, -\frac{1}{2r_{n+1}}\} \to \mathbb{R}$ given by $f(x) = \frac{1}{2r_{n+1} - \frac{1}{x}}$ is order reversing meaning that $x < y$ if and only if $f(x) > f(y)$ for $x, y \in \mathbb{R} \setminus \{0, -\frac{1}{2r_{n+1}}\}$. Therefore the ordering of the slopes on P^{φ}_{n+1} is precisely the reverse of the ordering of the slopes on P^0_{n+1} meaning that the path followed from $(0, \ldots, 0, 4^{n-1}, 4^n - 4^{n-1})$ will be precisely the shadow simplex path on P^0_{n+1} in reverse. In particular, the final step will be from $(1, 0, \ldots, 4^n - 2)$ to $(0, \ldots, 0, 4^n)$. The slope of that step is given by

$$\frac{4^n - (4^n - 2) - 1}{(4^n - (4^n - 2))r_{n+1} - r_1} = \frac{1}{2r_{n+1} - r_1}.$$

By our assumption that $r_{n+2} \geq 2r_{n+1} > 2r_{n+1} - r_1$, we therefore have that the final edge of the path has slope greater than $\frac{1}{r_{n+2}}$, which completes the induction and therefore the proof. □

Thus, for each $n \in \mathbb{N}$, there is a cone C_n of objectives given by the inequalities $\mathbf{r}_1 \geq 0$ and $2\mathbf{r}_i \leq \mathbf{r}_{i+1}$ for all $i \in [n-1]$ such that for any $\mathbf{r} \in C_n$, the $-\mathbf{r}$-shadow simplex path for solving the linear program for maximizing $\mathbf{1}^\intercal\mathbf{x}$ on P_n is always of length at least 2^n. All that remains is to ensure that we can modify P_n by adding new inequalities to create a new vertex whose normal cone is contained in $-C_n$. For this, we require the following lemma from polyhedral geometry:

Lemma 4. *Let $P \subseteq \mathbb{R}^n$ be an n-dimensional simple polytope with normal fan $\mathcal{N}$, and choose a cone $C \in \mathcal{N}$ generated by $\mathbf{x}_1, \mathbf{x}_2, \ldots, \mathbf{x}_n$. Let $\mathbf{w}$ be in the interior of C. Let r and s be the largest and second largest values of $\mathbf{w}^\intercal \mathbf{v}$ such that $\mathbf{v}$ is a vertex of P. Then consider the polytope P' given by $P \cap \{\mathbf{x} : \mathbf{w}^\intercal \mathbf{x} \leq \frac{r+s}{2}\}$. Then the normal fan $\mathcal{N}'$ of P' is given by $\mathcal{N}' = \mathcal{N} \setminus \{C\} \cup \{D_1, \ldots, D_n\}$, where D_i is the cone generated by $\mathbf{x}_1, \mathbf{x}_2, \ldots, \mathbf{x}_{i-1}, \mathbf{w}, \mathbf{x}_{i+1}, \ldots, \mathbf{x}_n$. In particular P' has precisely one more facet than P.*

Proof. Since $\mathbf{w}$ is on the interior of C, $\mathbf{w}$ is uniquely maximized at a vertex $\mathbf{v}$ of P corresponding to C. By assumption, $r > s$ meaning that $r > \frac{r+s}{2} > s$ and therefore $\mathbf{w}^\intercal \mathbf{u} < \frac{r+s}{2}$ for $\mathbf{u}$ a vertex of P if and only if $\mathbf{u} \neq \mathbf{v}$.

Hence, the half-space $\{\mathbf{x} : \mathbf{w}^\intercal \mathbf{x} = \frac{r+s}{2}\}$ is valid for all vertices other than $\mathbf{v}$, since $\frac{r+s}{2} > s$ and, in particular, intersects each edge incident to $\mathbf{v}$ on its interior. The intersection with each edge is a new vertex and these are exactly the new vertices created by intersecting P with that half-space. All other vertices and normal cones remain the same. The facet directions tight at each new vertex are given by $\mathbf{w}$ and the set of facets tight at the corresponding edge incident to $\mathbf{v}$. Hence, the set of cones for each of those vertices is precisely $D_1, D_2, \ldots, D_n$. All other cones remain unchanged.

This geometric observation tells us that we may add inequalities in such a way that we only ever subdivide precisely one cone in the normal fan. We will use this lemma to create a new vertex whose normal cone is contained in $-C_n$, which suffices to prove the result.

Proof (Proof of Theorem 1). Consider the sequence of vectors $\mathbf{a}^1, \mathbf{a}^2, \ldots, \mathbf{a}^n$ given by the recurrence $\mathbf{a}^1 = (1, 2, \ldots, 2^n)$ and $\mathbf{a}^k = \sum_{i=1}^{k-1} \mathbf{a}^i + \sum_{i=k}^{n} 2^i e_i$. Note that $\mathbf{a}^1 \in C_n$. Suppose for the sake of induction that $\mathbf{a}^j \in C_n$ for all $j < k$. Then, since $\sum_{i=k}^{n} 2^i e_i \in C_n$, $\mathbf{a}^k \in C_n$. Hence, $\mathbf{a}^1, \ldots, \mathbf{a}^n \in C_n$.

Note that by applying row operations we can replace each $\mathbf{a}^k$ with $\mathbf{a}^k - \sum_{i=1}^{k-1} \mathbf{a}^i$. Then the resulting sequence of vectors is

$$\left(\sum_{i=1}^{n} 2^i e_i, \sum_{i=2}^{n} 2^i e_i, \ldots, \sum_{i=k}^{n} 2^i e_i, e_{k+1}, \ldots, e_n \right).$$

The corresponding matrix is upper triangular with the diagonal all nonzero meaning that

$$\{\mathbf{a}^1, \ldots, \mathbf{a}^k, e_{k+1}, \ldots, e_n\}$$

is linearly independent for each $k \in [n]$. Hence the cone generated by them is full dimensional, and, since $\mathbf{a}^k$ is a positive combination of the rays given by $(\mathbf{a}^1, \ldots, \mathbf{a}^{k-1}, e_k, \ldots, e_n)$, $\mathbf{a}^k$ lies on the interior of that cone.

Now we may add inequalities to P_n to construct the exponential example. Each inequality is given by $(\mathbf{a}^k)^\intercal \mathbf{x} \geq b_k$ for some $b_k \in \mathbb{R}$. We build these inequalities inductively. Let $P_{k,n}$ denote P_n intersected with first k of those inequalities.

Let $c_k = \min((\mathbf{a}^k)^\intercal \mathbf{x} : \mathbf{x} \in P_{k-1,n})$, and let d_k be the second smallest value of $(\mathbf{a}^k)^\intercal \mathbf{x}$ on $P_{k-1,n}$. Then we define $b_k = c_k + (d_k - c_k)/2$. Note that since $\mathbf{a}^k$ is polynomial in size and $P_{k-1,n}$ may be assumed by induction to have polynomial encoding length, $P_{k,n}$ then has polynomial encoding length. Then, by construction, each new inequality adds precisely one new facet with normal given by $\mathbf{a}^k$. Furthermore, $\mathbf{a}^k$ is in the cone of $(\mathbf{a}^1, \dots, \mathbf{a}^{k-1}, e_k, \dots, e_n)$, which are the rays of the normal cone of a vertex in $P_{k-1,n}$ by induction. Hence, there is a new vertex of $P_{k,n}$ with normal cone with rays given by $(\mathbf{a}^1, \dots, \mathbf{a}^k, e_{k+1}, \dots, e_n)$ by Lemma 4. Therefore, by completing this induction, we arrive at $P_{n,n}$, which has a vertex $\mathbf{v}$ with normal cone with rays $(\mathbf{a}^1, \dots, \mathbf{a}^n)$. Since $\mathbf{a}^1, \dots, \mathbf{a}^n \in C_n$, any shadow simplex method for solving the linear program given by maximizing $\mathbf{1}^\intercal \mathbf{x}$ for $\mathbf{x} \in P_{n,n}$ from $\mathbf{v}$ must be of exponential length by Lemma 3. □

4 Steepest Edge Lower Bounds for Every Norm

In this section, our goal is to prove Theorems 2 and 3. As in the previous section, we modify Murty's cube P_n, but in this case, we shall do so by applying an affine transformation to it. In particular, we flatten P_n in the auxiliary direction $\mathbf{w}$ for choosing an exponential length shadow path, so that $\eta(\mathbf{x})$ is approximately the same as $\mathbf{w}^\intercal \mathbf{x}$ forcing the η-steepest edge rule to take the same path.

Proof (Proofs of Theorem 2). Let P_n be the polytope from Sect. 3. Then, by Lemmas 1 and 3, P_n is a combinatorial cube with an exponential sized shadow and, in particular, a shadow simplex path walking through all vertices of P_n for optimizing $\mathbf{1}^\intercal \mathbf{x}$. Let $\mathbf{w}$ be any auxiliary vector that chooses the shadow simplex path walking through all vertices.

We may assume without loss of generality that $\|\mathbf{w}\|_2 = 1$, since rescaling does not change the choice of shadow. Construct an orthonormal basis for $\mathbb{R}^n$ with respect to the typical 2-norm via the Gram Schmidt process given by $\mathbf{w}, \mathbf{a}^1, \dots, \mathbf{a}^{n-1}$. Then define a family of linear maps for all $k > 0$

$$A_k \left(\lambda \mathbf{w} + \sum_{i=1}^{n-1} \lambda_i \mathbf{a}^i \right) = \lambda \mathbf{w} + \sum_{i=1}^{n-1} \frac{\lambda_i}{k} \mathbf{a}^i.$$

By construction, A_k is invertible for all $k > 0$ and $A_k(\mathbf{w}) = \mathbf{w}$. Our goal is to show that for all k sufficiently large and vertices $\mathbf{v}$ of P_n, we have

$$\operatorname*{argmax}_{\mathbf{u} \in N_{\mathbf{c}}(\mathbf{v})} \left(\frac{\mathbf{c}^\intercal A_k^{-1}(A_k(\mathbf{u}) - A_k(\mathbf{v}))}{\mathbf{w}^\intercal A_k^{-1}(A_k(\mathbf{u}) - A_k(\mathbf{v}))} \right) = \operatorname*{argmax}_{\mathbf{u} \in N_{\mathbf{c}}(\mathbf{v})} \left(\frac{\mathbf{c}^\intercal A_k^{-1}(A_k \mathbf{u} - A_k \mathbf{v})}{\eta(A_k \mathbf{u} - A_k \mathbf{v})} \right), \quad (1)$$

for that readily implies that the η-steepest edge path on $A_k(P_n)$ must be exponentially long. Observe that Eq. 1 may be rewritten as

$$\operatorname*{argmax}_{\mathbf{u} \in N_{\mathbf{c}}(\mathbf{v})} \left(\frac{\mathbf{c}^\intercal (\mathbf{u} - \mathbf{v})}{\mathbf{w}^\intercal (\mathbf{u} - \mathbf{v})} \right) = \operatorname*{argmax}_{\mathbf{u} \in N_{\mathbf{c}}(\mathbf{v})} \left(\frac{\mathbf{c}^\intercal (\mathbf{u} - \mathbf{v})}{\eta(A_k(\mathbf{u} - \mathbf{v}))} \right).$$

In fact, rescaling $\mathbf{w}$ does not change the chosen vertex. Thus, the following is also equivalent to Eq. 1:

$$\operatorname*{argmax}_{\mathbf{u}\in N_{\mathbf{c}}(\mathbf{v})}\left(\frac{\mathbf{c}^{\intercal}(\mathbf{u}-\mathbf{v})}{\eta(\mathbf{w})\mathbf{w}^{\intercal}(\mathbf{u}-\mathbf{v})}\right)=\operatorname*{argmax}_{\mathbf{u}\in N_{\mathbf{c}}(\mathbf{v})}\left(\frac{\mathbf{c}^{\intercal}(\mathbf{u}-\mathbf{v})}{\eta(A_k(\mathbf{u}-\mathbf{v}))}\right).$$

Note that $\mathbf{c}$ and $\mathbf{w}$ may be perturbed slightly to be sufficiently generic so that $\frac{\mathbf{c}^{\intercal}(\mathbf{u}-\mathbf{v})}{\mathbf{w}^{\intercal}(\mathbf{u}-\mathbf{v})}$ is distinct for all pairs of vertices $\mathbf{u}$ and $\mathbf{v}$ of P such that $\mathbf{u}$ is a $\mathbf{c}$-improving neighbor of $\mathbf{v}$. In particular, there exists $\varepsilon > 0$ sufficiently small such that if

$$|\eta(A_k(\mathbf{u}-\mathbf{v}))-\eta(\mathbf{w})\mathbf{w}^{\intercal}(\mathbf{u}-\mathbf{v})|<\varepsilon,$$

for all $\mathbf{u},\mathbf{v}\in P_n$ with $\mathbf{u}$ a $\mathbf{c}$-improving neighbor of $\mathbf{v}$, then Eq. 1 must hold.

Let $S=\{\mathbf{u}-\mathbf{v} : \mathbf{u}\in N_{\mathbf{c}}(\mathbf{v}), \mathbf{v}\in V(P_n)\}$, where $V(P_n)$ is the set of vertices of P_n. To prove that Eq. 1 holds, it suffices to show there exists k sufficiently large that

$$\max_{\mathbf{s}\in S}(|\eta(A_k(\mathbf{s}))-\eta(\mathbf{w})\mathbf{w}^{\intercal}\mathbf{s}|)<\varepsilon.$$

In fact, it suffices to show for any $\mathbf{s}\in S$ one can find $k_{\mathbf{s}}$ sufficiently large so that

$$|\eta(A_k(\mathbf{s}))-\eta(\mathbf{w})\mathbf{w}^{\intercal}\mathbf{s}|<\varepsilon$$

for all $k > k_{\mathbf{s}}$, since one can take the maximum of all choices of $k_{\mathbf{s}}$ to achieve the uniform bound. This will be our strategy.

Let $\mathbf{t}\in S$. We now setup the technical bounds for our computation. The first steps are purely algebraic taking advantage of η being positive homogeneous:

$$\begin{aligned}|\eta(A_k(\mathbf{t}))-\eta(\mathbf{w})\mathbf{w}^{\intercal}\mathbf{t}| &= |\eta(A_k((\mathbf{w}^{\intercal}\mathbf{t})\mathbf{w}+\mathbf{t}-(\mathbf{w}^{\intercal}\mathbf{t})\mathbf{w}))-\eta(\mathbf{w})\mathbf{w}^{\intercal}\mathbf{t}|,\\ &=\left||\mathbf{w}^{\intercal}\mathbf{t}|\eta\left(\mathbf{w}+A_k\left(\frac{\mathbf{t}-\mathbf{w}^{\intercal}\mathbf{t}\mathbf{w}}{\mathbf{w}^{\intercal}\mathbf{t}}\right)\right)-\eta(\mathbf{w})\mathbf{w}^{\intercal}\mathbf{t}\right|.\end{aligned}$$

Lemma 2 of [7] states any shadow simplex path for vectors $\mathbf{c}$ and $\mathbf{w}$ is both strictly $\mathbf{c}$-increasing and $\mathbf{w}$-increasing. Since the path on P_n contains all the vertices of P, $\mathbf{c}$ and $\mathbf{w}$ must induce the same ordering on the vertices of P. In particular, an edge is $\mathbf{c}$-improving if and only if it is $\mathbf{w}$-improving. It follows that $\mathbf{w}^{\intercal}(\mathbf{u}-\mathbf{v})>0$ for all $\mathbf{u}$ and $\mathbf{v}$ such that $\mathbf{u}$ is a $\mathbf{c}$-improving neighbor of $\mathbf{v}$ meaning that $\mathbf{w}^{\intercal}\mathbf{s}>0$ for all $\mathbf{s}\in S$. This observation ensures that the sign of $\mathbf{w}^{\intercal}\mathbf{t}$ and $\eta(\mathbf{t})$ coincide. In particular, this allows us to factor out $\mathbf{w}^{\intercal}\mathbf{t}$ to find

$$|\eta(A_k(\mathbf{t}))-\eta(\mathbf{w})\mathbf{w}^{\intercal}\mathbf{t}|=\mathbf{w}^{\intercal}\mathbf{t}\left|\eta\left(\mathbf{w}+A_k\left(\frac{\mathbf{t}-\mathbf{w}^{\intercal}\mathbf{t}\mathbf{w}}{\mathbf{w}^{\intercal}\mathbf{t}}\right)\right)-\eta(\mathbf{w})\right|.$$

By continuity of η, there exists $\delta > 0$ sufficiently small and $\mathbf{y}\in\mathbb{R}^n$, such that whenever $\|\mathbf{y}\|_1<\delta$,

$$|\eta(\mathbf{w}+\mathbf{y})-\eta(\mathbf{w})|<\frac{\varepsilon}{\mathbf{w}^{\intercal}\mathbf{t}}.$$

Furthermore, there exists $k_{\mathbf{t}}$ so large that for all $k > k_{\mathbf{t}}$, by the definition of A_k,

$$\left\| A_k \left(\frac{\mathbf{t} - (\mathbf{w}^\intercal \mathbf{t} \mathbf{w})}{\mathbf{w}^\intercal \mathbf{t}} \right) \right\|_1 = \left\| \frac{1}{k \mathbf{w}^\intercal \mathbf{t}} (\mathbf{t} - \mathbf{w}^\intercal \mathbf{t} \mathbf{w}) \right\|_1 < \delta.$$

Hence, for all $k > k_{\mathbf{t}}$,

$$|\eta(A_k(\mathbf{t})) - \eta(\mathbf{w})\mathbf{w}^\intercal \mathbf{t}| = \mathbf{w}^\intercal \mathbf{t} \left| \eta \left(\mathbf{w} + A_k \left(\frac{\mathbf{t} - \mathbf{w}^\intercal \mathbf{t} \mathbf{w}}{\mathbf{w}^\intercal \mathbf{t}} \right) \right) - \eta(\mathbf{w}) \right| < (\mathbf{w}^\intercal \mathbf{t}) \frac{\varepsilon}{\mathbf{w}^\intercal \mathbf{t}} = \varepsilon,$$

which completes the proof of Theorem 2. □

To prove Theorem 3, we refine this argument under the assumption that η is a regular norm.

Proof (Proof of Theorem 3). Note that there are two choices we make in the proof of Theorem 2 that depend on the norm. The first is the value of ε, which depends on the value of $\eta(\mathbf{w})$. Without loss of generality, up to an invertible linear transformation, we may assume that $\mathbf{w} = e_1$. Then, since η is regular, $\eta(\mathbf{w}) = 1$. Hence, the choice of ε will be the same for all regular norms.

The choice of $k_{\mathbf{t}}$ depended solely on the choice of δ independent of the norm. Namely, we need δ to be made so small that whenever $\|\mathbf{y}\|_1 < \delta$,

$$|\eta(\mathbf{w} + \mathbf{y}) - \eta(\mathbf{w})| < \frac{\varepsilon}{\mathbf{w}^\intercal \mathbf{t}}.$$

Thus, it suffices to show that a singular choice of δ works for all regular norms. Suppose first that $\eta(\mathbf{w} + \mathbf{y}) \geq \eta(\mathbf{w})$. Then, by the triangle inequality,

$$\eta(\mathbf{w} + \mathbf{y}) - \eta(\mathbf{w}) \leq \eta(\mathbf{w}) + \eta(\mathbf{y}) - \eta(\mathbf{w}) = \eta(\mathbf{y}).$$

Similarly, if $\eta(\mathbf{w} + \mathbf{y}) < \eta(\mathbf{w})$, by the triangle inequality

$$\eta(\mathbf{w}) - \eta(\mathbf{w} + \mathbf{y}) = \eta(\mathbf{w} + \mathbf{y} - \mathbf{y}) - \eta(\mathbf{w} + \mathbf{y}) \leq \eta(-\mathbf{y}) + \eta(\mathbf{w} + \mathbf{y}) - \eta(\mathbf{w} + \mathbf{y}) = \eta(\mathbf{y}).$$

Hence, $|\eta(\mathbf{w} + \mathbf{y}) - \eta(\mathbf{w})| \leq \eta(\mathbf{y})$. Then, for any regular norm, whenever $\|\mathbf{y}\|_1 < \frac{\varepsilon}{\mathbf{w}^\intercal \mathbf{t}}$, by the triangle inequality

$$|\eta(\mathbf{w} + \mathbf{y}) - \eta(\mathbf{w})| \leq \eta(\mathbf{y}) \leq \sum_{i=1}^{n} \eta(\mathbf{y}_i e_i) = \sum_{i=1}^{n} |y_i| \eta(e_i) = \sum_{i=1}^{n} |y_i| = \|\mathbf{y}\|_1 < \frac{\varepsilon}{\mathbf{w}^\intercal \mathbf{t}}.$$

Hence, $\delta = \frac{\varepsilon}{\mathbf{w}^\intercal \mathbf{t}}$ works simultaneously for any regular norm meaning that there exists a combinatorial cube for which the path through all the vertices is chosen for the steepest edge for every regular norm simultaneously. □

Acknowledgments. This research was benefited by discussions with Eleon Bach, Jesús De Loera, Sophie Huiberts, and Sean Kafer, feedback from anonymous reviewers, and financial support from the NSF GRFP.

Disclosure of Interests. The author has no competing interests to declare.

References

1. Amenta, N., Ziegler, G.: Deformed products and maximal shadows of polytopes. Contemp. Math. **223**, 57–90 (1999)
2. Avis, D., Chvátal, V.: Notes on Bland's pivoting rule. Polyhedral Combinatorics: Dedicated to the memory of DR Fulkerson, pp. 24–34 (1978)
3. Ben-Tal, A., Nemirovski, A.: On polyhedral approximations of the second-order cone. Math. Oper. Res. **26**(2), 193–205 (2001)
4. Bertsimas, D., Tsitsiklis, J.: Introduction to Linear Optimization, vol. 6. Athena Scientific, Belmont (1997)
5. Bixby, R.: A brief history of linear and mixed-integer programming computation. Doc. Math. **2012**, 107–121 (2012)
6. Black, A.: Small shadows of lattice polytopes. In: Proceedings of the 2023 Annual ACM-SIAM Symposium on Discrete Algorithms (SODA), pp. 1669–1679. SIAM (2023)
7. Black, A., De Loera, J., Kafer, S., Sanità, L.: On the simplex method for 0/1-polytopes. Math. Oper. Res. (2024)
8. Black, A., De Loera, J., Lütjeharms, N., Sanyal, R.: The polyhedral geometry of pivot rules and monotone paths. SIAM J. Appl. Algebra Geom. **7**(3), 623–650 (2023)
9. Bonifas, N., Di Summa, M., Eisenbrand, F., Hähnle, N., Niemeier, M.: On subdeterminants and the diameter of polyhedra. Discret. Comput. Geom. **52**(1), 102–115 (2014)
10. Borgwardt, K.: Erratum: a sharp upper bound for the expected number of shadow vertices in LP-polyhedra under orthogonal projection on two-dimensional planes. Math. Oper. Res. **24**(4), 925–984 (1999)
11. Borgwardt, K.: The Simplex Method: A Probabilistic Analysis, vol. 1. Springer, Cham (2012)
12. Brunsch, T., Röglin, H.: Finding short paths on polytopes by the shadow vertex algorithm. In: Automata, Languages, and Programming: 40th International Colloquium (ICALP) Proceedings, pp. 279–290. Springer, Cham (2013)
13. Carstensen, P.: Complexity of some parametric integer and network programming problems. Math. Program. **26**(1), 64–75 (1983)
14. Dadush, D., Hähnle, N.: On the shadow simplex method for curved polyhedra. Discret. Comput. Geom. **56**(4), 882–909 (2016)
15. Dadush, D., Huiberts, S.: A friendly smoothed analysis of the simplex method. SIAM J. Comput. **49**(5), 18–449 (2019)
16. De Loera, J., Kafer, S., Sanità, L.: Pivot rules for circuit-augmentation algorithms in linear optimization. SIAM J. Optim. **32**(3), 2156–2179 (2022)
17. Deshpande, A., Spielman, D.: Improved smoothed analysis of the shadow vertex simplex method. In: 46th Annual IEEE Symposium on Foundations of Computer Science (FOCS), pp. 349–356. IEEE (2005)
18. Disser, Y., Friedmann, O., Hopp, A.: An exponential lower bound for Zadeh's pivot rule. Math. Program. **199**(1–2), 865–936 (2023)
19. Disser, Y., Mosis, N.: A unified worst case for classical simplex and policy iteration pivot rules. In: Proceedings of the 34th International Symposium on Algorithms and Computation (ISAAC), vol. 27, no. 17 (2023)
20. Forrest, J., Goldfarb, D.: Steepest-edge simplex algorithms for linear programming. Math. Program. **57**(1), 341–374 (1992)

21. Friedmann, O.: A subexponential lower bound for Zadeh's pivoting rule for solving linear programs and games. In: International Conference on Integer Programming and Combinatorial Optimization (IPCO), pp. 192–206. Springer, Cham (2011)
22. Friedmann, O., Hansen, T., Zwick, U.: Subexponential lower bounds for randomized pivoting rules for the simplex algorithm. In: Proceedings of the Forty-Third Annual ACM Symposium on Theory of Computing (STOC), pp. 283–292 (2011)
23. Gajjar, K., Radhakrishnan, J.: Parametric shortest paths in planar graphs. In: 2019 IEEE 60th Annual Symposium on Foundations of Computer Science (FOCS), pp. 876–895. IEEE (2019)
24. Gass, S., Saaty, T.: The computational algorithm for the parametric objective function. Naval Res. Logist. Quart. **2**, 39–45 (1955)
25. Goldfarb, D.: Worst case complexity of the shadow vertex simplex algorithm. Technical report (1983)
26. Goldfarb, D., Sit, W.: Worst case behavior of the steepest edge simplex method. Discret. Appl. Math. **1**(4), 277–285 (1979)
27. Huiberts, S., Lee, Y.T., Zhang, X.: Upper and lower bounds on the smoothed complexity of the simplex method. In: Proceedings of the 55th Annual ACM Symposium on Theory of Computing (STOC), pp. 1904–1917 (2023)
28. Jeroslow, R.: The simplex algorithm with the pivot rule of maximizing criterion improvement. Discret. Math. **4**(4), 367–377 (1973)
29. Kelner, J., Spielman, D.: A randomized polynomial-time simplex algorithm for linear programming. In: Proceedings of the Thirty-Eighth Annual ACM Symposium on Theory of Computing (STOC), pp. 51–60 (2006)
30. Klee, V., Minty, G.: How good is the simplex algorithm. Inequalities **3**(3), 159–175 (1972)
31. Kortenkamp, U., Richter-Gebert, J., Sarangarajan, A., Ziegler, G.: Extremal properties of 0/1-polytopes. Discret. Comput. Geom. **17**, 439–448 (1997)
32. Mulmuley, K., Shah, P.: A lower bound for the shortest path problem. In: Proceedings 15th Annual IEEE Conference on Computational Complexity (CCC), pp. 14–21. IEEE (2000)
33. Murty, K.: Computational complexity of parametric linear programming. Math. Program. **19**(1), 213–219 (1980)
34. Roos, C.: An exponential example for Terlaky's pivoting rule for the criss-cross simplex method. Math. Program. **46**, 79–84 (1990)
35. Shamir, R.: The efficiency of the simplex method: a survey. Manag. Sci. **33**(3), 301–334 (1987)
36. Spielman, D., Teng, S.: Smoothed analysis of algorithms: why the simplex algorithm usually takes polynomial time. J. ACM (JACM) **51**(3), 385–463 (2004)
37. Tano, M., Miyashiro, R., Kitahara, T.: Steepest-edge rule and its number of simplex iterations for a nondegenerate LP. Oper. Res. Lett. **47**(3), 151–156 (2019)
38. Terlaky, T., Zhang, S.: Pivot rules for linear programming: a survey on recent theoretical developments. Ann. Oper. Res. **46**, 203–233 (1993)
39. Vershynin, R.: Beyond Hirsch conjecture: walks on random polytopes and smoothed complexity of the simplex method. SIAM J. Comput. **39**(2), 646–678 (2009)
40. Zadeh, N.: A bad network problem for the simplex method and other minimum cost flow algorithms. Math. Program. **5**, 255–266 (1973)
41. Zadeh, N.: What is the worst case behavior of the simplex algorithm. Technical report (1980)

Online Matching on 3-Uniform Hypergraphs

Sander Borst[1(✉)], Danish Kashaev[2], and Zhuan Khye Koh[3]

[1] Max-Planck-Institut für Informatik, Saarbrücken, Germany
sborst@mpi-inf.mpg.de
[2] Centrum Wiskunde and Informatica (CWI), Amsterdam, The Netherlands
danish.kashaev@cwi.nl
[3] Boston University, Boston, USA
zkkoh@bu.edu

Abstract. The online matching problem was introduced by Karp, Vazirani and Vazirani (STOC 1990) on bipartite graphs with vertex arrivals. It is well-known that the optimal competitive ratio is $1-1/e$ for both integral and fractional versions of the problem. Since then, there has been considerable effort to find optimal competitive ratios for other related settings.

In this work, we go beyond the graph case and study the online matching problem on k-uniform hypergraphs. For $k = 3$, we provide an optimal primal-dual fractional algorithm, which achieves a competitive ratio of $(e-1)/(e+1) \approx 0.4621$. As our main technical contribution, we present a carefully constructed adversarial instance, which shows that this ratio is in fact optimal. It combines ideas from known hard instances for bipartite graphs under the edge-arrival and vertex-arrival models.

For $k \geq 3$, we give a simple integral algorithm which performs better than greedy when the online nodes have bounded degree. As a corollary, it achieves the optimal competitive ratio of $1/2$ on 3-uniform hypergraphs when every online node has degree at most 2. This is because the special case where every online node has degree 1 is equivalent to the edge-arrival model on graphs, for which an upper bound of $1/2$ is known.

1 Introduction

Online matching is a classic problem in the field of online algorithms. It was first introduced in the seminal work of Karp, Vazirani and Vazirani [22], who considered the bipartite version with one-sided vertex arrivals. In this setting, we are given a bipartite graph where vertices on one side are known in advance (offline), and vertices on the other side arrive sequentially (online). When an online vertex arrives, it reveals its incident edges, at which point the algorithm must decide how to match it (or not) irrevocably. The goal is to maximize the

The full version is available at arXiv:2402.13227. This project has received funding from the European Research Council (ERC) under the European Union's Horizon 2020 research and innovation programme (grant agreement no. 805241–QIP).
S. Borst and Z. K. Koh—Research was done while the authors were at CWI.

N. Megow and A. Basu (Eds.): IPCO 2025, LNCS 15620, pp. 100–113, 2025.
https://doi.org/10.1007/978-3-031-93112-3_8

cardinality of the resulting matching. Karp et al. [22] gave an elegant randomized algorithm RANKING, which achieves the optimal competitive ratio of $1 - 1/e$.

In certain applications, each offline vertex may be matched more than once. Examples include matching online jobs to servers, or matching online impressions to advertisers. This is the online b-matching model of Kalyanasundaram and Pruhs [20], where $b \geq 1$ is the maximum number of times an offline vertex can be matched. As b and the number of online vertices tend to infinity, it in turn captures the fractional relaxation of the Karp et al. [22] model. This means that the algorithm is allowed to match an online node fractionally to multiple neighbours, as long as the total load on every vertex does not exceed 1. For this problem, it is known that the deterministic algorithm BALANCE (or WATER-FILLING) achieves the optimal competitive ratio of $1 - 1/e$.

Online Hypergraph Matching. The online bipartite matching problem can be naturally generalized to hypergraphs as follows. For $k \geq 2$, let $\mathcal{H} = (V, W, H)$ be a k-uniform hypergraph with offline vertices V, online vertices W and hyperedges H. Every hyperedge $h \in H$ contains $k - 1$ elements from V and 1 element from W. Just like before, the online vertices arrive sequentially with their incident hyperedges, and the goal is to select a large matching, i.e., a set of disjoint hyperedges. The greedy algorithm is $1/k$-competitive. On the other hand, no integral algorithm can be $2/k$-competitive[1].

For the *fractional* version of the problem, Buchbinder and Naor [5] gave a deterministic algorithm which is $\Omega(1/\log k)$-competitive. They also constructed an instance showing that any algorithm is $O(1/\log k)$-competitive. In fact, their results apply to the more general setting of online packing linear program (LP), in which variables arrive sequentially. In the context of hypergraphs, this means that the hyperedges arrive sequentially. Note that for k-uniform hypergraphs, there is a trivial reduction from this *edge-arrival* model to our *vertex-arrival* model on $(k + 1)$-uniform hypergraphs, by adding degree 1 online nodes.

The aforementioned results show that asymptotically, both integral and fractional versions of the online matching problem on k-uniform hypergraphs are essentially settled (up to constant factors). However, our understanding of the problem for small values of k (other than $k = 2$) remains poor. Many applications of online hypergraph matching in practice have small values of k. For instance, in ride-sharing and on-demand delivery services [30], $k - 1$ represents the capacity of service vehicles, which is often small. Another example is network revenue management problems [25]. In this setting, given a collection of limited resources, a sequence of product requests arrive over time. When a product request arrives, we have to decide whether to accept it irrevocably. Accepting a product request generates profit, but also consumes a certain amount of each resource. The goal is to devise a policy which maximizes profit. In this context, $k - 1$ represents the maximum number of resources used by a product. As Ma

[1] In [31], it is shown that no algorithm can be $(2 + f(k))/k$-competitive for some positive function f with $f(k) = o(1)$. In the full version, we give a simple construction showing that no integral algorithm can be $2/k$-competitive.

et al. [25] noted, many of these problems have small values of k. In airlines, for example, $k-1$ corresponds to the maximum number of flight legs included in an itinerary, which usually does not exceed two or three.

Our Contributions. Motivated by the importance of online hypergraph matching for small values of k, we focus on 3-uniform hypergraphs, with the goal of obtaining tighter bounds. Our main result is a tight competitive ratio for the fractional version of this problem.

Theorem 1. *For the online fractional matching problem on 3-uniform hypergraphs, there is a deterministic $(e-1)/(e+1)$-competitive algorithm. Furthermore, every algorithm is at most $(e-1)/(e+1)$-competitive.*

The deterministic algorithm in Theorem 1 belongs to the class of WATER-FILLING algorithms. It uses the function $f(x) := e^x/(e+1)$ to decide which hyperedges receive load. In particular, for every online vertex w, the incident hyperedges $h = \{u, v, w\} \in \delta(w)$ which minimize $\phi(h) := f(x(\delta(u))) + f(x(\delta(v)))$ receive load until $\phi(h) \geq 1$.

Our main contribution is proving a matching upper bound in Theorem 1. For this, it suffices to consider deterministic algorithms because every randomized algorithm induces a deterministic fractional algorithm with the same expected value. This, in turn, allows us to construct an instance which is adaptive to the actions of the algorithm. The key idea is to combine two hard instances for online matching on *bipartite graphs* [13,22].

We start with the instance in [13], designed for the edge-arrival model. In this instance, edge arrivals are grouped into phases, such that the size of an online maximum matching increases by one per phase. At the end of every phase, as long as the total fractional value on the revealed edges exceeds a certain threshold, the next phase begins. Otherwise, the instance terminates. For our purpose, we want a more fine-grained control over the actions of the algorithm. So, we apply thresholding at the node level instead, based on fractional degrees, to determine which nodes become incident to the edges arriving in the next phase.

In our construction, we will have multiple copies of this modified edge-arrival instance. The edges in these instances are connected to the online nodes to form hyperedges. The way in which they are connected is inspired by the instance in [22], originally designed for the vertex-arrival model. The idea behind this vertex-arrival instance is to obfuscate the partners of the online nodes in an offline maximum matching, which is also applicable in our setting.

Our next result concerns the online integral matching problem on k-uniform hypergraphs. We show that one can do better than the greedy algorithm if the online nodes have bounded degree. It is achieved by the simple algorithm RANDOM: for every online vertex w, uniformly select a hyperedge among all the hyperedges incident to w which are disjoint from the current matching.

Theorem 2. *For the online matching problem on k-uniform hypergraphs where online vertices have maximum degree d, the competitive ratio of* RANDOM *is at least*

$$\min\left(\frac{1}{k-1}, \frac{d}{(d-1)k+1}\right).$$

Note that in Theorem 2, the first term is at most the second term if and only if $d \leq k-1$. Moreover, RANDOM is at least as good as the greedy algorithm, since the latter is $1/k$-competitive. For 3-uniform hypergraphs, the bound becomes $1/2$ for $d \leq 2$ and $1/(3-2/d)$ otherwise, thus interpolating between $1/3$ and $1/2$. Note that for $d \leq 2$, the bound is optimal, since the online matching problem on graphs under edge arrivals is a special case of this setting (with $k = 3, d = 1$), for which an upper bound of $1/2$ is known even against fractional algorithms on bipartite graphs [13].

Since every randomized algorithm for integral matching induces a deterministic algorithm for fractional matching, the upper bound of $(e-1)/(e+1) \approx 0.4621$ in Theorem 1 also applies to the integral problem on 3-uniform hypergraphs. However, the best known lower bound is $1/3$, given by the greedy algorithm. An interesting question for future research is whether there exists an integral algorithm better than greedy on 3-uniform hypergraphs.

Related Work. Since the online matching problem was introduced in [22], it has garnered a lot of interest, leading to extensive follow-up work. We refer the reader to the excellent survey by Mehta [28] for navigating this rich literature. The original analysis of RANKING [22] was simplified in a series of papers [3, 10, 11, 14]. Many variants of the problem have been studied, such as the online b-matching problem [20], and its extension to the AdWords problem [4,9,19,29]. Weighted generalizations have been considered, e.g., vertex weights [1,17] and edge weights [12]. Weakening the adversary by requiring that online nodes arrive in a random order has also been of interest [21,23,26]. Another line of research explored more general arrival models such as two-sided vertex arrival [32], general vertex arrival [13], edge arrival [6,13], and general vertex arrival with departure times [2,16,18].

In contrast, the literature on the online hypergraph matching problem is relatively sparse. Most work has focused on stochastic models, such as the random-order model. Korula and Pal [24] first studied the edge-weighted version under this model. For k-uniform hypergraphs, they gave an $\Omega(1/k^2)$-competitive algorithm. This was subsequently improved to $\Omega(1/k)$ by Kesselheim et al. [23]. Ma et al. [25] gave a $1/k$-competitive algorithm under 'nonstationary' arrivals. Pavone et al. [30] studied online hypergraph matching with delays under the adversarial model. At each time step, a vertex arrives, and it will depart after d time steps. A hyperedge is revealed once all of its vertices have arrived. Note that their model is incomparable to ours because every vertex has the same delay d.

In the prophet IID setting, every online node has a weight function which assigns weights to its incident hyperedges, and these functions are independently sampled from the same distribution. For this problem, [27] gave a $O(\log(k)/k)$ upper bound on the competitive ratio. We refer to [27] for an overview of known results in related settings.

Hypergraph matching on k-uniform hypergraphs is a well-studied problem in the offline setting. It is NP-hard to approximate within a factor of $\Omega(\log(k)/k)$ [15]. Moreover, the factor between the optimal solution and the optimal value of the natural LP relaxation is at least $1/(k-1+1/k)$ [7].

A special case that has also been studied is the restriction to k-partite graphs, where the vertices are partitioned into k sets and every hyperedge contains exactly one vertex from each set. This setting is called k-*dimensional matching*, and the optimal solution is known to be at least $1/(k-1)$ times the optimal value of the standard LP relaxation [7]. For $k = 3$, the best known polynomial time approximation algorithm gives a $(3/4 - \varepsilon)$-approximation [8].

Paper Organization. In Sect. 2, we give the necessary preliminaries and discuss notation. Section 3 presents the optimal primal-dual fractional algorithm for 3-uniform hypergraphs, which shows the first part of Theorem 1. Section 4 complements this with a tight upper bound, proving the second part of Theorem 1. The proof of Theorem 2 and other missing proofs are in the full version.

2 Preliminaries

Given a hypergraph $\mathcal{H} = (V, H)$ with vertex set V and hyperedge set H, the maximum matching problem involves finding a maximum cardinality subset of disjoint hyperedges. The canonical primal and dual LP relaxations for this problem are respectively given by:

$$\begin{aligned} \max \sum_{h \in H} x_h & \\ \sum_{h \in \delta(v)} x_h \leq 1 & \quad \forall v \in V \\ x_h \geq 0 & \quad \forall h \in H \end{aligned} \qquad\qquad \begin{aligned} \min \sum_{v \in V} y_v & \\ \sum_{v \in h} y_v \geq 1 & \quad \forall h \in H \\ y_v \geq 0 & \quad \forall v \in V. \end{aligned}$$

We denote by $\mathsf{OPT}_{\mathsf{LP}}(\mathcal{H})$ the offline optimal value of these two LPs. We denote by $\mathsf{OPT}(\mathcal{H})$ the objective value of an offline optimal integral solution to the primal LP, which clearly satisfies $\mathsf{OPT}(\mathcal{H}) \leq \mathsf{OPT}_{\mathsf{LP}}(\mathcal{H})$.

The online matching problem on k-uniform hypergraphs under vertex arrivals is defined as follows. An instance consists of a k-uniform hypergraph $\mathcal{H} = (V, W, H)$, where V is the set of offline nodes and $W = (w_1, w_2, \ldots)$ is the sequence of online nodes. The ordering of W corresponds to the arrival order of the online nodes. Every hyperedge $h \in H$ has exactly one node in W and $k - 1$ nodes in V. When an online node $w \in W$ arrives, its incident hyperedges $\delta(w)$ are revealed. A fractional algorithm is allowed to irrevocably increase x_h for every $h \in \delta(w)$, whereas an integral algorithm is allowed to irrevocably pick one of these hyperedges, i.e., setting $x_h = 1$ for some $h \in \delta(w)$.

Given an algorithm $\mathcal{A}$ and an instance $\mathcal{H}$, we denote by $\mathcal{V}(\mathcal{A}, \mathcal{H}) := \sum_{h \in H} x_h$ the value of the (fractional) matching obtained by $\mathcal{A}$ on $\mathcal{H}$. An integral algorithm is ρ-*competitive* if for any instance $\mathcal{H}$, $\mathcal{V}(\mathcal{A}, \mathcal{H}) \geq \rho\, \mathsf{OPT}(\mathcal{H})$. Similarly, a fractional algorithm is ρ-*competitive* if for any instance $\mathcal{H}$, $\mathcal{V}(\mathcal{A}, \mathcal{H}) \geq \rho\, \mathsf{OPT}_{\mathsf{LP}}(\mathcal{H})$.

In this paper, we focus on 3-uniform hypergraphs. For a 3-uniform instance $\mathcal{H} = (V, W, H)$, we denote by $\Gamma(\mathcal{H}) = (V, E)$ the graph on the offline nodes with edge set

$$E := \Big\{(u,v) \in V \times V, \quad \exists w \in W \text{ s.t. } \{u,v,w\} \in H\Big\}. \tag{1}$$

We remark that $\Gamma(\mathcal{H})$ is not a multigraph. In particular, an edge $(u,v) \in E$ can have several hyperedges in H containing it. A fractional matching x on the hyperedges H naturally induces a fractional matching x' on the edges E, i.e., $x'_e = \sum_{h:e \subseteq h} x_h$ for every $e \in E$. The value obtained by an algorithm $\mathcal{A}$ can thus also be counted as $\mathcal{V}(\mathcal{A},\mathcal{H}) = \sum_{h \in H} x_h = \sum_{e \in E} x'_e$. For an offline node $u \in V$, we denote its *load* (or *fractional degree*) as $\ell_u = x(\delta(u)) \in [0,1]$.

3 Optimal Fractional Algorithm for 3-Uniform Hypergraphs

In this section, we present a primal-dual algorithm for the online fractional matching problem on 3-uniform hypergraphs under vertex arrivals. This algorithm will turn out to be optimal with a tight competitive ratio of $(e-1)/(e+1) \approx 0.4621$. We define the following distribution function $f : [0,1] \to [0,1]$:

$$f(x) := \frac{e^x}{e+1}. \tag{2}$$

When an online node w arrives, our algorithm chooses to uniformly increase the primal variables of the hyperedges $\{u,v,w\}$ for which $f(x(\delta(u))) + f(x(\delta(v)))$ is minimal. We note that this belongs to the class of *water-filling* algorithms [20]. For this reason, we define the *priority* of a hyperedge $h = \{u,v,w\}$ as:

$$\phi(h) := f(x(\delta(u))) + f(x(\delta(v))). \tag{3}$$

Algorithm 3.1 Water-filling fractional algorithm for 3-uniform hypergraphs

Input : 3-uniform hypergraph $\mathcal{H} = (V, W, H)$ with online nodes W.
Output : Fractional matching $x \in [0,1]^H$

when $w \in W$ **arrives with** $\delta(w) \subseteq H$**:**
  set $x_h = 0$ for every $h \in \delta(w)$
  increase x_h for every $h = \{u,v,w\} \in \arg\min_{h \in \delta(w)}\{\phi(h)\}$ at rate 1
  increase y_u and y_v at rates $f(x(\delta(u)))$ and $f(x(\delta(v)))$
  increase y_w at rate $1 - f(x(\delta(u))) - f(x(\delta(v)))$
    until $x(\delta(w)) = 1$ **or** $\phi(h) \geq 1$ for every $h \in \delta(w)$.
return x

Theorem 3. *Algorithm 3.1 is* $(e-1)/(e+1)$*-competitive for the online fractional matching problem on* 3*-uniform hypergraphs.*

The proof is based on a primal-dual analysis. By construction, the primal and the dual objective values are equal at any point during the execution. In addition, $\sum_{v \in h} y_v \geq (e-1)/(e+1)$ holds for every hyperedge h at the end of the execution, meaning that the vector $(e+1)/(e-1)\,y$ is a feasible dual solution. Details are given in the full version.

4 Tight Upper Bound for 3-Uniform Hypergraphs

We now prove the second part of Theorem 1, i.e., every algorithm is at most $(e-1)/(e+1)$-competitive for the online fractional matching problem on 3-uniform hypergraphs under vertex arrivals.

4.1 Overview of the Construction

We construct an adversarial instance that is adaptive to the behaviour of the algorithm. The main idea is to combine the vertex-arrival instance of Karp et al. [22] and the edge-arrival instance of Gamlath et al. [13] for bipartite graphs.

We start by giving a high-level overview of the construction. The offline vertices of the hypergraph are partitioned into m sets $C_1, \ldots, C_m$, which we call *components*. Each component will induce a bipartite graph with bipartition $C_i = U_i \cup V_i$, where $|U_i| = |V_i| = T$.

The instance consists of T phases. In each phase $t \in \{1, \ldots, T\}$, the adversary first selects a bipartite matching $\mathcal{M}_i^{(t)}$ on each component C_i. Taking the union of these matchings gives a larger matching on the offline nodes: $\mathcal{M}^{(t)} := \bigcup_{i=1}^m \mathcal{M}_i^{(t)}$.

After selecting the matching $\mathcal{M}^{(t)}$ at phase t, the adversary selects the online nodes, with their incident hyperedges, arriving in that phase. The set of online nodes arriving in phase t is denoted by $W^{(t)}$. Each node $w \in W^{(t)}$ connects to a subset of edges $E(w) \subseteq \mathcal{M}^{(t)}$, meaning that the hyperedges incident to w are $\{\{w\} \cup e : e \in E(w)\}$.

We briefly explain how the matchings $\mathcal{M}_i^{(t)}$ are constructed and how the edges $E(w)$ are picked:

1. On each component C_i, the matching $\mathcal{M}_i^{(t)}$ is constructed based on the behaviour of the algorithm in phase $t-1$. It draws inspiration from the edge-arrival instance in [13], together with the function $f(x) = e^x/(e+1)$ defined in (2). The exact construction is described in Sect. 4.3.
2. For every online node $w \in W^{(t)}$, the edge set $E(w) \subseteq \mathcal{M}^{(t)}$ is selected based on the behaviour of the algorithm during phase t. This part can be seen as incorporating the vertex-arrival instance in [22]. The exact construction is described in Sect. 4.4.

To summarize, the instance is a hypergraph $\mathcal{H} = (V, W, H)$ with offline nodes V, online nodes W and hyperedges H given by

$$V := \bigcup_{i=1}^m C_i = \bigcup_{i=1}^m U_i \cup V_i \quad W := \bigcup_{t=1}^T W^{(t)} \quad H := \bigcup_{t=1}^T \bigcup_{w \in W^{(t)}} \{\{w\} \cup e : e \in E(w)\}.$$

4.2 Assumptions on the Algorithm

To simplify the construction and analysis of our instance, we will make two assumptions on the algorithm. First, we need the following definition, which relates the behaviour of an algorithm to the priority function ϕ defined in (3).

Definition 1. *Fix $\varepsilon \geq 0$. Let x be the fractional solution given by an algorithm $\mathcal{A}$ after the arrival of an online node w. We say that $\mathcal{A}$ is* ε-threshold respecting *on w if $\phi(h) = \sum_{v \in h \setminus \{w\}} f(x(\delta(v))) \leq 1 + \varepsilon$ for all incident hyperedges $h \in \delta(w)$ with $x_h > 0$. We also call $\mathcal{A}$* threshold respecting *if $\varepsilon = 0$.*

Remark 1. We emphasize that the property in Definition 1 only needs to hold for the fractional solution x after w has arrived, and before the arrival of the next online node. In particular, it is possible that $\phi(h) > 1 + \varepsilon$ in later iterations.

The two assumptions that we make are the following. In the full version, we show that they can be made without loss of generality.

1. The algorithm is *ε-threshold respecting* on all online nodes in the first $T - 1$ phases for some arbitrarily small $\varepsilon > 0$.
2. The algorithm is *symmetric* on each component $C_i = U_i \cup V_i$. In particular, for every $t \in \{1, \ldots, T\}$, the tth vertices of U_i and V_i have the same fractional degrees throughout the execution of the algorithm.

4.3 Constructing the Matching $\mathcal{M}^{(t)}$

In this section, we construct the matching $\mathcal{M}_i^{(t)}$ for every component $i \in [m]$ and phase $t \in [T]$. We will only describe the matchings for a single component C_i, i.e., $\mathcal{M}_i^{(1)}, \mathcal{M}_i^{(2)}, \ldots, \mathcal{M}_i^{(T)}$, because the same construction applies to other components. Intuitively, the value obtained by the algorithm in the first $T - 1$ phases is already limited by Assumption 1. So, the goal of this construction is to prevent the algorithm from gaining too much value in the last phase T. In particular, we will show that it can only obtain $O(\sqrt{T}) + \varepsilon\, O(T^2)$ in the last phase on every component C_i.

The matchings $\mathcal{M}_i^{(1)}, \ldots, \mathcal{M}_i^{(T)}$ are adaptive to the behaviour of the algorithm in every phase. It is essentially the instance of [13] with our threshold function incorporated. The vertex set of these matchings is on a bipartite graph, with T nodes on both sides of the bipartition. Let us denote this bipartition as $U_i = \{1, \ldots, T\}$ and $V_i = \{1, \ldots, T\}$. We index them the same way due to the symmetry assumption of the algorithm (Assumption 2). Each matching $\mathcal{M}_i^{(t)}$ satisfies the invariant $(u, v) \in \mathcal{M}_i^{(t)}$ if and only if $(v, u) \in \mathcal{M}_i^{(t)}$.

For an offline node $u \in V$, we denote its *load* (or *fractional degree*) at the end of phase t as $\ell_u^{(t)} = x^{(t)}(\delta(u)) \in [0, 1]$, where $x^{(t)}$ is the fractional matching generated by the algorithm at the end of phase t.

- $\mathcal{M}_i^{(1)}$ is a matching of size one that consists of the single edge $(1, 1)$.
- At the end of phase t, we will call a node *active* if it is incident to an edge $e = (u, v) \in \mathcal{M}_i^{(t)}$ satisfying $\phi(e) = f(\ell_u^{(t)}) + f(\ell_v^{(t)}) \geq 1$. All other nodes are said to be *inactive* and will not be used in any of the matchings of later phases. Let $\sigma_t(1) < \sigma_t(2) < \ldots < \sigma_t(r_t)$ be the active nodes in U_i at the end of phase t, where r_t denotes the number of such active nodes. By the aforementioned invariant and Assumption 2, the active nodes in V_i are also $\sigma_t(1) < \sigma_t(2) < \ldots < \sigma_t(r_t)$.

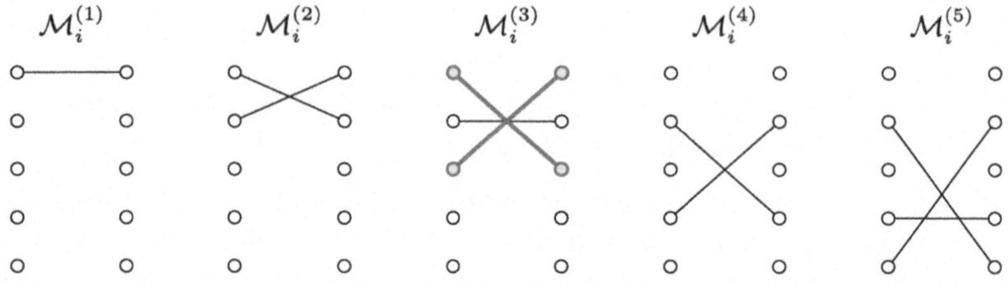

Fig. 1. In this example, the algorithm does not increase the edge $(1,3)$, and thus by symmetry the edge $(3,1)$, up to the threshold during phase $t = 3$. Hence, $f(\ell_1^{(3)}) + f(\ell_3^{(3)}) < 1$ and nodes 1 and 3 become inactive from that point on. The maximum matching at the end of phase 5 still has size five and consists of $\mathcal{M}_i^{(5)}$, in addition to the two edges $(1,3)$ and $(3,1)$ that are below the threshold.

- The matching at phase $t+1$ is then of size $r_t + 1$ and is defined as:

$$\mathcal{M}_i^{(t+1)} := \Big\{ \Big(\sigma_t(k), \sigma_t(r_t + 2 - k)\Big), k \in \{1, \ldots, r_t + 1\} \Big\},$$

  where we define $\sigma_t(r_t + 1) := t + 1$ for convenience. In particular, note that $t + 1 \in U_i$ and $t + 1 \in V_i$ are two fresh nodes with zero load, which are always part of the matching $\mathcal{M}_i^{(t+1)}$, but not part of any matching from a previous phase. Clearly, the invariant is maintained. Figure 1 illustrates the construction.

Let us denote $q_t := (r_t + 1)/2$. Observe that the nodes $\sigma_t(k) \in U_i$ and $\sigma_t(k) \in V_i$ for every $k \in \{1, \ldots, \lceil q_t \rceil\}$ form a vertex cover of the matching $\mathcal{M}_i^{(t+1)}$, meaning that every edge of the matching in phase $t+1$ is covered by one of these active nodes at phase t. Intuitively, this construction ensures that as t gets large, these nodes have a high fractional degree. Consequently, the algorithm does not have a lot of room to increase the fractional value on any edge of $\mathcal{M}_i^{(t+1)}$, due to the degree constraints. In order to upper bound the value that the algorithm can get in phase $t+1$, we will thus lower bound the fractional degree of the active nodes $\sigma_t(i)$ for $i \in \{1, \ldots, \lceil q_t \rceil\}$. For this reason, we define:

$$\ell(t,i) := x^{(t)}\Big(\delta(\sigma_t(i))\Big) = \sum_{e \in \delta(\sigma_t(i))} x_e^{(t)}.$$

In words, this is the fractional degree of the ith active node at the end of phase t. One can now see $\{\ell(t,i)\}_{t,i}$ as a process with two parameters, which depends on the behaviour of the algorithm. To analyze this process, we relate it to the CDF of the binomial distribution $B(t, 1/2)$. We show that

$$\sum_{i=1}^{\lceil q_{T-1} \rceil} 2(1 - \ell(T-1, i)) = O(\sqrt{T}) + \varepsilon\, O(T^2).$$

Since the left-hand side is the residual capacity of the vertex cover of $\mathcal{M}_i^{(T)}$, this yields an upper bound on the value obtained by the algorithm in component C_i during the last phase.

Theorem 4. *During the last phase, the value gained by the algorithm in each component C_i is at most $O(\sqrt{T}) + \varepsilon\, O(T^2)$.*

4.4 Connecting the Matching $\mathcal{M}^{(t)}$ to the Online Nodes

In this section, we connect the matching $\mathcal{M}^{(t)} = \cup_{i=1}^{m} \mathcal{M}_i^{(t)}$ to the online vertices to form hyperedges, for every phase $t \in [T]$. The way in which they are connected is similar to the vertex-arrival instance of [22] for bipartite graphs. The main idea is to obfuscate the partners of the online nodes in the optimal matching.

The following construction is an adaptation of the vertex-arrival instance for bipartite graphs [22] to tripartite hypergraphs. Given a graph matching $\mathcal{M}$ on the offline nodes, the first online node connects to every edge in $\mathcal{M}$. After the algorithm sets fractional values on every edge of $\mathcal{M}$, the second online node connects to $\mathcal{M} \setminus \{e_1\}$, where e_1 is the edge in the matching with the lowest fractional value. More generally, for every $k \in \{1, \ldots, |\mathcal{M}|\}$, the kth online node connects to the $|\mathcal{M}|-k+1$ edges $\mathcal{M} \setminus \{e_1, \ldots, e_{k-1}\}$, and e_k is defined as the edge having the lowest fractional value among them at the end of the kth iteration. This instance is illustrated in Fig. 2.

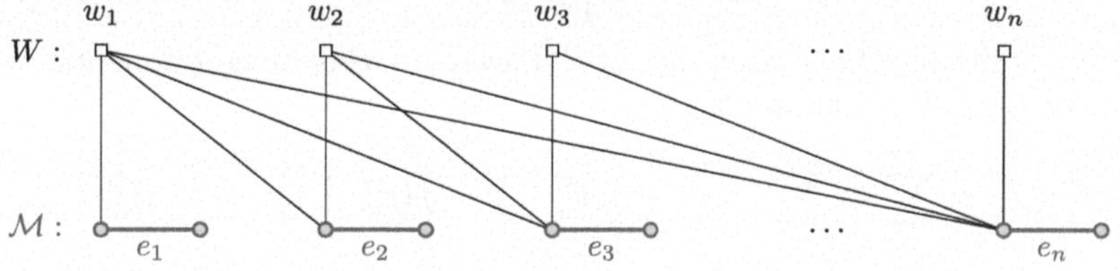

Fig. 2. An illustration of the instance constructed in Lemma 1

We now state the guarantee obtained by this construction, parametrized by the maximum (fractional) degree $\Delta \in [0, 1]$ attained by an offline node.

Lemma 1. *For any graph matching $\mathcal{M} = (V, E)$, there exists an online tripartite hypergraph instance $\mathcal{H} = (V, W, H)$ such that $\Gamma(\mathcal{H}) = \mathcal{M}$ and $\mathsf{OPT}(\mathcal{H}) = |\mathcal{M}|$. Moreover, for any fractional algorithm $\mathcal{A}$ whose returned solution x satisfies $x(\delta(v)) \leq \Delta$ for all offline nodes $v \in V$, we have*

$$\mathcal{V}(\mathcal{A}, \mathcal{H}) \leq (1 - e^{-\Delta})|\mathcal{M}| + 3/2.$$

The way now in which we apply this construction is by partitioning the matching $\mathcal{M}^{(t)}$ into submatchings based on the load (or fractional degree) of the vertices, and applying Lemma 1 on each submatching separately. More precisely, let us fix $N := \lceil 2|\mathcal{M}^{(t)}|^{1/3} \rceil$. We partition the edges of the matching $\mathcal{M}^{(t)}$ into N^2 submatchings as follows

$$\mathcal{M}^{(t)}(i,j) := \left\{ (u,v) \in \mathcal{M}^{(t)} : \ell_u \in \left[\frac{i-1}{N}, \frac{i}{N} \right], \ell_v \in \left[\frac{j-1}{N}, \frac{j}{N} \right] \right\}$$

for all $i, j \in [N]$. Then, we apply the construction illustrated in Fig. 2 to each submatching $\mathcal{M}^{(t)}(i,j)$. This finishes the description of our instance.

Recall from Sect. 4.1 that our instance consists of T phases. For ease of analysis, we will split the total value gained by the algorithm into the value gained in each phase. For an algorithm $\mathcal{A}$, let $\mathcal{V}^{(t)}(\mathcal{A})$ denote the value obtained by $\mathcal{A}$ in phase t. The next lemma upper bounds $\mathcal{V}^{(t)}(\mathcal{A})$ for a threshold-respecting algorithm $\mathcal{A}$, in terms of the loads of the offline nodes at the end of phase $t-1$. Recall that $W^{(t)}$ is the set of online nodes which arrive during phase t.

Lemma 2. *If $\mathcal{A}$ is threshold-respecting on $W^{(t)}$, then*

$$\mathcal{V}^{(t)}(\mathcal{A}) \leq \sum_{(u,v)\in\mathcal{M}^{(t)}} \left(1 - f(\ell_u^{(t-1)}) - f(\ell_v^{(t-1)})\right) + 15\,|\mathcal{M}^{(t)}|^{2/3}.$$

From the definition of f and the construction of the matching $\mathcal{M}^{(t)}$, we can convert the previous bound into the following expression. We remark that the threshold-respecting property is only used in the proof of Lemma 2.

Lemma 3. *If $\mathcal{A}$ is threshold-respecting on $W^{(t)}$, then*

$$\mathcal{V}^{(t)}(\mathcal{A}) \leq \frac{e-1}{e+1} m + 15\, t^{2/3} m^{2/3}.$$

Proof. By splitting the matching $\mathcal{M}^{(t)}$ based on the m components, we can rewrite the bound in Lemma 2 as

$$\mathcal{V}^{(t)}(\mathcal{A}) \leq \sum_{i=1}^{m} \sum_{(u,v)\in\mathcal{M}_i^{(t)}} \left(1 - f(\ell_u^{(t-1)}) - f(\ell_v^{(t-1)})\right) + 15\,|\mathcal{M}^{(t)}|^{2/3}. \quad (4)$$

Fix a component $i \in [m]$, and let $\mathcal{E}_i := \left\{ e \in \mathcal{M}_i^{(t-1)} \mid f(\ell_u^{(t-1)}) + f(\ell_v^{(t-1)}) \geq 1 \right\}$ be the subset of edges in the matching $\mathcal{M}_i^{(t-1)}$ which exceed the threshold at the end of phase $t-1$. By the construction of $\mathcal{M}_i^{(t)}$ in Sect. 4.3, we know that its node set consists of the nodes incident to $\mathcal{E}_i$, in addition to two new fresh nodes whose load is 0 at the end of phase $t-1$. This allows us to expand the inner sum in (4) as:

$$\begin{aligned}\sum_{(u,v)\in\mathcal{M}_i^{(t)}} 1 - f(\ell_u^{(t-1)}) - f(\ell_v^{(t-1)}) &= 1 - 2f(0) + \sum_{(u,v)\in\mathcal{E}_i} 1 - f(\ell_u^{(t-1)}) - f(\ell_v^{(t-1)}) \\ &\leq 1 - 2f(0) = \frac{e-1}{e+1},\end{aligned}$$

where the inequality follows from definition of $\mathcal{E}_i$. Plugging this into (4) with the bound $|\mathcal{M}^{(t)}| \leq tm$ (which is immediate by the construction in Sect. 4.3) yields the desired result. □

For an ε-threshold-respecting algorithm, we pick up an extra εtm term.

Corollary 1. *If $\mathcal{A}$ is ε-threshold-respecting on $W^{(t)}$ for some $\varepsilon \geq 0$, then*

$$\mathcal{V}^{(t)}(A) \leq \frac{e-1}{e+1}m + 15\,(tm)^{2/3} + \varepsilon tm.$$

Proof. Fix an edge $e \in \mathcal{M}^{(t)}$. Let $h_1, h_2, \ldots, h_k$ be the hyperedges arriving in phase t which contain e, denoted such that h_i arrives before h_j if and only if $i < j$. Let $j \in [k]$ be the smallest index such that $\phi(h_j) > 1$ immediately after $\mathcal{A}$ assigns x_{h_j} to h_j. Let $z_{h_j} \geq 0$ be the largest value such that $\phi(h_j) \geq 1$ if $\mathcal{A}$ were to assign $x_{h_j} - z_{h_j}$ to h_j instead. Define $z_{h_i} := 0$ for all $i < j$, and $z_{h_i} := x_{h_i}$ for all $i > j$. Since $\mathcal{A}$ is ε-threshold-respecting on $W^{(t)}$, we have $\sum_{i=1}^{k} z_{h_i} \leq \varepsilon$ because f is convex and $f' = f$.

Let z be the vector obtained by repeating this procedure on every edge $e \in \mathcal{M}^{(t)}$. Then, $1^\top z \leq \varepsilon tm$ as $|\mathcal{M}^{(t)}| \leq tm$. Moreover, observe that the algorithm which assigns $x - z$ in phase t is threshold-respecting on $W^{(t)}$. Thus, we can apply Lemma 3 to obtain the desired upper bound on $1^\top(x - z)$. □

4.5 Putting Everything Together

In this section, we complete the proof of Theorem 1. Since we assumed that the algorithm is ε-threshold respecting in the first $T-1$ phases, we can apply Corollary 1 to upper bound the value obtained in the first $T-1$ phases as

$$\sum_{t=1}^{T-1}\left(\frac{e-1}{e+1}m + \varepsilon tm + O\left((tm)^{2/3}\right)\right) \leq \frac{e-1}{e+1}Tm + \varepsilon T^2\,m + O\left(T^{5/3}m^{2/3}\right).$$

By Theorem 4, the value gained by the algorithm on each component C_i during the last phase T is at most $O(\sqrt{T} + \varepsilon T^2)$. Hence, the algorithm gains at most $O(\sqrt{T}m + \varepsilon T^2 m)$ in the last phase.

We now argue that our instance $\mathcal{H} = (V, W, H)$ has a perfect matching.

Lemma 4. *Our adversarial instance $\mathcal{H} = (V, W, H)$ satisfies $\mathsf{OPT}(\mathcal{H}) = Tm$.*

Proof. We prove that for every $t \in [T]$, there exists a hypergraph matching of size tm at the end of phase t. Let C_i be a component with bipartition $U_i = [T]$ and $V_i = [T]$. It suffices to show that there exists a graph matching $\widetilde{\mathcal{M}}_i^{(t)}$ with vertex set $[t]$ on each side. This is because $\widetilde{\mathcal{M}}^{(t)} := \cup_{i=1}^{m} \widetilde{\mathcal{M}}_i^{(t)}$ can be extended to a hypergraph matching in $\mathcal{H}$ by our construction (see Lemma 1). Let $E_i^{(t)} \subseteq \mathcal{M}_i^{(t)}$ be the edges whose endpoints are not active at the end of phase t. Then, a simple inductive argument on $t \geq 1$ shows that $\cup_{s=1}^{t-1} E_i^{(s)} \cup \mathcal{M}_i^{(t)}$ is a graph matching with vertex set $[t]$ on each side (see Fig. 1 for an example). □

By Lemma 4, the competitive ratio of the algorithm is at most

$$\frac{e-1}{e+1} + O(\varepsilon T + T^{2/3}m^{-1/3} + T^{-1/2}).$$

Hence, letting $m \to \infty$, picking $T = o(\sqrt{m})$ such that $T \to \infty$ and setting $\varepsilon = o(1/T)$, we conclude that the competitive ratio is upper bounded by $(e - 1)/(e + 1)$, thus finishing the proof of Theorem 1.

References

1. Aggarwal, G., Goel, G., Karande, C., Mehta, A.: Online vertex-weighted bipartite matching and single-bid budgeted allocations. In: Proceedings of the Twenty-Second Annual ACM-SIAM Symposium on Discrete Algorithms, pp. 1253–1264. SIAM (2011)
2. Ashlagi, I., Burq, M., Dutta, C., Jaillet, P., Saberi, A., Sholley, C.: Edge-weighted online windowed matching. Math. Oper. Res. **48**(2), 999–1016 (2023)
3. Birnbaum, B., Mathieu, C.: On-line bipartite matching made simple. ACM SIGACT News **39**(1), 80–87 (2008)
4. Buchbinder, N., Jain, K., Naor, J.S.: Online primal-dual algorithms for maximizing ad-auctions revenue. In: Arge, L., Hoffmann, M., Welzl, E. (eds.) ESA 2007. LNCS, vol. 4698, pp. 253–264. Springer, Heidelberg (2007). https://doi.org/10.1007/978-3-540-75520-3_24
5. Buchbinder, N., Naor, J.: Online primal-dual algorithms for covering and packing. Math. Oper. Res. **34**(2), 270–286 (2009)
6. Buchbinder, N., Segev, D., Tkach, Y.: Online algorithms for maximum cardinality matching with edge arrivals. Algorithmica **81**(5), 1781–1799 (2019)
7. Chan, Y.H., Lau, L.C.: On linear and semidefinite programming relaxations for hypergraph matching. Math. Program. **135**(1–2), 123–148 (2012)
8. Cygan, M.: Improved approximation for 3-dimensional matching via bounded pathwidth local search. In: 54th Annual IEEE Symposium on Foundations of Computer Science (FOCS), pp. 509–518. IEEE Computer Society (2013)
9. Devanur, N.R., Jain, K.: Online matching with concave returns. In: Proceedings of the Forty-Fourth Annual ACM Symposium on Theory of Computing, pp. 137–144 (2012)
10. Devanur, N.R., Jain, K., Kleinberg, R.D.: Randomized primal-dual analysis of ranking for online bipartite matching. In: Proceedings of the Twenty-Fourth Annual ACM-SIAM Symposium on Discrete Algorithms, pp. 101–107. SIAM (2013)
11. Eden, A., Feldman, M., Fiat, A., Segal, K.: An economics-based analysis of ranking for online bipartite matching. In: Symposium on Simplicity in Algorithms (SOSA), pp. 107–110. SIAM (2021)
12. Fahrbach, M., Huang, Z., Tao, R., Zadimoghaddam, M.: Edge-weighted online bipartite matching. J. ACM **69**(6), 1–35 (2022)
13. Gamlath, B., Kapralov, M., Maggiori, A., Svensson, O., Wajc, D.: Online matching with general arrivals. In: 60th IEEE Annual Symposium on Foundations of Computer Science (FOCS), pp. 26–37. IEEE (2019)
14. Goel, G., Mehta, A.: Online budgeted matching in random input models with applications to adwords. In: SODA, vol. 8, pp. 982–991 (2008)
15. Hazan, E., Safra, S., Schwartz, O.: On the complexity of approximating k-set packing. Comput. Complex. **15**, 20–39 (2006)
16. Huang, Z., et al.: Fully online matching. J. ACM **67**(3):17:1–17:25 (2020)
17. Huang, Z., Tang, Z.G., Wu, X., Zhang, Y.: Online vertex-weighted bipartite matching: beating 1–1/e with random arrivals. ACM Trans. Algorithms (TALG) **15**(3), 1–15 (2019)
18. Huang, Z., Tang, Z.G., Wu, X., Zhang, Y.: Fully online matching II: beating ranking and water-filling. In: 61st IEEE Annual Symposium on Foundations of Computer Science (FOCS), pp. 1380–1391. IEEE (2020)
19. Huang, Z., Zhang, Q., Zhang, Y.: Adwords in a panorama. In: 2020 IEEE 61st Annual Symposium on Foundations of Computer Science (FOCS), pp. 1416–1426. IEEE (2020)

20. Kalyanasundaram, B., Pruhs, K.R.: An optimal deterministic algorithm for online b-matching. Theor. Comput. Sci. **233**(1–2), 319–325 (2000)
21. Karande, C., Mehta, A., Tripathi, P.: Online bipartite matching with unknown distributions. In: Proceedings of the Forty-Third Annual ACM Symposium on Theory of Computing, pp. 587–596 (2011)
22. Karp, R.M., Vazirani, U.V., Vazirani, V.V.: An optimal algorithm for on-line bipartite matching. In: Proceedings of the Twenty-Second Annual ACM Symposium on Theory of Computing, pp. 352–358 (1990)
23. Kesselheim, T., Radke, K., Tönnis, A., Vöcking, B.: An optimal online algorithm for weighted bipartite matching and extensions to combinatorial auctions. In: Bodlaender, H.L., Italiano, G.F. (eds.) ESA 2013. LNCS, vol. 8125, pp. 589–600. Springer, Heidelberg (2013). https://doi.org/10.1007/978-3-642-40450-4_50
24. Korula, N., Pál, M.: Algorithms for secretary problems on graphs and hypergraphs. In: Albers, S., Marchetti-Spaccamela, A., Matias, Y., Nikoletseas, S., Thomas, W. (eds.) ICALP 2009. LNCS, vol. 5556, pp. 508–520. Springer, Heidelberg (2009). https://doi.org/10.1007/978-3-642-02930-1_42
25. Ma, Y., Rusmevichientong, P., Sumida, M., Topaloglu, H.: An approximation algorithm for network revenue management under nonstationary arrivals. Oper. Res. **68**(3), 834–855 (2020)
26. Mahdian, M., Yan, Q.: Online bipartite matching with random arrivals: an approach based on strongly factor-revealing LPs. In: Proceedings of the Forty-Third Annual ACM Symposium on Theory of Computing, pp. 597–606 (2011)
27. Marinkovic, J., Soto, J.A., Verdugo, V.: Online combinatorial assignment in independence systems. In: Integer Programming and Combinatorial Optimization - 25th International Conference (IPCO). LNCS, vol. 14679, pp. 294–308. Springer, Cham (2024)
28. Mehta, A.: Online matching and ad allocation. Found. Trends Theor. Comput. Sci. **8**(4), 265–368 (2013)
29. Mehta, A., Saberi, A., Vazirani, U., Vazirani, V.: Adwords and generalized online matching. J. ACM (JACM) **54**(5), 22–es (2007)
30. Pavone, M., Saberi, A., Schiffer, M., Tsao, M.W.: Technical note - online hypergraph matching with delays. Oper. Res. **70**(4), 2194–2212 (2022)
31. Tröbst, T., Udwani, R.: Almost tight bounds for online hypergraph matching. Oper. Res. Lett. **55**, 107143 (2024)
32. Wang, Y., Wong, S.C.: Two-sided online bipartite matching and vertex cover: beating the greedy algorithm. In: Halldórsson, M.M., Iwama, K., Kobayashi, N., Speckmann, B. (eds.) ICALP 2015. LNCS, vol. 9134, pp. 1070–1081. Springer, Heidelberg (2015). https://doi.org/10.1007/978-3-662-47672-7_87

On the Bidirected Cut Relaxation for Steiner Forest

Jarosław Byrka[1], Fabrizio Grandoni[2], and Vera Traub[3](✉)

[1] University of Wrocław, Wrocław, Poland
jaroslaw.byrka@cs.uni.wroc.pl
[2] IDSIA, USI-SUPSI, Lugano, Switzerland
fabrizio.grandoni@idsia.ch
[3] Research Institute for Discrete Mathematics and Hausdorff Center for Mathematics, University of Bonn, Bonn, Germany
traub@dm.uni-bonn.de

Abstract. The Steiner Forest problem is an important generalization of the Steiner Tree problem. We are given an undirected graph with non-negative edge costs and a collection of pairs of vertices. The task is to compute a cheapest forest with the property that the elements of each pair belong to the same connected component of the forest. The current best approximation factor for Steiner Forest is 2, which is achieved by the classical primal-dual algorithm; improving on this factor is a big open problem in the area.

Motivated by this open problem, we study an LP relaxation for Steiner Forest that generalizes the well-studied Bidirected Cut Relaxation for Steiner Tree. We prove that this relaxation has several promising properties. Among them, it is possible to round any half-integral LP solution to a Steiner Forest instance while increasing the cost by at most a factor $\frac{16}{9}$. To prove this result we introduce a novel recursive densest-subgraph contraction algorithm.

Keywords: network design · Steiner forest · bidirected cut relaxation

1 Introduction

In the *Steiner Forest* problem we are given an undirected graph $G = (V, E)$ with nonnegative edge costs $c : E \to \mathbb{R}_{\geq 0}$. Furthermore, we are given a collection $\mathcal{P}$ of pairs of vertices. The task is to compute a subset of edges $F \subseteq E$ of minimum total cost $c(F) := \sum_{e \in F} c(e)$ such that, for each pair $P = \{s, t\} \in \mathcal{P}$, the graph (V, F) contains an s-t path. There is always an optimal solution F where (V, F) is a forest. We will refer to vertices $R \subseteq V$ that are present in at least one pair from $\mathcal{P}$ as *terminals*, and call the remaining vertices *Steiner vertices.*

J. Byrka—Supported by NCN grant number 2020/39/B/ST6/01641.
F. Grandoni—Partially supported by the SNF Grant 200021_200731/1
V. Traub—Supported by the Deutsche Forschungsgemeinschaft (DFG, German Research Foundation) – 537750605.

N. Megow and A. Basu (Eds.): IPCO 2025, LNCS 15620, pp. 114–127, 2025.
https://doi.org/10.1007/978-3-031-93112-3_9

Steiner Forest generalizes the famous Steiner tree problem, where all terminals need to be connected to each other. Because Steiner Tree is NP-hard, indeed APX-hard [19], the same applies to Steiner Forest. The current-best approximation factor for Steiner Forest is 2, achieved in a classical paper by Agrawal, Klein, and Ravi [1] using the primal-dual method (see also [24]). Incidentally, this result is also one of the earliest and most famous examples of the primal-dual technique. Improving on the 2-approximation for Steiner Forest is probably among the most prominent open problems in the area of approximation algorithms (see e.g. [39]).

The main motivation for our work is a very recent result [10] about the integrality gap of the Bidirected Cut Relaxation for Steiner Tree (following a long line of research [14,15,21,27]). To define this relaxation, we first *bidirect* all edges, i.e., we replace each undirected edge $\{u, v\}$ with two directed edges (u, v) and (v, u), inheriting the same cost as $\{u, v\}$. We let $\overrightarrow{E}$ be the resulting set of directed edges. We also select an arbitrary terminal $r \in R$ as a root.

The idea behind the relaxation is to consider the equivalent problem of finding a directed Steiner Tree solution $F \subseteq \overrightarrow{E}$ where F is the edge set of a tree oriented towards the root r. Any such directed solution $F \subseteq \overrightarrow{E}$ has to contain a directed path from each terminal to r. Equivalently, for every vertex set U containing some terminal but not r, the solution F must contain at least one outgoing edge of U. We write $\delta^+(U)$ to denote the set of of outgoing edges of U. This leads to the Bidirected Cut Relaxation for Steiner Tree, which is defined as follows:

$$\min \sum_{e\in\overrightarrow{E}} c(e)\cdot x_e \qquad \text{(Tree-BCR)}$$

$$\text{s.t.} \sum_{e\in\delta^+(U)} x_e \geq 1 \qquad \forall U \subseteq V \setminus \{r\}, U\cap R \neq \emptyset.$$

Given a solution x to Tree-BCR, we denote its cost by $c(x) := \sum_{e\in\overrightarrow{E}} c(e)\cdot x_e$. Tree-BCR is one of the oldest and best-studied relaxations for Steiner Tree (see e.g. [22]). A classical result by Edmonds [20] shows that it is integral in the special case when $R = V$. In this case, Steiner Tree is equivalent to the Minimum Spanning Tree problem. The best-known lower bound on the integrality gap of Tree-BCR is only $\frac{6}{5}$ (see [38], improving on [9]). This gives hope that Tree-BCR might have a quite small integrality gap. Until very recently however, the best-known upper bound was only 2. In [10] this was improved for the first time, by showing an upper bound of 1.9988 on the integrality gap of Tree-BCR.

The latter result motivated us to study the Bidirected Cut Relaxation for Steiner Forest (Forest-BCR). To obtain this relaxation, we bidirect the edges as before. Consider any Steiner Forest solution F where each connected component of (V, F) is a tree. Then we can orient the edges of each such component towards an arbitrary root vertex. We consider the linear programming relaxation with variables z_P^r for $r \in V$ and $P \in \mathcal{P}$ specifying whether the vertices of the pair P belong to a connected component with root r. Moreover, we have variables x_e^r specifying whether a directed edge $e \in \overrightarrow{E}$ is used as part of a connected component with root r. This leads to the following relaxation:

$$\min \sum_{r \in V} \sum_{e \in \overrightarrow{E}} c(e) \cdot x_e^r \qquad \text{(Forest-BCR)}$$

$$\text{s.t.} \quad \sum_{r \in V} z_P^r = 1 \qquad \text{for all } P \in \mathcal{P}$$

$$\sum_{e \in \delta^+(U)} x_e^r \geq z_P^r \quad \text{for all } r \in V, P \in \mathcal{P}, \text{ and } U \subseteq V \setminus \{r\} \text{ with } P \cap U \neq \emptyset$$

$$x_e^r \geq 0 \qquad \text{for all } r \in V \text{ and } e \in \overrightarrow{E}$$

$$z_P^r \geq 0 \qquad \text{for all } r \in V \text{ and } P \in \mathcal{P}.$$

Given a solution (x, z) to (Forest-BCR), we use $c(x) := \sum_{r \in V} \sum_{e \in \overrightarrow{E}} c(e) \cdot x_e^r$ as a shortcut for its cost.

The integer program corresponding to (Forest-BCR) turned out to perform well in experimental studies [34],[1] but to the best of our knowledge, from a theoretical viewpoint (Forest-BCR) has not been studied before. In this work, we investigate the properties of this LP relaxation, which in the long run might help to improve on the longstanding 2-approximation algorithm for Steiner Forest.

1.1 Our Contributions

One of our main results is an algorithm to round a half-integral solution[2] to (Forest-BCR) while loosing a factor smaller than 2 in the cost. Half-integral solutions often play an important role in developing rounding algorithms for network design problems, see e.g. [7,12,29]. While it remains open whether the integrality gap of (Forest-BCR) is less than 2, the result from the next theorem (proof in Sect. 2) gives some supporting evidence that this might be the case.

Theorem 1. *Given any half-integral solution (x, z) to* (Forest-BCR), *we can compute in polynomial time a feasible solution to the associated Steiner Forest instance of cost at most $\frac{16}{9} \cdot c(x)$.*

To prove the above result, we provide a simple rounding algorithm which is applicable beyond half-integral solutions. The half-integrality of (x, z) will be used only in the analysis of the algorithm.

The basic idea of our algorithm is a recursive contraction of densest subgraphs with respect to x. First, we turn the given LP solution (x, z) into a more structured one, applying splitting-off operations as long as this maintains feasibility of the solution. (Splitting-off means that we reduce the value of variables $x_{(u,v)}^r$ and $x_{(v,w)}^r$ while increasing the value of $x_{(u,w)}^r$ by the same amount.) Next, we identify a densest subgraph with respect to x. More precisely, we compute a subset of vertices W that maximizes the sum of the values $x_{(u,v)}^r$ with $u, v \in W$,

[1] (Forest-BCR) is equivalent to the LP underlying the best performing IP in [34].
[2] I.e., a solution where the value of each variable is an integer multiple of $\frac{1}{2}$.

divided by $|W| - 1$[3]. Then we add a minimum spanning tree MST(W) on vertex set W to the solution under construction, contract W (updating $\mathcal{P}$ and (x, z) in the natural way), and continue recursively until each pair is connected.

We prove that our algorithm yields a solution of cost less than $2 \cdot c(x)$ if the maximum density of a set W with respect to x is always larger than $\frac{1}{2}$. Moreover, we prove that in the half-integral case, we can achieve a density of at least $\frac{9}{16}$, leading to Theorem 1.

Next, we study further basic properties of (Forest-BCR). Recall that the pairs $\mathcal{P}$ do not need to be disjoint and a vertex can appear in multiple pairs. Often the same Steiner Forest instance can be represented in multiple equivalent ways. For example, the pairs $\{\{a, b\}, \{b, c\}\}$ can be replaced by the pairs $\{\{a, b\}, \{a, c\}\}$ without changing the set of feasible solutions. We investigate the impact of such representation changes on (Forest-BCR).

Define the *demand graph* of an instance to be the graph with vertex set V that contains an edge between two vertices s and t if and only if the pair $\{s, t\}$ belongs to $\mathcal{P}$. All vertices that belong to the same connected component of the demand graph must be connected in any feasible Steiner Forest solution. Thus, if for two instances $((V, E), c, \mathcal{P}_1)$ and $((V, E), c, \mathcal{P}_2)$ the vertex sets of the connected components of the demand graphs are the same, then the two instances are equivalent. In this case, we also say that $\mathcal{P}_1$ is a different *representation* of $\mathcal{P}_2$ and vice versa.

Even though the feasible solutions of the two instances are identical (and the cost function is the same), it turns out that the value of (Forest-BCR) can depend on the representation.

Theorem 2. *There exists an instance $((V, E), c, \mathcal{P}_1)$ of the Steiner Forest problem and a different representation $\mathcal{P}_2$ of $\mathcal{P}_1$ such that the value of* (Forest-BCR) *is not the same for the two instances $((V, E), c, \mathcal{P}_1)$ and $((V, E), c, \mathcal{P}_2)$.*

We remark that, on the instances we consider in the proof of Theorem 2 (see full version [11]), (Forest-BCR) has optimal half-integral solutions. Hence already in this case, where we show that a good rounding algorithm exists, the representation matters.

Given Theorem 2, one might ask what the best way is to represent a given instance. For a given fixed instance, one could simply add all pairs $\{s, t\}$ for which s and t are in the same connected component of the demand graph. Alternatively, one could also generalize (Forest-BCR) by allowing $\mathcal{P}$ to contain arbitrary subsets of V rather than only subsets of size two. Then one can simply include the vertex set of every connected component of the demand graph in $\mathcal{P}$. The latter construction yields a relaxation that is at least as strong as any representation using only pairs.[4]

[3] Notice that we do not divide by $|W|$ as it is done in other definitions of density. This difference is crucial for us.

[4] Suppose $\mathcal{P}$ contains a subset A, where possibly $|A| > 2$, then we can turn any LP solution (x, z) into a feasible LP solution for the instance containing all pairs $P \subseteq A$ by setting $z_P^r := z_A^r$ for every such pair and each root $r \in V$. Hence, the LP relaxation where A is part of $\mathcal{P}$ is at least as strong as any equivalent representation containing pairs $P \subseteq A$ instead of A.

However, to prove an upper bound on the integrality gap for general instances of Steiner Forest, we need to prove an upper bound for any representation of the instance. To see this, note that whenever a vertex appears in multiple pairs, we can replace it by several collocated copies. This changes neither the value of (Forest-BCR) nor the cost of an optimum Steiner Forest solution. Of course, one could contract vertices at distance zero upfront, but a similar situation arises if the copies have a very small, but positive distance from each other. In this new instance, there is only one possible representation of the pairs but the integrality gap of the LP is essentially the same as for the original instance.

We say that a connected component of the demand graph is *trivial* if it contains only a single vertex and we call it *nontrivial* otherwise. Observe that Steiner Tree instances are instances of Steiner Forest where the demand graph has only one nontrivial connected component. In the full version of this paper [11], we prove that for Steiner Tree instances, the value of (Forest-BCR) does not depend on the representation of the pairs $\mathcal{P}$. Even more, this value is always equal to the value of Tree-BCR for the corresponding Steiner Tree instance. In particular, the result from [10] implies that such instances have integrality gap strictly smaller than 2, namely at most 1.9988.

Theorem 3. *Let $(G, c, \mathcal{P})$ be an instance of Steiner Forest where the demand graph has only one nontrivial connected component (with vertex set R). Then the value of* (Forest-BCR) *is equal to the value of Tree-BCR for the Steiner Tree instance (G, c, R).*

Finally, we provide a family of instances showing that the integrality gap of (Forest-BCR) is at least $\frac{3}{2}$ (see Sect. 3).

Theorem 4. *The integrality gap of* (Forest-BCR) *is at least $\frac{3}{2}$.*

Because (Forest-BCR) is a strengthening of the classical undirected LP relaxation of Steiner Forest,[5] which has an integrality gap of 2 [1,24,28], the integrality gap of (Forest-BCR) is at most 2. Whether the integrality gap of (Forest-BCR) is strictly less than 2, remains an interesting question for future work. We remark that for the instances we use to prove Theorem 4, the optimal solutions to (Forest-BCR) are not half-integral. Nevertheless, the rounding algorithm we use to prove Theorem 1 yields optimal integral solutions for these instances.

1.2 Related Work

The *Steiner Network* problem, also known as *Survivable Network Design*, is a generalization of Steiner Forest where we are given pairwise vertex connectivity requirements $\lambda_{u,v} \geq 0$, and the task is to compute a cheapest subgraph of G such that each such pair of vertices u, v is $\lambda_{u,v}$-edge connected. In a celebrated

[5] To see this, observe that for every feasible solution (x, z) to (Forest-BCR), the vector $\tilde{x} \in \mathbb{R}^E$ defined by $\tilde{x}_{\{u,v\}} := \sum_{r \in V}(x^r_{(u,v)} + x^r_{(v,u)})$ is a feasible solution to the undirected LP relaxation.

result, Jain [28] obtained a 2-approximation for this problem using the iterative rounding technique. Since then, improving on the 2-approximation barrier, even just for special cases of Steiner Network, became an important open problem. This was recently achieved for some problems in this family, such as Connectivity Augmentation [8,13,35–37] and Forest Augmentation [25]. Among the special cases for which 2 is still the best-known factor, we already mentioned the Steiner Forest problem. Another interesting special case is the Minimum-Weight 2-Edge Connected Spanning Subgraph problem.

For the Steiner Tree problem several better than 2 approximation algorithms are known. A series of works based on the relative-greedy approach [31,32,40, 41] culminated in a 1.55 approximation [33]. The current-best approximation algorithm achieves an approximation guarantee of $(\ln 4 + \varepsilon) \approx 1.39$ [9] and is based on an iterative randomized rounding approach, applied to the so-called Hypergraphic Cut Relaxation for Steiner tree. This LP relaxation has integrality gap at most $\ln 4$ [23]. Recently, the same approximation factor and a new proof of this integrality gap upper bound has been obtained using local search [36].

For Steiner Forest, better-than-2 approximation algorithms are only known in special cases. In particular, approximation schemes are known for planar and bounded tree-width graphs [4], and for Euclidean plane instances [6]. A better-than-2 approximation is also known for very dense unit weight graphs [30].

For the prize-collecting generalizations of Steiner Tree and Steiner Forest, where some terminals or terminal pairs might be left disconnected by paying a given associated cost, the currently best-known approximation factors are 1.79 for prize-collecting Steiner Tree [3] and 2 for prize-collecting Steiner Forest [2].

For the directed analogue of the Steiner Forest problem, approximation factors of $O(|R|^{1/2+\varepsilon})$ [17] and $O(|V|^{2/3+\varepsilon})$ [5] are known for every fixed $\varepsilon > 0$. For the special case of planar graphs, polylogarithmic approximation factors have been obtained in [18]. For the directed analogue of Steiner Tree, the current best approximation algorithms for it are $|R|^{\varepsilon}$ in polynomial time [16] and $O(\log^2 |R| / \log\log |R|)$ in quasi-polynomial time [26].

1.3 Notation

For a vertex v, we denote by $\delta^-(v)$ and $\delta^+(v)$ the sets of incoming and outgoing edges of v, respectively. Similarly, for a vertex set U, we denote by $\delta^-(U)$ and $\delta^+(U)$ the incoming and outgoing edges of U, respectively.

For a solution (x, z) to (Forest-BCR) and a vertex $r \in V$, we write $c(x^r) := \sum_{e \in \overrightarrow{E}} c(e) \cdot x^r_e$. Moreover, for an edge set $F \subseteq E$, we write $\overrightarrow{F}$ to denote the directed edge set obtained by bidirecting F and define $x^r(F) := \sum_{e \in \overrightarrow{F}} x^r_e$ and $x(F) := \sum_{r \in V} x^r(F)$.

Finally, for a graph $G = (V, E)$ and a set $W \subseteq V$, we write $G[W]$ to denote the subgraph of G induced by W, and we write $E[W]$ to denote the edge set of this subgraph, i.e., the set of all edges from E with both endpoints in W.

2 Rounding Half-Integral LP Solutions

In this section we prove Theorem 1, which we restate here for convenience.

Theorem 5. *Given any half-integral solution* (x, z) *to* (Forest-BCR), *we can compute in polynomial time a feasible solution to the associated Steiner Forest instance of cost at most* $\frac{16}{9} \cdot c(x)$.

To prove Theorem 1 we describe an algorithm that could be applied to round arbitrary solutions (x, z) to (Forest-BCR) and then prove that it yields a Steiner Forest solution of cost at most $\frac{16}{9}c(x)$ if the LP solution (x, z) is half-integral.

2.1 Description of the Algorithm

We now describe our recursive algorithm, which is summarized in Algorithm 1. See Fig. 1 for an example. If $\mathcal{P} = \emptyset$, we return the optimum solution, which is $F = \emptyset$.

Otherwise, we compute the metric closure $(\overline{G}, \overline{x})$ of the weighted graph (G, c). Next, we modify the given LP solution (x, z) into a more structured LP solution for the instance $(\overline{G}, \overline{c}, \mathcal{P})$. First, we ensure that whenever $z_P^r > 0$ for some vertex $r \in V$ and some pair $P \in \mathcal{P}$, then r is a terminal. This modification can be done without increasing the cost of the LP solution by a simple rerouting step (see full version [11]). We remark that we could have required this property in (Forest-BCR), in which case it would be automatically satisfied at the beginning of the algorithm. However, the property will not be automatically satisfied for the instances on which we call the algorithm recursively.

Then we do splitting-off steps, as long as possible: If for a vertex $v \in V$, a root $r \in V$, and edges (u, v) and (v, w) we can reduce $x^r_{(u,v)}$ and $x^r_{(v,w)}$ by $\varepsilon > 0$ and increase $x^r_{(u,w)}$ by ε while maintaining the feasibility of our LP solution, then we do so for the largest possible such ε. Because the cost function $\overline{c}$ is metric, this does not increase the cost of our LP solution. We then obtain an LP solution (x, z) with the following two properties:

(a) We have $z_P^r = 0$ for all $r \in V \setminus R$ and all $P \in \mathcal{P}$.
(b) No splitting-off step is possible while maintaining the feasibility of (x, z).

We call an LP solution with these two properties *well structured.*

Having a well structured LP solution for the instance $(\overline{G}, \overline{c}, \mathcal{P})$, we compute a densest subgraph of $\overline{G} = (V, E)$ with respect to the following notion of density:

Definition 1. *For* $W \subseteq V$ *with* $|W| \geq 2$, *we define the density of* W *with respect to* x *as*

$$\text{density}_x(W) := \frac{x(E[W])}{|W| - 1}.$$

We compute $W \subseteq V$ with $|W| \geq 2$ maximizing $\text{density}_x(W)$. Such a set W can be found in polynomial time by a reduction to submodular function minimization; details in the full version of the paper [11].

Algorithm 1. Recursive algorithm to round a solution (x, z) to (Forest-BCR) for a given instance $(G, c, \mathcal{P})$.

If $\mathcal{P} = \emptyset$, **return** $\emptyset$.
Let $(\overline{G}, \overline{c})$ be the metric closure of (G, c).
Replace (x, z) with a well structured solution for the instance $(\overline{G}, \overline{c}, \mathcal{P})$ without increasing the cost.
Compute a densest subgraph of G with respect to edge weights x, i.e., compute a set $W \subseteq V$ maximizing $\text{density}_x(W) := \frac{x(E[W])}{|W|-1}$.
Compute a minimum spanning tree $\text{MST}(W)$ of $\overline{G}[W]$ with respect to the costs $\overline{c}$.
Call the algorithm recursively on the instance and LP solution that arise by contraction of W and let F_W be the resulting edge set.
Let F result from $F_W \cup \text{MST}(W)$ by replacing every edge $\{v, w\}$ by a shortest v-w path in (G, c).
Return F.

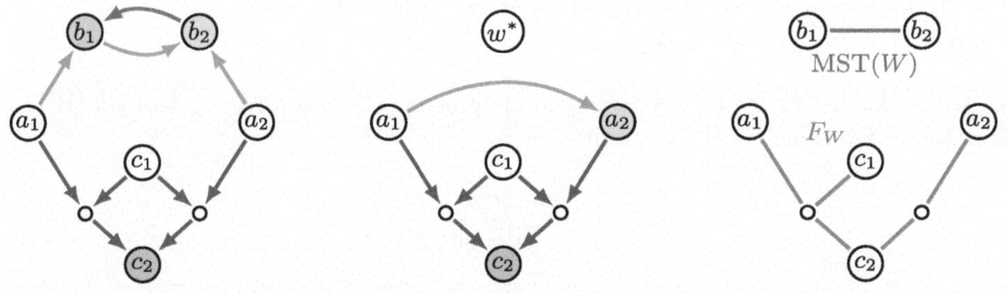

Fig. 1. The left part of figure illustrates an LP solution for an instance with pairs $\{a_1, a_2\}$, $\{b_1, b_2\}$, and $\{c_1, c_2\}$. Every edge e with $x_e^r = \frac{1}{2}$ is drawn in the same color as the vertex r. In the middle, we see the LP solution in the recursive call of Algorithm 1 (applied to the instance arising from the contraction of $W = \{b_1, b_2\}$) after rerouting and splitting-off. The solution returned by the algorithm is shown on the right.

Next we compute a minimum cost spanning tree $\text{MST}(W)$ in the graph $\overline{G}[W]$ induced by W with respect to edge costs $\overline{c}$. Then we call the algorithm recursively on the instance obtained by the contraction of W into a vertex w^*. We identify the edges in the contracted instance with the corresponding edges in the original instance. If parallel edges would arise by the contraction, we keep only one edge with minimal cost. Edges with both endpoints in W are no longer present after contraction.

When contracting the set W, we adapt the set of pairs in the natural way: we remove every pair $P \in \mathcal{P}$ with $P \subseteq W$; in every pair $\{s, t\}$ with $s \in W$ and $t \notin W$, we replace s with w^*. We also transform the LP solution (x, z) in the canonical way as follows. We set $z_P^{w^*} := \sum_{w \in W} z_P^w$ for each $P \in \mathcal{P}$. For all vertices $r, v \in V \setminus W$, we set $x^r_{(v,w^*)} := \sum_{w \in W} x^r_{(v,w)}$ and similarly $x^r_{(w^*,v)} = \sum_{w \in W} x^r_{(w,v)}$. Moreover, for each vertex $v \in V \setminus W$, we set $x^{w^*}_{(v,w^*)} := \sum_{w \in W} \sum_{u \in W} x^w_{(v,u)}$ and similarly $x^{w^*}_{(w^*,v)} := \sum_{w \in W} \sum_{u \in W} x^w_{(u,v)}$.

Then we take the union of MST(W) and the edge set F_W obtained from the recursive call to obtain a solution to the instance $(\overline{G}, \overline{c}, \mathcal{P})$. Finally, we replace every edge $\{v, w\}$ by a shortest v-w path in the original graph G with respect to the cost function c. In this way, we obtain a Steiner Forest solution to the original instance $(G, c, \mathcal{P})$.

2.2 Analysis of the Algorithm

For a proof that Algorithm 1 can be implemented to run in polynomial time we refer to the full version of the paper [11]. We next analyze the approximation ratio of the algorithm and prove that we obtain a solution of cost at most $\frac{16}{9} \cdot c(x)$ if (x, z) is half-integral.

First, we discuss how we can bound the cost $\text{mst}(W)$ of the minimum spanning tree MST(W) in $\overline{G}[W]$. We denote by $x|_{E[W]}$ the vector x restricted to directed edges with both endpoints in W and therefore write $c(x|_{E[W]}) = \sum_{v,w \in W} c(e) \cdot \sum_{r \in V} x^r_{(v,w)}$.

Lemma 1. *Let* (x, z) *be a solution to* (Forest-BCR) *and let* $W \subseteq V$ *with* $|W| \geq 2$ *maximizing* $\text{density}_x(W)$. *Then*

$$\text{mst}(W) \leq \frac{c(x|_{E[W]})}{\text{density}_x(W)}.$$

To prove Lemma 1, we consider the vector $y \in \mathbb{R}^{E[W]}_{\geq 0}$ that is given by $y_{\{v,w\}} := \frac{1}{\text{density}_x(W)} \cdot \sum_{r \in R} \big(x^r_{(v,w)} + x^r_{(w,v)}\big)$ for every edge $\{v, w\} \in E[W]$. One can show that y is contained in the spanning tree polytope, using the description of the spanning tree polytope by partition constraints. Here, we crucially exploit the fact that W maximizes the density with respect to x. Then we can bound $\text{mst}(W)$ by $c(y) = \frac{c(x|_{E[W]})}{\text{density}_x(W)}$. For details of the proof, see full version [11].

The next step our analysis is to prove that for a well structured half-integral LP solution (x, z), a sufficiently dense subgraph with respect to x does always exist:

Lemma 2. *Let* (x, z) *be a well structured solution to* (Forest-BCR) *for an instance with* $|\mathcal{P}| \geq 1$. *If* (x, z) *is half-integral, then there exists a set* $W \subseteq V$ *with* $|W| \geq 2$ *and* $\text{density}_x(W) \geq \frac{9}{16}$.

The proof of Lemma 2 is the only part of our analysis where we use the half-integrality of the LP solution. We first observe that we may assume that decreasing the value of any variable x^r_e will make the LP solution infeasible: otherwise, we can decrease x^r_e by $\frac{1}{2}$ and this can only decrease the density of any set W with respect to x. (The decrease of x^r_e will also maintain the property that (x, z) is well structured.)

Next, we consider the *projection multi-graph* $\hat{G} = (V, \hat{E})$ obtained from a half-integral solution (x, z) as follows. The graph $\hat{G}$ has vertex set V and its edge set contains $2 \cdot \sum_{r \in V} \big(x^r_{(v,w)} + x^r_{(w,v)}\big)$ copies of each edge $\{v, w\}$. We say that a vertex is a *high-degree vertex* if it has degree at least three in $\hat{G}$ and call it a *low-degree vertex*, otherwise.

We now use the fact that (x, z) is well structured to show that every Steiner vertex has either degree zero in $\hat{G}$ or it is a high-degree vertex. Consider a Steiner vertex v of nonzero degree in $\hat{G}$. Because of our assumption that decreasing the value of any variable x_e^r makes the LP solution infeasible and because $z_P^v = 0$ for every pair $P \in \mathcal{P}$, we have $x(\delta^+(v)) > 0$ and $x(\delta^-(v)) > 0$, implying that v has degree at least two in $\hat{G}$. If $x(\delta^+(v)) = \frac{1}{2}$ and $x(\delta^-(v)) = \frac{1}{2}$, we could have applied a splitting-off step at v, contradicting the fact that (x, z) is well structured. Hence, at least one of $x(\delta^+(v))$ and $x(\delta^-(v))$ must be at least 1, implying that v has degree at least three in $\hat{G}$.

Then, using the feasibility of the LP solution (x, z) for (Forest-BCR), we can show that every vertex v of nonzero degree in $\hat{G}$, which we will also call a *support vertex*, has degree at least two in $\hat{G}$ and satisfies in addition at least one of the following two properties:

(i) there is a vertex $w \in V$ such that $\hat{G}$ contains two parallel edges $\{v, w\}$;
(ii) either v itself is a high-degree vertex, or one of its neighbors in $\hat{G}$ is a high-degree vertex.

See Lemma 2.7 in the full version [11] for a proof of this fact.

To prove Lemma 2, we next lower bound the maximum density. If at least one terminal v satisfies (i), then $W = \{v, w\}$ satisfies $\text{density}_x(W) = 1 \geq \frac{9}{16}$. Otherwise every support vertex satisfies (ii) and we consider the set W consisting of all support vertices. In order to prove $\text{density}_x(W) \geq \frac{9}{16}$, we apply a token counting argument. Initially we distribute $2|\hat{E}|$ tokens by giving to every support vertex one token per incident edge in $\hat{G}$. Then we redistribute the tokens: every low-degree vertex in W keeps its 2 tokens and every high-degree vertex of degree $k \geq 3$ keeps $2 + \frac{k-2}{k+1} \geq \frac{9}{4}$ tokens for itself and gives $\frac{k-2}{k+1} \geq \frac{1}{4}$ tokens to each of its neighbors. Then (ii) implies that every support vertex has at least $\frac{9}{4}$ tokens, implying $2|\hat{E}[W]| = 2|\hat{E}| \geq \frac{9}{4} \cdot |W|$. From this we conclude $\text{density}_x(W) \geq \frac{9}{16}$. For full details of the proof of Lemma 2 we refer to the full version [11].

Having shown Lemma 1 and Lemma 2, we can prove that Algorithm 1 returns a Steiner Forest solution of cost at most $\frac{16}{9} \cdot c(x)$ whenever (x, z) is half-integral. To this end, we observe that the modifications of the solution (x, z) turning it into a well structured solution can only decrease the cost $c(x)$. Then we let (x_W, z_W) denote the LP solution that arises by contraction of W, i.e., the LP solution which we round to the edge set F_W by calling the algorithms recursively. Because the number of vertices decreases by at least one in every recursive call, we may assume by induction on the number of vertices, that $c(F_W) \leq \frac{16}{9} \cdot c(x_W)$. Moreover, by Lemma 1, we have $\text{mst}(W) \leq \frac{c(x|_{E[W]})}{\text{density}_x(W)}$ and thus by Lemma 2, $\text{mst}(W) \leq \frac{16}{9} \cdot c(x|_{E[W]})$. Using $c(x_W) + (x|_{E[W]}) \leq c(x)$, this implies the claimed upper bound on the cost of the returned edge set F.

3 Lower Bound on the Integrality Gap

In this section we prove that the integrality gap of (Forest-BCR) is at least $\frac{3}{2}$. We consider for every $q \in \mathbb{Z}_{>0}$, the following instance of Steiner Forest. The graph G has vertices s_i, v_i, and t_i for $i \in \{1, \ldots, q\}$. The edges of G consist of all edges between vertices s_i and v_j, and all edges between vertices v_i and t_j, i.e.,

$$E := \{\{s_i, v_j\} : i, j \in \{1, \ldots, q\}\} \cup \{\{v_i, t_j\} : i, j \in \{1, \ldots, q\}\}.$$

See Fig. 2. The set $\mathcal{P}$ contains the pairs $\{s_i, t_i\}$ for $i \in \{1, \ldots, q\}$ and the pairs $\{v_i, v_{i+1}\}$ for $i \in \{1, \ldots, q-1\}$. The cost of every edge is 1.

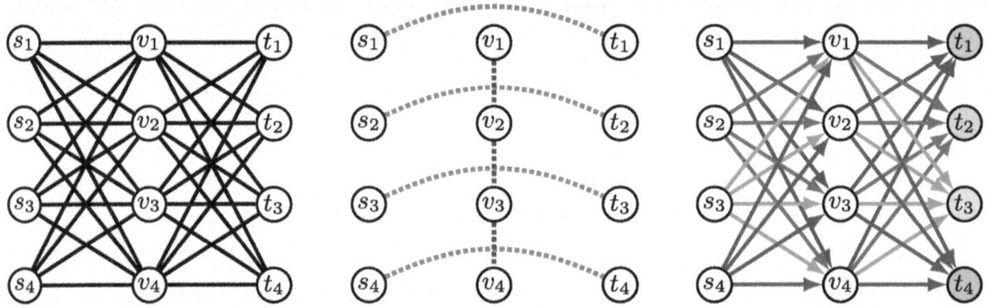

Fig. 2. The instance and LP solution we use to establish an integrality gap lower bound for $q = 4$. The figure shows the graph $G = (V, E)$ (left), the demand graph of the instance (middle), and an illustration of the values of x_e^r in the LP solution (right). For an edge $e \in \overrightarrow{E}$ we have $x_e^r = \frac{1}{q} = \frac{1}{4}$ if e is drawn in the same color as the vertex r. All other variables x_e^r are zero. In particular, for every non-colored vertex r, we have $x_e^r = 0$ for all $e \in \overrightarrow{E}$.

In any feasible solution F of this instance, the vertices $v_1, \ldots, v_q$ must belong to the same connected component of (V, F). Moreover, every path in G from s_i to t_i contains some vertex v_j and thus the vertices s_i and t_i must also belong to this connected component. We conclude that (V, F) has only a single connected component and is therefore a spanning tree of cost $|V| - 1 = 3q - 1$.

We now construct a feasible LP solution of cost $2q$. For $i \in \{1, \ldots, q\}$, we set

$$z^r_{\{s_i, t_i\}} := \begin{cases} 1 & \text{if } r = t_i \\ 0 & \text{otherwise} \end{cases}$$

and for $i \in \{1, \ldots, q-1\}$, we set

$$z^r_{\{v_i, v_{i+1}\}} := \begin{cases} \frac{1}{q} & \text{if } r = t_j \text{ for } j \in \{1, \ldots, q\} \\ 0 & \text{otherwise.} \end{cases}$$

For $i, j \in \{1, \ldots, q\}$, we set $x^{t_i}_{(s_i,v_j)} = x^{t_i}_{(v_j,t_i)} = \frac{1}{q}$. All other variables x^r_e with $r \in V$ and $e \in \overrightarrow{E}$ are 0. This defines a feasible solution to (Forest-BCR) of value $2q$. Therefore, the integrality gap of the instance is at least $\frac{3q-1}{2q}$, which becomes arbitrarily close to $\frac{3}{2}$ for q sufficiently large. This proves Theorem 4.

References

1. Agrawal, A., Klein, P.N., Ravi, R.: When trees collide: an approximation algorithm for the generalized Steiner problem on networks. SIAM J. Comput. **24**(3), 440–456 (1995). https://doi.org/10.1137/S0097539792236237
2. Ahmadi, A., Gholami, I., Hajiaghayi, M., Jabbarzade, P., Mahdavi, M.: 2-approximation for prize-collecting Steiner forest. In: Proceedings of the 2024 ACM-SIAM Symposium on Discrete Algorithms (SODA), pp. 669–693 (2024). https://doi.org/10.1137/1.9781611977912.25
3. Ahmadi, A., Gholami, I., Hajiaghayi, M., Jabbarzade, P., Mahdavi, M.: Prize-collecting Steiner tree: A 1.79 approximation. In: Proceedings of the 56th Annual ACM Symposium on Theory of Computing (STOC), pp. 1641–1652 (2024). https://doi.org/10.1145/3618260.3649789
4. Bateni, M., Hajiaghayi, M., Marx, D.: Approximation schemes for Steiner forest on planar graphs and graphs of bounded treewidth. J. ACM **58**(5) (2011). https://doi.org/10.1145/2027216.2027219
5. Berman, P., Bhattacharyya, A., Makarychev, K., Raskhodnikova, S., Yaroslavtsev, G.: Approximation algorithms for spanner problems and directed Steiner forest. Inf. Comput. **222**, 93–107 (2013)
6. Borradaile, G., Klein, P.N., Mathieu, C.: A polynomial-time approximation scheme for Euclidean Steiner forest. ACM Trans. Algorithms (TALG) **11**(3), 1–20 (2015)
7. Boyd, S., et al.: A 4/3-approximation algorithm for the minimum 2-edge connected multisubgraph problem in the half-integral case. SIAM J. Discret. Math. **36**(3), 1730–1747 (2022)
8. Byrka, J., Grandoni, F., Ameli, A.J.: Breaching the 2-approximation barrier for connectivity augmentation: a reduction to Steiner tree. SIAM J. Comput. **52**(3), 718–739 (2023). https://doi.org/10.1137/21M1421143
9. Byrka, J., Grandoni, F., Rothvoß, T., Sanità, L.: Steiner tree approximation via iterative randomized rounding. J. ACM **60**(1), 6:1–6:33 (2013). https://doi.org/10.1145/2432622.2432628
10. Byrka, J., Grandoni, F., Traub, V.: The bidirected cut relaxation for Steiner tree has integrality gap smaller than 2. In: Proceedings of the 65th IEEE Annual Symposium on Foundations of Computer Science (FOCS) (2024)
11. Byrka, J., Grandoni, F., Traub, V.: On the bidirected cut relaxation for Steiner forest. CoRR abs/2412.06518 (2024). https://doi.org/10.48550/ARXIV.2412.06518
12. Carr, R., Ravi, R.: A new bound for the 2-edge connected subgraph problem. In: International Conference on Integer Programming and Combinatorial Optimization, pp. 112–125. Springer, Cham (1998)
13. Cecchetto, F., Traub, V., Zenklusen, R.: Bridging the gap between tree and connectivity augmentation: unified and stronger approaches. In: Proceedings of the 53rd Annual ACM Symposium on Theory of Computing (STOC), pp. 370–383 (2021). https://doi.org/10.1145/3406325.3451086

14. Chakrabarty, D., Devanur, N.R., Vazirani, V.V.: New geometry-inspired relaxations and algorithms for the metric Steiner tree problem. Math. Program. **130**(1), 1–32 (2011). https://doi.org/10.1007/S10107-009-0299-0
15. Chakrabarty, D., Könemann, J., Pritchard, D.: Hypergraphic LP relaxations for Steiner trees. In: Eisenbrand, F., Shepherd, F.B. (eds.) IPCO 2010. LNCS, vol. 6080, pp. 383–396. Springer, Heidelberg (2010). https://doi.org/10.1007/978-3-642-13036-6_29
16. Charikar, M., et al.: Approximation algorithms for directed Steiner problems. J. Algorithms **33**(1), 73–91 (1999). https://doi.org/10.1006/jagm.1999.1042
17. Chekuri, C., Even, G., Gupta, A., Segev, D.: Set connectivity problems in undirected graphs and the directed Steiner network problem. ACM Trans. Algorithms (TALG) **7**(2), 1–17 (2011)
18. Chekuri, C., Jain, R.: A polylogarithmic approximation for directed Steiner forest in planar digraphs. arXiv preprint arXiv:2410.17403 (2024)
19. Chlebík, M., Chlebíková, J.: The Steiner tree problem on graphs: inapproximability results. Theor. Comput. Sci. **406**(3), 207–214 (2008). https://doi.org/10.1016/J.TCS.2008.06.046
20. Edmonds, J.: Optimum branchings. J. Res. Natl. Bur. Stand. **B71**, 233–240 (1967)
21. Feldmann, A.E., Könemann, J., Olver, N., Sanità, L.: On the equivalence of the bidirected and hypergraphic relaxations for Steiner tree. Math. Program. **160**(1–2), 379–406 (2016). https://doi.org/10.1007/S10107-016-0987-5
22. Goemans, M.X., Myung, Y.: A catalog of Steiner tree formulations. Networks **23**(1), 19–28 (1993). https://doi.org/10.1002/NET.3230230104
23. Goemans, M.X., Olver, N., Rothvoß, T., Zenklusen, R.: Matroids and integrality gaps for hypergraphic Steiner tree relaxations. In: Proceedings of the 44th ACM Symposium on Theory of Computing Conference (STOC), pp. 1161–1176 (2012). https://doi.org/10.1145/2213977.2214081
24. Goemans, M.X., Williamson, D.P.: A general approximation technique for constrained forest problems. SIAM J. Comput. **24**(2), 296–317 (1995). https://doi.org/10.1137/S0097539793242618
25. Grandoni, F., Ameli, A.J., Traub, V.: Breaching the 2-approximation barrier for the forest augmentation problem. In: Proceedings of the 54th Annual ACM Symposium on Theory of Computing (STOC), pp. 1598–1611 (2022). https://doi.org/10.1145/3519935.3520035
26. Grandoni, F., Laekhanukit, B., Li, S.: $O(\log^2 k/\log\log k)$-approximation algorithm for directed Steiner tree: a tight quasi-polynomial-time algorithm. In: Proceedings of the 51st Annual ACM Symposium on Theory of Computing (STOC), pp. 253–264 (2019). https://doi.org/10.1145/3313276.3316349
27. Hyatt-Denesik, D., Jabal Ameli, A., Sanità, L.: Finding almost tight witness trees. In: 50th International Colloquium on Automata, Languages, and Programming (ICALP 2023), pp. 79:1–79:16 (2023)
28. Jain, K.: A factor 2 approximation algorithm for the generalized Steiner network problem. Combinatorica **21**(1), 39–60 (2001). https://doi.org/10.1007/S004930170004
29. Karlin, A.R., Klein, N., Oveis Gharan, S.: An improved approximation algorithm for TSP in the half integral case. In: Proceedings of the 52nd Annual ACM SIGACT Symposium on Theory of Computing, STOC 2020, pp. 28–39. Association for Computing Machinery, New York (2020). https://doi.org/10.1145/3357713.3384273
30. Karpinski, M., Lewandowski, M., Meesum, S.M., Mnich, M.: Dense Steiner problems: approximation algorithms and inapproximability. CoRR abs/2004.14102 (2020). https://arxiv.org/abs/2004.14102

31. Karpinski, M., Zelikovsky, A.: New approximation algorithms for the Steiner tree problems. J. Comb. Optim. **1**(1), 47–65 (1997). https://doi.org/10.1023/A:1009758919736
32. Prömel, H.J., Steger, A.: A new approximation algorithm for the Steiner tree problem with performance ratio 5/3. J. Algorithms **36**(1), 89–101 (2000). https://doi.org/10.1006/JAGM.2000.1086
33. Robins, G., Zelikovsky, A.: Tighter bounds for graph Steiner tree approximation. SIAM J. Discret. Math. **19**(1), 122–134 (2005). https://doi.org/10.1137/S0895480101393155
34. Schmidt, D.R., Zey, B., Margot, F.: Stronger MIP formulations for the Steiner forest problem. Math. Program. **186**(1), 373–407 (2021). https://doi.org/10.1007/S10107-019-01460-6
35. Traub, V., Zenklusen, R.: A better-than-2 approximation for weighted tree augmentation. In: Proceedings of the 62nd IEEE Annual Symposium on Foundations of Computer Science (FOCS), pp. 1–12 (2021). https://doi.org/10.1109/FOCS52979.2021.00010
36. Traub, V., Zenklusen, R.: Local search for weighted tree augmentation and Steiner tree. In: Proceedings of the 2022 ACM-SIAM Symposium on Discrete Algorithms (SODA), pp. 3253–3272 (2022). https://doi.org/10.1137/1.9781611977073.128
37. Traub, V., Zenklusen, R.: A $(1.5+\varepsilon)$-approximation algorithm for weighted connectivity augmentation. In: Proceedings of the 55th Annual ACM Symposium on Theory of Computing (STOC), pp. 1820–1833 (2023). https://doi.org/10.1145/3564246.3585122
38. Vicari, R.: Simplex based Steiner tree instances yield large integrality gaps for the bidirected cut relaxation. CoRR abs/2002.07912 (2020). https://arxiv.org/abs/2002.07912
39. Williamson, D.P., Shmoys, D.B.: The Design of Approximation Algorithms. Cambridge University Press, Cambridge (2011)
40. Zelikovsky, A.: Better approximation bounds for the network and Euclidean Steiner tree problems. Technical report, University of Virginia, cS-96-06 (1996)
41. Zelikovsky, A.: An 11/6-approximation algorithm for the network Steiner problem. Algorithmica **9**(5), 463–470 (1993). https://doi.org/10.1007/BF01187035

Matroid Secretary via Labeling Schemes

Kristóf Bérczi[1], Vasilis Livanos[2,3], José A. Soto[3], and Victor Verdugo[4(✉)]

[1] MTA-ELTE Matroid Optimization Research Group and HUN-REN–ELTE Egerváry Research Group, Department of Operations Research, Eötvös Loránd University, Budapest, Hungary
kristof.berczi@ttk.elte.hu

[2] EPFL, Lausanne, Switzerland and Archimedes AI, Athens, Greece
vasileios.livanos@epfl.ch

[3] Department of Mathematical Engineering and Center for Mathematical Modeling, Universidad de Chile, Santiago, Chile
jsoto@dim.uchile.cl

[4] Institute for Mathematical and Computational Engineering, and Department of Industrial and Systems Engineering, Pontificia Universidad Católica de Chile, Santiago, Chile
victor.verdugo@uc.cl

Abstract. The matroid secretary problem (MSP) is one of the most prominent settings for online resource allocation and optimal stopping. A decision-maker is presented with a ground set of elements E revealed sequentially and in random order. Upon arrival, an irrevocable decision is made in a take-it-or-leave-it fashion, subject to a feasibility constraint on the set of selected elements captured by a matroid defined over E. The decision-maker only has ordinal access to compare the elements, and the goal is to design an algorithm that selects every element of the optimal basis with probability at least α (i.e., α-probability-competitive). While the existence of a constant probability-competitive algorithm for MSP remains a major open question, simple greedy policies are at the core of state-of-the-art algorithms for several matroid classes.

We introduce a flexible and general algorithmic framework to analyze greedy-like algorithms for MSP based on constructing a language associated with the matroid. Via this language, we establish a lower bound on the probability-competitiveness of the algorithm by studying a corresponding Poisson point process that governs the words' distribution in the language. Using our framework, we break the state-of-the-art guarantee for laminar matroids by settling the probability-competitiveness of the greedy-improving algorithm to be exactly $1 - \ln(2) \approx 0.3068$. For graphic matroids, we show a probability-competitiveness of 0.2693 when the underlying graph has no parallel edges and a guarantee of 0.2504 for general graphs, also breaking the state-of-the-art factor of 0.25.

Keywords: Matroid Secretary · Labeling Schemes · Poisson Processes

N. Megow and A. Basu (Eds.): IPCO 2025, LNCS 15620, pp. 128–141, 2025.
https://doi.org/10.1007/978-3-031-93112-3_10

1 Introduction

One of the most celebrated problems in online decision-making is the *secretary problem*, which aims to hire the best out of n candidates who arrive in a uniformly random order. The decision-maker can only rank the candidates observed up to that point and, upon observing a candidate, must make an immediate and irrevocable decision to accept or reject them. The famous optimal strategy of rejecting the first $1/e$ fraction of candidates and then selecting the first one who is better than all previously seen candidates guarantees at least a $1/e$ probability of hiring the best candidate (see, e.g., Lindley [17], Dynkin [9]).

Arguably, one of the most interesting generalizations is the *matroid secretary problem* (MSP) introduced by Babaioff et al. [2]. In the MSP, the decision-maker is given a matroid and can hire any set of candidates, forming an independent set in the matroid. An algorithm that, for a fixed optimal basis, selects a set in which every candidate from the optimal basis appears with probability at least α is called *α-probability-competitive.* For the remainder of this paper, we assume, without loss of generality, that a unique optimal basis exists. When the decision-maker observes a weight associated with each candidate and the selected set weights at least an α fraction of the optimal weight basis, the algorithm is *α-utility-competitive.* Over the last two decades, there has been a series of works studying the competitiveness for general matroids, resulting in a $O(\log\log r)$ utility-competitive and $O(\log r)$ probability-competitive algorithms [2,4,10,16,20] for rank r matroids. Babaioff et al. [2] conjectured the existence of a (utility) constant-competitive algorithm for every matroid. The *matroid secretary conjecture*, as it has been known since, has been resolved for several sub-classes of matroids [6,13–15,19,20], but it remains open in general and even for fundamental sub-classes such as gammoids or binary matroids.

1.1 Our Contributions and Techniques

In this work, we develop new tools for analyzing algorithms for the MSP that are more effective at managing the complex correlations involved in determining whether an element arriving at a specific time can be selected. Conceptually, our main contribution is introducing a new framework to analyze greedy-improving algorithms for the MSP using labeling schemes. Using this, we improve the state-of-the-art probability-competitiveness for the fundamental laminar and graphic matroid classes and match the best-known probability-competitiveness guarantees for several other matroids since, as we show in the full version of this paper, we extend the capabilities of the forbidden sets framework by Soto, Turkieltaub, and Verdugo [20].

The Labeling Scheme Framework. The algorithms we analyze initially skip the first p fraction of elements. Afterward, they mark every element that improves upon the current best set, i.e., an *improving element.* Some marked elements will be added to the output ALG when they arrive, according to an internal property P specific to the matroid. Property P filters the improving elements and guarantees the feasibility of the resulting selection, i.e., if P holds when an element e arrives, then it must be that ALG $+e$ is independent. To analyze the algorithm, we consider a separate procedure, called a *labeling scheme* that visits the marked elements in reverse order and assigns a label to each. At the end of the procedure, we obtain a word z by concatenating all labels assigned by our scheme. The core idea is that, for every matroid $\mathcal{M}$, we can assign a language $\mathcal{L}(\mathcal{M})$ such that, for any fixed element e^* in the optimal set, $\Pr[e^* \in \text{ALG}] \geq \Pr[z \in \mathcal{L}(\mathcal{M})]$, where ALG is the final set of our algorithm. Then, the competitiveness guarantee follows by computing $\Pr[z \in \mathcal{L}(\mathcal{M})]$, and we show that the uniformly random arrival order of the elements implies the characters (i.e., the labels) of z are chosen uniformly at random from the label set and the length of z follows a Poisson distribution.

Laminar Matroids. One of the first algorithms proposed for the MSP was the *greedy-improving* algorithm, which greedily selects every improving element, subject to feasibility. Despite its simplicity, it has been remarkably successful for many matroid classes. For laminar matroids, the best-known guarantee before our work was the very recent ≈ 0.2105-probability-competitiveness, achieved by Huang, Parsaeian, and Zhu [12] using the greedy-improving algorithm. Using our labeling scheme framework, we settle the probability-competitiveness of the greedy-improving algorithm for laminar matroids and show a tight guarantee of $1 - \ln(2) \approx 0.3068$ (Theorem 1). In fact, for every natural number r, we characterize the optimal competitiveness attained by the greedy-improving algorithm on laminar matroids of rank r. The competitiveness decreases as $r \to \infty$, with a limit of $1 - \ln(2)$. We also obtain a full characterization of the greedy-improving algorithm on uniform matroids for any fixed rank r and show that it grows from $1/e$ when $r = 1$ to $1 - 1/e$ as $r \to \infty$ (Proposition 1). Towards an answer for the matroid secretary conjecture, we study the simplest matroids for which the existence of a $1/e$-probability-competitive algorithm is still open – rank-2 matroids. As they are laminar matroids, we can specialize our results to this class, obtaining a probability-competitiveness of 0.3341. We show one can go beyond this via an alternative mixed algorithm to get a 0.3462-probability-competitive algorithm for the MSP on rank-2 matroids. The analysis of laminar MSP via our labeling schemes framework can be found in Sect. 3.

Graphic Matroids. In the graphic case, we adjust our labeling scheme to account for an algorithm that is not greedy-improving but avoids selecting edges that seem like they will induce cycles in the final set. The previous best algorithm, by Soto, Turkieltaub, and Verdugo [20], is 1/4-probability-competitive as it satisfies the 2-forbidden property. We improve over this guarantee via our labeling framework, showing that one can go beyond the forbidden sets technique.

We begin with a basic algorithm that orients the improving edges and greedily constructs an auxiliary digraph AUX with a maximum in-degree of one. Consequently, AUX has at most one cycle per connected component. The algorithm in [20] selects a subset of AUX that includes only edges connecting nodes with in-degree 0 when they appear. Hence, the resulting set does not contain the last edge of any cycle and is always feasible. However, this procedure may drop too many arcs from AUX. Our labeling schemes allow a more careful study of which edges we must delete to ensure that the graph is acyclic. For that, we assign a *non-negative generation* to each arc in AUX according to the order in which arcs were added. The basic algorithm outputs only edges of generation 0. We show that, in fact, by keeping edges with generation different from 1, we also obtain an acyclic graph. This new *generation algorithm* is 0.2693-probability-competitive for simple graphs (i.e., without parallel edges) – note that even for simple graphs, the previous algorithm by [20] is still 1/4-probability-competitive. By randomizing between the generation algorithm and an oblivious algorithm that is good for those optimal edges e^* that are likely to belong to some 2-cycles in AUX, we obtain our final algorithm for graphic matroids that achieves a 0.2504-probability-competitiveness for general graphs (Theorem 2). The analysis of graphic MSP via our labeling schemes framework can be found in Sect. 4.

1.2 Related Work

The MSP was introduced by Babaioff, Immorlica, Kempe, and Kleinberg [2], in which the authors gave a $\Omega(1/\log r)$-utility-competitive algorithm for general matroids. Later, Chakraborty and Lachish [4] gave an improved $\Omega(1/\sqrt{\log r})$-utility-competitive algorithm, before the current best bound of $\Omega(1/\log\log r)$ obtained first by Lachish [16] and then by Feldman, Svensson, and Zenklusen [10]. To the best of our knowledge, the only probability-competitive algorithms for general matroids are the $\Omega(1/\log^2 r)$-competitive algorithm by Bateni, Hajiaghayi, and Zadimoghaddam [3] and a $\Omega(1/\log r)$-competitive algorithm by Soto, Turkieltaub, and Verdugo [20].

The matroid secretary conjecture has been shown to hold for the *random assignment and random order* model by Soto [19], a setting with a slightly weaker adversary in which the weights of the elements are still chosen adversarially, but the assignment of weights to elements is done uniformly at random. Oveis-Gharan and Vondrák [11] extend this result to the setting where the arrival order of the elements is adversarial instead of uniformly random. Santiago, Sergeev, and Zenklusen [18] circumvent this problem by obtaining an algorithm that only requires access to an independence oracle in the random-assignment random-order model. Recently, Cristi, Dütting, Kleinberg, and Paes Leme [5] gave reductions from the unknown-matroid to the known-matroid variants of MSP for several sub-classes of matroids and showed an impossibility result for such a reduction in the case of a general matroid. Very recently, a series of works by Dughmi [7,8] uses duality to establish an equivalence between the matroid secretary conjecture and the existence of random-order contention resolution schemes. Regarding sub-classes of matroids, Babaioff, Dinitz, Gupta, Immorlica, and Talwar [1] unified several results via the α-partition property technique. Soto, Turkieltaub, and Verdugo [20] introduced the forbidden-sets technique and obtained many of the current best competitive ratios for several sub-classes of matroids.

2 The Labeling Scheme Framework

2.1 Improving Elements and Times

In this section, we describe our new approach to the matroid secretary problem using labeling schemes. Given a ground set E together with a subset $F \subseteq E$ and $e \in E$, the sets $F \setminus \{e\}$ and $F \cup \{e\}$ are abbreviated as $F - e$ and $F + e$, respectively. We use $\mathcal{M} = (E, \mathcal{I})$ to denote the matroid and r to indicate its rank. The elements of E can be compared (there is an underlying total order), or equivalently, we assume that all the elements have different weights. For a subset $X \subseteq E$, we denote by $\mathrm{OPT}(X)$ the unique optimal basis of X and refer to the elements in $\mathrm{OPT}(E)$ as *optimal elements*. An equivalent way to model a uniformly random arrival order is to assume that every element $e \in E$ picks an arrival time t_e from the uniform distribution over $[0,1]$, and the algorithm observes elements one by one in increasing arrival time. Although arrival times are not part of the input, our algorithms for matroids with n elements can simulate this process by first generating n independent, uniformly distributed arrival times in $[0,1]$. These *arrival times* are sorted as $s_1 < \cdots < s_n$ and assigned to elements as they are being presented; if e is the i-th arriving element, then $t_e = s_i$. For every positive $t \in (0,1]$, let E_t denote the set of elements arriving in the interval $[0,t)$ and E_t^+ the set of elements arriving in the interval $[0,t]$.

Definition 1 (Improving Elements and Improving Times). *An element $e \in E$ is improving if $e \in \mathrm{OPT}(E_{t_e}^+) = \mathrm{OPT}(E_{t_e} + e)$. The times when improving elements arrive are called improving times.*

Contrary to arrival times, we usually assume they are sorted in decreasing order, i.e., for a set of improving times $\{t_1, t_2, \ldots, t_k\}$, we have $t_k < t_{k-1} < \cdots < t_1$. For $0 \le a < b \le 1$, we denote the *last improving time* in increasing order in $[0,b)$ by $S(b)$, and the *number of improving times* in $[a,b)$ or $[a,b]$ by $N[a,b)$ and $N[a,b]$, respectively. The following is a folklore lemma for matroids (see [20, Lemma 2] for a proof). It allows us to restrict our analysis to algorithms for the matroid secretary problem that always skip non-improving elements.

Lemma 1. *If $X \subseteq Y$ and $e \in \mathrm{OPT}(Y+e)$, then $e \in \mathrm{OPT}(X+e)$. In particular, every optimal element $e \in \mathrm{OPT}(E)$ is improving.*

Our algorithms skip all elements with arrival time before a certain parameter $p \in (0,1)$, which has the same effect as choosing a value $n_0 \sim \mathrm{Bin}(n,p)$ from the binomial distribution of parameters n and p and then skipping the first n_0 elements. Among the elements that arrive after time p, our algorithms skip all non-improving elements. In particular, the output of all our algorithms will be a subset of the improving elements arriving in the interval $[p,1]$. The following assumption is unnecessary for our algorithms, but it simplifies their analysis.

Assumption 1. The input matroid is augmented in such a way that, at every time t, the set E_t has full rank. Every new element added in this way has an infinitesimal weight compared to its original counterpart.

We remark that this is only used to analyze our algorithms, and we refer to the full version of this paper for more details about the construction. One of the main tools we use is the following lemma, which characterizes the distribution of the arrival times of improving elements. Its proof can be found in the full version of this paper.

Lemma 2. *Under Assumption 1, the following hold.*

(i) *For every* $b \in \mathbb{R}_+$, $S(b)$ *has cumulative distribution function* $\Pr[S(b) \leq x] = 0$ *if* $x < 0$, $\Pr[S(b) \leq x] = \left(\frac{x}{b}\right)^r$ *if* $x \in [0, b)$, *and* $\Pr[S(b) \leq x] = 0$ *if* $x \geq b$.

(ii) *For all* $k \in \mathbb{N}$, *let* $x_k = -\ln(y_k)$ *where* $y_0 = 1$ *and* $y_k = S(y_{k-1})$ *otherwise.*

  (a) *The arrival times* $x_k - x_{k-1}$ *follow an exponential distribution with mean* $1/r$. *In particular,* $\{x_k\}_{k \in \mathbb{N}}$ *is a Poisson point process restricted to the positive half-line with rate* r.

  (b) $N[a, b)$ *distributes as a Poisson random variable with rate* $\lambda = r \ln(b/a)$.

  (c) *For any finite family of disjoint intervals* $\{[a_i, b_i)\}_i$, *the random variables* $\{N[a_i, b_i)\}_i$ *are mutually independent.*

2.2 Labeling Schemes and Improving Words

Our main contribution is a new way of analyzing algorithms based on assigning labels from $[r]$ to the elements of OPT at improving times and then analyzing the *word* formed by the labels of the improving elements.

Definition 2 (Labeling Scheme). *A labeling scheme* Λ *for an interval* $[a, b]$ *assigns, for each improving time* $t \in [a, b]$, *a unique label from* $[r]$ *to each element of the set* $X = \mathrm{OPT}(E_t^+)$.

It is worth emphasizing that the labels are assigned independently of the arrival order of E_t. Also note that an element's label may change from improving time to improving time and can depend on the set of improving elements and times in $(t, 1]$. We often use the following labeling scheme.

Definition 3 (Labeling Scheme Induced by a Total Order). *Let* π *be any total order on* E. *For every* $t \in [0, 1]$, *the labeling scheme* Λ_π *induced by* π *assigns a label to each* $e \in \mathrm{OPT}(E_t^+)$ *equal to the relative rank of* e *in* $\mathrm{OPT}(E_t^+)$ *with respect to* π.

In other words, for $X = \mathrm{OPT}(E_t^+)$, the labeling scheme assigns label 1 to the element of X that appears first in π, label 2 to the element of X that appears next, and so on. Note that even for this simple labeling scheme, the same element may have different labels at different times. Given a labeling scheme Λ, we say that an *improving time* t *has label* i, denoted as $z(t) = i$ if the improving element arriving at time t receives label i. We define the following.

Definition 4 (Improving Word in Time Interval $[a, b]$). *Let* $0 \leq a < b \leq 1$ *and* $\{e_k, \ldots, e_1\}$ *be the set of improving elements that arrive in time interval* $[a, b]$ *with improving times* $t_j = t_{e_j}$, *where the elements are indexed in a decreasing order with respect to their arrival times, i.e.,* $t_k < \cdots < t_1$. *The improving word in* $[a, b]$ *is* $z = z_1 \ldots z_k \in [r]^k$, *where* $z_i = z(t_i)$ *is the label of* e_i *in* $\mathrm{OPT}(E_{t_i}^+)$.

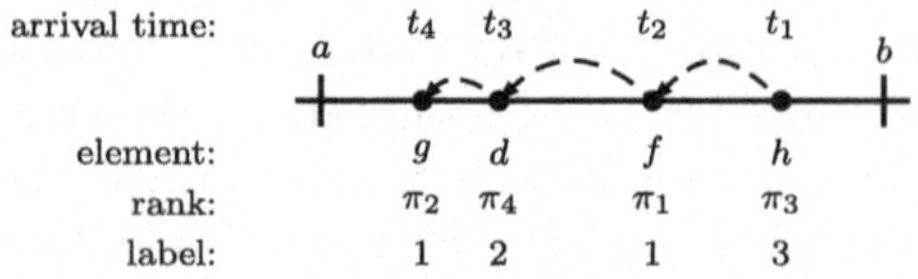

Fig. 1. A sequence of improving element arrivals in $[a, b]$. Assume that $E_{t_4} = \{g\}$, $E_{t_3} = \{g, d\}$, $E_{t_2} = \{g, d, f\}$, and $E_{t_1} = \{g, d, f, h\}$. If the elements are ranked as $f >_\pi g >_\pi > h >_\pi d$, the improving word induced by π in $[a, b]$ is $z = 3121$.

For an illustration of the previous definition, see Fig. 1. Whenever we apply a labeling scheme Λ and the resulting improving word in an interval $[a, b]$ is z, we say that Λ *produces* z *in* $[a, b]$. The uniformly random arrival order of the elements implies the labels of z are chosen uniformly at random from $[r]$, and the length of z follows a Poisson distribution. The main idea of our analysis is to design a language $\mathcal{L}$, depending on the matroid class and rank, such that, for every optimal element $e^* \in \mathrm{OPT}(E)$, we can find a labeling scheme Λ which ensures that e^* is selected by the algorithm whenever the improving word produced by Λ in $[p, 1]$ belongs to $\mathcal{L}$. We call such a language an *improving language*.

2.3 Algorithms via Improving Languages

In our framework, after skipping the elements arriving before p, we select an improving element if the sequence of different optimum sets we have seen (which correspond to the optimum sets at improving times) satisfies a property P, where P is tailored to the specific matroid class and rank r. If P holds when an element e arrives, P must guarantee that $\mathrm{ALG} + e \in \mathcal{I}$. To analyze such an algorithm, our main goal is to find, for every $e^* \in \mathrm{OPT}(E)$, a labeling scheme $\Lambda := \Lambda(e^*)$ and a language $\mathcal{L} \subseteq [r]^*$ such that, if $z \in [r]^*$ is the improving word in $[p, 1]$ induced by the labeling scheme Λ, then $z \in \mathcal{L} \implies e^* \in \mathrm{ALG}$. The significance of this goal is based on the fact that $\Pr[z \in \mathcal{L}]$ is much easier to compute due to z coming from a Poisson point process. In the following sections, we exploit this approach for laminar matroids (Sect. 3) and graphic matroids (Sect. 4).

3 Laminar Matroids

In this section, we apply the labeling framework to obtain an improved analysis of the well-known Greedy-Improving algorithm for laminar matroids. In other words, for laminar matroids, we let the property P simply be that $\mathrm{ALG} + e \in \mathcal{I}$.

Algorithm 1. Greedy-Improving Algorithm

```
Initialize ALG_0 ← ∅.
Skip all elements with arrival time in [0, p).
for each element e with t_e ∈ [p, 1], in arrival order do
    if e ∈ OPT(t_e) and ALG_0 + e ∈ I then
        ALG_0 ← ALG_0 + e
Return ALG_0.
```

As an illustration of our techniques, we first consider, in Sect. 3.1, the special case of uniform matroids and characterize the probability-competitiveness of the algorithm (Proposition 1). Then, we generalize our analysis for the broader class of laminar matroids in Sect. 3.2, showing that Greedy-Improving is $(1 - \ln(2))$-probability-competitive for every laminar matroid (Theorem 1); we prove this bound is tight, thereby settling the probability-competitiveness of Greedy-Improving completely. We design a new algorithm for rank-2 matroids (a subclass of the laminar family) that outperforms Greedy-Improving and provides a novel approach. We note that, in the context of laminar matroids, Assumption 1 corresponds to augmenting the matroid with a large set F of parallel copies of a single basis B_0 (i.e., each element of B_0 has a countably infinite number of parallel copies). All missing proofs can be found in the full version of this paper.

3.1 Uniform Matroids

As a warm-up, we present a new analysis of Greedy-Improving for the special class of uniform matroids using labeling schemes. It is worth noting that the algorithm is not optimal for uniform matroids; we describe it to aid the reader and illustrate our analysis. When $(E, \mathcal{I})$ is a rank-r uniform matroid on n elements, the condition $\mathrm{ALG}_0 + e \in \mathcal{I}$ in Step 4 is equivalent to $|\,\mathrm{ALG}_0\,| \leq r - 1$.

Definition 5 (Labeling Scheme for Uniform Matroids). *Let $e^* \in \mathrm{OPT}(E)$. Consider a total order π of E in which e^* appears first. We set $\Lambda_{uniform}(e^*)$ to be the labeling scheme induced by π as in Definition 3.*

Lemma 3. *Let z be the improving word in $[p, 1]$ produced by the labeling scheme $\Lambda_{uniform}(e^*)$, ALG_0 be the output of* Greedy-Improving, *and $\mathcal{L}_{uniform} = \{x1y \in [r]^* \colon x \in ([r] \setminus \{1\})^* \text{ and } |y| \leq r - 1\}$. Then $e^* \in \mathrm{ALG}_0$ if and only if $z \in \mathcal{L}_{uniform}$.*

Let $c(r, p) = -p \ln(p)$ if $r = 1$ and

$$c(r,p) = \frac{1-\left(1-\frac{1}{r}\right)^r}{\left(1-\frac{1}{r}\right)^r} p + p^r \sum_{k=0}^{r-1} \frac{\left(r\ln\left(\frac{1}{p}\right)\right)^k}{k!} - \frac{p^r}{\left(1-\frac{1}{r}\right)^r} \sum_{k=0}^{r-1} \frac{\left((r-1)\ln\left(\frac{1}{p}\right)\right)^k}{k!},$$

for $r \geq 2$. In the full version of this paper we show that $\Pr[z \in \mathcal{L}_{\text{uniform}}]$ is exactly $c(r, p)$. Using that, we obtain the following proposition

Proposition 1. *Algorithm 1 is exactly $c(r, p)$-probability-competitive for rank-r uniform matroids. Furthermore, as $r \to \infty$, the best sample size in Algorithm 1 for uniform matroids is $1/e$, achieving a probability-competitiveness of $1 - 1/e$.*

3.2 General Laminar Matroids

A *laminar matroid* is defined on a ground set with a laminar family of subsets. This family is organized into layers, where the sets in the i-th layer are pairwise disjoint, have rank i and sets in lower layers are nested within sets in higher layers. In this section, we show the following theorem.

Theorem 1. *The greedy-improving algorithm is* $1 - \ln(2) \approx 0.3068$*-probability-competitive for the MSP on laminar matroids. Furthermore, for every* $\varepsilon > 0$*, there exists a laminar matroid* $\mathcal{M}(\varepsilon)$ *such that the probability-competitiveness of the greedy-improving algorithm on* $\mathcal{M}(\varepsilon)$ *is at most* $1 - \ln(2) + \varepsilon$*.*

Lower Bound in Theorem 1. For uniform matroids, our analysis works for total order π of the elements on which e^* appears first. In the laminar case, however, we need to consider specific orders that depend on the laminar structure.

Definition 6 (Chain Order). *Let* $e^* \in \mathrm{OPT}(E)$ *and* $C_1 \subseteq \cdots \subseteq C_q = E$ *be the members of the laminar family containing* e^**. We refer to these sets as the chain induced by* e^**. We say that a total order* $(\pi_1, \ldots, \pi_n)$ *of* E *is a chain order with respect to* e^* *if* $\pi_1 = e^*$ *and for* $1 \leq i < j \leq r$*, we have* $\{\ell \in [q] \colon \pi_i \in C_\ell\} \subseteq \{\ell \in [q] \colon \pi_j \in C_\ell\}$*. In other words, in the enumeration* π*,* e^* *appears first, then the rest of the elements from* $E \cap C_1$*, then all the elements from* $(E \cap C_2) \setminus C_1$*, then all the elements from* $(E \cap C_3) \setminus C_2$ *and so on.*

Definition 7 (Chain Labeling Scheme for Laminar Matroids). *Let* $e^* \in \mathrm{OPT}(E)$ *be an element of the optimal basis and* π *be a chain order with respect to* e^**. We set* $\Lambda^{\pi}_{chain}(e^*)$ *as the labeling scheme induced by* π*. Since the choice of* π *is arbitrary, we write* $\Lambda_{chain}(e^*)$ *to denote a labeling scheme induced by an arbitrary but fixed chain order* π *with respect to* e^**.*

Definition 8 (Well-indexed Word). *We say that a word* z *is* well-indexed *if it satisfies the following:*

1. *There exists a unique symbol 1 in* z*; i.e.* $z = x1y$ *with* $x, y \in \{[r] \setminus \{1\}\}^*$*.*
2. *For each* $c \in [|y|]$*, it holds that*

$$|\{i \in [|y|] \colon y_i \leq c\}| \leq c - 1. \tag{1}$$

In particular, since $1 \leq y_i \leq r$ *for all* i*, this condition implies* $|y| \leq r - 1$*.*

Denote by $\mathcal{L}_{laminar} = \{z \in [r]^* \colon z \text{ is well-indexed}\}$*.*

Lemma 4. *Let* z *be the improving word of* $e^* \in \mathrm{OPT}(E)$ *produced by the labeling scheme* $\Lambda_{chain}(e^*)$*. If* $z \in \mathcal{L}_{laminar}$ *then* $e^* \in \mathrm{ALG}_0$*.*

Thus, to lower bound the competitiveness of Algorithm 1, it suffices to compute the probability that the improving word z produced in $[p, 1]$ is well-indexed.

Lemma 5. *Let* $m \leq r - 1$ *and* $y \in [r]^m$ *be a word chosen uniformly at random from* $[r]^m$*. Then, the probability that condition* (1) *holds is equal to* $1 - m/r$*.*

For any $p \in (0, 1)$, let us define $a(r, p) = -p\ln(p)$ if $r = 1$ and

$$\begin{aligned} a(r,p) = & -2p + (2 + \ln p)\Pr[P(-r\ln p) < r - 1] \\ & + 2\Pr[P(-r\ln p) = r - 1] + \frac{p}{\left(1 - \frac{1}{r}\right)^r}\Pr[P(-(r-1)\ln p) \geq r], \end{aligned}$$

for $r \geq 2$, where $P(\mu)$ is a Poisson random variable with rate μ.

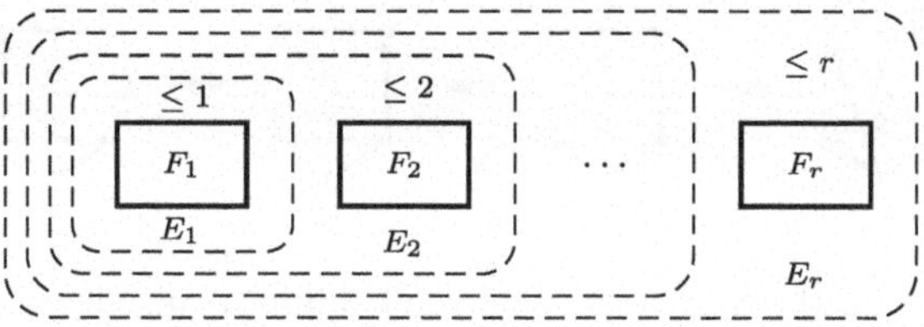

Fig. 2. A tight example for Algorithm 1. Sets $F_1, \ldots, F_r$ are presented with solid lines; $E_1, \ldots, E_r$ are presented with dashed lines. The numbers denote the bounds on the E_i's.

Lemma 6. *Algorithm 1 is at least $a(r, p)$-probability-competitive for rank-r laminar matroids.*

Upper Bound in Theorem 1. We show that the analysis of Algorithm 1 is tight for laminar matroids. For $q, r \in \mathbb{N}$, we define a laminar matroid $\mathcal{M}_{q,r} = (E_{q,r}, \mathcal{I}_{q,r})$ together with a weight function as follows. The ground set of the matroid is $E_{q,r} = F_1 \sqcup \cdots \sqcup F_r$, where each F_i has cardinality q. Let $E_i = \bigcup_{j=1}^{i} F_j$. The independent sets are $\mathcal{I}_{q,r} = \{I \subseteq E_{q,r} : |I \cap E_i| \leq i \text{ for all } i \in [r]\}$. Finally, for each $i \in [r]$, we set the weight of each element $e \in F_i$ to be $w_e \sim U[r - i, r - i + 1]$ independently from all other elements (see Fig. 2). Choosing q and r carefully results in an instance that provides an upper bound on the probability-competitiveness of GREEDY-IMPROVING for laminar matroids.

4 Graphic Matroids

Given a graph with edges as the ground set, the independent sets in the *graphic matroid* are edge sets that do not contain any cycles. In this section, we use our labeling schemes framework to improve the state-of-the-art competitive ratio for graphic matroid secretary. Our main result is the following.

Theorem 2. *There is a* 0.2504*-probability-competitive algorithm for the graphic MSP. For simple graphs, there is a* 0.2693*-probability-competitive algorithm.*

To show the theorem above, in Sect. 4.1 we first describe a basic algorithm that uses a simple labeling scheme and can be considered a variant of the 2-forbidden algorithm of [20]. In Sect. 4.2, we introduce one of the key notions of the paper, the *generation* of an edge, which suggests a new algorithm that we analyze through our labeling scheme framework to show the improved 0.2693-probability-competitiveness for simple graphs, i.e., with no parallel edges. To tackle the case of general graphs, we combine this algorithm with another that performs well on different types of instances, breaking the 1/4 barrier. All missing proofs can be found in the full version of this paper.

4.1 Basic Algorithm and Auxiliary Digraph

Let $G' = (V', E')$ denote the input graph of the graphic MSP. For graphic matroids, we can implement Assumption 1 with the following construction.

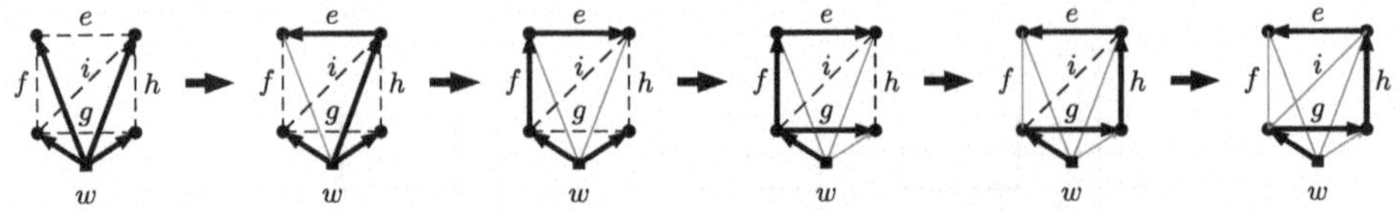

Fig. 3. An example for the canonical orientation of improving edges. Vertex w is chosen as the root, the original edges of the canonical orientation are drawn with thick lines, and a set $\{e, f, g, h, i\}$ arrives in the order e, f, g, h, i, with ranking $e > g > h > i > f$. Note that e, f, g and h are improving edges whereas i is not, the orientation of e changes several times, and f leaves the optimal tree upon the arrival of h.

Assumption 2. The input graph of our algorithm is the augmented graph $G = (V + w, E)$ with $E = E' \cup F$, where w is a new node considered a *root* and F is a collection of dummy edges consisting of an infinite number of parallel copies of the edges $\{wv\colon v \in V\}$. These dummy edges have infinitesimal weight.

Recall that every edge e in $E = E' \cup F$ picks its arrival time t_e from the uniform distribution over $[0, 1]$. Under Assumption 2, for every $t > 0$, the set E_t of all edges in $E = E' \cup F$ arriving in the interval $[0, t)$ has full rank $r = (|V + w| - 1) = |V|$, and thus $\mathrm{OPT}(E_t)$ is a tree with vertex set $V + w$.

Definition 9 (Canonical Orientation of an Improving Edge). *The arborescence $A(t)$ (resp. $A(t^+)$) obtained by orienting the tree* $\mathrm{OPT}(E_t)$ *(resp.* $\mathrm{OPT}(E_t^+)$*) away from the root w is called the* canonical orientation *of the optimum at t (resp. t^+). For an edge $e \in E$, we say that e* is oriented from u to v *at time t if $e \in \mathrm{OPT}(E_t^+)$ and e is oriented as (u, v) in $A(t^+)$.*

Note that the orientation of an edge changes from improving time to improving time and that only improving elements are oriented. Indeed, if $e = \{u, v\}$ is an edge arriving at time t_e, then, at that time, it receives an orientation (u, v) according to the arborescence $A(t_e^+)$. When the next improving edge arrives, say f, the optimum tree $\mathrm{OPT}(E_{t_f}^+)$ is different from $\mathrm{OPT}(E_{t_e}^+)$ since f enters and some other edge leaves the tree. If e is still part of the optimum tree, then its orientation may have changed; see Fig. 3 for an illustration. The following algorithm is a variant of the 2-forbidden algorithm of [20] and is the foundation of our algorithms for graphic matroids. Algorithm 2 is $p(1-p)$-probability-competitive; see the full version for details. In particular, 1/4-probability-competitive for $p = 1/2$.

Algorithm 2. BASIC ALGORITHM

```
Initialize ALG_0 ← ∅.
Initialize AUX ← digraph (V + w, ∅).
Skip all edges with arrival time in [0, p).
for each edge e with t_e ∈ [p, 1], in arrival order do
    if e ∈ OPT(E_{t_e}^+) is oriented from u to v at time t_e and deg^-_AUX(v) = 0 then
        AUX ← AUX + (u, v)
        if deg^-_AUX(u) = 0 then
            ALG_0 ← ALG_0 + e
Return ALG_0.
```

4.2 Cycles and Generations of the Auxiliary Digraph

Let $G' = (V, E')$ be an undirected, not necessarily simple graph, and $G = (V + w, E)$ with $E = E' \cup F$ be defined as before. In the following, we study the cycles of the auxiliary digraph AUX in Algorithm 2 when applied to the augmented graph G. For that, enumerate all arcs in AUX in the order in which they entered as $(u_1, v_1), \dots (u_k, v_k)$. Note that $\{v_1, \dots, v_k\} \subseteq V$ and they are all different, while $\{u_1, \dots, u_k\} \subseteq V + w$ and they might repeat. For every time $t \in [0, 1]$, let AUX(t) be the sub-digraph of AUX consisting of all arcs that entered ALG_0 in the interval $[0, t]$.

Definition 10 (Generation of an Arc in AUX). *Let e be an edge such that its orientation $\mathbf{e} = (u, v)$ entered* AUX *at time t. The* generation $\text{Gen}(e) \in \mathbb{N}$ *is defined as follows. If $e \in F$ is a dummy edge, we set* $\text{Gen}(e) = 1$. *If $e \in E'$ and* $\deg^{-}_{\text{AUX}(t)}(u) = 0$, *we set* $\text{Gen}(e) = 0$. *Otherwise, there is a unique arc $\boldsymbol{f}$ with head u that entered* AUX *at a time $s < t$, and we set* $\text{Gen}(e) = \text{Gen}(f) + 1$.

Let $C \subseteq E' \cup F$ be a cycle in AUX. No edge in C can be in F, since the edges of F are oriented away from w by construction, and no arc in AUX is directed toward w. Figure 4 shows an example of a cycle and the generations of its edges. Also, the edges e that enter ALG_0 are exactly those with $\text{Gen}(e) = 0$.

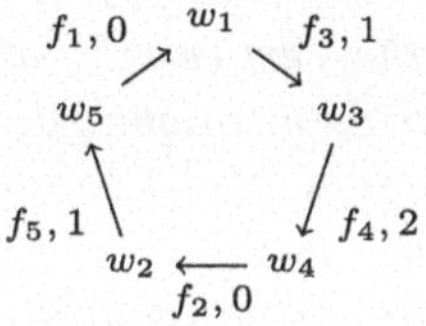

Fig. 4. A directed 5-cycle in AUX arriving in the order f_1, f_2, f_3, f_4, f_5; each arc is marked as $f_i, \text{Gen}(f_i)$.

Lemma 7. *Let* AUX' *be the sub-digraph of* AUX *obtained by deleting all edges with generation equal exactly to one. Then* AUX' *is acyclic.*

The previous discussion and the fact that the generation of each arc can be computed easily at its inclusion to AUX suggest the following algorithm.

Lemma 8. *Algorithm 3 outputs a set* ALG_1 *that is independent.*

We show that if G does not contain parallel edges, the competitive ratio of GENERATION is strictly better than $1/4$. Let $G' = (V, E')$ be a simple graph. Since the case $|V| = 2$ is straightforward, we assume that $|V| \geq 3$. Let AUX and ALG_1 be obtained by applying Algorithm 3 to $G = (V + w, E' \cup F)$, and let $e^* \in \text{OPT}(E)$ be an optimal edge arriving at a time $t^* = t_{e^*}$.

Algorithm 3. GENERATION ALGORITHM

$\mathrm{ALG}_1 \leftarrow \emptyset$
Initialize AUX $\leftarrow$ digraph $(V + w, \emptyset)$.
Skip all edges with arrival time in $[0, p)$.
for each edge e with $t_e \in [p, 1]$, in arrival order **do**
  if $e \in \mathrm{OPT}(E^+_{t_e})$ is oriented from u to v at time t_e and $\deg^-_{\mathrm{AUX}}(v) = 0$ **then**
    AUX $\leftarrow$ AUX $+(u, v)$
    if $u \neq w$ **then**
      if $\deg^-_{\mathrm{AUX}}(u) = 0$ **then**
        $\mathrm{ALG}_1 \leftarrow \mathrm{ALG}_1 + e$
      else let $f = (u', u)$ be the unique arc in AUX pointing toward u
        if $u' = w$ or $\deg^-_{\mathrm{AUX}(t_f)}(u') = 1$ **then**
          $\mathrm{ALG}_1 \leftarrow \mathrm{ALG}_1 + e$
Return ALG_1.

Definition 11 (Labeling Scheme $\Lambda_1(e^*)$). *On the time interval $[t^*, 1]$, we use the usual labeling scheme Λ_π induced by any total order of E such that $\pi_1 = e^*$. Suppose that e^* is oriented from a vertex u^* to a vertex v^* at time t^*, and let w_0^* be an arbitrary vertex in $V \setminus \{u^*, v^*\}$. For $s < t^*$, the labeling scheme is as follows: let $\boldsymbol{f} = (u', u^*)$ be the earliest improving arc in $(s, 1]$ having head u^*. If $\boldsymbol{f}$ exists and $u' \neq w$, then we set $w_s^* := u'$; since the original graph $G' = (V, E')$ is simple, f is not parallel to e^*, showing that $u' \neq v^*$ holds as well. Otherwise, we set $w_s^* = w_0^*$. In any case, the vertices v^*, u^*, and w_s^* are different. Recall that for any vertex $v \in V$, there is a unique edge in* $\mathrm{OPT}(E_s)$ *that is oriented toward v. Then, at time s, we assign label 1 to the edge oriented toward v^*, label 2 to the edge oriented toward u^*, and label 3 to the edge oriented toward w_s^*. The remaining $(r-3)$ labels are assigned arbitrarily to the other elements of* $\mathrm{OPT}(E_s)$.

Denote by $\mathcal{L}_{\text{generation}} = \{z \in [r]^* \colon z_{\{1,2,3\}} = x1y$, where x and y are words not containing the symbol 1 and y is a word that does not end with symbol $2\}$. Here, $z_{\{1,2,3\}}$ denotes the subsequence of z restricted to the alphabet $\{1, 2, 3\}$.

Lemma 9. *Let z be the improving word of $e^* \in$* $\mathrm{OPT}(E)$ *produced by $\Lambda_1(e^*)$. If $z \in \mathcal{L}_{generation}$ then $e^* \in$* AUX *and* $\mathrm{Gen}(e^*) \neq 1$. *In particular, $e^* \in$* ALG_1.

Using this lemma, we show that Algorithm 3 is $(\frac{1}{4}p(1-p^2) - \frac{1}{2}p\ln(p))$-probability-competitive for simple graphs. In particular, it is 0.2693-probability-competitive when $p \approx 0.4485$. Algorithm 3 may not achieve a good competitive ratio on general graphs with parallel edges, and the reason is that the probability that a given edge $e \in \mathrm{OPT}(E)$ has generation 1 may be much larger in this case. We remedy this issue by considering an algorithm suitable for those elements $e \in \mathrm{OPT}(E)$ for which $\mathrm{OPT}(E_q)$ is likely to contain a parallel copy of e when q is a constant. We combine this algorithm with GENERATION, leading to an algorithm called MIXTURE, which is 0.2504-probability-competitive; we refer to the full version for all the details.

References

1. Babaioff, M., Dinitz, M., Gupta, A., Immorlica, N., Talwar, K.: Secretary problems: weights and discounts. In: SODA, pp. 1245–1254 (2009)
2. Babaioff, M., Immorlica, N., Kempe, D., Kleinberg, R.: Matroid secretary problems. J. ACM **65**(6) (2018)
3. Bateni, M., Hajiaghayi, M., Zadimoghaddam, M.: Submodular secretary problem and extensions. ACM Trans. Algorithms **9**(4), 1–23 (2013)
4. Chakraborty, S., Lachish, O.: Improved competitive ratio for the matroid secretary problem. In: SODA, pp. 1702–1712 (2012)
5. Cristi, A., Dütting, P., Kleinberg, R., Leme, R.P.: Online matroid embeddings (2024). https://arxiv.org/abs/2407.10316
6. Dinitz, M., Kortsarz, G.: Matroid secretary for regular and decomposable matroids. SIAM J. Comput. **43**(5), 1807–1830 (2014)
7. Dughmi, S.: The outer limits of contention resolution on matroids and connections to the secretary problem. In: ICALP, pp. 42:1–42:18 (2020)
8. Dughmi, S.: Matroid secretary is equivalent to contention resolution. In: ITCS, pp. 58:1–58:23 (2022)
9. Dynkin, E.B.: The optimum choice of the instant for stopping a Markov process. Soviet Math. Dokl **4** (1963)
10. Feldman, M., Svensson, O., Zenklusen, R.: A simple O(log log(rank))-competitive algorithm for the matroid secretary problem. Math. Oper. Res. **43**(2), 638–650 (2018)
11. Gharan, S.O., Vondrák, J.: On variants of the matroid secretary problem. Algorithmica **67**(4), 472–497 (2013). https://doi.org/10.1007/S00453-013-9795-Y
12. Huang, Z., Parsaeian, Z., Zhu, Z.: Laminar matroid secretary: greedy strikes back. To appear in ESA (2024)
13. Im, S., Wang, Y.: Secretary problems: laminar matroid and interval scheduling. In: SODA, pp. 1265–1274 (2011)
14. Kesselheim, T., Radke, K., Tönnis, A., Vöcking, B.: An optimal online algorithm for weighted bipartite matching and extensions to combinatorial auctions. In: ESA, pp. 589–600 (2013)
15. Korula, N., Pál, M.: Algorithms for secretary problems on graphs and hypergraphs. In: ICALP (2009)
16. Lachish, O.: O(log log rank) competitive ratio for the matroid secretary problem. In: FOCS, pp. 326–335 (2014)
17. Lindley, D.V.: Dynamic programming and decision theory. J. R. Stat. Soc. Ser. C (Appl. Stat.) **10**(1), 39–51 (1961)
18. Santiago, R., Sergeev, I., Zenklusen, R.: Constant-competitiveness for random assignment matroid secretary without knowing the matroid. In: IPCO, pp. 423–437 (2023). https://doi.org/10.1007/978-3-031-32726-1_30
19. Soto, J.A.: Matroid secretary problem in the random-assignment model. SIAM J. Comput. **42**(1), 178–211 (2013)
20. Soto, J.A., Turkieltaub, A., Verdugo, V.: Strong algorithms for the ordinal matroid secretary problem. Math. Oper. Res. **46**(2), 642–673 (2021)

Valid Cuts for the Design of Potential-Based Flow Networks

Pascal Börner[1], Max Klimm[2], Annette Lutz[1], Marc E. Pfetsch[1], Martin Skutella[2], and Lea Strubberg[2](✉)

[1] Department of Mathematics, TU Darmstadt, Darmstadt, Germany
{boerner,lutz,pfetsch}@mathematik.tu-darmstadt.de
[2] Institute of Mathematics, TU Berlin, Berlin, Germany
{klimm,skutella,strubberg}@math.tu-berlin.de

Abstract. The construction of a cost minimal network for flows obeying physical laws is an important problem for the design of electricity, water, hydrogen, and natural gas infrastructures. We formulate this problem as a mixed-integer non-linear program with potential-based flows. The non-convexity of the constraints stemming from the potential-based flow model together with the binary variables indicating the decision to build a connection make these programs challenging to solve. We develop a novel class of valid inequalities on the fractional relaxations of the binary variables. Further, we show that this class of inequalities can be separated in polynomial time for solutions to a fractional relaxation. This makes it possible to incorporate these inequalities into a branch-and-cut framework. The advantage of these inequalities is lastly demonstrated in a computational study on the design of real-world gas transport networks.

1 Introduction

The availability of networks for electricity, water, hydrogen, natural gas, and traffic is a main driving force for human and economic development. Yet, the installation and maintenance of these infrastructure networks requires long term efforts and vast financial and environmental resources. Thus, for an economical usage of resources, mathematical optimization methods are very important.

This paper studies the mathematical optimization problem of designing a cost-minimal network that satisfies given demands. This problem—sometimes coined the *network synthesis problem*—has been studied extensively for a classic flow model [1,10,11,13,28]. In this problem, we are given an underlying multigraph $G = (V, A)$ of possible network connections. Every potential arc $a \in A$ has linear costs for installing capacities and the goal is to minimize costs such that a given set of demands can be satisfied.

While obviously important, these models have shortcomings as a model for the construction of physical infrastructure networks for electricity, water, hydrogen, and natural gas since they are based on a simplistic flow model where flow

N. Megow and A. Basu (Eds.): IPCO 2025, LNCS 15620, pp. 142–156, 2025.
https://doi.org/10.1007/978-3-031-93112-3_11

can be routed arbitrarily along the network as long as flow conservation constraints at the nodes are satisfied. In reality, physical flows such as electricity, water, hydrogen, natural gas, and (to a lesser and not immediate extent) individual traffic cannot be routed at will in the network, but instead is governed by physical principles such as resistances, pipe pressures, and minimal travel times.

To address this issue, we adopt the potential-based flow model introduced by Birkhoff and Diaz [6] and further discussed by Rockafellar [26]. In this model, we are given a directed graph $G = (V, A)$ where each arc is characterized by a *resistance* $\beta_a > 0$ that depends on the physical properties of the connection such as electrical resistances of wires, or the roughness of pipes. A flow is a vector $f \in \mathbb{R}^A$ where, for an arc a, a positive flow value $f_a > 0$ indicates a physical (mass) flow in the direction of the orientation of the arc, and a negative flow $f_a < 0$ indicates a (mass) flow against the direction of the orientation of the arc. The arc flow is governed by a vertex potential vector $\pi \in \mathbb{R}^V$ in the following way. For each arc $a = (u, v)$, the flow on a satisfies the equation

$$\pi_u - \pi_v = \beta_a \operatorname{sign}(f_a)|f_a|^r, \tag{1}$$

where the parameter $r > 0$ is called the *degree* of the potential-based flow network. We have $r = 1$ for DC power networks, $r = 1.852$ for water networks, and $r = 2$ for hydrogen or natural gas networks [12]. The potential π_v on node v corresponds to voltages in power networks, hydraulic heads in water networks, or squared pressures in hydrogen or natural gas networks. Due to safety reasons, it is assumed that the node potentials are within given bounds, i.e., that $\pi \in [0, \bar{\pi}]^V$ for a given global maximal potential value $\bar{\pi} > 0$.

For the cost-minimal construction of potential-based flow networks, we are given a directed multi-graph $G = (V, A)$ that represents possible connections. Parallel arcs can be used to allow for connections between two nodes with different properties like diameters, resistances and costs. We are further given a balance vector $b \in \mathbb{R}^V$ with the property that $\sum_{v \in V} b_v = 0$. Our goal is to select a cost-minimal subset of arcs such that there is a potential-based flow satisfying the given balances. That is, we are interested in solving the following mixed-integer non-linear program (MINLP):

$$\min \quad \sum_{a \in A} c_a x_a \tag{2a}$$

$$\text{s.t.} \quad \beta_a \operatorname{sign}(f_a)\,|f_a|^r = x_a\,(\pi_u - \pi_v) \qquad \text{for all } a = (u, v) \in A, \tag{2b}$$

$$\sum_{a \in \delta^+(v)} f_a - \sum_{a \in \delta^-(v)} f_a = b_v \qquad \text{for all } v \in V, \tag{2c}$$

$$f \in \mathbb{R}^A,\ x \in \{0,1\}^A, \tag{2d}$$

$$\pi \in [0, \bar{\pi}]^V. \tag{2e}$$

Our Results. The MINLP (2) involves binary decision variables $x \in \{0,1\}^A$ that govern which arcs are constructed, and also features non-linear constraints (2b). This makes the problem difficult to solve in practice. One key challenge when solving (2) in a branch-and-cut framework is to obtain good linear inequalities.

Our main result is Theorem 5, which yields a family of inequalities that are valid for all feasible solutions (π, f, x) to (2). These inequalities depend on the global maximum pressure bound $\bar{\pi}$ in the network and take the resistances and the degree of the network into account. We further show that these inequalities can be separated in polynomial time, i.e., for a point (π, f, x), we can decide in polynomial time whether all exponentially many inequalities of this type are satisfied, or we can construct one of the inequalities from the family that is violated by (π, f, x). For the proof of the result, we reduce the problem of finding one of the violated inequalities to the problem of finding disjoint cuts in the network with a suitably chosen total cut value. Further using the submodularity of the cut function, we can optimize over this set in polynomial time by using submodular function minimization routines.

Finally, we employ these inequalities and their efficient separation in a spatial branch-and-cut framework. Our computational results show that the inclusion of these inequalities in the branch-and-bound process improves the overall computation time.

Some of the previous literature (e.g., Thürauf et al. [29]) deal with individual pressure bounds

$$\pi_v \in [\bar{\pi}_v, \underline{\pi}_v] \quad \text{for all } v \in V \tag{2e$'$}$$

instead of the global pressure bound (2e). Setting $\bar{\pi} := \max_{v \in V} \bar{\pi}_v - \min_{v \in V} \underline{\pi}_v$, it is straightforward to verify that the MINLP with global pressure bounds $[0, \bar{\pi}]$ is a relaxation of the MINLP with the individual pressure bounds (2e′). Thus, all valid inequalities on (π, f, x) for the relaxation remain valid for the restriction.

Related Work. Topology optimization has been independently considered for the different applications of potential-based flows. Moreover, for physical networks like gas and water networks, there are mainly two different approaches.

The first approach and also the approach followed in this paper is to construct arcs with minimum cost from a given set of possible connections. This problem has been studied for gas networks [8,16–18,21,27], water networks [9,22], electrical networks [5] and general potential-based flow networks [29]. The NP-hardness of Problem (2) has been shown explicitly in [16] for the degree of $r = 2$; for electrical networks (with a degree of $r = 1$) it follows from the NP-hardness result of the switching problem for DC-networks shown in [20].

The second approach is to determine optimal capacities of the installed arcs corresponding, e.g., to diameters of pipes [2–4,15,32].

2 Preliminaries

Let $G = (V, A)$ be a directed and weakly connected graph with possibly parallel arcs but without loops. To clarify notation, we define $\delta^+(v) \subseteq A$ as the set of arcs leaving node $v \in V$, and $\delta^-(v) \subseteq A$ as the set of arcs entering v. For a set $S \subseteq V$, we denote by $\delta(S) \subseteq A$ the set of arcs that connect a node $u \in S$ with a node $v \notin S$ in either direction.

A *flow* in G is a vector $f \in \mathbb{R}^A$, where negative flow over an arc indicates a positive flow in the opposite direction. Let $B := \{b \in \mathbb{R}^V : \sum_{v \in V} b_v = 0\}$ be the set of *balanced vectors*. We call a flow $f \in \mathbb{R}^A$ a *b-transhipment* or the *flow induced by* b with $b \in B$ if it satisfies the flow conservation constraints (2c). For $s, t \in V$ with $s \neq t$, a b-transhipment is an *(s,t)-flow of value* $d \in \mathbb{R}$ if and only if $b = d \cdot \chi_{s,t}$. Here, we define the vector $\chi_{s,t} := \chi_s - \chi_t \in B$, where χ_v is the standard unit vector in $\mathbb{R}^V$ corresponding to component v.

We call the tuple $\mathcal{N} = (G, \beta, r)$ with resistances $\beta \in \mathbb{R}^A_{>0}$ and degree $r > 0$ a *potential-based flow network*, or simply *network*. The *conductance* $\mu \in \mathbb{R}^A_{>0}$ is defined by $\mu_a = 1/\sqrt[r]{\beta_a}$ for $a \in A$. A *potential-based flow* in a network $\mathcal{N}$ is a pair (π, f) of a potential vector $\pi \in \mathbb{R}^V$ and flow vector f if (1) holds. It is an easy observation that the potential vector is shift-invariant, i.e., for a potential-based flow (π, f), the shifted potentials $\pi + \lambda \mathbf{1}$ also fulfill (1) for all $\lambda \in \mathbb{R}$, where $\mathbf{1} \in \mathbb{R}^V$ is the all ones vector. Thus, we will henceforth assume without loss of generality that all potential-based flows (π, f) have their smallest node potential normalized to 0. We call a potential-based flow (π, f) in a network $\mathcal{N}$ a *potential-based b-transhipment* or a *potential-based (s,t)-flow* if f is a b-transhipment or an (s,t)-flow in G, respectively.

Birkhoff and Diaz [6] showed that for every $b \in B$ there exists a unique potential-based b-transhipment (π, f). This establishes the existence of a function $\pi^{\mathcal{N}} : B \to \mathbb{R}^V$ that maps a balanced vector $b \in B$ to potentials $\pi^{\mathcal{N}}(b) \in \mathbb{R}^V$, with a smallest node potential of 0, and a function $f^{\mathcal{N}} : B \to \mathbb{R}^A$, mapping $b \in B$ to the flow $f^{\mathcal{N}}(b) \in \mathbb{R}^A$, so that the pair $(\pi^{\mathcal{N}}(b), f^{\mathcal{N}}(b))$ fulfills (2c) and (1). Note that $\pi^{\mathcal{N}}(\lambda b) = \lambda^r \pi^{\mathcal{N}}(b)$ for all $\lambda \in \mathbb{R}$; see also [12, Observation 3.1]. In the context of infrastructure networks, the *entry* nodes $T^+ \subseteq V$ and *exit* nodes $T^- \subseteq V$, with $T^+ \cap T^- = \emptyset$, are typically fixed. We refer to their union $T := T^+ \cup T^-$ collectively as *terminals*. As a demand on these terminals we consider only those that are balanced. Therefore, let

$$B(T) := \{b \in B : b_v \geq 0 \,\forall v \in T^+,\; b_v \leq 0 \,\forall v \in T^-,\; b_v = 0 \,\forall v \notin T\}.$$

For $X \subseteq T$ and $b \in B(T)$, we define $b(X) := \sum_{v \in X} b_v$.

2.1 Effective Resistance and Conductance

For a potential-based flow network and two nodes s and t, the *effective resistance* $R^{\mathcal{N}}_{s,t}$ is the potential difference required to send one unit of flow from s to t, i.e.,

$$R^{\mathcal{N}}_{s,t} := \pi^{\mathcal{N}}_s(\chi_{s,t}) - \pi^{\mathcal{N}}_t(\chi_{s,t}).$$

We define the *effective conductance* between s and t as $U^{\mathcal{N}}_{s,t} = (R^{\mathcal{N}}_{s,t})^{-1/r}$. Two potential-based flow networks $\mathcal{N}$ on $G = (V, A)$ and $\mathcal{N}'$ on $G' = (V', A')$ of degree r are called *equivalent* with respect to a set of terminal nodes $T \subseteq V \cap V'$ if $\pi^{\mathcal{N}}_v(b) = \pi^{\mathcal{N}'}_v(b)$ for all $v \in T$ and $b \in B(T)$. The following result shows that every two-terminal network is equivalent to a network consisting of a single arc.

Theorem 1 ([12,19]). *A network $\mathcal{N} = (G, \beta, r)$ with terminal nodes $T = \{s,t\} \subseteq V$ is equivalent to the network $((T, \{(s,t)\}), \beta', r)$ with $\beta'_{(s,t)} = R^{\mathcal{N}}_{s,t} > 0$.*

The following lemma states another useful property of two-terminal networks.

Lemma 1 ([19, Lemma 9]). *Let s, $t \in V$ be two distinct nodes in a network $\mathcal{N}$ and $d > 0$. Let $\pi := \pi^{\mathcal{N}}(d \cdot \chi_{s,t}) \in \mathbb{R}^V$ be the induced potentials of the potential-based (s,t)-flow of value d in $\mathcal{N}$. Then, $\pi_s \geq \pi_v \geq \pi_t$ for all $v \in V$. Furthermore, $\pi_s - \pi_t = d^r \cdot R^{\mathcal{N}}_{s,t}$.*

The following results show how effective resistance and conductance behave under different network operations.

Proposition 1 ([19,24]). *Let $\mathcal{N} = (G, \beta, r)$ be a potential-based flow network and let s, $t \in V$. Then, the following hold:*

(i) If $\mathcal{N}'$ is a network resulting from $\mathcal{N}$ by either increasing the resistance of an arbitrary arc or by deleting an arc, then

$$R^{\mathcal{N}'}_{s,t} \geq R^{\mathcal{N}}_{s,t} \quad \text{and} \quad U^{\mathcal{N}'}_{s,t} \leq U^{\mathcal{N}}_{s,t}.$$

(ii) If $\mathcal{N}''$ is a network resulting from $\mathcal{N}$ by either decreasing the resistance of an arbitrary arc or by contracting two nodes, then

$$R^{\mathcal{N}''}_{s,t} \leq R^{\mathcal{N}}_{s,t} \quad \text{and} \quad U^{\mathcal{N}''}_{s,t} \geq U^{\mathcal{N}}_{s,t}.$$

2.2 Graph Cuts

In this paper, we present valid inequalities for (2), which are based on cuts inside the network. A *cut* in a graph G is a partition of the node set V into two disjoint subsets $S \subseteq V$ and $V \setminus S \subseteq V$. We call the arc set $\delta(S) \subseteq A$ the set of *arcs crossing cut* S. A cut S is an (X,Y)*-cut* if $X \subseteq S$ and $Y \subseteq V \setminus S$. If $X = \{s\}$ and $Y = \{t\}$, we change the notation slightly and call it an (s,t)-cut.

For a weighted graph $G = (V, A, w)$ with non-negative arc weights $w \in \mathbb{R}^A_{\geq 0}$, we define the *cut function* z that maps a cut to the total weight of arcs crossing the cut, i.e., $z \colon 2^V \to \mathbb{R}_{\geq 0}$, $S \mapsto \sum_{a \in \delta(S)} w_a$, where 2^V is the power set of V. The function z is submodular (see, e.g., [23]), that is,

$$z(A \cup B) + z(A \cap B) \leq z(A) + z(B) \quad \text{for all } A, B \subseteq V. \tag{3}$$

Submodular functions can be minimized in polynomial time; see, e.g., [23]. More generally, a minimum of a submodular function defined on a subset $\mathcal{X} \subseteq 2^V$ closed under union and intersection can be computed in polynomial time.

We say that $S_1, \ldots, S_k \in 2^V$ are k disjoint (X,Y)-cuts if S_i is an (X,Y)-cut and $\delta(S_i) \cap \delta(S_j) = \emptyset$ for all $i \neq j$. The k disjoint (X,Y)-cut problem is the problem of finding k disjoint (X,Y)-cuts such that the sum of the weight of the arcs crossing the cuts is minimal over all such cuts. This problem is a slight generalization of the k disjoint (s,t)-cut problem studied in [31]. In particular, for non-negative arc weights, it is shown in [31] that the problem has a compact LP formulation and can thus be solved in polynomial time.

3 Inequalities for Potential-Based Flows

In the following we state valid inequalities that must hold if there exists a potential-based flow satisfying global pressure bounds. We start with an observation on rather simple networks defined on a path with parallel arcs.

3.1 (s, t)-Flows on Multi-paths

In this subsection, we consider a multi-graph $G = (V, A)$ with $k + 1$ nodes, $V = \{v_0, v_1, \ldots, v_k\}$, where only arcs between consecutive nodes v_i and v_{i+1} are allowed. We denote by A_i the set of arcs connecting node v_{i-1} to v_i such that $A = \cup_{i=1}^k A_i$. We call such a graph *multi-path of length k*. Additionally, we define a *multi-path network of length k* as a network with an underlying multi-path of corresponding length. Given such a multi-path network $\mathcal{N}$ with conductance vector $\mu \in \mathbb{R}^A_{>0}$, we introduce variables $u_i \in \mathbb{R}^k_{>0}$ for the sum of the conductances of the arcs between the nodes v_{i-1} and v_i for $i = 1, \ldots, k$, so that

$$u_i := \sum_{a \in A_i} \mu_a. \tag{4}$$

This first lemma gives a closed form expression for the effective conductance of multi-path network.

Lemma 2. *Let $\mathcal{N}$ be a multi-path network of length k and degree $r > 0$. The effective conductance between $s = v_0$ and $t = v_k$ is given by $U^{\mathcal{N}}_{s,t} = (\sum_{i=1}^k u_i^{-r})^{-1/r}$.*

Proof. It is shown in [12, Lemmas 6 & 7] that the conductance of parallel arcs sums up in the same way as the resistance of arcs in series. We can therefore reduce all parallel arcs in A_i to one arc $a_i = (v_{i-1}, v_i)$ with conductance u_i without changing the behavior of the potentials at s and t. The resistance of these arcs is then given by $\beta_{a_i} = u_i^{-r}$. We further reduce the network to one arc $a = (s, t)$ with resistance $\beta_a = \sum_{i=1}^k \beta_{a_i}$ again without changing the behavior of the potentials at s and t. Let $\mathcal{N}' = ((\{s, t\}, \{(s, t)\}), (\beta_a), r)$ be the resulting network. Because $\mathcal{N}$ and $\mathcal{N}'$ are equivalent with respect to the node set $\{s, t\}$, we have $R^{\mathcal{N}}_{s,t} = R^{\mathcal{N}'}_{s,t} = \beta_a$ according to Theorem 1. Using the definition of the effective conductance, we obtain

$$U^{\mathcal{N}}_{s,t} = \frac{1}{\sqrt[r]{\beta_a}} = (\textstyle\sum_{i=1}^k \beta_{a_i})^{-1/r} = (\sum_{i=1}^k u_i^{-r})^{-1/r}.$$

□

We proceed to use this closed formula to obtain a lower bound on the effective conductance of a network that supports an (s, t)-flow of value d.

Lemma 3. *Let $\mathcal{N}$ be a multi-path network of degree $r > 0$ and length k, let $s = v_0$ and $t = v_k$. Let (π, f) be a potential-based (s, t)-flow of value $d > 0$. If $\pi \in [0, \bar{\pi}]^V$, then $U^{\mathcal{N}}_{s,t} \geq \frac{d}{\sqrt[r]{\bar{\pi}}}$.*

Proof. Because (π, f) is the potential-based flow in $\mathcal{N}$ induced by $b = d \cdot \chi_{s,t}$, it follows from Lemma 1 that

$$\pi_s - \pi_t = \pi_s^{\mathcal{N}}(d \cdot \chi_{s,t}) - \pi_t^{\mathcal{N}}(d \cdot \chi_{s,t}) = d^r R_{s,t}^{\mathcal{N}},$$

which is equivalent to $\sqrt[r]{\pi_s - \pi_t} \cdot U_{s,t}^{\mathcal{N}} = d$. Since $\bar{\pi} \geq \pi_s - \pi_t$, we obtain

$$\sqrt[r]{\bar{\pi}} \cdot U_{s,t}^{\mathcal{N}} \geq \sqrt[r]{\pi_s - \pi_t} \cdot U_{s,t}^{\mathcal{N}} = d.$$

□

In the following we state an inequality for potential-based (s,t)-flows with global pressure bounds $[0, \bar{\pi}]$ which is linear in the conductance μ. Intuitively, it states that the average flow that can be sent from node v_i to node v_{i+1}, when giving these nodes a potential difference of $\bar{\pi}/k$, must be at least the value of a maximum potential-based (s,t)-flow; see Fig. 1 for an example.

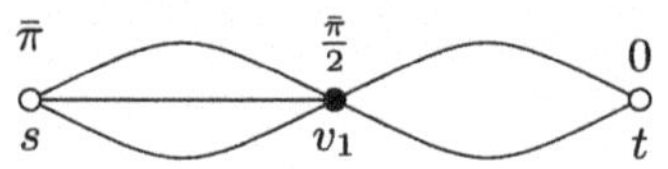

Fig. 1. Assuming $\mu = \mathbf{1}$, we can send a flow of $3\sqrt[r]{\bar{\pi}/2}$ over A_1, and $2\sqrt[r]{\bar{\pi}/2}$ over A_2. Thus, by Lemma 2 a feasible potential-based (s,t)-flow in the depicted network has at most a value of $5/2\sqrt[r]{\bar{\pi}/2}$.

Theorem 2. *Let $\mathcal{N}$ be a multi-path network of length k and degree $r > 0$, let $s = v_0$ and $t = v_k$. Let (π, f) be a potential-based (s,t)-flow of value $d > 0$. If $\pi \in [0, \bar{\pi}]^V$, then*

$$\frac{1}{k\sqrt[r]{k}} \sum_{i=1}^{k} \sum_{a \in A_i} \mu_a \geq \frac{d}{\sqrt[r]{\bar{\pi}}}.$$

Proof. For the proof we use Lemma 3 and claim that

$$\frac{1}{k\sqrt[r]{k}} \sum_{i=1}^{k} \sum_{a \in A_i} \mu_a \geq U_{s,t}^{\mathcal{N}}. \tag{5}$$

Note that (5) is by Eqs. (4) and Lemma 2 equivalent to

$$\frac{1}{k\sqrt[r]{k}} \sum_{i=1}^{k} u_i \geq \Big(\sum_{i=1}^{k} u_i^{-r}\Big)^{-1/r}. \tag{6}$$

We show that (6) holds by applying the inequality of arithmetic and geometric mean twice, once for $u_1, \ldots, u_k$ and once for $u_1^{-r}, \ldots, u_k^{-r}$, so that

$$\begin{aligned}\Big(\textstyle\sum_{i=1}^{k} u_i\Big)\Big(\textstyle\sum_{i=1}^{k} u_i^{-r}\Big)^{1/r} &\geq k\Big(\textstyle\prod_{i=1}^{k} u_i\Big)^{1/k}\Big(k\Big(\textstyle\prod_{i=1}^{k} u_i^{-r}\Big)^{1/k}\Big)^{1/r}\\ &= k\sqrt[r]{k}\Big(\textstyle\prod_{i=1}^{k} u_i(u_i^{-r})^{1/r}\Big)^{1/k} = k\sqrt[r]{k}.\end{aligned}$$

□

3.2 General (s, t)-Flows

For (s, t)-flows in general networks, Lemma 2 extends in the following way.

Theorem 3. *Let $\mathcal{N} = (G, \beta, r)$ be a potential-based flow network and let s, $t \in V$. If there exists a potential-based (s, t)-flow of value $d > 0$ in $\mathcal{N}$ with potentials $\pi \in [0, \bar{\pi}]^V$, then for all k disjoint (s, t)-cuts $S_1 \subseteq \cdots \subseteq S_k \subseteq V$ it is*

$$\frac{1}{k\sqrt[r]{k}} \sum_{i=1}^{k} \sum_{a \in \delta(S_i)} \mu_a \geq \frac{d}{\sqrt[r]{\bar{\pi}}}. \tag{7}$$

Proof. By contracting the nodes in $S_i \setminus S_{i-1}$ for each $i = 1, \ldots, k+1$, where we set $S_0 := \emptyset$ and $S_k := V$, we obtain a multi-path network $\mathcal{N}'$ of length k, so that $A_i = \delta(S_i)$. Note that the effective resistance between s and t does not increase in $\mathcal{N}'$ by Proposition 1(ii), i.e., $R_{s,t}^{\mathcal{N}'} \leq R_{s,t}^{\mathcal{N}}$.

Consider the potential $\pi' := \pi^{\mathcal{N}}(d \cdot \chi_{s,t})$ of the potential-based (s, t)-flow of value d in $\mathcal{N}'$, with a fixed potential of 0 at node t. Since the highest potential is attained at s and the lowest at t, according to Lemma 1 it holds for every node v_i in the multi-path that

$$0 \leq \pi'_{v_i} \leq \pi'_s - \pi'_t = d^r R_{s,t}^{\mathcal{N}'} \leq d^r R_{s,t}^{\mathcal{N}} = \pi_s - \pi_t \leq \bar{\pi}.$$

Applying Lemma 2 to $\mathcal{N}'$ yields the desired inequality. □

3.3 b-Transshipments

Before we state our main result, we note that every potential-based b-transshipment satisfying global pressure bounds yields, for any subset $X \subseteq T$ of terminal nodes, a valid potential-based flow with sources in X and sinks not in X.

Theorem 4. *Let $\mathcal{N} = (G, \beta, r)$ be a potential-based flow network with balance vector $b \in B(T)$, and f a potential-based b-transshipment with corresponding potentials $\pi \in [0, \bar{\pi}]^V$. Then, for any $X \subseteq T$ with $b(X) \geq 0$, there is a potential-based b'-transshipment f' with corresponding potentials $\pi' \in [0, \bar{\pi}]^V$, and*

$$\{v : b'_v > 0\} \subseteq T^+ \cap X, \quad \{v : b'_v < 0\} \subseteq T^- \setminus X, \quad \sum_{v \in T^+ \cap X} b'_v = b(X). \tag{8}$$

Proof. We assume w.l.o.g. that the arcs in A are oriented such that $f_a \geq 0$ for all $a \in A$. Flow conservation implies that

$$\sum_{a \in \delta^+(S)} f_a \geq b(X) \qquad \text{for each } S \subseteq V \text{ with } X \cap T^+ \subseteq S \subseteq V \setminus (T^- \setminus X). \tag{9}$$

If we interpret the flow value f_a as a capacity of arc $a \in A$, then (9) and the max-flow min-cut theorem imply that there exists a feasible $\hat{b}$-transshipment $\hat{f}$ in G with $0 \leq \hat{f}_a \leq f_a$ for all $a \in A$ and $\hat{b} \in \mathbb{R}^V$ satisfying the exact same conditions stated for b' in (8).

We can now interpret $\hat{f}$ as a potential-based flow with the same node potentials $\pi \in [0, \bar{\pi}]^V$ as f in network $\hat{\mathcal{N}} := (G, \hat{\beta}, r)$ with increased arc resistances $\hat{\beta} \geq \beta$: simply let

$$\hat{\beta}_a := \frac{\pi_v - \pi_w}{\hat{f}_a^{\,r}} \geq \frac{\pi_v - \pi_w}{f_a^{\,r}} = \beta_a \qquad \text{for all } a = (v, w) \in A.$$

In the case that $\hat{f}_a = 0$ for some arc $a \in A$, the resistance $\hat{\beta}_a$ is infinite, which is equivalent to removing the arc from the network.

We further modify the network by adding a super-source node s with arcs from s to all source nodes $v \in V$ with $\hat{b}_v > 0$, as well as a super-sink node t with arcs from all sink nodes $v \in V$ with $\hat{b}_v < 0$ to t. We extend potentials $\pi \in [0, \bar{\pi}]^V$ to potentials on $V \cup \{s, t\}$ by setting $\pi_s := \bar{\pi}$ and $\pi_t := 0$. Moreover, we set the resistances of arcs leaving s and of arcs entering t in such a way that the node potentials π induce an (s, t)-flow of value $b(X)$ in the extended network, i.e.,

$$\hat{\beta}_{(s,v)} := \frac{\pi_s - \pi_v}{\hat{b}_v^r} \qquad \text{for all } v \in V \text{ with } \hat{b}_v > 0, \quad \text{and}$$

$$\hat{\beta}_{(v,t)} := \frac{\pi_v - \pi_t}{-\hat{b}_v^r} \qquad \text{for all } v \in V \text{ with } \hat{b}_v < 0.$$

If the resistance of an arc (s, v) or (v, t) is set to zero, this is equivalent to merging node v with s or t, respectively, as both have the same potential. If we now decrease the arc resistances of arcs $a \in A$ from $\hat{\beta}_a$ to their original values β_a, then, by Prposition 1, there still exists a potential-based (s, t)-flow f' of the same value $b(X)$ with corresponding node potentials $\pi' \in [0, \bar{\pi}]^{V \cup \{s,t\}}$. Restricting f' to the original network, that is, deleting nodes s and t together with their incident arcs, yields the desired b'-transshipment satisfying (8). □

4 Inequalities for the Topology Optimization Problem

After having established several inequalities for potential-based flows with global pressure bound $\bar{\pi}$, we are now returning to the topology optimization problem 2 and present our main result. Given an instance $(\mathcal{N}, r, b, \bar{\pi}, c)$ of (2), we consider a feasible solution (π, f, x). We state a linear inequality in x, which can then directly be added in a spatial branch-and-bound algorithm to solve (2), see also Sect. 5.

Definition 1. *Let $\mathcal{N} = (G, \beta, r)$ be a potential-based flow network. Let $x \in [0, 1]^A$ be given. We call $\mathcal{N}^x := (G^x, \beta^x, r)$ the network induced by x, where we define the graph $G^x := (V, A^x)$ with arc set $A^x := \{a \in A : x_a \neq 0\}$. For the resistance, we set $\beta_a^x := \beta_a / x_a$ for $a \in A^x$ and denote the corresponding conductance with μ^x.*

Note that by the definition of conductance, we obtain:

$$\mu_a^x = \sqrt[r]{x_a}\, \mu_a \quad \text{for } a \in A^x. \tag{10}$$

Remark 1. There is a one-to-one correspondence between a feasible solution to (2) and the potential-based flow in $\mathcal{N}^x$. If (π, f, x) is feasible for (2), then (π, f^x), with $f^x_a = f_a$ for $a \in A^x$, is exactly the potential-based flow induced by b in the network $\mathcal{N}^x$. Conversely, if (π, f') is a potential-based b-transhipment in $\mathcal{N}^x$ with $x \in [0,1]^A$, then (π, f, x), with $f_a = f'_a$ for $a \in A^x$ and $f_a = 0$ otherwise, is feasible for (2) if and only if $\pi_v \in [0, \bar{\pi}]$.

Theorem 5. *Let $(\mathcal{N}, r, b, \bar{\pi}, c)$ be an instance of (2). If (π, f, x) is feasible, then*

$$\frac{1}{k\sqrt[r]{k}} \sum_{i=1}^{k} \sum_{a \in \delta(S_i)} \mu_a \, x_a \geq \frac{b(X)}{\sqrt[r]{\bar{\pi}}} \tag{11}$$

for all subsets $X \subseteq T$ and all k disjoint $(X \cap T^+, T^- \setminus X)$-cuts $S_1 \subseteq \cdots \subseteq S_k \subseteq V$.

Proof. Since all variables are non-negative, the inequality always holds if $b(X)$ is non-positive. We therefore assume, without loss of generality, that $b(X) > 0$.

Consider the network $\mathcal{N}^x$ induced by x from $\mathcal{N}$ (Definition 1). According to Remark 1, (π, f) is a potential-based b-transshipment in $\mathcal{N}^x$, satisfying the given pressure bounds if we ignore the flow over arcs outside the support of x. By Theorem 4, network $\mathcal{N}^x$ admits a potential-based b'-transhipment with corresponding potentials $\pi' \in [0, \bar{\pi}]^V$ such that all source nodes of b' are contained in $T^+ \cap X$ and the sinks contained in $T^- \setminus X$. As shown in the theorem's proof, we can modify the network $\mathcal{N}^x$ by adding a super-source s with arcs to all sources of b' and a super-sink node t with arcs connecting all sinks of b' to t.

The resistances of the arcs added can be chosen so that the potentials π' extend to $V \cup \{s, t\}$. This can be done in such a way that they induce a potential-based (s,t)-flow of value $b(X)$, while still satisfying the given pressure bounds. Note that the arcs crossing the k disjoint (s,t)-cuts $S_1 \cup \{s\} \subseteq \cdots \subseteq S_k \cup \{s\}$ in the modified network are exactly those crossing the $(X \cap T^+, T^- \setminus X)$-cuts $S_1, \ldots, S_k$ in $\mathcal{N}^x$. Applying Theorem 3 together with (10) yields:

$$\frac{1}{k\sqrt[r]{k}} \sum_{i=1}^{k} \sum_{a \in \delta(S_i) \cap A^x} \mu_a \sqrt[r]{x_a} \geq \frac{d}{\sqrt[r]{\bar{\pi}}}, \tag{12}$$

To conclude the proof, note that $\sqrt[r]{x_a}$ can be replaced by $x_a \in \{0, 1\}$. □

Inequality (12) still holds for non-binary $x \in [0,1]^A$. However, expression (11) is stronger if $r > 1$.

Theorem 6. *Inequalities of type (11) can be separated for given $x \in [0,1]^A$ in polynomial time.*

Proof. To simplify notation, we say that $S_1, \ldots, S_k \in 2^V$ are *disjoint X-cuts* for a set $X \subseteq T$ if $S_1 \subseteq \cdots \subseteq S_k$ are k disjoint $(X \cap T^+, T^- \setminus X)$-cuts. We show that a set $X \subseteq T$ minimizing the following expression

$$\frac{1}{k\sqrt[r]{k}} \sum_{i=1}^{k} \sum_{a \in \delta(S_i)} \mu_a x_a - \frac{b(X)}{\sqrt[r]{\bar{\pi}}} \tag{13}$$

for all k disjoint X-cuts can be found in polynomial time. We define the family $\mathcal{X}_k := \{X \subseteq T : \text{there are disjoint } X\text{-cuts } S_1, \ldots, S_k \in 2^V\}$ for every $k \in \mathbb{N}$.

In the full version of this paper, we show that $\mathcal{X}_k \subseteq 2^T$ is closed under set union and intersection. On $\mathcal{X}_k$ we then define the function $\sigma_k\colon \mathcal{X}_k \to \mathbb{R}_{\geq 0}$ which maps $X \in \mathcal{X}_k$ to the value

$$\sigma_k(X) := \min\left\{\frac{1}{k\sqrt[r]{k}} \sum_{i=1}^{k} \sum_{a \in \delta(S_i)} \mu_a x_a \,:\, \text{disjoint } X\text{-cuts } S_1, \ldots, S_k \subseteq V\right\}.$$

Claim 1. The set function σ_k is submodular.

Proof (of Claim 1). Let $X, Y \in \mathcal{X}_k$, and $S_1^X, \ldots, S_k^X$ and $S_1^Y, \ldots, S_k^Y$ the respective minimal, k disjoint X-cuts and Y-cuts. For the submodularity of σ_k, we first use the definition of σ_k and then the submodularity of the cut function, i.e., Inequality (3). The final step follows from the fact that $S_1^X \cup S_1^Y, \ldots, S_k^X \cup S_k^Y$ are disjoint $X \cup Y$-cuts and $S_1^X \cap S_1^Y, \ldots, S_k^X \cap S_k^Y$ disjoint $X \cap Y$-cuts. Both statements follow from the proof of the closedness of $\mathcal{X}_k$. Formally,

$$\begin{aligned}
\sigma_k(X) + \sigma_k(Y) &= \frac{1}{k\sqrt[r]{k}} \sum_{i=1}^{k} \left(\sum_{a \in \delta(S_i^X)} \mu_a x_a + \sum_{a \in \delta(S_i^Y)} \mu_a x_a\right) \\
&\geq \frac{1}{k\sqrt[r]{k}} \sum_{i=1}^{k} \left(\sum_{a \in \delta(S_i^X \cup S_i^Y)} \mu_a x_a + \sum_{a \in \delta(S_i^X \cap S_i^Y)} \mu_a x_a\right) \\
&\geq \sigma_k(X \cup Y) + \sigma_k(X \cap Y).
\end{aligned}$$

◁

Claim 2. The set function σ_k can be evaluated in polynomial time.

Proof (of Claim 2). We show that we can decide if $X \in \mathcal{X}_k$ and, if this is the case, also evaluate $\sigma_k(X)$ in polynomial time. We consider the underlying undirected graph $G = (V, E, w)$ of the given network $\mathcal{N}$ with edge weights given by $w_e = \mu_e x_e \geq 0$. We then add nodes s and t and connect s to all nodes in $X \cap T^+$ and t to all nodes in $T^- \setminus X$ where we set the weight of all these edges to infinity. We solve the k disjoint (s,t)-cut problem by solving an LP formulation stated in the full version of this paper to obtain the value $k\sqrt[r]{k}\,\sigma_k(X)$. ◁

Note that b as a function on 2^T is modular. This implies together with Claim 1 that the function $g_k\colon \mathcal{X}_k \to \mathbb{R}_{\geq 0}$, $X \mapsto \sigma_k(X) - b(X)/\sqrt[r]{\pi}$ is submodular. Thus, the expression (13) can be minimized by computing the minimum of g_k for every $k \in \mathbb{N}$. There exists at most $|V| - 1$ disjoint $(X \cap T^+, T^- \setminus X)$-cuts. □

5 Computational Results

We performed computational experiments with the derived inequalities on gas networks with $r = 2$, concentrating on the single source, single sink case. The code was implemented in C and uses SCIP 9.1.1 [7], which performs linear programming based spatial branch-and-bound. We use CPLEX 12.10 to solve the LPs and IPOPT 3.14.17 [30] for the NLPs. We include the acyclicity constraints from [14]. The computations were obtained single-threaded in exclusive mode on a Linux cluster with Intel Xeon E5-1620 CPUs running at 3.5 Ghz with 32 GB memory. Our results are shown in Table 1; TL refers to reaching the time limit of 8 h.

Table 1. Computational results with (11) for the (s, t)-case

instance	(11)	$\overline{\pi}$	time (s)	# nodes	gap %	k
Path27	yes	70	0.7	1	0	26
Path27	no	70	21789	2056431	0	–
Path27	yes	100	TL	1321823	8.6	26
Path27	no	100	TL	837490	40.9	–
Path27	yes	200	6758	1701253	0	26
Path27	no	200	TL	1700438	0.45	–
Path52	yes	98	0.1	1	0	51
Path52	no	98	TL	539108	94	–
GasLib40	yes	100	18	453	0	12
GasLib40	no	100	17	854	0	–
GasLib40large	yes	100	19735	2058996	0	12
GasLib40large	no	100	280	8484	0	–
GasLib134	yes	70	TL	469862	41.5	40
GasLib134	no	70	TL	386046	114.7	–

We tested the impact of the inequality (11) on several instances with different parameters for the maximum pressure and the cut depth with the (s, t)-balance d being fixed. We created path instances Path27 and Path52 consisting of 27 and 52 nodes respectively. Between each pair of nodes there are three different pipe options to be built. Furthermore, we introduced a super source and sink to GasLib40 and GasLib134 to use them for our tests. Moreover, GasLib40large contains more diameter options, whereas for GasLib40 only the original arcs can be built. We considered the GasLib instances as passive instances by deleting active elements, like valves and switches, and by considering compressors as normal pipes. The costs $c \in \mathbb{R}^A_{\geq 0}$ depend on the diameter and length of the arc and are computed according to Reuß et al. [25].

The results align with the structure of the inequalities, meaning that (11) is especially strong if the length k of the disjoint cuts and the balance d are high as well as the maximum pressure $\bar{\pi}$ is low. We found that the maximum possible value for k performed best on path instances, which is why we do not include results with lower k. On the other instances, usually a higher k performed better. For PATH27 and PATH52 with the tested settings, the inequality describes the integral hull resulting in the instances being solved at the root node. Overall, we conclude that the addition of the inequalities boosts performance. It is especially helpful on the large GASLIB134 and PATH52, where the gap is huge without it. Note that for GASLIB40LARGE the version with the inequalities performs worse because an optimal solution is found very late. In future experiments, we will investigate the behavior of primal heuristics and adapting k to the instance.

Acknowledgments. This work is funded by the DFG via the SFB TRR 154 subprojects A01, A007, and A09.

Disclosure of Interests.. The authors have no competing interests to declare that are relevant to the content of this article.

References

1. Àlvarez-Miranda, E., Cacchiani, V., Lodi, A., Parriani, T., Schmidt, D.R.: Single-commodity robust network design problem: complexity, instances and heuristic solutions. Eur. J. Oper. Res. **238**, 711–723 (2014). https://doi.org/10.1016/j.ejor.2014.04.023
2. André, J., Bonnans, F., Cornibert, L.: Optimization of capacity expansion planning for gas transportation networks. Eur. J. Oper. Res. **197**(3), 1019–1027 (2009). https://doi.org/10.1016/j.ejor.2007.12.045
3. André, J., Bonnans, J.F.: Optimal structure of gas transmission trunklines. Optim. Eng. **12**(1), 175–198 (2011). https://doi.org/10.1007/s11081-009-9094-2
4. Babonneau, F., Nesterov, Y., Vial, J.P.: Design and operations of gas transmission networks. Oper. Res. **60**(1), 34–47 (2012). https://doi.org/10.1287/opre.1110.1001
5. Farias de Barros, H., Alvarez-Herault, M.C., Raison, B., Tran, Q.T.: Optimal AC/DC distribution systems expansion planning from DSO's perspective considering topological constraints. IEEE Trans. Power Deliv. **38**(5), 3417–3428 (2023). https://doi.org/10.1109/TPWRD.2023.3277089
6. Birkhoff, G., Diaz, J.B.: Non-linear network problems. Quart. Appl. Math. **13**(4), 431–443 (1956)
7. Bolusani, S., et al.: The SCIP Optimization Suite 9.0. Technical report, Optimization Online (2024). https://optimization-online.org/2024/02/the-scip-optimization-suite-9-0/
8. Borraz-Sánchez, C., Bent, R., Backhaus, S., Hijazi, H., Hentenryck, P.V.: Convex relaxations for gas expansion planning. INFORMS J. Comput. **28**(4), 645–656 (2016). https://doi.org/10.1287/ijoc.2016.0697
9. Bragalli, C., D'Ambrosio, C., L.J., Lodi, A., Toth, P.: On the optimal design of water distribution networks: a practical MINLP approach. Optim Eng. **13**, 219–246 (2012). https://doi.org/10.1007/s11081-011-9141-7

10. Buchheim, C., Liers, F., Sanità, L.: An exact algorithm for robust network design. In: Pahl, J., Reiners, T., Voß, S. (eds.) INOC 2011. LNCS, vol. 6701, pp. 7–17. Springer, Heidelberg (2011). https://doi.org/10.1007/978-3-642-21527-8_2
11. Gomory, R.E., Hu, T.C.: An application of generalized linear programming to network flows. J. Soc. Ind. Appl. Math. **10**, 260–283 (1962)
12. Groß, M., Pfetsch, M.E., Schewe, L., Schmidt, M., Skutella, M.: Algorithmic results for potential-based flows: easy and hard cases. Networks **73**(3), 306–324 (2019). https://doi.org/10.1002/net.21865
13. Gusfield, D.: Simple constructions for the multi-terminal network flow synthesis. SIAM J. Comput. **12**, 157–165 (1983). https://doi.org/10.1137/0212010
14. Habeck, O., Pfetsch, M.E.: Combinatorial acyclicity models for potential-based flows. Networks **79**(1), 83–104 (2021). https://doi.org/10.1002/net.22038
15. Hansen, C.T., Madsen, K., Nielsen, H.B.: Optimization of pipe networks. Math. Program. **52**(1–3), 45–58 (1991). https://doi.org/10.1007/BF01582879
16. Humpola, J.: Gas Network Optimization by MINLP. Ph.D. thesis, TU Berlin (2014). https://doi.org/10.14279/depositonce-4255
17. Humpola, J., Fügenschuh, A., Koch, T.: Valid inequalities for the topology optimization problem in gas network design. OR Spect. **38**(3), 597–631 (2015). https://doi.org/10.1007/s00291-015-0390-2
18. Humpola, J., Serrano, F.: Sufficient pruning conditions for MINLP in gas network design. EURO J. Comput. Optim. (3), 239–261 (2016). https://doi.org/10.1007/s13675-016-0077-8
19. Klimm, M., Pfetsch, M.E., Raber, R., Skutella, M.: Reduction of potential-based flow networks. Math. Oper. Res. **48**(4), 2287–2303 (2023)
20. Lehmann, K., Grastien, A., Van Hentenryck, P.: The Complexity of DC-Switching Problems. Preprint 1411.4369, arXiv (2014). https://arxiv.org/abs/1411.4369
21. Li, Y., Dey, S.S., Sahinidis, N.V.: A reformulation-enumeration MINLP algorithm for gas network design. J. Glob. Optim. **90**(4), 931–963 (2024). https://doi.org/10.1007/s10898-024-01424-x
22. Li, Y., Dey, S.S., Sahinidis, N.V., Susarla, N., Zamarripa, M.A., Drouven, M.G.: Optimizing the design and operation of water networks: Two decomposition approaches. Comput. Chem. Eng. **192**, 108897 (2025). https://doi.org/10.1016/j.compchemeng.2024.108897
23. McCormick, S.T.: Submodular function minimization. In: Aardal, K., Nemhauser, G., Weismantel, R. (eds.) Discrete Optimization, Handbooks in Operations Research and Management Science, vol. 12, pp. 321–391. Elsevier (2005). https://doi.org/10.1016/S0927-0507(05)12007-6
24. Raber, R.: Optimization, reduction, and robustness of potential-based flow networks. Dissertation, Technische Universität Berlin (2022)
25. Reuß, M.: Modeling hydrogen networks for future energy systems: A comparison of linear and nonlinear approaches. Int. J. Hydrog. Energy **44**(60), 32136–32150 (2019). https://doi.org/10.1016/j.ijhydene.2019.10.080
26. Rockafellar, R.T.: Network Flows and Monotropic Optimization. Wiley-Interscience (1984)
27. Schweiger, J., Liers, F.: A decomposition approach for optimal gas network extension with a finite set of demand scenarios. Optim. Eng. **19**(2), 297–326 (2018). https://doi.org/10.1007/s11081-017-9371-4
28. Talluri, K.T.: Network synthesis with few edges. Networks **27**, 109–115 (1996). https://doi.org/10.1002/(SICI)1097-0037(199603)27:2⟨109::AID-NET2⟩3.0.CO;2-O

29. Thürauf, J., Grübel, J., Schmidt, M.: Adjustable robust nonlinear network design under demand uncertainties. Preprint, Optimization Online (2024). https://optimization-online.org/?p=26035
30. Wächter, A., Biegler, L.T.: On the implementation of an interior-point filter line-search algorithm for large-scale nonlinear programming. Math. Program. **106**(1), 25–57 (2006). https://doi.org/10.1007/s10107-004-0559-y
31. Wagner, D.K.: Disjoint (s, t)-cuts in a network. Networks **20**(4), 361–371 (1990). https://doi.org/10.1002/net.3230200402
32. de Wolf, D., Smeers, Y.: Optimal dimensioning of pipe networks with application to gas transmission networks. Oper. Res. **44**(4), 596–608 (1996). https://doi.org/10.1287/opre.44.4.596

The Differentiable Feasibility Pump

Matteo Cacciola[1], Alexandre Forel[1(✉)], Antonio Frangioni[2], and Andrea Lodi[3]

[1] Polytechnique Montréal, Montréal, Canada
{matteo.cacciola,alexandre.forel}@polymtl.ca
[2] Dipartimento di Informatica, Universita di Pisa, Pisa, Italy
frangio@di.unipi.it
[3] Jacobs Technion-Cornell Institute, Cornell Tech and Technion–IIT, New York, USA
andrea.lodi@cornell.edu

Abstract. Although nearly 20 years have passed since its conception, the feasibility pump algorithm remains a widely used heuristic to find feasible primal solutions to mixed-integer linear problems. Many extensions of the initial algorithm have been proposed. Yet, its core algorithm remains centered around two key steps: solving the linear relaxation of the original problem to obtain a solution that respects the constraints, and rounding it to obtain an integer solution. This paper shows that the traditional feasibility pump and many of its follow-ups can be seen as gradient-descent algorithms with specific parameters. A central aspect of this reinterpretation is observing that the traditional algorithm differentiates the solution of the linear relaxation with respect to its cost. This reinterpretation opens many opportunities for improving the performance of the original algorithm. We study how to modify the gradient-update step as well as extending its loss function. We perform extensive experiments on MIPLIB instances and show that these modifications can substantially reduce the number of iterations needed to find a solution.

Keywords: mixed-integer programming · differentiable optimization

1 Introduction

Despite significant improvements in open-source and commercial solvers in recent years, solving large-scale mixed-integer linear optimization problems remains a computational challenge. Primal heuristics, which return a solution that satisfies all constraints of the problem, are a cornerstone of the overall solving process. Among them, the feasibility pump, originally proposed by [24], remains a popular and efficient method.

The feasibility pump is based on two main operations: (i) solving the linear relaxation of the original problem to obtain a solution that is not necessarily integer but respects the linear constraints, and (ii) rounding the solution to respect the integrality constraints, at the risk of violating the linear constraints. The algorithm iterates over these two steps, changing the objective of the problem at

N. Megow and A. Basu (Eds.): IPCO 2025, LNCS 15620, pp. 157–171, 2025.
https://doi.org/10.1007/978-3-031-93112-3_12

each iteration. When it gets stuck in a cycle, restart operations are performed to perturb the objective and resume the algorithm.

In this paper, we propose a novel interpretation of the feasibility pump. We show that the traditional algorithm can be seen as a gradient-descent with a specific loss function, gradient estimation, and gradient update steps. This perspective builds upon the recent works on differentiating through optimization models and integrating them as layers into deep learning [3]. Differentiable optimization has seen extensive applications in structured learning and contextual optimization [31,35].

This reinterpretation allows us to derive a new generalized feasibility pump. Because of the central role of differentiable optimization, we call this generalization the *differentiable feasibility pump*. It extends the original algorithm in multiple ways. First, it allows simple modifications of the gradient-update parameters as well as applying more advanced gradient-descent algorithms. Second, it allows adapting the loss function to take advantage of the gradient-descent setting. Specifically, we show that a piecewise convex loss function leads to better performance than the original one, which is piecewise linear. Third, it allows extending the loss function by adding differentiable loss terms. In this paper, we introduce a loss term to penalize the infeasibility of the rounded solution.

Related Works. Several extensions have been proposed since the work of [24]. While the original algorithm was proposed for binary mixed-integer linear problems, it was quickly extended to general linear mixed-integer problems [7] and non-linear ones (see, e.g., [6,15,21,33]). We present below several extensions and refer an interested reader to [9] and [10] for an extensive overview.

Generalizations. [13] interpret the feasibility pump as a proximal point algorithm. [18] and [17] cast feasibility pumps as a Frank-Wolfe method and introduce concave penalty functions for measuring solution integrality. [28] frame the feasibility pump as an alternating direction method applied to a specific reformulation of the original problem. They then study modifications of the restarts to escape cycles. [19] focus on the restart components and develop extensions based on the WalkSAT local-search algorithm.

Objective-Informed. A limitation of the original algorithm is that it has no explicit component to emphasize the quality of the solution in terms of cost. Ideally, a primal heuristic would return a feasible solution that also has a small cost for minimization problems. [1] introduce an objective component to guide the pump towards feasible solutions with a low cost. Recently, [32] introduce objective scaling terms.

Constraint-Informed. A second often-discussed limitation is that the two key steps of the feasibility pump are performed "blind" to each other. That is, the integrality constraints are ignored when solving the linear relaxation, and the linear constraints are ignored when rounding the relaxed solution. Several approaches have been proposed to remedy this issue, mostly focusing on making the rounding component aware of the problem's constraints. [25] investigate a modified rounding operation to encourage constraint propagation. [4] modify the rounding step to round on the segment between the computed feasible point and the analytic center of the relaxed problem. Other modifications include introducing an integer line search [14], with a barrier term to take into account active constraints [26], or multiple reference vectors [32].

Contributions. In this paper, we make the following contributions:

(i) We reinterpret the feasibility pump algorithm as a gradient-descent with specific loss function, gradient estimation, and gradient-update parameters.
(ii) We present the differentiable feasibility pump, a generalized feasibility pump algorithm based on gradient descent, and show how this general form encompasses many existing algorithms.
(iii) We investigate several variations of the differentiable feasibility pump. In particular, we introduce a differentiable rounding operation that allows us to include a loss term to measure the infeasibility of the rounded solution. This provides a new solution to the "blindness" of the feasibility pump.
(iv) We show the value of the differentiable pump in extensive experiments on MIPLIB instances. Our experiments show that the original feasibility pump is remarkably stable with respect to its hyperparameters. However, introducing a feasibility loss together with changes in the hyper-parameters can significantly improve performance.

2 Background

We consider binary mixed-integer linear optimization problems of the form

$$\min_{x} \left\{ c^\top x \mid Ax \geq b,\, x \in \{0,1\}^n \right\}, \tag{1}$$

where $A \in \mathbb{R}^{m\times n}$, $b \in \mathbb{R}^m$, and $c \in \mathbb{R}^n$ is the cost vector. We denote the feasible region of Problem (1) as $\mathcal{X} = \{x \in \{0,1\}^n \mid Ax \geq b\}$ and the feasible region of the relaxed problem as $\hat{\mathcal{X}} = \{x \in [0,1]^n \mid Ax \geq b\}$. We say that an element of $\hat{\mathcal{X}}$ is feasible and an element of $\{0,1\}^n$ is integer. A solution of Problem (1) is both feasible and integer. While continuous variables can be easily included in Problem (1), we do not consider them to simplify our presentation.

2.1 Feasibility Pump

The main goal of the feasibility pump is to find a feasible solution of Problem (1). To achieve this, it alternates between solving the linear relaxation of the problem for a given objective, and rounding its solution until a solution that is both feasible and integer is found. This approach is motivated by the fact that the extreme points of $\mathcal{X}$ are also extreme points of $\hat{\mathcal{X}}$ when Problem (1) has only binary variables. Therefore, there always exists a cost vector $\bar{\theta}$ such that $\bar{x} \in \operatorname{argmin}\left\{\bar{\theta}^\top x \mid x \in \hat{\mathcal{X}}\right\}$ is a feasible integer solution, provided that $\mathcal{X} \neq \varnothing$.

Algorithm. We denote by $x^{(k)}$ a solution obtained at the k-th iteration, and by x_i the i-th component of solution x. The algorithm starts by solving the linear relaxation of Problem (1) to obtain the feasible solution $\hat{x}^{(0)}$. If $\hat{x}^{(0)}$ is integer, it belongs to $\mathcal{X}$ and the algorithm terminates. Otherwise, the integer solution $\lfloor x \rceil^{(0)}$ is computed by rounding x^* component-wise to the closest integer. If $\lfloor x \rceil^{(0)}$ is feasible then it belongs to $\mathcal{X}$ and, again, the algorithm terminates. In any iteration, if the rounded solution $\lfloor x \rceil^{(k)}$ is not feasible, the algorithm projects it onto the feasible domain $\hat{\mathcal{X}}$ and finds the feasible point $\hat{x}^{(k+1)}$. If $\hat{x}^{(k+1)}$ is integral the algorithm terminates, otherwise $\lfloor x \rceil^{(k+1)}$ is computed as the rounding of $\hat{x}^{(k+1)}$. If $\lfloor x \rceil^{(k)}$ is feasible the algorithm terminates, otherwise another iteration is performed.

Distance Function. Projecting the integer solution onto the feasible domain requires a distance measure. The feasibility pump uses the ℓ_1 norm, also called Hamming distance. It is defined as $\Delta(x, \lfloor x \rceil) = \sum_i |x_i - \lfloor x \rceil_i|$. Given a fixed vector $\lfloor x \rceil$, $\Delta(x, \lfloor x \rceil)$ is an affine function in x and, since $\lfloor x \rceil$ is a binary vector, it can be computed as

$$\Delta(x, \lfloor x \rceil) = \sum_{i:\lfloor x \rceil_i = 1} (1 - x_i) + \sum_{i:\lfloor x \rceil_i = 0} x_i = \sum_{i=1}^{n} \lfloor x \rceil_i - \sum_{i:\lfloor x \rceil_i = 1} x_i + \sum_{i:\lfloor x \rceil_i = 0} x_i. \quad (2)$$

Hence, in iteration k, the feasible solution that is closest to $\lfloor x \rceil^{(k-1)}$ is found by solving the linear program given by $\min\{\Delta(x, \lfloor x \rceil^{(k-1)}) \mid x \in \hat{\mathcal{X}}\}$ or, equivalently, by solving $\min\{\theta^\top x \mid x \in \hat{\mathcal{X}}\}$, where $\theta \in \mathbb{R}^n$ is such that $\theta_i = -1$ if $\lfloor x \rceil_i^{(k-1)} = 1$, and $\theta_i = 1$ otherwise. This shows that the feasibility pump solves a sequence of linear problems that differ only in their cost parameters.

Restarts. A limitation of the feasibility pump is that it can get stuck in cycles when the objective given by Eq. (2) has already been obtained in the past. Several strategies have been introduced to escape these cycles. The first strategy of [24] is to flip a random number of entries of the rounded solution $\lfloor x \rceil$ at the end of an iteration when a cycle of size one is detected. The second is to apply random perturbations to $\lfloor x \rceil$ when a cycle of size larger than two is detected. These restart strategies, called *flip* and *perturbation* respectively, are not the focus of this paper. To simplify our presentation, we do not discuss them in this paper and assume thereafter that our feasibility pump algorithms implement them exactly as the original one.

2.2 Differentiable Optimization

This work is motivated by the novel perspective brought by recent works on differentiable optimization. This stream studies how to integrate mixed-integer linear optimization models as differential layers in structured learning [3,11,12] and contextual stochastic optimization [22,35]. This allows training a machine-learning pipeline to predict a set of parameters that respect the constraints of a mixed-integer linear problem. Further, the machine-learning model can be trained in an end-to-end fashion: minimizing the cost of a downstream optimization task (see, e.g., [20,22,23,39]. To achieve this, a meaningful gradient must be propagated from the loss back to the machine-learning model. This can be done in several ways. Early works focus on quadratic programs and use an implicit differentiation approach based on the Karush–Kuhn–Tucker conditions of optimality [2,3].

While several types of optimization models have been considered, the most studied form is the linear Problem (1) parameterized by its cost coefficients. Let $x^*(\theta)$ be a solution of the combinatorial linear problem $\min\{\theta^\top x \mid x \in \mathcal{X}\}$ parameterized by θ. A key challenge is that the gradient of the argmin of a (combinatorial) linear program with respect to its cost vector is non-informative. Formally, the Jacobian $J_\theta x^*(\theta)$ is equal to 0 almost everywhere since the solution $x^*(\theta)$ is constant for small variations of θ, and undefined as the solution jumps from one vertex to another for a large enough change. Differentiable optimization studies how to obtain surrogate gradients.

A common technique is to consider a surrogate optimization model and using its gradient in place of the ones of the original model. The perturbation method of [8] regularizes the optimization model by perturbing its cost vector [16]. With an additive perturbation, the solution of the regularized problem is given by

$$x^*_\varepsilon(\theta) = \mathbb{E}_Z \left[\operatorname*{argmin}_{x \in \mathcal{X}} (\theta + \varepsilon Z)^\top x \right],$$

where $\varepsilon > 0$ is a scaling parameter and Z follows an exponential distribution with positive differentiable density. The Jacobian of $x^*_\varepsilon(\theta)$ is then given by

$$J_\theta x^*_\varepsilon(\theta) = 1/\varepsilon \, \mathbb{E}_Z \left[x^*_\varepsilon(\theta + Z) Z^\top \right]. \tag{3}$$

This gradient can be evaluated using a Monte Carlo approximation, by sampling perturbation vectors and solving the corresponding optimization problems. Other methods to obtain a differentiable surrogate optimization model include adding a quadratic regularization term [39], using a linear interpolation between solutions [38], or linking gradient computations with interior point methods [30].

A limitation of the perturbed optimizers is that they need to solve several problems for each gradient computation, which can be computationally expensive. An efficient alternative is to take minus the identity matrix as surrogate Jacobian [27,36], i.e.,

$$J_\theta x^*(\theta) = -I_n, \tag{4}$$

with I_n the identity matrix of size n.

3 Differentiable Feasibility Pump

Differentiable optimization underlies our reinterpretation of the feasibility pump algorithm as a gradient-descent algorithm. Further, it is based on two key observations: (i) the feasibility pump solves a sequence of problems that differ only in their cost parameters, and (ii) the feasibility pump focuses precisely on the class of problems studied in differentiable optimization.

3.1 Regularized Gradient-Descent Interpretation

We now present the differentiable feasibility pump, an interpretation of the feasibility pump as a regularized gradient-descent algorithm. This algorithm minimizes a loss function $\mathcal{L} : \mathbb{R}^n \to \mathbb{R}$ over the cost vector $\theta \in \mathbb{R}^n$. First, we present the differentiable feasibility pump algorithm shown in Algorithm 1, where $\eta > 0$ is the step size. Second, we show how a restricted version of this generalized pump recovers the original feasibility pump.

Algorithm 1. Differentiable feasibility pump

```
Initialize: θ^(0) = c, k = 0
while k < N_max do
    Compute the feasible solution x̂^(k) = argmin{θ^(k)ᵀ x : x ∈ X̃}
    Round it to obtain the integer solution ⌊x̂⌉^(k)
    if ⌊x̂⌉^(k) is feasible then
        Return ⌊x̂⌉^(k)
    else
        Update the cost vector θ^(k+1) = θ^(k) − η∇_θ L(θ)
    end if
end while
```

Integrality Loss. The first success condition of the original feasibility pump is finding a cost vector $\bar{\theta}$ such that $\hat{x}(\bar{\theta}) = \operatorname{argmin}_x\{\bar{\theta}^\top x : x \in \hat{\mathcal{X}}\}$ is integer. We introduce the loss function $f : \mathbb{R}^n \to \mathbb{R}$ to measure the non-integrality of a feasible solution $\hat{x}$. In this work, we consider the class of integrality losses

$$f(x) = \sum_{i=1}^{n} \left(\min\left\{x_i, 1 - x_i\right\}\right)^p, \tag{5}$$

where $p > 0$ is the loss order. Based on the value of p, the loss f has local convexity/concavity properties. Formally, consider the partition of the n-dimensional hypercube $\mathcal{D} = \left\{\prod_{j=1}^{n} D_j \mid D_j \in \{[0, 0.5), [0.5, 1]\}\right\}$. The restriction of $f(x)$ to any element of $\mathcal{D}$ is concave if $p \leq 1$ and convex if $p \geq 1$.

Regularized Loss. Consider now the regularized loss function given by

$$\mathcal{L}(\theta) = f(\hat{x}(\theta)) + \gamma \frac{\|\theta\|_2^2}{2}, \tag{6}$$

where $\gamma > 0$ is the regularization weight. The second component of this loss adds a regularization term that controls the magnitude of the cost vector θ. Minimizing over this regularized loss is equivalent to minimizing the integrality loss f with a constraint on the norm of θ. Notice that the norm of θ does not influence the solution of the linear relaxation. That is $\hat{x}(\theta) = \hat{x}(\theta/r)$ for any scalar $r > 0$, with a slight abuse of notation in case the argmin is set-valued.

Gradient Update. Using this regularized loss, the differentiable feasibility pump minimizes the optimization problem $\min_\theta\{\mathcal{L}(\theta) \mid \theta \in \mathbb{R}^n\}$. Solving this problem with a gradient-descent algorithm requires computing the gradient of the loss $\mathcal{L}$ with respect to θ, that is

$$\nabla_\theta \mathcal{L}(\theta) = \nabla_\theta \left(f(\hat{x}(\theta)) + \gamma \frac{\|\theta\|_2^2}{2} \right), \tag{7}$$

and updating the cost vector θ as $\theta \Longleftarrow \theta - \eta \nabla_\theta \mathcal{L}(\theta)$. Using the chain rule, the gradient of the loss can be computed as

$$\nabla_\theta \mathcal{L}(\theta) = (J_\theta \hat{x}(\theta))^\top \nabla_{\hat{x}(\theta)} f(\hat{x}(\theta)) + \gamma\theta. \tag{8}$$

As discussed in Sect. 2.2, several techniques can be applied to obtain a surrogate Jacobian $J_\theta \hat{x}(\theta)$.

3.2 Equivalence with the Original Algorithm

We now show that the original feasibility pump is a special case of the differentiable pump with specific parameters, loss function, and gradient estimation.

Proposition 1. *When* $f(x) = \sum_{i=1}^n \min\{x_i, 1 - x_i\}$, $\eta = 1$, $\gamma = 1$, *and the Jacobian of the* argmin *is computed as* $J_\theta \hat{x}(\theta) = -I_n$, *Algorithm 1 generate the same sequence of solutions* $\{\hat{x}^{(k)}\}$ *as the original feasibility pump algorithm.*

Proof. Consider the regularized loss given in Eq. (6) with integrality loss as in Eq. (5) with $p = 1$. The gradient update is now given by

$$\theta \Longleftarrow \theta - \eta \nabla_\theta \left(f(\hat{x}(\theta)) \right) - \eta\gamma\theta.$$

Choosing a specific set of parameters such that $\eta\gamma = 1$, for example $\eta = 1$ and $\gamma = 1$, the gradient update becomes $\theta \Longleftarrow -\nabla_\theta \left(f(\hat{x}(\theta)) \right)$. The updated cost vector is now equal to minus the gradient of the integrality loss. Using the chain rule, we obtain

$$\theta \Longleftarrow -(J_\theta \hat{x}(\theta))^\top \cdot \nabla_{\hat{x}(\theta)} f(\hat{x}(\theta)), \tag{9}$$

which simplifies to $\theta \Longleftarrow \nabla_{\hat{x}(\theta)} f(\hat{x}(\theta))$ when taking the Jacobian as minus the identity matrix as in Eq. (4). The integrality loss with $p = 1$ is piecewise linear. Its gradient is given by

$$\nabla_{\hat{x}(\theta)} f(\hat{x}(\theta)) = \begin{cases} -1 & \text{if } \hat{x}(\theta) > 0.5, \\ 1 & \text{otherwise.} \end{cases} \tag{10}$$

In Sect. 2.1, we noticed that the original pump solves a sequence of problems $\operatorname{argmin}\{\theta^\top x \mid x \in \hat{\mathcal{X}}\}$, where $\theta_i = -1$ if $\lfloor x \rceil_i = 1$ and $\theta_i = 1$ otherwise. Hence, the sequences of problems solved by the two algorithms are equivalent. □

Remark 1. For conciseness, we do not discuss the cycle detection and restarts operations in the proof of Proposition 1. The cycle detection can be kept the same as the original algorithm since the same information is available. It is direct to define the restart operations by flipping or perturbing θ rather than $\lfloor x \rceil$.

Proposition 1 shows that the feasibility pump can be seen as a gradient descent with a very specific set of parameters. In particular, the algorithm parameters are such that $\eta\gamma = 1$. An interesting consequence of these parameters is that the new cost vector obtained following the gradient update depends only on the gradient of the loss.

Novel Perspective. The differentiable feasibility pump interpretation opens many possibilities to alter, extend, and improve the algorithm. For instance, without any change to the algorithm, it allows tuning the main hyper-parameters such as the step size η and the regularization weight γ.

Further, it allows reassessing the loss function chosen to measure the non-integrality of a feasible solution. The loss $f(x) = \sum_i \min\{x_i, 1 - x_i\}$ (i.e., with order $p = 1$) is piecewise linear. Therefore, it has a piecewise constant gradient. This means that the gradient of the integrality loss is independent of the magnitude of the integrality loss. This might lead to an unstable behavior. Intuitively, we would like the gradient of the integrality loss to be small when the loss is small, and large when the loss is large. This can be achieved by using a convex integrality loss, for instance by taking $p > 1$. Finally, it is notable that the Jacobian surrogate $J_\theta \hat{x}(\theta) = -I_n$ is used implicitly even though it was formalized around twenty years after the feasibility pump algorithm.

4 Extending the Differentiable Pump

In this section, we extend the differentiable pump beyond a simple reinterpretation. We present a generalized loss function and introduce a novel loss component to take into account the infeasibility of the rounded solution.

4.1 General Loss Function

We consider the general loss given by

$$\mathcal{L}(\theta) = \alpha C(\theta) + \beta f(\hat{x}(\theta)) + \lambda g(\lfloor \hat{x}(\theta) \rceil) + \gamma \Omega(\theta), \tag{11}$$

where $(\alpha, \beta, \lambda, \gamma)$ is a vector of positive weights. This general loss has four components. The first component C measures the cost of the solutions of the feasibility pump with respect to the objective of the problem. It is motivated by the idea that a feasibility pump should provide a primal solution that is not only feasible but also of good quality, i.e., with a low cost [1]. It can be taken for instance as $C(\theta) = c^\top \hat{x}(\theta)$, $C(\theta) = c^\top \lfloor \hat{x}(\theta) \rceil$, or a convex combination of them. The second component f aims to minimize the non-integrality of the solution. Hence, it depends only on the feasible solution $\hat{x}(\theta)$. The third component g is a novel loss function that depends only on the rounded solution $\lfloor \hat{x}(\theta) \rceil$. It minimizes the infeasibility of the rounded solution, thus mirroring the previous term. Finally, the fourth term is the regularization of the cost vector. The remainder of this section presents a feasibility loss and discusses how to adapt the rounding operation to propagate a meaningful gradient.

4.2 Feasibility Loss

The main novelty of this generalized loss is taking into account the infeasibility of the rounded solution $\lfloor \hat{x}(\theta) \rceil$. As previously discussed, most existing approaches only consider the non-integrality of $\hat{x}(\theta)$, and the update of the cost vector at each iteration is agnostic to the feasibility of the rounded solution. A notable exception is the approach of [26] that encourages the satisfaction of the constraints by modifying the integrality loss to take into account the slack and active constraints of the feasible solution.

Again, many loss functions could be used to measure the infeasibility of the rounded solution. A suitable loss function should be: (i) equal to zero when all constraints are satisfied, (ii) easily differentiable, and (iii) not heavily impacted by the number of constraints or magnitude of the constraint matrices. To achieve this, we measure the infeasibility of the rounded solution element-wise, normalize the infeasibility by the constraint coefficients, and average over the constraints.

Let $s_j(\tilde{x}) : \{0,1\}^n \to \mathbb{R}$ measure the slackness of the j-th constraint as $s_j(\tilde{x}) = (b_j - A_j\tilde{x})/\|[A_j\, b_j]\|_2$, where A_j is the j−th row of the constraint matrix, b_j is the j−th right-hand side coefficient, and $[A_j\, b_j]$ is their concatenation. We define $g_j : \mathcal{X} \to \mathbb{R}$ to measure the infeasibility of the j-th constraint as

$$g_j(\tilde{x}) = \max(s_j(\tilde{x}) - \epsilon, 0) = \begin{cases} s_j(\tilde{x}), & \text{if } \tilde{x} \text{ violates the } j\text{-th constraint,} \\ 0, & \text{otherwise,} \end{cases}$$

where $\epsilon > 0$ is a small constant introduced to avoid numerical instability. This function is known in the machine-learning community as the rectified linear unit (ReLU). Hence, it can be compactly rewritten as $g_j(\tilde{x}) = \text{ReLU}(s_j(\tilde{x}) - \epsilon)$. Finally, we average the infeasibility over the constraints by taking

$$g(\tilde{x}) = \frac{\sum_{j=1}^{m} g_j(\tilde{x})}{m}. \tag{12}$$

4.3 Differentiable Rounding

The last step needed to obtain a fully differentiable feasibility pump algorithm is to find a surrogate gradient for the rounding operation. Here, we apply again techniques of differentiable optimization. Using perturbation, we show that this gradient can be expressed in closed-form and is especially efficient to compute.

Rounding is applied element-wise. Hence, we consider that x is single dimensional. Our approach is based on the observation that rounding can be seen as a linear program, i.e., $\lfloor x \rceil$ is the solution of the optimization problem given by

$$\max_{y}\{(x-0.5)^\top y \mid y \in [0,1]\}. \tag{13}$$

This problem depends only on x in its objective coefficients. Thus, it fits the class of problems studied extensively in the differentiable optimization literature. The rounding problem can be differentiated by applying the techniques discussed in Sect. 2.2 such as taking the identity matrix as Jacobian. A more informative gradient can be obtained using perturbation.

Define the perturbed rounding operation as $\lfloor x \rceil_\varepsilon = \mathbb{E}_Z[\lfloor x + \varepsilon Z \rceil]$, where Z is a random variable with positive and differentiable density on $\mathbb{R}$. Using (3), the Jacobian of the perturbed rounded solution can be computed as

$$J_x \lfloor x \rceil_\varepsilon = \frac{1}{\varepsilon} \mathbb{E}_Z[Z \mid x + \varepsilon Z > 0.5] \, \mathbb{P}(x + \varepsilon Z > 0.5). \tag{14}$$

This expression is obtained by Bayes's law. The expectation is also known as the inverse Mills ratio. The gradient of the perturbed rounded solution can be obtained in closed-form as

$$J_x \lfloor x \rceil_\varepsilon = \frac{1}{\varepsilon}\phi\left(\frac{0.5-x}{\varepsilon}\right), \tag{15}$$

with ϕ the probability density function (p.d.f.) of the standard Gaussian. Remarkably, we do not need to evaluate the expectation in Eq. (14) using a Monte Carlo approximation: it is sufficient to compute the p.d.f. of a standard Gaussian, which is very efficient. Hence, introducing a feasibility loss based on perturbed rounding does not increase the complexity of the feasibility pump. The computational bottleneck remains solving the relaxed problem $\hat{x}(\theta)$.

Remark 2. This expression can be also obtained by observing that the perturbed rounding can be equivalently expressed as $\lfloor x \rceil_\varepsilon = \mathbb{E}_Z[\mathbb{I}(x + \varepsilon Z \geq 0.5)]$ so that $\lfloor x \rceil_\varepsilon = \Phi_Z(\frac{x-0.5}{\varepsilon})$ with Φ_Z the cumulative distribution function of Z. This expression holds for any perturbation distribution in the exponential family. Its gradient can thus be directly deduced.

We call this differentiable rounding operator a *soft* rounding, akin to softmax being a differentiable surrogate to the max operator. Other soft operators commonly used in machine learning include the differentiable top-k layer [40] or differentiable sorting [34]. Soft rounding is also closely linked to straight-through estimators [5].

5 Numerical Study

In this section, we investigate the performance of the differentiable feasibility pump. We show that substantial performance improvements can be obtained when modifying the parameters of the original algorithm. Next, we investigate the value of the feasibility loss introduced in Sect. 4.

We run experiments on a set of 125 instances from MIPLIB 2017 [29]. We select instances with continuous or binary variables and exclude exceptionally large instances. All linear problems are solved using the commercial solver `Gurobi`. To ensure consistent and reliable results, we tune the solver's parameters to use a single thread with the smallest possible tolerance for optimality and feasibility. The differentiable pump is implemented in `python` using the `pytorch` package for automatic differentiation. The differentiable optimization models are adapted from the `PyEPO` package of [37]. The code used to run all the experiments in this paper is openly accessible at https://github.com/MCacciola94/DiffPump.

5.1 Experimental Setting

We investigate several variants of the differentiable feasibility pump, with increasing flexibility compared to the original algorithm. All methods use the general loss function given in Eq. (11) with a specific set of weights (β, λ, γ) and step size η. We do not consider the cost component of the loss (i.e., we set $\alpha = 0$) in order to better compare the results of the differentiable pump with the ones of the original algorithm.

The integrality loss f is taken as in Equation (5) with p denoting its order. The feasibility loss g is taken as in Eq. (12). The regularization is given by $\Omega(\theta) = \|\theta\|^2/2$. All methods use minus the identity matrix as a surrogate for the Jacobian of the argmin of the linear relaxation, and all variants using the feasibility loss use soft rounding with $\varepsilon = 0.15$. All methods use the same cycle detection method, restarts, and stopping criteria. Finally, all variants use a limit of $N_{\max} = 1000$ iterations. If a pump cannot find a feasible solution within the iteration limit, the experiment is considered failed.

The original algorithm is denoted as `FP`. We investigate four variants of the differentiable feasibility pump with increasing flexibility compared to `FP`. The first variant of the differentiable pump is denoted by `DP1`. It changes the weight of the regularization loss γ and the step size η. The second variant `DP2` additionally changes the order p of the integrality loss. The third variant `DP3` uses only the feasibility loss with soft rounding as described in Sect. 4.3. The last variant `DP4` combines both loss functions with positive weights β and λ respectively, and is allowed to vary all hyperparameters. For each method, the hyperparameters are tuned using a grid search.

We consider three main performance metrics: (i) the number of instances for which the algorithm did not find a solution within the limited number of iterations, (ii) the total number of iterations of the algorithm over all instances, and (iii) the percentage of iterations in which a restart operation was performed.

For instance, a ratio of 50% means that the algorithm performed a perturbation or flip operation every two iterations on average.

5.2 Main Results

Our main experimental results are presented in Table 1. The table shows the number of fails, the total number of iterations, and the restart ratio for all feasibility pump variants. It shows that several variants of the differentiable feasibility pump achieve better performance than the original algorithm.

Table 1. Number of fails, iterations, and randomization ratio over all instances.

	FP	DP1	DP2	DP3	DP4
Number of fails	28	28	21	21	20
Total number of iterations	30071	30037	24084	25684	23383
Restart ratio	43.71%	42.79%	24.61%	3.9%	25.98%

A first observation is that simply changing the gradient-update step size η and the weight of the regularization loss γ has minimal impact: the performance of DP1 stays very close to the one of the original pump. The fact that changing the values of the parameters does not bring any significant improvement is remarkable. It suggests that the parameters of the original algorithm are a local optimum.

However, Table 1 shows that changing the order of the integrality loss to $p = 2$ and jointly adapting the step size and regularization weight leads to significant improvements both in the number of iterations and the number of instances solved. Indeed, DP2 solves 7 more instances than FP and uses more than 5000 fewer iterations, a reduction of around 20%. Further, it leads to fewer flipping and restarts, having almost half the restart ratio of the original algorithm. This suggests that the new loss paired with $\eta\gamma \neq 1$ make the algorithm less likely to get stuck in a cycle.

The third variant, DP3, also outperforms the original algorithm FP. Table 1 shows that it achieves almost the same performance of DP2 but is substantially more stable. Indeed, its restart ratio is one order of magnitude smaller than that of FP. DP3 behaves more closely to a pure gradient-descent algorithm as it only rarely uses the heuristic restart operations. This suggests that the loss landscape induced by the feasibility loss has fewer local minima.

Finally, Table 1 shows that DP4, which uses a combination of feasibility loss and integrality loss can provide a further improvement, although it is not substantial. The algorithm can find the feasible solution of one more instance and slightly reduce the total number of iterations. Interestingly, its restart ratio is similar to the one of DP2.

6 Conclusion

This paper reinterprets the feasibility pump algorithm of [24] as a gradient-descent algorithm, building upon recent works on differentiable optimization. This interpretation generalizes many existing feasibility pump algorithms and allows novel extensions in terms of designing a general loss function or gradient-update algorithms. For instance, our experiments show that using a locally convex integrality loss or a differentiable feasibility loss based on soft rounding can provide substantial performance improvements.

Acknowledgement. The authors thank Youssouf Emine and David E. Bernal Neira for the momentous discussions.

References

1. Achterberg, T., Berthold, T.: Improving the feasibility pump. Disc. Optim. **4**(1), 77–86 (2007)
2. Amos, B.: Differentiable optimization-based modeling for machine learning. Ph. D. thesis (2019)
3. Amos, B., Kolter, J.Z.: OptNet: differentiable optimization as a layer in neural networks. In: International Conference on Machine Learning, pp. 136–145. PMLR (2017)
4. Baena, D., Castro, J.: Using the analytic center in the feasibility pump. Oper. Res. Lett. **39**(5), 310–317 (2011)
5. Bengio, Y., Léonard, N., Courville, A.: Estimating or propagating gradients through stochastic neurons for conditional computation. arXiv preprint arXiv:1308.3432 (2013)
6. Bernal, D.E., Vigerske, S., Trespalacios, F., Grossmann, I.E.: Improving the performance of DICOPT in convex MINLP problems using a feasibility pump. Optim. Methods Softw. **35**(1), 171–190 (2020)
7. Bertacco, L., Fischetti, M., Lodi, A.: A feasibility pump heuristic for general mixed-integer problems. Disc. Optim. **4**(1), 63–76 (2007)
8. Berthet, Q., Blondel, M., Teboul, O., Cuturi, M., Vert, J.P., Bach, F.: Learning with differentiable perturbed optimizers. In: Advances in Neural Information Processing Systems, vol. 33, pp. 9508–9519. Curran Associates, Inc. (2020)
9. Berthold, T., Lodi, A., Salvagnin, D.: Ten years of feasibility pump, and counting. EURO J. Comput. Optim. **7**(1), 1–14 (2019)
10. Berthold, T., Lodi, A., Salvagnin, D.: Primal Heuristics in Integer Programming. Cambridge University Press, Cambridge (2025)
11. Blondel, M., Martins, A.F., Niculae, V.: Learning with Fenchel-Young losses. J. Mach. Learn. Res. **21**(35), 1–69 (2020)
12. Blondel, M., Roulet, V.: The elements of differentiable programming. arXiv preprint arXiv:2403.14606 (2024)
13. Boland, N.L., Eberhard, A.C., Engineer, F., Tsoukalas, A.: A new approach to the feasibility pump in mixed integer programming. SIAM J. Optim. **22**(3), 831–861 (2012)

14. Boland, N.L., Eberhard, A.C., Engineer, F.G., Fischetti, M., Savelsbergh, M., Tsoukalas, A.: Boosting the feasibility pump. Math. Program. Comput. **6**(3), 255–279 (2014). https://doi.org/10.1007/s12532-014-0068-9
15. Bonami, P., Cornuéjols, G., Lodi, A., Margot, F.: A feasibility pump for mixed integer nonlinear programs. Math. Program. **119**(2), 331–352 (2009)
16. Dalle, G., Baty, L., Bouvier, L., Parmentier, A.: Learning with combinatorial optimization layers: a probabilistic approach. arXiv preprint arXiv:2207.13513 (2022)
17. De Santis, M., Lucidi, S., Rinaldi, F.: A new class of functions for measuring solution integrality in the feasibility pump approach. SIAM J. Optim. **23**(3), 1575–1606 (2013)
18. De Santis, M., Lucidi, S., Rinaldi, F.: Feasibility pump-like heuristics for mixed integer problems. Disc. Appl. Math. **165**, 152–167 (2014)
19. Dey, S.S., Iroume, A., Molinaro, M., Salvagnin, D.: Improving the randomization step in feasibility pump. SIAM J. Optim. **28**(1), 355–378 (2018)
20. Donti, P., Amos, B., Kolter, J.Z.: Task-based end-to-end model learning in stochastic optimization. In: Advances in Neural Information Processing Systems, vol. 30 (2017)
21. D'Ambrosio, C., Frangioni, A., Liberti, L., Lodi, A.: A storm of feasibility pumps for nonconvex MINLP. Math. Program. **136**, 375–402 (2012)
22. Elmachtoub, A.N., Grigas, P.: Smart "predict, then optimize." Manag. Sci. **68**(1), 9–26 (2022)
23. Ferber, A., Wilder, B., Dilkina, B., Tambe, M.: MIPaaL: mixed integer program as a layer. In: Proceedings of the AAAI Conference on Artificial Intelligence, vol. 34, no. 02, pp. 1504–1511 (2020)
24. Fischetti, M., Glover, F., Lodi, A.: The feasibility pump. Math. Program. **104**, 91–104 (2005)
25. Fischetti, M., Salvagnin, D.: Feasibility pump 2.0. Math. Program. Comput. **1**(2-3), 201–222 (2009)
26. Frangioni, A., Pan, S., Traversi, E., Calvo, R.W.: A constraints-aware reweighted feasibility pump approach. Oper. Res. Lett. **49**(5), 671–675 (2021)
27. Fung, S.W., Heaton, H., Li, Q., Mckenzie, D., Osher, S., Yin, W.: JFB: Jacobian-free backpropagation for implicit networks. In: Proceedings of the AAAI Conference on Artificial Intelligence, vol. 36, no. 66, pp. 6648–6656 (2022)
28. Geissler, B., Morsi, A., Schewe, L., Schmidt, M.: Penalty alternating direction methods for mixed-integer optimization: a new view on feasibility pumps. SIAM J. Optim. **27**(3), 1611–1636 (2017)
29. Gleixner, A., et al.: MIPLIB 2017: Data-driven compilation of the 6th mixed-integer programming library. Math. Program. Comput. (2021)
30. Mandi, J., Guns, T.: Interior point solving for LP-based prediction+optimisation. Adv. Neural. Inf. Process. Syst. **33**, 7272–7282 (2020)
31. Mandi, J., et al.: Decision-focused learning: foundations, state of the art, benchmark and future opportunities. J. Artif. Intell. Res. **80**, 1623–1701 (2024)
32. Mexi, G., Berthold, T., Salvagnin, D.: Using multiple reference vectors and objective scaling in the feasibility pump. EURO J. Comput. Optim. **11**, 100066 (2023)
33. Misener, R., Floudas, C.A.: Global optimization of mixed-integer quadratically-constrained quadratic programs (MIQCQP) through piecewise-linear and edge-concave relaxations. Math. Program. **136**(1), 155–182 (2012)
34. Prillo, S., Eisenschlos, J.: Softsort: a continuous relaxation for the argsort operator. In: International Conference on Machine Learning, pp. 7793–7802. PMLR (2020)

35. Sadana, U., Chenreddy, A., Delage, E., Forel, A., Frejinger, E., Vidal, T.: A survey of contextual optimization methods for decision making under uncertainty. Eur. J. Oper. Res. (2024)
36. Sahoo, S.S., Paulus, A., Vlastelica, M., Musil, V., Kuleshov, V., Martius, G.: Backpropagation through combinatorial algorithms: identity with projection works. In: International Conference on Learning Representations (2023)
37. Tang, B., Khalil, E.B.: PyEPO: a pytorch-based end-to-end predict-then-optimize library for linear and integer programming. arXiv preprint arXiv:2206.14234 (2022)
38. Vlastelica, M., Paulus, A., Musil, V., Martius, G., Rolinek, M.: Differentiation of blackbox combinatorial solvers. In: International Conference on Learning Representations (2019)
39. Wilder, B., Dilkina, B., Tambe, M.: Melding the data-decisions pipeline: Decision-focused learning for combinatorial optimization. In: Proceedings of the AAAI Conference on Artificial Intelligence, vol. 33, no. 01, pp. 1658–1665 (2019)
40. Xie, Y., et al.: Differentiable top-k with optimal transport. Adv. Neural. Inf. Process. Syst. **33**, 20520–20531 (2020)

Approximate Cut & Packing Ratios for Multi-commodity Arborescences

Parinya Chalermsook[1](✉) and Chien-Chung Huang[2]

[1] The University of Sheffield, Sheffield, UK
chalermsook@gmail.com
[2] École normale supérieure, Université PSL, Paris, France

Abstract. Packing arborescences in directed graphs is a central concept in combinatorial optimization. A seminal work of Edmonds (1973) presents a min-max (duality) relation between maximum r-arborescence packing and minimum r-cut (a cut that separates some vertex from the root r). Quite a few researchers (Frank (1978), Cai (1983), and Kamiyama, Katoh, and Takizawa (2009)) have generalized Edmonds' min-max result to characterize packings of arborescences at multiple roots. Nonetheless, these min-max relations, unlike Edmonds' result, cannot be seen as linear programming (LP) duality between cuts and packings.

In this paper, we initiate the study of two min-max relations of multi-commodity arborescence packings based on the LP duality view. We show tight relations between fractional arborescence packings and cuts as well as efficient algorithms that achieve such ratios. Our cut problems are natural and rich in modeling power, e.g., they capture cornerstone optimization problems such as set cover, densest sub (hyper) graphs. Therefore, our algorithmic results generalize and subsume quite a few existing algorithms. Our techniques are based on a simple and illustrative primal-dual analysis, that allows us to reduce the task of upper bounding the min-max ratios for our problems to that of set covering problems. As a bonus, these techniques offer new primal-dual insights to the original result of Edmonds and Kamiyama, Katoh and Takizawa.

1 Introduction

We are given a directed graph $G = (V, E)$ and a specific root $r \in V$. An *r-arborescence* is a subset of edges $E' \subseteq E(G)$ so that E' contains no cycle, and r can reach, using only the edges in E', every vertex reachable from r in G; if an arborescence spans every vertex in $V(G)$ (which happens when r can reach all vertices in $V(G)$), we call it a *strict* r-arborescence.[1] An *r-arborescence packing* is a collection of edge-disjoint r-arborescences $T_1, \ldots, T_M$. The goal of

[1] In most papers, an arborescence would correspond to our notion of strict arborescence. For simplicity of exposition, an arborescence in this paper spans only the vertices reachable from r, instead of all vertices.

Supported by the European Research Council (ERC) under the European Union's Horizon 2020 research and innovation programme (grant agreement No. 759557).

N. Megow and A. Basu (Eds.): IPCO 2025, LNCS 15620, pp. 172–183, 2025.
https://doi.org/10.1007/978-3-031-93112-3_13

the r-arborescence packing problem is to compute a packing that maximizes the value of M (that is, to pack as many arborescences as possible). The question can be generalized/extended to capacitated graphs (where graph G is associated with capacities $c(e)$ on the edges $e \in E$) and fractional arborescence packings (instead of integral ones).[2] A natural upper bound on the value of maximum r-arborescence packing is the size of the r-*cut*: A set of edges $E' \subseteq E$ whose removal disconnects r from some vertex reachable from r. The r-cut problem asks for the r-cut with the minimum capacity. It should be emphasized that r-cut not only upper bounds the value of r-arborescence packing, but also represents the dual problem (in the sense of LP duality) of the latter. In a seminal result [15], Edmonds proves the following mini-max theorem, thus giving a duality characterization of arborescence packings.

Theorem 1. *Suppose that G has integer capacity on all edges. Then there is an integral r-arborescence packing of value C if and only if every r-cut has value at least C.*

Edmonds' pioneering result has led to a large body of follow-up works. We will discuss some of the most relevant results in the related work section. For readers interested in a more systematic interview, we suggest a recent survey by Bérczi and Frank [5].

Generalization to Multiple Roots: The focus of this paper is on packings of arborescences at multiple roots (that we call *commodities*). The study of multi-commodity arborescences perhaps originated from Frank already in 1978 [18] and a follow-up by Cai [35] (see also more recent extensions [6]), where they present necessary and sufficient partitions conditions for packing disjoint arborescences. Note that these results consider *strict* arborescences.

A different approach to multicommodity arborescence packings was initiated by Kamiyama, Katoh and Takizawa (KKT) [26], where they ask for packing of arborescences that only span the reachable vertices.[3] They show a characterization of arborescence packing in terms of a simple and natural cut condition.

The generalization of KKT and Frank-Cai rely on cut/partition conditions to characterize the maximum packing of arborescences. It should be emphasized that, unlike Edmonds' result, they are not based on LP duality: The r-cut and r-arborescence packing problems correspond to a pair of LP primal and dual problems, but the cut/partition conditions in KKT and Frank-Cai are not dual LP problems.

In this work, we initiate the study of primal-dual (cut & packing ratio) perspectives of multicommodity arborescence packing. We define two cut problems

[2] In a fractional packing, each T_i is associated with value $\tau_i \in [0,1]$ such that the capacity is not violated on any edge, that is, $\sum_{i\in[M]:e\in T_i} \tau_i \leq c(e)$. An arborescence packing is *integral* if $\tau_i \in \{0,1\}$ for all i. The value of an arborescence packing is $\sum_{i\in[M]} \tau_i$.

[3] They call this object an *in-tree* instead of an arborescence, while Szigeti called it a *reachability arborescence* [40].

that generalize the original r-cut problem, prove tight bounds on the min-max ratios, present efficient algorithms achieving such ratios, and derive various applications and connections to other fundamental optimization problems. We remark that understanding the primal-dual view can often provide insights that may not be visible in other forms of min-max characterizations and algorithmic results of this type often transfer very well across computational settings (e.g., online algorithms [7] and fast algorithm design [2]).

1.1 Our New Cut Problems

There are two natural ways to define optimization problems that generalize a single commodity arborescence packing to multicommodity ones. For readers familiar with maximum flow literature, such generalizations are analogous to the extension of single-commodity maximum (s,t)-flow to maximum concurrent flow [33] and maximum multicommodity flow [22]. We show that our extensions lead to natural cut problems that capture several well-known problems and applications. Before we start, we should emphasize that in all the following problems, we allow the capacities $c : E \to \mathbb{R}_{\geq 0}$ to take *real* values. This generalization is significant from an algorithmic perspective.

Max-sum: Given a capacitated digraph G together with terminals $\mathcal{R} \subseteq V(G)$, our goal is to find a maximum-cardinality collection of edge-disjoint arborescences whose roots lie in $\mathcal{R}$. This generalization mimics the generalization from maximum (s,t)-flow to maximum multi-commodity flow [22]. We call this problem the *maximum multicommodity arborescence packing (*MCAR*)*.

Next, we define a dual cut problem (determined by the LP duality) whose optimal value upper bounds the above packing problem: Given a capacitated digraph G together with k terminals $\mathcal{R} \subseteq V(G)$, we say a terminal $r \in \mathcal{R}$ is *separated* by edge set $F \subseteq E(G)$ if r cannot reach some of its (previously) reachable vertices using the remaining edge sets $E(G)\backslash F$. Here the objective is to find a minimum-capacity cut $F \subseteq E(G)$ that separates all terminals $r \in \mathcal{R}$. We call this optimization problem *Minimum Directed Multicut with Rooted Demands (*DIRMCR*)*. Observe that the value of a feasible cut provides an upper bound on the value of every packing: given an arborescence packing, $T_1, \ldots, T_M$, every T_i must contain at least one edge in F. Figure 1 shows an illustration of an input instance of our problems.

When $|\mathcal{R}| = 1$ and all vertices can be reached by the unique root, DIRMCR is exactly Edmonds' r-cut problem, and when $\mathcal{R} = V(G)$, the problem can be seen as a counter-part of the global min-cut problem in a undirected graph, i.e., we are interested in a cut $F \subseteq E(G)$ so that every vertex $v \in V(G)$ cannot reach some of its reachable vertices after we remove F from $E(G)$.[4] Moreover, DIRMCR easily

[4] The standard global minimum cut in directed graphs is defined as finding the min-cost set of edges whose removal makes the graph no longer strongly connected. Our way of viewing the global min-cut in the directed graph is thus different, but we believe ours is natural enough in its own right.

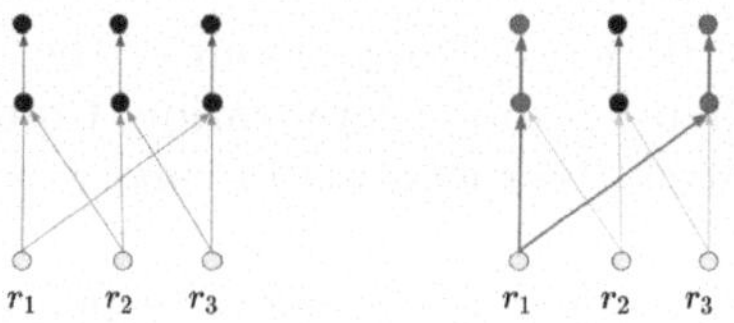

Fig. 1. Assume that all the edges have unit cost. The terminal set is $\mathcal{R} = \{r_1, r_2, r_3\}$. On the right, we show in red the vertices and edges that are reachable from r_1. The optimal solution for DIRMCR has value 2 (cutting two edges on the top row). The optimal integral value for MCAR is 1, while fractionally we can pack 1.5 (assigning 1/2 to each arborescence rooted at r_i). As an instance of DIRSPR (where each root has unit demand), the optimal DIRSPR solution has value 0.5 (by removing an arbitrary edge from the top row, we can disconnect two demands), which is the same as the optimal CONCAR where we can pack 1/2 arborescence at each root r_i.

captures the SETCOVER problem. We are interested in the following analogue of Edmond's result:

> Assume that the optimal DIRMCR is at least C, can we fractionally pack $C/f(k)$ arborescences for some function f? Is there an efficient algorithm for approximating DIRMCR?

A familiar reader may have observed that $f(k)$ here corresponds to the integrality gap of a natural LP relaxation for DIRMCR. As DIRMCR captures SETCOVER as a special case, we cannot expect function f to be asymptotically better than logarithmic. Therefore, the main question is whether $f(k) = \Theta(\log k)$.

Max-min: Instead of maximizing the total number of arborescences, another natural goal is to guarantee a certain sense of fairness between the demands. In particular, we are given a capacitated digraph G together with k roots/terminals $\mathcal{R}$ and the demands $d : \mathcal{R} \to \mathbb{Z}_{\geq 0}$. Our goal is to compute a collection of arborescences $T_1, \ldots, T_M$ so that, for each $r \in \mathcal{R}$, at least $\gamma \cdot d(r)$ such arborescences are rooted at r, in order to maximizing the value of γ, while respecting the capacities. In other words, we aim at maximizing the mininum ratio of the demands that have been satisfied. We call this problem the *maximum concurrent arborescence packing* (CONCAR). This generalization mimics the generalization from maximum (s, t)-flow to the maximum concurrent flow [33] (hence the term concurrent in our problem).

Now we define the corresponding cut problem. Given a capacitated digraph G together with roots $\mathcal{R}$ and demands d, our goal is to find a cut $F \subseteq E(G)$ so as to minimize the ratio between the capacity of F and the total demands separated by F.[5] This problem is called Minimum Directed Sparsest Cut with Rooted Demands (DIRSPR). Besides capturing the Edmonds' r-cut problem

[5] That is, let $\mathcal{R}_F = \{r \in \mathcal{R} : r \text{ is separated by removing } F\}$. Then, we minimize $c(F)/d(\mathcal{R}_F)$.

(when $\mathcal{R} = \{r\}$), the DIRSPR problem generalizes the Densest Subgraph Problem (DSP) as well as its hypergraph generalization. Computing dense subgraphs or subhypergraphs has been a cornerstone in theory [8] and practice [23] (see, e.g., the surveys [31,32]).

We are interested in the following question (again, an answer to which would generalize Edmonds' theorem):

> Assume that the optimal DIRSPR is at least C, can we fractionally pack $C/g(k)$ arborescences for some function g? Is there an efficient algorithm for computing the optimal DIRSPR value?

As DSP has integrality gap equal to one [8], we do not have any prior knowledge about g for this problem.

1.2 Our Results and Techniques

In this paper, we provide tight answers to all these questions.

Theorem 2. *There exists a fractional concurrent arborescence packing of value C if and only if the value of minimum directed sparsest cut with rooted demands is at least C. Moreover, there exists a polynomial-time algorithm that solves* DIRSPR *optimally.*

Theorem 3. *There exists a fractional multicommodity arborescence packing of value $C/H(k)$ (where $k = |\mathcal{R}|$ and $H(k) = \sum_{i\in[k]}(1/i)$) if the optimal value of minimum directed multicut with rooted demands is C. Moreover, there exists a polynomial-time $H(k)$-approximation algorithm for* DIRMCR.

To prove these theorems, we start by observing that the non-constructive proofs (i.e., they do not lead to efficient algorithms for DIRSPR and DIRMCR) can be obtained from the result of Kamiyama, Katoh, and Takizawa [26] in a black-box fashion. We briefly summarize their result and explain why it does not lead to efficient algorithms for our cut problems. We say that the **flow condition** is satisfied if, for all $S \subseteq V$, and for every v that is reachable from all terminals in S, the value of maximum flow from S to v is at least $d(S)$.

Theorem 4. (KKT, informal and rephrased). *Given a graph G, terminals $\mathcal{R}$ and demands δ on $\mathcal{R}$, if the flow condition is satisfied, then there exist edge-disjoint arborescences that "realize" the demand δ (i.e., at least $\delta(v)$ arboresences are rooted at v for all $v \in \mathcal{R}$) that can be computed in time $\Delta \cdot poly(n)$ where $\Delta = \sum_{v\in\mathcal{R}} \delta(v)$.*

To prove Theorem 2, assume the DIRSPR on the instance $(G, \mathcal{R}, d)$ has value at least C. It is easy to verify that the flow condition is satisfied on $(G, \mathcal{R}, C \cdot d)$, which allows us to invoke KKT to say that an arborescence packing exists, hence proving the theorem. This argument does not lead to an efficient algorithm because the total demands in d (and hence the value of Δ) can be exponentially large. With Theorem 2, a folklore argument now can be used to derive Theorem 3:

In particular, any α-approximation for DIRSPR can be turned into $\alpha \cdot H(k)$ approximation for DIRMCR (similarly to the reduction from multicut to sparsest cut.)

We emphasize that the above argument only leads to pseudo-polynomial time algorithms (assuming that the capacities c are integers or rationals). One might think that the whole issue is just how to turn the proof of KKT into a more efficient algorithm. However, this seems a challenging task. The original proof of KKT (generalized from the proof of Lovász [34] on Edmonds' theorem) proceeds by constructing the arborescences one by one carefully, in order to preserve the cut condition. It seems difficult to construct the required arborescences in a "batch" fashion. Let us mention two more known proofs of KKT, and unfortunately, they suffer the same issue for our purpose. Hörsch and Szigeti [24] gave a significantly simpler proof of KKT, using the stronger form of Edmonds' packing theorem. Their proof only works on unit-demand termianls. To apply their proof, one needs to create Δ auxiliary terminal vertices. Another result by Király, Szigeti, and Tanigawa [30] shows that the packing of edge-disjoint arborescences of KKT can be formulated as a matroid intersection problem and they provide polyhedral descriptions for all packings promised by KKT. This allows them to even find a "min-weight" packing of KKT.[6] This approach relies on defining certain matroids over "uncapacited" edge sets. To apply their approach with a capacitated graph, again we need to make copies of the edges, making the algorithm inefficient.

Our constructive proofs take a totally different approach, based on developing new tools exploiting LP-duality. We proceed by decomposing the optimal fractional cut into a combination of integral "structured" cuts, with which we associate a (variant of) SETCOVER instance. This approach essentially turns the problem of bounding the min-max ratio into that of bounding the SETCOVER integrality gap. An additional minor advantage of this second approach is that when SETCOVER has smaller integrality gap, the ratio in our Theorem 3 gets improved as well (e.g., while $H(2) > 1$, the set cover integrality gap is 1 when there are two elements).

To give a pre-taste of our technique, in Sect. 2, we give an LP-rounding proof that a fractional arborescence packing of value C exists when $|\mathcal{R}| = 1$, where C is optimal value of DIRMCR; in particular, this is a fractional variant of Theorem 1. The alternative proof of Gabow and Manu [20] involves a primal-dual schema to construct a fractional packing of r-arboresences and a r-cut with the same objective, while ours is by simply rounding the optimal fractional r-cut directly. We believe our proof is significantly simpler. We note that this rounding technique also leads to a polynomial time algorithm to solve DIRMCR when $|\mathcal{R}| = 2$, and show that there is a fractional packing of value C, when $|\mathcal{R}| = 2$. At the end of Sect. 2, we then explain, on a high-level, how to generalize this LP-rounding idea for general $\mathcal{R}$ of any cardinality.

[6] In fact, the result of [30] holds for a further generalization of KKT, where extra matroid constraints are imposed.

Application 1: Modeling Known Optimization Problems. We illustrate the modeling power of our cut problems. The key idea is that our cut problems can encode AND and OR constraints as well as a natural combination of these constraints. Therefore, we can capture optimization problems whose corresponding constraint satisfaction formula is somewhat "simple". To illustrate this idea, given an OR formula $\mathsf{OR}(x_1, \dots, x_k) = \vee_{i \in [k]} x_i$, we can encode the formula as a directed graph G and $r \in V(G)$ so that there is a one-to-one correspondence between finite-cost r-cut solutions in G_ϕ and Boolean assignments to variables x_i. The idea is shown in Fig. 2. The AND formula $\mathsf{AND}(x_1, \dots, x_k) = \bigwedge_{i \in [k]} x_i$ can be encoded similarly.

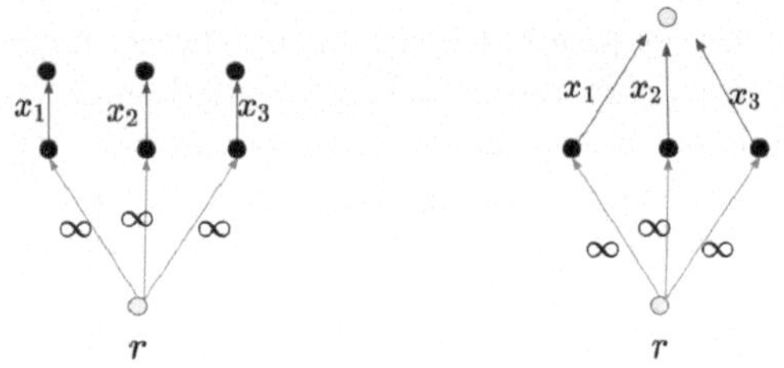

Fig. 2. Gadgets for encoding OR and AND constraints. On the left-hand-side, in order to separate the root r, at least one of the edges corresponding to x_1, x_2, x_3 must be cut; intuitively choosing the edges x_1 and x_2 corresponds to the assignment $x_1 = 1, x_2 = 1, x_3 = 0$. So this gadget represents the constraint $(x_1 \vee x_2 \vee x_3)$ in the sense that r is separated by a solution if and only if the OR formula is satisfied. Similarly, the gadget on the right-hand-side represents the AND constraint $(x_1 \wedge x_2 \wedge x_3)$.

Once we have the gadgets for the AND and OR constraints, it is a relatively simple task to show how our problem captures an optimization problem.

- **Set Cover**: We are given the ground set U, a collection of sets $\mathcal{F} \subseteq 2^U$, and the weight function $w : \mathcal{F} \to \mathbb{R}_{\geq 0}$. The goal is to choose a minimum-weight subset $\mathcal{F}' \subseteq \mathcal{F}$ so as to cover all the ground elements, i.e., $U \subseteq \bigcup \mathcal{F}'$. We have variables $\{x_F\}_{F \in \mathcal{F}}$ and write, for each $u \in U$, a constraint $\phi_u(\mathbf{x}) = \bigvee_{u \in F} x_F$. Since we want to satisfy all constraints, we can capture the problem using DIRMCR by creating a root for each u and gluing the gadgets together.
- **Densest Subgraphs**: We are given a graph $H = (V, E)$, and the objective is to choose a subset $S \subseteq V$ so as to maximize the ratio $\frac{|E(H[S])|}{|S|}$, i.e., the ratio between the number of edges and vertices in the induced subgraph. This ratio is the *density* of set S, and we want to find the set with largest density. A CSP formulation is as follows: We have variables $\{x_u\}_{u \in V}$. For each edge $e = (u, v)$ we have constraint $\psi_e(\mathbf{x}) = (x_u \wedge x_v)$. Notice that, in this CSP, we would like to minimize the ratio between the support $|\{u : x_u = 1\}|$ and the number of satisfied constraints.

Application 2: LP Rounding Proofs for Existing and New Results

Theorem 5. *For* $|\mathcal{R}| \leq 2$, *there exists of a packing of value* C *for* MCAR *if and only if the optimal value of* DIRMCR *is* C.

As mentioned earlier, when $|\mathcal{R}| = 1$, this gives the fractional counterpart of Edmonds' theorem. When $|\mathcal{R}| = 2$, our result is new. In particular, it is a neat parallel to Hu's "max-biflow-min-cut" theorem [25][7]. For arbitrary $\mathcal{R}$, our techniques can derive a fractional counterpart of KKT. We state the result formally here. We are given an input graph G and demands $d : V(G) \to \mathbb{R}_{\geq 0}$.

Theorem 6. (G, d) *admits a perfect (fractional) packing (where at least* $d(v)$ *arborescences are rooted at* v*) if and only if the flow condition is satisfied.*

Notice that by continuity argument, the above theorem also follows from KKT. We nonetheless provide an independent proof.

Application 3: Generalized Nash-Williams and Tutte and Parameterization. As discussed above, the DIRMCR problem can be seen as a certain interpretation of global minimum cut. In undirected graphs, a classical theorem on tree packing given by Nash-Williams [37] and Tutte [42] implies that there exists an integral tree packing in undirected graph G of value $C/2$ where C is the global minimum cut. We prove the following (fractional and directed) analogue of this theorem, which gives integrality gap bound in terms of the number of strongly-connected components.

Theorem 7. *There exists a (fractional) arborescence packing of value* $C/2$ *in a strongly connected graph where* C *is the optimal value of* DIRMCR. *More generally, in any digraph* G, *there exists a packing of value* $\frac{C}{2\mathsf{scc}(G)}$.

When $\mathsf{scc}(G) = 1$, this theorem subsumes a fractional directed analogue of the Nash-Williams-Tutte theorem (see also another proof [10]).

Further Related Works: Edmonds' result on packing arborescences has motivated a host of follow-up works. His original proof is simplified [34] and his algorithm is made more efficient [20,41]. In terms of generalization, besides those of Frank [18], Cai [35], and KKT [26] already mentioned, researchers have considered imposing matroid constraints [13,17,27] and the mixed graphs [21,36]. Regarding KKT, right after [26], Fujishige [19] considered a further generalization, where he asks for packing of arborescences that span certain subsets of reachable vertices from given terminals. Király [29] gave a simpler proof of [19].

Set cover admits $\ln n$ approximation algorithm and is NP-hard to approximate to within a factor of $(1 - o(1)) \ln n$ [14,16]. Integrality gap of the densest subgraph problem has been shown in [8], extended in [4], and generalized in many

[7] Hu's theorem states that in an undirected capacitated graph with terminal pairs $\{s_i, t_i\}_{i=1}^{2}$. The maximum total flow from s_i to t_i, $i \in \{1, 2\}$ is equivalent to the minimum cut separating s_i and t_i, $i \in \{1, 2\}$, simultaneously.

directions. Recently, Chekuri et al. [9] considers a generalization, that they call the *densest supermodular subset* problem. The integrality gap questions for flow and cut problems have received a lot of attention. For the directed multicut problem, the integrality gap is at most $O(n^{11/23})$ [1] (improving upon the previous works [1,11]) and at least $\Omega(n^{1/7})$ [12]. When the number of terminals $k = O(\log n)$, the tight gap of $k - o(1)$ is shown by Saks et al. [38]. Flow-cut gaps in undirected graphs have been studied extensively (e.g., [3,22,28,33]).

2 Overview of Techniques

As a warm-up, we show that, for integer C, the value of r-cut is at least C iff we can fractionally pack C arborescences. We are given digraph $G = (V, E)$, capacity function $c : E \to \mathbb{R}_{\geq 0}$ and a root vertex $r \in V$. Assume that every vertex in V is reachable from r. Denote by $\mathcal{T}$ be the set of all r-arborescences. We write the following primal-dual LPs.

$$\begin{aligned} \text{(CUT) min} & \sum_{e \in E(G)} c(e)x(e) \\ \text{s.t.} & \sum_{e \in E(T)} x(e) \geq 1 \ \forall T \in \mathcal{T} \\ & x(e) \geq 0 \text{ for all } e \in E \end{aligned} \qquad \begin{aligned} \text{(PACK) max} & \sum_{T \in \mathcal{T}} y(T) \\ \text{s.t.} & \sum_{T : e \in T} y(T) \leq c(e) \forall e \in E(G) \\ & y(T) \geq 0 \text{ for all } T \in \mathcal{T} \end{aligned}$$

When $x(e) \in \{0, 1\}$, the value of $x(e)$ tentatively indicates whether an edge e is cut or not. Each constraint $\sum_{e \in E(T)} x(e) \geq 1$ ensures that the cut solution x removes at least one edge from T. By strong duality, the optimal objective values of these two LPs are equal. Denote by λ^* the optimal objective value (so there exists a fractional arborescence packing of value λ^*). Our goal is to show that there exists a feasible r-cut solution F such that $c(F) = \lambda^*$, thus implying the fractional counterpart of Edmonds' theorem.

Let x^* and y^* be optimal solutions to (CUT) and (PACK) respectively. Let $\mathsf{supp}(y^*) = \{T : y^*(T) > 0\}$ denote the set of arborescences in the support of y^*. Our algorithm and analysis rely heavily on the complementary slackness property, which we recall here: (i) If $x^*(e) > 0$, then $\sum_{T : e \in E(T)} y^*(T) = c(e)$; (ii) If $y^*(T) > 0$, then $x^*(E(T)) = \sum_{e \in E(T)} x^*(e) = 1$.

Our algorithm will find a cut $Z \subseteq V$ with two properties: (P1) All edges going into this subset have positive LP values, i.e., for all $e \in F = \delta^{in}(Z)$, we have $x^*(e) > 0$, and (P2) every arborescence T in the support intersects F in exactly one edge, i.e., $|T \cap F| = 1$ for all $T : y^*(T) > 0$. We call the set Z that has these properties an *excellent* set. We claim that if such a subset exists, we are immediately done: Notice that removing F disconnects r from vertices in Z, so F is a feasible r-cut, and that $\sum_{e \in F} c(e) = \sum_{e \in F} \sum_{T : e \in E(T)} y^*(T) = \sum_{T \in \mathsf{supp}(y^*)} y^*(T)|T \cap F| = \sum_{T \in \mathsf{supp}(y^*)} y^*(T) = \lambda^*$, where the first equality follows from complementary slackness and (P1) and the third from (P2).

Finding an Excellent Set: Now we show how to find an excellent set. Denote by E_0 the set of edges e with $x^*(e) = 0$ (i.e., the non-support edges). From the properties of excellent sets, the edges in E_0 should not enter the set Z. For any arborescence T, we use the notation T_v to denote the subtree of T rooted at v and, for $E' \subseteq E(G)$ and $U \subseteq V(G)$, $E'[U]$, to denote the subset of edges of E' both of whose endpoints are in U. For instance, $E_0[V(T_v^*)]$ consists of the non-support edges that are induced by the vertices in T_v^*

Algorithm:

1. Let T^* be a minimum x^*-weight arborescence, so $x^*(E(T^*)) = 1$.
2. Let $e = uv \in E(T^*)$ be the edge farthest from r such that $x^*(e) > 0$.
3. Compute strongly connected components in $(V(T_v^*), E_0[V(T_v^*)])$. Define Z to be the component that contains v.

It is well-known that we can choose any vertex v in a strongly connected component S and build a v-arborescence in S (see, e.g. [39].)

To argue (P1), suppose $x^*(f) = 0$ for some $f = ab \in F = \delta^{in}(Z)$ (so $b \in Z$ and $a \notin Z$). Notice that the unique path in T^* from r to a does not intersect Z: Otherwise, we would have $Z \cup \{a\} \subseteq V(T_v^*)$ and therefore there is a path from v to a using only non-support edges, and vice versa, which would have implied that a and Z are in the same strongly connected component (a contradiction). We modify T^* into another r-arborescence T' as follows. Firstly, we remove all edges of $E(T^*) \cap F$ and then add edge f. Next we remove all edges of T^* in Z and add an b-arborescence in Z that uses only edges in E_0. It is easy to verify that T' is still an r-arborescence and $x^*(E(T')) < x^*(E(T^*)) = 1$ (we remove at least one support edge and add only edges in E_0), a contradiction.

To argue (P2), consider any arborescence $T \in \mathsf{supp}(y^*)$. By the complementary slackness, $T \in \mathsf{supp}(y^*)$ implies $x^*(T) = 1$. Suppose $|T \cap F| \geq 2$. Let $e = ab \in T \cap F$ be the edge such that a is closest to the root r in the arborescence T. Therefore the path from r to a in T does not contain any vertex in Z. We modify T in almost the same way as before: Remove all edges of T into Z except for edge $e = ab$, and remove all edges of T in Z then add a b-arborescence in Z using only edges in E_0. This would give us an r-arborescence T' such that $x^*(E(T')) < x^*(E(T)) = 1$, again a contradiction to the feasibility of x^*.

References

1. Agarwal, A., Alon, N., Charikar, M.S.: Improved approximation for directed cut problems. In: STOC (2007)
2. Arora, S., Hazan, E., Kale, S.: The multiplicative weights update method: a meta-algorithm and applications. ToC **8**, 121–164 (2012)
3. Arora, S., Rao, S., Vazirani, U.: Expander flows, geometric embeddings and graph partitioning. JACM **56**, 1–37 (2009)
4. Balalau, O.D., Bonchi, F., Chan, T.H., Gullo, F., Sozio, M.: Finding subgraphs with maximum total density and limited overlap. In: WSDM (2015)
5. Bérczi, K., Frank, A.: Packing arborescences. Technical report, Egerváry Research Group. TR-2009-04 (2009)

6. Bérczi, K., Frank, A.: Supermodularity in unweighted graph optimization i: Branchings and matchings. Math, OR (2018)
7. Buchbinder, N., Naor, J.: Online primal-dual algorithms for covering and packing. Math, OR (2009)
8. Charikar, M.: Greedy approximation algorithms for finding dense components in a graph. In: APPROX (2000)
9. Chekuri, C., Quanrud, K., Torres, M.R.: Densest subgraph: supermodularity, iterative peeling, and flow. In: SODA (2022)
10. Chekuri, C., Quanrud, K., Chao, X.: LP relaxation and tree packing for minimum k-cut. SIDMA **34**, 1334–1353 (2020)
11. Cheriyan, J., Karloff, H., Rabani, Y.: Approximating directed multicuts. In: FOCS (2001)
12. Chuzhoy, J., Khanna, S.: Polynomial flow-cut gaps and hardness of directed cut problems. JACM **56**(2), 1–28 (2009)
13. de Gevigney, O.D., Nguyen, V.H., Szigeti, Z.: Matroid-based packing of arborescences. SIDMA **27**(1), 567–574 (2013)
14. Dinur, I., Steurer, D.: Analytical approach to parallel repetition. In: STOC (2014)
15. Edmonds, J.: Edge-disjoint branchings. Comb. Algor. (1973)
16. Feige, U.: A threshold of ln n for approximating set cover. JACM **45**, 634–652 (1998)
17. Fortier, Q., Király, C., Léonard, M., Szigeti, Z., Talon, A.: Old and new results on packing arborescences in directed hypergraphs. DAM **242**, 26–33 (2018)
18. Frank, A.: On disjoint trees and arborescences. In: Algebraic Methods in Graph Theory (1978)
19. Fujishige, S.: A note on disjoint arborescences. Combinatorica **30**, 247–252 (2010)
20. Gabow, H.N., Manu, K.S.: Packing algorithms for arborescences (and spanning trees) in capacitated graphs. Math. Prog. **82**, 83–109 (1998)
21. Gao, H., Yang, D.: Packing of maximal independent mixed arborescences. DAM **289**, 313–319 (2021)
22. Garg, N., Vazirani, V.V., Yannakakis, M.: Approximate max-flow min-(multi) cut theorems and their applications. SICOMP **25**, 235–251 (1996)
23. Gionis, A., Tsourakakis, C.E.: Dense subgraph discovery: Kdd 2015 tutorial. In: KDD (2015)
24. Hörsch, F., Szigeti, Z.: Reachability in arborescence packings. DAM **320**, 170–183 (2022)
25. Hu, T.C.: Multi-commodity network flows. OR (1963)
26. Kamiyama, N., Katoh, N., Takizawa, A.: Arc-disjoint in-trees in directed graphs. Combinatorica **29**, 197–214 (2009)
27. Katoh, N., Tanigawa, S.: Rooted-tree decompositions with matroid constraints and the infinitesimal rigidity of frameworks with boundaries. SIDMA **27**, 155–185 (2013)
28. Khandekar, R., Rao, S., Vazirani, U.: Graph partitioning using single commodity flows. JACM **56**, 1–15 (2009)
29. Király, C.: On maximal independent arborescence packing. SIDMA **30**, 2107–2114 (2016)
30. Király, C., Szigeti, Z., Tanigawa, S.: Packing of arborescences with matroid constraints via matroid intersection. Math. Prog. **181**, 85–117 (2020)
31. Lanciano, T., Miyauchi, A., Fazzone, A., Bonchi, F.: A survey on the densest subgraph problem and its variants. ACM Comput. Surv. **56**, 1–14 (2024)
32. Lee, V.E., Ruan, N., Jin, R., Aggarwal, C.: A survey of algorithms for dense subgraph discovery. Manag. Mining Graph Data, 303–336 (2010)

33. Leighton, T., Rao, S.: Multicommodity max-flow min-cut theorems and their use in designing approximation algorithms. JACM **46**, 787–832 (1999)
34. Lovász, L.: On two minimax theorems in graph. JCTB **21**, 96–103 (1976)
35. Mao-cheng, C.: Arc-disjoint arborescences of digraphs. J. Graph Theory **7**, 235–240 (1983)
36. Matsuoka, T., Tanigawa, S.: On reachability mixed arborescence packing. Discret, OPT (2019)
37. Nash-Williams, C.: Edge-disjoint spanning trees of finite graphs. J. Lond. Math. Soc. **1**(1), 445–450 (1961)
38. Saks, M., Samorodnitsky, A., Zosin, L.: A lower bound on the integrality gap for minimum multicut in directed networks. Combinatorica **24**, 525–530 (2004)
39. Schrijver, A.: Combinatorial Optimization: Polyhedra and Efficiency. Springer, Cham (2003)
40. Szigeti, Z.: Packing mixed hyperarborescences. Discrete OPT (2023)
41. Tarjan, R.: A good algorithm for edge-disjoint branching. IPL (1974)
42. Tutte, W.T.: On the problem of decomposing a graph into n connected factors. J. Lond. Math. Soc. **1**, 221–230 (1961)

Lagrangian Dual for Integer Optimization with Zero Duality Gap that Admits Decomposition

Diego Cifuentes, Santanu S. Dey(✉), and Jingye Xu

H. Milton Stewart School of Industrial and Systems Engineering, Georgia Institute of Technology, Atlanta, USA
{dfc3,sdey30,jxu673}@gatech.edu

Abstract. For mixed integer programs (MIPs) with block structures and coupling constraints, on dualizing the coupling constraints the resulting Lagrangian relaxation becomes decomposable into blocks which allows for the use of parallel computing. However, the resulting Lagrangian dual can have non-zero duality gap due to the inherent non-convexity of MIPs. In this paper, we propose two reformulations of such MIPs by adding redundant constraints, such that the Lagrangian dual obtained by dualizing the coupling constraints and the redundant constraints have zero duality gap while still remaining decomposable. One of these reformulations is similar, although not the same as the RLT hierarchy. In this case, we present multiplicative bounds on the quality of the dual bound at each level of the hierarchy for packing and covering MIPs. We show our results are applicable to general sparse MIPs, where decomposability is revealed via the tree-decomposition of the intersection graph of the constraint matrix. In preliminary experiments, we observe that the proposed Lagrangian duals give better dual bounds than classical Lagrangian dual and Gurobi in equal time, where Gurobi is not exploiting decomposability.

Keywords: Distributed Computing · Lagrangian Dual · Strong Duality

1 Introduction

Consider a two-block mixed integer program (MIP) with coupling constraints of the following form:

$$\text{OPT} := \min_{(\mathbf{x},\mathbf{y})} \sum_{i\in\{1,2\}} \left\langle \mathbf{c}^{(i)}, \mathbf{x}^{(i)} \right\rangle + \left\langle \mathbf{d}^{(i)}, \mathbf{y}^{(i)} \right\rangle \tag{1a}$$

$$\text{s.t.} (\mathbf{x}^{(i)}, \mathbf{y}^{(i)}) \in \mathcal{X}^{(i)}, \forall i \in \{1,2\}, \tag{1b}$$

$$\mathbf{x}^{(1)} = \mathbf{x}^{(2)} \in \{0,1\}^n. \tag{1c}$$

N. Megow and A. Basu (Eds.): IPCO 2025, LNCS 15620, pp. 184–198, 2025.
https://doi.org/10.1007/978-3-031-93112-3_14

where $\mathcal{X}^{(i)} := \left\{ (\mathbf{x}^{(i)}, \mathbf{y}^{(i)}) \;\middle|\; \begin{array}{l} A^{(i)}\mathbf{x}^{(i)} + B^{(i)}\mathbf{y}^{(i)} \leq \mathbf{b}^{(i)}, \\ \mathbf{y}^{(i)} \text{ is nonnegative and mixed-integer}, \\ \mathbf{x}^{(i)} \in \{0,1\}^n \end{array} \right\}$ with $A^{(i)}$, $B^{(i)}$, $\mathbf{b}^{(i)}$, $\mathbf{c}^{(i)}$, $\mathbf{d}^{(i)}$ being rational data of suitable dimension for each $i \in \{1, 2\}$.

If the coupling constraints (1c) are ignored, then the remaining problem can be decomposed into independent optimization tasks over each $\mathcal{X}^{(i)}$. One classic approach that exploits this structure to obtain dual bounds for (1) is that of Lagrangian relaxation [19]. Specifically, by dualizing (1c), we obtain:

$$\begin{aligned} L(\lambda) := \min_{(\mathbf{x},\mathbf{y})} & \left(\sum_{i\in\{1,2\}} \left\langle \mathbf{c}^{(i)}, \mathbf{x}^{(i)} \right\rangle + \left\langle \mathbf{d}^{(i)}, \mathbf{y}^{(i)} \right\rangle \right) + \left\langle \lambda, \mathbf{x}^{(1)} - \mathbf{x}^{(2)} \right\rangle \\ \text{s.t. } & (\mathbf{x}^{(i)}, \mathbf{y}^{(i)}) \in \mathcal{X}^{(i)}, \forall i \in \{1, 2\}, \end{aligned} \tag{2}$$

and

$$\text{DUAL} := \max_{\lambda} L(\lambda). \tag{3}$$

It is well-known that $L(\lambda)$ is a non-smooth concave function and sub-gradients for $L(\lambda)$ can be obtained by solving (2), which is a collection of independent optimization tasks over $\mathcal{X}^{(i)}$ *that could be solved in parallel.* Using these sub-gradients, one can solve (3) via non-smooth optimization methods [6].

Even though weak duality always holds, due to non-convexity, strong duality generally fails, that is, OPT > DUAL. On the other hand, one can solve (1) directly without exploiting decomposability, trivially obtaining zero duality gap. This motivates the main question of this paper: *Can we obtain zero duality gap using Lagrangian-style relaxations while simultaneously allowing decomposition into sub-problems?* Our main contributions are:

(1.) Obtaining zero duality gap and decomposability simultaneously: We design two reformulations of (1) whose Lagrangian duals, namely M-Lagrangian dual (Sect. 2.1, Sect. 2.2) and V-Lagrangian dual (Sect. 2.1, Sect. 2.4), achieve the twin goal of zero duality gap and decomposability. The M-Lagrangian method is a hierarchy of reformulations of (1) similar to the Reformulation-Linearization-Technique (RLT) [1] but not the same. M-Lagrangian duals achieve zero duality gap in the last step of the hierarchy, while simultaneously each level admits decomposition into sub-problems. We compare the difference between M-Lagrangian and RLT in Remark 1.

(2.) Analysis of bounds: We present multiplicative bounds on the duality gap at different levels of the M-Lagrangian hierarchy for packing and covering problems; see Sect. 2.3.

(3.) Run-time for solving V-Lagrangian dual: We present bounds on the number of iterations needed to solve the V-Lagrangian dual via sub-gradient methods in comparison to an algorithm in [2], see Sect. 2.4.

(4.) Generalization to arbitrary MIPs: Consider a loosely coupled general MIP where the block structure is revealed using a tree-decomposition [16] of the intersection graph [18] of the constraint matrix. Here, the blocks correspond to smaller problems defined over variables in the bags of the tree-decomposition. We show how to generalize the above results: simultaneously achieving decomposability and strong duality, and multiplicative bounds for packing and covering instances, for the M-Lagrangian dual to this setting, see Sect. 3.

(5.) Preliminary computational results: We illustrate how the proposed Lagrangian duals can outperform classical Lagrangian relaxation and Gurobi in terms of dual bounds achieved, where Gurobi solves the whole problem without exploiting decomposition; see Sect. 4.

Finally, in Sect. 5 we make concluding remarks.

One another exact algorithm that admits decomposition for (1) (and more generally via tree-decomposition of intersection graph) is the Dynamic Programming (DP) algorithm [5,9,10,26,29]. However, DP suffers from the disadvantage that we do not obtain any information on bounds until the algorithm terminates. On the other hand, the advantage of Lagrangian duals is that even if one terminates without reaching optimality, it is possible to retrieve dual bounds.

Literature Survey. There are very few exact algorithms for integer optimization that also achieve decomposability. Apart from DP discussed above, in the special case of two-stage stochastic programming, there are specialized algorithms [2,33] and the integer L-shaped method [3,25]. Attempts to achieve the above two goals simultaneously fail for Augmented Lagrangian method (AL) [17,22] and Semi-Lagrangian (SL) [7] which achieve strong duality, but not decomposability. Recently, the paper [32], develops a variant of AL method that achieves decomposability.

Notation. For a positive integer n, we use $[n]$ to denote the set $\{1,\dots,n\}$. Given a finite set S we represent its power set as 2^S. We denote the collection of subsets of $[n]$ containing all elements of cardinality k or less as $\binom{[n]}{\leq k}$. Given a graph G, we refer to its vertices by $V(G)$ and its edges by $E(G)$.

2 Decomposable Reformulation and its Lagrangian Dual for Two-Block Problem.

2.1 Two Reformulations

The main idea of the reformulations is to add redundant constraints to (1) that are separable after dualizing. Consider a separable constraint implied by $\mathbf{x}^{(1)} = \mathbf{x}^{(2)} \in \{0,1\}^n$, that is,

$$\mathbf{x}^{(1)} = \mathbf{x}^{(2)} \in \{0,1\}^n \implies F^{(1)}(\mathbf{x}^{(1)}) + F^{(2)}(\mathbf{x}^{(2)}) \geq 0. \tag{4}$$

We extend $\mathcal{X}^{(i)}$ to $\mathcal{X}^{(i)}_{ex}$ in the following fashion: $\mathcal{X}^{(i)}_{ex} := \{(\mathbf{x}^{(i)}, \mathbf{y}^{(i)}, w^{(i)}) : (\mathbf{x}^{(i)}, \mathbf{y}^{(i)}) \in \mathcal{X}^{(i)}, w^{(i)} = F^{(i)}(\mathbf{x}^{(i)})\}, \forall i \in \{1,2\}$, and obtain the following reformulation of (1):

$$\begin{aligned}\text{OPT} := \min_{(\mathbf{x},\mathbf{y},\mathbf{w})} & \sum_{i\in\{1,2\}} \left\langle \mathbf{c}^{(i)}, \mathbf{x}^{(i)} \right\rangle + \left\langle \mathbf{d}^{(i)}, \mathbf{y}^{(i)} \right\rangle && \text{(5a)}\\ \text{s.t.} & (\mathbf{x}^{(i)}, \mathbf{y}^{(i)}, w^{(i)}) \in \mathcal{X}_{ex}^{(i)}, \forall i \in \{1,2\}, && \text{(5b)}\\ & \mathbf{x}^{(1)} = \mathbf{x}^{(2)}, && \text{(5c)}\\ & w^{(1)} + w^{(2)} \geq 0. && \text{(5d)}\end{aligned}$$

Since the constraint (5d) is also a coupling constraint, we introduce additional multipliers and dualize it as follows:

$$\begin{aligned} L^{ex}(\lambda,\mu) := \min_{(\mathbf{x},\mathbf{y},\mathbf{w})} & \left(\sum_{i\in\{1,2\}} \left\langle \mathbf{c}^{(i)}, \mathbf{x}^{(i)} \right\rangle + \left\langle \mathbf{d}^{(i)}, \mathbf{y}^{(i)} \right\rangle \right) \\ & + \left\langle \lambda, \mathbf{x}^{(1)} - \mathbf{x}^{(2)} \right\rangle + \left\langle \mu, w^{(1)} + w^{(2)} \right\rangle \\ & \text{s.t.} (\mathbf{x}^{(i)}, \mathbf{y}^{(i)}, w^{(i)}) \in \mathcal{X}_{ex}^{(i)}, \forall i \in \{1,2\}, \end{aligned} \tag{6}$$

and

$$\text{DUAL}^{ex} := \max_{\mu\geq 0,\lambda} L^{ex}(\lambda,\mu). \tag{7}$$

Although the addition of (4) has no effect on (1), that is on the primal side the optimal value trivially remains the same; on the dual side they can indeed improve the bound, that is it is possible that DUAL $\neq$ DUAL^{ex}.

We first establish a result to identify when DUAL $=$ DUAL^{ex}. The following result is a strengthening on a result from [27].

Proposition 1. *If either $F^{(1)}(\cdot)$ or $F^{(2)}(\cdot)$ is affine, then DUAL $=$ DUAL^{ex}.*

Proposition 1 motivates us to consider the following two different choices of redundant non-affine constraints:

1. *Monomial-based reformulation*: Let $\mathcal{S} \subseteq 2^{[n]}$ be a collection of subsets of $[n]$, then the primal-redundant constraints take the form of

$$\prod_{j\in S} x_j^{(1)} = \prod_{j\in S} x_j^{(2)}, \forall S \in \mathcal{S}. \tag{8}$$

2. *Vertex-based reformulation*: The primal-redundant constraints take the form

$$\sigma_{\mathbf{v}}(\mathbf{x}^{(1)}) = \sigma_{\mathbf{v}}(\mathbf{x}^{(2)}), \text{ for each vertex } \mathbf{v} \text{ of } [0,1]^n, \tag{9}$$

where $\sigma_{\mathbf{v}}(\mathbf{x}) := \begin{cases} 1 & \text{if } \mathbf{x} = \mathbf{v} \\ 0 & \text{otherwise} \end{cases}$.

We refer to the Lagrangian dual of monomial-based reformulation and vertex-based reformulation as *M-Lagrangian dual* and *V-Lagrangian dual* respectively.

Remark 1. Difference between traditional RLT-type constraints and (8): For the constraints of the form $x_j^{(1)} = x_j^{(2)}$ for say $j \in \{1, 2\}$, traditional RLT would involve multiplying these constraints, that is, adding new constraints of the form $\left(x_1^{(1)} - x_1^{(2)}\right)\left(x_2^{(1)} - x_2^{(2)}\right) = 0$ in the reformulation step of RLT. Such a constraint would not be separable after dualizing. Constraints of the form of (8) are more similar to those used in the reformulation in [12,14]. Also see [28].

We note here that the monomial-based reformulation and the vertex-based reformulation are quite different. As we will see in Sect. 2.2, one has to add (8) corresponding to all monomials to obtain zero duality gap in general. In practice, in Sect. 4, we will see significant dual bound improvement over the classical Lagrangian dual with just the quadratic monomials.

On the other hand, V-Lagrangian dual is more like a "column-generation" approach. For strong duality to be attained, we need to add (9) for all vertices of $[0, 1]^n$ (Sect. 2.4). We will show that the inner problem (6) in this case can be emulated in the original space $\mathcal{X}^{(i)}$ using "no-good cuts" [3]. The number of extra cuts is initially zero and increases with each iteration, until (7) is solved. The resulting method is similar to the algorithm described in [2], where adding weights to vertices is replaced by cutting off the vertices. We will show in Sect. 2.4 that sub-gradient method for V-Lagrangian dual has a convergence rate that is at least as good as that of [2].

2.2 M-Lagrangian Dual for Two-Block Problems

Fix any $\mathcal{S} \subseteq 2^{[n]}$, we let

$$\mathcal{X}_M^{(i)}(\mathcal{S}) := \left\{ (\mathbf{x}^{(i)}, \mathbf{y}^{(i)}, \mathbf{w}^{(i)}) : (\mathbf{x}^{(i)}, \mathbf{y}^{(i)}) \in \mathcal{X}^{(i)}, w_S^{(i)} = \prod_{j \in S} x_j^{(i)}, \forall S \in \mathcal{S} \right\}.$$

We assume that $\{j\} \in \mathcal{S}, \forall j \in [n]$. In this case, (5c) is included within (5d) and so the monomial reformulation takes the form:

$$\text{OPT} = \min_{(\mathbf{x},\mathbf{y},\mathbf{w})} \sum_{i \in \{1,2\}} \left\langle \mathbf{c}^{(i)}, \mathbf{x}^{(i)} \right\rangle + \left\langle \mathbf{d}^{(i)}, \mathbf{y}^{(i)} \right\rangle \tag{10a}$$

$$\text{s.t.} (\mathbf{x}^{(i)}, \mathbf{y}^{(i)}, \mathbf{w}^{(i)}) \in \mathcal{X}_M^{(i)}(\mathcal{S}), \forall i \in \{1, 2\}, \tag{10b}$$

$$\mathbf{w}_S^{(1)} = \mathbf{w}_S^{(2)}, \forall S \in \mathcal{S}. \tag{10c}$$

If we apply Lagrangian decomposition to (10) by dualizing (10c), then by Geoffrion's Theorem (also called primal characterization [11]) the optimal value of the M-Lagrangian dual is equal to:

$$\begin{aligned} \text{DUAL}^M := & \min_{(\mathbf{x},\mathbf{y},\mathbf{w})} \sum_{i \in \{1,2\}} \left\langle \mathbf{c}^{(i)}, \mathbf{x}^{(i)} \right\rangle + \left\langle \mathbf{d}^{(i)}, \mathbf{y}^{(i)} \right\rangle \\ & \text{s.t.} (\mathbf{x}^{(i)}, \mathbf{y}^{(i)}, \mathbf{w}^{(i)}) \in \mathbf{conv}\{\mathcal{X}_M^{(i)}(\mathcal{S})\}, \forall i \in \{1, 2\}, \\ & \mathbf{w}_S^{(1)} = \mathbf{w}_S^{(2)}, \forall S \in \mathcal{S}. \end{aligned} \tag{11}$$

Throughout the section, we make the following assumption:

Assumption 1. *$\mathcal{S}$ is down-closed. Namely, if $S \in \mathcal{S}$, then $S' \in \mathcal{S}$ for all $S' \subseteq S$.*

As the objective function of (11) only involves $(\mathbf{x}, \mathbf{y})$, we are interested in the projection of the feasible region of (11) onto $(\mathbf{x}, \mathbf{y})$:

$$\mathcal{A}(\mathcal{S}) := \left\{ (\mathbf{x}, \mathbf{y}) \;\middle|\; \begin{array}{l} \exists \mathbf{w}, (\mathbf{x}^{(i)}, \mathbf{y}^{(i)}, \mathbf{w}^{(i)}) \in \mathbf{conv}\{\mathcal{X}_M^{(i)}(\mathcal{S})\}, \forall i \in \{1, 2\}, \\ \mathbf{w}_S^{(1)} = \mathbf{w}_S^{(2)}, \forall S \in \mathcal{S} \end{array} \right\}. \tag{12}$$

We further define

$$\mathcal{B}(\mathcal{S}) := \bigcap_{\mathcal{U} \in \mathcal{S}} \mathbf{conv} \left\{ (\mathbf{x}, \mathbf{y}) \;\middle|\; \begin{array}{l} (\mathbf{x}^{(i)}, \mathbf{y}^{(i)}) \in \mathcal{X}^{(i)}, \forall i \in \{1, 2\}, \\ x_j^{(1)} = x_j^{(2)}, \forall j \in \mathcal{U} \end{array} \right\}. \tag{13}$$

Theorem 1. *Under Assumption 1 on $\mathcal{S}$, we have $\mathcal{A}(\mathcal{S}) \subseteq \mathcal{B}(\mathcal{S})$. As a consequence, when $\mathcal{S} = 2^{[n]}$, strong duality holds between (10) and (11).*

2.3 Bounds for Two-Block Packing and Covering Problem

In this subsection, we consider the M-Lagrangian dual when $\mathcal{S} = \binom{[n]}{\leq k}$ for some fixed number $k \in [n]$. We refer to the corresponding dual optimal value by DUAL_k^M. The goal of this subsection is to present a multiplicative bound on DUAL_k^M for packing and covering instances. Since we use a "min" objective and "$\leq$" constraints, (1) is called *packing* (resp. *covering*) if each $\mathbf{c}^{(i)}$, $\mathbf{d}^{(i)}$ are non-positive (resp. non-negative) and $A^{(i)}$, $B^{(i)}$, $\mathbf{b}^{(i)}$ are entry-wise non-negative (resp. non-positive). Therefore, in our notation, OPT and DUAL_k^M are non-positive (resp. non-negative) for packing (resp. covering) problems.

For packing problems, we require the following assumption for our analysis, which is commonly made in the stochastic programming literature (see, for example [13,23,31]).

Assumption 2. *(Relatively complete recourse) For any $(\mathbf{x}^{(1)}, \mathbf{y}^{(1)}) \in \mathcal{X}^{(1)}$, there exists some $\mathbf{y}^{(2)}$ that $(\mathbf{x}^{(1)}, \mathbf{y}^{(2)}) \in \mathcal{X}^{(2)}$. Symmetrically, for any $(\mathbf{x}^{(2)}, \mathbf{y}^{(2)}) \in \mathcal{X}^{(2)}$, there exists some $\mathbf{y}^{(1)}$ that $(\mathbf{x}^{(2)}, \mathbf{y}^{(1)}) \in \mathcal{X}^{(1)}$.*

Note that if this assumption is violated, then one can simply add $A^{(2)}\mathbf{x}^{(1)} \leq \mathbf{b}^{(2)}$ and $A^{(1)}\mathbf{x}^{(2)} \leq \mathbf{b}^{(1)}$ into $\mathcal{X}^{(1)}$ and $\mathcal{X}^{(2)}$ respectively and it is straightforward to see that this assumption is then satisfied for packing problems.

Theorem 2. *For any packing instance (1), under Assumption 2, we have that $\left(2 + \dfrac{1}{t-2}\right) \cdot OPT \leq DUAL_k^M \leq OPT$, where $t = k/n$.*

Theorem 3. *For any covering instance (1), we have $\left(\dfrac{1}{2-t}\right) \cdot OPT \leq DUAL_k^M \leq OPT$ where $t = k/n$.*

The proof of the above results uses techniques similar to those presented in [15].

2.4 V-Lagrangian Dual for Two-Block Problem

$$\text{Let } \mathcal{X}_V^{(i)} := \left\{ (\mathbf{x}^{(i)}, \mathbf{y}^{(i)}, \mathbf{w}^{(i)}) \;\middle|\; \begin{array}{l} (\mathbf{x}^{(i)}, \mathbf{y}^{(i)}, \mathbf{w}^{(i)}) \in \mathcal{X}^{(i)}, \\ w_{\mathbf{v}}^{(i)} = \sigma_{\mathbf{v}}(\mathbf{x}^{(i)}), \forall \mathbf{v} \text{ vertex of } [0,1]^n \} \end{array} \right\}.$$

By replacing $\mathcal{X}_V^{(i)}$ in place of $\mathcal{X}_{ex}^{(i)}$ in (5) and dualizing, we obtain:

$$\begin{aligned} L^V(\lambda, \mu) := \min_{(\mathbf{x},\mathbf{y},\mathbf{w})} & \left(\sum_{i \in \{1,2\}} \left\langle \mathbf{c}^{(i)}, \mathbf{x}^{(i)} \right\rangle + \left\langle \mathbf{d}^{(i)}, \mathbf{y}^{(i)} \right\rangle \right) \\ & + \left\langle \lambda, \mathbf{x}^{(1)} - \mathbf{x}^{(2)} \right\rangle + \left\langle \mu, \mathbf{w}^{(1)} - \mathbf{w}^{(2)} \right\rangle \\ & \text{s.t. } (\mathbf{x}^{(i)}, \mathbf{y}^{(i)}, \mathbf{w}^{(i)}) \in \mathcal{X}_V^{(i)}, \forall i \in \{1, 2\}, \end{aligned} \tag{14}$$

and

$$\text{DUAL}^V := \max_{\lambda,\mu} L^V(\lambda, \mu). \tag{15}$$

Theorem 4. *Strong duality holds for V-Lagrangian dual, that is OPT = DUAL*^{V}.

Now we show how to solve (15) in the original space. The inner problem (14) is decomposed into sub-problems taking the following form for some values of λ, μ:

$$\min_{\mathbf{x}^{(i)},\mathbf{y}^{(i)}} \left\{ \left\langle \mathbf{c}^{(i)} + \lambda, \mathbf{x}^{(i)} \right\rangle + \left\langle \mathbf{d}^{(i)}, \mathbf{y}^{(i)} \right\rangle + \left\langle \mu, \mathbf{w}^{(i)} \right\rangle : (\mathbf{x}^{(i)}, \mathbf{y}^{(i)}, \mathbf{w}^{(i)}) \in \mathcal{X}_V^{(i)} \right\}. \tag{16}$$

Given $\mathcal{V} \subseteq \{0,1\}^n$, one can model $\mathbf{x} \in \{0,1\}^n \setminus \mathcal{V}$ by no-good cuts or more general methods [4,25]. Thus, solving (16) can be accomplished in the original space by dividing the optimization task into two parts as shown in Algorithm 1.

Algorithm 1: Solve (16)

Input i, λ, μ **Output** the optimal solution of (16)

$\mathcal{V} \leftarrow \{\mathbf{v} : \mu_{\mathbf{v}} \neq 0\}$

$(\mathbf{x}_*, \mathbf{y}_*) \leftarrow \text{argmin}_{(\mathbf{x}^{(i)},\mathbf{y}^{(i)}) \in \mathcal{X}^{(i)}} \left\langle \mathbf{c}^{(i)} + \lambda, \mathbf{x}^{(i)} \right\rangle + \left\langle \mathbf{d}^{(i)}, \mathbf{y}^{(i)} \right\rangle : \mathbf{x}^{(i)} \notin \mathcal{V}$

$p_* \leftarrow \left\langle \mathbf{c}^{(i)} + \lambda, \mathbf{x}_* \right\rangle + \left\langle \mathbf{d}^{(i)}, \mathbf{y}_* \right\rangle$

foreach $\mathbf{v} \in \mathcal{V}$ **do**

  $(\mathbf{v}, \mathbf{y}') \leftarrow \text{argmin}_{(\mathbf{v},\mathbf{y}^{(i)}) \in \mathcal{X}^{(i)}} \left\langle \mathbf{c}^{(i)} + \lambda, \mathbf{v} \right\rangle + \left\langle \mathbf{d}^{(i)}, \mathbf{y}^{(i)} \right\rangle$

  $p' \leftarrow \left\langle \mathbf{c}^{(i)} + \lambda, \mathbf{v} \right\rangle + \left\langle \mathbf{d}^{(i)}, \mathbf{y}' \right\rangle + \mu_{\mathbf{v}}$

  if $p' < p_*$ **then**

    $\mathbf{x}_* \leftarrow \mathbf{v}, \mathbf{y}_* \leftarrow \mathbf{y}'\ p_* \leftarrow p'$

return $\mathbf{x}^*, \mathbf{y}_*, p_*$

Algorithm 1 uses the fact that any feasible solution $(\mathbf{w}^{(i)})_{\mathbf{v}}$ is non-zero if and only if $\mathbf{x}^{(i)} = \mathbf{v}$. Another consequence of this fact is that, since there are two sub-problems of the type (16), the support of the sub-gradient (on the μ variables) is at most 2. Therefore, if one updates μ, λ by the sub-gradient method or classical bundle(-level) method, then $\|\mu\|_0$ increases by at most 2 for each iteration. Also, observe that $\text{argmin}_{(\mathbf{v},\mathbf{y}^{(i)})\in\mathcal{X}^{(i)}} \langle \mathbf{c}^{(i)} + \lambda, \mathbf{v}\rangle + \langle \mathbf{d}^{(i)}, \mathbf{y}^{(i)}\rangle$ is independent of λ, μ and therefore one may store the choice of $\mathbf{y}'$ for each $\mathbf{v}$ to avoid repeated evaluation in Algorithm 1.

We next compare the number of iterations to solve (15) using a sub-gradient method against the number of iterations for the scenario decomposition algorithm (SDA) of [2]. At a high level, the main difference between our approach here and the SDA is that instead of updating the dual values corresponding to the solutions found in previous iteration, the SDA simply eliminates the corresponding solutions from each sub-problem in the next iteration. In particular, we show that using a sub-gradient method to solve (15) cannot be much worse than SDA with $\lambda = 0$ setting (this setting is the main focus of [2]) under the assumption of non-negativity of objective function value.

Proposition 2. *Suppose that $c^{(i)}, d^{(i)}$ are non-negative for $i \in \{1, 2\}$. Suppose that SDA with $\lambda = 0$ terminates in t_{SDA} iterations. Let $\phi := \max_{i\in\{1,2\},(\mathbf{x}^{(i)},\mathbf{y}^{(i)})\in\mathcal{X}^{(i)}} \left|\langle \mathbf{c}^{(i)}, \mathbf{x}^{(i)}\rangle + \langle \mathbf{d}^{(i)}, \mathbf{y}^{(i)}\rangle\right|$. Then there exists a sub-gradient algorithm that solves (15) to ϵ-optimality (additive error) within $O\left(\left(\frac{\sqrt{n}\phi}{\epsilon}\right)^2\right) \cdot t_{SDA}$ iterations.*

We also show that there are simple instances where SDA can be exponentially worse than a sub-gradient algorithm to solve (15).

Proposition 3. *There exists an instance where SDA with $\lambda = 0$ setting requires $O(2^n)$ iterations to achieve ϵ-optimality, while (15) can be solved to ϵ-optimality by classic sub-gradient method in $O(n^2(n+2))$ iterations.*

3 Lagrangian Dual for General MIPs

In this section, we aim to generalize the ideas applied to the two-block case to the case of general MIPs, which can be thought of as multiple blocks of constraints and each pair of blocks of constraints share some common variables.

V-Lagrangian Dual. We first note that the Algorithm 1 that allows for conducting the computations in the original space, does not generalize very easily to the case of multiple-blocks of constraints, where different pairs of blocks share different sets of common variables. Therefore, the most natural generalization that can be tackled by the V-Lagrangian dual is the case of two-stage stochastic programming type instances (also called star instances) where the blocks are: (i) the constraints involving the first stage variables only, (ii) the constraints corresponding to a particular scenario in second stage. If the same set of first-stage binary variables appears in the constraints for every scenario, then it is easy to generalize Algorithm 1 in this setting, and it is straightforward to see that the V-Lagrangian dual has zero duality gap.

M-Lagrangian for Multi-block Problems. For the case of general MIPs, we only pursue the M-Lagrangian dual approach. We first observe that strong duality may fail when naively constructing the M-Lagrangian dual, even if all constraints corresponding to all monomials are included. Consider the 3-block example:

$$\min t^{(1)} + t^{(2)} + t^{(3)} \tag{17a}$$

$$\text{s.t. } (\mathbf{x}^{(1)}, t^{(1)}) \in \{(\mathbf{x}, t) : \mathbf{x} \in \{0,1\}^2, t = x_1 x_2\}, \tag{17b}$$

$$(\mathbf{x}^{(2)}, t^{(2)}) \in \{(\mathbf{x}, t) : \mathbf{x} \in \{0,1\}^2, t = (1 - x_1) x_2\}, \tag{17c}$$

$$(\mathbf{x}^{(3)}, t^{(3)}) \in \{(\mathbf{x}, t) : \mathbf{x} \in \{0,1\}^2, t = (1 - x_1)(1 - x_2)\}, \tag{17d}$$

$$x_1^{(1)} = x_1^{(2)}, x_2^{(1)} = x_1^{(3)}, x_2^{(2)} = x_2^{(3)}. \tag{17e}$$

The optimal value (17) is 1. On the other hand, each coupling constraint between any two blocks in (17e) only involves one variable. Therefore, the classical Lagrangian of (17) by dualizing (17e) coincides with M-Lagrangian with all monomials. The optimal value of classical Lagrangian dual is 0 for this example.

Example (17) illustrates that strong duality may not be achieved when naively applying the reformulations discussed in Sect. 2.1. Such issue arises because the coupling constraints (17e) form a cycle and the generalization of Theorem 1 does not go through. Therefore, it is natural to explore the tree-decomposition approach.

Consider a general MIP:

$$\min\{\mathbf{c}^\top \mathbf{x} : A\mathbf{x} \leq \mathbf{b}, \mathbf{x} \text{ is mixed-binary}\}, \tag{18}$$

where $\mathbf{c}, \mathbf{b}, A$ are all rational.

Definition 1. *The intersection graph [18] of (18) is a simple undirected graph that has a vertex for each variable in (18) and two vertices are adjacent if and only if their associated variables appear in any common constraint of $Ax \leq b$.*

Using the intersection graph of (18), one can reformulate (18) into "a tree-structure" via the *tree decomposition* [30].

Definition 2. *Let G be a simple undirected graph. A tree decomposition of G is a pair of $(\mathcal{T}, \mathcal{Q})$ where $\mathcal{T}$ is a tree and $\mathcal{Q} = \{\mathcal{Q}_t : t \in V(\mathcal{T})\}$ is a collection of vertices of $V(G)$ such that the following holds:*

1. *For each $v \in V(G)$, the set $\{t \in V(\mathcal{T}) : v \in \mathcal{Q}_t\}$ forms a subtree of $\mathcal{T}$,*
2. *If $(u, v) \in E(G)$, then there exists $t \in V(\mathcal{T})$ such that $u, v \in \mathcal{Q}_t$,*
3. *$\bigcup_{t \in V(\mathcal{T})} \mathcal{Q}_t = V(G)$.*

Given a general MIP, its intersection graph $\mathcal{G}$, and a corresponding tree decomposition $(\mathcal{T}, \mathcal{Q})$, one can reformulate (18) as:

$$\begin{aligned} &\min \sum_{i \in V(\mathcal{T})} \left\langle \mathbf{c}^{(i)}, \mathbf{x}^{(i)} \right\rangle \\ &\text{s.t. } \mathbf{x}^{(i)} \in \mathcal{X}^{(i)}, \forall i \in V(\mathcal{T}),\ x_v^{(i)} = x_v^{(j)}, \forall (i,j) \in E(\mathcal{T}) \text{ and } v \in \mathcal{Q}_i \cap \mathcal{Q}_j, \end{aligned} \tag{19}$$

where each $\mathcal{X}^{(i)}$ makes local copies of variables in $\mathcal{Q}_i$ and each constraint in $A\mathbf{x} \leq \mathbf{b}$ is absorbed in some $\mathcal{X}^{(i)}$ using a copy of the variables in $\mathcal{X}^{(i)}$. Henceforth, we assume that coupling constraints involve only binary variables.

3.1 Strong Duality for M-Lagrangian Dual for Multi-block Problems

For each $i \in V(\mathcal{T})$, we define

$$\mathcal{X}_M^{(i)} := \left\{ (\mathbf{x}^{(i)}, \mathbf{w}^{(i)}) : \mathbf{x}^{(i)} \in \mathcal{X}^{(i)}, w_S^{(i)} = \prod_{i \in S} x_S^{(i)}, \forall S \subseteq \mathcal{Q}_i \right\}.$$

Let $\mathcal{Q}_{ij} \subseteq V(G)$ denote $\mathcal{Q}_i \cap \mathcal{Q}_j$. For every edge $(i,j) \in E(\mathcal{T})$, let $\mathcal{S}_{ij} \subseteq 2^{\mathcal{Q}_{ij}}$ be a collection of subsets of $\mathcal{Q}_{ij}$. The extended formulation of (19) parameterized by $\mathcal{S} := \{\mathcal{S}_{ij}\}_{(i,j) \in E(\mathcal{T})}$takes the form of

$$\begin{aligned} &\min \sum_{i \in V(\mathcal{T})} \left\langle \mathbf{c}^{(i)}, \mathbf{x}^{(i)} \right\rangle \\ &\text{s.t. } (\mathbf{x}^{(i)}, \mathbf{w}^{(i)}) \in \mathcal{X}_M^{(i)}, \forall i \in V(\mathcal{T}),\ \mathbf{w}_S^{(i)} = \mathbf{w}_S^{(j)}, \forall (i,j) \in E(\mathcal{T}), S \in \mathcal{S}_{ij}. \end{aligned} \tag{20}$$

Assumption 3. *$\mathcal{S}_{ij}$ is down-closed for all $(i,j) \in E(\mathcal{T})$.*

Similar to arguments in Sect. 2.2, we consider the projection of the feasible region of the primal characterization of M-Lagrangian dual onto $\mathbf{x}$:

$$\mathcal{A}(\mathcal{S}, \mathcal{T}) := \left\{ \mathbf{x} \;\middle|\; \begin{array}{l} \exists \mathbf{w}, (\mathbf{x}^{(i)}, \mathbf{w}^{(i)}) \in \mathbf{conv}\left\{\mathcal{X}_M^{(i)}\right\}, \forall i \in V(\mathcal{T}), \\ \mathbf{w}_S^{(i)} = \mathbf{w}_S^{(j)}, \forall (i,j) \in E(\mathcal{T}), S \in \mathcal{S}_{ij} \end{array} \right\}, \tag{21}$$

and

$$\mathcal{B}(\mathcal{S}, \mathcal{T}) := \bigcap_{\mathcal{U} \trianglelefteq \mathcal{S}} \mathbf{conv} \left\{ \begin{array}{l} \mathbf{x}^{(i)} \in \mathcal{X}^{(i)}, \forall i \in V(\mathcal{T}), \\ x_v^{(i)} = x_v^{(j)}, \forall (i,j) \in E(\mathcal{T}), \forall v \in \mathcal{U}_{ij}. \end{array} \right\}, \tag{22}$$

where $\mathcal{U} \trianglelefteq \mathcal{S}$ implies $\mathcal{U} = \{\mathcal{U}_{ij}\}_{(i,j) \in E(\mathcal{T})}$ and $\mathcal{U}_{ij} \in \mathcal{S}_{ij}$ for all $(i,j) \in \mathcal{T}$.

Theorem 5. *Under the assumption 3, $\mathcal{A}(\mathcal{S}, \mathcal{T}) \subseteq \mathcal{B}(\mathcal{S}, \mathcal{T})$. Therefore, when $\mathcal{S}_{ij} = 2^{\mathcal{Q}_{ij}}, \forall (i,j) \in E(\mathcal{T})$, then strong duality holds between (19) and its M-Lagrangian relaxation.*

3.2 Hierarchical Bounds for Multi-block Packing and Covering Problems

In this subsection, we consider the M-Lagrangian dual when $\mathcal{S}_{ij} = \binom{\mathcal{Q}_{ij}}{\leq k}$ for some fixed number $k \in [n]$. We refer to $\text{DUAL}^M, \mathcal{A}(\mathcal{S},\mathcal{T}), \mathcal{B}(\mathcal{S},\mathcal{T})$ in this setting by $\text{DUAL}_k^M, \mathcal{A}_k(\mathcal{T}), \mathcal{B}_k(\mathcal{T})$. The goal of this subsection is again to analyze DUAL_k^M as a function of k for packing and covering instances through the lens of bound implied by $\mathcal{B}_k$. Similar to assumption 2 in the two-block case, we need an additional assumption to analyze packing problems:

Assumption 4. *(Relatively complete recourse - general) For any $i \in V(\mathcal{T})$ and any $\mathbf{x}^{(i)} \in \mathcal{X}^{(i)}$, there exists a feasible solution $\mathbf{x}_*$ in (19) such that $\mathbf{x}_*^{(i)} = \mathbf{x}^{(i)}$.*

Like before, this assumption is easy to achieve by addition of suitable constraints. To present our result, we require the following definitions.

Definition 3 (Good and k-good). *Given a subset $\mathcal{W}$ of variables $\mathbf{x}$ in (18), let $\mathcal{V} := \{i \in V(\mathcal{T}) : \mathcal{Q}_i \cap \mathcal{W} \neq \emptyset\}$. Consider the sub-graph $\mathcal{T}(\mathcal{W})$ of $\mathcal{T}$ induced by $\mathcal{V}$. We say $\mathcal{W}$ is good if every connected component $\mathcal{C}$ of $\mathcal{T}(\mathcal{W})$ satisfies either*

1. *For any $(i,j) \in E(\mathcal{C}), |\mathcal{Q}_i \cap \mathcal{Q}_j \cap \mathcal{W}| \leq k$.*
2. *There exists $i \in V(\mathcal{C})$ such that $(\mathcal{Q}_j \cap \mathcal{W}) \subseteq (\mathcal{Q}_i \cap \mathcal{W}), \forall j \in V(\mathcal{C})$.*

When every connected component $\mathcal{C}$ of $\mathcal{T}(\mathcal{W})$ satisfies (1.), we call $\mathcal{W}$ k-good.

For a subset $\mathcal{W}$ of variables in (18), we let $\chi_{\mathcal{W}}$ to be the indicator vector of $\mathcal{W}$, that is we let $(\chi_{\mathcal{W}})_x = 1$ if $x \in \mathcal{W}$ and 0 otherwise. Consider the following two linear programs:

$$\eta_k := \left\{ \min \sum_{\mathcal{W} \text{ is good}} \alpha_{\mathcal{W}} : \sum_{\mathcal{W} \text{ is good}} \alpha_{\mathcal{W}} \chi_{\mathcal{W}} \geq \mathbf{1} \text{ and } \alpha_{\mathcal{W}} \geq 0. \right\} \tag{23}$$

$$\theta_k := \left\{ \min \sum_{\mathcal{W} \text{ is } k\text{-good}} \alpha_{\mathcal{W}} : \sum_{\mathcal{W} \text{ is } k\text{-good}} \alpha_{\mathcal{W}} \chi_{\mathcal{W}} \geq \mathbf{1} \text{ and } \alpha_{\mathcal{W}} \geq 0. \right\} \tag{24}$$

Theorem 6. *For any packing instance, under Assumption 4, we have that $\eta_k \cdot OPT \leq DUAL_k^M \leq OPT$.*

Theorem 7. *For any covering instance (18), let $\tau := \max_{v \in V(\mathcal{G})} |\{i \in V(\mathcal{T}) : v \in \mathcal{Q}_i\}|$. Then we have that $\dfrac{\theta_k}{1 - \tau + \tau \cdot \theta_k} \cdot OPT \leq DUAL_k^M \leq OPT$.*

In fact, Theorem 6 and Theorem 7 are generalizations of respectively Theorem 2 and Theorem 3.

Corollary 1. *Let $t := \dfrac{k}{n}$. For two-stage packing problem with Z scenarios, under Assumption 4, we have $\left(2 + \dfrac{2t - Z \cdot t - 1}{Z - t}\right) \cdot OPT \leq DUAL_k^M \leq OPT$. For two-stage covering problem with Z scenarios, we have $\dfrac{1}{(1-Z) \cdot t + Z} \cdot OPT \leq DUAL_k^M \leq OPT$.*

4 Preliminary Numerical Experimental

Our experiments are run on a Windows PC with a 12th Gen Intel(R) Core(TM) i7 processor and 16 GB RAM. We use Julia 1.9, relying on Gurobi version 9.0.2 as MIP solver and Clarabel [21] as QP solver (for bundle method implementation).

We generated two classes of instances which we refer to as (STAR-STAB) and (PATH-STAB). For both problems we construct 10 blocks, each block solving a maximum cardinality stable set problem on a random graph on 100 nodes. For (STAR-STAB) the blocks are coupled by $\mathbf{x}_j^{(1)} = \mathbf{x}_j^{(i)}$, $\forall i \in \{2, \ldots, 10\}$, $\forall j \in \{1, \ldots, 33\}$. For (PATH-STAB) the blocks are coupled by $\mathbf{x}_{67+j}^{(i)} = \mathbf{x}_j^{(i+1)}, \forall i \in \{1, \ldots, 9\}, \forall j \in \{1, \ldots, 33\}$. For each class of problems, we generate 10 instances.

For (PATH-STAB) we consider classic Lagrangian dual (L) and M-Lagrangian dual with only quadratic terms (QL). For (STAR-STAB), we additionally consider V-Lagrangian dual (VL) and SDA [2]. We solve the Lagrangian dual by bundle method. For all Lagrangian duals we first start by solving classical Lagrangian dual for 600 s. Then we take this dual solution, append zeros for the additional dual variables and use this solution as the starting point for solving the more sophisticated Lagrangian dual in the remaining time. In our naive implementation, we solved the sub-problems sequentially.

We set 1200 s as time limit for all methods. In our experiments, none of the methods terminated. In particular, the initial upper bound required by bundle methods was derived using the feasible solution of Gurobi. During the course of the run of the bundle methods, these upper bounds did not change.

Table 1. STAR-STAB

Methods	Primal-dual gap	Time	Iterations
Gurobi	10.0%	1200	–
L	6.0%	1200	127
QL	4.4%	1200	96
VL	3.9%	1200	103
SDA	8.6%	1200	151

Table 2. PATH-STAB

Methods	Primal-dual gap	Time	Iterations
Gurobi	10.8%	1200	–
L	3.5%	1200	369
QL	1.2%	1200	254

The results of our computation are present in Table 1. The primal-dual gap is computed as 1 minus the ratio of the dual bound from the Lagrangian relaxation and the primal bound given by Gurobi's feasible solution. We observe that in the case of (STAR-STAB), the V-Lagrangian dual has the best performance, closely followed by M-Lagrangian dual with quadratic terms. The dual bounds obtained from classical Lagrangian dual and SDA are worse. For the (PATH-STAB) instances (Table 2), the M-Lagrangian dual with quadratic terms outperforms the classical Lagrangian dual. For both instance classes, Gurobi's performance in obtaining dual bounds is significantly worse as it does not exploit decomposability.

5 Concluding Remarks

We show that constructing Lagrangian duals which achieve the twin goal of zero duality gap and maintaining decomposability of the sub-problems, comes at the cost of solving more challenging sub-problems with non-linear objective functions. Thus, this approach is perhaps most suitable when we have a large number of sub-problems, but each subproblem is not too large. An open question is whether one can achieve Lagrangian duals with decomposability and zero duality gap while solving an "easier" subproblem in each iteration.

On the computational side, we see that even implementing the sub-problems sequentially leads to significant improvements over both Gurobi (that cannot exploit decomposability) or the classical Lagrangian dual. We caution the reader that there are significantly more sophisticated methods for star (2-stage stochastic) problems [24]. While our methods may not necessarily be competitive against sophisticated methods, the preliminary results are encouraging. Significant more engineering in the implementation of our methods and also applying our methods to instances with more general tree-decomposition structures (see for example [8] for identification of sparsity structures of MIPLIB instances [20]) remain future research directions.

References

1. Adams, W.P., Sherali, H.D.: Linearization strategies for a class of zero-one mixed integer programming problems. Oper. Res. **38**(2), 217–226 (1990)
2. Ahmed, S.: A scenario decomposition algorithm for 0–1 stochastic programs. Oper. Res. Lett. **41**(6), 565–569 (2013)
3. Angulo, G., Ahmed, S., Dey, S.S.: Improving the integer l-shaped method. INFORMS J. Comput. **28**(3), 483–499 (2016)
4. Angulo, G., Ahmed, S., Dey, S.S., Kaibel, V.: Forbidden vertices. Math. Oper. Res. **40**(2), 350–360 (2015)
5. Aris, R., Nemhauser, G.L., Wilde, D.J.: Optimization of multistage cyclic and braching systems by serial procedures. AIChE J. **10**(6), 913–919 (1964)
6. Bagirov, A., Karmitsa, N., Mäkelä, M.M.: Introduction to Nonsmooth Optimization: Theory, Practice and Software, vol. 12. Springer, Heidelberg (2014)

7. Beltran, C., Tadonki, C., Vial, J.P.: Semi-lagrangian relaxation. HEC Genève (2004)
8. Bergner, M., et al.: Automatic dantzig-wolfe reformulation of mixed integer programs. Math. Program. **149**(1), 391–424 (2015)
9. Bertele, U., Brioschi, F.: On non-serial dynamic programming. J. Comb. Theory, Ser. A **14**(2), 137–148 (1973)
10. Bienstock, D., Munoz, G.: Lp formulations for polynomial optimization problems. SIAM J. Optim. **28**(2), 1121–1150 (2018)
11. Boland, N., et al.: Combining progressive hedging with a frank-wolfe method to compute lagrangian dual bounds in stochastic mixed-integer programming. SIAM J. Optim. **28**(2), 1312–1336 (2018)
12. Burer, S.: On the copositive representation of binary and continuous nonconvex quadratic programs. Math. Program. **120**(2), 479–495 (2009)
13. Chen, R., Luedtke, J.: On sample average approximation for two-stage stochastic programs without relatively complete recourse. Math. Program. **196**(1), 719–754 (2022)
14. Cifuentes, D., Dey, S.S., Xu, J.: Sensitivity analysis for mixed binary quadratic programming. In: International Conference on Integer Programming and Combinatorial Optimization, pp. 446–459. Springer, Heidelberg (2024). https://doi.org/10.1007/978-3-031-59835-7_33
15. Dey, S.S., Molinaro, M., Wang, Q.: Analysis of sparse cutting planes for sparse milps with applications to stochastic milps. Math. Oper. Res. **43**(1), 304–332 (2018)
16. Diestel, R.: Graph theory. Springer (print edition); Reinhard Diestel (eBooks) (2024)
17. Feizollahi, M.J., Ahmed, S., Sun, A.: Exact augmented lagrangian duality for mixed integer linear programming. Math. Program. **161**, 365–387 (2017)
18. Fulkerson, D., Gross, O.: Incidence matrices and interval graphs. Pac. J. Math. **15**(3), 835–855 (1965)
19. Geoffrion, A.M.: Lagrangean relaxation for integer programming. In: Approaches to Integer Programming, pp. 82–114. Springer, Heidelberg (2009)
20. Gleixner, A., et al.: Miplib 2017: data-driven compilation of the 6th mixed-integer programming library. Math. Program. Comput. **13**(3), 443–490 (2021)
21. Goulart, P.J., Chen, Y.: Clarabel: An interior-point solver for conic programs with quadratic objectives. arXiv preprint arXiv:2405.12762 (2024)
22. Gu, X., Ahmed, S., Dey, S.S.: Exact augmented lagrangian duality for mixed integer quadratic programming. SIAM J. Optim. **30**(1), 781–797 (2020)
23. Higle, J.L.: Stochastic programming: optimization when uncertainty matters. In: Emerging theory, methods, and applications, pp. 30–53. Informs (2005)
24. Küçükyavuz, S., Sen, S.: An introduction to two-stage stochastic mixed-integer programming. In: Leading Developments from INFORMS Communities, pp. 1–27. INFORMS (2017)
25. Laporte, G., Louveaux, F.V.: The integer l-shaped method for stochastic integer programs with complete recourse. Oper. Res. Lett. **13**(3), 133–142 (1993)
26. Laurent, M.: Sums of squares, moment matrices and optimization over polynomials. In: Emerging Applications of Algebraic Geometry, pp. 157–270. Springer, Heidelberg (2009). https://doi.org/10.1007/978-0-387-09686-5_7
27. Lemaréchal, C., Renaud, A.: A geometric study of duality gaps, with applications. Math. Program. **90**, 399–427 (2001)
28. Minoux, M., Ouzia, H.: Drl*: a hierarchy of strong block-decomposable linear relaxations for 0–1 mips. Disc. Appl. Math. **158**(18), 2031–2048 (2010)

29. Nemhauser, G.: Introduction to Dynamic Programming. John Wiley, New York (1966)
30. Robertson, N., Seymour, P.D.: Graph minors. ii. Algorithmic aspects of tree-width. J. Algor. **7**(3), 309–322 (1986)
31. Rockafellar, R.T., Wets, R.J.: Stochastic convex programming: relatively complete recourse and induced feasibility. SIAM J. Control. Optim. **14**(3), 574–589 (1976)
32. Sun, K., Sun, M., Yin, W.: Decomposition methods for global solution of mixed-integer linear programs. SIAM J. Optim. **34**(2), 1206–1235 (2024)
33. Zhang, M., Kucukyavuz, S.: Finitely convergent decomposition algorithms for two-stage stochastic pure integer programs. SIAM J. Optim. **24**(4), 1933–1951 (2014)

Reducing the Large Set Threshold for Oertel's Conjecture on the Mixed-Integer Volume

Andrés Cristi[1](✉) and David Salas[2](✉)

[1] EPFL, Lausanne, Switzerland
andres.cristi@epfl.ch
[2] Universidad de O'Higgins, Rancagua, Chile
david.salas@uoh.cl

Abstract. In 1960, Grünbaum proved that for any convex body $C \subset \mathbb{R}^d$ and every halfspace H containing the centroid of C, one has that the volume of $H \cap C$ is at least a $\frac{1}{e}$-fraction of the volume of C. Recently, in 2014, Oertel conjectured that a similar result holds for mixed-integer convex sets. Concretely, he proposed that for any convex body $C \subset \mathbb{R}^{n+d}$, there should exist a point $\mathbf{x} \in S = C \cap (\mathbb{Z}^n \times \mathbb{R}^d)$ such that for every halfspace H containing $\mathbf{x}$, one has that

$$\mathcal{H}_d(H \cap S) \geq \frac{1}{2^n}\frac{1}{e}\,\mathcal{H}_d(S),$$

where $\mathcal{H}_d$ denotes the d-dimensional Hausdorff measure. While the conjecture remains open, Basu and Oertel proved in 2017 that the above inequality holds true for sufficiently large sets, in terms of a measure known as the *lattice width* of a set. In this work, by following a geometric approach, we improve this result by substantially reducing the threshold at which a set can be considered large. We reduce this threshold from an exponential to a polynomial dependency on the dimension, therefore significantly enlarging the family of mixed-integer convex sets over which Oertel's conjecture holds true.

1 Introduction

A classic result by Grünbaum [13] states that for every convex body $C \subset \mathbb{R}^d$, and for every halfspace H of $\mathbb{R}^d$, one has that

$$H \text{ contains the centroid of } C \implies \frac{\mathrm{vol}_d(H \cap C)}{\mathrm{vol}_d(C)} \geq \left(\frac{d}{d+1}\right)^d \geq \frac{1}{e}, \tag{1.1}$$

where $\mathrm{vol}_d(\cdot)$ stands for the usual d-dimensional volume (given by the Lebesgue measure over $\mathbb{R}^d$). This result, nowadays known as Grünbaum's inequality (together with the study of general measures over convex sets, see [13,14]) is at the core of many advances in convex geometry, see, e.g., [6,8–10,17,18,21,22].

N. Megow and A. Basu (Eds.): IPCO 2025, LNCS 15620, pp. 199–212, 2025.
https://doi.org/10.1007/978-3-031-93112-3_15

It has also been used in the field of optimization in the study of cutting plane algorithms [19] and randomized algorithms [7,11,15]. A particularly interesting application is the study of the information complexity of mixed-integer convex optimization [3–5]: if we want to optimize a linear objective function over a convex body that we can access only via a separation oracle, Grünbaum's inequality gives a bound on the minimum number of queries to the oracle required to find the solution.

Motivated by its application to cutting plane methods in convex optimization, Oertel [20] studied how to extend Grünbaum's inequality to *mixed-integer convex bodies*, that is, sets of the form $S := C \cap (\mathbb{Z}^n \times \mathbb{R}^d)$, where C is a convex body. His analysis followed three steps. First, the relative volume of a set $A \subset S$ should be given by

$$\mu(A) = \frac{\sum_{z \in \mathbb{Z}^n} \mathrm{vol}_d\left(A \cap (\{z\} \times \mathbb{R}^d)\right)}{\sum_{z \in \mathbb{Z}^n} \mathrm{vol}_d\left(S \cap (\{z\} \times \mathbb{R}^d)\right)}, \tag{1.2}$$

This idea of volume for mixed-integer convex sets can be formalized using Hausdorff measures, see Sect. 2 below. Second, since the centroid of S can be outside S, he proposed to replace it by a *centerpoint*: a point $\bar{\mathbf{x}} \in S$ is a centerpoint of S if it is one of the points that attains the maximum in

$$\mathcal{F}(S) := \max_{\mathbf{x} \in S} \left(\inf\{\mu(H \cap S) \,:\, H \text{ is a halfspace with } \mathbf{x} \in H\} \right). \tag{1.3}$$

We call $\mathcal{F}(S)$ the Oertel radius of S. Note that S always admits centerpoints, which explains the max operation in the definition of $\mathcal{F}(S)$ (see [5]). Finally, he studied lower bounds for $\mathcal{F}(S)$, in the spirit of Grünbaum's inequality (1.1). The centerpoint approach was then extended by Basu and Oertel in [5], and used in a series of papers [3–5] to study information complexity in convex mixed-integer optimization.

A Worst-Case-Like Example and Oertel's Conjecture on the Mixed-Integer Volume: Let $K \subset \mathbb{R}^d$ be a d-dimensional compact cone, $C = [0,1]^n \times K$ and $S = C \cap (\mathbb{Z}^n \times \mathbb{R}^d) = \{0,1\}^n \times K$. In this example, one can easily prove the following facts:

- All centerpoints are given by the points (z, c) with $z \in \{0,1\}^n$ and c being the centroid of K.
- For any halspace H containing a centerpoint (z, c), one has that

$$\mu(H \cap S) \geq \frac{1}{2^n} \left(\frac{d}{d+1} \right)^d$$

- By taking a halfspace $H = H_n \times H_d$ such that H_n separates z from the other points in $[0,1]^n$ and H_d is the halfspace of $\mathbb{R}^d$ that contains the vertex of K defined by the hyperplane parallel to the base of K that passes through the centroid of K (exactly the case where Grünbaum's inequality is tight), one has that

$$\mu(H \cap S) = \frac{\mathrm{vol}_d(H_d \cap K)}{2^n \mathrm{vol}_d(K)} = \frac{1}{2^n} \left(\frac{d}{d+1} \right)^d.$$

Thus, in this example, $\mathcal{F}(S) = \frac{1}{2^n}\left(\frac{d}{d+1}\right)^d$. This is, however, a kind of worst-case example: since one can isolate the integer variables of $\{0,1\}^n$ and the volume of all the fibers is the same, one basically is forced to apply Grünbaum's inequality at one fiber that has a fraction of $\frac{1}{2^n}$ of the total volume. Then, *Oertel's conjecture* is that this is, in fact, the worst scenario.

Conjecture 1 (Oertel's conjecture [20, *Conjecture 4.1.20*]*).* Let $C \subset \mathbb{R}^{n+d}$ be a convex body and $S = C \cap (\mathbb{Z}^n \times \mathbb{R}^d)$. Then,

$$\mathcal{F}(S) \geq \frac{1}{2^n}\frac{1}{e}.$$

Partial Results on Oertel's conjecture: Prior to our work, the main contributions towards settling Conjecture 1, can be found in [20] and in the following-up article [5].[1] First, following an approach based on Helly numbers (see, e.g., [3] and the references therein), Oertel showed that

$$\forall S \text{ mixed-integer convex body on } \mathbb{Z}^n \times \mathbb{R}^d, \quad \mathcal{F}(S) \geq \frac{1}{2^n(d+1)}. \tag{1.4}$$

(See [20, Theorem 4.1.19] or [5, Corollary 3.4].) While the factor $\frac{1}{2^n}$ is expected after the previous discussion, the factor $\frac{1}{d+1}$ is only better than the conjectured constant factor $1/e$ for the case $d = 1$.

Basu and Oertel [5] showed that if the set of integer points in $\mathbb{Z}^n$ belonging to the projection of S is sufficiently large, then Conjecture 1 holds true. The concept of "large set" was measured in terms of the *lattice width*, which is the minimum width of the set along integral directions. Formally, for a set $D \subset \mathbb{R}^n$, it is defined as

$$\omega(D) = \min_{u \in \mathbb{Z}^n \setminus \{0\}} \left(\max_{z \in D} u^\top z - \min_{z \in D} u^\top z \right). \tag{1.5}$$

For a set C in $\mathbb{R}^{n+d}$, let us denote $\text{proj}_{\mathbb{R}^n}(C)$ as the (orthogonal) projection of C onto $\mathbb{R}^n$ with the natural identification $\mathbb{R}^{n+d} = \mathbb{R}^n \times \mathbb{R}^d$. Then, the positive result of Basu and Oertel can be summarized in the following result.

Theorem 1 ([5, Theorem 3.6]). *There exists a universal constant $\alpha > 0$ such that for every $n, d \in \mathbb{N}$ and every convex body $C \subset \mathbb{R}^{n+d}$ with $\omega(\text{proj}_{\mathbb{R}^n}(C)) > 2cn(n+d)^{5/2}\alpha n^{n+1}$ for some $c \in \mathbb{R}_+$, then*

$$\mathcal{F}(C \cap (\mathbb{Z}^n \times \mathbb{R}^d)) \geq e^{-\frac{1}{c}-1} + e^{-\frac{2}{c}} - 1.$$

In particular, if $c \in \mathbb{R}_+$ is such that $e^{-\frac{1}{c}-1} + e^{-\frac{2}{c}} - 1 \geq 2^{-(n+1)}$, then Conjecture 1 holds true for C.

[1] The works [5,20] and the subsequent works [3,4] are much richer than what we are describing here. To reduce the exposition, we focus our attention only on the elements that are pertinent to Conjecture 1.

To get an idea of how large $\omega(\mathrm{proj}_{\mathbb{R}^n}(C))$ needs to be in the above theorem, we can look at the values of c at which the bound $e^{-\frac{1}{c}-1}+e^{-\frac{2}{c}}-1$ is non-negative. That is, we can write

$$e^{-\frac{1}{c}-1}+e^{-\frac{2}{c}}-1 \geq 0 \implies e^{-2/c}\left(e^{1/c-1}+1\right) \geq 1$$
$$\implies -\frac{2}{c}+\log(e^{1/c-1}+1) \geq 0 \implies c \geq 1.$$

Thus, the threshold given in Theorem 1 is quite large: For the bound to be meaningful, it requires the lattice width $\omega(\mathrm{proj}_{\mathbb{R}^n}(C))$ to be at least $\Omega((n+d)^{5/2}n^{n+2})$.

Our Contributions: To the best of our knowledge, Conjecture 1 is still open and the results of [20] and [5] described above are the state-of-the-art for it. In our work, we focus on reducing the threshold at which a set is large enough so that Conjecture 1 holds. Our main results are Theorem 2 and Corollary 1:

- We prove that there exists a universal constant $\alpha > 0$ such that if $\mathrm{proj}_{\mathbb{R}^n}(C)$ contains a ball of radius $k \geq \alpha d^2 n^{3/2}$, then Conjecture 1 holds.
- By passing through a unimodular transformation, we deduce a new threshold in terms of the lattice width of the set. We prove that there exists a universal constant $\alpha' > 0$ such that if $\omega(\mathrm{proj}_{\mathbb{R}^n}(C)) \geq \alpha' d^2 n^6$, then Conjecture 1 holds true.

These results improve Theorem 1 by reducing the exponential dependency on n to an explicit polynomial one, and also by reducing the dependency on d from $d^{5/2}$ to d^2. This enlarges considerably the family of sets where Conjecture 1 holds true and provides further indications that it might be true in general.

Due to space constraints, we do not provide all details in this extended abstract and defer them to the full version of the paper. In particular:

- We omit an alternative proof of the case $n = 1$ based on the geometry of cones, which provides several complementary insights on the geometry of the problem.
- We omit some of the proofs, providing instead sketches of proofs. This gives an idea of how geometrical properties of convex sets allow to derive the results.

2 Preliminaries

In what follows, we consider $n, d \in \mathbb{N}$ and we set $C \subset \mathbb{R}^{n+d}$ to be a convex body, that is, a convex compact set with nonempty interior. We denote by $S = C \cap (\mathbb{Z}^n \times \mathbb{R}^d)$ as the mixed-integer set induced by C.

It will be useful to adopt the following convention: We identify $\mathbb{R}^{n+d}$ with $\mathbb{R}^n \times \mathbb{R}^d$. We use the letters z, w to denote elements of $\mathbb{R}^n$ and letters x, y to denote elements of $\mathbb{R}^d$. We use bold characters $\mathbf{x}, \mathbf{y}$ to denote elements of $\mathbb{R}^n \times \mathbb{R}^d$. Then, for every $\mathbf{x} \in \mathbb{R}^{n+d}$, we write $\mathbf{x} = (z, x) \in \mathbb{R}^n \times \mathbb{R}^d$.

We formalize the notion of volume for general sets using Hausdorff measures. For every $r \geq 0$, we denote by $\mathcal{H}_r$ the r-dimensional Hausdorff measure (see, e.g., [12, Definition 2.1]). When r is an integer, $\mathcal{H}_r$ coincides with the r-dimensional Lebesgue measure over each affine space of dimension r. For any set $K \subset \mathbb{R}^{n+d}$ we denote by $\dim K$ its Hausdorff dimension (see, e.g., [12, Definition 2.2]) which is given as

$$\dim K = \inf\{r > 0 \;:\; \mathcal{H}_r(K) = 0\}.$$

When K is a convex set, the $\dim K$ coincides with the dimension of the affine hull of K. So, if K is a convex body and $\dim K = q$, then the usual q-dimensional volume of K coincides $\mathcal{H}_q(K)$. When K is a disjoint union of convex sets (as it is the case of mixed-integer convex sets), then the Hausdorff dimension of K is given by the maximal Hausdorff dimension of its convex components. In particular, if the convex body $C \subset \mathbb{R}^{n+d}$ has at least one interior point $\mathbf{x} = (z, x)$ with $z \in \mathbb{Z}^n$, then $\dim S = d$. Thus, in this case, one has that the measure $\mu(\cdot)$ defined in Eq. 1.2 coincides with the normalization of $\mathcal{H}_d$, that is,

$$\forall A \subset S, \quad \mu(A) = \frac{\sum_{z\in\mathbb{Z}^n} \mathrm{vol}_d(A \cap (\{z\} \times \mathbb{R}^d))}{\sum_{z\in\mathbb{Z}^n} \mathrm{vol}_d(S \cap (\{z\} \times \mathbb{R}^d))} = \frac{\mathcal{H}_d(A)}{\mathcal{H}_d(S)}.$$

In general, for every compact set K one has that

$$\mathcal{H}_q(K) = \begin{cases} +\infty & \text{if } q < \dim K, \\ V \in [0, +\infty) & \text{if } q = \dim K, \\ 0 & \text{if } q > \dim K. \end{cases}$$

Thus, the Hausdorff measure $\mathcal{H}_q$ provides a good extension of the q-dimensional volume of an arbitrary compact set. Moreover, the "intrinsic" volume of a set K is given by $\mathcal{H}_{\dim K}(K)$. For further information on Hausdorff measures, we refer the reader to [12].

For a convex set K of dimension $q = \dim K$, we define the centroid of K

$$\mathfrak{c}(K) = \frac{1}{\mathcal{H}_q(K)} \int_K x \, d\mathcal{H}_q(x). \tag{2.1}$$

Note that the centroid of K is contained in K, thanks to convexity.

Let $K \subseteq \mathbb{R}^p$. For $x \in K$ and $u \in \mathbb{R}^p \setminus \{0\}$ we define $H(u, x) = \{y \in \mathbb{R}^p : u^\top(y - x) \geq 0\}$. Let $q = \dim K$. We define the *Oertel radius* of K as

$$\mathcal{F}(K) = \sup_{x\in K} \inf_{u\in\mathbb{R}^p\setminus\{0\}} \frac{\mathcal{H}_q(H(u, x) \cap K)}{\mathcal{H}_q(K)}. \tag{2.2}$$

Note that we consider the $(\dim K)$-Hausdorff measure of $H(u, x) \cap K$ rather than its volume in its own dimension, to put zero measure on the extreme cases where $H(u, x) \cap K$ has lower dimension. Also note that when K is a mixed-integer convex body, the Oertel radius is attained by some elements of K, as we discussed in the introduction (see [5]).

3 Main Result: A Polynomial Large Set Threshold

We first need to define what "large set" means in this context. We will consider C to be large in $\mathbb{R}^{n+d}$ if its projection $\text{proj}_{\mathbb{R}^n}(C)$ contains a ball of large radius. Assuming without loss of generality that $0 \in C$, and denoting by $\mathbb{B}_n$ the Euclidean unit ball in $\mathbb{R}^n$, let k be such that

$$k\mathbb{B}_n \subset \text{proj}_{\mathbb{R}^n}(C).$$

Our goal is to prove that there exists a universal constant $\alpha > 0$ such that

$$k \geq \alpha d^2 n^{3/2} \implies \mathcal{F}(S) \geq \frac{1}{2^n}\left(\frac{d}{d+1}\right)^d. \tag{3.1}$$

3.1 Outline of the Proof

The main idea is to show that slices approximate the volume of boxes, is some sense emulating the idea of Riemann sums in Riemann integration. We start the following key observation: "away from the boundary" of C, any point $z \in \text{proj}_{\mathbb{R}^n}(C)$ verifies that

$$\mathcal{H}_d(C \cap (\{z\} \times \mathbb{R}^d)) \approx \mathcal{H}_{n+d}(C \cap (\{w \,:\, \|z - w\|_\infty \leq 1/2\} \times \mathbb{R}^d)) \tag{3.2}$$

In what follows, we will consider the following definitions, illustrated in Fig. 1:

- For $z \in \mathbb{R}^n$, we denote the **slice** of C induced by z as $S_z(C) = C \cap (\{z\} \times \mathbb{R}^d)$ (solid blue line inside the set, see Fig. 1).
- For $z \in \mathbb{R}^n$, we denote the n-dimensional **box** of z as $\text{Box}_z^n = \{w \in \mathbb{R}^n : \|w - z\|_\infty \leq 1/2\}$ (solid red square over the grid, see Fig. 1).
- We denote the **rectangular cut** of C induced by z as $B_z(C) = C \cap (\text{Box}_z^n \times \mathbb{R}^d)$ (volumetric section of the convex body in blue, see Fig. 1).

With this notation, Eq. 3.2 can be written as

$$\mathcal{H}_d(S_z(C)) \approx \mathcal{H}_{n+d}(B_z(C)).$$

Our strategy can be split in the following steps:

1. **Approximate volume with slices.** The volume of a convex body that contains a large ball in the projection is well approximated by the measure of its slices. More precisely, if $\text{proj}_{\mathbb{R}^n}(C)$ contains a ball of radius $\Omega(k)$, then $\mathcal{H}_{n+d}(C) = \mathcal{H}_d(C \cap (\mathbb{Z}^n \times \mathbb{R}^d)) \cdot (1 \pm O(dn^{3/4}/k^{1/2}))$ (see Lemma 3). We prove this in two steps.
   (a) Take the set $C_{-\varepsilon} = (1-\varepsilon) \cdot C$. We have that $\mathcal{H}_{n+d}(C_{-\varepsilon}) = (1-\varepsilon)^{n+d}\mathcal{H}_{n+d}(C)$, and for every point $z \in \text{proj}_{\mathbb{R}^n}(C_{-\varepsilon})$ the ball of radius εr centered at z is contained in $\text{proj}_{\mathbb{R}^n}(C)$ (a simple consequence of Lemma 1).

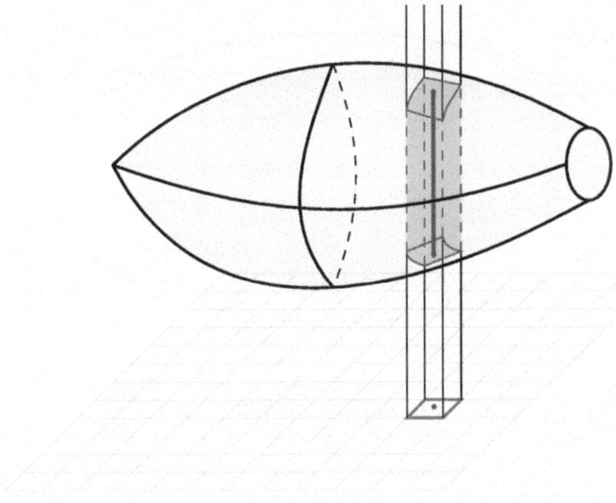

Fig. 1. Illustration of slices, boxes and rectangular cut with $n = 2$ and $d = 1$. (Color figure online)

(b) If a point in the projection has a large ball around, then the volume of the corresponding d-dimensional slice approximates well the volume of the $(n + d)$-dimensional rectangular cut. More precisely, for a point $z \in \text{proj}_{\mathbb{R}^n}(C)$, if a ball of radius r centered at z is contained in $\text{proj}_{\mathbb{R}^n}(C)$, then $\mathcal{H}_{n+d}(B_z(D)) = \mathcal{H}_d(S_z(D)) \cdot (1 \pm O(d\sqrt{n}/r))$. See Lemma 2.

2. **Move the centroid.** we can assume the centroid of C is in $\mathbb{Z}^n \times \mathbb{R}^d$ by losing a small fraction of the volume. More precisely, we can shift C and obtain a new set C' with centroid in $\mathbb{Z}^n \times \mathbb{R}^d$, so that $\mathcal{H}_{n+d}(C \Delta C') \leq \mathcal{H}_{n+d}(C) \cdot O(\sqrt{n}(n + d)/k)$. See Lemma 4.
3. **Cut and use the continuous bound.** If we cut C' with an arbitrary hyperplane passing through its centroid and denote the two sides by A' and B', then both $\mathcal{H}_{n+d}(A')$ and $\mathcal{H}_{n+d}(B')$ are at least $\mathcal{H}_{n+d}(C')/e$.
4. **Approximate volumes of sides by the slices.** Either $\text{proj}_{\mathbb{R}^n}(A')$ or $\text{proj}_{\mathbb{R}^n}(B')$ must contain a ball of radius $k/2$. In general, if the union of two convex sets contains a ball of radius k, one of them contains a ball of radius $k/2$ [16]. Assume without loss of generality that A' contains the ball. The volumes of C' and A' are well approximated by the slices, so the volume of B' is also well approximated by the slices.

3.2 Technical Lemmas

For a convex set D, we have that for any pair of elements $z, w \in D$, $u = \frac{1}{1+\varepsilon}z + \frac{\varepsilon}{1+\varepsilon}w \in D$, and then $z + \varepsilon w = (1 + \varepsilon)u \in (1 + \varepsilon)D$, so we can write the following lemma (see Fig. 2).

Lemma 1. *Let $D \subseteq \mathbb{R}^n$ be a convex set. For all $z, w \in D$, and $\varepsilon > 0$, we have that $z + \varepsilon w \in (1 + \varepsilon) \cdot D$.*

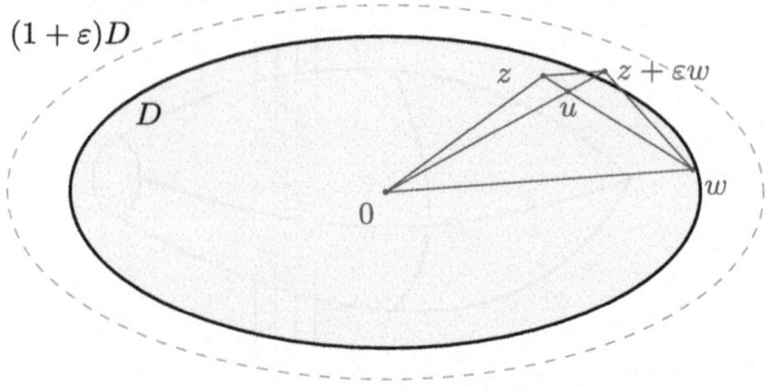

Fig. 2. A graphic representation of Lemma 1

Lemma 2. *For a convex body $C \subseteq \mathbb{R}^{n+d}$ and a point $z \in \text{proj}_{\mathbb{R}^n}(C)$ such that $\text{proj}_{\mathbb{R}^n}(C)$ contains a ball of radius $r > \sqrt{n}/2$ centered at z. We have that*

$$\mathcal{H}_d(S_z(C)) \cdot \left(1 - \frac{\sqrt{n}}{2r}\right)^d \leq \mathcal{H}_{n+d}(B_z(C)) \leq \mathcal{H}_d(S_z(C)) \cdot \left(1 + \frac{\sqrt{n}}{2r}\right)^d .$$

Sketch of Proof of Lemma 2. The volume of $B_z(C)$ can be written as the integral of the (d-dimensional) volume of the slices defined by points $w \in \text{Box}_z^n$. Thus, we compare the volume of each of these slices $S_w(C)$ with the volume of the slice $S_z(C)$. Any point w in the n-dimensional hypercube is at a distance at most $\sqrt{n}/2$ of z. Because we assumed there is a ball of radius r around z, there is a point $y \in C$ in the same direction as w from z, but at distance r. The stereographic projection of $S_z(C)$ from y on $\{w\} \times \mathbb{R}^d$ is contained in $S_w(C)$. The projection is a rescaled copy of $S_z(C)$, by a factor no smaller than $(1-(\sqrt{n}/2)/r)$, so $\mathcal{H}_d(S_z(C)) \cdot (1-(\sqrt{n}/2)/r)^d \leq \mathcal{H}_d(S_w(C))$. Similarly, we can project $S_w(C)$ on $\{z\} \times \mathbb{R}^d$ from the other side, and obtain that $\mathcal{H}_d(S_w(C)) \leq \mathcal{H}_d(S_z(C)) \cdot (1+(\sqrt{n}/2)/r)^d$. See Fig. 3. □

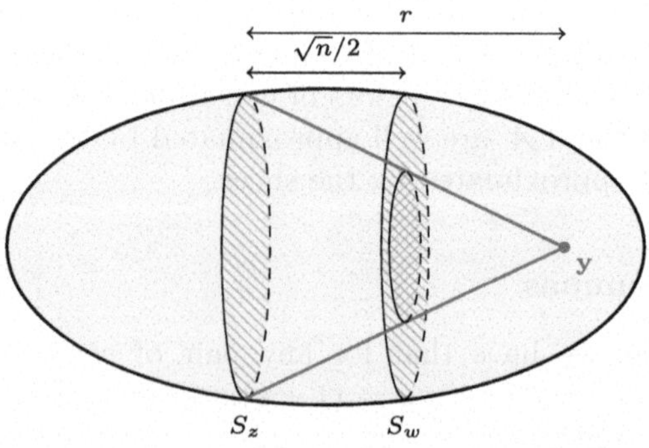

Fig. 3. A graphic representation of the procedure used in Lemma 2 to compare the volume of two slices.

Lemma 3. *Let* $C \subseteq \mathbb{R}^{n+d}$ *be a convex body such that* $\text{proj}_{\mathbb{R}^n}(C)$ *contains a ball of radius* k*, and let* $S = C \cap (\mathbb{Z}^n \times \mathbb{R}^d)$*. If* $5\frac{dn^{3/4}}{k^{1/2}} \leq 1$*, we have that*

$$\mathcal{H}_{n+d}(C) \cdot \left(1 - 5\frac{dn^{3/4}}{k^{1/2}}\right) \leq \mathcal{H}_d(S) \leq \mathcal{H}_{n+d}(C) \cdot \left(1 + 5\frac{dn^{3/4}}{k^{1/2}}\right).$$

Sketch of Proof of Lemma 3. Move the origin to the center of the ball of radius k in $\text{proj}_{\mathbb{R}^n}(C)$, and for an appropriate $\varepsilon > 0$, define $C_{-\varepsilon} = C \cdot (1 - \varepsilon)$ and $C_{+\varepsilon} = C \cdot (1 + \varepsilon)$. By Lemma 1, for every $z \in \text{proj}_{\mathbb{R}^n}(C_{-\varepsilon})$, the ball of radius $k\varepsilon$ centered at z is contained in $\text{proj}_{\mathbb{R}^n}(C)$. Similarly, for every $z \in \text{proj}_{\mathbb{R}^n}(C)$, the ball of radius $k\varepsilon$ centered at z is contained in $\text{proj}_{\mathbb{R}^n}(C_{+\varepsilon})$. For the lower bound, we apply Lemma 2 on the slices of $C_{-\varepsilon}$, taking $r = k\varepsilon$ to approximate the slices of C. From this approximation, we lose a factor of $(1 - O(\frac{d\sqrt{n}}{k\varepsilon}))$. On top of that, $\mathcal{H}_{n+d}(C_{-\varepsilon}) = (1 - \varepsilon)^{n+d} \cdot \mathcal{H}_{n+d}(C)$, so in total, we lose a factor $(1 - O(\frac{d\sqrt{n}}{k\varepsilon} + \varepsilon(n+d)))$. Analogously, using C and $C_{+\varepsilon}$ this time, we get an upper bound where we lose a factor $(1 + O(\frac{d\sqrt{n}}{k\varepsilon} + \varepsilon(n + d)))$. Taking $\varepsilon = 1/(n^{1/4}k^{1/2})$, we get the bounds of the lemma. □

Lemma 4. *Let* $C \subseteq \mathbb{R}^{n+d}$ *be a convex body such that* $\text{proj}_{\mathbb{R}^n}(C)$ *contains a ball of radius* k*. There exists a vector* $\mathbf{x}$ *such that the centroid of* $C + \mathbf{x}$ *is in* $\mathbb{Z}^n \times \mathbb{R}^d$ *and such that*

$$\mathcal{H}_{n+d}\left((C + \mathbf{x}) \setminus C\right) \leq \left(\left(1 + \frac{\sqrt{n}}{2k}\right)^{n+d} - 1\right) \cdot \mathcal{H}_{n+d}(C).$$

Proof. Without losing any generality, assume that the ball of radius k contained in $\text{proj}_{\mathbb{R}^n}(C)$ is centered in the origin. Let (z, x) be the centroid of C. There must be a point $z' \in \mathbb{Z}^n$ at distance at most $\sqrt{n}/2$ of z. Thus, $z'' = \frac{2k}{\sqrt{n}}(z' - z)$ is at a distance at most k of the origin, and therefore, it is in $\text{proj}_{\mathbb{R}^n}(C)$. That means there is $x'' \in \mathbb{R}^d$ such that $\mathbf{x}'' = (z'', x'') \in C$. We shift C by $\frac{\sqrt{n}}{2k}\mathbf{x}''$, so the projection of the centroid of the shifted set is z'.

Since $\mathbf{x}''$ is in C, Lemma 1 implies that $C + \frac{\sqrt{n}}{2k}\mathbf{x}'' \subseteq (1 + \frac{\sqrt{n}}{2k}) \cdot C$, and therefore, $(C + \frac{\sqrt{n}}{2k}\mathbf{x}'') \setminus C \subseteq \left((1 + \frac{\sqrt{n}}{2k}) \cdot C\right) \setminus C$. The result follows. □

3.3 Statement and Proof of the Main Result

We now present and prove our main result.

Theorem 2. *Let* $C \subseteq \mathbb{R}^{n+d}$ *be a convex body such that* $\text{proj}_{\mathbb{R}^n}(C)$ *contains a ball of radius* k*. There is a point* $\mathbf{x} \in S = C \cap (\mathbb{Z}^n \times \mathbb{R}^d)$ *such that for every halfspace* H *that contains* $\mathbf{x}$*,*

$$\mathcal{H}_d(S \cap H) \geq \left(\frac{1}{e} - 16\frac{dn^{3/4}}{k^{1/2}}\right) \cdot \mathcal{H}_d(S).$$

In particular, there exists a universal constant $\alpha > 0$ such that if $k \geq \alpha d^2 n^{3/2}$, then

$$\mathcal{F}(S) \geq \frac{1}{2^n}\left(\frac{d}{d+1}\right)^d.$$

Sketch of proof of Theorem 2. We take the set $C' = C + \mathbf{x}$ with centroid in $\mathbb{Z}^n \times \mathbb{R}^d$ guaranteed by Lemma 4, and a halfspace H defined by a hyperplane that passes through the centroid. Applying Grünbaum's inequality, both sides $A' = C' \cap H$ and $B' = C' \setminus H$ have a volume of at least $(1/e) \cdot \mathcal{H}_{n+d}(C') = (1/e) \cdot \mathcal{H}_{n+d}(C)$. By Lemma 4, both $A = C \cap H$ and $B = C \setminus H$ have a volume of at least $((1/e) - O(dn^{3/2}/k)) \cdot \mathcal{H}_{n+d}(C)$. Necessarily, one of the two sides, A or B, contains a ball of radius $k/2$. Assume this side is A. By Lemma 3,

$$\mathcal{H}_d(A \cap (\mathbb{Z}^n \times \mathbb{R}^d)) = \left(1 \pm O\left(\frac{dn^{3/4}}{k^{1/2}}\right)\right) \cdot \mathcal{H}_{n+d}(A).$$

The same holds for C, and since $B = C \setminus A$, the same holds for B.

Proof. Let $C' = C + \mathbf{x}$, where $\mathbf{x}$ is the vector guaranteed to exist by Lemma 4 so that the centroid of C' is in $S' = C' \cap (\mathbb{Z}^n \times \mathbb{R}^d)$. Let H be an arbitrary affine halfspace defined by affine hyperplane that passes through the centroid of C'. Let $A' = C' \cap H$, $A = C \cap H$, $B' = C' \setminus H$, and $B = C \setminus H$. Without losing any generality, we assume that all these sets are full dimensional. We have that $\mathcal{H}_{n+d}(A') \geq (1/e) \cdot \mathcal{H}_{n+d}(C')$ and $\mathcal{H}_{n+d}(B') \geq (1/e) \cdot \mathcal{H}_{n+d}(C')$. Either $\mathrm{proj}_{\mathbb{R}^n}(A)$ or $\mathrm{proj}_{\mathbb{R}^n}(B)$ contains a ball of radius $k/2$. Since we will prove the same lower bound for both sides, let us assume that $\mathrm{proj}_{\mathbb{R}^n}(A)$ contains a ball of radius $k/2$. By Lemma 3 and the fact that $(1-a)(1-b) \geq 1-a-b$ for every $a, b \geq 0$, we have that

$$\begin{aligned}
\mathcal{H}_d(A \cap (\mathbb{Z}^n \times \mathbb{R}^d)) &\geq \mathcal{H}_{n+d}(A) \cdot \left(1 - 5\sqrt{2}\frac{dn^{3/4}}{k^{1/2}}\right) \\
&\geq (\mathcal{H}_{n+d}(A') - \mathcal{H}_{n+d}(C' \setminus C)) \cdot \left(1 - 5\sqrt{2}\frac{dn^{3/4}}{k^{1/2}}\right) \\
&\geq \mathcal{H}_{n+d}(C)\left(\frac{1}{e} - \frac{2\sqrt{n}(n+d)}{k}\right) \cdot \left(1 - 5\sqrt{2}\frac{dn^{3/4}}{k^{1/2}}\right) \\
&\geq \frac{1}{e}\mathcal{H}_{n+d}(C)\left(1 - \frac{2e}{11} \cdot d\frac{n^{3/4}}{k^{1/2}}\right) \cdot \left(1 - 5\sqrt{2}\frac{dn^{3/4}}{k^{1/2}}\right) \\
&\geq \mathcal{H}_{n+d}(C) \cdot \frac{1}{e} \cdot \left(1 - 8\frac{dn^{3/4}}{k^{1/2}}\right),
\end{aligned}$$

where in the third inequality we applied Lemma 4, and then the assumption that $16\frac{dn^{3/4}}{k^{1/2}} \leq 1$, so $\frac{\sqrt{n}(n+d)}{k} \leq d\frac{n^{3/4}}{k^{1/2}} \cdot \frac{n^{3/4}}{k^{1/2}} \leq \frac{1}{11} \cdot d\frac{n^{3/4}}{k^{1/2}}$. In the last inequality, we used the fact that $2e/11 + 5\sqrt{2} \leq 8$.

On the other hand,

$$\begin{aligned}
\mathcal{H}_d(B \cap (\mathbb{Z}^n \times \mathbb{R}^d)) &= \mathcal{H}_d(S) - \mathcal{H}_d(A \cap (\mathbb{Z}^n \times \mathbb{R}^d)) \\
&\geq (\mathcal{H}_{n+d}(C) - \mathcal{H}_{n+d}(A)) \cdot \left(1 - 5\sqrt{2}\frac{dn^{3/4}}{k^{1/2}}\right) \\
&= \mathcal{H}_{n+d}(B) \cdot \left(1 - 5\sqrt{2}\frac{dn^{3/4}}{k^{1/2}}\right) \\
&\geq (\mathcal{H}_{n+d}(B') - \mathcal{H}_{n+d}(C' \setminus C)) \cdot \left(1 - 5\sqrt{2}\frac{dn^{3/4}}{k^{1/2}}\right) \\
&\geq \mathcal{H}_{n+d}(C) \cdot \frac{1}{e} \cdot \left(1 - 8\frac{dn^{3/4}}{k^{1/2}}\right).
\end{aligned}$$

Now, directly by Lemma 3,

$$\mathcal{H}_d(S) \leq \mathcal{H}_{n+d}(C) \cdot \left(1 + 5\sqrt{2}\frac{dn^{3/4}}{k^{1/2}}\right).$$

Therefore,

$$\begin{aligned}
\frac{\min\{\mathcal{H}_d(A \cap (\mathbb{Z}^n \times \mathbb{R}^d)), \mathcal{H}_d(B \cap (\mathbb{Z}^n \times \mathbb{R}^d))\}}{\mathcal{H}_d(S)} &\geq \frac{1}{e} \cdot \frac{1 - 8\frac{dn^{3/4}}{k^{1/2}}}{1 + 8\frac{dn^{3/4}}{k^{1/2}}} \\
&\geq \frac{1}{e} \cdot \left(1 - 16\frac{dn^{3/4}}{k^{1/2}}\right).
\end{aligned}$$

Finally, if $k \geq \alpha d^2 n^{3/2}$, we get that

$$\mathcal{F}(S) \geq \frac{1}{e}\left(1 - \frac{16}{\sqrt{\alpha}}\right).$$

We can choose $\alpha = \left(\frac{16 \cdot 4}{4-e}\right)^2$ so $\frac{1}{e}\left(1 - \frac{11}{\sqrt{\alpha}}\right) \geq \frac{1}{4} \geq \frac{1}{2^n}\left(\frac{d}{d+1}\right)^d$. □

3.4 A New Threshold for the Lattice Width

Here we discuss the relation between the condition of our result, that the projection contains a ball of large radius, and the condition of large lattice width in the result of Basu and Oertel. We show that if the lattice width of a convex set is large, after an appropriate transformation, its projection in $\mathbb{R}^n$ contains a large ball, and therefore, we can reinterpret our bound in terms of the lattice width.

Here, $\mathrm{Flt}(n)$ is the flatness constant in dimension n, which is defined as the supremum of the lattice width of convex sets contained in $\mathbb{R}^n \setminus \mathbb{Z}^n$. It is known that $\mathrm{Flt}(n) \leq n^{5/2}$ [2, Section 7.4, Theorem 8.3].

Proposition 1. *Given a convex body $K \subseteq \mathbb{R}^n$ there is an unimodular linear transformation $L : \mathbb{R}^n \to \mathbb{R}^n$ such that $L(K)$ contains a ball of radius $\omega(K)/(2n^2\mathrm{Flt}(n)) \geq \omega(K)/(2n^{9/2})$.*

Proof. Let Δ_n be the n-dimensional standard simplex in $\mathbb{R}^n$, that is, $\Delta_n = \operatorname{conv}\{\mathbf{0}, \mathbf{e}_1, \dots, \mathbf{e}_n\}$. Averkov et al. [1, Theorem 2.1] showed that a convex set $K \subseteq \mathbb{R}^n$ always contains a unimodular copy of $\frac{\omega(K)}{2n\mathrm{Flt}(n)} \cdot \Delta_n$, i.e., there is an affine unimodular transformation such that the image of C contains $\frac{\omega(K)}{2n\mathrm{Flt}(n)} \cdot \Delta_n$. It is a well-known fact that the standard simplex Δ_n contains a ball of radius $1/n$, and therefore, the image of K contains a ball of radius $\frac{\omega(K)}{2n^2\mathrm{Flt}(n)}$. □

A unimodular linear transformation in $\mathbb{R}^n$ is a linear transformation that is invertible over the integers, i.e., it identifies points in $\mathbb{Z}^n$ with points in $\mathbb{Z}^n$. If we extend L, the linear transformation of the proposition, to $\mathbb{R}^{n+d}$ using the identity in $\mathbb{R}^d$, we obtain an invertible linear transformation $\bar{L} = (L, \mathrm{Id})$ that preserves all the volumes in $\mathbb{Z}^n \times \mathbb{R}^d$. This means that we can effectively work with $\bar{L}(C \cap (\mathbb{Z}^n \times \mathbb{R}^d))$ instead of $C \cap (\mathbb{Z}^n \times \mathbb{R}^d)$, and any conclusion about the volume also holds for $C \cap (\mathbb{Z}^n \times \mathbb{R}^d)$. Thus, we obtain the following as a consequence of our theorem.

Corollary 1. *Let $C \subseteq \mathbb{R}^{n+d}$ be a convex body. There is a point $\mathbf{x} \in S = C \cap (\mathbb{Z}^n \times \mathbb{R}^d)$ such that for every halfspace H that contains $\mathbf{x}$,*

$$\mathcal{H}_d(H \cap S) \geq \left(\frac{1}{e} - 11\sqrt{2} \frac{dn^{7/4}\sqrt{Flt(n)}}{\sqrt{\omega(\mathrm{proj}_{\mathbb{R}^n}(C))}} \right) \cdot \mathcal{H}_d(S).$$

In particular, there is a universal constant $\alpha > 0$ such that if $\omega(\mathrm{proj}_{\mathbb{R}^n}(C)) \geq \alpha d^2 n^6$, then

$$\mathcal{F}(S) \geq \frac{1}{2^n} \left(\frac{d}{d+1} \right)^d.$$

4 Final Comments and Perspectives

We provided new thresholds for large sets such that Oertel's Conjecture (Conjecture 1) holds true. The first one is based on the idea that a large set should contain a large ball, and the second one is based on the concept of lattice width, as in Theorem 1. In both cases, our thresholds are polynomial: we require that the radius of the inscribed ball is $\Omega(d^2 n^{3/2})$, and alternatively, that the lattice width (of the projection to the integer variables) is $\Omega(d^2 n^6)$.

We believe that Oertel's Conjecture holds in general, and not only for large sets. However, this has proven to be a very challenging problem.

Acknowledgments. The first and second author were partially funded by the Center of Mathematical Modeling (CMM), FB210005, BASAL funds for centers of excellence (ANID-Chile). The second author was partially funded by the project FONDECYT 1251159 (ANID-Chile). The second author whishes to thank A. Basu for his lectures and useful discussions during the conference IPCO 2023, that motivated this work. Part of this research was conducted during the post-doctoral fellowship of the first author at the Center of Mathematical Modeling (CMM).

References

1. Averkov, G., Hofscheier, J., Nill, B.: Generalized flatness constants, spanning lattice polytopes, and the Gromov width. Manuscripta Math. **170**(1–2), 147–165 (2023). https://doi.org/10.1007/s00229-021-01363-x
2. Barvinok, A.: A course in convexity, Graduate Studies in Mathematics, vol. 54. American Mathematical Society, Providence (2002). https://doi.org/10.1090/gsm/054
3. Basu, A.: Complexity of optimizing over the integers. Math. Program. **200**(2), 739–780 (2023). https://doi.org/10.1007/s10107-022-01862-z
4. Basu, A., Jiang, H., Kerger, P., Molinaro, M.: Information complexity of mixed-integer convex optimization. Math. Program. (2024). https://doi.org/10.1007/s10107-024-02099-8
5. Basu, A., Oertel, T.: Centerpoints: a link between optimization and convex geometry. SIAM J. Optim. **27**(2), 866–889 (2017). https://doi.org/10.1137/16M1092908
6. Belloni, A., Freund, R.M.: On the symmetry function of a convex set. Math. Program. **111**(1–2), 57–93 (2008). https://doi.org/10.1007/s10107-006-0074-4
7. Bertsimas, D., Vempala, S.: Solving convex programs by random walks. J. ACM **51**(4), 540–556 (2004). https://doi.org/10.1145/1008731.1008733
8. Bobkov, S.G.: Convex bodies and norms associated to convex measures. Probab. Theory Related Fields **147**(1–2), 303–332 (2010). https://doi.org/10.1007/s00440-009-0209-7
9. Böröczky, K.J., Henk, M.: Cone-volume measure and stability. Adv. Math. **306**, 24–50 (2017). https://doi.org/10.1016/j.aim.2016.10.005
10. Brzezinski, P.: Volume estimates for sections of certain convex bodies. Math. Nachr. **286**(17–18), 1726–1743 (2013). https://doi.org/10.1002/mana.201200119
11. Dabbene, F., Shcherbakov, P.S., Polyak, B.T.: A randomized cutting plane method with probabilistic geometric convergence. SIAM J. Optim. **20**(6), 3185–3207 (2010). https://doi.org/10.1137/080742506
12. Evans, L.C., Gariepy, R.F.: Measure Theory and Fine Properties of Functions. Textbooks in Mathematics, revised edn. CRC Press, Boca Raton (2015)
13. Grünbaum, B.: Partitions of mass-distributions and of convex bodies by hyperplanes. Pacific J. Math. **10**, 1257–1261 (1960)
14. Grünbaum, B.: Measures of symmetry for convex sets. In: Proceedings of Symposium Pure Mathematics, Vol. VII, pp. 233–270. Amer. Math. Soc., Providence (1963)
15. He, S., Luo, Z.Q., Nie, J., Zhang, S.: Semidefinite relaxation bounds for indefinite homogeneous quadratic optimization. SIAM J. Optim. **19**(2), 503–523 (2008). https://doi.org/10.1137/070679041
16. Kadets, V.: Coverings by convex bodies and inscribed balls. Proc. Am. Math. Soc. **133**(5), 1491–1495 (2005). https://doi.org/10.1090/S0002-9939-04-07650-6
17. Klartag, B.: A central limit theorem for convex sets. Invent. Math. **168**(1), 91–131 (2007). https://doi.org/10.1007/s00222-006-0028-8
18. Lovász, L., Vempala, S.: The geometry of logconcave functions and sampling algorithms. Random Struct. Algor. **30** (2007). https://doi.org/10.1002/rsa.20135
19. Nemirovsky, A.S., Yudin, D.B.: Problem complexity and method efficiency in optimization. Wiley-Interscience Series in Discrete Mathematics. John Wiley & Sons, Inc., New York (1983)
20. Oertel, T.: Integer convex minimization in low dimensions. Ph.D. thesis, ETH Zurich (2014)

21. Stephen, M., Zhang, N.: Grünbaum's inequality for projections. J. Funct. Anal. **272**(6), 2628–2640 (2017). https://doi.org/10.1016/j.jfa.2016.09.009
22. Stephen, M., Yaskin, V.: Applications of Grünbaum-type inequalities. Trans. Am. Math. Soc. **372**(9), 6755–6769 (2019). https://doi.org/10.1090/tran/7879

An Unconditional Lower Bound for the Active-Set Method on the Hypercube

Yann Disser and Nils Mosis(✉)

TU Darmstadt, Darmstadt, Germany
{disser,mosis}@mathematik.tu-darmstadt.de

Abstract. The existence of a polynomial-time pivot rule for the simplex method is a fundamental open question in optimization. While many super-polynomial lower bounds exist for individual or very restricted classes of pivot rules, there currently is little hope for an unconditional lower bound that addresses all pivot rules. We approach this question by considering the active-set method as a natural generalization of the simplex method to non-linear objectives. This generalization allows us to prove the first unconditional lower bound *for all pivot rules*. More precisely, we construct a multivariate polynomial of degree linear in the number of dimensions such that the active-set method started in the origin visits all vertices of the hypercube. We hope that our framework serves as a starting point for a new angle of approach to understanding the complexity of the simplex method.

Keywords: simplex method · active-set method · lower bound · linear programming · non-linear programming

1 Introduction

The simplex method [12] is widely regarded as one of the most natural and important practical algorithms for solving linear programming problems. However, its theoretical complexity remains an open question to this day. This has to do with the fact that the simplex method's behavior is highly sensitive to the choice of pivot rule, which determines how ties between improving edges at a vertex are broken and significantly influences the method's behavior, such as whether it guarantees termination.

The discovery of a polynomial-time pivot rule for the simplex method would have profound implications. In particular, it would most likely provide a strongly polynomial algorithm for linear programming, thereby resolving the ninth open problem on Smale's list of mathematical problems for the 21st century [35]. While weakly polynomial algorithms for linear optimization are known, namely the ellipsoid and the interior-point method [27,38], and while the latter has recently been adapted to a strongly polynomial algorithm for linear programs

N. Megow and A. Basu (Eds.): IPCO 2025, LNCS 15620, pp. 213–227, 2025.
https://doi.org/10.1007/978-3-031-93112-3_16

with at most two variables per constraint [2,10], barrier methods cannot yield strongly polynomial guarantees [3]. For linear programs with zero-one vertices strongly polynomial simplex pivot rules are known [8]. A general polynomial-time pivot rule for the simplex method would also imply a polynomial bound on the combinatorial diameter of polytopes, thus resolving the long-standing polynomial Hirsch conjecture [11,33].

Over the years, super-polynomial lower running time bounds were established for various natural pivot rules. Bounds for straightforward rules were found early-on by utilizing distorted hypercubes [5,7,24,26,29,31]. More recently, linear programs derived from Markov Decision Processes have been used to handle more sophisticated, memory-based rules [6,14,20–22]. In some cases, the exponential worst-case behavior of the simplex method could be attributed to its mightiness [1,17,18], which can be seen as a conditional lower bound. Nevertheless, results from smoothed complexity [9,13,25,36] suggest that the constructions underlying the above bounds are fragile, which indicates that fundamentally different approaches may be necessary to achieve an unconditional lower bound that applies universally across all pivot rules. While existing hypercube constructions were generalized to small classes of pivot rules [4,15], the existence of an unconditional super-polynomial lower bound remains wide open.

We approach this question by considering the active-set method [19,32] as a natural generalization of the simplex method to non-linear programs. A case of particular importance for the active-set method are quadratic, concave programs for which, much like in the linear setting, only weakly polynomial algorithms are known [38,39], and the active-set method is a promising candidate for the first strongly polynomial algorithm. Similarly to the simplex method, the active-set method is governed by a pivot rule that determines the improving direction in each step in case there is more than one option. Again, no general running time bounds for all pivot rules are known.

Our Results. We prove the first unconditional super-polynomial lower bounds on the running time of the active-set method *for all pivot rules.* More precisely, we show the following.

Theorem 1. *For all $n \in \mathbb{N}_{>2}$, there is a multivariate polynomial $\mathcal{F}_n$ of degree n such that the active-set method started in $\mathbf{0}$ needs $2^n - 1$ iterations to optimize $\mathcal{F}_n$ over the n-dimensional hypercube $[0,1]^n$, irrespective of the pivot rule.*

In particular, our construction forces the active-set method to visit all vertices of the hypercube while traversing along its edges in a simplex-like fashion.

This result yields a new method of approach for unconditional lower bounds on the running time of the simplex method: If we could lower the degree of the polynomial to 1 by considering more general polytopes, we would obtain a super-polynomial (even exponential) lower bound for the simplex method using any pivot rule.

As a first step in this direction, we can lower the degree of the polynomial, while maintaining a super-polynomial bound, simply by neglecting most of the input dimensions.

Corollary 1. *For every $d(n) = \omega(\log n)$ with $d(n) \leq n$, there are polynomials $(\mathcal{F}_n)_{n\in\mathbb{N}}$ of degrees $\mathcal{O}(d(n))$ such that the active-set method started in* $\mathbf{0}$ *needs $2^{\omega(\log n)}$ iterations to optimize $\mathcal{F}_n$ over the n-dimensional hypercube $[0,1]^n$, irrespective of the pivot rule.*

Importantly, we provide a single construction that addresses all pivot rules simultaneously (there is a unique improving direction in each step). This means that lowering the degree of the polynomial for general polytopes would constitute progress towards disproving the monotone polynomial Hirsch conjecture [40].

We observe that it is NP-complete to optimize polynomials of degree 3 over the hypercube (Proposition 5). This already implies a super-polynomial lower bound on the running time of the active-set method, assuming P $\neq$ NP. The value of our contribution lies in the fact that we provide an *unconditional* lower bound.

In fact, it turns out that the polynomials of Theorem 1 and Corollary 1 induce decomposable unique sink orientations of the hypercube (Proposition 6) that can be solved in linear time [34]. In particular, the polynomials in our construction do not give rise to NP-hard optimization problems (unless P = NP).

Notation. We write all vectors in boldface and denote the i-th unit vector by $\boldsymbol{e}^i$. For $x, y \in \mathbb{N} \cup \{0\}$, we write $x \equiv_2 y$ if both x and y are either even or odd. For $n \in \mathbb{N}$, we write $[n] =: \{1, \ldots, n\}$. We denote the i-th row of a matrix $A \in \mathbb{R}^{m\times n}$ by $\boldsymbol{A}_{i\cdot}$ and, for an index set $\mathcal{A} \subseteq [m]$, we write $A_{\mathcal{A}\cdot}$ to denote the matrix that consists of the rows $\boldsymbol{A}_{i\cdot}$ with $i \in \mathcal{A}$.

2 Simplex and Active-Set Method

The aim of this section is to illustrate that the active-set method is a natural generalization of the simplex method. This connection is well-known.

Proposition 1 (Exercise 8.17 in [19]). *The active-set method, when applied to linear objectives, is* equivalent *to the simplex method, i.e., the same intermediate solutions are computed during the application of both algorithms.*

Consider Algorithms 1 & 2 for a side-by-side comparison of Simplex and ActiveSet. Both algorithms are formulated in a way that makes their connection evident. We now describe the behavior of both algorithms in more detail.

Algorithm 1: $\textsc{Simplex}(\boldsymbol{c}, A, \boldsymbol{b}, \boldsymbol{x})$

input: $\boldsymbol{c} \in \mathbb{Q}^n$, vertex $\boldsymbol{x}$ of polytope $\mathcal{P} := \{\boldsymbol{y} \colon A\boldsymbol{y} \leq \boldsymbol{b}\}$,
  system $A\boldsymbol{y} \leq \boldsymbol{b}$ non-degenerate
output: optimum solution $\boldsymbol{x}$ of $\max\{\boldsymbol{c}^\top \boldsymbol{y} \colon \boldsymbol{y} \in \mathcal{P}\}$

$\mathcal{B} \leftarrow \mathrm{eq}(\boldsymbol{x})$ `(basis)`
while $\mathcal{D} := \{\boldsymbol{d} \colon A_{\mathrm{eq}(\boldsymbol{x}).}\boldsymbol{d} \leq \boldsymbol{0},\ \boldsymbol{c}^\top \boldsymbol{d} > 0\} \neq \emptyset$**:**
  take $\boldsymbol{d} \in \mathcal{D}$ with $|\{i \in \mathcal{B} \colon \boldsymbol{A_{i.}d} = 0\}| = n - 1$
  if $A_{\mathcal{B}.}\boldsymbol{d} \neq \boldsymbol{0}$: `(always true)`
    $\mathcal{B} \leftarrow \mathcal{B} \setminus \{\ell\}$ for the unique $\ell \in \mathcal{B}$ with $\boldsymbol{A_{\ell.}d} < 0$
  if $A_{\mathcal{B}.}\boldsymbol{d} = \boldsymbol{0}$: `(always true)`
    $\mu \leftarrow \inf\{\mu \geq 0 \colon \boldsymbol{x} + \mu\boldsymbol{d} \notin \mathcal{P}\}$
    $\boldsymbol{x} \leftarrow \boldsymbol{x} + \mu\boldsymbol{d}$
    if $\boldsymbol{c}^\top \boldsymbol{d} > 0$: `(always true)`
      $\mathcal{B} \leftarrow \mathcal{B} \cup \{j\}$ for the unique $j \in \mathrm{eq}(\boldsymbol{x}) \setminus \mathcal{B}$

Algorithm 2: $\textsc{ActiveSet}(f, A, \boldsymbol{b}, \boldsymbol{x})$

input: $f \colon \mathbb{R}^n \to \mathbb{R}$ continuously differentiable,
  point $\boldsymbol{x}$ of polytope $\mathcal{P} := \{\boldsymbol{y} \colon A\boldsymbol{y} \leq \boldsymbol{b}\}$
output: critical point $\boldsymbol{x}$ of $\max\{f(\boldsymbol{y}) \colon \boldsymbol{y} \in \mathcal{P}\}$

$\mathcal{A} \leftarrow \mathrm{eq}(\boldsymbol{x})$ `(active set)`
while $\mathcal{D} := \{\boldsymbol{d} \colon A_{\mathrm{eq}(\boldsymbol{x}).}\boldsymbol{d} \leq \boldsymbol{0},\ \boldsymbol{\nabla f(x)}^\top \boldsymbol{d} > 0\} \neq \emptyset$**:**
  take $\boldsymbol{d} \in \mathcal{D}$ maximizing $|\{i \in \mathcal{A} \colon \boldsymbol{A_{i.}d} = 0\}|$
  if $A_{\mathcal{A}.}\boldsymbol{d} \neq \boldsymbol{0}$**:**
    $\mathcal{A} \leftarrow \mathcal{A} \setminus \{i\}$ for some $i \in \mathcal{A}$ with $\boldsymbol{A_{i.}d} < 0$
  if $A_{\mathcal{A}.}\boldsymbol{d} = \boldsymbol{0}$**:**
    $\mu \leftarrow \inf\{\mu \geq 0 \colon \boldsymbol{x} + \mu\boldsymbol{d} \notin \mathcal{P}$ or $\boldsymbol{\nabla f(x + \mu d)}^\top \boldsymbol{d} \leq 0\}$
    $\boldsymbol{x} \leftarrow \boldsymbol{x} + \mu\boldsymbol{d}$
    if $\boldsymbol{\nabla f(x)}^\top \boldsymbol{d} > 0$**:**
      $\mathcal{A} \leftarrow \mathcal{A} \cup \{j\}$ for some $j \in \mathrm{eq}(\boldsymbol{x}) \setminus \mathcal{A}$

The simplex method solves arbitrary linear programs of the form

$$\begin{aligned} \max \quad & \boldsymbol{c}^\top \boldsymbol{x} \\ \text{s.t.} \quad & A\boldsymbol{x} \leq \boldsymbol{b}, \end{aligned} \tag{LP}$$

where $\boldsymbol{c} \in \mathbb{Q}^n$, $A \in \mathbb{Q}^{m \times n}$, and $\boldsymbol{b} \in \mathbb{Q}^m$. For simplicity, we assume that the feasible region $\mathcal{P} := \{\boldsymbol{x} \colon A\boldsymbol{x} \leq \boldsymbol{b}\}$ is bounded, and that the constraints are non-degenerate, i.e., we assume that every vertex $\boldsymbol{x}$ of $\mathcal{P}$ satisfies $|\operatorname{eq}(\boldsymbol{x})| = n$, where we write $\operatorname{eq}(\boldsymbol{x}) := \{i \colon \boldsymbol{A}_{i.}\boldsymbol{x} = b_i\}$. The simplex method can be understood combinatorially as a traversal of $\mathcal{P}$ along its edges.

More precisely, it maintains a subset $\mathcal{B}$, called a *basis*, of n constraint indices such that the system $A_{\mathcal{B}.}\boldsymbol{x} = \boldsymbol{b}_{\mathcal{B}}$ uniquely determines the current vertex $\boldsymbol{x}$. By non-degeneracy, this is equivalent to $\mathcal{B} = \operatorname{eq}(\boldsymbol{x})$. As long as $\boldsymbol{x}$ is not already an optimum, the simplex method moves along edges of $\mathcal{P}$ to neighboring vertices with better objective function values. We now describe how our formulation of SIMPLEX reflects this behavior.

A vertex $\boldsymbol{x}$ of $\mathcal{P}$ is not optimal if and only if there is a direction $\boldsymbol{d}$ that is feasible and improving, i.e., that satisfies $A_{\operatorname{eq}(\boldsymbol{x}).}\boldsymbol{d} \leq \boldsymbol{0}$ and $\boldsymbol{c}^\top \boldsymbol{d} > 0$. In particular, there is at least one improving edge direction[1] in each vertex $\boldsymbol{x}$ that is not optimal. Due to non-degeneracy, every edge direction $\boldsymbol{d}$ corresponds to a unique index $\ell \in \mathcal{B}$ with $\boldsymbol{A}_{\ell.}\boldsymbol{d} < 0$ and $A_{\mathcal{B}\setminus\{\ell\}.}\boldsymbol{d} = \boldsymbol{0}$. In fact, a feasible direction $\boldsymbol{d}$ is an edge direction in $\boldsymbol{x}$ if and only if

$$|\{i \in \mathcal{B} \colon \boldsymbol{A}_{i.}\boldsymbol{d} = 0\}| = n - 1 = |\mathcal{B}| - 1.$$

Now, if the current vertex $\boldsymbol{x}$ is not optimal, SIMPLEX chooses an improving edge direction $\boldsymbol{d}$, deletes the corresponding index ℓ from $\mathcal{B}$, and moves along $\boldsymbol{d}$ as far as possible without becoming infeasible. Note that, by boundedness of $\mathcal{P}$, this transitions to another vertex $\boldsymbol{y}$ of $\mathcal{P}$. The method then adds a new active constraint index $j \in \operatorname{eq}(\boldsymbol{y})$ to $\mathcal{B}$ to obtain a basis uniquely identifying the new vertex. By non-degeneracy, this index is uniquely determined by $j \in \operatorname{eq}(\boldsymbol{y}) \setminus \mathcal{B}$.

The number of iterations of the simplex method depends on the *pivot rule* that determines the improving edge direction $\boldsymbol{d}$ chosen in each step in case there is more than one. The question of whether or not there exists a pivot rule guaranteeing a polynomial running time of the simplex method is arguably one of the most famous open problems in (linear) optimization. We refer to [11] for more details on the simplex method.

The active-set method can be applied to non-linear programs with linear constraints of the form

$$\begin{aligned} \max \quad & f(\boldsymbol{x}) \\ \text{s.t.} \quad & A\boldsymbol{x} \leq \boldsymbol{b}, \end{aligned} \tag{NLP}$$

where $f \colon \mathbb{R}^n \to \mathbb{R}$ is continously differentiable, $A \in \mathbb{Q}^{m \times n}$, and $\boldsymbol{b} \in \mathbb{Q}^m$. As before, we denote the feasible region by $\mathcal{P}$, and the set of indices of the constraints

[1] An edge direction in vertex $\boldsymbol{x}$ is a vector of the form $\boldsymbol{d} := \lambda(\boldsymbol{y} - \boldsymbol{x})$ where $\boldsymbol{y}$ is a neighboring vertex of $\boldsymbol{x}$ and $\lambda > 0$.

that are active in some $\boldsymbol{x} \in \mathcal{P}$ by $\text{eq}(\boldsymbol{x})$. For simplicity, we again assume that $\mathcal{P}$ is bounded.

If $\boldsymbol{x}^*$ is a local optimum of (NLP), then there are no feasible improving directions in $\boldsymbol{x}^*$. That is, for all $\boldsymbol{d}$ with $A_{\text{eq}(\boldsymbol{x}^*)}.\boldsymbol{d} \leq \boldsymbol{0}$, we have $\boldsymbol{\nabla f}(\boldsymbol{x}^*)^\top \boldsymbol{d} \leq 0$. The active-set method aims to find a *critical point*, i.e., a point satisfying this necessary optimality condition, by determining the set of constraints that are active in such a point.

More precisely, the active-set method maintains a subset $\mathcal{A} \subseteq \text{eq}(\boldsymbol{x})$, called *active set*, of constraints that are active in the current solution $\boldsymbol{x} \in \mathcal{P}$. As long as $\boldsymbol{x}$ is not already a critical point, the method tries to move along an improving feasible direction $\boldsymbol{d}$ that respects the active set, i.e., that satisfies $A_{\mathcal{A}}.\boldsymbol{d} = \boldsymbol{0}$. If there is no such direction, the method removes as few indices as possible from $\mathcal{A}$ until such a direction $\boldsymbol{d}$ can be found. Now it moves as far as possible along $\boldsymbol{d}$ without becoming infeasible, or until reaching a point $\boldsymbol{y}$ with $\boldsymbol{\nabla f}(\boldsymbol{y})^\top \boldsymbol{d} \leq 0$. If this movement is stopped by hitting the boundary of $\mathcal{P}$, the method adds a new active constraint to $\mathcal{A}$. This procedure is iterated until a critical point for (NLP) is found. We now describe how our formulation of ACTIVESET allows this behavior.

A feasible point $\boldsymbol{x}$ is *not critical* if and only if there is a feasible improving direction $\boldsymbol{d}$, i.e., a direction $\boldsymbol{d}$ with $A_{\text{eq}(\boldsymbol{x})}.\boldsymbol{d} \leq \boldsymbol{0}$ and $\boldsymbol{\nabla f}(\boldsymbol{x})^\top \boldsymbol{d} > 0$. In ACTIVESET, we take such a direction $\boldsymbol{d}$ which also maximizes the number of indices $j \in \mathcal{A}$ with $\boldsymbol{A}_j.\boldsymbol{d} = 0$, and can then remove indices i from $\mathcal{A}$ with $\boldsymbol{A}_i.\boldsymbol{d} < 0$ (in separate iterations). This corresponds to deleting a smallest set $\mathcal{I} \subseteq \mathcal{A}$ from $\mathcal{A}$ that ensures the existence of a feasible improving direction $\boldsymbol{d}$ with $A_{\mathcal{A}\setminus\mathcal{I}}.\boldsymbol{d} = \boldsymbol{0}$. Now the method moves along the improving direction $\boldsymbol{d}$ as far as possible without becoming infeasible or reaching a point where the derivative of the objective in direction $\boldsymbol{d}$ is non-positive. If the directional derivative is still positive in the new point, the movement was stopped by hitting the boundary of $\mathcal{P}$, so the method adds a new active constraint to $\mathcal{A}$.

Implementations of the active-set method are usually formulated for strictly concave, quadratic objective functions (see e.g. [19,32]), where they compute Karush-Kuhn-Tucker (KKT) points [28,30] of equality constrained subproblems until reaching a KKT point, that is, the unique global optimum solution, of (NLP). More precisely, given a feasible solution $\boldsymbol{x}$ of (NLP) and an active set $\mathcal{A} \subseteq \{i \colon \boldsymbol{A}_i.\boldsymbol{x} = b_i\}$ of constraint indices, the method computes a KKT point $\boldsymbol{x}^*$ of

$$\begin{aligned} &\max f(\boldsymbol{y}) \\ &\text{s.t. } A_{\mathcal{A}}.\boldsymbol{y} = \boldsymbol{b}_{\mathcal{A}}. \end{aligned} \qquad (\text{NLP}_{\mathcal{A}})$$

Then, it moves from $\boldsymbol{x}$ towards $\boldsymbol{x}^*$ as far as possible without becoming infeasible. If it reaches $\boldsymbol{x}^*$, the method removes a constraint from the active set to enable further progress. Otherwise, it adds a new constraint to the active set.

Note that our formulation of ACTIVESET allows this behavior for strictly concave, quadratic functions. If the current solution $\boldsymbol{x}$ is not already the KKT point of ($\text{NLP}_{\mathcal{A}}$), the direction $\boldsymbol{d} := \boldsymbol{x}^* - \boldsymbol{x}$ is improving (by concavity of the objective) and satisfies $A_{\mathcal{A}}.\boldsymbol{d} = \mathbf{0}$, i.e., maximizes $|\{i \in \mathcal{A}\colon \boldsymbol{A}_i.\boldsymbol{d} = 0\}|$. Further, we have $\boldsymbol{x}^* = \boldsymbol{x} + \bar{\mu}\boldsymbol{d}$ for

$$\bar{\mu} := \inf\{\mu \geq 0\colon \boldsymbol{\nabla f}(\boldsymbol{x} + \mu\boldsymbol{d})^\top \boldsymbol{d} \leq 0\}.$$

If $\boldsymbol{x}$ is the KKT point of (NLP$_{\mathcal{A}}$), then there is no feasible improving direction $\boldsymbol{d}$ with $A_{\mathcal{A}}.\boldsymbol{d} = \mathbf{0}$, so ActiveSet removes constraints from $\mathcal{A}$ preventing progress.

While concave quadratic functions allow for an efficient implementation, the computations in an iteration of ActiveSet are non-trivial for general functions. In the following, we assume that ActiveSet has access to an oracle, which finds $\boldsymbol{d}$ and computes μ in each step.

The running time of the algorithm, i.e., the number of iterations, highly depends on the *pivot rule* that determines the direction $\boldsymbol{d}$ and the indices $i \in \mathcal{A}$ and $j \in \text{eq}(\boldsymbol{x}) \setminus \mathcal{A}$ in each iteration in case there is more than one choice. It is an open problem whether there is a polynomial time pivot rule for the active-set method for (strictly) concave, quadratic objective functions.

3 Unconditional Super-Polynomial Bounds for Active-Set

In Sect. 3.1, we define multivariate polynomials $\mathcal{F}_n\colon \mathbb{R}^n \to \mathbb{R}$ of degree n for all $n > 2$. We then see in Sect. 3.2 that there is exactly one improving edge in every vertex of the boolean hypercube $[0,1]^n$ which is not optimal; and that the value of $\mathcal{F}_n$ increases when moving along this edge to the neighboring vertex. The path π obtained by starting in $\mathbf{0}$ and iteratively moving along unique improving edges to neighboring vertices visits all 2^n vertices of the hypercube, see Sect. 3.3. In Sect. 3.4, we obtain our main results by showing that ActiveSet follows the path π. Finally, we make some remarks on the complexity of the optimization problem induced by our polynomials in Sect. 3.5.

3.1 Polynomials of Linear Degree

For each $n \in \mathbb{N}$, we define a multivariate polynomial $\mathcal{F}_n\colon \mathbb{R}^n \to \mathbb{R}$ by

$$\mathcal{F}_n(\boldsymbol{x}) = \sum_{i=1}^{n} \left(2^{i-1}\alpha_{n,i}(\boldsymbol{x}) - \beta_{n,i}(\boldsymbol{x})\right),$$

with $\boldsymbol{x} = (x_1, \ldots, x_n)^\top \in \mathbb{R}^n$ and $\alpha_{n,i}, \beta_{n,i}\colon \mathbb{R}^n \to \mathbb{R}$ given by $\alpha_{n,n+1}(\boldsymbol{x}) = 0$ and

$$\alpha_{n,i}(\boldsymbol{x}) = x_i + (1 - 2x_i)\alpha_{n,i+1}(\boldsymbol{x}), \qquad \forall i \in [n], \tag{1}$$

$$\beta_{n,i}(\boldsymbol{x}) = 2^i\left(x_i - x_i^2\right)\left(1 - x_{i-1} + \sum\nolimits_{j=1}^{i-2} x_j\right), \qquad \forall i \in [n],$$

where we write $x_0 := 1$, i.e., we have $\beta_{n,1}(\boldsymbol{x}) = 0$ for all $\boldsymbol{x}$. Whenever $\boldsymbol{x}$ and its dimension are clear from the context, we write $\alpha_i := \alpha_{n,i}(\boldsymbol{x})$ and $\beta_i := \beta_{n,i}(\boldsymbol{x})$. Note that $\mathcal{F}_n$ is a multivariate polynomial of degree n – except for $\mathcal{F}_2$ which is of degree 3.

We will analyze the behavior of ActiveSet optimizing $\mathcal{F}_n$ on the hypercube $[0,1]^n$ starting in $\mathbf{0}$. As a preparation, we now compute the maximum of $\mathcal{F}_n$ on $[0,1]^n$ as well as the gradients of $\mathcal{F}_n$ in the vertices of the hypercube.[2]

Lemma 1. *For all $n \in \mathbb{N}$, the optimum solution of*

$$\begin{aligned} \max \quad & \mathcal{F}_n(\boldsymbol{x}) \\ s.t. \quad & \boldsymbol{x} \in [0,1]^n \end{aligned}$$

is attained by the unit vector $\mathbf{e}^n$.

Lemma 2. *For all $n \in \mathbb{N}$, $k \in [n]$, and $\boldsymbol{x} \in \{0,1\}^n$, the k-th partial derivative of $\mathcal{F}_n$ is given by*

$$\partial_k \mathcal{F}_n(\boldsymbol{x}) = (1 - 2\alpha_{k+1}) \sum_{i=1}^{k} 2^{i-1} \prod_{j=i}^{k-1} (1 - 2x_j) - 2^k (1 - 2x_k) \left(1 - x_{k-1} + \sum_{i=1}^{k-2} x_i \right),$$

where we write $x_0 := 1$, i.e., we have $\partial_1 \mathcal{F}_n(\boldsymbol{x}) = 1 - 2\alpha_2$.

3.2 Uniqueness of Improving Edges

We now show that in every vertex of the hypercube that is not optimal, there is exactly one improving edge direction.

Proposition 2. *For all $n \in \mathbb{N}$ and all $\boldsymbol{x} \in \{0,1\}^n \setminus \{\mathbf{e}^n\}$, there is exactly one $k \in [n]$ such that $\partial_k \mathcal{F}_n(\boldsymbol{x}) > 0$ and $x_k = 0$, or $\partial_k \mathcal{F}_n(\boldsymbol{x}) < 0$ and $x_k = 1$.*

Before giving the proof, we introduce notation tightening our further analysis.

Definition 1. *Given some $\boldsymbol{x} = (x_1, \ldots, x_n)^\top \in \mathbb{R}^n$ and $k \in [n]$, we write*

$$\mathcal{P}_n(\boldsymbol{x}, k) = x_{k-1} \prod_{j=1}^{k-2} (1 - x_j) \quad \text{and} \quad \mathcal{S}_n(\boldsymbol{x}, k) = \sum_{j=k+1}^{n} x_j,$$

where we set $x_0 := 1$, i.e., we have $\mathcal{P}_n(\boldsymbol{x}, 1) = 1$ for all $\boldsymbol{x}$. Whenever n and $\boldsymbol{x}$ are clear from the context, we write $\mathcal{P}(k) := \mathcal{P}_n(\boldsymbol{x}, k)$ and $\mathcal{S}(k) := \mathcal{S}_n(\boldsymbol{x}, k)$.

We now observe how the dimension k of Proposition 2 is determined by the values of $\mathcal{P}_n$ and $\mathcal{S}_n$.(see footnote 2)

Lemma 3. *The following are equivalent for all $\boldsymbol{x} \in \{0,1\}^n$ and $k \in [n]$:*

(i) $\partial_k \mathcal{F}_n(\boldsymbol{x}) > 0$ *and* $x_k = 0$, *or* $\partial_k \mathcal{F}_n(\boldsymbol{x}) < 0$ *and* $x_k = 1$,
(ii) $\mathcal{S}(k) \equiv_2 x_k$ *and* $\mathcal{P}(k) = 1$.

[2] All missing proofs are provided in the full version [16].

We obtain our proposition by showing that there is a unique dimension k that satisfies condition (ii) from Lemma 3 in each vertex that is not optimal.

Proof of Proposition 2. We assume $n \geq 3$, the cases $n \in \{1,2\}$ can be checked directly. We start with two technical observations. First, fix some $\boldsymbol{x} \in \{0,1\}^n$ and $1 < k_1 < k_2 \leq n$. We can easily see that $\mathcal{P}(k_1) = 1$ implies $x_{k_1-1} = 1$, while $\mathcal{P}(k_2) = 1$ implies $x_{k_1-1} = 0$. Since $\mathcal{P}(k) \in \{0,1\}$ for all $k \in [n]$, this yields the following:

$$\forall \boldsymbol{x} \in \{0,1\}^n, 1 < k_1 < k_2 \leq n \colon \mathcal{P}(k_1) = 0 \vee \mathcal{P}(k_2) = 0. \tag{2}$$

Now fix some $\boldsymbol{x} \in \{0,1\}^n$ with $\mathcal{S}(1) \equiv_2 x_1$. Assume that $\mathcal{P}(k) = 1$ holds for some fixed $k \in \{2,\dots,n\}$. Then, we have $x_{k-1} = 1$ and $x_j = 0$ holds for all $j \in [k-2]$. Thus, $\mathcal{S}(1) \equiv_2 x_1$ implies $\mathcal{S}(k) \equiv_2 x_k + 1$, which gives us the following:

$$\forall \boldsymbol{x} \in \{0,1\}^n \text{ with } \mathcal{S}(1) \equiv_2 x_1,\ \forall k \in \{2,\dots,n\} \colon \mathcal{P}(k) = 0 \vee \mathcal{S}(k) \equiv_2 x_k + 1. \tag{3}$$

For all $\boldsymbol{x} \in \{0,1\}^n$, Eqs. (2) & (3) yield that we have $\mathcal{S}(k) \equiv_2 x_k$ and $\mathcal{P}(k) = 1$ for at most one $k \in [n]$.

Now consider some $\boldsymbol{x} \in \{0,1\}^n \setminus \{\boldsymbol{e}^n\}$. It remains to argue that there exists some $k \in [n]$ with $\mathcal{S}(k) \equiv_2 x_k$ and $\mathcal{P}(k) = 1$. We may assume $\mathcal{S}(1) \equiv_2 x_1 + 1$ since $\mathcal{P}(1) = 1$. This yields the existence of some $j \in [n]$ with $x_j = 1$. Further, we have $k := \min\{j \in [n] \colon x_j = 1\} < n$ as $\boldsymbol{x} \neq \boldsymbol{e}^n$, hence $\mathcal{P}(k+1) = 1$ and $\mathcal{S}(k+1) \equiv_2 x_{k+1}$. The statement now follows directly by Lemma 3. □

Let k be the unique dimension from Proposition 2 for some vertex $\boldsymbol{x}$ that is not optimal. We show that the value of $\partial_k \mathcal{F}_n(\boldsymbol{x})$ does not depend on x_k.(see footnote 2)

Proposition 3. *Let $n \in \mathbb{N}$ and $\boldsymbol{x} \in \{0,1\}^n \setminus \{\boldsymbol{e}^n\}$. Let $k \in [n]$ be the unique dimension from Proposition 2. Then, we have*

$$\partial_k \mathcal{F}_n(\boldsymbol{x}) = \partial_k \mathcal{F}_n(\boldsymbol{x} + \mu \boldsymbol{e}^k)$$

for all $\mu \in \mathbb{R}$.

3.3 Visiting all Vertices of the Hypercube

We will see that ActiveSet applied to $\mathcal{F}_n$ over the hypercube and starting in $\mathbf{0}$ moves from one vertex to another along the unique improving edges we identified in Sect. 3.2. This trajectory results in a Hamiltonian path in the polyhedral graph of $[0,1]^n$, i.e., in a path along the edges of the hypercube that visits all of its 2^n vertices exactly once.

Proposition 4. *Let $n \in \mathbb{N}$. The sequence $(\boldsymbol{x}^i)_{i\in[2^n]} \subseteq \{0,1\}^n$ given by $\boldsymbol{x}^1 := \boldsymbol{0}$ and*

$$\boldsymbol{x}^{i+1} := \boldsymbol{x}^i + (1 - 2(x^i)_{k_i})\boldsymbol{e}^{k_i}, \quad \forall i \in [2^n - 1], \tag{4}$$

where $k_i \in [n]$ is the unique dimension of Proposition 2 in $\boldsymbol{x}^i$, forms a Hamiltonian path in the polyhedral graph of $[0,1]^n$ ending in $\boldsymbol{e}^n$.

Proof. By Lemma 3 and Proposition 2, the choice of $k_i \in [n]$ is indeed unique for $\boldsymbol{x}^i \in \{0,1\}^n \setminus \{\boldsymbol{e}^n\}$ and we have $\mathcal{P}(k_i) = 1$ and $\mathcal{S}(k_i) \equiv_2 (x^i)_{k_i}$. We prove the statement by induction on $n \in \mathbb{N}$. It obviously holds for $n = 1$, so assume it holds for some fixed $n \in \mathbb{N}$.

Consider the sequence $(\boldsymbol{x}^i)_{i\in[2^{n+1}]} \subseteq \{0,1\}^{n+1}$ as defined in the statement. By induction, the subsequence $(\boldsymbol{x}^i)_{i\in[2^n]}$ forms a Hamiltonian path in the polyhedral graph of $[0,1]^n$ ending in $\boldsymbol{x}^{2^n} = \boldsymbol{e}^n$, since the additional entry $(x^i)_{n+1} = 0$ does not change the value of $\mathcal{P}(k)$ or $\mathcal{S}(k)$ for any $k \in [n]$. Further, since we have $\mathcal{P}_{n+1}(\boldsymbol{e}^n, n+1) = 1$ and $\mathcal{S}_{n+1}(\boldsymbol{e}^n, n+1) = 0 = (\boldsymbol{e}^n)_{n+1}$, it holds that $k_{2^n} = n+1$ and thus $\boldsymbol{x}^{2^n+1} = \boldsymbol{e}^n + \boldsymbol{e}^{n+1}$.

For the remaining part of the sequence, note that having $(x^i)_{n+1} = 1$ instead of $(x^i)_{n+1} = 0$ in the additional dimension does not change $\mathcal{P}_{n+1}(\boldsymbol{x}^i, k)$ and flips the value of $(\mathcal{S}_{n+1}(\boldsymbol{x}^i, k) \mod 2) \in \{0,1\}$ for all $k \in [n]$.

By Eq. (4), for all $i \in [2^n - 1]$, we have $\mathcal{P}_{n+1}(\boldsymbol{x}^i, k_i) = \mathcal{P}_{n+1}(\boldsymbol{x}^{i+1}, k_i)$ and $\mathcal{S}_{n+1}(\boldsymbol{x}^i, k_i) = \mathcal{S}_{n+1}(\boldsymbol{x}^{i+1}, k_i)$. Therefore, the second half of the sequence is the inverse of the first half; more precisely, we have

$$(\boldsymbol{x}^i)_{i\in\{2^n+1, 2^n+2, \ldots, 2^{n+1}\}} = (\boldsymbol{y}^k)_{k\in[2^n]},$$

where $\boldsymbol{y}^k := \boldsymbol{x}^{2^n+1-k} + \boldsymbol{e}^{n+1}$ for all $k \in [2^n]$. In particular, $(\boldsymbol{x}^i)_{i\in[2^{n+1}]}$ forms a Hamiltonian path in the polyhedral graph of $[0,1]^{n+1}$ ending in $\boldsymbol{e}^{n+1}$. □

3.4 Proofs of the Super-Polynomial Bounds

We define a matrix $C \in \{-1,0,1\}^{2n\times n}$ by $\boldsymbol{C}_{i\cdot} = (\boldsymbol{e}^i)^\top$ and $\boldsymbol{C}_{(i+n)\cdot} = -(\boldsymbol{e}^i)^\top$ for all $i \in [n]$, and a vector $\boldsymbol{c} \in \{0,1\}^{2n}$ by $c_i = 1$ and $c_{i+n} = 0$ for all $i \in [n]$. So we have

$$[0,1]^n = \{\boldsymbol{x} \colon C\boldsymbol{x} \le \boldsymbol{c}\}.$$

Fixing this implementation of the hypercube, we can now state our main result formally. For the proof, we combine the insights of the previous subsections.

Theorem 1. *For all $n \in \mathbb{N}$, $\textsc{ActiveSet}(\mathcal{F}_n, C, \boldsymbol{c}, \boldsymbol{0})$ solves*

$$\begin{aligned} \max \quad & \mathcal{F}_n(\boldsymbol{x}) \\ \text{s.t.} \quad & \boldsymbol{x} \in [0,1]^n \end{aligned}$$

in $2^n - 1$ iterations, irrespective of the pivot rule.

Proof. We show that ActiveSet visits all vertices of the hypercube $[0,1]^n$. Let $(\boldsymbol{x}^i)_{i\in[2^n]}$ denote the sequence of Proposition 4. The initial solution is given by $\mathbf{0} = \boldsymbol{x}^1$. Now fix some $i \in [2^n - 1]$ and assume that $\boldsymbol{x}^i$ is the intermediate solution at the start of iteration i of ActiveSet$(\mathcal{F}_n, C, \boldsymbol{c}, \mathbf{0})$ and that the current active set is $\mathcal{A} = \mathrm{eq}(\boldsymbol{x}^i) := \{j\colon \boldsymbol{C}_j.\boldsymbol{x}^i = c_j\}$. Note that, according to Proposition 4, we have $\boldsymbol{x}^i \neq \boldsymbol{e}^n$ so we can fix $k \in [n]$ as the unique dimension of Proposition 2 in the point $\boldsymbol{x}^i$.

All edge directions in $\boldsymbol{x}^i$ are of the form

$$\boldsymbol{d}^{j,\lambda} := \lambda(1 - 2(x^i)_j)\boldsymbol{e}^j$$

for some $j \in [n]$ and some $\lambda > 0$. Note that $\partial_j \mathcal{F}_n = \boldsymbol{\nabla}\boldsymbol{\mathcal{F}}_n{}^\top \boldsymbol{e}^j = -\boldsymbol{\nabla}\boldsymbol{\mathcal{F}}_n{}^\top(-\boldsymbol{e}^j)$ for all $j \in [n]$, so we have $\boldsymbol{\nabla}\boldsymbol{\mathcal{F}}_n(\boldsymbol{x}^i)^\top \boldsymbol{d}^{j,\lambda} > 0$ if and only if $\partial_j \mathcal{F}_n(\boldsymbol{x}^i) > 0$ and $(x^i)_j = 0$, or $\partial_j \mathcal{F}_n(\boldsymbol{x}^i) < 0$ and $(x^i)_j = 1$. Hence, by Proposition 2, the improving edge directions in $\boldsymbol{x}^i$ are given by $\boldsymbol{d}^{k,\lambda} = \lambda(1 - 2(x^i)_k)\boldsymbol{e}^k$ for $\lambda > 0$.

Since the system $C\boldsymbol{x} \leq \boldsymbol{c}$ is non-degenerate and $\mathcal{A} = \mathrm{eq}(\boldsymbol{x}^i)$, every feasible direction $\boldsymbol{d}$ in the vertex $\boldsymbol{x}^i$ satisfies

$$|\{i \in \mathcal{A}\colon \boldsymbol{A}_i.\boldsymbol{d} = 0\}| = |\mathcal{A}| - 1$$

if and only if it is an edge direction in $\boldsymbol{x}^i$. In particular, for every edge direction $\boldsymbol{d}$ in $\boldsymbol{x}^i$, there is a unique index $\ell \in A$ with $A_\ell.\boldsymbol{d} < 0$ and $A_{\mathcal{A}\setminus\{\ell\}}.\boldsymbol{d} = \mathbf{0}$. Further, the system $A_{\mathcal{A}}.\boldsymbol{d} = \mathbf{0}$ only has the trivial solution. Hence, taking an arbitrary feasible improving direction maximizing $|\{i \in \mathcal{A}\colon \boldsymbol{A}_i.\boldsymbol{d} = 0\}|$ is equivalent to taking an arbitrary improving edge direction.

This yields that ActiveSet chooses the improving edge direction $\boldsymbol{d} := \boldsymbol{d}^{k,\lambda}$ for some fixed $\lambda > 0$ in iteration i and deletes the unique $\ell \in \mathcal{A}$ with $A_\ell.\boldsymbol{d} < 0$. We have $A_{\mathcal{A}\setminus\{\ell\}}.\boldsymbol{d} = \mathbf{0}$ and, by Proposition 3,

$$\boldsymbol{\nabla}\boldsymbol{\mathcal{F}}_n(\boldsymbol{x}^i + \mu\boldsymbol{d})^\top \boldsymbol{d} = \boldsymbol{\nabla}\boldsymbol{\mathcal{F}}_n(\boldsymbol{x}^i)^\top \boldsymbol{d} > 0$$

for all $\mu \in \mathbb{R}$. Hence, ActiveSet moves from $\boldsymbol{x}^i$ along $\boldsymbol{d}$ in iteration i until reaching the next vertex

$$\boldsymbol{x}^i + \boldsymbol{d} = \boldsymbol{x}^i + (1 - 2(x^i)_k)\boldsymbol{e}^k = \boldsymbol{x}^{i+1}.$$

Finally, since $\boldsymbol{\nabla}\boldsymbol{\mathcal{F}}_n(\boldsymbol{x}^{i+1})^\top \boldsymbol{d} > 0$, an index $j \in \mathrm{eq}(\boldsymbol{x}^{i+1}) \setminus \mathcal{A}$ is added to $\mathcal{A}$. In fact, since the system $C\boldsymbol{x} \leq \boldsymbol{c}$ is non-degenerate, this choice of j is unique, i.e., ActiveSet starts iteration $i+1$ in the point $\boldsymbol{x}^{i+1}$ with $\mathcal{A} = \mathrm{eq}(\boldsymbol{x}^{i+1})$.

We have seen that ActiveSet transitions from $\boldsymbol{x}^i$ to $\boldsymbol{x}^{i+1}$ in iteration i, where $i \in [2^n - 1]$ was chosen arbitrarily. This proves the statement since the optimum solution of the given program is $\boldsymbol{x}^{2^n} = \boldsymbol{e}^n$ by Lemma 1. □

If we are willing to relax the exponential bound from Theorem 1, we can lower the degrees of the polynomials. In Corollary 1, the degrees are chosen just large enough to yield a super-polynomial bound.

Proof of Corollary 1. Let $d(n) = \omega(\log n)$ with $d(n) \leq n$ for all $n \in \mathbb{N}$. Since the additional dimensions do not affect its behavior, by Theorem 1, ActiveSet started in $\mathbf{0}$ solves

$$\begin{aligned} \max \quad & \mathcal{F}_{\lfloor d(n) \rfloor}(x_1, \ldots, x_{\lfloor d(n) \rfloor}) \\ \text{s.t.} \quad & \boldsymbol{x} \in [0,1]^n \end{aligned}$$

in $2^{\lfloor d(n) \rfloor} - 1$ iterations for all $n \in \mathbb{N}_{>2}$. Since $2^{\lfloor d(n) \rfloor} = 2^{\omega(\log n)}$ and the polynomials $\left(\mathcal{F}_{\lfloor d(n) \rfloor}\right)_{n \in \mathbb{N}}$ are of degrees $\lfloor d(n) \rfloor = \mathcal{O}(d(n))$, this concludes the proof. □

3.5 The Complexity of Maximizing Over the Hypercube

The lower bounds in Theorem 1 and Corollary 1 are unconditional, meaning they show that ActiveSet is inefficient on the hypercube even if P = NP. It is folklore that the underlying decision problem is already NP-hard for polynomials of degree 3. We include a proof to make our presentation self-contained.(see footnote 2)

Proposition 5. *Given a multivariate polynomial $f\colon \mathbb{R}^n \to \mathbb{R}$ of degree* 3, *it is* NP*-complete to decide whether there is $\boldsymbol{x} \in \{0,1\}^n$ with $f(\boldsymbol{x}) \geq 0$.*

While the underlying problem is hard, the next statements show that our polynomial $\mathcal{F}_n$ belongs to a class of functions for which linear time algorithms exist that find the optimum vertex of the hypercube. We need some definitions before we can state this formally.

Every function $f\colon \mathbb{R}^n \to \mathbb{R}$ attaining unique values on the vertices of $[0,1]^n$ induces an orientation σ_f of the edges of the hypercube if we direct every edge $\{\boldsymbol{x}, \boldsymbol{y}\}$ from $\boldsymbol{x}$ to $\boldsymbol{y}$ if $f(\boldsymbol{x}) < f(\boldsymbol{y})$, and from $\boldsymbol{y}$ to $\boldsymbol{x}$ if $f(\boldsymbol{y}) < f(\boldsymbol{x})$. We write $\boldsymbol{x} \to_\sigma \boldsymbol{y}$ if edge $\{\boldsymbol{x}, \boldsymbol{y}\}$ is directed from $\boldsymbol{x}$ to $\boldsymbol{y}$ in σ.

An orientation σ of the edges of the hypercube is called *unique sink orientation* (USO) if there is exactly one *sink*, i.e., a vertex without outgoing edges, in every face of the hypercube. For more details on USOs, see e.g. [23,37].

We say that an orientation σ is *combed* if there is a dimension $i \in [n]$ such that all edges along dimension i point in the same direction, i.e., if we have

$$\boldsymbol{x} \to_\sigma \boldsymbol{x} + (1 - 2x_i)\boldsymbol{e}^i$$

for all $\boldsymbol{x} \in \{0,1\}^n$ with $x_i = \delta$, where $\delta \in \{0,1\}$ is fixed. An orientation is called *decomposable* if it is combed on every nonzero-dimensional subcube of $[0,1]^n$.(see footnote 2)

Proposition 6. *The function $\mathcal{F}_n$ induces a decomposable orientation of* $\{0,1\}^n$.

Every decomposable orientation is also a USO, and can be solved in linear time [34]. Hence, the polynomials of Theorem 1 and Corollary 1 do not induce NP-hard optimization problems (unless P = NP).

4 Future Research

In this paper, we considered the active-set method as a natural generalization of the simplex method in order to derive the first unconditional running time bounds for all pivot rules. A canonical next step towards a general bound for the simplex method would be to lower the degree of our polynomial.

Question 1. Is there a polynomial of degree $\mathcal{O}(\log n)$ over the hypercube for which active-set takes super-polynomially many iterations for all pivot rules?

It might be possible to reduce the complexity of the objective function even more drastically by increasing the complexity of the feasible region.

Question 2. Is there a polynomial of constant degree over some polytope for which active-set takes super-polynomially many iterations for all pivot rules?

While a general bound for linear objectives, i.e., for the simplex method, would be a massive breakthrough, reaching quadratic and concave objectives would already be very interesting: Much like in the linear setting, only weakly polynomial algorithms are known, namely ellipsoid and interior-point methods [38,39], and the active-set method is a promising candidate for a strongly polynomial algorithm, due to its combinatorial nature.

Question 3. Is there a concave, quadratic polynomial over some polytope for which active-set takes super-polynomially many iterations for all pivot rules?

References

1. Adler, I., Papadimitriou, C., Rubinstein, A.: On simplex pivoting rules and complexity theory. In: Proceedings of the 17th Conference on Integer Programming and Combinatorial Optimization (IPCO), pp. 13–24 (2014)
2. Allamigeon, X., Dadush, D., Loho, G., Natura, B., Végh, L.A.: Interior point methods are not worse than simplex. In: Proceedings of the 63rd Annual IEEE Symposium on Foundations of Computer Science (FOCS), pp. 267–277 (2022)
3. Allamigeon, X., Gaubert, S., Vandame, N.: No self-concordant barrier interior point method is strongly polynomial. In: Proceedings of the 54th Annual ACM Symposium on Theory of Computing (STOC), pp. 515–528 (2022)
4. Amenta, N., Ziegler, G.M.: Deformed products and maximal shadows of polytopes. Contemp. Math. **223**, 57–90 (1999)
5. Avis, D., Chvátal, V.: Notes on Bland's pivoting rule. Polyhedral Combinatorics: Dedicated to the memory of D.R. Fulkerson, pp. 24–34 (1978)
6. Avis, D., Friedmann, O.: An exponential lower bound for Cunningham's rule. Math. Program. **161**, 271–305 (2017)
7. Black, A.E.: Exponential lower bounds for many pivot rules for the simplex method (2024). https://arxiv.org/abs/2403.04886, to appear in IPCO 2025
8. Black, A.E., De Loera, J.A., Kafer, S., Sanità, L.: On the simplex method for 0/1-polytopes. Math. Oper. Res. (2024)

9. Dadush, D., Huiberts, S.: A friendly smoothed analysis of the simplex method. In: Proceedings of the 50th Annual ACM Symposium on Theory of Computing (STOC), pp. 390–403 (2018)
10. Dadush, D., Koh, Z.K., Natura, B., Olver, N., Végh, L.A.: A strongly polynomial algorithm for linear programs with at most two nonzero entries per row or column. In: Proceedings of the 56th Annual ACM Symposium on Theory of Computing (STOC), pp. 1561–1572 (2024)
11. Dantzig, G.: Linear Programming and Extensions. Princeton University Press, Princeton (1963)
12. Dantzig, G., Orden, A., Wolfe, P.: The generalized simplex method for minimizing a linear form under linear inequality restraints. Pac. J. Math. **5**(2), 183–195 (1955)
13. Deshpande, A., Spielman, D.A.: Improved smoothed analysis of the shadow vertex simplex method. In: Proceedings of the 46th Annual IEEE Symposium on Foundations of Computer Science (FOCS), pp. 349–356 (2005)
14. Disser, Y., Friedmann, O., Hopp, A.V.: An exponential lower bound for Zadeh's pivot rule. Math. Program. **199**(1–2), 865–936 (2023)
15. Disser, Y., Mosis, N.: A unified worst case for classical simplex and policy iteration pivot rules. In: Proceedings of the 34th International Symposium on Algorithms and Computation (ISAAC), pp. 27:1–27:17 (2023)
16. Disser, Y., Mosis, N.: An unconditional lower bound for the active-set method on the hypercube (2025). https://arxiv.org/abs/2502.18019
17. Disser, Y., Skutella, M.: The simplex algorithm is NP-mighty. ACM Trans. Algorithms **15**(1), 1–19 (2018)
18. Fearnley, J., Savani, R.: The complexity of the simplex method. In: Proceedings of the 47th Annual ACM Symposium on Theory of Computing (STOC), pp. 201–208 (2015)
19. Fletcher, R.: Practical Methods of Optimization. John Wiley & Sons, Hoboken (2000)
20. Friedmann, O.: A subexponential lower bound for Zadeh's pivoting rule for solving linear programs and games. In: Proceedings of the 15th Conference on Integer Programming and Combinatorial Optimization (IPCO), pp. 192–206 (2011)
21. Friedmann, O., Hansen, T.D., Zwick, U.: Subexponential lower bounds for randomized pivoting rules for the simplex algorithm. In: Proceedings of the 43rd Annual ACM Symposium on Theory of Computing (STOC), pp. 283–292 (2011)
22. Friedmann, O., Hansen, T.D., Zwick, U.: A subexponential lower bound for the random facet algorithm for parity games. In: Proceedings of the 22nd Annual ACM-SIAM Symposium on Discrete Algorithms (SODA), pp. 202–216 (2011)
23. Gärtner, B., Schurr, I.: Linear programming and unique sink orientations. In: Proceedings of the 17th Annual ACM-SIAM Symposium on Discrete Algorithms (SODA), pp. 749–757. No. 26 (2006)
24. Goldfarb, D., Sit, W.Y.: Worst case behavior of the steepest edge simplex method. Discret. Appl. Math. **1**(4), 277–285 (1979)
25. Huiberts, S., Lee, Y.T., Zhang, X.: Upper and lower bounds on the smoothed complexity of the simplex method. In: Proceedings of the 55th Annual ACM Symposium on Theory of Computing (STOC), pp. 1904–1917 (2023)
26. Jeroslow, R.G.: The simplex algorithm with the pivot rule of maximizing criterion improvement. Discret. Math. **4**(4), 367–377 (1973)
27. Karmarkar, N.: A new polynomial-time algorithm for linear programming. In: Proceedings of the 16th Annual ACM Symposium on Theory of Computing (STOC), pp. 302–311 (1984)

28. Karush, W.: Minima of functions of several variables with inequalities as side conditions. In: Traces and emergence of nonlinear programming, pp. 217–245. Springer (2013)
29. Klee, V., Minty, G.J.: How good is the simplex algorithm? Inequalities **3**(3), 159–175 (1972)
30. Kuhn, H., Tucker, A.: Nonlinear programming. In: Proceedings of the 2nd Berkeley Symposium, pp. 481–492 (1951)
31. Murty, K.G.: Computational complexity of parametric linear programming. Math. Program. **19**(1), 213–219 (1980)
32. Nocedal, J., Wright, S.J.: Numerical optimization. Springer (1999)
33. Santos, F.: A counterexample to the Hirsch conjecture. Ann. Math. 383–412 (2012)
34. Schurr, I., Szabó, T.: Finding the sink takes some time: an almost quadratic lower bound for finding the sink of unique sink oriented cubes. Discret. Comput. Geom. **31**(4), 627–642 (2004)
35. Smale, S.: Mathematical problems for the next century. Math. Front. Perspect. 271–294 (2000)
36. Spielman, D.A., Teng, S.H.: Smoothed analysis of algorithms: why the simplex algorithm usually takes polynomial time. J. ACM **51**(3), 385–463 (2004)
37. Szabó, T., Welzl, E.: Unique sink orientations of cubes. In: Proceedings of the 42nd IEEE Symposium on Foundations of Computer Science (FOCS), pp. 547–555 (2001)
38. Tarasov, S.P., Khachiyan, L.G.: Polynomial solvability of convex quadratic programming. In: Doklady Akademii Nauk, vol. 248, pp. 1049–1051. Russian Academy of Sciences (1979)
39. Ye, Y., Tse, E.: An extension of Karmarkar's projective algorithm for convex quadratic programming. Math. Program. **44**, 157–179 (1989)
40. Ziegler, G.M.: Lectures on polytopes, vol. 152. Springer (2012)

Non-distributive Lattices, Stable Matchings, and Linear Optimization

Christopher En(✉) and Yuri Faenza

IEOR, Columbia University, New York, USA
{ce2456,yf2414}@columbia.edu

Abstract. We show that all finite lattices, including non-distributive lattices, arise as stable matching lattices under standard assumptions on choice functions. In the process, we introduce new tools to reason on general lattices for optimization purposes: the *partial representation* of a lattice, which partially extends Birkhoff's representation theorem to non-distributive lattices; the *distributive closure* of a lattice, which gives such a partial representation; and *join constraints*, which can be added to the distributive closure to obtain a representation for the original lattice. Then, we use these techniques to show that the minimum cost stable matching problem under the same standard assumptions on choice functions is NP-hard, by establishing a connection with antimatroid theory.

Keywords: Stable matchings · Lattices · Antimatroids

1 Introduction

Since their introduction by Gale and Shapley [19], stable matchings have been studied extensively by the economics, computer science, and optimization communities. They are today applied in areas as different as school choice [1], online dating [23] and ride-sharing [39], and have been the subject of extensive theoretical research, often collected in monographs or large portions of them [21,24,31,35]. One reason for such popularity is their rich mathematical structure, which has been leveraged to design efficient algorithms. In particular, stable matchings in Gale and Shapley's classical one-to-one model (the *marriage* setting) form a distributive lattice under a natural partial order [28]. This fact has been used to devise an efficient combinatorial algorithms for various problems over the set of stable matchings (see, e.g., [13,14]) including the problem of finding a stable matching of minimum cost, when costs are defined over possible pairs [25]. The latter algorithm has been extended to more general models of many-to-many two-sided matchings [4,15], still exploiting the underlying distributive lattice structure. Moreover, [15,27] showed that even the linear description of the stable matching polytope [36,38] and its properties can be understood, or even deduced, starting from the distributive lattice structure of stable matchings.

N. Megow and A. Basu (Eds.): IPCO 2025, LNCS 15620, pp. 228–241, 2025.
https://doi.org/10.1007/978-3-031-93112-3_17

Interestingly, [7] (and [21] with a simpler construction) proved the converse implication of [28]: every finite distributive lattice arises as the stable matching lattice of some one-to-one instance (all lattices, and more generally all sets, from this paper are assumed to be finite). Thus, the class of stable matchings in the one-to-one model is universal enough to model any distributive lattice structure, making the theory of stable matchings relevant beyond its original scope.

Considerable attention has been devoted to matching markets more general than the ones investigated in [4,15,19]. In particular, Roth [34] proved that, in two-sided markets where agents' choice functions are consistent and substitutable (see Sect. 2 for definitions), a stable matching exists and can be found efficiently. Interestingly, stable matchings in such models are also known to form a lattice, which is however non-distributive [8]. This is a fundamental difference with respect to simpler models, since the absence of a distributive lattice structure prevents the application of Birkhoff's Representation Theorem, on which many of the algorithms, implicitly or explicitly, rely (see Sect. 2 for definitions and [15] for a discussion). As a consequence, beyond Roth's result mentioned above, no algorithm is known for the matching markets studied in [34], despite their relevance for applications (see, e.g., [11,29,37]). More generally, while [15] explored the algorithmic implications of associating a distributive lattice structure to the feasible region of any discrete optimization problem (not necessarily a matching problem), to the best of our knowledge nothing is known on the role of (non-distributive) lattice structures for devising algorithms. This seems to be a relevant question, since general lattices appear much more frequently than distributive lattices, in settings such as matching with contracts [22], formal concept analysis [5,20], and symbolic computation [33], among others.

Our Contributions. Our first main contribution is a universality theorem for stable matchings: We show that every lattice arises as the lattice of stable matchings for a *standard matching market instance* - that is, a matching market instance with substitutable and consistent choice functions, given as its set of agents together with an algorithm that computes the choice functions of the agents in time polynomial in the number of agents[1].

Theorem 1 (All finite lattices are stable matching lattices). *Let $\mathcal{L} = (X, \succeq)$ be a finite lattice. Then, there is a standard matching market instance I^* such that $\mathcal{L}$ is order-isomorphic to the lattice of stable matchings of I^*. Furthermore, I^* has $O(|X|^4)$ agents, and can be constructed in $O(|X|^6)$ time.*

Similarly to the results from [7,28] for distributive stable matching lattices in the one-to-one model, Theorem 1 together with [8] establishes an equivalence between stable matching lattices and general lattices. It also shows that standard matching markets are substantially more general than matching markets from [4, 15,19], suggesting that algorithmic results from [4,15,19] may not extend to this setting. Indeed, for our second main contribution, we show that the minimum

[1] Note that the number of choice functions that are substitutable and consistent is doubly exponential in the number of agents of the market [10].

cost stable matching problem in standard matching market instances is NP-Hard by establishing a novel connection between stable matchings and antimatroids.

Theorem 2 (Linear optimization over matching markets is NP-Hard). *The following minimum cost stable matching problem is NP-Hard:*

Given: a standard matching market instance I with agents $F \uplus W$ and a cost function $c : F \times W \to \mathbb{Z}$.
Find: a stable matching μ of I that minimizes $c(\mu)$.

While both our main results only pertain to stable matchings, we prove them by introducing new tools that apply to general lattices. These tools include the *partial representation* of a lattice $\mathcal{L}$, which partially extends Birkhoff's representation theorem to non-distributive lattices; the *distributive closure* of a lattice, which gives such a partial representation by defining a distributive lattice that "contains" $\mathcal{L}$; and *join constraints*, which can be added to the distributive closure to obtain a representation for the original lattice $\mathcal{L}$. We further show that a join constraint can be enforced on a stable matching lattice via a *join constraint augmentation*, a construction that involves adding auxiliary agents and editing certain choice functions. These tools and the idea of iteratively pruning the distributive closure via join constraints may be useful in dealing more generally with optimization problems over non-distributive lattices.

The Techniques: Theorem 1. The known fact [7,21] that every distributive lattice $\mathcal{L} = (X, \succeq)$ is the lattice of stable matchings in some one-to-one matching market can be proved as follows. Birkhoff's representation theorem [6] implies that there is an order-isomorphism between $\mathcal{L}$ and the lattice $(\mathcal{D}(X_j), \supseteq)$ of the lower closed sets $\mathcal{D}(X_j)$ of an associated poset $(X_j, \succeq)$, called the *representation poset.* Then, a construction by Gusfield and Irving [21] gives a one-to-one matching market instance $I_{(X_j,\succeq)}$ whose stable matching lattice is order-isomorphic to $(\mathcal{D}(X_j), \supseteq)$ (see Theorem 6), hence to $\mathcal{L}$, concluding the proof.

While Birkhoff's theorem fails for a non-distributive lattice $\mathcal{L} = (X, \succeq)$, a relaxed version holds. In particular, in Sect. 3 we define an injective order-embedding $\psi : X \to \mathcal{D}(X_j)$ (see Lemma 1 and Fig. 1). However, some elements of $\mathcal{D}(X_j)$ do not correspond to elements of $\mathcal{L}$. We call therefore $(X_j, \succeq)$ a *partial representation of* $\mathcal{L}$, and call ψ the *partial representation function* (see Definition 6). Then, the Gusfield-Irving construction produces a matching market $I_{(X_j,\succeq)}$ with stable matching lattice order-isomorphic to $(\mathcal{D}(X_j), \supseteq)$, which in turn contains the image of the original lattice $(X, \succeq)$ via ψ. Our goal is now to modify the matching market instance $I_{(X_j,\succeq)}$ to an instance I^*, so that the lattice of stable matchings of I^* is order-isomorphic to $\mathcal{L}$. To do this, we show that it suffices to apply a polynomial number of *join constraint augmentations* to $I_{(X_j,\succeq)}$. Roughly speaking, a join constraint augmentation edits the matching market instance so that certain matchings lose stability, without adding new stable matchings. In this way, we can "prune" the stable matching lattice until the only remaining stable matchings form a lattice order-isomorphic to the original lattice $(X, \succeq)$. Join constraint augmentations can be seen as an implementation for the stable matching lattice of the more general concept of *join constraints* (see Definition 7).

To enforce join constraint augmentations, we introduce the concept of *extendable matching markets*. These markets are edited versions of the original instance $I_{(X_j,\succeq)}$ whose stable matchings, when projected to the agents of $I_{(X_j,\succeq)}$, are contained in the stable matchings of $I_{(X_j,\succeq)}$. We then construct a sequence of extendable matching markets, where each successive market enforces an additional join constraint on the stable matchings of $I_{(X_j,\succeq)}$.

The Techniques: Theorem 2. Our starting point is a connection between lattices and antimatroids, first explored in [3,9,12]: one can univocally define an antimatroid $(V, \mathcal{G})$ via its *path poset* $(Q, \supseteq)$, see Definition 10. With this encoding, the problem of finding an element of the antimatroid minimizing a linear function over the ground set V given $(Q, \supseteq)$ is well-posed. [32] showed that this problem is NP-Hard. After presenting some building blocks in Sect. 5, in Sect. 6 we prove Theorem 2 by constructing, for any path poset $(Q, \supseteq)$ and function $c : V \to \mathbb{Z}$, a standard matching market instance I and cost function c' over pairs of agents, so that the lattice of stable matchings of I is order-isomorphic to the corresponding antimatroid $(V, \mathcal{G})$, and the minimizers of c over $\mathcal{G}$ biject with stable matchings of I minimizing c'. Our construction relies on techniques developed in the first part of the paper.

Additional Related Work. The relationship between stable matching lattices and distributive lattices has been studied in [2,7,8,21,28]. [6] first showed a connection between distributive lattices and the poset of join-irreducible elements. [4,15,25] leveraged this relationship to propose algorithms for finding stable matchings of minimum cost. Different extensions of stability in two-sided markets using concepts from matroid theory have been explored, e.g., in [17,18,26]. [16] studies another generalization of the stable marriage problem involving stable common antichains, which also display a lattice structure, and shows that the minimum cost stable common antichain problem is NP-Hard.

2 Preliminaries

Throughout the paper, we often define functions $\sigma : S_1 \to S_2$ between sets S_1, S_2. We also extend these functions to the power sets $\sigma : 2^{S_1} \to 2^{S_2}$ in a manner similar to valued functions, so that for $T \subseteq S_1$, $\sigma(T) = \bigcup_{s \in T} \sigma(s)$. We denote the symmetric difference operator by $\triangle$. We next introduce the basic structures and tools used throughout the paper.

2.1 Lattices, Matching Markets, and Stability

A poset $\mathcal{L} = (X, \succeq)$ is a *lattice* if any pair of elements $a, b \in X$ has a unique *join*, or least upper bound, denoted $a \vee b$, and a unique *meet*, or greatest lower bound, denoted $a \wedge b$. It also follows by induction that for any $Y \subseteq X$, there exists a unique least upper bound $\vee Y$ (again called join) and greatest lower bound $\wedge Y$ (again called meet), where $\vee\emptyset$ (resp., $\wedge\emptyset$) is the minimal (resp., maximal) element of $\mathcal{L}$. A lattice is *distributive* if the join and meet operations distribute over each other: $a \vee (b \wedge c) = (a \vee b) \wedge (a \vee c)$ and $a \wedge (b \vee c) = (a \wedge b) \vee (a \wedge c)$ for all $a, b, c \in X$. An element $a \in X$ is *join-irreducible* if there does not exist $Y \subseteq X \setminus \{a\}$ such that $\vee Y = a$. The set of join-irreducible elements is denoted by X_j. Note that $(X_j, \succeq)$ forms a poset, with the order induced by the original lattice order.

Given $x, y \in X$ with $x \prec y$, if there does not exist $z \in X$ such that $x \prec z \prec y$, then x is an *immediate predecessor* of y, and y is an *immediate successor* of x.

Throughout the paper, we assume that a lattice $(X, \succeq)$ is given as the set X of elements, plus an oracle which returns $x \wedge y$ and $x \vee y$ in constant time. This is polynomially equivalent to an oracle which decides in constant time if $x \succeq y$.

Next, we define the main objects of study of this paper, matching markets, and related notions. Fix two disjoint finite sets of *agents*, called *firms* F and *workers* W. Each firm f has a choice function $\mathcal{C}_f : 2^W \to 2^W$, such that $\mathcal{C}_f(S) \subseteq S$ for all $S \subseteq W$. Similarly, each worker w has a choice function $\mathcal{C}_w : 2^F \to 2^F$, such that $\mathcal{C}_w(S) \subseteq S$ for all $S \subseteq F$. Then, a matching market instance is given by $I = (F, W, (\mathcal{C}_f)_{f\in F}, (\mathcal{C}_w)_{w\in W})$.

Given a matching market $I = (F, W, (\mathcal{C}_f)_{f\in F}, (\mathcal{C}_w)_{w\in W})$, a *matching* μ is a mapping from $F \cup W$ to $2^{F\cup W}$ such that for all $w \in W$ and all $f \in F$, we have $\mu(w) \subseteq F$, $\mu(f) \subseteq W$, and $f \in \mu(w)$ if and only if $w \in \mu(f)$. A matching can also be viewed as a set of firm-worker pairs, so we write $(f, w) \in \mu$, $f \in \mu(w)$, and $w \in \mu(f)$ interchangeably. A matching is *individually rational* if for every agent a, we have $\mathcal{C}_a(\mu(a)) = \mu(a)$. A firm-worker pair $(f, w) \notin \mu$ is called a *blocking pair* if $f \in \mathcal{C}_w(\mu(w) \cup \{f\})$ and $w \in \mathcal{C}_f(\mu(f) \cup \{w\})$. The solution concept we are concerned with is *stability*: a matching is *stable* if it is individually rational and admits no blocking pairs. The set of stable matchings of I is denoted $\mathcal{S}(I)$ and, when I is clear from context, simply by $\mathcal{S}$.

Next, we define two common assumptions on choice functions, *substitutability* and *consistency*, which we also assume throughout the paper.

Definition 1 (Substitutability and Consistency). *A choice function $\mathcal{C}$ is substitutable if for any set of partners S, $a \in \mathcal{C}(S)$ implies that for all $T \subseteq S$, $a \in \mathcal{C}(T \cup \{a\})$. It is consistent if for any sets of partners S, T, $\mathcal{C}(S) \subseteq T \subseteq S$ implies $\mathcal{C}(S) = \mathcal{C}(T)$. If $\mathcal{C}$ is substitutable and consistent, we say it is* path-independent.

[8] showed that under substitutability and consistency, the set of stable matchings is nonempty and forms a lattice, when ordered following firm preferences.

Theorem 3 (Stable Matching Lattice [8]). *Let $(F, W, (\mathcal{C}_f)_{f\in F}, (\mathcal{C}_w)_{w\in W})$ be an instance of the stable matching problem, such that all choice functions satisfy path-independence. Then, the set of stable matchings $\mathcal{S}$ is nonempty and $(\mathcal{S}, \succeq)$ forms a lattice, called the* Stable Matching Lattice, *where $\mu \succeq \mu'$ if and only if $\mathcal{C}_f(\mu(f) \cup \mu'(f)) = \mu(f)$ for all $f \in F$. Hence, there exist $\mu_F, \mu_W \in \mathcal{S}$,* *called the* firm-optimal *and* worker-optimal *stable matchings respectively, such that* $\mu_F \succeq \mu \succeq \mu_W$ *for all* $\mu \in \mathcal{S}$.

A matching market instance $(F, W, (\mathcal{C}_f)_{f \in F}, (\mathcal{C}_w)_{w \in W})$ is called *standard* if all choice functions are path-independent, and the instance is given in input as F, W, plus an algorithm that runs in time polynomial in $|F|, |W|$ and, when given in input an agent a and set T of agents from the side of the market opposite to a, outputs $\mathcal{C}_a(T)$. If moreover, for any agent a and any $T \subseteq F \cup W$, we have $|\mathcal{C}_a(T)| \leq 1$, we call it a *one-to-one* instance.

2.2 Birkhoff's Representation Theorem and the Poset of Rotations

We next recall basic tools from order theory used throughout the paper.

Definition 2 (Lower closed set). *Given a poset* $(X, \succeq)$, *a* lower closed set *of* X *is any set* $S \subseteq X$ *such that* $x \in S$ *and* $x' \preceq x$ *implies* $x' \in S$. *The set of all lower closed sets of* X *is denoted* $\mathcal{D}(X)$. *Note that* $(\mathcal{D}(X), \supseteq)$ *is a lattice.*

Order-embeddings and order-isomorphisms are monotone functions between posets which preserve the partial order structure.

Definition 3 (Order-embeddings and order-isomorphisms). *Let* $(X, \succeq_X)$ *and* $(Y, \succeq_Y)$ *be posets. A function* $\psi : X \to Y$ *is an* order-embedding *of* $(X, \succeq_X)$ *into* $(Y, \succeq_Y)$, *if for any* $x, x' \in X$, *we have* $x \succeq_X x' \iff \psi(x) \succeq_Y \psi(x')$. *Note that such a function must be injective, as* $\psi(x) = \psi(x')$ *implies* $x \succeq_X x'$ *and* $x \preceq_X x'$. *An order-embedding is an* order-isomorphism *if it is also bijective.*

A representation is an order-isomorphism between a distributive lattice and the lower closed sets of a partial order.

Definition 4 (Representation). *Let* $\mathcal{L} = (X, \succeq)$ *be a distributive lattice and* $(B, \succeq_B)$ *be a poset.* $(B, \succeq_B)$ *is a* representation *of* $\mathcal{L}$ *if there exists an order-isomorphism* $\psi : (X, \succeq) \to (\mathcal{D}(B), \supseteq)$. ψ *is called a* representation function. *By definition, a representation function is an order-isomorphism, hence invertible.*

Next, we introduce Birkhoff's representation theorem, which allows us to represent a distributive lattice using the poset of join-irreducible elements.

Theorem 4 (Birkhoff's Representation Theorem [6]). *Let* $\mathcal{L} = (X, \succeq)$ *be a distributive lattice. Then,* $(X_j, \succeq)$ *is a representation of* $\mathcal{L}$, *with Birkhoff representation function* $\psi_j : X \to \mathcal{D}(X_j)$ *given by* $\psi_j(x) = \{x' \in X_j \mid x' \preceq x\}$.

When a matching market is one-to-one, the stable matching lattice is distributive [28], and a representation is given by its poset of rotations [25]. Rotations are minimal differences between stable matchings.

Definition 5 (Rotations). *Let* $\mathcal{L} = (\mathcal{S}, \succeq)$ *be a stable matching lattice of a one-to-one market. Given two matchings* $\mu, \mu' \in \mathcal{S}$ *such that* μ *is an immediate predecessor of* μ' *in* $\mathcal{L}$, *the corresponding* rotation ρ *is given by* (ρ^+, ρ^-) *where* $\rho^+ = \mu' \setminus \mu$ *and* $\rho^- = \mu \setminus \mu'$. *The set of all rotations of* $\mathcal{S}$ *is denoted* $\Pi(\mathcal{S})$.

Theorem 5 **([21]).** *Let $(F, W, (\mathcal{C}_f)_{f \in F}, (\mathcal{C}_w)_{w \in W})$ be a one-to-one stable matching instance, and let $(\mathcal{S}, \succeq)$ be the stable matching lattice, with join-irreducible elements $\mathcal{S}_j$. Then, $(\mathcal{S}, \succeq)$ is distributive. Furthermore, there is a bijection $\phi_j : \Pi(\mathcal{S}) \to \mathcal{S}_j$, which induces a partial order $\succeq$ on $\Pi(\mathcal{S})$, such that $(\Pi(\mathcal{S}), \succeq)$ is a representation for $\mathcal{S}$ with representation function $\psi_{\mathcal{S}} : \mathcal{S} \to \mathcal{D}(\Pi(\mathcal{S}))$ so that, for $R \in \mathcal{D}(\Pi(\mathcal{S}))$, we have*

$$\psi_{\mathcal{S}}^{-1}(R) = (\triangle_{\rho \in R}(\rho^- \triangle \rho^+)) \triangle \mu_W = \mu_W \cup \left(\bigcup_{\rho \in R} \rho^+ \right) \setminus \left(\bigcup_{\rho \in R} \rho^- \right).$$

Hence, $\psi_{\mathcal{S}}^{-1}(R)$ is obtained from μ_W by adding all pairs in $\cup_{\rho \in R} \rho^+$ and removing all those in $\cup_{\rho \in R} \rho^-$. We refer to $\psi_{\mathcal{S}}$ as the rotation representation function.

Note that we define the order $\succeq$ on $\mathcal{S}$ from the firm-perspective, while lower closed sets of rotations are defined so that the empty set corresponds to the worker-optimal stable matching μ_W.

In essence, in the one-to-one case each matching can be seen as a unique set of rotations applied to the worker-optimal matching. If $\rho \in \psi_{\mathcal{S}}(\mu)$, we say that ρ *occurs* in μ. Gusfield and Irving [21] also showed constructively that all partial orders are order-isomorphic to the rotation poset of a stable matching instance.

Theorem 6 **([21]).** *Let $(X, \succeq)$ be a poset. Then, there exists a stable matching instance $I_{(X,\succeq)}$ such that $(X, \succeq)$ is order-isomorphic to the poset of rotations $(\Pi(\mathcal{S}), \succeq)$ of the stable matching lattice $(\mathcal{S}, \succeq)$ for $I_{(X,\succeq)}$, with order-isomorphism given by $\phi : X \to \Pi(\mathcal{S})$. $I_{(X,\succeq)}$ is called the GI instance of $(X, \succeq)$. Given $(X, \succeq)$, we can compute $I_{(X,\succeq)}$, $\Pi(\mathcal{S})$, and ϕ in $O(|X|^2)$ running time.*

From the previous two results, the following immediately holds.

Corollary 1. *Let $(X, \succeq)$ be a poset, and let $(\mathcal{S}, \succeq)$ be the stable matching lattice for the instance $I_{(X,\succeq)}$. Then $(X, \succeq)$ is a representation for $(\mathcal{S}, \succeq)$, with representation function $\phi^{-1} \circ \psi_{\mathcal{S}}$.*

3 Representing Non-Distributive Lattices

In this section, we introduce new tools for representing general non-distributive lattices, and show how to implement them on stable matching lattices. Figure 1 gives an example of such concepts. We will then use these tools in the next section to prove our universality theorem, i.e., Theorem 1.

3.1 Partial Representation and Distributive Closure

To extend Birkhoff's theorem to any (possibly non-distributive) lattice $\mathcal{L}$, we introduce the notion of partial representation. Intuitively, we relax the conditions on ψ from order-embeddings to certain order-isomorphisms, so that some lower closed sets of the representation poset may not correspond to elements of $\mathcal{L}$.

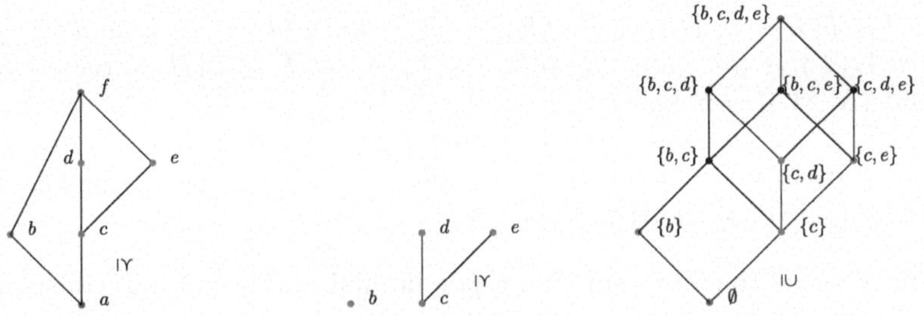

(a) $(X, \succeq)$ (b) $(X_j, \succeq)$ (c) $(\mathcal{D}(X_j), \supseteq)$

Fig. 1. Hasse diagrams of a non-distributive lattice $(X, \succeq)$, the corresponding poset of join-irreducible elements $(X_j, \succeq)$, and the distributive closure $(\mathcal{D}(X_j), \supseteq)$ of $(X, \succeq)$. Note that, e.g., $\{b, c\} \notin \psi_X(X)$. Imposing the join constraints $\Omega = \{(\{b, c\}, \{d, e\}), (\{c, d, e\}, \{b\})\}$ gives an order-isomorphism from $(X, \succeq)$ to $(\{T \in \mathcal{D}(X_j) \mid T \text{ satisfies } (\alpha, \beta)\ \forall (\alpha, \beta) \in \Omega\}, \supseteq)$.

Definition 6 (Partial representation). *Let $\mathcal{L} = (X, \succeq)$ be a lattice and $\mathcal{P} = (B, \succeq_B)$ be a poset. Then $\mathcal{P}$ is a* partial representation *of $\mathcal{L}$ if there exists an order-embedding ψ of $\mathcal{L}$ into $(\mathcal{D}(B), \supseteq)$ such that $\psi(\overline{x}) = B$ (resp., $\psi(\underline{x}) = \emptyset$), where $\overline{x} \in X$ (resp., $\underline{x} \in X$) is the maximal (resp., minimal) element of $\mathcal{L}$. ψ is called a* partial representation function.

Note that a representation function for a distributive lattice is also a partial representation function. For a lattice $\mathcal{L} = (X, \succeq)$, we define its *distributive closure* to be the lattice $(\mathcal{D}(X_j), \supseteq)$. $(\mathcal{D}(X_j), \supseteq)$ can be used to define a "canonical" partial representation, as discussed next.

Lemma 1 (Partial representation via the distributive closure). *Let $\mathcal{L} = (X, \succeq)$ be a lattice. Then, $(X_j, \succeq)$ is a partial representation of $\mathcal{L}$, with partial representation function $\psi_X : X \to \mathcal{D}(X_j)$ given by*

$$\psi_X(x) = \{x' \in X_j \mid x' \preceq x\}.$$

We refer to ψ_X as the canonical *partial representation function for $(X, \succeq)$.*

3.2 Join Constraints on Lattices

The main new tool we use throughout the paper is the *join constraint*, which is a constraint on a lattice that excludes certain elements while allowing others.

Definition 7 (Join constraint). *Let $(X, \succeq)$ be a lattice with partial representation $(B, \succeq)$ and partial representation function ψ_X. A* join constraint *on B is a pair (α, β) where $\alpha, \beta : 2^B \to \{0, 1\}$ are functions of the form*

$$\alpha(T) = (\mathbf{1}_{b_{11}}(T) \vee \mathbf{1}_{b_{12}}(T) \vee \ldots) \wedge (\mathbf{1}_{b_{21}}(T) \vee \mathbf{1}_{b_{22}}(T) \vee \ldots) \wedge \ldots,$$

$$\beta(T) = \mathbf{1}_{b_1}(T) \wedge \mathbf{1}_{b_2}(T) \wedge \ldots$$

for some $b_{11}, b_{12}, \cdots \in B$ *such that* $b_{i_1 i_2} \not\succ b_{i_3 i_4}$ *for all* i_1, i_2, i_3, i_4 *and* $b_1, b_2, \cdots \in B$*, where* $\mathbf{1}_b(T)$ *is the indicator that* $b \in T$*. A set* $T \in \mathcal{D}(B)$ satisfies *the join constraint* (α, β) *if, whenever* $\alpha(T) = 1$*, we also have* $\beta(T) = 1$*. Similarly, an element* $x \in X$ *satisfies the join constraint if* $\psi_X(x)$ *satisfies the join constraint. The* arguments *of the join constraint are the elements* $b_{11}, b_{22}, \ldots$ *and* $b_1, b_2, \ldots$*. We sometimes denote* (α, β) *by* $(\{b_{11}, b_{12}, \ldots\}, \{b_1, b_2, \ldots\})$.

We next argue that, by applying a polynomial number of join constraints to the distributive closure, we can obtain the original lattice. Note that the running time of the lemma below does not depend on $|\mathcal{D}(X_j)|$, which may be exponential in $|X|$.

Lemma 2 (Polynomially many join constraints suffice to describe a lattice). *Let* $(X, \succeq)$ *be a lattice, with canonical partial representation* ψ_X*, and distributive closure* $(\mathcal{D}(X_j), \supseteq)$*. Then there exists a set* Ω *of* $O(|X|^2)$ *join constraints, such that* ψ_X *is an order-isomorphism from* $(X, \succeq)$ *to*

$$(\{T \in \mathcal{D}(X_j) \mid T \text{ satisfies } (\alpha, \beta)\ \forall (\alpha, \beta) \in \Omega\}, \supseteq). \tag{1}$$

Furthermore, the number of arguments $|(\alpha, \beta)|$ *of each join constraint is at most* $O(|X|)$*, and given* $(X, \succeq)$*, the set* Ω *can be computed in* $O(|X|^3)$ *time.*

3.3 Implementing Join Constraints for the Stable Matching Lattice

Following Lemma 2, to prove Theorem 1, we would like to create, given a lattice $(X, \succ)$, a standard matching market instance whose stable matchings admit an order-isomorphism with (1). In order to do that, we introduce the following definition. Throughout the section, fix a GI instance $I = (F, W, (\mathcal{C}_f)_{f \in F}, (\mathcal{C}_w)_{w \in W})$ with stable matching lattice $(\mathcal{S}, \succeq)$ and representation function $\psi_{\mathcal{S}} : \mathcal{S} \to \mathcal{D}(\Pi(\mathcal{S}))$. We also let Ω be a set of join constraints defined over $\Pi(\mathcal{S})$.

Definition 8 (Ω-Extension of a matching market). *Let* $I^* = (F', W', (\mathcal{C}'_f)_{f \in F'}, (\mathcal{C}'_w)_{w \in W'})$ *be a matching market with stable matching lattice* $(\mathcal{S}', \succeq')$ *and such that* $F \subseteq F'$*,* $W \subseteq W'$*. We say that* I^* *is an* Ω-extension *of* I *if there exists a projection* $\xi_{\mathcal{S}'} : 2^{F' \times W'} \to 2^{F \times W}$ *that acts as the identity on* $2^{F \times W}$ *and is an order-isomorphism from* $(\mathcal{S}', \succeq')$ *to*

$$(\{\mu \in \mathcal{S} : \mu \text{ satisfies } (\alpha, \beta), \text{ for all } (\alpha, \beta) \in \Omega)\}, \succeq).$$

We next show that a standard Ω-extension of I always exists.

Theorem 7 (GI instances are Ω-extendable). *There exists an* Ω*-extension* I^* *for* I *with projection* $\xi_{\mathcal{S}^*}$ *that has* $O(|\Omega||W|)$ *additional agents and is a standard matching market instance. Given* I *as an input,* I^* *can be computed in* $O(|\Omega||F|(|W| + |(\alpha, \beta)^*|))$ *time, where* $|(\alpha, \beta)^*|$ *is the maximum number of arguments in a join constraint in* Ω*. Additionally,* $\psi_{\mathcal{S}^*} = \psi_{\mathcal{S}} \circ \xi_{\mathcal{S}^*}$ *is a partial representation function for the stable matching lattice* $(\mathcal{S}^*, \succeq)$ *of* I^**.*

We present here an overview of the proof. We extend the matching market instance by a single join constraint at the time, using a novel construction we call a *join constraint augmentation*, which produces an $\{(\alpha,\beta)\}$-extension of the current extension by adding additional agents and editing certain choice functions. Although it is unclear how to perform a join constraint augmentation on a generic instance, we define a class of *extendable instances*, to which we can apply a join constraint augmentation, and show that join constraint augmentations preserve extendability. We show that a GI instance is extendable. In this way, we can iteratively apply join constraint augmentations to the original instance, until the final instance is an Ω-extension.

4 Proof of Theorem 1

Let $(X,\succeq)$ be a lattice. By Theorem 6 applied to $(X_j,\succeq)$, in $O(|X_j|^2) \subseteq O(|X|^2)$ time we can construct a GI instance $I := I_{(X_j,\succeq)}$ whose rotation poset $(\Pi(\mathcal{S}),\succeq)$ is order-isomorphic to the poset of join-irreducible elements $(X_j,\succeq)$ via $\phi : X_j \to \Pi(\mathcal{S})$. By Corollary 1, $\phi^{-1} \circ \psi_{\mathcal{S}} : (\mathcal{S},\succeq) \to (\mathcal{D}(X_j),\supseteq)$ is an order-isomorphism, where $(\mathcal{S},\succeq)$ is the lattice of stable matchings of I. By Lemma 2, we can compute in $O(|X|^3)$ time a set Ω of $O(|X|^2)$ join constraints on $(X_j,\succeq)$, such that the canonical partial representation $\psi_X : X \to \mathcal{D}(X_j)$ is an order-isomorphism from $(X,\succeq)$ to $(\{T \in \mathcal{D}(X_j) \mid T \text{ satisfies } (\alpha,\beta)\ \forall(\alpha,\beta) \in \Omega\},\supseteq)$. Setting $\Omega_{\mathcal{S}} = \{(\alpha \circ \phi^{-1}, \beta \circ \phi^{-1}) \mid (\alpha,\beta) \in \Omega\}$ as a set of join constraints over $(\mathcal{S},\succeq)$, we deduce that $\psi_{\mathcal{S}}^{-1} \circ \phi \circ \psi_X$ is an order-isomorphism from $(X,\succeq)$ to

$$(\{\mu \in \mathcal{S} \mid \mu \text{ satisfies } \omega,\ \forall \omega \in \Omega_{\mathcal{S}}\},\succeq). \tag{2}$$

By Theorem 7, we can construct a standard $\Omega_{\mathcal{S}}$-extension I^* of I such that (2) is order-isomorphic to $(\mathcal{S}^*,\succeq)$, where $(\mathcal{S}^*,\succeq)$ is the lattice of stable matchings of I^*. Thus, $(\mathcal{S}^*,\succeq)$ is order-isomorphic to $(X,\succeq)$, as desired.

For the runtime, first observe that, by Theorem 6, I has $O(|X|^2)$ agents and can be constructed in time $O(|X|^2)$. By Lemma 2 we have $|\Omega| = O(|X|^2)$, and by Theorem 7 and Lemma 2, it takes $O(|\Omega||F|(|W| + |(\alpha,\beta)^*|)) = O(|X|^6)$ time to construct I^*, which then has $O(|\Omega||W|) = O(|X|^4)$ agents. □

5 Antimatroids and Complements of Join Constraints

In this section, we introduce tools that will allow us to show that optimizing a linear function over the set of stable matchings is NP-hard via a reduction from the problem of finding a feasible set of minimum cost in an antimatroid.

Definition 9 (Antimatroid). *Let V be a finite ground set and let $\mathcal{G}$ be a nonempty family of subsets of V, called* feasible. *$(V,\mathcal{G})$ is an* antimatroid *if the following conditions hold:*

1. (**Feasibility of the ground set**) $V \in \mathcal{G}$.
2. (**Closure under union**) $G_1, G_2 \in \mathcal{G} \implies G_1 \cup G_2 \in \mathcal{G}$.

3. **(Accessibility)** $G \in \mathcal{G} \setminus \emptyset \implies \exists g \in G : G \setminus \{g\} \in \mathcal{G}$.

It is well-known that the feasible sets of an antimatroid, when ordered by inclusion, form a lattice $(\mathcal{G}, \supseteq)$ (see, e.g., [30]).

Definition 10 (Paths and endpoints). *Let $(V, \mathcal{G})$ be an antimatroid. An* endpoint *of $G \in \mathcal{G}$ is an element $g \in G$ such that $G \setminus \{g\} \in \mathcal{G}$. A* path *is a feasible set $G \in \mathcal{G}$ with exactly one endpoint, denoted by $e(G)$. We let Q be the collection of all paths of $(V, \mathcal{G})$. Given any $x \in V$, we define $Q(x) := \{G \in Q \mid e(G) = x\}$. The set of paths Q can be partially ordered as to form the* path poset $(Q, \supseteq)$.

[30] also showed that, similarly to Birkhoff's representation theorem for lattices, $(Q, \supseteq)$ generates all feasible sets of an antimatroid.

Proposition 1 (Feasible sets as unions of paths [30]). *Let $(V, \mathcal{G})$ be an antimatroid with path poset $(Q, \supseteq)$. Define $\psi_{\mathcal{G}} : \mathcal{G} \to 2^Q$ by $\psi_{\mathcal{G}}(G) = \{G' \in Q \mid G' \subseteq G\}$. Then, for any $G \in \mathcal{G}$, we have $G = \bigcup \psi_{\mathcal{G}}(G)$. Conversely, for any $Q' \subseteq Q$, we have $\cup Q' \in \mathcal{G}$.*

Thus, a natural way to encode an antimatroid is by its path poset. This encoding can be used to formulate the following optimization problem.

Definition 11 (Minimum cost feasible set in an antimatroid). *Let $(V, \mathcal{G})$ be an antimatroid encoded by its path poset $(Q, \supseteq)$. Let $c : V \to \mathbb{Z}$. The goal of the minimum cost feasible set problem is to find an element in* $\arg\min_{G \in \mathcal{G}} \sum_{g \in G} c(g)$.

Theorem 8 ([32]). *The minimum cost feasible set problem is NP-hard.*

In the following, fix a lattice $\mathcal{L} = (X, \succeq)$, its *complement lattice* $\mathcal{L}^c = (X, \preceq)$, a poset $(B, \succeq)$ and define an *upper closed set* of B to be $S \subseteq B$ such that $x \in S$ and $x' \succeq x$ implies $x' \in S$. The set of upper closed sets of B is $\mathcal{D}^c(B)$. Note that a set is upper closed if and only if its complement is lower closed.

Definition 12 (Complement of a join constraint). *Given a join constraint (α, β) defined over $\mathcal{L}$, the complement (β^c, α^c) of (α, β) is defined as*

$$\beta^c(T) = \mathbf{1}_{b_1}(T) \vee \mathbf{1}_{b_2}(T) \vee \cdots, \; \alpha^c(T) = (\mathbf{1}_{b_{11}}(T) \wedge \mathbf{1}_{b_{12}}(T) \wedge \ldots) \vee (\mathbf{1}_{b_{21}}(T) \wedge \ldots) \vee \ldots$$

For a partial representation function $\psi_X : X \to \mathcal{D}(B)$ for $\mathcal{L}$, we define $\psi_X^c : X \to \mathcal{D}^c(B)$ as $\psi_X^c(x) = B \setminus \psi_X(x)$ for all $x \in \mathcal{L}$.

Lemma 3 (Complement lattice satisfies complement join constraint). *Assume that the elements of X, with partial representation function $\psi_X : X \to \mathcal{D}(B)$, satisfy a join constraint (α, β) on B as in Definition 7. Then $\psi_X^c(x) : X \to \mathcal{D}^c(B)$ is a partial representation function for $\mathcal{L}^c$. Furthermore, elements $x \in X$ satisfy $\beta^c(\psi_X^c(x)) = 1$ only if $\alpha^c(\psi_X^c(x)) = 1$.*

If $T \subseteq B$ satisfies $\beta^c(T) = 1$ only if $\alpha^c(T) = 1$, we say that T satisfies the *complement join constraint* (β^c, α^c). If $x \in \mathcal{L}$ satisfies $\beta^c(\psi(x)) = 1$ only if $\alpha^c(\psi(x)) = 1$, we say that x satisfies (β^c, α^c) via ψ. We next show that, analogously to Lemma 2, we can use polynomially many complement join constraints to describe the path poset of an antimatroid, starting from the poset $(V, \sim)$ where $x, y \in V$ satisfy $x \sim y$ if and only if $x = y$. Note that $(\mathcal{D}(V), \supseteq) = (\mathcal{D}^c(V), \supseteq)$, and $(\mathcal{D}(V), \supseteq)$ has canonical representation function $\text{id} : \mathcal{D}(V) \to \mathcal{D}(V)$.

Lemma 4 (Polynomially many complement join constraints describe an antimatroid). *Let $(V, \mathcal{G})$ be an antimatroid with path poset $(Q, \supseteq)$. There is a set Ω of $O(|V|)$ complement join constraints on $(V, \sim)$ with $O(|Q|^2)$ arguments each so that $\mathcal{G} = \{T \in \mathcal{D}^c(V) \mid T$ satisfies (β^c, α^c) via $\text{id}\ \forall (\beta^c, \alpha^c) \in \Omega\}$.*

Last, we show how to translate costs between rotations and matchings.

Lemma 5 (Cost function). *Let $I := I_{(V,\sim)}$ be the GI instance for $(V, \sim)$ with stable matching lattice $(\mathcal{S}, \succeq)$ and representation function $\psi_\mathcal{S} : \mathcal{S} \to \mathcal{D}(\Pi(\mathcal{S}))$. Let Ω be a set of join constraints over $(\mathcal{S}, \succeq)$, and let $I^* = (F^*, W^*, (\mathcal{C}_f^*)_{f \in F^*}, (\mathcal{C}_w^*)_{w \in W^*})$ be an Ω-extension of I with stable matching lattice $(\mathcal{S}^*, \succeq)$, and partial representation function $\psi_{\mathcal{S}^*} = \psi_\mathcal{S} \circ \xi_{\mathcal{S}^*}$. Let $c : \Pi(\mathcal{S}) \to \mathbb{R}$. Define $c' : F^* \times W^* \to \mathbb{R}$ as follows. For each $\rho \in \Pi$ and each $(f, w) \in \rho^-$, define $c'(f, w) = c(\rho)/|\rho^-|$. For all other (f, w), set $c'(f, w) = 0$. Then, $c'(\mu) = c(\psi^c_{\mathcal{S}^*}(\mu))$ for all $\mu \in \mathcal{S}^*$.*

In Lemma 5, c' is well-defined since for every pair (f, w) of agents of a one-to-one instance, $(f, w) \in \rho^-$ for at most one rotation ρ (see, e.g., [21]).

6 Proof of Theorem 2

We reduce from minimum cost feasible set in an antimatroid. Let $(V, \mathcal{G})$ be an antimatroid, $(Q, \supseteq)$ its path poset, and $c : V \to \mathbb{Z}$ a cost function. Let $I := I_{(V,\sim)}$ be the GI instance for $(V, \sim)$ defined as in Theorem 6 with stable matching lattice $(\mathcal{S}, \succeq)$, and bijection $\phi : V \to \Pi(\mathcal{S})$. By Lemma 4, there is a set Ω of $O(|V|)$ complement join constraints on $(V, \sim)$ each with $O(|Q|^2)$ arguments such that

$$\mathcal{G} = \{T \in \mathcal{D}^c(V) \mid T \text{ satisfies } (\beta^c, \alpha^c) \text{ via id } \forall (\beta^c, \alpha^c) \in \Omega\}.$$

Let $\Omega' = \{(\alpha, \beta) \mid (\beta^c, \alpha^c) \in \Omega\}$ be the join constraints corresponding to the complement join constraints in Ω. For $\psi_{\mathcal{S}^*}$ defined as in Theorem 7, construct a standard Ω'-extension I^* of I, with matching lattice $(\mathcal{S}^*, \succeq)$ order-isomorphic to $\xi_{\mathcal{S}^*}(\mathcal{S}^*) = \{\mu \in \mathcal{S} \mid \mu \text{ satisfies } (\alpha, \beta),\ \forall (\alpha, \beta) \in \Omega'\}$, and partial representation function $\phi^{-1} \circ \psi_{\mathcal{S}^*} : \mathcal{S}^* \to \mathcal{D}(V)$. By Lemma 3, we know that the complement lattice $(\mathcal{S}^*, \preceq)$ has partial representation $(\phi^{-1} \circ \psi_{\mathcal{S}^*})^c = \phi^{-1} \circ \psi^c_{\mathcal{S}^*}$ and $\xi_{\mathcal{S}^*}(\mathcal{S}^*) = \{\mu \in \mathcal{S} \mid \mu \text{ satisfies } (\beta^c, \alpha^c) \text{ via } \phi^{-1} \circ \psi^c_{\mathcal{S}^*},\ \forall (\beta^c, \alpha^c) \in \Omega\}$. Thus,

$$\begin{aligned}
\phi^{-1}(\psi^c_{\mathcal{S}^*}(\mathcal{S}^*)) &= \{T \in \mathcal{D}^c(V) \mid T = \phi^{-1}(\psi^c_{\mathcal{S}^*}(\mu)),\ \mu \in \mathcal{S}^*\} \\
&= \{T \in \mathcal{D}^c(V) \mid T = \phi^{-1}(\psi^c_{\mathcal{S}}(\mu)),\ \mu \in \mathcal{S},\ \mu \text{ satisfies} \\
&\qquad (\beta^c, \alpha^c) \text{ via } \phi^{-1} \circ \psi^c_{\mathcal{S}^*},\ \forall (\beta^c, \alpha^c) \in \Omega\} \\
&= \{T \in \mathcal{D}^c(V) \mid T \text{ satisfies } (\beta^c, \alpha^c) \text{ via id},\ \forall (\beta^c, \alpha^c) \in \Omega\} = \mathcal{G}.
\end{aligned}$$

Extend c to $\Pi(\mathcal{S})$ by setting $c(\rho) = c(\phi^{-1}(\rho))$ for each $\rho \in \Pi(\mathcal{S})$, and define c' as in Lemma 5. Then we know that for each $\mu \in \mathcal{S}^*$ we have $c'(\mu) = c(\psi^c_{\mathcal{S}^*}(\mu)) = c(\phi^{-1}(\psi^c_{\mathcal{S}^*}(\mu)))$. Then, because $\phi^{-1}(\psi^c_{\mathcal{S}^*}(\mathcal{S}^*)) = \mathcal{G}$, the optimal value of c' over $\mathcal{S}^*$ is equal to the optimal value of c over $\mathcal{G}$. Since I^* can be obtained using $O(|V|)$ join constraint augmentations, each with $O(|Q|^2)$ arguments, by Theorem 7 we have a polynomial reduction from minimum cost feasible set in an antimatroid to minimum cost stable matching in a standard instance. □

Acknowledgments. Supported by the grants NSF 2046146 and AFOSR FA9550-23-1-0697, and by a Cheunk-Keung Innovation Doctoral Fellowship. The authors wish to thank Samuel Fiorini and a reviewer for suggesting papers [32] and [16], respectively.

Disclosure of Interests. The authors have no competing interests to declare that are relevant to the content of this article.

References

1. Abdulkadiroğlu, A., Sönmez, T.: School choice: a mechanism design approach. Am. Econ. Rev. **93**(3), 729–747 (2003)
2. Alkan, A.: On preferences over subsets and the lattice structure of stable matchings. Rev. Econ. Des. **6**, 99–111 (2001)
3. Avann, S.: Application of the join-irreducible excess function to semi-modular lattices. Math. Ann. **142**(4), 345–354 (1961)
4. Bansal, V., Agrawal, A., Malhotra, V.S.: Polynomial time algorithm for an optimal stable assignment with multiple partners. Theor. Comput. Sci. **379**(3), 317–328 (2007)
5. Bělohlávek, R.: Concept lattices and order in fuzzy logic. Ann. Pure Appl. Log. **128**(1–3), 277–298 (2004)
6. Birkhoff, G.: Rings of sets (1937)
7. Blair, C.: Every finite distributive lattice is a set of stable matchings. J. Comb. Theory Ser. A **37**(3), 353–356 (1984)
8. Blair, C.: The lattice structure of the set of stable matchings with multiple partners. Math. Oper. Res. **13**(4), 619–628 (1988)
9. Dilworth, R.P.: Lattices with unique irreducible decompositions. The Dilworth Theorems: Selected Papers of Robert P. Dilworth, pp. 93–99 (1990)
10. Echenique, F.: Counting combinatorial choice rules. Games Econ. Behav. **58**(2), 231–245 (2007)
11. Echenique, F., Oviedo, J.: A theory of stability in many-to-many matching markets. Theor. Econ. **1**(2), 233–273 (2006)
12. Edelman, P.H., Jamison, R.E.: The theory of convex geometries. Geom. Dedicata. **19**(3), 247–270 (1985)
13. Faenza, Y., Foussoul, A., He, C.: Two-stage stochastic stable matching. In: Proceedings of IPCO 2024, pp. 154–167. Springer (2024)
14. Faenza, Y., Mourtos, I., Samaris, M., Sethuraman, J.: (un) stable matchings with blocking costs. Oper. Res. Lett. **49**(5), 655–662 (2021)
15. Faenza, Y., Zhang, X.: Affinely representable lattices, stable matchings, and choice functions. Math. Program. **197**(2), 721–760 (2023)
16. Fleiner, T.: Stable and crossing structures. Ph. D. thesis, Tecnische Universiteit Eindhoven (2000)

17. Fleiner, T.: A matroid generalization of the stable matching polytope. In: Aardal, K., Gerards, B. (eds.) IPCO 2001. LNCS, vol. 2081, pp. 105–114. Springer, Heidelberg (2001). https://doi.org/10.1007/3-540-45535-3_9
18. Fleiner, T., Kamiyama, N.: A matroid approach to stable matchings with lower quotas. Math. Oper. Res. **41**(2), 734–744 (2016)
19. Gale, D., Shapley, L.S.: College admissions and the stability of marriage. Am. Math. Mon. **69**(1), 9–15 (1962)
20. Garg, V.K.: Introduction to Lattice Theory with Computer Science Applications. John Wiley & Sons, Hoboken (2015)
21. Gusfield, D., Irving, R.W.: The Stable Marriage Problem: Structure and Algorithms. MIT Press, Cambridge (1989)
22. Hatfield, J.W., Milgrom, P.R.: Matching with contracts. Am. Econ. Rev. **95**(4), 913–935 (2005)
23. Hitsch, G.J., Hortaçsu, A., Ariely, D.: Matching and sorting in online dating. Am. Econ. Rev. **100**(1), 130–63 (2010)
24. Immorlica, N., Echenique, F., Vazirani, V.V.: Online and Matching-based Market Design. Cambridge University Press, Cambridge (2023)
25. Irving, R.W., Leather, P., Gusfield, D.: An efficient algorithm for the "optimal" stable marriage. J. ACM (JACM) **34**(3), 532–543 (1987)
26. Iwata, S., Yokoi, Y.: Finding a stable allocation in polymatroid intersection. Math. Oper. Res. **45**(1), 63–85 (2020). https://doi.org/10.1287/MOOR.2018.0976
27. Király, T., Pap, J.: Total dual integrality of rothblum's description of the stable-marriage polyhedron. Math. Oper. Res. **33**(2), 283–290 (2008)
28. Knuth, D.E.: Marriages stables. Technical report (1976)
29. Konishi, H., Ünver, M.U.: Credible group stability in many-to-many matching problems. J. Econ. Theory **129**(1), 57–80 (2006)
30. Korte, B., Lovász, L., Schrader, R.: Greedoids, vol. 4. Springer Science & Business Media (2012)
31. Manlove, D.: Algorithmics of matching under preferences, vol. 2. World Scientific (2013)
32. Merckx, K.: Optimization and Realizability Problems for Convex Geometries. Ph.D. thesis, Université Libre de Bruxelles (2019)
33. Reynolds, J.C.: Transformational systems and the algebraic structure of atomic formulas. Mach. Intell. **5**, 135–152 (1970)
34. Roth, A.E.: Stability and polarization of interests in job matching. Econom.: J. Econom. Soc. 47–57 (1984)
35. Roth, A.E., Sotomayor, M.: Two-sided matching. Handb. Game Theory Econ. Appl. **1**, 485–541 (1992)
36. Rothblum, U.G.: Characterization of stable matchings as extreme points of a polytope. Math. Program. **54**, 57–67 (1992)
37. Sönmez, T., Ünver, M.U.: Course bidding at business schools. Int. Econ. Rev. **51**(1), 99–123 (2010)
38. Vate, J.: Linear programming brings marital bliss. Oper. Res. Lett. **8**(3), 147–153 (1989)
39. Wang, X., Agatz, N., Erera, A.: Stable matching for dynamic ride-sharing systems. Transp. Sci. **52**(4), 850–867 (2018)

Separating Coverage and Submodular: Maximization Subject to a Cardinality Constraint

Yuval Filmus[1(✉)], Roy Schwartz[1], and Alexander V. Smal[2]

[1] Technion Israel Institute of Technology, Technion City, Haifa 3200003, Israel
{yuvalfi,schwartz}@cs.technion.ac.il
[2] JetBrains Research, Paphos, Cyprus
avsmal@gmail.com

Abstract. We consider two classic problems: maximum coverage and monotone submodular maximization subject to a cardinality constraint. [Nemhauser–Wolsey–Fisher '78] proved that the greedy algorithm provides an approximation of $1 - 1/e$ for both problems, and it is known that this guarantee is tight ([Nemhauser–Wolsey '78; Feige '98]). Thus, one would naturally assume that everything is resolved when considering the approximation guarantees of these two problems, as both exhibit the same tight approximation and hardness.

In this work we show that this is not the case, and study both problems when the cardinality constraint is a constant fraction $c \in (0, 1]$ of the ground set. We prove that monotone submodular maximization subject to a cardinality constraint admits an approximation of $1 - (1 - c)^{1/c}$; This approximation equals 1 when $c = 1$ and it gracefully degrades to $1 - 1/e$ when c approaches 0. Moreover, for every $c = 1/s$ (for any integer $s \in \mathbb{N}$) we present a matching hardness.

Surprisingly, for $c = 1/2$ we prove that Maximum Coverage admits an approximation of 0.7533, thus separating the two problems. To the best of our knowledge, this is the first known example of a well-studied maximization problem for which coverage and monotone submodular objectives exhibit a different best possible approximation.

Keywords: coverage · submodular · approximation · linear programming · semi-definite programming

1 Introduction

In this work we consider two related classic maximization problems. First, in Maximum Coverage (MC) we are given a (weighted) collection of elements E equipped with $w: E \to \mathbb{R}_+$, sets $S_1, \dots, S_n \subseteq E$, and a cardinality bound $k \in \mathbb{N}$. The goal is to find a collection of at most k sets $X \subseteq [n]$ that maximizes the total weight of covered elements: $\sum_{e \in \bigcup_{i \in X} S_i} w_e$.

MC is captured by the well-studied problem of maximizing a nonnegative monotone submodular function subject to a cardinality constraint. Given a

N. Megow and A. Basu (Eds.): IPCO 2025, LNCS 15620, pp. 242–255, 2025.
https://doi.org/10.1007/978-3-031-93112-3_18

universe U of size n, a set function $f\colon 2^U \rightarrow \mathbb{R}_+$ is *submodular* if it satisfies $f(A)+f(B) \geq f(A\cup B)+f(A\cap B)$ $\forall A, B \subseteq U$,[1] and is monotone if $f(A) \leq f(B)$ $\forall A \subseteq B$. In this problem, denoted by SM, we are given U, a monotone submodular function f represented by a value oracle,[2] and a cardinality bound $k \in \mathbb{N}$. The goal is to find $S \subseteq U$ of size at most k that maximizes $f(S)$. Clearly, SM captures MC since coverage functions are submodular and monotone.

SM was studied as early as the late 70's, when in their celebrated work, Nemhauser, Wolsey, and Fisher [37] proved that the simple greedy algorithm provides an approximation of $1 - 1/e$. Moreover, Nemhauser and Wolsey [36] proved that this guarantee is tight, i.e., any algorithm that achieves an approximation better than $1 - 1/e$ is required to perform exponentially many value queries. Later, Feige [25] proved that assuming $P \neq NP$ the bound of $1 - 1/e$ is also tight for the special case of MC.

Both MC and SM have numerous theoretical and practical applications, e.g., the spread of influence in networks [29,30], social graph analysis [38], dictionary selection [20], data summarization [35], and many more. Thus, it is no surprise that these problems have been studied extensively in different settings, e.g., fast and implementable algorithms [6,14,34], non-monotone submodular objectives [9,10,13,24], randomized vs. deterministic algorithms [8,11,12], and online algorithms [15,18], to name a few.

One would naturally assume that with respect to the approximation of both these classic and well-studied problems everything is resolved, for two main reasons. First, matching upper and lower bounds are known. Second, there is no known difference (to the best of our knowledge) between the best possible approximation when considering maximization, with any constraint, of coverage and monotone submodular objectives. Another classic example of the latter is a knapsack constraint [31,40].

In this work we prove that this is not the case. We study the behavior of the best possible approximation for both MC and SM as a function of the cardinality bound k. Specifically, we assume that $k = cn$ for some absolute constant $0 < c \leq 1$, and aim to prove both upper and lower bounds on the possible approximation as a function of c. To the best of our knowledge, this behavior has never been previously studied until this work. The aforementioned works imply that regardless of the value of c, the approximation is always at least $1 - 1/e$ for both MC and SM. Moreover, for both problems a uniform random solution provides an approximation of at least c (see, e.g., [26, Lemma 2.2]).

A surprising result of our work is that MC and SM admit different approximations when studied in this regime. For example, for $c = 1/2$ we prove that MC admits an approximation that is strictly better than the hardness of SM. This separates maximization of a coverage function when compared to maximization of a general monotone submodular function. We are not aware of any previously known natural example of this unexpected phenomenon in the context of combi-

[1] An equivalent definition: $f(A\cup\{e\}) - f(A) \geq f(B\cup\{e\}) - f(B)$, $\forall A \subseteq B$ and $e \notin B$.

[2] A value oracle is a black box that allows the algorithm to access f by querying the oracle with some $S \subseteq U$ and receiving from the oracle $f(S)$.

natorial optimization. The situation can be compared to the difference between the optimal approximation ratios of MAX-CUT and (symmetric or arbitrary) non-monotone submodular maximization.

1.1 Our Results

Algorithms for MC and SM. First, for every constant $0 < c \leq 1$ we present a simple LP-based approximation algorithm for MC whose approximation ratio depends on c.

Theorem 1. *For every constant $0 < c \leq 1$ there exists a polynomial time algorithm for Maximum Coverage with $k = cn$ that achieves an approximation of $\rho(c)$. If $c = 1/s$ for some integer $s \in \mathbb{N}$, then $\rho(c) = 1 - (1-c)^{1/c}$. Otherwise, $1/(s+1) < c < 1/s$ for some integer $s \in \mathbb{N}$ and $\rho(c) = 1 - \sigma(\alpha^*, s)$, where:*

(1) $\sigma(\alpha, m) = (1 - \alpha c - (1-\alpha)/m)^m$.
(2) $\alpha^* \in (0,1)$ is the unique solution of the equation $\sigma(\alpha^*, s) = \sigma(\alpha^*, s+1)$.

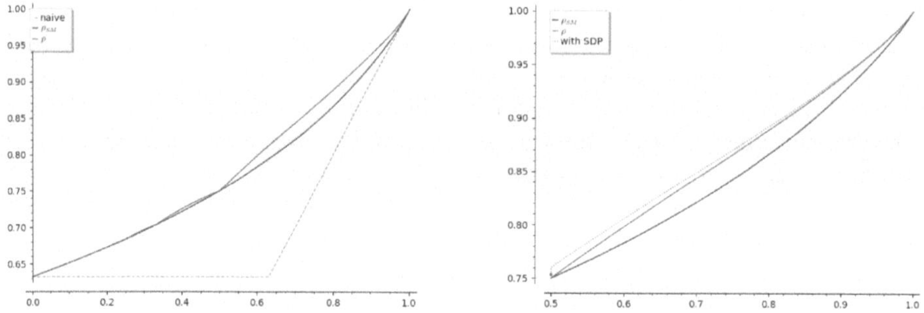

(a) Entire curve (b) Zoom-in on $c \in [1/2, 1]$

Fig. 1. $1-(1-c)^{1/c}$ (blue) is approximation for SM, $\rho(c)$ (green) is hardness for SM and approximation for MC, $\max\{1 - 1/e, c\}$ (dashed gray) is naive algorithm, and red point (proven) extended by dotted red line (numerical under a generalization of Austrin's simplicity conjecture [3]) is approximation for MC. Both red dot and dotted red line separate MC and SM.

Figure 1 depicts $\rho(c)$. Some notable remarks regarding $\rho(c)$:

(1) $\rho(c)$ is an increasing function in c.
(2) If $c = 1/s$ for some integer $s \in \mathbb{N}$ then $\rho(c) = 1 - (1-c)^{1/c}$.
(3) $\rho(1) = 1$, i.e., if $c = 1$ our algorithm gives an optimal solution.
(4) $\lim_{c \to 0} \rho(c) = (1 - 1/e)$.
(5) $\rho(c) > \max\{c, 1 - 1/e\}$ for all $0 < c < 1$.

An immediate consequence of the above is that our approximation is strictly better than the naive algorithm that takes the best of the two previously known algorithms. It is worth noting that for every c we present a matching integrality gap of $\rho(c)$ for the LP that is used by our algorithm, thus proving the tightness of our analysis.

The algorithm underlying Theorem 1 relies on the explicit description of the MC instance. It can be modified to work even given only value oracle access to the instance.

Theorem 2. *For every constant $0 < c \leq 1$ there exists a polynomial time algorithm for Maximum Coverage with $k = cn$ that achieves an approximation of $\rho(c)$, where the input is provided by a value oracle.*

Second, for every constant $0 < c \leq 1$ we present a simple approximation algorithm for SM whose approximation ratio depends on c. The algorithm is implicit in [27], and we include it here for completeness.

Theorem 3. *For every constant $0 < c \leq 1$ there exists a polynomial time algorithm for monotone submodular maximization subject to a cardinality constraint $k = cn$ that achieves an approximation of $1 - (1 - c)^{1/c}$.*

Figure 1 depicts the curve $1 - (1 - c)^{1/c}$. Some notable remarks regarding this approximation guarantee:

(1) For every $0 < c < 1$, $1 - (1 - c)^{1/c} > \max\{c, 1 - 1/e\}$, i.e., our algorithm is strictly better than the naive algorithm that takes the best of the two previously known algorithms.
(2) If $c = 1/s$ for some integer $s \in \mathbb{N}$ then $1 - (1 - c)^{1/c} = \rho(c)$, thus Theorems 1 and 3 provide the same guarantee for these values of c.
(3) $\rho(c) > 1 - (1 - c)^{1/c}$ for all c such that there does not exist an integer $s \in \mathbb{N}$ for which $c = 1/s$, i.e., for these values of c, Theorem 1 provides an approximation strictly better than that of Theorem 3.

One should recall that Theorem 3 is for SM, whereas Theorem 1 is for MC, which is a special case of SM.

Value Oracle Hardness for SM. Third, we present a hardness result for SM that equals $\rho(c)$ and (almost) matches the guarantee of Theorem 3 for every c. The case $c = 1/s$ of the theorem essentially appears already in [41].

Theorem 4. *For every rational constant $0 < c \leq 1$ and every $\varepsilon > 0$, any (possibly randomized) algorithm for monotone submodular maximization subject to a cardinality constraint $k = cn$ that achieves an approximation of $(1 + \varepsilon)\rho(c)$ requires exponentially many value queries.*

It is worth nothing that Theorems 3 and 4 resolve the approximability of SM when $c = 1/s$ for $s \in \mathbb{N}$ as both the algorithm and the hardness match (recall that in this case $\rho(c) = 1 - (1 - c)^{1/c}$).

We do not prove similar hardness results for MC, as we actually prove that MC admits a better approximation. Moreover, the results of Austrin and Stanković [5] imply that MC is APX-hard for all $c \in (0,1)$, even in the special case of Max k-Vertex-Cover, in which each element is covered by exactly two sets.

Separating Coverage and Submodular Maximization. Perhaps the most surprising result is that we are able to *separate* MC and SM, even when in both cases the instance is given via a value oracle, as the following theorem summarizes.

Theorem 5. *There exists a polynomial time algorithm for Maximum Coverage with* $k = n/2$ *that achieves an approximation of at least* $0.7533 - o(1)$*. The algorithm only requires value oracle access to the instance.*

It is important to note that the above theorem implies that MC, when $c = 1/2$, has an approximation that is strictly better than $\rho(1/2) = 3/4$. Recall that the latter is the hardness for SM, when $c = 1/2$, as Theorem 4 states. Refer to Fig. 1 for this result.

Moreover, assuming a generalization of the simplicity conjecture of Austrin [3], this separation extends (numerically) to all constants $1/2 < c < 1$, showing that MC with $k = cn$ admits an approximation that is *strictly* better than $\rho(c)$ by an additive absolute constant.

1.2 Our Techniques

A main ingredient in both algorithms for MC, i.e., Theorems 1 and 5, is that we round fractional solutions that are *not* optimal with respect to our relaxations. We average a fractional optimal solution, LP in the case of Theorem 1 and both LP and SDP in the case of Theorem 5, with a fractional solution that corresponds to a uniform random solution. The mixing probability needs to be carefully chosen, as it depends on both c and the rounding algorithm we use for the LP or SDP. The SDP is used to handle the elements in the instance which are covered exactly twice. These elements form an instance of Max k-Vertex-Cover, a problem for which Austrin, Benabbas and Georgiou [4] gave an SDP-based approximation algorithm.

Moreover, in order to obtain (numerically) better algorithms for MC when $1/2 < c < 1$ (conditional on a generalization of Austrin's simplicity conjecture), our algorithm generalizes Austrin's algorithm for the case $c = 1/2$ by handling arbitrary c and by allowing elements to be covered only once.

Our results for SM are based on the measured continuous greedy algorithm [17,27]. This algorithm simulates a continuous process. The longer the process runs, the higher the value of the output is. However, if the process runs for too long, we might end up with an infeasible solution. We exploit the simple structure of the cardinality constraint to show that there exists a good stopping time for this process.

Our hardness results for SM are based on the symmetry gap machinery of Vondrák [41], in which a fixed instance with a gap between symmetric and arbitrary solutions is lifted to value oracle hardness. This lifting process maintains the ratio k/n. Curiously, the instances used to prove hardness for SM are identical to those that are used to show a tight integrality gap for the LP of Theorem 1.

1.3 Related Work

While we are unaware of previous work that studies either MC or SM in the regime $k = cn$, Buchbinder et al. [13] study *non-monotone* submodular maximization subject to a cardinality constraint $k \leq cn$ or $k = cn$ (in contrast to the monotone setting, in the non-monotone setting the two constraints are different). In both cases, their algorithm gives an approximation ratio of $(1 + {}^1\!/\!{}_2\sqrt{c(1-c)})^{-1} - o(1)$ for all $c \leq {}^1\!/\!{}_2$, and ${}^1\!/\!{}_2 - o(1)$ for all $c > {}^1\!/\!{}_2$.

Another related work in the same vein due to Li et al. [33] studies the problem of maximizing a monotone linear function subject to the constraint $k \leq cn$, when the linear function is given as a value oracle. The paper, which focuses on algorithms making very few queries, shows that any algorithm whose approximation ratio beats c must make $\Omega({}^n\!/\!{}_{\log n})$ value oracle queries.

Separations between maximum coverage and monotone submodular functions do appear in other settings. One example is mechanism design: Dughmi, Roughgarden and Yan [22] give a $1 - 1/e$-approximation mechanism for welfare maximization when the valuations of the bidders are matroid rank sums, a class of functions that includes coverage functions. In contrast, Dughmi and Vondrák [23] and Dobzinski and Vondrák [21] rule out any constant approximation for general monotone submodular valuations.

1.4 Paper Organization

Sect. 3 contains the algorithms for proving Theorem 1 to 3. Due to lack of space, the remaining parts of the paper are relegated to the full version. These include Theorem 4, the tight hardness for SM; a tight integrality gap for the LP underlying Theorem 1; and the separation of MC and SM for $c = {}^1\!/\!{}_2$ (provably) and for $c \in ({}^1\!/\!{}_2, 1)$ (conditionally).

2 Preliminaries

Given a set function $f\colon 2^U \to \mathbb{R}_+$, the *multilinear* extension $F\colon [0,1]^U \to \mathbb{R}_+$ is defined as follows: $F(x) \triangleq \sum_{T \subseteq E} f(T) \prod_{e \in T} x_e \prod_{e \notin T}(1 - x_e)$, $\forall x \in [0,1]^U$. Intuitively, $F(x)$ has a probabilistic meaning: it is the expected value of f on a random set which includes each element $e \in E$ independently with probability x_e.

We require an algorithm for rounding a fractional point with respect to the multilinear extension.

Theorem 6 (Pipage Rounding [1,16]). *There exists a polynomial time algorithm that, given a matroid $M = (U, \mathcal{I})$ and a point $\mathbf{z} \in \mathcal{P}$ (where $\mathcal{P}$ is the independent sets polytope associated with M), finds $\mathbf{r} \in \mathcal{P}$ satisfying:*

(1) $\mathbf{r} \in \mathcal{P} \cap \{0,1\}^U$.
(2) $F(\mathbf{r}) \geq F(\mathbf{z})$.

We note that instead of pipage rounding one can use, e.g., a deterministic version of swap rounding [19].

The Max k-Vertex-Cover problem is a special case of MC. Given an undirected graph G on n vertices, the elements E are the edges of the graph, and each vertex v corresponds to a set that contains all edges touching v. Thus, the goal (as in MC) is to choose $X \subseteq V$ of size at most k that maximize the total weight of edges covered by X, i.e., edges that have at least one endpoint in X. We utilize the following result of Austrin, Benabbas, and Georgiou [4] for the case that $c = 1/2$ (this result applies to the more general problem of Max 2-CNF where exactly half the variables are required to be true). The result of [4] is based on semi-definite programming and on the works of Lewin, Livant, and Zwick [32] and Raghavendra and Tan [39]. It is based on a certain *simplicity conjecture*, recently proved by Brakensiek, Huang and Zwick [7].

Theorem 7 (Max k-Vertex-Cover [4]). *There is a polynomial time algorithm for Max k-Vertex-Cover with $k = n/2$ that achieves an approximation of at least* 0.9401.

3 Algorithms for MC and SM

3.1 Algorithm for MC

Our algorithm is based on a simple "non-oblivious" rounding of the natural LP, i.e., it rounds an altered solution and not the optimal solution for the LP. We start by stating the classic LP for MC (see, e.g., [2]):

$$\max \left\{ \sum_{e \in E} w_e x_e : x_e \leq \min \left\{ 1, \sum_{i:\, e \in S_i} y_i \right\} \forall e \in E,\ \sum_{i=1}^{n} y_i \leq cn,\ y_i \geq 0\ \forall i \in [n] \right\}.$$

In the above, every y_i indicates whether S_i is chosen, and x_e indicates whether e is covered. Clearly, for an optimal solution to this LP we have that $x_e \geq 0$, $\forall e \in E$, and $y_i \leq 1$, $\forall i \in [n]$. From this point onwards we assume that $\{x_e\}_{e \in E}$ and $\{y_i\}_{i \in [n]}$ are an optimal solution to the above LP.

We define a new set of variables z_i, $\forall i \in [n]$, by averaging the LP's solution with a uniform random fractional solution:

$$z_i \triangleq \alpha c + (1 - \alpha) y_i,$$

where α depends on c and will be chosen later. Our main observation is summarized in the following lemma. It states that if every set S_i is chosen independently with a probability of z_i then the probability that element $e \in E$ is covered is at least $\rho(c)x_e$. We state the lemma via the multilinear extension F of the set function $f\colon 2^{[n]} \to \mathbb{R}_+$ which is the weighted coverage function of the given instance of MC, i.e., $f(T) \triangleq \sum_{e\in E} w_e \mathbf{1}_{\{e\in\cup_{i\in T}S_i\}}$, $\forall T \subseteq [n]$.

Lemma 1. *$F(\mathbf{z}) \geq \rho(c)\sum_{e\in E} x_e$, where $\rho(c)$ is defined as in Theorem 1.*

Proof. Recalling the probabilistic interpretation of F (see Sect. 2), it suffices to fix an element $e \in E$ and prove that: $1 - \prod_{i:e\in S_i}(1 - z_i) \geq \rho(c)x_e$. For simplicity of presentation assume that e belongs to $S_1, \ldots, S_m$. Hence,

$$1 - \prod_{i=1}^{m}(1 - z_i) = 1 - \prod_{i=1}^{m}(1 - \alpha c - (1 - \alpha)y_i) \tag{1}$$
$$\geq 1 - (1 - \alpha c - (1 - \alpha)(y_1 + \cdots + y_m)/m)^m \tag{2}$$
$$\geq 1 - (1 - \alpha c - (1 - \alpha)x_e/m)^m \tag{3}$$
$$\geq x_e(1 - (1 - \alpha c - (1 - \alpha)/m)^m). \tag{4}$$

In the above, (1) is by the definition of z_i, (2) is just the AM-GM inequality, (3) follows from the constraint upper bounding x_e in the LP, and (4) follows from the concavity of $1 - (1 - \alpha c - (1 - \alpha)x_e/m)^m$ in x_e when $m \geq 1$ (using $x_e \leq 1$).

Given c, all that is left to conclude the proof is to choose α and prove that:

$$\min_{m\geq 1}\{1 - (1 - \alpha c - (1 - \alpha)/m)^m\} \geq \rho(c).$$

We denote $\sigma(\alpha, m) \triangleq (1 - \alpha c - (1 - \alpha)/m)^m = (1 - {}^1/_m - (c - {}^1/_m)\alpha)^m$ and $\sigma(\alpha) \triangleq \max_{m\geq 1}\{\sigma(\alpha, m)\}$, where the maximum is taken over integer m. Note that $\sigma(\alpha, m)$ is decreasing in α when $m > {}^1/_c$, increasing in α when $m < {}^1/_c$, and equals $(1 - {}^1/_m)^m$ when $m = {}^1/_c$. Moreover, we note that $\sigma(\alpha, m)$ is log-concave in m since $\frac{\partial^2 \log\sigma(\alpha,m)}{\partial m^2} = -\frac{(1-\alpha)^2/m}{[(1-\alpha c)m-(1-\alpha)]^2} < 0$.

Our proof considers two cases, according to whether $c = {}^1/_s$ for some integer $s \in \mathbb{N}$ or not. First, assume that $c = {}^1/_s$ for some $s \in \mathbb{N}$, where $s \geq 2$ (the case $c = 1$ is trivial). For any fixed α, denote by $m^*(\alpha)$ the (not necessarily integer) maximizer of $\sigma(\alpha, m)$, i.e., $m^*(\alpha) \triangleq \arg\max_{m\geq 1}\{\sigma(\alpha, m)\}$. We prove that in this case, choosing $\alpha^* = 1 - (s - 1)\ln({}^s/_{(s-1)})$ implies that $m^*(\alpha^*) = s$; note $\alpha^* \in (0, 1)$ since $\ln({}^s/_{(s-1)}) < {}^1/_{(s-1)}$. Thus, $\sigma(\alpha^*, m^*(\alpha)) = (1 - c)^{{}^1/_c}$, since $m^*(\alpha^*) = s$, $s = {}^1/_c$, and $\sigma(\alpha^*, {}^1/_c) = (1 - c)^{{}^1/_c}$.

Recall that $\sigma(\alpha, m)$ is log-concave in m, thus to find $m^*(\alpha^*)$ it suffices to find the (unique) m for which $\partial(\log\sigma(\alpha^*, m))/\partial m = 0$. Direct calculation of the latter shows that it is equivalent to:

$$\ln(1 - \alpha^* c - (1 - \alpha^*)/m) + \frac{(1 - \alpha^*)/m}{1 - \alpha^* c - (1 - \alpha^*)/m} = 0.$$

Plugging $c = {}^1/_s$, $m = s$, and $\alpha^* = 1 - (s - 1)\ln({}^s/_{(s-1)})$, we have $1 - \alpha^* c - (1 - \alpha^*)/m = {}^{(s-1)}/_s$, from which the equality is easy to verify.

Second, we assume that c satisfies ${}^1/_{(s+1)} < c < {}^1/_s$ for some integer $s \in \mathbb{N}$. Define the following function $\delta(\alpha) \triangleq \sigma(\alpha, s + 1) - \sigma(\alpha, s)$. Note that $\delta(\alpha)$ is a decreasing function in α since:

(1) $\sigma(\alpha, m)$ is increasing when $m < 1/c$ (and $s < 1/c$).
(2) $\sigma(\alpha, m)$ is decreasing when $m > 1/c$ (and $s+1 > 1/c$).

Moreover:

$$\delta(0) = (1 - 1/(s+1))^{s+1} - (1 - 1/s)^s > 0 \text{ and } \delta(1) = (1-c)^{s+1} - (1-c)^s < 0.$$

Therefore, there exists a (unique) α^* for which $\delta(\alpha^*) = 0$. This α^* satisfies: $\sigma(\alpha^*, s+1) = \sigma(\alpha^*, s)$. Moreover, since $\sigma(\alpha^*, m)$ is log-concave in m, it must be the case that $s < m^*(\alpha^*) < s+1$. Thus, we proved that $\sigma(\alpha^*) = \sigma(\alpha^*, s) = \sigma(\alpha^*, s+1) < \sigma(\alpha^*, m^*(\alpha^*))$. This concludes the proof of the lemma.

Note that in the second case, $\rho(c) > 1 - \sigma(\alpha^*, 1/c) = 1 - (1-c)^{1/c}$. We give a matching integrality gap for the LP in the full version of the paper.

Proof (Proof of Theorem 1*).* Our algorithm performs the following steps:

(1) Find an optimal solution $\{x_e\}_{e \in E}$ and $\{y_i\}_{i \in [n]}$ to the LP.
(2) Compute $z_i \triangleq \alpha c + (1-\alpha) y_i \ \forall i \in [n]$, where α is chosen as in the proof of Lemma 1.
(3) Apply Theorem 6 on $\mathbf{z}$ and $\mathcal{P} \triangleq \{\mathbf{x} \in [0,1]^{[n]} : \sum_{i=1}^n x_i \le cn\}$ (note that $\mathcal{P}$ is the independent set polytope of the uniform matroid defined on $[n]$ with $k = cn$).

Step (3) can be applied since $\mathbf{z} \in \mathcal{P}$. Thus, Lemma 1, alongside the observation that $\sum_{e \in E} w_e x_e$ is an upper bound on the value of an optimal solution, suffice to conclude the proof.

It is worth noting that the approximation ratio $\rho(c)$ has an elegant closed form solution when $1/2 < c < 1$:

$$\rho(c) = 1 - \frac{(1-c)\left(1 - 2\sqrt{c(1-c)}\right)}{(2c-1)^2}, \qquad \forall c \in (1/2, 1).$$

To see this, we first determine α^* by solving the equation $\sigma(\alpha^*, 1) = \sigma(\alpha^*, 2)$. This is a quadratic equation $(1-c)\alpha^* = (1/2 - (c - 1/2)\alpha^*)^2$ whose solutions are $\frac{1 \pm 2\sqrt{c(1-c)}}{(2c-1)^2}$. Since $1 > (1-2c)^2$, the only solution in $[0,1]$ is $\alpha^* = \frac{1-2\sqrt{c(1-c)}}{(2c-1)^2}$. We then compute $\rho(c) = 1 - \sigma(\alpha^*, 1) = 1 - (1-c)\alpha^*$.

3.2 Algorithm for MC in the Value Oracle Model

The algorithm underlying Sect. 3.1 seems to rely on knowing the set system explicitly. However, using simple considerations, we can make do given only value oracle access to f. The essential observation is that for a given set $I \subseteq [n]$, the total weight of elements belonging to all sets S_i for $i \in I$ and to no set S_j for $j \notin I$ is

$$w_I := \sum_{J \subseteq I} (-1)^{|I \setminus J|} f(J).$$

This formula follows easily from Möbius inversion.

In particular, for every M we can determine, using $O(n^M)$ value oracle queries, the value of w_I for all $|I| \le M$, as well as

$$w' := \sum_{|I|>M} w_I = f([n]) - \sum_{|I|\le M} w_I.$$

We now solve the following LP:

$$\max\left\{ \sum_{|I|\le M} w_I x_I + w'x' : x_I \le \min\left\{1, \sum_{i\in I} y_i\right\} \ \forall |I| \le M, \right.$$
$$\left. x' \le \min\left\{1, \sum_{i=1}^{n} y_i\right\}, \sum_{i=1}^{n} y_i \le cn, \ y_i \ge 0 \ \forall i \in [n] \right\}.$$

This LP corresponds to the instance in which all elements belonging to more than M sets are assumed to belong to all sets. In particular, if O is an optimal solution to the original instance, then it is a feasible solution to the LP, and its objective value is at least $f(O)$.

It remains to prove an analog of Lemma 1. The proof of the lemma boils down to the following two inequalities:

$$1 - \prod_{i\in I}(1 - z_i) \ge \rho(c)x_I,$$
$$\sum_{|I|>M} w_I \left(1 - \prod_{i\in I}(1 - z_i)\right) \ge \rho(c)w'x'.$$

The first inequality is proved in Lemma 1. Since $x' \le 1$, the second inequality will follow from

$$|I| > M \Longrightarrow 1 - \prod_{i\in I}(1 - z_i) \ge \rho(c).$$

To this end, we observe that $\alpha > 0$ (this is proved as part of the proof of Lemma 1), and so $1 - \prod_{i\in I}(1 - z_i) \ge 1 - (1 - \alpha c)^{M+1} \ge \rho(c)$ for some constant $M = M(c)$, completing the proof of the generalized version of Lemma 1 and so of Theorem 2.

One drawback of this algorithm is that its running time $O(n^M)$ depends on c, where M blows up as $c \to 1$. In contrast, the algorithm behind Theorem 1 has the same running time for all values of c.

3.3 Algorithm for SM

Our algorithm for SM follows the same type of analysis used by Feldman, Naor, and Schwartz [27] when applying the measured continuous greedy to a partition matroid independence constraint. The measured continuous greedy algorithm of

Feldman, Naor, and Schwartz [27] is based on the celebrated continuous greedy algorithm of Calinescu, Chekuri, Pál, and Vondrák [17].

Both algorithms (approximately) solve the following problem:

$$\max\{F(\mathbf{x}) : \mathbf{x} \in \mathcal{P}\},$$

where $\mathcal{P} \subseteq [0,1]^U$ is a down-monotone polytope.[3] The measured continuous greedy algorithm can be viewed as a continuous time process whose location at time t is $\mathbf{y}(t)$.[4] It is defined as follows: $\mathbf{y}(0) \leftarrow 0$ and

$$\frac{\partial y_i(t)}{\partial t} \leftarrow (1 - y_i(t))x_i(t) \quad \forall i \in U, \forall t \geq 0, \tag{5}$$

where $\mathbf{x}(t) \leftarrow \arg\max\{\mathbf{x} \cdot \nabla F(\mathbf{y}(t)) : \mathbf{x} \in \mathcal{P}\}$. It is proved in [27] that:

$$F(\mathbf{y}(t)) \geq (1 - e^{-t})f(S^*) \quad \forall t \geq 0, \tag{6}$$

where $S^* \subseteq U$ is an optimal solution to the problem.

In our case $\mathcal{P}$ is just the uniform matroid polytope, i.e., $\mathcal{P} = \{\mathbf{x} \in [0,1]^U : \sum_{i \in U} x_i \leq k\}$ (recall that $k = cn$), and S^* is an integral optimal solution, i.e., $S^* = \arg\max\{f(S) : S \subseteq U, |S| \leq k\}$. Clearly, (6) implies that one would like to terminate the process as late as possible since this would result in a solution with higher value. However, (5) implies that the longer the process runs the larger $\mathbf{y}(t)$ becomes, and at some point it might move outside of $\mathcal{P}$. Thus, our goal is to find the largest stopping time T_c for which we are guaranteed that $\mathbf{y}(T_c) \in \mathcal{P}$.

We are now ready to prove Theorem 3.

Proof (Proof of Theorem 3). We note that at time $t = 0$: $\sum_{i \in U} y_i(0) = 0$. Moreover, one can verify that for every $t \geq 0$ and $i \in U$:

$$y_i(t) \leq 1 - e^{-\int_{s=0}^{t} x_i(s)ds}.$$

Since the function $1 - e^{-t}$ is concave in t, $\sum_{i \in U} y_i(t)$ is maximized when every $y_i(s)$, for every $0 \leq s \leq t$, is increased at the same rate of $k/n = c$ (recall that $\mathbf{x}(s) \in \mathcal{P}$, i.e., $\sum_{i \in U} x_i(s) \leq k$). Thus, we can conclude that $\sum_{i \in U} y_i(t) \leq n(1 - e^{-ct})$. Choose T_c to be the time for which this upper bound is tight, i.e., $1 - e^{-cT_c} = k/n = c$, implying that $T_c = \ln(1/(1-c))/c$. Plugging this T_c into (6) yeilds that $F(\mathbf{y}(T_c)) \geq (1 - (1-c)^{1/c})f(S^*)$. Applying pipage rounding (Theorem 6) concludes the proof.

4 Open Questions

What is the optimal approximation ratio for SM? Theorem 3 gives an approximation ratio of $1 - (1-c)^{1/c}$, which is matched by Theorem 4 when $1/c \in \mathbb{N}$. We conjecture that $1 - (1-c)^{1/c}$ is the optimal approximation ratio for all c.

[3] $\mathcal{P}$ is down-monotone if $\mathbf{x} \in \mathcal{P}$ and $0 \leq \mathbf{y} \leq \mathbf{x}$ imply $\mathbf{y} \in \mathcal{P}$.

[4] This continuous time process can be implemented as an algorithm by choosing a small step size and discretizing the movement of the process. This incurs a small additive $o(1)$ loss in the approximation (we refer the reader to [17,27] for the details).

What is the optimal approximation ratio for MC? We conjecture that the SDP-based algorithm for Max k-Vertex-Cover, used to prove Theorem 5, can be extended to give a $1 - 1/e$ approximation for MC for arbitrary k, as well as to yield the optimal approximation ratio for $k = cn$ for all c.

We thank StackExchange user John for proposing the problem [28], and Zach Chase for useful discussions. We thank the anonymous IPCO reviewers for many helpful suggestions, including one that led to Theorem 2. Yuval Filmus has received funding from the European Union's Horizon 2020 research and innovation program under grant agreement no. 802020-ERC-HARMONIC, and by ISF grant 507/24. Roy Schwartz and Alexander V. Smal have received funding from the European Union's Horizon 2020 research and innovation program under grant agreement no. 852870-ERC-SUBMODULAR.

References

1. Ageev, A.A., Sviridenko, M.I.: Pipage rounding: a new method of constructing algorithms with proven performance guarantee. J. Comb. Optim. **8**(3), 307–328 (2004)
2. Ageev, A.A., Sviridenko, M.I.: Approximation algorithms for maximum coverage and max cut with given sizes of parts. In: Integer programming and combinatorial optimization (Graz, 1999), LNCS, vol. 1610, pp. 17–30. Springer, Berlin (1999)
3. Austrin, P.: Balanced Max 2-Sat might not be the hardest. In: STOC'07—Proceedings of the 39th Annual ACM Symposium on Theory of Computing, pp. 189–197. ACM, New York (2007)
4. Austrin, P., Benabbas, S., Georgiou, K.: Better balance by being biased: a 0.8776-approximation for max bisection. ACM Trans. Algorithms **13**(1) (2016)
5. Austrin, P., Stanković, A.: Global cardinality constraints make approximating some Max-2-CSPs harder. In: Approximation, randomization, and combinatorial optimization. Algorithms and techniques, LIPIcs. Leibniz Int. Proc. Inform., vol. 145, pp. Art. No. 24, 17. Schloss Dagstuhl. Leibniz-Zent. Inform., Wadern (2019)
6. Badanidiyuru, A., Vondrák, J.: Fast algorithms for maximizing submodular functions. In: Proceedings of the Twenty-Fifth Annual ACM-SIAM Symposium on Discrete Algorithms, pp. 1497–1514. SODA '14, Society for Industrial and Applied Mathematics, USA (2014)
7. Brakensiek, J., Huang, N., Zwick, U.: Tight approximability of MAX 2-SAT and relatives, under UGC. In: Proceedings of the 2024 Annual ACM-SIAM Symposium on Discrete Algorithms (SODA), pp. 1328–1344. SIAM, Philadelphia, PA (2024)
8. Buchbinder, N., Feldman, M.: Deterministic algorithms for submodular maximization problems. ACM Trans. Algorithms **14**(3) (2018)
9. Buchbinder, N., Feldman, M.: Constrained submodular maximization via a non-symmetric technique. Math. Oper. Res. **44**(3), 988–1005 (2019)
10. Buchbinder, N., Feldman, M.: Constrained submodular maximization via new bounds for dr-submodular functions. In: Proceedings of the 56th Annual ACM Symposium on Theory of Computing, STOC 2024, Vancouver, BC, Canada, 24–28 June 2024, pp. 1820–1831. ACM (2024)
11. Buchbinder, N., Feldman, M.: Deterministic algorithm and faster algorithm for submodular maximization subject to a matroid constraint. In: 2024 IEEE 65th Annual Symposium on Foundations of Computer Science—FOCS 2024, pp. 700–712. IEEE Computer Soc., Los Alamitos, CA ([2024] 2024)

12. Buchbinder, N., Feldman, M.: Extending the extension: Deterministic algorithm for non-monotone submodular maximization. In: 2025 IEEE 66th Annual Symposium on Foundations of Computer Science—FOCS 2025. IEEE Computer Soc., Los Alamitos, CA ([2025] 2025)
13. Buchbinder, N., Feldman, M., Naor, J., Schwartz, R.: Submodular maximization with cardinality constraints. In: Proceedings of the Twenty-Fifth Annual ACM-SIAM Symposium on Discrete Algorithms, pp. 1433–1452. ACM, New York (2014)
14. Buchbinder, N., Feldman, M., Schwartz, R.: Comparing apples and oranges: query trade-off in submodular maximization. Math. Oper. Res. **42**(2), 308–329 (2017)
15. Buchbinder, N., Feldman, M., Schwartz, R.: Online submodular maximization with preemption. ACM Trans. Algorithms **15**(3) (2019)
16. Calinescu, G., Chekuri, C., Pál, M., Vondrák, J.: Maximizing a submodular set function subject to a matroid constraint (extended abstract). In: Fischetti, M., Williamson, D.P. (eds.) IPCO 2007. LNCS, vol. 4513, pp. 182–196. Springer, Heidelberg (2007). https://doi.org/10.1007/978-3-540-72792-7_15
17. Calinescu, G., Chekuri, C., Pál, M., Vondrák, J.: Maximizing a monotone submodular function subject to a matroid constraint. SIAM J. Comput. **40**(6), 1740–1766 (2011)
18. Chan, T.H.H., Huang, Z., Jiang, S.H.C., Kang, N., Tang, Z.G.: Online submodular maximization with free disposal. ACM Trans. Algorithms **14**(4), Art. 56, 29 (2018). https://doi.org/10.1145/3242770
19. Chekuri, C., Vondrák, J., Zenklusen, R.: Dependent randomized rounding via exchange properties of combinatorial structures. In: 2010 IEEE 51st Annual Symposium on Foundations of Computer Science—FOCS 2010, pp. 575–584. IEEE Computer Soc., Los Alamitos, CA (2010)
20. Das, A., Kempe, D.: Approximate submodularity and its applications: subset selection, sparse approximation and dictionary selection. J. Mach. Learn. Res. **19**(3), 1–34 (2018)
21. Dobzinski, S., Vondrák, J.: The computational complexity of truthfulness in combinatorial auctions. ArXiv **abs/1202.2789** (2012). https://api.semanticscholar.org/CorpusID:149760
22. Dughmi, S., Roughgarden, T., Yan, Q.: From convex optimization to randomized mechanisms: toward optimal combinatorial auctions. In: Proceedings of the forty-third Annual ACM Symposium on Theory of Computing (2011). https://api.semanticscholar.org/CorpusID:9192953
23. Dughmi, S., Vondrák, J.: Limitations of randomized mechanisms for combinatorial auctions. In: 2011 IEEE 52nd Annual Symposium on Foundations of Computer Science, pp. 502–511 (2011). https://api.semanticscholar.org/CorpusID:7549214
24. Ene, A., Nguyen, H.L.: Constrained submodular maximization: Beyond $1/e$. In: IEEE 57th Annual Symposium on Foundations of Computer Science, FOCS 2016, 9–11 October 2016, Hyatt Regency, New Brunswick, New Jersey, USA, pp. 248–257. IEEE Computer Society (2016)
25. Feige, U.: A threshold of $\ln n$ for approximating set cover. J. ACM **45**(4), 634–652 (1998)
26. Feige, U., Mirrokni, V.S., Vondrák, J.: Maximizing non-monotone submodular functions. SIAM J. Comput. **40**(4), 1133–1153 (2011)
27. Feldman, M., Naor, J., Schwartz, R.: A unified continuous greedy algorithm for submodular maximization. In: 2011 IEEE 52nd Annual Symposium on Foundations of Computer Science, pp. 570–579 (2011)

28. John: Is the maximum coverage problem remains as hard when taking most sets? Theoretical Computer Science Stack Exchange. https://cstheory.stackexchange.com/q/54170. Accessed 11 Apr 2024
29. Kempe, D., Kleinberg, J., Tardos, E.: Influential nodes in a diffusion model for social networks. In: Proceedings of the 32nd International Conference on Automata, Languages and Programming, pp. 1127–1138. ICALP'05, Springer-Verlag, Berlin, Heidelberg (2005)
30. Kempe, D., Kleinberg, J., Tardos, E.: Maximizing the spread of influence through a social network. Theory Comput. **11**(4), 105–147 (2015)
31. Khuller, S., Moss, A., Naor, J.S.: The budgeted maximum coverage problem. Inf. Process. Lett. **70**(1), 39–45 (1999)
32. Lewin, M., Livnat, D., Zwick, U.: Improved rounding techniques for the max 2-sat and max di-cut problems. In: Proceedings of the 9th International IPCO Conference on Integer Programming and Combinatorial Optimization, pp. 67–82. Springer-Verlag (2002)
33. Li, W., Feldman, M., Kazemi, E., Karbasi, A.: Submodular maximization in clean linear time. In: Koyejo, S., Mohamed, S., Agarwal, A., Belgrave, D., Cho, K., Oh, A. (eds.) Advances in Neural Information Processing Systems, vol. 35, pp. 17473–17487. Curran Associates, Inc. (2022). https://proceedings.neurips.cc/paper_files/paper/2022/file/6faf3b8ed0df532c14d0fc009e451b6d-Paper-Conference.pdf
34. Mirzasoleiman, B., Badanidiyuru, A., Karbasi, A., Vondrák, J., Krause, A.: Lazier than lazy greedy. In: Proceedings of the Twenty-Ninth AAAI Conference on Artificial Intelligence, pp. 1812–1818. AAAI'15, AAAI Press (2015)
35. Mitrovic, M., Kazemi, E., Zadimoghaddam, M., Karbasi, A.: Data summarization at scale: a two-stage submodular approach. In: International Conference on Machine Learning (2018)
36. Nemhauser, G.L., Wolsey, L.A.: Best algorithms for approximating the maximum of a submodular set function. Math. Oper. Res. **3**(3), 177–188 (1978)
37. Nemhauser, G.L., Wolsey, L.A., Fisher, M.L.: An analysis of approximations for maximizing submodular set functions-I. Math. Program. **14**(1), 265–294 (1978)
38. Norouzi-Fard, A., Tarnawski, J., Mitrovic, S., Zandieh, A., Mousavifar, A., Svensson, O.: Beyond 1/2-approximation for submodular maximization on massive data streams. In: Proceedings of the 35th International Conference on Machine Learning. Proceedings of Machine Learning Research, vol. 80, pp. 3829–3838. PMLR, 10–15 July 2018
39. Raghavendra, P., Tan, N.: Approximating csps with global cardinality constraints using sdp hierarchies. In: Proceedings of the Twenty-Third Annual ACM-SIAM Symposium on Discrete Algorithms, pp. 373–387. SODA '12 (2012)
40. Sviridenko, M.: A note on maximizing a submodular set function subject to a knapsack constraint. Oper. Res. Lett. **32**(1), 41–43 (2004)
41. Vondrák, J.: Symmetry and approximability of submodular maximization problems. SIAM J. Comput. **42**(1), 265–304 (2013)

Completeness in the Polynomial Hierarchy for Many Natural Problems in Bilevel and Robust Optimization

Christoph Grüne[1(✉)] and Lasse Wulf[2]

[1] Department of Computer Science, RWTH Aachen University, Aachen, Germany
gruene@algo.rwth-aachen.de
[2] Section of Algorithms, Logic, and Graphs, Technical University of Denmark, Kongens Lyngby, Denmark
lawu@dtu.dk

Abstract. In bilevel and robust optimization we are concerned with combinatorial min-max problems, for example from the areas of min-max regret robust optimization, network interdiction, and two-stage adjustable robust optimization. Even though these areas are well-researched for over two decades and one would naturally expect many (if not most) of the problems occurring in these areas to be complete for the classes Σ_2^p or Σ_3^p from the polynomial hierarchy, almost no hardness results in this regime are currently known. However, such complexity insights are important, since they imply that no polynomial-sized integer program for these min-max problems exist, and hence conventional IP-based approaches fail. We address this lack of knowledge by introducing at least 72 new Σ_2^p-complete and Σ_3^p-complete problems. The majority of all earlier publications on Σ_2^p- and Σ_3^p-completeness in said areas are special cases of our meta-theorem. Precisely, we introduce a large list of problems for which the meta-theorem is applicable (including clique, vertex cover, knapsack, TSP, facility location, and many more). We show that for each of these problems, the corresponding min-max (i.e. interdiction/regret) variant is Σ_2^p- and the min-max-min (i.e. two-stage) variant is Σ_3^p-complete.

Keywords: Computational Complexity · Bilevel Optimization · Robust Optimization · Stackelberg Games · Min-Max Regret · Most Vital Nodes · Most Vital Vertex · Most Vital Edges · Blocker Problems · Interdiction Problems · Two-Stage Problems · Polynomial Hierarchy

1 Introduction

In recent years, there has been enormous interest in the areas of Robust Optimization [7], Network Interdiction [34], Stackelberg Games [27], Attacker-

C. Grüne—Funded by the German Research Foundation (DFG) – GRK 2236/2.
L. Wulf—Funded by the Austrian Science Fund (FWF): W1230 and Carlsberg Foundation CF21-0302 "Graph Algorithms with Geometric Applications".

N. Megow and A. Basu (Eds.): IPCO 2025, LNCS 15620, pp. 256–269, 2025.
https://doi.org/10.1007/978-3-031-93112-3_19

Defender games [21], and many other bilevel problems [12]. These research areas are vital in helping us to understand, which parts of a network are most vulnerable, prevent terrorist attacks, understand economic processes, and to understand and improve the robustness properties of many systems.

The common property of all these areas of research is that they study min-max optimization problems. Equivalently, these problems can be characterized as an abstract two-turn game between two players. The first player (from now on called Alice) starts the game and takes an action with the goal of minimizing some objective. Afterwards, the second player (from now on called Bob) responds to Alice. Typically (but not always), Alice's and Bob's goals are opposite of each other.

Let us take a look at the maximum clique interdiction problem [31] as a prototypical example. In this problem, we are given a graph $G = (V, E)$ and some budget $k \in \mathbb{N}$. The goal of Bob is to find a clique of largest possible size in the graph. However, before Bob's turn, Alice can delete up to k vertices from the graph in order to impair Bob's objective. Hence the game is a min-max optimization problem described by

$$\min_{W \subseteq V, |W| \leq k} \max\{|C| : C \text{ is a clique in the graph } G \text{ with } W \cap C = \emptyset\}.$$

We remark that this problem follows a natural pattern, in which researchers often come up with new problems: First, some *nominal problem* is taken (in this case, the maximum clique problem), and afterwards it is *modified* into a more complicated min-max problem by adding an additional component (in this case, the possibility of Alice to interdict). The goal of this paper is to consider this very general pattern and shed some light on the understanding of its computational complexity. Roughly speaking, we show that whenever the nominal problem is already an NP-complete problem, then under some mild assumptions the complexity of the min-max variant will be significantly larger than the complexity of the nominal variant. As a first example of this behavior, consider the following: It was proven by Rutenburg [32] that the maximum clique interdiction problem is complete for the complexity class Σ_2^p.

The Natural Complexity Class for Bilevel Problems. In the year 1976, Stockmeyer [35] introduced the *polynomial hierarchy*, containing the complexity classes Σ_k^p for all $k \in \mathbb{N}$. While the complexity class $\Sigma_1^p = NP$ is very well known even to outsiders of theoretical computer science, the complexity class Σ_2^p seems to be less known to non-specialists. However, for min-max optimization problems, it turns out that Σ_2^p is the natural class to describe them. A formal definition of the classes Σ_k^p for all $k \in \mathbb{N}$ is given in Sect. 2. Roughly speaking, the class Σ_2^p contains all the problems of the form: Does there EXIST some object x, such that FOR ALL objects y some easy-to-check property $P(x, y)$ holds?

Researchers are interested in the question which problems are complete for the class Σ_2^p, since this has important consequences. Similarly to the widely believed conjecture $P \neq NP$, it is also widely believed that $NP \neq \Sigma_2^p$. If this conjecture is true, it means that a Σ_2^p-complete problem can not be described by a mixed integer program of polynomial size (because this would imply a polynomial-time reduction to MIP, which is NP-complete). This means that current existing integer program solvers, which have in some cases been very successful in tackling NP-complete problems, are not very successful at solving the harder case of Σ_2^p-problems. As for example observed in [36], today the community is at a threshold, where new and more powerful tools and techniques need to be developed in order to tackle Σ_2^p-hard problems.

1.1 Literature Overview

In this paper, we restrict our attention to the three areas of *network interdiction*, *min-max regret robust optimization* (with interval uncertainty), and *two-stage adjustable robust optimization* (with budgeted uncertainty). We choose these areas, since they represent three highly popular sub-areas of bilevel optimization, and since we can showcase a diverse set of techniques in the three different cases, while staying concise at the same time. In the area of *network interdiction*, one is concerned with the question, which parts of a network are most vulnerable to an attack or failure. Other names for slight variants of the network interdiction problem are the most vital nodes/most vital edges problem and the vertex/edge blocker problem. Network interdiction has been considered for a vast amount of standard optimization problems. Among these are problems in P such as shortest path [29], matching [38], minimum spanning tree [28], or maximum flow [37]. Furthermore, algorithms for interdiction problems that are NP-complete were developed, for example for vertex cover [5], independent set [5], coloring [4], clique [31], 1- and p-center [6], or 1- and p-median [6]. Typically, the complexity of these problems is not analyzed beyond NP-hardness.

In the area of *min-max regret robust optimization*, one is faced with an uncertain cost scenario. The decision-maker seeks to minimize the maximum deviation between incurred cost and optimal cost. The min-max regret criterion is popular in robust optimization and has been considered for many standard optimization problems. Among the problems, where the nominal problem is contained in P and are analyzed from a complexity viewpoint, are for example shortest path [3], minimum spanning tree [3], min cut [1], min s-t cut [1], 1-center and 1-median with uncertainty in the node or edge weights [2]. In the realm of interval min-max regret problems, for which the nominal problem is NP-complete, there are publications considering the complexity of the knapsack problem [11], the set cover problem [10], and the TSP and Steiner tree problem [15].

In the area of *two-stage adjustable robust optimization*, one is faced with an uncertain cost scenario. The decision-maker has to make a two-step decision, where in the first step a partial decision has to be made without full knowledge of the scenario (here-and-now), and in a second step a partial corrective decision can be made with full knowledge of the scenario (wait-and-see). Two-stage adjustable optimization has been considered for a large amount of standard optimization problems, for example in [9,17,24–26].

Closely Related Work. Despite a tremendous interest in bilevel optimization, and despite the complexity classes Σ_k^p being the natural complexity classes for this type of decision problems, we are aware of only a handful of publications on the matter of Σ_k^p-completeness with respect to these areas. Rutenburg [32] shows Σ_2^p-completeness of the maximum clique interdiction problem. Deineko and Woeginger [11] show Σ_2^p-completeness of the min-max regret interval knapsack problem. Caprara, Carvalho, Lodi, and Woeginger [8] show Σ_2^p-completeness of an interdiction variant of knapsack that was originally introduced by DeNegre [13]. Fröhlich and Ruzika [14] show the Σ_2^p-completeness of two versions of a location-interdiction problem. Nabli, Carvalho, and Hosteins [30] study the multilevel critical node problem and prove its Σ_3^p-completeness. Coco, Santos, and Noronha [10] show Σ_2^p-completeness of the min-max regret maximum benefit set covering problem. Goerigk, Lendl, and Wulf [18] show Σ_3^p-completeness of two-stage adjustable variants of TSP, independent set, and vertex cover. Grüne [19] introduces a reduction framework for so-called universe gadget reductions to show Σ_3^p-completeness of recoverable robust versions of typical optimization problems. The compendium by Umans and Schaefer [33] contains many Σ_2^p-complete problems, but few of them are related to bilevel optimization. A seminal paper by Jeroslow [22] shows Σ_k^p-completeness for various classes of general multi-level integer programs.

The research that is most closely connected to this paper is the Ph.D. thesis of Johannes [23]. She proves Σ_2^p-hardness for a variety of problems including "adversarial problems", "partial inverse optimization problems" and several other types of classes. She also discusses a transitive transformation for NP-completeness results that she calls "value preserving" [23, Theorem 2.1.1]. She shows that a wide range of adversarial problems and partial preprocessing problems can be proven to be Σ_2^p-hard by relying on their corresponding NP-completeness proofs, providing the NP-completeness transformation was "value preserving." In our paper, we establish a similar meta-theorem for other classes of problems, including interdiction, min-max regret, and two-stage adjustable optimization.

Recently, we have learned that Johannes and Orlin [2020] have written an unpublished manuscript that extended the research in Johannes's thesis. They refer to certain NP-completeness transformations as "ratcheting transformations" because they can be used to prove hardness for extensions of these problems to problems higher in the polynomial time hierarchy. There are additional overlaps of their work with our meta-theorem.

1.2 Our Contribution

We make a large step towards understanding the complexity of bilevel combinatorial optimization, by introducing a general and powerful meta-theorem to prove Σ_k^p-completeness. Our meta-theorem can be applied to such problems which are already NP-complete and have an additional property (explained below). Using the meta-theorem, we can "upgrade" an existing NP-completeness proof to a Σ_k^p-completeness proof with very little work. In other words, this means that for a lot of NP-complete problems, the Σ_2^p-completeness of its min-max variant follows essentially "for free" from its NP-completeness. An analogous statement holds for min-max-min problems and Σ_3^p-completeness. We apply our meta-theorem to the areas of bilevel optimization, in particular network interdiction, min-max regret robust optimization, and two-stage adjustable optimization. This way, we obtain at least 72 natural Σ_2^p- or Σ_3^p-complete problems relevant to these areas. We remark that earlier papers showed Σ_2^p- or Σ_3^p-completeness only for one problem at a time, usually in a tedious fashion. In contrast, our meta-theorem contains essentially all known Σ_2^p- or Σ_3^p-completeness results in the areas of network interdiction, min-max regret, and two-stage adjustable robust optimization as a special case (namely, min-max-regret knapsack [11], min-max regret maximum benefit set cover [10], interdiction maximum clique [32], and interdiction knapsack [8], as well as two-stage adjustable TSP, vertex cover, independent set [18]). While each of these earlier works proves Σ_2^p-hardness of only one specific problem, our meta-theorem proves Σ_2^p-completeness simultaneously for a large class of problems, including all problems mentioned in Sect. 7, as well as possible further problems which are added in the future.

We call this class of problems for which the meta-theorem is applicable the class SSP-NP-complete (SSP-NPc), for reasons which are explained in Sect. 3. We show that over 20 classical problems are contained in the class SSP-NPc. A complete list is provided in Sect. 7. Furthermore, one can add new problems to the class SSP-NPc with very little work by finding a so-called *SSP reduction* starting from any problem which is already contained in the class. This offers future researchers a convenient way to prove Σ_k^p-completeness of relevant problems.

Since the class SSP-NPc contains so many well-known problems, our work shows that, in a certain sense, the "standard" or "normal" behavior of a NP-complete problem is to become Σ_2^p-complete when modified to be a min-max problem. While this behavior is very intuitive, we are the first to be able to prove this intuition to be true for a very broad range of problems in robust and bilevel optimization. Now we are ready to describe our concrete main results.

Network Interdiction/Blocker Problems/Most Vital Nodes or Edges. In Sect. 4, we are concerned with the *minimum cost blocker problem.* In the minimum cost blocker problem one is given a nominal problem Π and additionally a cost function. The question is for Alice to find a *blocker* of minimum cost. A blocker is a set which intersects every solution of the nominal problem. (For example, every Hamiltonian cycle, every minimum vertex cover, every maximum clique, etc.) Our main result is that for every problem $\Pi \in$ SSP-NPc, the corresponding minimum cost blocker problem is Σ_2^p-complete.

Min-max Regret Robust Optimization with Interval Uncertainty. In Sect. 5, we are concerned with the *min-max regret* robust optimization problem with interval uncertainty. Faced with an uncertain cost function, the goal is to minimize the maximum deviation between incurred cost and optimal cost. The uncertain cost is modelled by assuming that each cost coefficient stems from a pre-specified interval. Our main result in this section is that for every problem $\Pi \in$ SSP-NPc, the corresponding min-max regret robust optimization problem with interval uncertainty is Σ_2^p-complete.

Two-stage Adjustable Robust Optimization. In Sect. 6, we are concerned with *two-stage adjustable robust optimization* with discrete budgeted uncertainty (also called discrete Γ-uncertainty). Roughly speaking, the class of two-stage problems models problems that are divisible into two stages: The decision on the first stage has to be made without full information on the real instance (here-and-now), and the decision in the second stage is made after the uncertainty is revealed (wait-and-see). Two-stage adjustable problems can be formulated via min-max-min expressions. Therefore, the natural class for them is the class Σ_3^p. Consequently, in Sect. 6, we demonstrate how our main idea can also be used to show completeness for the third stage (and potentially higher stages) of the polynomial hierarchy.

1.3 Technical Overview

Our main idea is to show that many well-known NP-completeness proofs from some problem Π_1 to some other problem Π_2 have a special property, which we call the *SSP property*. On an intuitive level, this property states that the universe of Π_1 can be injectively embedded into the universe of Π_2 in such a way that the following two properties hold: (P1) every solution of Π_1 corresponds to a partial solution of Π_2, and (P2) every solution of Π_2 when restricted to the image of the embedding corresponds to a solution of Π_1.

We show that a surprisingly large amount of NP-completeness reductions which are known from the literature actually have the SSP property (Sect. 7). Even the historically first NP-completeness reduction, i.e. the reduction used by Cook and Levin to show that SATISFIABILITY (SAT) is NP-complete, has the SSP property (Theorem 1).

Our ideas are best explained with an example. Consider the classic 3SAT problem. Given an instance of 3SAT, let $L := \{\ell_1, \dots, \ell_n\} \cup \{\overline{\ell}_1, \dots, \overline{\ell}_n\}$ be the set of literals used by this formula. In this paper, we use the following perspective: We imagine that the literal set is a *universe* $\mathcal{U} = L$ associated to the instance. Similarly given an instance of VERTEX COVER, we associate the universe $\mathcal{U}' = V$ (the vertices) to it. We recall the classical NP-hardness reduction from 3SAT to VERTEX COVER, from the book of Garey and Johnson [16]. Given a 3SAT instance consisting out of literals $\{\ell_1, \dots, \ell_n\} \cup \{\overline{\ell}_1, \dots, \overline{\ell}_n\}$ and clauses $C \subseteq L^3$, the reduction constructs a graph $G = (V, E)$ the following way: The graph contains vertices $W := \{v_{\ell_1}, \dots, v_{\ell_n}\} \cup \{v_{\overline{\ell}_1}, \dots, v_{\overline{\ell}_n}\}$ such that each vertex v_{ℓ_i} is connected to vertex $v_{\overline{\ell}_i}$ with an edge. Furthermore, for each clause, we add a new triangle to the graph, such that the three vertices of the triangle are connected to the corresponding vertices of the literals appearing in the clause. The following is easily verified: Every vertex cover of G has size at least $|L|/2 + 2|C|$ and G has a vertex cover of size $|L|/2 + 2|C|$ if and only if the 3SAT instance is a Yes-instance.

Consider the set W. For every single solution α of the 3SAT instance, we can find a small vertex cover S' (of size $|L|/2 + 2|C|$) such that $S' \cap W$ encodes α.

This is property (P1). Conversely, every small vertex cover (of size $|L|/2+2|C|$) restricted to the set W directly encodes a possible solution of the 3SAT instance. This is property (P2).

Let $\mathcal{S}(\varphi) \subseteq 2^{\mathcal{U}}$ denote the solutions of 3SAT. Let $k := |L|/2 + 2|C|$ and $\mathcal{S}'(G,k) \subseteq 2^{\mathcal{U}'}$ denote the solutions of VERTEX COVER. We consider the injective function $f : \mathcal{U} \to \mathcal{U}'$ which maps the literal ℓ to its representative vertex v_ℓ. This function f can be interpreted as a function which embeds the universe $\mathcal{U}$ into the universe $\mathcal{U}'$. It describes which literals in $\mathcal{U}$ correspond to which vertices in $\mathcal{U}'$. We have $W = f(\mathcal{U})$. With this definition, (P1) $\wedge$ (P2) is equivalent to

$$\{f(S) : S \in \mathcal{S}(\varphi)\} = \{S' \cap f(\mathcal{U}) : S' \in \mathcal{S}'(G,k)\}. \tag{1}$$

The above equation defines the SSP property. We call reductions fulfilling this property SSP reductions. We note that another well-known variant of reductions exists in the literature, so-called *parsimonious* reductions. On the first glance, these seem similar to our reductions. However, these two concepts are not the same, because parsimonious reductions map solutions to solutions bijectively, while our reductions map elements to elements injectively.

Finally, we explain how the idea of SSP reductions is used to obtain a meta-theorem. The main idea is that the existence of an SSP reduction $\Pi_1 \leq_{\text{SSP}} \Pi_2$ tells us that one can imagine Π_1 as a "sub-instance" of Π_2. Therefore, it suffices to prove Σ_2^p-completeness for the min-max variant of only one single problem in SSP-NPc (say SAT, for example), and then invest a little extra work to show that this Σ_2^p-completeness actually carries over to the min-max variant of all other problems in SSP-NPc.

2 Preliminaries

A *language* is a set $L \subseteq \{0,1\}^*$. A language L is contained in Σ_k^p iff there exists some polynomial-time computable function V (verifier), and $m_1, m_2, \ldots, m_k \leq \text{poly}(|w|)$ such that for all $w \in \{0,1\}^*$

$$w \in L \Leftrightarrow \exists y_1 \in \{0,1\}^{m_1} \forall y_2 \in \{0,1\}^{m_2} \ldots Q y_k \in \{0,1\}^{m_k} : V(w, y_1, y_2, \ldots, y_k),$$

where $Q = \exists$, if k is odd, and $Q = \forall$, if k is even.

A *many-one-reduction* from a language L to a language L' is a map $f : \{0,1\}^* \to \{0,1\}^*$ such that $w \in L$ iff $f(w) \in L'$ for all $w \in \{0,1\}^*$. A language L is Σ_k^p-hard, if every $L' \in \Sigma_k^p$ can be reduced to L with a polynomial-time many-one reduction. If L is both Σ_k^p-hard and contained in Σ_k^p, it is Σ_k^p-complete. For a map $f : A \to B$ and some subset $A' \subseteq A$, we define the image of the subset A' as $f(A') = \{f(a) : a \in A'\}$. For space reasons, we defer the proofs to the full version of the paper [20].

3 Framework

The goal of this section is to introduce the class of *SSP-NP-complete* problems, i.e. the class of all problems for which our meta-theorem is applicable. We first consider linear optimization problems (LOP problems). An LOP problem for us is a tuple $(\mathcal{I}, \mathcal{U}, \mathcal{F}, d, t)$, where $\mathcal{I}$ is the set of instances. Associated to each instance I, there is a universe $\mathcal{U}(I)$, a set of feasible solutions $\mathcal{F}(I) \subseteq 2^{\mathcal{U}(I)}$, some linear cost function $d^{(I)} : \mathcal{U}(I) \to \mathbb{Z}$, and a threshold $t^{(I)}$. The goal is to find a feasible solution $F \in \mathcal{F}(I)$ of small cost $d^{(I)}(F) \leq t^{(I)}$. With Π = VERTEX COVER, we obtain:

VERTEX COVER **Instances:** Graph $G = (V, E)$, number $k \in \mathbb{N}$. **Universe:** Vertex set $V =: \mathcal{U}$. **Feasible solution set:** The set of all vertex covers. **Solution set:** The set of all vertex covers of size at most k.

Next, we consider a simplifying abstraction of LOP problems, which we call SSP problems. An SSP problem (short for "subset search problem") is a tuple $(\mathcal{I}, \mathcal{U}, \mathcal{S})$ of input instances $\mathcal{I}$, as well as a universe $\mathcal{U}(I)$, and a solution set $\mathcal{S}(I) \subseteq 2^{\mathcal{U}(I)}$ associated to each input instance $I \in \mathcal{I}$. Every LOP problem can be viewed as SSP problem with the definition $\mathcal{S}(I) = \{F \in \mathcal{F}(I) : d^{(I)}(F) \leq t^{(I)}\}$. The formal definition of these concepts is given in the full version of this paper.

Given an SSP problem, we say that an instance $I \in \mathcal{I}$ is called Yes-instance, if $\mathcal{S}(I) \neq \emptyset$. The decision problem associated to the SSP problem Π is the language $\{I \in \mathcal{I} : \mathcal{S}(I) \neq \emptyset\}$ of all Yes-instances.

An *SSP reduction* is a usual many-one reduction which additionally has the SSP property. By formalizing the intuition gained in Sect. 1.3, we obtain the following definition. Note that in this definition the function g corresponds to the standard many-one reduction, while the functions $(f_I)_{I \in \mathcal{I}}$ are the injective embedding functions corresponding to the SSP property (analogous to Eq. (1)). Note that since $\mathcal{U}(I)$ can be different for every instance $I \in \mathcal{I}$, we have that $(f_I)_{I \in \mathcal{I}}$ is a family of functions, and not a single function.

Definition 1 (SSP Reduction). *Let $\Pi = (\mathcal{I}, \mathcal{U}, \mathcal{S})$ and $\Pi' = (\mathcal{I}', \mathcal{U}', \mathcal{S}')$ be two SSP problems. We say that there is an SSP reduction from Π to Π', and write $\Pi \leq_{SSP} \Pi'$, if*

1. *There exists a function $g : \{0,1\}^* \to \{0,1\}^*$ computable in polynomial time in the input size $|I|$, such that I is a Yes-instance iff $g(I)$ is a Yes-instance (i.e. $\mathcal{S}(I) \neq \emptyset$ iff $\mathcal{S}'(g(I)) \neq \emptyset$).*
2. *There exist functions $(f_I)_{I \in \mathcal{I}}$ computable in polynomial time in $|I|$ such that for all instances $I \in \mathcal{I}$, we have that $f_I : \mathcal{U}(I) \to \mathcal{U}'(g(I))$ is an injective function mapping from the universe of the instance I to the universe of the instance $g(I)$ such that*

$$\{f_I(S) : S \in \mathcal{S}(I)\} = \{S' \cap f_I(\mathcal{U}(I)) : S' \in \mathcal{S}'(g(I))\}.$$

An example of an SSP reduction from 3SAT to VERTEX COVER was shown in Sect. 1.3. Next, we show that SSP reductions are transitive, which enables us to easily show reductions between a multitude of problems.

Lemma 1. *SSP reductions are transitive, i.e. for SSP problems Π_1, Π_2, Π_3 with $\Pi_1 \leq_{SSP} \Pi_2$ and $\Pi_2 \leq_{SSP} \Pi_3$, it holds that $\Pi_1 \leq_{SSP} \Pi_3$.*

Further, we define the class SSP-NP, which is the analogue of NP restricted to SSP problems.

Definition 2 (SSP-NP). *The class SSP-NP consists of all the SSP problems which are polynomial-time verifiable. Formally, an SSP problem $\Pi = (\mathcal{I}, \mathcal{U}, \mathcal{S})$ belongs to SSP-NP, if $|\mathcal{U}(I)| = poly(|I|)$ and if there is an algorithm receiving tuples of an instance $I \in \mathcal{I}$ and a subset $S \subseteq \mathcal{U}(I)$ as input and decides in time polynomial in $|I|$, whether $S \in \mathcal{S}(I)$.*

Because of the analogous definition of the class SSP-NP to NP and the natural SSP adaptation of SAT, we are able to adapt the theorem of Cook and Levin to the class SSP-NP and show that SAT is the canonical SSP-NP-complete problem.

Theorem 1 (Cook-Levin Theorem Adapted to SSPs). *SATISFIABILITY is SSP-NP-complete with respect to polynomial-time SSP reductions, i.e. for every SSP problem Π contained in SSP-NP, we have $\Pi \leq_{SSP}$ SATISFIABILITY.*

With Theorem 1 in mind, we define the class of *SSP-NP-complete problems* (SSP-NPc) as the set of all SSP-NP problems that are complete for the class SSP-NP with respect to SSP reductions.

4 Interdiction Problems

In this section, we consider the closely related topics of *minimum cost blocker problems*, *most vital vertex/edge problems*, and *interdiction problems*. All of these problems are slight variants of each other. Formally, we consider the following problem.

Definition 3 (Interdiction Problem). *Let an SSP problem $\Pi = (\mathcal{I}, \mathcal{U}, \mathcal{S})$ be given. The interdiction problem associated to Π is denoted by INTERDICTION-Π and defined as follows: The input is an instance $I \in \mathcal{I}$ together with a cost function $c : \mathcal{U}(I) \to \mathbb{N}$ and a threshold $t \in \mathbb{N}$. The question is whether $\exists B \subseteq \mathcal{U}(I)$ with $c(B) \leq t : \forall S \in \mathcal{S}(I) : B \cap S \neq \emptyset$.*

The main result of this section is then stated as follows.

Theorem 2. *For all SSP-NP-complete problems Π, the interdiction variant INTERDICTION- Π is Σ_2^p-complete.*

The proof is given in the full version of the paper. We quickly sketch the proof: First, we consider Definition 3 only for the single problem $\Pi =$ SAT. We show using traditional techniques that INTERDICTION-SAT is Σ_2^p-complete (by a reduction from $\exists\forall$-3DNF-SAT). Next, we consider an arbitrary SSP-NP-complete problem Π'. Since Π' is SSP-NP-complete, there is a reduction SAT $\leq_{\text{SSP}} \Pi'$.

This SSP reduction implies that given a SAT instance I, we can find a Π' instance I', such that I can be imagined as a "sub-instance" of I'. Specifically, there is an injective function f mapping the universe of $\mathcal{U}(I)$ into the universe $\mathcal{U}'(I')$. Furthermore, the topology of solutions is maintained (by Eq. (1), or equivalently properties (P1) and (P2)). Hence the SAT instance I can be imagined as a Π' sub-instance of I'. We show that this relation extends in such a way that the corresponding INTERDICTION-SAT instance can be imagined as a sub-instance of INTERDICTION-Π'. Namely we can modify the costs of interdiction such that all newly added elements that are part of I', but not part of I receive infinite costs. On the other hand, all universe elements that are part of I receive the same costs in I'. This has the following consequence: If some "interdictor" tries to interdict all solutions of the Π' instance I', i.e. find a set $B' \subseteq \mathcal{U}'(I')$ with $B' \cap S' \neq \emptyset$ for all $S' \in \mathcal{S}'(I')$ of minimal cost, then w.l.o.g. the blocker is contained in the subinstance of I' that corresponds to the SAT-instance. Therefore all solutions of I' are blocked by some blocker B' if and only if the sub-instance of SAT is blocked by the corresponding blocker $B := f^{-1}(B')$. Since INTERDICTION-SAT is Σ_2^p-hard, it follows that INTERDICTION-Π' is Σ_2^p-hard.

5 Min-Max Regret Problems with Interval Uncertainty

In this section, we consider min-max regret robust optimization problems with interval uncertainty. Our main result in this section is that for every single LOP problem with the property that the SSP problem derived from it is in SSP-NPc, the corresponding min-max regret problem is Σ_2^p-complete.

In order to define the min-max regret version of Π, we use the following definitions, which are standard in the area of min-max regret robust optimization [11]: For some cost function $c : \mathcal{U}(I) \to \mathbb{Z}$, we define $S_c^\star \in \arg\min_{S' \in \mathcal{F}(I)} c(S')$ to be an optimal feasible solution of Π with respect to cost function c. The *regret* of some feasible solution $S \in \mathcal{F}(I)$ is defined as $\text{reg}(S, c) := c(S) - c(S_c^\star)$. Given numbers $\underline{c}_e, \overline{c}_e \in \mathbb{Z}$ for every universe element $e \in \mathcal{U}(I)$ such that $\underline{c}_e \leq \overline{c}_e$, the *interval uncertainty set* defined by the coefficient sequence $(\underline{c}_e, \overline{c}_e)_{e \in \mathcal{U}(I)}$ is defined as $C := \{c \mid c : \mathcal{U}(I) \to \mathbb{R} \text{ is a function s.t. } c(e) \in [\underline{c}_e, \overline{c}_e]\ \forall e \in \mathcal{U}(I)\}$.

Definition 4 (Min-Max Regret Problem with Interval Uncertainty). *Let an LOP problem $\Pi = (\mathcal{I}, \mathcal{U}, \mathcal{F}, d, t)$ be given. The min-max regret problem with interval uncertainty associated to Π is denoted by* INTERVAL MIN-MAX REGRET- Π *and defined as follows: The input is an instance $I \in \mathcal{I}$ together with integers $\underline{c}_e \leq \overline{c}_e$ for all $e \in \mathcal{U}(I)$ and a threshold $t_R \in \mathbb{Z}$. The question is whether $\min_{S \in \mathcal{F}(I)} \max_{c \in C} \text{reg}(S, c) \leq t_R$.*

We define the *maximum regret* $r_{\max}(S) := \max_{c \in C} \text{reg}(S, c)$. For a fixed feasible set $S \in \mathcal{F}(I)$, it is a well-known fact [11] that the maximum regret is caused by the so-called *canonical cost scenario* $c_S : \mathcal{U}(I) \to \mathbb{Z}$, defined by $c_S(e) := \overline{c}_e$ if $e \in S$ and $\underline{c}_e$ if $e \notin S$. This implies that $r_{\max}(S) = \text{reg}(S, c_S) = c_S(S) - c_S(S_{c_S}^\star) = c_S(S) - \min_{S' \in \mathcal{F}(I)} c_S(S')$.

With this observation on the canonical cost scenario, we can view interval min-max regret problems as a two-player game of an $\exists$-player that wants to find a feasible solution against an adversary (the $\forall$-player). The $\exists$-player is in control of the *min* and thus is able to choose the solution S. On the other hand, the adversary is in control of the *max* and is thus able to choose the cost function c from the set C as well as the solution S'. Consequently, we can reformulate the question to $\exists S \in \mathcal{F}(I) : \forall S' \in \mathcal{F}(I) : c_S(S) - c_S(S') \leq t_R$. Comparable to interdiction problems, we are able to prove the following theorem.

Theorem 3. *For all LOP problems Π with the property that the SSP problem derived from them is contained in SSP-NPc, the interval min-max regret variant* Interval Min-Max Regret- Π *is Σ_2^p-complete.*

6 Two-Stage Adjustable Problems

In this section, we consider two-stage adjustable robust optimization problems with discrete budgeted uncertainty. We show for every single LOP problem, whose derived SSP problem is SSP-NP-complete, that the corresponding two-stage adjustable problem is Σ_3^p-complete. We choose to make the following definition of a two-stage adjustable problem, where in the first stage an arbitrary set S_1 can be selected, but needs to be completed to a feasible set in the second stage (i.e. $S_1 \cup S_2 \in \mathcal{F}(I)$).

Definition 5 (Two-Stage Adjustable Problem). *Let an LOP problem $\Pi = (\mathcal{I}, \mathcal{U}, \mathcal{F}, d, t)$ be given. The two-stage adjustable problem associated to Π is denoted by* Two-Stage Adjustable- Π *and defined as follows: The input is an instance $I \in \mathcal{I}$ together with three cost functions: the first stage cost function $c_1 : \mathcal{U}(I) \to \mathbb{Z}$ and the second stage cost functions $\underline{c}_2 : \mathcal{U}(I) \to \mathbb{Z}$ and $\overline{c}_2 : \mathcal{U}(I) \to \mathbb{Z}$, a threshold $t_{TS} \in \mathbb{Z}$, and an uncertainty parameter $\Gamma \in \mathbb{Z}$. Then, the uncertainty set is defined by all cost functions such that at most Γ elements $u \in \mathcal{U}(I)$ have costs of $\overline{c}_2(u)$, while all others have costs $\underline{c}_2(u)$:*

$$C_\Gamma := \{c_2 \mid \forall u \in \mathcal{U} : c_2(u) = \underline{c}_2(u) + \delta_u(\overline{c}_2(u) - \underline{c}_2(u)),\ \delta_u \in \{0,1\}, \sum_{u \in \mathcal{U}} \delta_u \leq \Gamma\}.$$

The question is whether

$$\min_{S_1 \subseteq \mathcal{U}(I)} \max_{c_2 \in C_\Gamma} \min_{\substack{S_2 \subseteq \mathcal{U}(I) \setminus S_1 \\ S_1 \cup S_2 \in \mathcal{F}(I)}} c_1(S_1) + c_2(S_2) \leq t_{TS}.$$

Comparable to interdiction problems and min-max regret problems, we are able to show the following theorem.

Theorem 4. *For all LOP problems Π with the property that the SSP problem derived from them is contained in SSP-NPc, the two-stage adjustable variant* Two-Stage Adjustable- Π *is Σ_3^p-complete.*

7 SSP Reductions for Various Problems

In this section, we present a multitude of SSP-NP-complete problems. In the full version of the paper, we show SSP reductions between all the problems, hence proving that all considered problems are contained in the class SSP-NPc. All reductions are already known and can be found in the literature. In Fig. 1, the tree of all presented reductions can be found, beginning at SATISFIABILITY. We heavily use the transitivity of SSP reductions.

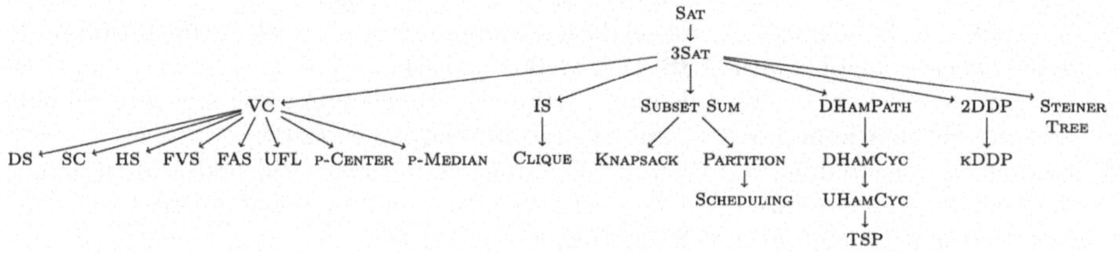

Fig. 1. The tree of SSP reductions for all considered problems.

With the tree of SSP reductions shown in Fig. 1, we derive the following theorem. We remark that the formal definition of all problems mentioned can be found in the full version of this paper.

Theorem 5. *The following problems are SSP-NP-complete: SATISFIABILITY, 3-SATISFIABILITY, VERTEX COVER, DOMINATING SET, SET COVER, HITTING SET, FEEDBACK VERTEX SET, FEEDBACK ARC SET, UNCAPACITATED FACILITY LOCATION, P-CENTER, P-MEDIAN, INDEPENDENT SET, CLIQUE, SUBSET SUM, KNAPSACK, PARTITION, SCHEDULING, DIRECTED HAMILTONIAN PATH, DIRECTED HAMILTONIAN CYCLE, UNDIRECTED HAMILTONIAN CYCLE, TRAVELING SALESMAN PROBLEM, TWO DIRECTED VERTEX DISJOINT PATH, k-VERTEX DIRECTED DISJOINT PATH, STEINER TREE*

8 Conclusion

We have shown that for a large amount of NP-complete problems, their min-max variant (min-max-min variant, respectively) is automatically Σ_2^p-complete (Σ_3^p-complete, respectively). We have showcased this behavior in many different areas. However, many questions still remain unanswered. First, we would like to understand if our theorem can also be adapted to work with other popular areas of robust or multi-level optimization, for example recoverable robust optimization, Stackelberg games, or attacker-defender games. For space reasons, we restricted our focus in this paper mainly on multi-stage problems with two or three stages. It seems natural to extend our arguments to multi-stage problems with an arbitrary number of stages. Finally, our framework only applies to nominal problems which are NP-complete in the first place. However, researchers in the area often are interested in robust variants of nominal problems in P. In contrast to NP-completeness reductions, in this regime there are much less known reductions which can be easily upgraded.

References

1. Aissi, H., Bazgan, C., Vanderpooten, D.: Complexity of the min-max (regret) versions of min cut problems. Discret. Optim. **5**(1), 66–73 (2008)
2. Averbakh, I., Berman, O.: Minmax regret median location on a network under uncertainty. INFORMS J. Comput. **12**(2), 104–110 (2000)
3. Averbakh, I., Lebedev, V.: Interval data minmax regret network optimization problems. Discret. Appl. Math. **138**(3), 289–301 (2004)
4. Bazgan, C., Bentz, C., Picouleau, C., Ries, B.: Blockers for the stability number and the chromatic number. Graphs Comb. **31**(1), 73–90 (2015)
5. Bazgan, C., Toubaline, S., Tuza, Z.: Complexity of most vital nodes for independent set in graphs related to tree structures. In: Combinatorial Algorithms - 21st International Workshop, IWOCA 2010, pp. 154–166
6. Bazgan, C., Toubaline, S., Vanderpooten, D.: Complexity of determining the most vital elements for the 1-median and 1-center location problems. In: Combinatorial Optimization and Applications - 4th International Conference, COCOA 2010, Part I, pp. 237–251
7. Ben-Tal, A., Ghaoui, L.E., Nemirovski, A.: Robust Optimization, Princeton Series in Applied Mathematics, vol. 28. Princeton University Press, Princeton (2009)
8. Caprara, A., Carvalho, M., Lodi, A., Woeginger, G.J.: A study on the computational complexity of the bilevel knapsack problem. SIAM J. Optim. **24**(2), 823–838 (2014)
9. Chassein, A.B., Goerigk, M., Kasperski, A., Zielinski, P.: On recoverable and two-stage robust selection problems with budgeted uncertainty. Eur. J. Oper. Res. **265**(2), 423–436 (2018)
10. Coco, A.A., Santos, A.C., Noronha, T.F.: Robust min-max regret covering problems. Comput. Optim. Appl. **83**(1), 111–141 (2022)
11. Deineko, V.G., Woeginger, G.J.: Pinpointing the complexity of the interval min-max regret knapsack problem. Discret. Optim. **7**(4), 191–196 (2010)
12. Dempe, S., Zemkoho, A. (eds.): Bilevel Optimization. SOIA, vol. 161. Springer, Cham (2020). https://doi.org/10.1007/978-3-030-52119-6
13. Denegre, S.: Interdiction and Discrete Bilevel Linear Programming. Ph.D. thesis, USA (2011). aAI3456385
14. Fröhlich, N., Ruzika, S.: On the hardness of covering-interdiction problems. Theor. Comput. Sci. **871**, 1–15 (2021)
15. Ganesh, A., Maggs, B.M., Panigrahi, D.: Robust algorithms for TSP and steiner tree. ACM Trans. Algorithms **19**(2), 12:1–12:37 (2023)
16. Garey, M.R., Johnson, D.S.: Computers and Intractability: A Guide to the Theory of NP-Completeness. W. H. Freeman, New York (1979)
17. Goerigk, M., Lendl, S., Wulf, L.: Two-stage robust optimization problems with two-stage uncertainty. Eur. J. Oper. Res. **302**(1), 62–78 (2022)
18. Goerigk, M., Lendl, S., Wulf, L.: On the complexity of robust multi-stage problems with discrete recourse. Discret. Appl. Math. **343**, 355–370 (2024)
19. Grüne, C.: The complexity classes of hamming distance recoverable robust problems. In: LATIN 2024: Theoretical Informatics - 16th Latin American Symposium, Part I, pp. 321–335

20. Grüne, C., Wulf, L.: On the complexity of recoverable robust optimization in the polynomial hierarchy. CoRR **abs/2411.18590** (2024)
21. Hunt, K., Zhuang, J.: A review of attacker-defender games: current state and paths forward. Eur. J. Oper. Res. (2023)
22. Jeroslow, R.G.: The polynomial hierarchy and a simple model for competitive analysis. Math. Program. **32**(2), 146–164 (1985)
23. Johannes, B.: New classes of complete problems for the second level of the polynomial hierarchy (2011)
24. Kasperski, A., Zielinski, P.: Robust two-stage network problems. In: Operations Research Proceedings 2015, Selected Papers of the International Conference of the German, Austrian and Swiss OR Societies, pp. 35–40
25. Kasperski, A., Zieliński, P.: Robust discrete optimization under discrete and interval uncertainty: a survey. Robustness analysis in decision aiding, optimization, and analytics, pp. 113–143 (2016)
26. Kasperski, A., Zielinski, P.: Robust recoverable and two-stage selection problems. Discret. Appl. Math. **233**, 52–64 (2017)
27. Li, T., Sethi, S.P.: A review of dynamic stackelberg game models. Discret. Contin. Dyn. Syst.-B **22**(1), 125 (2017)
28. Lin, K., Chern, M.: The most vital edges in the minimum spanning tree problem. Inf. Process. Lett. **45**(1), 25–31 (1993)
29. Malik, K., Mittal, A.K., Gupta, S.K.: The k most vital arcs in the shortest path problem. Oper. Res. Lett. **8**(4), 223–227 (1989)
30. Nabli, A., Carvalho, M., Hosteins, P.: Complexity of the multilevel critical node problem. J. Comput. Syst. Sci. **127**, 122–145 (2022)
31. Pajouh, F.M., Boginski, V., Pasiliao, E.L.: Minimum vertex blocker clique problem. Networks **64**(1), 48–64 (2014)
32. Rutenburg, V.: Propositional truth maintenance systems: classification and complexity analysis. Ann. Math. Artif. Intell. **10**(3), 207–232 (1993)
33. Schaefer, M., Umans, C.: Completeness in the polynomial-time hierarchy a compendium (2008)
34. Smith, J.C., Prince, M., Geunes, J.: Modern network interdiction problems and algorithms. In: Handbook of Combinatorial Optimization, pp. 1949–1987. Springer, New York (2013)
35. Stockmeyer, L.J.: The polynomial-time hierarchy. Theor. Comput. Sci. **3**(1), 1–22 (1976)
36. Woeginger, G.J.: The trouble with the second quantifier. 4OR **19**(2), 157–181 (2021). https://doi.org/10.1007/s10288-021-00477-y
37. Wood, R.: Deterministic network interdiction. Math. Comput. Model. **17**(2), 1–18 (1993)
38. Zenklusen, R.: Matching interdiction. Discret. Appl. Math. **158**(15), 1676–1690 (2010)

Semirandom Planted Clique via 1-Norm Isometry Property

Venkatesan Guruswami[1,2] and Hsin-Po Wang[1,2](✉)

[1] University of California, Berkeley, USA
[2] National Taiwan University, Taipei, China
hsinpo@ntu.edu.tw

Abstract. We give a polynomial time algorithm that finds a planted clique of size $k \geqslant \sqrt{n \log n}$ in the semirandom model, improving the state-of-the-art $\sqrt{n}(\log n)^2$ bound. The *semirandom planted clique problem* concerns finding the planted subset S of k vertices V of a graph G, where the induced subgraph $G[S]$ is complete, the cut edges in $G[S; V \setminus S]$ are random, and the remaining edges in $G[V \setminus S]$ are adversarial.

An elegant greedy algorithm (Błasiok, Buhai, Kothari, and Steurer, FOCS 2024) finds S by sampling inner products of the columns of the adjacency matrix of G, and checking if they deviate significantly from typical inner products of random vectors. Their analysis uses a suitably random matrix that, with high probability, satisfies a certain restricted isometry property. Inspired by Wootters' work on list decoding, we put forth and implement the 1-norm analog of this argument, and quantitatively improve their analysis to work all the way up to the conjectured optimal $\sqrt{n \log n}$ bound on clique size, answering one of the main open questions posed in [BBKS24].

1 Introduction

An important and well-studied class of computational problems concerns finding an atypical structure amidst ambient noise. One such quintessential challenge is the fundamental *planted clique* problem from combinatorial optimization, where we are given a graph G on n vertices that contains a clique of size k (structure), while the remaining edges are randomly placed (noise). See Fig. 1 for an example.

A situation in which the clique is obvious is when $k \gg \sqrt{n \log n}$. In this situation, since the degree of every vertex in a uniformly random graph is in the range $n/2 \pm O(\sqrt{n \log n})$ with high probability, one can find the planted k-clique by simply picking out the k vertices with the highest degrees. A classic spectral algorithm improves this bound and succeeds in finding the planted clique when $k \gg \sqrt{n}$ [AKS98].

However, if k is much smaller than $\sqrt{n}$ but still[1] much greater than $\log n$, these simple algorithms cease to work. In fact, no efficient algorithm is known in

[1] A random graph will have cliques of size $2 \log_2 n$ with high probability, so we must assume $k \gg \log n$ for the planted clique to "stand out.".

N. Megow and A. Basu (Eds.): IPCO 2025, LNCS 15620, pp. 270–282, 2025.
https://doi.org/10.1007/978-3-031-93112-3_20

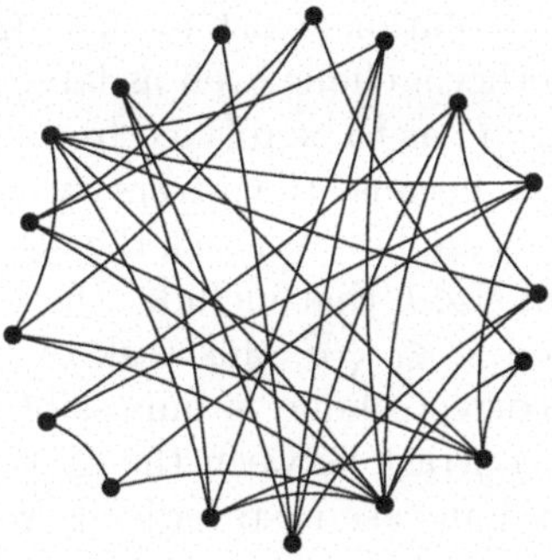

$$\begin{bmatrix}
0&0&0&0&0&0&0&0&0&0&1&0&1&0&1&0&1\\
0&1&1&1&0&0&0&1&0&0&1&0&0&1&0&0&0\\
0&0&1&0&0&0&0&0&1&1&0&1&0&0&0&0&1\\
0&0&1&0&0&0&1&0&0&1&0&1&0&1&0&1&0\\
0&0&0&0&1&1&0&0&0&0&0&0&0&0&0&0&1\\
0&1&1&0&1&0&1&0&0&0&0&1&0&1&1&0&0\\
0&1&0&0&0&0&0&1&0&0&0&0&0&0&0&1&1\\
0&0&0&0&0&0&0&0&0&0&0&0&0&1&1&0&0\\
0&0&0&0&0&0&0&1&0&0&0&0&0&0&1&0&0\\
1&1&0&0&0&1&0&1&1&0&1&0&0&0&0&1&0\\
0&0&1&0&0&0&1&0&0&0&0&1&0&1&0&0&0\\
0&0&0&0&0&0&0&1&0&0&0&0&1&0&0&0&0\\
0&0&0&1&0&0&0&0&0&0&0&1&1&0&0&0&0\\
1&0&0&0&1&0&0&0&0&0&0&0&0&0&0&1&0\\
0&1&1&0&0&0&1&0&0&0&0&1&0&1&1&1&0\\
0&0&1&0&0&0&0&1&0&0&1&1&0&0&0&1&0\\
1&0&0&1&0&0&0&1&0&0&0&0&0&0&0&0&0
\end{bmatrix}$$

Fig. 1. An instance of planted clique problem. Left is the graph and right is the adjacency matrix. Answer is in Fig. 2.

this regime, and the well-known planted clique hypothesis asserts that no such algorithm exists. In other words, there seems to be a *statistical vs. computational gap* [BB20] of $\log n$ vs $\sqrt{n}$, in which range certain tasks are information-theoretically possible but (seemingly) computationally intractable. The study of this phenomenon, and which computational problems with statistical inputs exhibit such gaps, is a central theme in average-case complexity that has received accelerated attention in recent years.

In addition to the large k needed for the degree-based algorithm to work, the algorithm is also fragile and easy to trick. If we add k new edges to a non-clique vertex, the top k high-degree vertices will contain a false positive. The algorithm is thus confounded by even modest fluctuations from a fully random base graph. To study the robustness of clique-finding algorithms, Feige and Kilian introduced the insightful and influential *semirandom* model [FK01]. In this model, S is the vertices of the planted clique, edges between S and $V \setminus S$ (where V the set of vertices) are placed randomly, and the edges within $V \setminus S$ can be configured adversarially with access to the random edges.

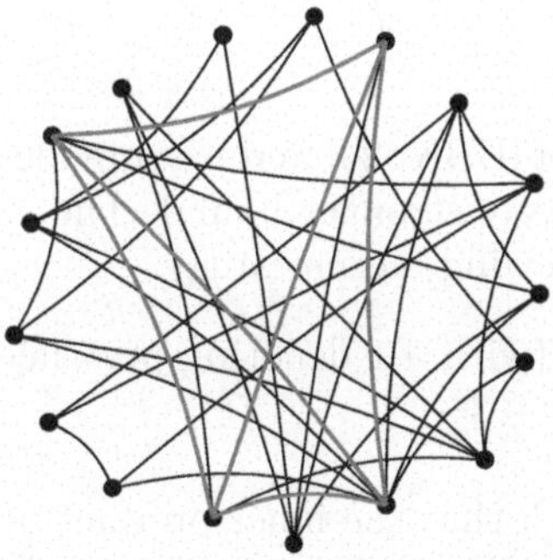

$$\begin{bmatrix}
0&0&0&0&0&0&0&0&0&0&1&0&1&0&1&0&1\\
0&1&1&1&0&0&0&1&0&0&1&0&0&1&0&0&0\\
0&0&1&0&0&0&0&0&1&1&0&1&0&0&0&0&1\\
0&0&1&0&0&0&1&0&0&1&0&1&0&1&0&1&0\\
0&0&0&0&1&1&0&0&0&0&0&0&0&0&0&0&1\\
0&1&1&0&1&0&1&0&0&0&0&1&0&1&1&0&0\\
0&1&0&0&0&0&0&1&0&0&0&0&0&0&0&1&1\\
0&0&0&0&0&0&0&0&0&0&0&0&0&1&1&0&0\\
0&0&0&0&0&0&0&1&0&0&0&0&0&0&1&0&0\\
1&1&0&0&0&1&0&1&1&0&1&0&0&0&0&1&0\\
0&0&1&0&0&0&1&0&0&0&0&1&0&1&0&0&0\\
0&0&0&0&0&0&0&1&0&0&0&0&1&0&0&0&0\\
0&0&0&1&0&0&0&0&0&0&0&1&1&0&0&0&0\\
1&0&0&0&1&0&0&0&0&0&0&0&0&0&0&1&0\\
0&1&1&0&0&0&1&0&0&0&0&1&0&1&1&1&0\\
0&0&1&0&0&0&0&1&0&0&1&1&0&0&0&1&0\\
1&0&0&1&0&0&0&1&0&0&0&0&0&0&0&0&0
\end{bmatrix}$$

Fig. 2. The planted clique is highlighted.

Unlike the random model, for which simple solutions exist for $k \geq \Omega(\sqrt{n})$, in the semirandom model it is not obvious whether a clique even as large as say $n^{2/3}$ can be easily identified. In fact, the goal itself has to be relaxed from finding the specific planted clique S, which is impossible because the adversary can add up to $(n-k)/k$ cliques of size k in $G[V \setminus S]$. One can, however, ask the algorithm to find a list of at most $(1+o(1))n/k$ cliques of size k that includes the planted clique S. Steinhardt [Ste18] showed that when $k \ll \sqrt{n}$, this task of finding a list of $(1+o(1))n/k$ cliques including the correct clique is impossible, *even information-theoretically*. This shows a sharp contrast between the random vs. semirandom planted clique problem, as the random case is statistically solvable for k as small as $O(\log n)$.

This raises the natural challenge of whether we can solve semirandom planted clique for k approaching the Steinhardt threshold of $\approx \sqrt{n}$. This question has received a fair bit of attention in recent years. Algorithms based on (different) semidefinite programming (SDP) relaxations were given to solve this problem when $k \gtrsim n^{2/3} \log^{1/3} n$ in [CSV17], and when $k \gtrsim n^{2/3}$ in [MMT20]. The latter work also suggested that their SDP should likely fail when $k = o(n^{2/3})$. More recently, Buhai, Kothari, and Steurer [BKS23] gave a complicated algorithm based on sum-of-squares SDP to solve the problem all the way to $k \geq n^{1/2+\varepsilon}$; the runtime of the algorithm was $n^{O(1/\varepsilon)}$ and thus very large as k approached Steinhardt's threshold of $n^{1/2}$.

Finally, Błasiok, Buhai, Kothari, and Steurer [BBKS24] gave a delightfully elegant greedy algorithm for the problem that works as long as $k \geq \sqrt{n}(\log n)^2$. It was also shown in these works [BKS23,BBKS24] that when $k \gtrsim \sqrt{n \log n}$, one can prune any polynomial-sized list containing the planted set S to a list of size $(1+o(1))n/k$.

Given this backdrop, a natural question is whether one can solve semirandom planted clique, i.e., output a polynomial-sized list including the planted set S, for k all the way down to $O(\sqrt{n \log n})$, which is believed to be the right threshold.

This is precisely what we achieve in this short paper.

Theorem 1 (main). *The semi-random planted clique problem on n vertices, where the planted clique has size $k \geqslant \mathcal{O}(\sqrt{n \log n})$ and random edges have density $1/2$, has a $\mathsf{poly}(n)$-time algorithm that outputs a list of $(1+o(1))n/k$ guesses of the clique, and is correct with probability 0.99.*

In fact, our algorithm is the same as that of [BBKS24] and we provide a simpler analysis that also leads to the quantitatively stronger bound. The starting point of the algorithm of [BBKS24] is the following base strategy.

- Consider A the ± 1 adjacency matrix of G (with $+1$ denoting presence of an edge). Let A^v be the vth column of A for $v \in V$.
- Randomly choose a vertex $u \in V$.
- Let $S_u \subset V$ be the set of vertices $v \in V$ such that the inner product between A^v and A^u is in the range $k \pm \mathcal{O}(\sqrt{n})$. This S_u will be one guess of S.
- Repeat the two steps above $\Theta(n/k)$ times so that with high probability, we are lucky at least once and u is in S. It remains to show that when u is in S,

the guess S_u contains at most $o(k)$ vertices outside S (the spurious vertices in $S_u \setminus S$ can then be easily pruned out).

This strategy can handle the region $k \geqslant \tilde{\mathcal{O}}(n^{3/4})$ [BBKS24, Section 2.1]. As it turns out, the obstacle is that there are only n choices of $u \in V$, so it is relatively easy[2] for the adversary to manipulate inner products. Błasiok *et al.* strengthened this strategy in [BBKS24] by comparing more inner products.

- Randomly choose *three* vertices $u_1, u_2, u_3 \in V$.
- Let $S_{u_1 u_2 u_3} \subset V$ be the set of vertices $v \in V$ such that the inner product between A^v and $A^{u_1} \odot A^{u_2} \odot A^{u_3}$ (where $\odot$ denotes entry-wise product of vectors) "looks right" (this will be made precise later). This $S_{u_1 u_2 u_3}$ will be one guess for S.
- Repeat the two steps above $\Theta((n/k)^3)$ times so that with high probability, we are lucky at least once and u_1, u_2, u_3 all belong to S. It remains to show that when u_1, u_2, u_3 are all in S, the guess $S_{u_1 u_2 u_3}$ has at most $o(k)$ vertices outside S (which can be easily pruned out).

The improved algorithm works in the region $k \geqslant \mathcal{O}(\sqrt{n}(\log n)^2)$. The improvement comes from the fact that it is harder for the adversary to manipulate inner products when triples $(u_1, u_2, u_3) \in V^3$ are sampled from a much larger sample space.[3]

The crux of the analysis of [BBKS24] is to control deviations of inner products via the restricted isometry property (RIP) of a certain random matrix. This RIP, which concerns ℓ_2-norms is hard to control. Our key insight is to realize that one can work with a simpler ℓ_1 version of the RIP, and to control it with a simpler and quantitatively tighter analysis. This aspect is directly inspired by a similar use of ℓ_1-RIP in the novel analysis of list-decoding properties of random linear codes by Wootters [Woo13].

The paper is organized as follows. In Sect. 2, we review the inner-product algorithms in greater detail. In Sect. 3, we explain how their performance analysis is related to whether certain random matrices have restricted isometry property. In Sect. 4, we discuss list decoding and how proof techniques from there can be applied here. In Sect. 5, we wrap up the proof of the main theorem.

2 Check Inner Products

In this section, we review [BBKS24]'s greedy algorithms that are based on sampling inner products between the columns of the adjacency matrix of G.

[2] To clarify, it is not proved that the adversary has a way to fool the algorithm when $k < \tilde{\mathcal{O}}(n^{3/4})$; it is just that the analysis given in [BBKS24, Section 2.1] needs $k \geqslant \tilde{\mathcal{O}}(n^{3/4})$.

[3] A natural question is what the bounds on k are if we sample pairs (u_1, u_2), quadruples (u_1, u_2, u_3, u_4), and so on. That is, why stop at triples? A rule of thumb is one can work with $k > n^a$, where a is $3/(3 + \text{the length of tuples})$. But we also know that $\sqrt{n}$ cannot be beaten. Hence, stopping at triples is the best option. See [BBKS24, Section 2.2] for a discussion of the tradeoff when choosing pairs.

Since it is easier to talk about deviations when the expectation is zero, we assume that the adjacency matrix $A \in \{+1, -1\}^{n \times n}$ records an edge as $+1$ and a non-edge as -1.

Let u and v be two vertices. Consider the inner product between A^u and A^v, the uth and the vth columns of A. There are three cases.

- If both u and v are in the clique, then the inner product looks like
$$k + (\text{sum of } n - k \text{ random signs}),$$
which is about $k \pm \mathcal{O}(\sqrt{n-k})$ by the central limit theorem, or $k \pm \sqrt{n}$ for short.
- If u is in the clique but v is not, then the inner product looks like the sum of n random signs, which is about $\pm\sqrt{n}$.
- If neither of them is in the clique, then the inner product is the sum of n adversarial numbers, which can be anything.

Accordingly, if u happens to be in the clique, we can separate $v \in S$ from $v \notin S$ by checking if the inner product is $k \pm \sqrt{n}$ or $\pm\sqrt{n}$. We can, for instance, define $S_u \subset V$ to be those $v \in V$ such that the inner product is at least $k/2$.

Now for the improved algorithm that samples three vertices $u_1, u_2, u_3 \in V$ instead of one, we first compute the entry-wise product of A^{u_1}, A^{u_2}, and A^{u_3}. That is, we let $T^{u_1u_2u_3} \in \{+1, -1\}^{n \times 1}$ be the column vector whose wth row is the product $A^{u_1}_w \cdot A^{u_2}_w \cdot A^{u_3}_w \in \{+1, -1\}$. Then we can compute the inner products of $T^{u_1u_2u_3}$ with A^v for all vertices $v \in V$.

Note that, when u_1, u_2, u_3 are in the clique, the wth entry of $T^{u_1u_2u_3}$ is

$$T^{u_1u_2u_3}_w = \begin{cases} 1 & \text{if } w \in S, \\ \text{product of three random signs} & \text{if } w \notin S. \end{cases}$$

So $T^{u_1u_2u_3}$ plays the same role as A^u in the earlier algorithm in the sense that the inner product between $T^{u_1u_2u_3}$ and A^v would be $k \pm \sqrt{n}$ if $v \in S$, and $\sqrt{n}$ if $v \notin S$. This is what "looks right" means in the introduction. We define $S_{u_1u_2u_3} \subset V$ to be those $v \in V$ such that the inner product is at least $k/2$.

We claim, while referring readers to [BBKS24], that all we want now is that S_u is very close to S for a considerable fraction of $u \in S$. And for the latter algorithm, we want that $S_{u_1u_2u_3}$ is very close to S for a considerable fraction of triples $(u_1, u_2, u_3) \in S^3$. We do not need all u (and all triples) to be good because we can simply sample more times. We do not need S_u (and $S_{u_1u_2u_3}$) to be exactly S because we can post-process a close enough S_u (and $S_{u_1u_2u_3}$) to get S.

3 The Reduction to Restricted Isometry Property

In this section, we explain how [BBKS24] reduces the performance of the inner-product–based algorithm to whether a certain random matrix has a form of restricted isometry property (RIP) with high probability. A matrix $M \in \mathbb{R}^{m \times n}$

is said to have the (t, ϵ)-RIP if for all t-sparse vectors $z \in \mathbb{R}^n$, $(1-\epsilon)\|z\|_2 \leq \|Mz\|_2 \leq (1+\epsilon)\|z\|_2$. This property is useful to recover an unknown sparse "signal" z using only $m \ll n$ "measurements" Mz.

The reader will see (a) the randomness of the "measurement" matrix M corresponds to the random edges between S and $V \setminus S$, (b) the z in the RIP statement

$$\|Mz\|_2 \approx (1 \pm \varepsilon)\|z\|_2, \tag{1}$$

corresponds to the set of triples $(u_1, u_2, u_3) \in S^3$ where $S_{u_1 u_2 u_3}$ misclassifies some vertex $v \notin S$, and (c) the sparsity condition on x corresponds to the success rate of the inner-product–based algorithm.

3.1 The Simple Version

In order to explain the reduction better, let us first focus on the simpler version (the one that samples u, not triples). From the perspective of the adversary, there are two options that can fool the inner-product algorithm.

- The adversary can manipulate edges such that, when u is not in the clique, S_u appears to be a clique different from S.
- The adversary can manipulate edges such that, when u is in the clique, a $v \notin S$ appears to be in S.

The first case does not pose any issue as the algorithm is allowed to output a list of candidate cliques. We only need to make sure that the correct clique appears in the list.

For the second option to be a threat, the adversary should realize that it is not enough to fabricate a bad inner product between only one pair of vertices; this is because if a guess S_u intersects S at $k \pm o(k)$ vertices, one can post-process S_u to get to S per se [BKS23, Claim 5.9]. As a result, the adversary should seek to manipulate the edges within $V \setminus S$ to maximize the number of pairs $(u, v) \in S \times (V \setminus S)$ such that v appears to be in S when we compute the inner product between u and v.

Now, suppose that we can show the following statement.

($\heartsuit$) For any fixed $v \in V \setminus S$, the number of $u \in S$ leading to bad inner products, i.e., at least $k/2$, is less than $\mathcal{O}(n^2/k^2)$.

Then by a double counting argument we can see that (a) there are $\mathcal{O}(n^3/k^2)$ bad pairs $(u, v) \in S \times (V \setminus S)$ and (b) for more than 50% of $u \in S$, the number of bad $v \in V \setminus S$ is less than $\mathcal{O}(n^3/k^3)$. Equate $\mathcal{O}(n^3/k^3)$ to $o(k)$, we conclude that the algorithm works when $k > \omega(n^{3/4})$.

It remains to reduce ($\heartsuit$) to an RIP inequality of the form (1). For a fixed $v \in V \setminus S$, let $B \subset S$ be the collection of bad u's, that is, $\langle A^u, A^v \rangle \geqslant k/2$ and we want to show that $|B| \leqslant \mathcal{O}(n^2/k^2)$. Summing the inner products over B, we obtain

$$|B|\frac{k}{2} \leqslant \sum_{u \in B} \langle A^u, A^v \rangle = \Big\langle \sum_{u \in B} A^u, A^v \Big\rangle = \Big\langle \sum_{u \in B} A^u_S, A^v_S \Big\rangle + \Big\langle \sum_{u \in B} A^u_{V \setminus S}, A^v_{V \setminus S} \Big\rangle.$$

Here, A^u_S is the sub-vector of A^u where we keep the rows corresponding to vertices in S. The same goes for $A^u_{V\setminus S}$, A^v_S, and $A^v_{V\setminus S}$. Note that $\langle \sum_{u\in B} A^u_S, A^v_S \rangle$ is the boring part as the randomness in A^v_S makes the expectation zero. The other term, the interesting part, can be bounded with the Cauchy–Schwarz inequality

$$\Big\langle \sum_{u\in B} A^u_{V\setminus S}, A^v_{V\setminus S} \Big\rangle \leqslant \Big\| \sum_{u\in B} A^u_{V\setminus S} \Big\|_2 \cdot \big\| A^v_{V\setminus S} \big\|_2$$

All entries are ± 1, so $\|A^v_{V\setminus S}\|_2$ is just $\sqrt{n-k}$. The other 2-norm can be understood as

$$\Big\| \sum_{u\in B} A^u_{V\setminus S} \Big\|_2 = \big\| A^S_{V\setminus S} 1_B \big\|_2$$

where $A^S_{V\setminus S} \in \{+1,-1\}^{(n-k)\times k}$, the sub-matrix of A where we only keep the rows corresponding to $V \setminus S$ and the columns corresponding to S, plays the role of the measurement matrix M and 1_B, the indicator vector of $B \subset S$, plays the role of the x with sparsity $|B|$.

However, one should not expect any meaningful bound[4] from the RIP literature because, as much as $A^S_{V\setminus S}$ is a random matrix, it is not a wide matrix but a tall one.

3.2 The Triple Version

RIP comes into play when we consider the triple version and the statement

($\diamondsuit$) for any fixed $v \in V \setminus S$, the number of $(u_1, u_2, u_3) \in S^3$ leading to bad inner products, i.e., at least $k/2$, is less than $\mathcal{O}(n^2/k^2)$.

Let $\mathbf{B}$ be the collection of bad triples $(u_1, u_2, u_3) \in S^3$ that misclassifies some fixed vertex $v \in V \setminus S$, $1_{\mathbf{B}} \in \{0,1\}^{k^3\times 1}$ is the indicator vector of $\mathbf{B}$, and $T \in \{+1,-1\}^{n\times k^3}$ is the matrix whose (u_1, u_2, u_3)th column is $T^{u_1u_2u_3}$, the entry-wise product of A^{u_1}, A^{u_2}, and A^{u_3}. To prove ($\diamondsuit$), we follow the same logic as above: summing inner products, linearity, splitting rows, and Cauchy–Schwarz.

[4] In [BBKS24], $\|A^S_{V\setminus S} 1_B\|_2$ is bounded using the fact that such a random matrix likely has an operator norm $\mathcal{O}(\sqrt{n})$, so $\|A^S_{V\setminus S} 1_B\|_2 \leqslant \mathcal{O}(\sqrt{n}\sqrt{|B|})$. Overall we have $|B|k/2 \leqslant \mathcal{O}(\sqrt{n}\sqrt{|B|}\sqrt{n})$ which leads to $|B| \leqslant \mathcal{O}(n^2/k^2)$, as desired.

$$\begin{aligned}
|\mathbf{B}| \cdot \frac{k}{2} &\leqslant \sum_{(u_1,u_2,u_3)\in\mathbf{B}} \left\langle T^{u_1u_2u_3}, A^v \right\rangle \\
&= \left\langle \sum_{(u_1,u_2,u_3)\in\mathbf{B}} T^{u_1u_2u_3}, A^v \right\rangle \\
&= \left\langle T\mathbb{1}_{\mathbf{B}}, A^v \right\rangle \\
&= \left\langle T_{V\setminus S}\mathbb{1}_{\mathbf{B}}, A^v_{V\setminus S} \right\rangle + \left\langle T_S\mathbb{1}_{\mathbf{B}}, A^v_S \right\rangle \\
&\leqslant \left\|T_{V\setminus S}\mathbb{1}_{\mathbf{B}}\right\|_2 \cdot \left\|A^v_{V\setminus S}\right\|_2 + (\text{zero-mean boring part}) \\
&= \left\|T_{V\setminus S}\mathbb{1}_{\mathbf{B}}\right\|_2 \cdot \sqrt{n-k} + (\text{zero-mean boring part}),
\end{aligned} \tag{2}$$

Now that $T_{V\setminus S}$ is a wide random matrix[5], we can expect

$$\left\|T_{V\setminus S}\mathbb{1}_{\mathbf{B}}\right\|_2 \leqslant (1+1)\sqrt{n-k}\left\|\mathbb{1}_{\mathbf{B}}\right\|_2$$

as $T_{V\setminus S}/\sqrt{n-k}$ likely has RIP with error term $\varepsilon := 1$.

It is not hard to see that the bound on k would not be tight if the Cauchy–Schwarz inequality (2)

$$\left\langle T_{V\setminus S}\mathbb{1}_{\mathbf{B}}, A^v_{V\setminus S} \right\rangle \leqslant \left\|T_{V\setminus S}\mathbb{1}_{\mathbf{B}}\right\|_2 \cdot \left\|A^v_{V\setminus S}\right\|_2$$

does not assume equality. The equality is achieved when $T_{V\setminus S}\mathbb{1}_{\mathbf{B}}$ is parallel to $A^v_{V\setminus S}$, which is a ± 1-vector. But it is unlikely that all the entries of $T_{V\setminus S}\mathbb{1}_{\mathbf{B}}$ assume similar absolute values. In the next section, we replace Cauchy–Schwarz with

$$\left\langle T_{V\setminus S}\mathbb{1}_{\mathbf{B}}, A^v_{V\setminus S} \right\rangle \leqslant \left\|T_{V\setminus S}\mathbb{1}_{\mathbf{B}}\right\|_1 \cdot \left\|A^v_{V\setminus S}\right\|_\infty. \tag{3}$$

The equality is achieved if the adversary constructs $A^v_{V\setminus S}$ by matching the signs of the entries of $T_{V\setminus S}\mathbb{1}_{\mathbf{B}}$. That is to say, it is always achievable. So hopefully we can get a better bound on k.

4 List Deccodability and the 1-Norm Analog of RIP

In this section, we show how to bound the 1-norm $\|T_{V\setminus S}\mathbb{1}_{\mathbf{B}}\|_1$ in (3). The key is to show an inequality of the form

$$\left\|T_{V\setminus S}\mathbb{1}_{\mathbf{B}}\right\|_1 \leqslant \mathcal{O}(n\sqrt{|\mathbf{B}|}) + (\text{deviation term}),$$

where $\sqrt{|\mathbf{B}|}$ reflects the norm of $\mathbb{1}_{\mathbf{B}}$ and the deviation term should be small.

Before we proceed to bounding $\|T_{V\setminus S}\mathbb{1}_{\mathbf{B}}\|_1$, we want to briefly elaborate on the context of Wootters' work [Woo13] as the proof of Theorem 1 is almost a replica of hers. Our proof is driven by the serendipitous discovery that the approach of [Woo13] can fill the gap left by [BBKS24].

[5] Note that the entries of T are not independent as they are algorithmically generated from a random source A. That being said, all that is needed for the restricted isometry property to hold with high probability is that $\mathbf{E}[T^\top T]$ is a multiple of the identity matrix, where the expectation is taking over the randomness of A.

4.1 List Decodability

In coding theory, there is a codebook $\mathcal{C} \subset \{+1,-1\}^n$ and we want that, whenever we observe a message $y \in \{+1,-1\}^n$, we should be able to decode the original message $x \in \mathcal{C}$ by minimizing the Hamming distance to the observation

$$\min_{x \in C} \mathrm{HDist}(x, y).$$

But sometimes the x that minimizes the distance is not the correct message. So in the theory of list coding, we give attention to a list of candidates

$$\mathcal{L} := \mathrm{HBall}(y, r) \cap \mathcal{C},$$

where $\mathrm{HBall}(y, r)$ is a Hamming ball centered at y and r is the *search radius* (which is our estimate of how distorted y can be). We hope that the list size $|\mathcal{L}|$ is small so that, given other clues (such as side channels), we can recover the correct message from $\mathcal{L}$. This leads to the question of how to construct a codebook $\mathcal{C}$ with specified alphabet size, block length, code rate, search radius, and list size.

As it turns out, determining the region of the five parameters where list-decodable codes exist is a difficult problem. One of the many difficulties is that the list size, $|\mathrm{HBall}(y, r) \cap \mathcal{C}|$ maximized over all y, is hard to control. In [Woo13], Wootters made the following observation[6]

$$\begin{gathered}\text{two codewords } x \text{ and } y \text{ are } r\text{-close}\\ \Updownarrow \\ \text{inner product between } x \text{ and } y \text{ is } \geqslant n - 2r\end{gathered}$$

and thus

$$\begin{gathered}\text{there is a large list } \mathcal{L} \text{ of } x\text{'s that are } r\text{-close to } y\\ \Updownarrow \\ \text{inner product between every } x \in \mathcal{L} \text{ and } y \text{ is large}\\ \Downarrow \\ \langle \text{the sum of the } x\text{'s in } \mathcal{L}, y\rangle \text{ is large}\\ \Downarrow \\ |\text{the sum of the } x\text{'s in } \mathcal{L}|_1 \text{ is large}\\ \Downarrow \\ |\Phi \cdot 1_{\mathcal{L}}|_1 \text{ is large, need to bound } |\Phi \cdot 1_{\mathcal{L}}|_1\end{gathered}$$

where $\Phi \in \{+1,-1\}^{n \times |\mathcal{C}|}$ is a matrix whose columns are codewords of $\mathcal{C}$.

---

6 Here we assume binary alphabet $\{+1,-1\}$, but Wootters's work covers larger alphabets.

The 1-norm Analog of RIP

Now that the readers can see the parallel between $T_{V\setminus S}1_{\mathbf{B}}$ and $\Phi 1_{\mathcal{L}}$, we move on to the actual bound. To begin, we center the random variable $\|T_{V\setminus S}1_{\mathbf{B}}\|_1$ by subtracting its expectation

$$\max_{\mathbf{B}}\left\{\left\|T_{V\setminus S}1_{\mathbf{B}}\right\|_1\right\} \leqslant \max_{\mathbf{B}}\left\{\mathop{\mathbf{E}}_{A}\left[\left\|T_{V\setminus S}1_{\mathbf{B}}\right\|_1\right]\right\} + \max_{\mathbf{B}}\left\{\left\|T_{V\setminus S}1_{\mathbf{B}}\right\|_1 - \mathop{\mathbf{E}}_{A}\left[\left\|T_{V\setminus S}1_{\mathbf{B}}\right\|_1\right]\right\}. \tag{4}$$

It suffices to upper bound the two terms on the right-hand side.

For the first term $\max_{\mathbf{B}}\{\mathbf{E}_A[\|T_{V\setminus S}1_{\mathbf{B}}\|_1]\}$, note that the first row of $T_{V\setminus S}1_{\mathbf{B}}$ is a sum of $|\mathbf{B}|$ random signs. If the entries of $T_{V\setminus S}$ are iid, the sum will be roughly $\pm\sqrt{|\mathbf{B}|}$. But they are not; they are only pairwisely independent. Let $X_1, X_2, \ldots, X_{|\mathbf{B}|}$ be the first row of $T_{V\setminus S}1_{\mathbf{B}}$, Jensen's inequality gives

$$\begin{aligned}\left(\mathop{\mathbf{E}}_{A}|X_1 + X_2 + \cdots + X_{|\mathbf{B}|}|\right)^2 &\leqslant \mathop{\mathbf{E}}_{A}(X_1 + X_2 + \cdots + X_{|\mathbf{B}|})^2\\ &= \mathop{\mathbf{E}}_{A}[X_1^2 + X_2^2 + \cdots + X_{|\mathbf{B}|}^2].\\ &= |\mathbf{B}|.\end{aligned}$$

So the same bound $\sqrt{|\mathbf{B}|}$ applies. There are $|V \setminus S| = n - k$ rows, so

$$\mathop{\mathbf{E}}_{A}\|T_{V\setminus S}1_{\mathbf{B}}\|_1 \leqslant (n-k)\cdot\sqrt{|\mathbf{B}|} \leqslant n\sqrt{|\mathbf{B}|}\ . \tag{5}$$

For the second term $\max_{\mathbf{B}}\{\|T_{V\setminus S}1_{\mathbf{B}}\|_1 - \mathbf{E}_A[\|T_{V\setminus S}1_{\mathbf{B}}\|_1]\}$ of (4), we want to apply Markov's inequality to obtain an upper bound that is valid 50% of time. Therefore, we take the expectation over A and attempts to bound

$$\mathop{\mathbf{E}}_{A}\left[\max_{\mathbf{B}}\left\{|t|_1 - \mathop{\mathbf{E}}_{A}[|t|_1]\right\}\right],$$

where t is short for $T_{V\setminus S}1_{\mathbf{B}}$. A standard symmetrization argument upper bounds this expectation with

$$\mathop{\mathbf{E}}_{A}\left[\max_{\mathbf{B}}\left\{|t|_1 - \mathop{\mathbf{E}}_{A}[|t|_1]\right\}\right] \leqslant \mathcal{O}\left(\mathop{\mathbf{E}}_{A}\mathop{\mathbf{E}}_{g}\left[\max_{\mathbf{B}}\left\{\sum_j g_j t_j\right\}\right]\right) \tag{6}$$

where t_j is the jth entry of t and $g_1, g_2, \ldots$ are iid standard Gaussian random variables [Woo13, (7)]. In greater details, (6) is

$$\begin{aligned}
\mathop{\mathbf{E}}_{A}\Big[\max_{\mathbf{B}}\Big\{|t|_1 - \mathop{\mathbf{E}}_{A}[|t|_1]\Big\}\Big] &= \mathop{\mathbf{E}}_{A}\Big[\max_{\mathbf{B}}\Big\{|t|_1 - \mathop{\mathbf{E}}_{A'}[|t'|_1]\Big\}\Big] \\
&= \mathop{\mathbf{E}}_{A}\Big[\max_{\mathbf{B}}\Big\{\sum_j |t_j| - \mathop{\mathbf{E}}_{A'} |t'_j|\Big\}\Big] \\
&\leqslant \mathop{\mathbf{E}}_{A}\mathop{\mathbf{E}}_{A'}\Big[\max_{\mathbf{B}}\Big\{\sum_j t_j - t'_j\Big\}\Big] \\
&= \mathop{\mathbf{E}}_{A}\mathop{\mathbf{E}}_{A'}\mathop{\mathbf{E}}_{\varepsilon}\Big[\max_{\mathbf{B}}\Big\{\sum_j \varepsilon_j(t_j - t'_j)\Big\}\Big] \\
&\leqslant \mathop{\mathbf{E}}_{A}\mathop{\mathbf{E}}_{A'}\mathop{\mathbf{E}}_{\varepsilon}\Big[\max_{\mathbf{B}}\Big\{\sum_j \varepsilon_j t_j\Big\} + \max_{\mathbf{B}}\Big\{\sum_j \varepsilon_j t'_j\Big\}\Big] \\
&= 2\mathop{\mathbf{E}}_{A}\mathop{\mathbf{E}}_{\varepsilon}\Big[\max_{\mathbf{B}}\Big\{\sum_j \varepsilon_j t_j\Big\}\Big] \\
&\leqslant \mathcal{O}\Big(\mathop{\mathbf{E}}_{A}\mathop{\mathbf{E}}_{g}\Big[\max_{\mathbf{B}}\Big\{\sum_j g_j t_j\Big\}\Big]\Big)
\end{aligned}$$

Here A' and t' are independent copies of A and t, respectively, and $\varepsilon_1, \varepsilon_2, \ldots$ are iid Rademacher random variables.

Note that the sum inside $\max_{\mathbf{B}}$ is a linear combination of Gaussian random variables, which are Gaussian themselves. Hence, the expectation $\mathbf{E}_g$ of the maximum $\max_{\mathbf{B}}$ is upper bounded by

$$\sqrt{\text{variance} \cdot \log(\#\text{ Gaussian random variables})}. \tag{7}$$

This bound is a consequence of the union bound, which does not care if the Gaussian variables are correlated. Clearly the number of Gaussian variables is the number of ways we can choose $\mathbf{B}$, which is $\binom{n}{|\mathbf{B}|}$. But there is a better bound: Instead of all subsets of $\{1, \ldots, n\}$, we let $\mathbf{B}$ enumerate over all *multi-sets* that consist of 1 to n. This only provides more choices to the maximum function, and increases the final bound. But this way, because the extreme values of $\langle g, t\rangle$ takes place when $\mathbf{B}$ is extreme, i.e., when $\mathbf{B}$ consists of only one number, (7) can be bounded by $|\mathbf{B}|\sqrt{n \log n}$ from above. Here, $|\mathbf{B}|\sqrt{n}$ comes from the variance part, and $\sqrt{\log n}$ comes from the number of extremal choices of multi-sets $\mathbf{B}$.

With everything above, we have

$$\begin{aligned}
|\mathbf{B}| \cdot \frac{k}{2} &\leqslant \sum_{(u_1,u_2,u_3)\in\mathbf{B}} \Big\langle T^{u_1u_2u_3}, A^v \Big\rangle \\
&= \langle T_{V\setminus S}1_{\mathbf{B}}, A^v_{V\setminus S}\rangle + \langle T_S 1_{\mathbf{B}}, A^v_S\rangle \\
&\leqslant \big\|T_{V\setminus S}1_{\mathbf{B}}\big\|_1 \cdot \big\|A^v_{V\setminus S}\big\|_\infty + \langle T_S 1_{\mathbf{B}}, A^v_S\rangle \\
&= \mathcal{O}\big(n\sqrt{|\mathbf{B}|}\big) + \mathcal{O}\big(|\mathbf{B}|\sqrt{n \log n}\big) + \langle T_S 1_{\mathbf{B}}, A^v_S\rangle
\end{aligned}$$

where the last step used $\left\|A^v_{V\setminus S}\right\|_\infty = 1$, and the upper bounds on the two terms (5) and (7) of (4). The term $\langle T_S 1_{\mathbf{B}}, A^v_S\rangle$ has expected value zero and, by Hoeffding's inequalities, deviates away from zero by $\Omega(|\mathbf{B}|\sqrt{n\log n})$ with probability at most $1/n^{10}$ [BBKS24, Lemma 3.2]. So we may assume that with high probability it contributes an amount $\mathcal{O}(|\mathbf{B}|\sqrt{n\log n})$. When $k \geq c\sqrt{n\log n}$ for some large enough absolute constant c, we can conclude from the above that

$$k|\mathbf{B}| \leq \mathcal{O}(n\sqrt{|\mathbf{B}|})$$

and thus $|\mathbf{B}| \leqslant \mathcal{O}(n^2/k^2)$, as desired in ($\diamondsuit$).

5 Putting Everything Together

In this section, we present a formal proof of Theorem 1. We begin with the algorithm.

- Sample $u_1, u_2, u_3 \in V$ randomly.
- Compute $T^{u_1u_2u_3}$ as the entry-wise product of A^{u_1}, A^{u_2}, and A^{u_3}.
- For each $v \in V$, compute the inner product $\langle T^{u_1u_2u_3}, A^v\rangle$. Let $S_{u_1u_2u_3}$ be those $v \in V$ such that the inner product is at least $k/2$.
- Post-process $S_{u_1u_2u_3}$ as described in [BKS23, Claim 5.9].
- Repeat the above steps $\mathcal{O}(n^3/k^3)$ times.
- Prune the $\mathcal{O}(n^3/k^3)$ guesses down to a list of $(1+o(1))n/k$ guesses using [BBKS24, Appendix B].

The validity of the algorithm follows from the argument below:

- Thanks to [BKS23, Claim 5.9], if any $S_{u_1u_2u_3}$ matches the clique S with up to $o(k)$ errors, the post-processing will recover S with high probability.
- Thanks to [BBKS24, Appendix B], we can reduce the number of guesses to the desired $(1+o(1))n/k$ with high probability.
- It remains to show that (a) with high probability, one of our choices of u_1, u_2, u_3 are all in S and (b) when (a) happens, with high probability, $S_{u_1u_2u_3}$ matches S with up to $o(k)$ errors.
- Statement (a) is obvious as the probability that u_1, u_2, u_3 are all in S is k^3/n^3, and we try $\mathcal{O}(n^3/k^3)$ times.
- Statement (b) can be rewritten as follows: for the majority of choices of $(u_1, u_2, u_3) \in S^3$, the number of bad $v \in V \setminus S$ is $o(k)$. By a double counting argument, (b) is a consequence of the diamond statement ($\diamondsuit$), that for any fixed $v \in V \setminus S$, the number of bad $(u_1, u_2, u_3) \in S^3$ (which is denoted $|\mathbf{B}|$) is $\mathcal{O}(n^2/k^2)$.
- Finally, to prove $|\mathbf{B}| \leqslant \mathcal{O}(n^2/k^2)$, follow the last paragraph of Sect. 4.

This finishes the proof of Theorem 1 and concludes our paper.

Acknowledgments. Supported in part by NSF CCF-2211972 and a Simons Investigator award.

References

AKS98. Alon, N., Krivelevich, M., Sudakov, B.: Finding a large hidden clique in a random graph. In: Proceedings of the Ninth Annual ACM-SIAM Symposium on Discrete Algorithms, SODA '98, pp. 594–598, USA, January 1998. Society for Industrial and Applied Mathematics, 1998

BB20. Brennan, M., Bresler, G.: Reducibility and statistical-computational gaps from secret leakage. in: Proceedings of Thirty Third Conference on Learning Theory, pp. 648–847. PMLR, July 2020. https://proceedings.mlr.press/v125/brennan20a.html

BBKS24. Błasiok, J., Buhai, R.D., Kothari, P.K., Steurer, D.: Semirandom planted clique and the restricted isometry property. In: Proceedings of the IEEE Symposium on Foundations of Computer Science, 2024. http://arxiv.org/abs/2404.14159

BKS23. Buhai, R.D., Kothari, P.K., Steurer, D.: Algorithms approaching the threshold for semi-random planted clique. In: Proceedings of the 55th Annual ACM Symposium on Theory of Computing, STOC 2023, pp. 1918–1926, New York, NY, USA, June 2023. Association for Computing Machinery, 2023. https://dl.acm.org/doi/10.1145/3564246.3585184, https://doi.org/10.1145/3564246.3585184

CSV17. Charikar, M., Steinhardt, J., Valiant, G.: Learning from untrusted data. In: Proceedings of the 49th Annual ACM SIGACT Symposium on Theory of Computing, pp. 47–60. ACM, 2017. https://doi.org/10.1145/3055399.3055491

FK01. Feige, U., Kilian, J.: Heuristics for semirandom graph problems. J. Comput. Syst. Sci. **63**(4), 639–671 (2001). https://www.sciencedirect.com/science/article/pii/S0022000001917738, https://doi.org/10.1006/jcss.2001.1773

MMT20. McKenzie, T., Mehta, H., Trevisan, L.: A new algorithm for the robust semi-random independent set problem. In: Proceedings of the 2020 ACM-SIAM Symposium on Discrete Algorithms, pp. 738–746. SIAM, 2020. https://doi.org/10.1137/1.9781611975994.45

Ste18. Steinhardt, J.: Does robustness imply tractability? A lower bound for planted clique in the semi-random model, September 2018. http://arxiv.org/abs/1704.05120, arXiv:1704.05120, https://doi.org/10.48550/arXiv.1704.05120

Woo13. Wootters, M.: On the list decodability of random linear codes with large error rates. In: Proceedings of the ACM Symposium on Theory of Computing, pp. 853–860. ACM, 2013. https://doi.org/10.1145/2488608.2488716

Sequential Testing with Subadditive Costs

Blake Harris, Viswanath Nagarajan, and Rayen Tan(✉)

University of Michigan, Ann Arbor, MI, USA
{blakehar,viswa,rayent}@umich.edu

Abstract. In the classic sequential testing problem, we are given a system with several components, each of which fails with some independent probability. The goal is to identify whether or not some component has failed. When the test costs are additive, it is well known that a greedy algorithm finds an optimal solution. We consider a much more general setting with *subadditive* cost functions and provide a $(4\rho + \gamma)$-approximation algorithm, assuming a γ-approximate value oracle (that computes the cost of any subset) and a ρ-approximate ratio oracle (that finds a subset with minimum ratio of cost to failure probability). While the natural greedy algorithm has a poor approximation ratio in the subadditive case, we show that a suitable truncation achieves the above guarantee. Our analysis is based on a connection to the minimum-sum set cover problem. As applications, we obtain the first approximation algorithms for sequential testing under various cost-structures: $(5 + \epsilon)$-approximation for tree-based costs, 9.5-approximation for routing costs and $(4 + \ln n)$-approximation for machine activation costs. We also show that sequential testing under submodular costs does not admit any polylogarithmic approximation (assuming the exponential time hypothesis).

Keywords: Sequential testing · Approximation algorithms · Stochastic optimization

1 Introduction

Consider a manufacturing facility that needs to check a product for defects. There are n components in the product, each of which is defective with independent probability. There is also a test for each component that incurs some cost: the test "fails" if the component is defective. The goal is to identify whether any of the components has a defect. Sequential testing involves performing tests one by one until either some defect is found or it is verified that there is no defect. The objective is to minimize the *expected* total cost of tests performed. It is well known that the sequence that orders tests in increasing order of cost to failure-probability is optimal [4,19]. In addition to the manufacturing application [9], sequential testing is applicable in healthcare [15] and job-screening [11] settings.

However, in many situations, the cost of testing is not additive (as in the setting above) because one might *batch* tests together and benefit from economies of scale. Motivated by such considerations, there has been some recent work on

N. Megow and A. Basu (Eds.): IPCO 2025, LNCS 15620, pp. 283–296, 2025.
https://doi.org/10.1007/978-3-031-93112-3_21

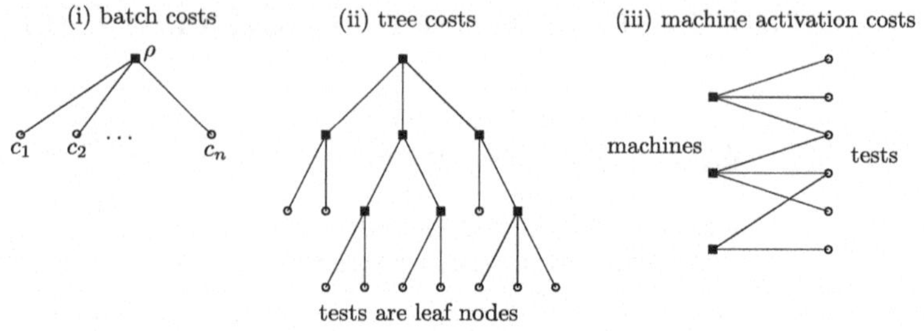

Fig. 1. Examples of subadditive cost structures.

sequential testing with *batch-costs*, where in addition to the individual testing costs (as above) there is a fixed setup cost ρ that is incurred for every batch. It has been observed that the batch-cost problem becomes much harder. [5] obtained an approximation ratio ≈ 6.83 and [21] improved this to a PTAS.

While the batch-cost setting is one way of modeling economies of scale, it is still quite restrictive. Our goal is to address more complex "joint" cost structures in sequential testing. We mention two such examples here (see also Fig. 1).

- *Hierarchical (tree-based) costs.* The components in many systems have a modular structure, where they are arranged hierarchically in "modules". In such cases, testing a component involves removing all modules that the component is contained in. See [17] for an application in aircraft maintenance. This cost structure can be modeled by a tree where each component is a leaf-node and each module is an internal-node. The cost of testing a subset S of components is the total cost of all nodes in the subtree induced by S. The previously-studied batch-cost setting corresponds to a tree of depth one.
- *Machine activation costs.* Suppose that the tests need to be performed on a set of machines, and each test has a specific subset of machines that can perform it. If a machine is "activated" then it can perform all the tests that are allowed on it. The cost of performing some subset of tests is the minimum number of machines that need to be activated for these tests.

In this paper, we address such problems systematically by considering sequential testing problems where the cost-structure is a *subadditive* function. We provide a generic approximation algorithm for subadditive sequential testing (SST) where the approximation ratio depends on certain properties of the cost-function. We also provide several applications of our result, which include tree-based and machine activation costs mentioned above. We obtain the first approximation algorithms for these complex cost-structures. We also show that some natural cost-structures cannot be approximated well. In particular, submodular-cost sequential testing is harder to approximate than the densest k-subgraph problem [18], which essentially rules out any sub-polynomial approximation ratio (under the exponential time hypothesis).

1.1 Problem Definition

This paper deals with *series systems*: systems that only function when all of its constituent components work. There is a test associated with every component and the goal is to diagnose the system by identifying a defective component (if any). Observe that it is not necessary to conduct all tests. It suffices to find one component (test) that fails, in which case we know the system is defective, regardless of the outcome of other tests. Testing is conducted until a failure is found (whereby we declare the system defective) or until all components are tested and deemed to work (whereby we declare the system working). The objective is to minimize the expected cost of testing.

Formally, an instance of sequential batch testing consists of n tests, given by $[n] := \{1, \dots, n\}$. Each test has a known independent probability p_i of passing. We let $q_i = 1 - p_i$ be the probability of failure for each $i \in [n]$.[1] We represent the outcome of test i as a random variable $X_i \in \{0, 1\}$ where $X_i = 1$ if tests i fails and $X_i = 0$ if test i passes.

Cost Structure. We consider a very general cost structure that allows us to model several applications involving batched tests. In particular, there is a family $\mathcal{F} \subseteq 2^{[n]}$ of subsets that represent allowed batches of tests. Each batch $B \in \mathcal{F}$ also has a non-negative cost c_B. This means that all tests in B can be performed simultaneously at cost c_B. Without loss of generality, the family $\mathcal{F}$ is "downward closed", i.e., if $B \in \mathcal{F}$ and $B' \subseteq B$ then $B' \in \mathcal{F}$. We note that the family $\mathcal{F}$ and its costs may be specified implicitly: so $|\mathcal{F}|$ may be exponentially large. It will be convenient to work with a cost-function $c : 2^{[n]} \to \mathbb{R}_+$ representing the cost to perform *any* subset S of tests, i.e.,

$$c(S) = \min \left\{ \sum_{B \in \mathcal{B}} c_B : S \subseteq \cup_{B \in \mathcal{B}} B \,,\, \mathcal{B} \subseteq \mathcal{F} \right\}, \quad \forall S \subseteq [n].$$

The cost function $c(S)$ is the minimum cost of a collection of batches that "covers" S: so this is defined for every subset S. Moreover, the cost-function c is monotone and subadditive. Recall that function c is *subadditive* if $c(A) + c(B) \geq c(A \cup B)$ for all $A, B \subseteq [n]$.

Note that for any allowed batch $B \in \mathcal{F}$ we have $c(B) \leq c_B$. Moreover, we can assume (without loss of generality) that $c(B) = c_B$ for all $B \in \mathcal{F}$. (If $c(B) < c_B$ then we just re-define the cost $c_B = c(B)$, which results in an equivalent instance.)

A solution to the *subadditive series testing* (SST) problem is given by a sequence $\mathcal{B} = \langle B_1, \dots, B_k \rangle$ of batches that form a partition of $[n]$. For any batch B_j let $P(B_j) := \prod_{i \in B_j} p_i$ denote the probability that all tests in B_j pass. Solution $\mathcal{B}$ performs batch B_j if and only if all tests in the preceding batches $B_1, \dots, B_{j-1}$ pass. So, the probability that B_j is performed is $\prod_{\ell=1}^{j-1} P(B_\ell)$ and

[1] We assume that all the probabilities are provided as rational numbers and the instance size is $n \log L$ where L is the largest integer in the rational representation.

the expected cost of the solution, $\text{Cost}(\mathcal{B}) = \sum_{j=1}^{k} \prod_{\ell=1}^{j-1} P(B_\ell) \cdot c(B_j)$. The objective is to find a sequence $\mathcal{B}$ to minimize $\text{Cost}(\mathcal{B})$.

We note that for most stochastic optimization problems, there is a distinction between adaptive solutions (where the observed outcomes influence subsequent decisions), and nonadaptive solutions (where a sequence of decisions is fixed a priori) [1,7,8,13]. However, for SST, the adaptivity gap is 1, so we focus on the nonadaptive case.

Some special cases of SST are listed below.

- In the classic series-testing problem, costs are additive, i.e., $c(S) = \sum_{i \in S} c_i$ where c_i is the individual cost for test i.
- In the batch series-testing problem, there is a setup cost ρ in addition to the individual costs $\{c_i\}_{i=1}^n$ and $c(S) = \rho + \sum_{i \in S} c_i$.
- In the tree-cost series-testing problem, there is a hierarchical structure on the tests given by a rooted tree with leaves corresponding to the n tests. Each node in the tree has a weight. For any subset S, the cost $c(S)$ equals the total weight of all nodes in the subtree containing the leaves in S.

We will discuss more applications of SST in Sect. 3.

Oracles for the Cost Function. Our algorithm relies on two oracles for (approximately) solving certain optimization problems related to the cost function c. The first oracle is a "value" oracle, which may itself be non-trivial as the cost structure is specified implicitly.

Definition 1 *(γ-**approximate value oracle**). Given any subset $S \subseteq [n]$ this oracle returns a collection $\mathcal{B} \subseteq \mathcal{F}$ of batches that covers S with total cost $\sum_{B \in \mathcal{B}} c_B \le \gamma \cdot c(S)$.*

In other words, we assume that there is a polynomial-time algorithm to obtain a γ-approximation to the cost of any subset. The second oracle corresponds to the natural step in designing greedy algorithms for SST.

Definition 2 *(ρ -**approximate ratio oracle**). Given any subset $U \subseteq [n]$ this oracle returns a batch B^* that ρ-approximately minimizes the following ratio problem:*

$$\min_{B \in \mathcal{F}, B \subseteq U} \quad \frac{c_B}{1 - P(B)}, \tag{1}$$

where $P(B) = \prod_{i \in B} p_i$ is the probability that all tests in B pass. In addition to the batch B^*, we assume that the oracle also returns an upper-bound C^* on the cost c_{B^*} such that $\frac{C^*}{1-P(B^*)}$ is at most ρ times the minimum ratio in (1).

Note that $1 - P(B)$ is exactly the probability that testing will stop after batch B: so this can be viewed as the "benefit" of selecting batch B, and the above ratio minimizes the cost-to-benefit ratio over all batches. We could also have defined the ratio in (1) by minimizing $\frac{c(B)}{1-P(B)}$ over *all* subsets $B \subseteq U$ (not just those in $\mathcal{F}$), which turns out to be equivalent. Finally, the additional assumption that the ratio oracle returns an upper-bound C^* on the cost is a mild assumption: this holds in all our applications (and is a direct consequence of the ρ-approximation to the min-ratio value).

1.2 Results and Techniques

We present a generic approximation algorithm for SST that relies on the approximate value and ratio oracles (Definitions 1 and 2).

Theorem 1. *There is a* $(4\rho + \gamma)$*-approximation algorithm for the subadditive sequential testing problem whenever there is a* γ*-approximate value oracle and a* ρ*-approximate ratio oracle.*

Our algorithm uses the greedy approach that iteratively selects batches with the minimum cost-to-benefit ratio. In fact, this "ratio" algorithm was already proposed in [6] where it was evaluated computationally on a special case of SST. We first show that the approximation ratio of this simple greedy algorithm is $\Omega(\sqrt{n})$ even when $\gamma = \rho = 1$. In order to bypass such bad examples, we modify the greedy algorithm by considering all possible "truncation" points where instead of continuing with the greedy algorithm we just perform all remaining tests in a final batch. Roughly speaking, our SST solution is the truncated greedy solution that has the minimum expected cost. In addition to the two oracles, our analysis relies on a connection to the *minimum-sum set cover* problem [10], for which a 4-approximation algorithm is known.

Next, we show that several specific cost-functions satisfy our assumptions, leading to good approximation algorithms for SST under these cost structures. Although the ratio problem (1) has a non-linear term in the denominator, we show that it can be linearized at the loss of a small approximation factor, which then allows us to use existing algorithms for "quota" versions of the respective covering problems. Some of our applications are:

- Hierarchical (tree) costs: we show that there is an FPTAS for the ratio oracle, which leads to a $(5 + \epsilon)$-approximation algorithm for SST.
- Routing costs: each test corresponds to a node in a metric and $c(S)$ is the minimum cost of a route that visits all nodes in S from a root. We obtain a 9.5-approximation algorithm for SST using the TSP as the value oracle and the k-TSP problem [12] as the ratio oracle.
- Machine activation costs: each test has an "allowed" subset of machines that may perform the test, and once a machine is activated it can perform all the tests that are allowed on it. $c(S)$ is the minimum number (or cost) of machines to "activate" to perform all tests in S. We obtain a $(4 + \ln n)$-approximation algorithm for SST using set-cover as the value oracle.

We also consider SST when the cost function is *submodular*, which is a natural and well-structured class of subadditive functions. Interestingly, this problem turns out to be very hard to approximate. By establishing a relation to the densest k-subgraph problem [18], we prove that the submodular-cost series testing problem cannot be approximated to a factor better than $n^{1/\operatorname{poly}(\log\log n)}$ (assuming the exponential time hypothesis).[2] The submodular function used in

[2] The exponential time hypothesis states that there is no $2^{o(n)}$ time algorithm for 3-SAT.

the hard instance is just a coverage function. For this instance, there is a $\gamma = 1$ approximate oracle for the value oracle and the hardness comes from the ratio oracle.

1.3 Related Work

Sequential testing has been applied to many settings as mentioned earlier, and the greedy algorithm is known to be optimal for additive costs [4,19]. The extension to batched cost structures was proposed by [6], where the authors obtained some efficient heuristics (without approximation guarantees). Later, [5] obtained a ≈ 6.28-approximation algorithm for batched cost. This bound was subsequently improved to a PTAS by [21]. To the best of our knowledge, ours is the first paper to consider more complex cost-structures.

The minimum-sum set cover problem (that is used in our analysis) is a central problem in approximation algorithms: it is well-known that a natural greedy algorithm is a 4-approximation [2,10] and that this ratio cannot be improved unless P = NP.

Sequential testing problems have also been considered for other systems (beyond series systems). Some examples are k-of-n systems (which determines if at least k components work) [3], linear threshold functions [8] and score classification (which involves classifying the system based on the number of defects) [13,14]. See also [20] and [23] for surveys. All these results involve the simple additive cost structure. A recent paper [22] obtains constant-factor approximation algorithms under batch-costs (with a fixed setup cost ρ and unconstrained batches) for many of these systems. However, this result relies heavily on this particular cost-structure and does not extend to the subadditive setting.

1.4 Preliminaries on Minimum-Sum Set Cover

Our algorithm relies on a connection to the well-known *minimum-sum set cover* (MSSC) problem. An instance of MSSC consists of a set E of elements with weights $\{w_e\}_{e\in E}$ and M subsets $\{S_i \subseteq E\}_{i=1}^{M}$ with costs $\{c_i\}_{i=1}^{M}$. An MSSC solution is a permutation $\sigma = \langle\sigma(1), \sigma(2), \ldots, \sigma(M)\rangle$ of the M sets. Given solution σ, the *cover-time* of any element $e \in E$, denoted $\mathrm{Cov}(\sigma, e)$, is the cost of the smallest prefix of σ that covers e. That is, if $e \in S_{\sigma(j)} \setminus \left(S_{\sigma(1)} \cup \cdots \cup S_{\sigma(j-1)}\right)$ then $\mathrm{Cov}(\sigma, e) = c_{\sigma(1)} + \cdots + c_{\sigma(j)}$. The objective in MSSC is to minimize the total weighted cover time $\sum_{e\in E} w_e \cdot \mathrm{Cov}(\sigma, e)$.

The greedy algorithm for MSSC works as follows. If R denotes the set of uncovered elements (initially $R = E$) then we select the set S_i that minimizes the ratio $\frac{c_i}{\sum_{e\in S_i\cap R} w_e}$.

We will use the (known) result that this algorithm is a 4-approximation [10]. In fact, we need a more robust version that allows for approximate greedy choices, which also follows from prior work.

2 Algorithm

The simple greedy algorithm iteratively chooses the batch $B \in \mathcal{F}$ (among the remaining tests) that minimizes the ratio $\frac{c_B}{1-P(B)}$. This is known to be optimal in the special case of additive testing costs. Moreover, this algorithm was proposed as a heuristic even for the batched setting in [6]: there was no approximation bound known for it. We first show that this greedy algorithm has approximation ratio $\Omega(\sqrt{n})$ for SST. Then, we present a modified greedy algorithm that truncates the greedy solution suitably, which leads to Theorem 1.

Bad Instance for Greedy. There are n tests with cost function $c(S) = \min(|S|, \sqrt{n})$, which is *subadditive.* Each test $i \in [n]$ has failure probability $q_i := \frac{1}{2^{i+1}}$. Recall that $p_i = 1 - q_i$; so $p_1 \leq p_2 \leq \cdots \leq p_n$.

We will show that the greedy solution performs tests in singleton batches, i.e., $\langle \{1\}, \{2\}, \ldots, \{n\} \rangle$. To this end, we show the following:

$$\text{For any } j \geq 1\text{, the min-ratio batch in } \{j, j+1, \ldots, n\} \text{ is } \{j\}. \tag{2}$$

To see this, let $U = \{j, j+1, \ldots, n\}$ denote the remaining tests. Fix any value $1 \leq k \leq n-j+1$ and consider all batches $B \subseteq U$ with $|B| = k$: it is clear that the minimum ratio among these batches is achieved by $B_k := \{j, j+1, \ldots, j+k-1\}$ because these are the tests with minimum p_is.

We now have:

$$\frac{1}{2^{j+1}} = q_j \leq \Pr[\text{some test in } B_k \text{ fails}] \leq q_j + \cdots + q_{j+k-1} < \frac{2}{2^{j+1}}.$$

So, the ratio of B_k is $\frac{c(B_k)}{1-P(B_k)} > 2^j \cdot \min(k, \sqrt{n})$. Moreover, the ratio of $B_1 = \{j\}$ is exactly 2^{j+1}. It now follows that $\{j\}$ minimizes the ratio, proving (2).

We now show that the expected cost of the greedy solution is at least $\frac{n}{2}$. Indeed, the probability that *all* tests pass is at least $1 - \sum_{i=1}^{n} q_i \geq \frac{1}{2}$: under this event the solution will have to incur cost of $\sum_{i=1}^{n} c(\{i\}) = n$.

By performing all tests in a single batch, we get an expected cost of $\min\{n, \sqrt{n}\} = \sqrt{n}$. So the optimal cost is at most $\sqrt{n}$. This implies that the greedy algorithm has approximation ratio at least $\frac{n/2}{\sqrt{n}} = \frac{\sqrt{n}}{2}$.

Our Algorithm. At a high level, the reason that greedy fails is that it does not utilize the property that testing *all* remaining components necessarily diagnoses the system (irrespective of the probabilities of failure). This motivates our modified greedy algorithm that considers all possible "truncation" points, where all remaining tests are performed in one final batch. The final solution is obtained by choosing the truncated greedy solution of minimum expected cost (see Algorithm 1).

Observe that our algorithm relies on both the ratio oracle (Definition 2) and the value oracle (Definition 1). We will show that this algorithm achieves a

Algorithm 1. Modified Greedy Algorithm

set of remaining tests $U \leftarrow [n]$; number of batches $\ell \leftarrow 0$
while $U \neq \varnothing$ **do**
    use the ratio-oracle to pick the batch $B_{\ell+1}$ that ρ-approximately minimizes

$$\min_{B \subseteq U, B \in \mathcal{F}} \quad \frac{c(B)}{1 - P(B)},$$

    along with an upper-bound $C_{\ell+1}$ on the cost $c(B_{\ell+1})$.
    $U \leftarrow U \setminus B_{\ell+1}$ and $\ell \leftarrow \ell + 1$.
end while
let $\pi = \langle B_1, \ldots, B_\ell \rangle$ be the greedy solution.
for $k = 0, 1, \ldots, \ell$ **do**
    let π_k be the truncated solution $\langle B_1, \ldots, B_k, [n] \setminus \cup_{j=1}^{k} B_j \rangle$.
    use the value-oracle to obtain a γ-approximate value D_k for batch $[n] \setminus \cup_{j=1}^{k} B_j$.
    define an upper-bound on the expected cost of solution π_k as

$$G_k = \sum_{j=1}^{k} P(\cup_{h=1}^{j-1} B_h) \cdot C_j + P(\cup_{h=1}^{k} B_h) \cdot D_k.$$

end for
Return the solution from $\{\pi_0, \pi_1, \ldots, \pi_\ell\}$ with minimum upper-bound $\min_{k=0}^{\ell} G_k$.

$(4\rho + \gamma)$ approximation, which would prove Theorem 1. The analysis proceeds in two steps. First, we bound the cost of the greedy solution $\pi = \langle B_1, B_2, \ldots, B_\ell \rangle$ by relating it to an MSSC instance. Next, we show that there exists a truncation of the greedy solution that achieves a good approximation ratio.

Minimum-Sum Set Cover Instance. Given any instance of SST, we create an instance of MSSC as follows:

- The elements are all non-zero vectors in $\{0,1\}^n$. Each element $x \in \{0,1\}^n$ represents a *realization* of all tests where test i has outcome x_i. The *weight* of any element x is $w_x = \prod_{i:x_i=1} q_i \cdot \prod_{i:x_i=0}(1 - q_i) = \Pr[X_i = x_i, \ \forall i \in [n]]$, the probability of realization x. Note that we exclude the all-zero realization (where all tests pass).
- There is a set S_B for each batch of tests $B \subseteq [n]$ in SST, where $S_B = \{x \in \{0,1\}^n : \exists i \in B \, x_i = 1\}$, which are all realizations that have a "fail" outcome for some test in B. The cost of set S_B is just the cost $c(B)$ of the corresponding batch.

Although this MSSC instance has an exponential number of elements and sets, it is not an issue because we only use this view in the analysis.

Recall that a solution to MSSC is a permutation σ of the sets, and $\text{Cov}(\sigma, x)$ denotes the cover time of any element x. We note that the solution σ need not be a full permutation: σ can be any sequence of sets such that every element is covered. The MSSC objective is $\sum_{x \neq \mathbf{0}} w_x \cdot \text{Cov}(\sigma, x)$. The greedy algorithm for

MSSC involves always selecting the set S_B that minimizes the ratio: $\frac{c(B)}{\sum_{x \in S_B \cap R} w_x}$, where R is the current set of un-covered elements. We will use the following result on the MSSC algorithm with approximate greedy choices.

Theorem 2 *([10]). The* MSSC *algorithm that always selects a set which is a ρ-approximation to the greedy criterion has approximation ratio 4ρ.*

We now observe that the greedy sequence $\mathsf{GRD} = \langle B_1, B_2, \ldots, B_\ell \rangle$ constructed in Steps 2–5 is exactly a ρ-approximate greedy solution to the above MSSC instance. Consider an iteration in the algorithm when $U \subseteq [n]$ is the set of remaining tests. Irrespective of the exact batches chosen so far, the elements in the MSSC instance that are still un-covered are $R = \{x \in \{0,1\}^n \setminus \mathbf{0} : x_j = 0\ \forall j \in [n] \setminus U\}$. For any batch $B \subseteq U$, the new elements covered would then be:

$$S_B \cap R = \{x \in \{0,1\}^n \ : \ x_j = 0\ \forall j \in [n] \setminus U \text{ and } \exists i \in B : x_i = 1\}$$

By definition of the weights w_x and the fact that the outcomes X_i are independent, we have

$$\sum_{x \in S_B \cap R} w_x = \Pr[X_j = 0\ \forall j \in [n] \backslash U] \cdot \Pr[\exists i \in B : X_i = 1] = P([n] \backslash U) \cdot (1 - P(B)),$$

where we used the definition of $P(B) = \Pr[X_i = 0\ \forall i \in B]$. Therefore, the greedy criterion for MSSC is:

$$\min_{B \subseteq U} \frac{c(B)}{P([n] \setminus U) \cdot (1 - P(B))} = \frac{1}{P([n] \setminus U)} \cdot \min_{B \subseteq U} \frac{c(B)}{1 - P(B)},$$

where the equality is because the term $P([n] \setminus U)$ is a fixed value (not dependent on B). So the greedy MSSC criterion is just a scaled version of our greedy criterion in Step 3. It follows that the greedy solution π in Step 6 is a ρ-approximate greedy solution to this MSSC instance.

Relating SST and MSSC Objectives. For any $x \in \{0,1\}^n$, let $\mathrm{Cov}'(\pi, x) = \sum_{j=1}^{k} C_j$ where B_k is the first batch that covers x. (For $x = \mathbf{0}$ which is never covered, we set $\mathrm{Cov}'(\pi, \mathbf{0}) = \sum_{j=1}^{\ell} C_j$.) Note that $\mathrm{Cov}'(\pi, x)$ is an upper-bound on the actual cover time $\mathrm{Cov}(\pi, x)$ as each C_j is an upper-bound on the cost of batch B_j. Viewing the solution π as a ρ-approximate greedy solution to MSSC and using Theorem 2, we have:

Lemma 1. *If π^* is an optimal solution to the* SST *instance then*

$$\sum_{\substack{x \in \{0,1\}^n \\ x \neq \mathbf{0}}} w_x \cdot \mathrm{Cov}'(\pi, x) \leq 4\rho \cdot \sum_{\substack{x \in \{0,1\}^n \\ x \neq \mathbf{0}}} w_x \cdot \mathrm{Cov}(\pi^*, x).$$

The objective of any solution σ to SST can be written as

$$\mathbb{E}[\mathrm{cost}(\sigma)] = \sum_{\substack{x \in \{0,1\}^n \\ x \neq \mathbf{0}}} w_x \cdot \mathrm{Cov}(\sigma, x) + w_{\mathbf{0}} \cdot \mathrm{Cov}(\sigma, \mathbf{0}), \tag{3}$$

where we explicitly separate the term for the all-zero realization $\mathbf{0}$, to highlight that MSSC does not account for it. Note that $\mathrm{Cov}(\sigma, \mathbf{0})$ is the maximum cost of running solution σ, which occurs when all tests pass.

Truncating the Greedy Solution. We define $C^* := \mathrm{Cov}(\pi^*, \mathbf{0})$ to be the maximum cost in any run of the optimal solution π^*. By subadditivity of the cost function, we have $C^* \geq c([n])$. Let $\alpha > 0$ be some parameter that will be set later. We now define a truncated greedy solution as $\pi_r = \langle B_1, \ldots, B_r, [n] \setminus \cup_{j=1}^r B_j \rangle$, where r is the maximum index such that $\sum_{j=1}^r C_r \leq \alpha C^*$. See Fig. 2. We will show that the upper-bound on $\mathbb{E}[\mathrm{cost}(\pi_r)]$ is $G_r \leq (4\rho + \gamma) \cdot \mathbb{E}[\mathrm{cost}(\pi^*)]$. This would complete the proof of Theorem 1.

Below, we work with the cost upper-bounds C_j for each batch B_j (for $j \in [\ell]$) and D_r for the last batch $[n] \setminus \cup_{j=1}^r B_j$ in π_r. Recall that $\mathrm{Cov}'(\pi, x)$ is the corresponding upper-bound on the cover-time of any $x \in \{0,1\}^n$ for the greedy solution π. Similarly, we define $\mathrm{Cov}'(\pi_r, x)$ to be the upper-bound on the cover-time of any $x \in \{0,1\}^n$ in the truncated solution π_r.

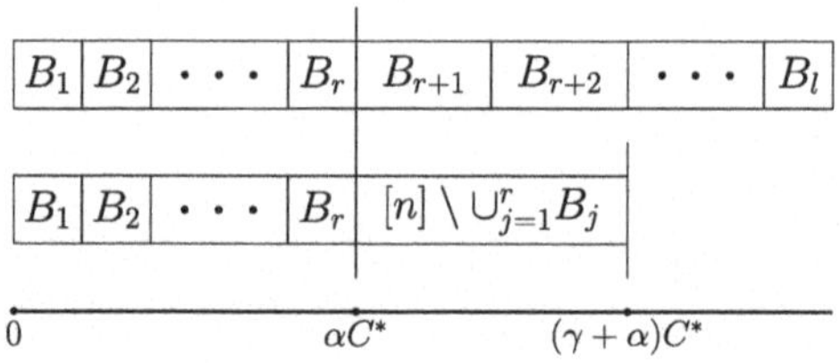

Fig. 2. Ordering of greedy solution (Top) and the truncated solution (Bottom).

Lemma 2. *For each* $x \in \{0,1\}^n$, $\mathrm{Cov}'(\pi_r, x) \leq \left(1 + \frac{\gamma}{\alpha}\right) \cdot \mathrm{Cov}'(\pi, x)$.

Proof. First, suppose that $\mathrm{Cov}'(\pi, x) \leq \alpha C^*$. Then we have $\mathrm{Cov}'(\pi_r, x) = \mathrm{Cov}'(\pi, x)$ as x gets covered at the same point in both π and π_r.

Next, we claim that $D_r \leq \gamma \cdot C^*$. Indeed, by monotonicity and subadditivity of the cost-function, we have $c([n] \setminus \cup_{j=1}^r B_j) \leq c([n]) \leq C^*$. So, using the γ-approximate value oracle, we get $D_r \leq \gamma \cdot C^*$.

Now suppose that $\mathrm{Cov}'(\pi, x) > \alpha C^*$. We have

$$\mathrm{Cov}'(\pi_r, x) \leq \mathrm{Cov}'(\pi_r, \mathbf{0}) = \sum_{j=1}^{r} C_j + D_r \leq \alpha C^* + \gamma C^* < \left(1 + \frac{\gamma}{\alpha}\right) \mathrm{Cov}'(\pi, x),$$

which completes the proof. □

We are now ready to bound G_r.

$$\begin{aligned}
G_r &= \sum_{\substack{x\in\{0,1\}^n \\ x\neq \mathbf{0}}} w_x \cdot \mathrm{Cov}'(\pi_r, x) + w_{\mathbf{0}} \cdot \mathrm{Cov}'(\pi_r, \mathbf{0}) \\
&\leq \sum_{\substack{x\in\{0,1\}^n \\ x\neq \mathbf{0}}} w_x \cdot \mathrm{Cov}'(\pi_r, x) + w_{\mathbf{0}} \cdot (\alpha+\gamma) C^* && (4)\\
&\leq \left(1+\frac{\gamma}{\alpha}\right) \sum_{\substack{x\in\{0,1\}^n \\ x\neq \mathbf{0}}} w_x \cdot \mathrm{Cov}'(\pi, x) + w_{\mathbf{0}} \cdot (\alpha+\gamma) C^* && (5)\\
&\leq 4\rho\left(1+\frac{\gamma}{\alpha}\right) \sum_{\substack{x\in\{0,1\}^n \\ x\neq \mathbf{0}}} w_x \cdot \mathrm{Cov}(\pi^*, x) + w_{\mathbf{0}} \cdot (\alpha+\gamma) C^* && (6)\\
&= 4\rho\left(1+\frac{\gamma}{\alpha}\right) \sum_{\substack{x\in\{0,1\}^n \\ x\neq \mathbf{0}}} w_x \cdot \mathrm{Cov}(\pi^*, x) + w_{\mathbf{0}} \cdot (\alpha+\gamma) \,\mathrm{Cov}(\pi^*, \mathbf{0}) && (7)\\
&\leq \max\left\{\frac{4\rho}{\alpha}(\alpha+\gamma),\, \alpha+\gamma\right\} \cdot \sum_{x\in\{0,1\}^n} w_x \cdot \mathrm{Cov}(\pi^*, x) \\
&= \max\left\{\frac{4\rho}{\alpha}(\alpha+\gamma),\, \alpha+\gamma\right\} \cdot \mathbb{E}[\mathrm{cost}(\pi^*)]
\end{aligned}$$

Above, (4) uses the fact that $\mathrm{Cov}'(\pi_r, \mathbf{0}) \leq \alpha C^* + \gamma C^*$, (5) uses Lemma 2, (6) uses Lemma 1 and (7) uses the definition of C^*. Setting $\alpha = 4\rho$, we obtain $G_r \leq (4\rho+\gamma)\cdot \mathbb{E}[\mathrm{cost}(\pi^*)]$ as claimed.

2.1 The Ratio Oracle

Here, we show how we can transform the ratio problem into a simpler constrained optimization problem known as the "quota problem". This turns out to be very useful in our applications. Recall that the ratio oracle requires finding the batch B that minimizes $\frac{c(B)}{1-P(B)}$. Towards simplifying the ratio problem, we define $r_i = -\log p_i$ for all $i \in [n]$ and let $r(S) = \sum_{i\in S} r_i$. Moreover, define the function $d(Q) = 1 - e^{-Q}$, which is monotone and concave. This allows us to write the ratio as $\mathrm{ratio}(B) = \frac{c(B)}{d(r(B))}$. Using the above properties of function d, we can reduce the ratio problem to a linear-constrained optimization problem.

Minimum Cost Subject to Quota (QP). Given a subadditive cost function $c : 2^{[n]} \to \mathbb{R}_+$ and non-negative rewards $\{r_i\}_{i=1}^n$, and a quota Q the goal is to find a subset $S \subseteq [n]$ of minimum cost such that the total reward is at least Q, i.e., $\min\left\{c(S) : S \subseteq [n],\, \sum_{i\in S} r_i \geq Q\right\}$.

Compared to the ratio oracle (which has a non-linear term in the objective), the quota problem has a familiar knapsack-type structure found in optimization problems. We rely on a bicriteria approximation for QP:

Definition 3. *An (α, β)-bicriteria approximation algorithm for* QP *finds a subset $\hat{S}$ such that the cost $c(\hat{S}) \leq \alpha \cdot c(S^*)$ and $r(\hat{S}) \geq Q/\beta$, where S^* denotes the optimal* QP *solution.*

The next lemma (proved in the full version of the paper [16]) shows how such an approximation for QP can be used for the ratio oracle.

Lemma 3. *If there is an (α, β)-bicriteria approximation for* QP*, then there is a $((1+\epsilon)\alpha\beta)$-approximation for the ratio oracle for any fixed parameter $\epsilon > 0$.*

3 Applications

In this section, we present applications of our results. These instances are special cases of SST that have good approximations for the value oracle and the ratio oracle. In each of these cases, we summarize the approximation ratio for the ratio and value oracle. In some cases, the ratio oracles require first finding an (α, β)-bicriteria approximation for QP (Defnition 3), then deriving the ratio oracle using Lemma 3. The approximation ratio for SST is then obtained using Theorem 1. A summary of our applications is presented in Table 1. We provide details for a setting where our cost function is concave. Details of the rest of the applications are found in the full paper [16].

Table 1. Approximation Ratio under different cost structure.

Cost Structure	ρ	γ	Approximation Ratio
Concave cardinality	1	1	5
Tree	$1+\epsilon$	1	$5+\epsilon$
Tree with batch capacity	$1+\epsilon$	$1+\epsilon$	$5+\epsilon$
Machine Activation	1	$\ln n$	$4+\ln n$
Routing (General Metric)	$2+\epsilon$	$1.5-\epsilon$	9.5
Routing (Euclidean Metric)	$1+\epsilon$	1	$5+\epsilon$

3.1 Concave Cardinality Costs

We consider settings where the cost of testing depends only on the *cardinality* of the batch (i.e., number of tests conducted), and the cost structure exhibits economies of scale. Formally, there is a monotone increasing and concave function $g : \mathbb{R} \to \mathbb{R}$ such that the cost of testing any subset $S \subseteq [n]$ is given by $c(S) = g(|S|)$. For this problem, we will show that the two oracles admit ratios $\gamma = \rho = 1$.

Value Oracle. As subset S is a valid batch, computing the cost just involves evaluating the function g. This sets $\gamma = 1$.

Ratio Oracle. We can solve the ratio oracle optimally by renumbering the tests by increasing probabilities $p_1 \leq p_2 \leq \cdots \leq p_n$. The algorithm then picks the prefix $B_i = \{1, 2, \ldots, i\}$ that minimizes the ratio $\frac{c(B_i)}{1-P(B_i)}$. Enumerating all $i = \{1, 2, \ldots, n\}$ only takes $\mathcal{O}(n)$ time. It is easy to see that the optimal batch is among the prefix set $\{B_i\}_{i=1}^n$. Indeed, as all batches of cardinality i have the same cost $g(i)$, testing the size i batch with the minimum p_is minimizes the ratio.

By Theorem 1, this gives us a 5-approximation.

It is worth noting that the bad instance for the natural greedy algorithm (in Sect. 2) involves concave cardinality costs. Our modified greedy algorithm achieves a constant approximation ratio in this setting: so it overcomes the limitation of the natural greedy algorithm. We note however that SST with concave cardinality costs can be solved exactly with a dynamic programming algorithm. The dynamic program relies on the symmetry in the cost structure, which implies that the batched solution is an ordered partition of the tests sorted by increasing probabilities p_i.

4 Hardness for Submodular Costs

Theorem 3 states the hardness result for approximating sequential testing with submodular costs (which is a natural special case of subadditive costs).

Theorem 3. *Assuming the exponential time hypothesis (ETH), there is no* $n^{1/\operatorname{poly}(\log\log n)}$*-approximation algorithm for submodular-cost sequential testing.*

In particular, this rules out any poly-logarithmic approximation ratio for SST in the special case of submodular costs (assuming ETH).

The setting for the hard instance consists of n tests, each of which requires multiple machines to be activated. Formally, there are m machines and each test $i \in [n]$ is associated with a subset $M_i \subseteq [m]$ of machines that need to be activated. For any batch of tests $B \subseteq [n]$, the cost $c(B) = |\bigcup_{i\in B} M_i|$ the number of machines required to conduct every test in B. Observe that c is a coverage function: so it is monotone and submodular.

In the full paper [16], we show Theorem 3 via a reduction from the densest k-subgraph problem.

Acknowledgments. Research supported in part by NSF grant CCF-2418495.

References

1. Bansal, N., Gupta, A., Li, J., Mestre, J., Nagarajan, V., Rudra, A.: When LP is the cure for your matching woes: improved bounds for stochastic matchings. Algorithmica **63**(4), 733–762 (2012)
2. Bar-Noy, A., Bellare, M., Halldórsson, M.M., Shachnai, H., Tamir, T.: On chromatic sums and distributed resource allocation. Inf. Comput. **140**(2), 183–202 (1998)

3. Ben-Dov, Y.: Optimal testing procedures for special structures of coherent systems. Manage. Sci. **27**(12), 1410–1420 (1981)
4. Butterworth, R.: Some reliability fault-testing models. Oper. Res. **20**(2), 335–343 (1972)
5. Daldal, R., Gamzu, I., Segev, D., Ünlüyurt, T.: Approximation algorithms for sequential batch-testing of series systems. Nav. Res. Logist. **63**(4), 275–286 (2016)
6. Daldal, R., Özlük, O., Selçuk, B., Shahmoradi, Z., Ünlüyurt, T.: Sequential testing in batches. Ann. Oper. Res. **253**(1), 97–116 (2017)
7. Dean, B.C., Goemans, M.X., Vondrák, J.: Approximating the stochastic knapsack problem: the benefit of adaptivity. Math. Oper. Res. **33**(4), 945–964 (2008)
8. Deshpande, A., Hellerstein, L., Kletenik, D.: Approximation algorithms for stochastic submodular set cover with applications to Boolean function evaluation and min-knapsack. ACM Trans. Algorithms **12**(3), 1–28 (2016)
9. Duffuaa, S., Raouf, A.: An optimal sequence in multicharacteristics inspection. J. Optim. Theory Appl. **67**, 79–86 (1990)
10. Feige, U., Lovász, L., Tetali, P.: Approximating min sum set cover. Algorithmica **40**, 219–234 (2004)
11. Garey, M.: Optimal task sequencing with precedence constraints. Discret. Math. **4**(1), 37–56 (1973)
12. Garg, N.: Saving an epsilon: a 2-approximation for the k-MST problem in graphs. In: Proceedings of the Thirty-seventh Annual ACM Symposium on Theory of Computing, pp. 396–402 (2005)
13. Ghuge, R., Gupta, A., Nagarajan, V.: Non-adaptive stochastic score classification and explainable halfspace evaluation. In: Integer Programming and Combinatorial Optimization - 23rd International Conference, pp. 277–290. Springer (2022)
14. Gkenosis, D., Grammel, N., Hellerstein, L., Kletenik, D.: The stochastic score classification problem. In: 26th Annual European Symposium on Algorithms. LIPIcs, vol. 112, pp. 1–14 (2018)
15. Greiner, R., Hayward, R., Jankowska, M., Molloy, M.: Finding optimal satisficing strategies for and-or trees. Artif. Intell. **170**(1), 19–58 (2006)
16. Harris, B., Nagarajan, V., Tan, R.: Sequential testing with subadditive costs (2025). https://arxiv.org/abs/2501.18010
17. Levi, R., Magnanti, T., Muckstadt, J., Segev, D., Zarybnisky, E.: Maintenance scheduling for modular systems: modeling and algorithms. Naval Res. Logistics (NRL) **61**(6), 472–488 (2014)
18. Manurangsi, P.: Almost-polynomial ratio eth-hardness of approximating densest k-subgraph. In: Proceedings of the 49th Annual ACM SIGACT Symposium on Theory of Computing, pp. 954–961 (2017)
19. Mitten, L.: An analytic solution to the least cost testing sequence problem. J. Ind. Eng. **11**(1), 17 (1960)
20. Moret, B.: Decision trees and diagrams. ACM Comput. Surv. **14**(4), 593–623 (1982)
21. Segev, D., Shaposhnik, Y.: A polynomial-time approximation scheme for sequential batch testing of series systems. Oper. Res. **70**(2), 1153–1165 (2022)
22. Tan, R., Xu, A., Nagarajan, V.: A general framework for sequential batch-testing. Preprint at Optimization Online (2024)
23. Ünlüyurt, T.: Sequential testing of complex systems: a review. Discret. Appl. Math. **142**(1–3), 189–205 (2004)

(Near)-Optimal Algorithms for Sparse Separable Convex Integer Programs

Christoph Hunkenschröder[1], Martin Koutecký[3], Asaf Levin[2], and Tung Anh Vu[3(✉)]

[1] TU Berlin, Berlin, Germany
hunkenschroeder@tu-berlin.de
[2] Technion–Israel Institute of Technology, Haifa, Israel
levinas@ie.technion.ac.il
[3] Charles University, Prague, Czech Republic
{koutecky,tung}@iuuk.mff.cuni.cz

Abstract. We study the general integer programming (IP) problem of optimizing a separable convex function over the integer points of a polytope: $\min\{f(\mathbf{x}) \mid A\mathbf{x} = \mathbf{b}, \mathbf{l} \leq \mathbf{x} \leq \mathbf{u}, \mathbf{x} \in \mathbb{Z}^n\}$. The number of variables n is a variable part of the input, and we consider the regime where the constraint matrix A has small coefficients $\|A\|_\infty$ and small primal or dual treedepth $\mathrm{td}_P(A)$ or $\mathrm{td}_D(A)$, respectively. Equivalently, we consider block-structured matrices, in particular n-fold, tree-fold, 2-stage and multi-stage matrices.

We ask about the possibility of near-linear algorithms in the general case of (non-linear) separable convex functions. The techniques of previous works for the linear case are inherently limited to it; in fact, no strongly-polynomial algorithm may exist due to a simple unconditional information-theoretic lower bound of $n \log \|\mathbf{u} - \mathbf{l}\|_\infty$, where $\mathbf{l}, \mathbf{u}$ are the vectors of lower and upper bounds. Our first result is that with parameters $\mathrm{td}_P(A)$ and $\|A\|_\infty$, this lower bound can be matched (up to dependency on the parameters). Second, with parameters $\mathrm{td}_D(A)$ and $\|A\|_\infty$, the situation is more involved, and we design an algorithm with complexity $g(\mathrm{td}_D(A), \|A\|_\infty) n \log n \log \|\mathbf{u} - \mathbf{l}\|_\infty$ where g is some computable function. We conjecture that a stronger lower bound is possible in this regime, and our algorithm is in fact optimal.

Our algorithms combine ideas from scaling, proximity, and sensitivity of integer programs, together with a new dynamic data structure allowing fast sparse updates.

Keywords: integer programming · parameterized complexity · Graver basis · treedepth · n-fold · tree-fold · 2-stage stochastic · multi-stage stochastic

N. Megow and A. Basu (Eds.): IPCO 2025, LNCS 15620, pp. 297–311, 2025.
https://doi.org/10.1007/978-3-031-93112-3_22

1 Introduction

Our focus is on the integer (linear) programming problem in standard form

$$\min\{f(\mathbf{x}) \mid A\mathbf{x} = \mathbf{b},\, \mathbf{l} \le \mathbf{x} \le \mathbf{u},\, \mathbf{x} \in \mathbb{Z}^n\}, \text{ and} \tag{IP}$$
$$\min\{\mathbf{w}\mathbf{x} \mid A\mathbf{x} = \mathbf{b},\, \mathbf{l} \le \mathbf{x} \le \mathbf{u},\, \mathbf{x} \in \mathbb{Z}^n\}, \tag{ILP}$$

with $A \in \mathbb{Z}^{m\times n}$, $\mathbf{b} \in \mathbb{Z}^m$, $\mathbf{l}, \mathbf{u}, \mathbf{w} \in \mathbb{Z}^n$, and $f : \mathbb{R}^n \to \mathbb{R}$ a separable convex function, that is, f can be expressed as $f(\mathbf{x}) = \sum_{i=1}^n f_i(x_i)$ and each $f_i : \mathbb{R} \to \mathbb{R}$ is convex. We write vectors in boldface (e.g., $\mathbf{x}, \mathbf{y}$) and their entries in normal font (e.g., the i-th entry of $\mathbf{x}$ is x_i). Throughout this paper we shall follow the notation of Grötschel, Lovász, and Schrijver [24] and write the dot product of vectors $\mathbf{w}$ and $\mathbf{x}$ as $\mathbf{w}\mathbf{x}$ instead of $\mathbf{w}^\intercal\mathbf{x}$. In this work the objective function f is be represented by a comparison oracle: given two points $\mathbf{x}, \mathbf{y} \in \mathbb{R}^n$, it returns the answer to the query $f(\mathbf{x}) > f(\mathbf{y})$ in constant time. IP is a fundamental optimization problem with rich theory and many practical applications [1,3,4,7, 21,44,47,53,54].

Because it is NP-hard [34], we study tractable subclasses corresponding to block-structured matrices, namely so-called multi-stage stochastic and tree-fold matrices, depicted in Figure 1. Both classes have been a subject of much study and have many theoretical and practical applications; see the survey of Schultz et al. [52] as well as an exposition article by Gavenčiak et al. [22] for examples of applications of multi-stage stochastic matrices, and see [10,15,22,32,38,39,41, 42] for examples of applications of IP with tree-fold matrices.

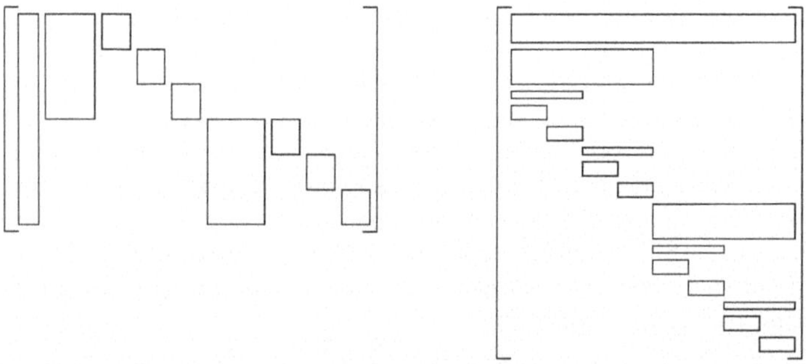

Fig. 1. On the left a schematic depiction of a multi-stage stochastic matrix with three levels. On the right a schematic tree-fold matrix with 4 layers. All entries outside of the rectangles must be zero. Entries within rectangles can be non-zero.

Equivalently, assuming $\|A\|_\infty$ is small, multi-stage stochastic matrices are exactly those with small primal treedepth $\mathrm{td}_P(A)$, and tree-fold matrices are those with small dual treedepth $\mathrm{td}_D(A)$ [45, Lemmas 25 and 26]; we will formally introduce these notions later. Note that both restrictions on $\|A\|_\infty$ and treedepth are necessary to obtain efficient algorithms: INDEPENDENT SET can be modelled as an ILP with $\|A\|_\infty = 1$, and KNAPSACK can be modelled as an ILP with one row, meaning that both cases are NP-hard.

A stream of papers [5,11,12,15,18,26,27,33,35] spanning the last 20+ years has brought gradually more and more general as well as faster algorithms for block-structured IPs, as well as provided complexity lower bounds [31,43]. The pinnacle of these developments is the fundamental result that IP is solvable in time $g(\|A\|_\infty, d)(n^\omega + \min(m,n)nm)\text{poly}(n)\log\|\mathbf{u}-\mathbf{l}\|_\infty \log f_{\text{gap}}$ [20] where $d = \min\{\text{td}_P(A), \text{td}_D(A)\}$, $f_{\text{gap}} = \max_{\mathbf{x},\mathbf{y}:\mathbf{l}\leq\mathbf{x},\mathbf{y}\leq\mathbf{u}} f(\mathbf{x}) - f(\mathbf{y})$, g is some computable function, and n^ω is time required to multiply two $n \times n$ matrices. Cslovjecsek et al. [11,12] have designed an algorithm for ILP with complexity $g(\|A\|_\infty, d)n\log^{2^{\mathcal{O}(d)}} n$ by coupling the multidimensional search technique [49] with new structural results. This algorithm is *strongly* near-linear in the sense that the number of arithmetic operations it performs does not depend on the numbers in $\mathbf{b}, \mathbf{l}, \mathbf{u}, \mathbf{w}$ on input. While the exponent over $\log n$ could conceivably be reduced, we consider the complexity of ILP in the regime of small $\text{td}_P(A)$ or $\text{td}_D(A)$ and with small coefficients to be essentially settled by these results.

1.1 Our Contribution

The situation is quite different for IP with a general separable convex function: no strongly polynomial algorithm is possible for IP in general, in fact, at least $n\log\|\mathbf{u}-\mathbf{l}\|_\infty$ comparisons are required to solve an IP instance by the following information-theoretic argument. For each $i \in [n]$, finding the integer minimum of a convex function $f_i : \mathbb{R} \to \mathbb{R}$ in the interval $[l_i, u_i]$ can be optimally done by binary search with $\log(u_i - l_i)$ comparisons (and cannot be done faster), and the integer minimum of $f(\mathbf{x}) = \sum_i f_i(x_i)$ is simply the vector of minima for each coordinate. Thus, minimizing $f(\mathbf{x})$ over the box $[\mathbf{l}, \mathbf{u}]$ takes at least $n\log\|\mathbf{u}-\mathbf{l}\|_\infty$ comparisons. It is then not surprising that the techniques of [11,12] are inherently limited to linear objectives and cannot be extended to the more general separable convex case. We focus on this question:

> *What is the complexity of* IP *with a separable convex objective given by a comparison oracle, parameterized by* $\text{td}_P(A)$ *or* $\text{td}_D(A)$ *and* $\|A\|_\infty$*?*

In the primal treedepth case, we design an algorithm matching the lower bound, up to the dependency on the parameters $\text{td}_P(A)$ and $\|A\|_\infty$.

Theorem 1. *There is a computable function g and an algorithm which solves* IP *in time* $g(\text{td}_P(A), \|A\|_\infty)n\log\max(\|\mathbf{u}-\mathbf{l}\|_\infty, \|\mathbf{b}\|_\infty)$.

Note that the dependence on $\log\|\mathbf{b}\|_\infty$ above (and below) is only necessary to find an initial feasible solution. In the dual treedepth case (i.e., tree-fold matrices), we devise an algorithm which misses the lower bound only by a $\log n$ factor:

Theorem 2 ($\star$). *There is a computable function g and an algorithm which solves* IP *in time* $g(\mathrm{td}_D(A), \|A\|_\infty) n \log(n) \log \max(\|\mathbf{u} - \mathbf{l}\|_\infty, \|\mathbf{b}\|_\infty)$.

We conjecture that the information-theoretic lower bound could be strengthened when A has small dual treedepth, and that our algorithm is actually optimal. We provide some indirect evidence in this direction in the full version of the paper.

Note that focusing on convex (non-linear) objective functions is well justified. For example, Bertsimas et al. note in their spectacular work [6] on the notorious subset selection problem in statistical learning that, over the past decades, algorithmic and hardware advances have elevated *convex* (i.e., non-linear) optimization to a level of relevance in applications comparable to its linear counterpart. Moreover, our Theorem 2 directly speeds up algorithms relying on n-fold IP with separable convex objectives [22,37,38].

Techniques. Both our algorithms are based on scaling, akin to the well-known capacity scaling algorithm for maximum flow [2, Section 7.3] or the scaling algorithm for IP of Hochbaum and Shanthikumar [29]. Theorem 1 utilizes a recent strong proximity result of Klein and Reuter [36]. Theorem 2 is much more delicate; to show it, we adopt an approach similar to iterative rounding used in approximation algorithms [46]. This approach is enabled by a new sensitivity result (Theorem 6) showing that "de-scaling" one variable at a time will lead to a sparse update; a novel dynamic data structure allowing fast sparse updates then completes the picture.

Note that our focus here is explicitly *not* on the parameter dependence since the dependence on g in our algorithm stems from a source common to most existing algorithms for those classes of IP. So any positive progress is likely to translate also to our algorithms.

1.2 Related Work

Unless stated otherwise, we omit the dependency on the parameters (i.e., $\|A\|_\infty$, block sizes, $\mathrm{td}_P(A)$, $\mathrm{td}_D(A)$, etc.). Jansen et al. [33] have designed the first near-linear algorithm for n-fold ILP, a subclass of ILP with small $\mathrm{td}_D(A)$, with complexity $n \log^{\mathcal{O}(1)} n \log f_{\text{gap}} \log \|\mathbf{u}-\mathbf{l}, \mathbf{b}\|_\infty$. Their algorithm only works for linear objectives, but could be modified to work also for separable convex functions. Like Theorem 2, it is also based on the idea of sparse updates, but uses a less efficient and less general color-coding-based approach. Jansen et al. [33] were also the first to study sensitivity in the context of n-fold ILP. Eisenbrand et al. [20] give a weakly-polynomial algorithm in the case of separable convex objectives, and, in another paper [19], show how to reduce the dependency on the objective function, if given the continuous optimum. Hunkenschröder et al. [30] show lower bounds on the parameter dependencies for algorithms for block-structured ILP, however, our focus here is on the polynomial dependency. Cslovjecsek et al. [13] give novel algorithms for n-fold and 2-stage ILP which avoid the usual approach based on the Graver basis, but also seem inherently limited to the linear case; in the case of 2-stage ILP even to the case of testing the feasibility.

A major inspiration for our work is the seminal paper of Hochbaum and Shanthikumar [29] which explores ideas such as proximity and scaling in the context of separable convex optimization. Compared to them, we focus on the case of small treedepth while they study programs with small subdeterminants. However, our iterative scaling approach, and the associated dynamic data structure, are new. Finally, various dynamic data structures allowing for sparse updates are a major focus of recent breakthroughs regarding LP [8,9,16,17].

We postpone some definitions, statements, and proofs to the full version of the paper. Statements whose proofs have been deferred are marked with ($\star$).

2 Preliminaries

If A is a matrix, $A_{i,j}$ denotes the j-th coordinate of the i-th row, $A_{i,\bullet}$ denotes the i-th row and $A_{\bullet,j}$ denotes the j-th column. We use $\log := \log_2$. For $a, b \in \mathbb{Z}$ we denote by $[a, b] = \{c \in \mathbb{Z} \mid a \leq c \leq b\}$, and $[a] := [1, a]$. We extend this notation to vectors $\mathbf{u}, \mathbf{v} \in \mathbb{Z}^a$ with $a \in \mathbb{N}$ where $[\mathbf{u}, \mathbf{v}] := \{\mathbf{w} \in \mathbb{Z}^a \mid u_i \leq w_i \leq v_i \, \forall i \in [a]\}$. For multiple real vectors $\mathbf{u}_1, \ldots, \mathbf{u}_n$ and any $1 \leq p \leq +\infty$ we denote by $\|\mathbf{u}_1, \ldots, \mathbf{u}_n\|_p$ the ℓ_p-norm of the vector obtained by concatenating vectors $\mathbf{u}_1, \ldots, \mathbf{u}_n$.

The *relaxation* of IP is the problem

$$\min \{f(\mathbf{x}) \mid \ A\mathbf{x} = \mathbf{b}, \ \mathbf{l} \leq \mathbf{x} \leq \mathbf{u}, \ \mathbf{x} \in \mathbb{R}^n\}. \tag{RP}$$

For an IP instance $\mathcal{I}$, let $\text{Sol}(\mathcal{I}) := \{\mathbf{x} \in \mathbb{Z}^n \mid A\mathbf{x} = \mathbf{b}, \mathbf{l} \leq \mathbf{x} \leq \mathbf{u}\}$ denote the set of feasible solutions of $\mathcal{I}$. We will use the notion of *centering an IP instance at a vector* $\mathbf{v} \in \mathbb{Z}^n$ which defines a new IP instance $\bar{\mathcal{I}} = (A, \bar{\mathbf{b}}, \bar{\mathbf{l}}, \bar{\mathbf{u}}, \bar{f})$ where $\bar{\mathbf{b}} = \mathbf{b} - A\mathbf{v}$, $\bar{\mathbf{l}} = \mathbf{l} - \mathbf{v}$, $\bar{\mathbf{u}} = \mathbf{u} - \mathbf{v}$, $\bar{f}(\mathbf{x}) = f(\mathbf{x} - \mathbf{v})$. Centering at $\mathbf{v}$ can be viewed as a translation $\tau(\mathbf{x}) = \mathbf{x} + \mathbf{v}$. Then τ is a bijection between $\text{Sol}(\mathcal{I})$ and $\text{Sol}(\bar{\mathcal{I}})$, and $\bar{f}$ only translates its input which it passes to the oracle for f. Thus given an optimum $\mathbf{v}^*$ of $\bar{\mathcal{I}}$ we can obtain an optimum of $\mathcal{I}$ in time $\mathcal{O}(n)$. If $\mathbf{v}$ is a feasible solution, i.e. we have $A\mathbf{v} = \mathbf{b}$, then we can perform the centering in time $\mathcal{O}(n)$ as we can skip the matrix multiplication $A\mathbf{v}$ and directly set $\bar{\mathbf{b}} = \mathbf{0}$.

We use the standard way to find an initial feasible solution to IP akin to Phase I of the simplex method. We formalize this in the full version of the paper. For now it suffices to say that one has to solve an auxiliary IP instance

$$\min\{\mathbf{w}'\mathbf{x}' \mid A_I(\mathbf{x}, \mathbf{x}') = \mathbf{b}, (\mathbf{l}, \mathbf{l}') \leq (\mathbf{x}, \mathbf{x}') \leq (\mathbf{u}, \mathbf{u}'), (\mathbf{x}, \mathbf{x}') \in \mathbb{Z}^{n+m}\}, \tag{A_I-feasible IP}$$

satisfying $\|\bar{\mathbf{l}}\|_\infty \leq \|\mathbf{l}, \mathbf{b}\|_\infty$ and $\|\bar{\mathbf{u}}\|_\infty \leq \|\mathbf{u}, \mathbf{b}\|_\infty$.

Graphs of A *and treedepth.* By $G_P(A)$ we denote the *primal graph of* A whose vertex set is $[n]$ (corresponding to the columns of A), and we connect two vertices $i \neq j \in [n]$ if there exists a row k of A such that $A_{k,i} \neq 0$ and $A_{k,j} \neq 0$. We denote $G_D(A) := G_P(A^T)$. From this point on we always assume that $G_P(A)$ and $G_D(A)$ are connected, otherwise A has (up to row and column permutations) a diagonal structure with some number $d \geq 2$ of blocks, and solving IP amounts to solving d smaller IP instances independently. Thus, the bounds we will show provide similar bounds for each such independent instance.

Definition 1 (Treedepth). *The* closurecl(F) *of an undirected rooted tree* F *is the undirected graph obtained from* F *by making every vertex adjacent to all of its ancestors. For a rooted tree* F its *height* height(F) *is the maximum number of vertices on any root-leaf path.* The *treedepth* td(G) *of a connected graph* G *is the minimum height of a tree* F *where* $V(F) = V(G)$ *such that* $G \subseteq \text{cl}(F)$, *namely* $E(G) \subseteq E(\text{cl}(F))$. A td-*decomposition of* G *is a tree* F *such that* $G \subseteq \text{cl}(F)$. *A* td-*decomposition* F of G is *optimal* if height$(F) = \text{td}(G)$.

Computing td(G) is NP-hard, but can be done in time $2^{\text{td}(G)^2} \cdot |V(G)|$ [51]. We define the *primal treedepth of* A to be $\text{td}_P(A) := \text{td}(G_P(A))$ and the *dual treedepth of* A to be $\text{td}_D(A) := \text{td}(G_D(A))$. We assume that an optimal td-decomposition is given since the time required to find it is dominated by other terms. Constructing $G_P(A)$ or $G_D(A)$ can be done in linear time if A is given in its sparse representation because there are at most $\text{td}(G)|V(G)|$ edges. Throughout we shall assume that $G_P(A)$ or $G_D(A)$ are given.
Graver Bases. Define a partial order $\sqsubseteq$ on $\mathbb{R}^n$ as follows: for $\mathbf{x}, \mathbf{y} \in \mathbb{R}^n$, write $\mathbf{x} \sqsubseteq \mathbf{y}$ and say that $\mathbf{x}$ is *conformal* to $\mathbf{y}$ if $\mathbf{x} \leq \mathbf{y}$, and for each $i \in [n]$, $x_i y_i \geq 0$ holds, i.e. $\mathbf{x}$ and $\mathbf{y}$ lie in the same orthant. Let $\ker_{\mathbb{Z}}(A) = \{\mathbf{x} \in \mathbb{Z}^n \mid A\mathbf{x} = \mathbf{0}\}$.

Definition 2 (Graver basis [23]**).** *The Graver basis of an integer* $m \times n$ *matrix* A *is the finite set* $\mathcal{G}(A) \subset \mathbb{Z}^n$ of $\sqsubseteq$*-minimal elements in* $\ker_{\mathbb{Z}}(A) \setminus \{\mathbf{0}\}$.

For a matrix $A \in \mathbb{Z}^{m \times n}$ and $1 \leq p \leq \infty$ we denote $g_p(A) := \max_{\mathbf{g} \in \mathcal{G}(A)} \|\mathbf{g}\|_p$. One important property of $\mathcal{G}(A)$ is as follows:

Proposition 1 (Positive Sum Property [50, Lemma 3.4] **).** *Let* $A \in \mathbb{Z}^{m \times n}$.*For any* $\mathbf{x} \in \ker_{\mathbb{Z}}(A)$, *there exists an* $n' \leq 2n - 2$ *and a decomposition* $\mathbf{x} = \sum_{j=1}^{n'} \lambda_j \mathbf{g}_j$ *with* $\lambda_j \in \mathbb{N}$ and $\mathbf{g}_j \in \mathcal{G}(A)$ *for each* $j \in [n']$, *and with* $\mathbf{g}_j \sqsubseteq \mathbf{x}$, i.e., all $\mathbf{g}_j$ *belonging to the same orthant as* $\mathbf{x}$.

3 Scaling Algorithm & Proximity Bounds

Our goal now is to show a scaling algorithm for IP with the following properties:

Theorem 3. *Let* $\mathcal{I} = (A, \mathbf{b}, \mathbf{l}, \mathbf{u}, f)$ *be an* IP *instance, and* $\beta = \lceil \log \|\mathbf{l}, \mathbf{u}\|_\infty \rceil + 1$. *There exists* $\rho \in \mathbb{R}^+$ *which only depends on* A *such that, given an initial feasible solution* $\mathbf{x}_0$ of $\mathcal{I}$, *we can find the optimum of* $\mathcal{I}$ *by solving* β IP *instances* $(\mathcal{I}^i = (A, \mathbf{0}, \mathbf{l}^i, \mathbf{u}^i, f^i))_{i=1}^{\beta}$, *where, for every* $i \in [\beta]$, f^i *is a separable convex function if* f *was, and* $\|\mathbf{u}^i - \mathbf{l}^i\|_\infty \leq 3\rho$.

Note that in the statement of Theorem 3, if for each of the instances $\mathcal{I}_i$ we have $\mathbf{l}^i \leq \mathbf{0} \leq \mathbf{u}^i$, then $\mathbf{0} \in \mathrm{Sol}(\mathcal{I}_i)$ due to the zero right hand side.

Proof Ideas for Theorem 3. Let $\mathcal{I}$ be an IP instance. For the rest of this section, we shall assume $\mathbf{0} \in \mathrm{Sol}(\mathcal{I})$: This is without loss of generality as we can, obtain a feasible solution $\mathbf{z}$ by the standard approach sketched in the preliminaries (Section 2) and then center $\mathcal{I}$ at $\mathbf{z}$. Note that this approach only increases $\mathrm{td}_P(A)$ and $\mathrm{td}_D(A)$ by at most 1 and preserves $\|A\|_\infty$.

For $s \in \mathbb{N}$, let the *s-scaled lattice* be $s\mathbb{Z}^n := \{s\mathbf{a} \mid \mathbf{a} \in \mathbb{Z}^n\}$. For any IP instance $\mathcal{J} = (A', \mathbf{b}', \mathbf{l}', \mathbf{u}', f')$ and any $s \in \mathbb{N}$, let its *s-scaled instance* be

$$\min\{f(\mathbf{x}) \mid A'\mathbf{x} = \mathbf{b}',\ \mathbf{l}' \leq \mathbf{x} \leq \mathbf{u}',\ \mathbf{x} \in s\mathbb{Z}^n\}\ , \tag{s-IP}$$

where $(A', \mathbf{b}', \mathbf{l}', \mathbf{u}', f', s)$ is the input. The only difference between IP and s-IP is that in the former we look for a solution in $\mathbb{Z}^n$ while in the latter in $s\mathbb{Z}^n$. For notational convenience, we will also define a "reverse version" of s-IP, called IP-scaled, where instead of searching on a coarser lattice, we shrink the feasible region, but "upscale" each point with respect to the objective function f:

$$\min\left\{f(s\mathbf{x}') \mid A\mathbf{x}' = \mathbf{b},\ \lceil \tfrac{1}{s} \rceil \mathbf{l} \leq \mathbf{x}' \leq \lfloor \tfrac{1}{s} \rfloor \mathbf{u},\ \mathbf{x}' \in \mathbb{Z}^n\right\}. \tag{IP-scaled}$$

where $(A, \mathbf{b}, \mathbf{l}, \mathbf{u}, f, s)$ is the input. For $s \in \mathbb{N}$ let $\mathcal{I}_s$ be the 2^s-scaled instance of $\mathcal{I}$. Observe that $\mathbf{0}$ is the only feasible solution in $\mathcal{I}_\beta$,thus it is the optimum of $\mathcal{I}_\beta$. In $\mathcal{I}_{\beta-1}$ it still holds that $\mathbf{0}$ is a feasible solution, so we may use it as an initial feasible solution from which we search for an optimum $\mathbf{x}^{\beta-1}$ of $\mathcal{I}_{\beta-1}$, and so on for $\beta - 2$ etc. That is, we let $\mathbf{x}^\beta = 0$, and then, for every $i = \beta - 1, \beta - 2, \ldots, 0$, we search for an optimum $\mathbf{x}^i$ of $\mathcal{I}_i$ with $\mathbf{x}^{i+1}$ as an initial feasible solution, and we center $\mathcal{I}_i$ at $\mathbf{x}^i$. We prove that for each instance $\mathcal{I}_i$, there exists an optimum $\mathbf{x}^i$ near the initial solution $\mathbf{x}^{i+1}$, that is, with $\left\|\mathbf{x}^{i+1} - \mathbf{x}^i\right\|_\infty$ small. So, we can restrict the original bounds to this smaller box, corresponding to $\|\mathbf{u} - \mathbf{l}\|_\infty$ being small. Each f^i is simply a translation and scaling of f, so it is separable convex.

3.1 A Scaling Proximity Bound

A *proximity bound* in optimization bounds the distance between optima of related formulations of the same problem, e.g., an integer program and its relaxation, etc. [11,25,29,40,48]. The notion of proximity bound we use is the following.

Definition 3 (Conformal proximity bound, [12]**).** Let $1 \leq p \leq +\infty$. *We say that $\mathcal{P}_p(A) \geq 0$ is the conformal ℓ_p-proximity bound of A if it is the infimum of reals $\rho \geq 0$ satisfying the following: for every* IP *with A as its constraint matrix, for every fractional solution* $\mathbf{x}$ of RP *and every integer solution* $\mathbf{z}$ of IP , *there is an integer solution* $\mathbf{z}'$ of IP *such that* $\|\mathbf{x} - \mathbf{z}'\|_p \leq \rho$ *and* $\mathbf{z}' \sqsubseteq \mathbf{x} - \mathbf{z}$.

An advantage of Definition 3 compared to prior notions of proximity bounds is that the bound ρ does not depend on f. Our main result here is:

Theorem 4 (⋆). *Let* $1 \leq p \leq +\infty$, $\mathcal{I}$ *be an* IP *instance,* $\mathbf{z}^*$ *be an optimum of* $\mathcal{I}$, *and* $\mathcal{I}'$ *be the* s*-scaled* $\mathcal{I}$ *with* $s \in \mathbb{N}$. *Then there exists an optimum* $\hat{\mathbf{z}}$ of $\mathcal{I}'$ *with*

$$\|\mathbf{z}^* - \hat{\mathbf{z}}\|_p \leq (s+1)\mathcal{P}_p(A), \tag{1}$$

and for every optimum $\hat{\mathbf{z}}$ of $\mathcal{I}'$ *there exists an optimum* $\mathbf{z}^*$ of $\mathcal{I}$ *satisfying* (1).

We need the following lemma which shows that, for separable convex functions, a conformal proximity bound is indeed useful to relate optima of IP and RP. The proof is a re-phrasing of known arguments [14,25,29].

Lemma 1 (⋆). *Let* $1 \leq p \leq +\infty$, $\hat{\mathbf{z}}$ *be an optimum of* IP , *and* $\hat{\mathbf{x}}$ *an optimum of the relaxation* RP *of* IP. *Then there exist* $\mathbf{x}^* \in \mathbb{R}^n$ *and* $\mathbf{z}^* \in \mathbb{Z}^n$ *optima of* RP *and* IP, *respectively, with* $\|\hat{\mathbf{x}} - \mathbf{z}^*\|_p = \|\mathbf{x}^* - \hat{\mathbf{z}}\|_p \leq \mathcal{P}_p(A)$.

Proof idea (Theorem 4). We use Lemma 1 to bound the distance between an optimum of IP and its relaxation RP, and the distance between an optimum of s-IP and RP, and then use triangle inequality on these two bounded distances. This is valid because the optima of the relaxations of IP and s-IP coincide.

3.2 Scaling Algorithm for IP

As a 2^k-scaled instance is also a 2-scaled instance of a 2^{k-1}-scaled instance, we get the following algorithm to solve a general IP. The algorithm is parameterized by some upper bound ρ on the proximity bound $\mathcal{P}_\infty(A)$ (see Theorem 4).

(1) Given an initial feasible solution $\mathbf{x}_0$, recenter the instance at $\mathbf{x}_0$, obtaining IP where $\mathbf{0}$ is feasible. Set $\mathbf{x}^\beta = \mathbf{0}$.
(2) For each $i = \beta - 1, \beta - 2, \ldots, 0$, perform the following steps: Intersect the box constraints of the 2^{i+1}-scaled instance with the proximity bounds $\|\mathbf{x}^{i+1} - \mathbf{x}\|_\infty \leq 3\rho$. Solve this instance with initial solution $\mathbf{0} = \mathbf{x}^{i+1}$ to optimality, and let $\mathbf{x}^i$ denote the optimum. Center the current instance at $\mathbf{x}^i$.
(3) Output $\mathbf{x}^0$.

The 3ρ bound in step (2) follows from the fact that we are viewing the 2^k-scaled instance as an s-scaled instance of the 2^{k-1}-scaled instance where $s = 2$, and using $3 = s + 1$ in the bound of Theorem 4.

Proof of Theorem 3 Given an initial feasible solution $\mathbf{x}_0$, we center the original instance at $\mathbf{x}_0$ which transforms its right hand side to $\mathbf{b} - A\mathbf{x}_0 = \mathbf{0}$ as required. It is clear that we need at most $\lceil \log \|\mathbf{l}, \mathbf{u}\|_\infty \rceil + 1$ iterations, in each of which we apply Theorem 4 with $s = 2$ to the bounds $\mathbf{l}^i, \mathbf{u}^i$. □

Remark 1 We can somewhat decrease the number of iterations of the scaling algorithm as follows. In each instance $\mathcal{I}^i$ we have $\|\mathbf{u}^i - \mathbf{l}^i\| \leq 3\rho$. In the first $\lceil \log \rho \rceil$ instances $\mathcal{I}^i$, we are searching in the $2^{\beta-i}$-lattice. Let us view these instances as IP-scaled instances: then each has $\mathbf{l}^i = 2^{-(\beta-i)}\mathbf{l}$ and $\mathbf{u}^i = 2^{-(\beta-i)}\mathbf{u}$. But then for each of them we already have $\|\mathbf{l}^i, \mathbf{u}^i\|_\infty \leq \rho$, so the guaranteed bound on $\|\mathbf{u} - \mathbf{l}\|_\infty$ is satisfied even without invoking the proximity theorem. Thus, we can skip the first $\lceil \log \rho \rceil$ iterations by initially setting $k = \beta - \lceil \log \rho \rceil$, reducing the number of iterations to $(\lceil \log \|\mathbf{l}, \mathbf{u}\|_\infty \rceil + 1)/\rho$. This neatly matches the $\log(B/n)$ bound on the number of scaling phases of Hochbaum [28] who has essentially $B = \|\mathbf{u} - \mathbf{l}\|_\infty$, and her proximity bound is $\mathcal{O}(n)$.

4 Primal Algorithm

Theorem 1 is proven by instantiating the scaling algorithm of Theorem 3 with a new strong proximity of Klein and Reuter [36], and the standard branching algorithm for each subinstance [20, Lemma 8].

The parameter dependency of the function g is essentially determined by $g_\infty(A)$ and the proximity bound $\mathcal{P}_\infty(A)$. The currently best known bounds for both are triple-exponential in terms of $\text{td}_P(A)$ [36, Corollary 2.1, Lemma 2.3], and this seems to be optimal for $g_\infty(A)$, see [30, Theorem 1]. Since the exact dependence of g on $\text{td}_P(A)$ and $\|A\|_\infty$ is not the focus of this work, we shall simply focus on proving the theorem in its stated form. Klein and Reuter [36] (improving on Cslovjecsek et al. [12]) give a bound on $\mathcal{P}_\infty(A)$ independent of n:

Proposition 2 ([36, Lemma 2.3]). *There is a computable function g' s.t.*

$$\mathcal{P}_\infty(A) \leq g'(\text{td}_P(A), \|A\|_\infty) \ . \tag{2}$$

Proof of Theorem 1 We instantiate the scaling algorithm from Theorem 1 as follows. We provide Proposition 2 as an upper bound ρ on $\mathcal{P}_\infty(A)$. We solve each of the $\lceil \log(\|\mathbf{l}, \mathbf{u}\|_\infty) \rceil + 1$ IP instances in the scaling algorithm using the algorithm of Eisenbrand et al. [20, Lemma 8] which has running time $\text{td}_P(A)^2(2\rho+1)^{\text{td}_P(A)}n$. Taking $g(\text{td}_P(A), \|A\|_\infty) = \text{td}_P(A)^2(2\rho+1)^{\text{td}_P(A)}$ and the fact that $\rho = g'(\text{td}_P(A), \|A\|_\infty)$ for some computable function g' concludes the proof. □

5 Scaling and Sensitivity

The situation is more delicate regarding the algorithm for dual treedepth (Theorem 2). Not only is there no analogue of the proximity result of Klein and Reuter [36], in fact, Cslovjecsek et al. [11, Proposition 4.1] show that there are instances of IP with small $\text{td}_D(A) + \|A\|_\infty$ where an integer optimum is $\Omega(n)$ far in the ℓ_∞-norm from any continuous optimum. The proximity results for stronger relaxations [11,40] are also of no obvious help. Their bounds which are independent of n only apply to vertex solutions, which are guaranteed to include an optimum only for linear objectives. Moreover, solving these stronger relaxations is either not near-linear time (reliant on the ellipsoid method [40]) or limited to linear objectives [11].

Let us re-examine the scaling algorithm. In each iteration, we go from an optimum in the lattice $2s\mathbb{Z}^n$ to an optimum in $s\mathbb{Z}^n$. From the point of view of IP-scaled considering $s\mathbb{Z}^n$ instead of $2s\mathbb{Z}^n$ means that for every $i \in [n]$ we modify the separable objective function from $f_i(2sx_i')$ to $f_i(sx_i')$, the lower bound from $\frac{1}{2s}l_i$ to $\frac{1}{s}l_i$, and the upper bound from $\frac{1}{2s}u_i$ to $\frac{1}{s}u_i$. Afterwards, we solve the new instance in the finer lattice $s\mathbb{Z}^n$ with an initial feasible solution which was the optimum in $2s\mathbb{Z}^n$.

Our approach for dual treedepth is to find the optimum solution in $\mathbb{Z}^n$ by changing the scaling factor one variable at a time, which means that once we modify f_i, l_i, u_i for some $i \in [n]$, we immediately solve the instance resulting from refining coordinate i. It is not clear why this should be advantageous – after all, in each iteration of the scaling algorithm we need to solve IP-scaled n times instead of just once. The advantage lies in the fact that whenever we refine a single variable, then an optimum of the refined instance is close (in terms of the ℓ_1-norm) to an optimum of the instance before refinement.

To show this proximity bound, we proceed in two steps. First, we show a sensitivity theorem for changes in the box constraints $\mathbf{l}, \mathbf{u}$. Then, we would like to use this theorem to show that refining few variables means only a small change in the optimal solution; however, this is not possible directly, because going from $l_i/2$ to l_i may be a big change in the absolute sense. Thus, we use an encoding trick to circumvent this issue. Let us start with the first step.

Theorem 5 ($\star$). Let $\mathbf{x}$ *be an optimum of* IP , $\delta \in \mathbb{Z}^n$, *and* $\mathbf{l}' := \mathbf{l} + \delta$. *If the* IP *with* $\mathbf{l}$ *replaced by* $\mathbf{l}'$ *is feasible, then it has an optimum* $\mathbf{x}'$ *with* $\|\mathbf{x} - \mathbf{x}'\|_1 \leq \|\delta\|_1 g_1(A)$.

Proof Ideas. Consider the special case when δ is a unit vector, i.e. $\delta \in \{\mathbf{e}_i, -\mathbf{e}_i\}$, and let $\mathbf{x}'$ be an optimum of the modified instance which is closest to $\mathbf{x}$ in ℓ_1-norm. As $\mathbf{x} - \mathbf{x}' \in \ker_{\mathbb{Z}}(A)$, it has a conformal decomposition $\mathbf{x} - \mathbf{x}' = \sum_j \mathbf{g}_j$ with $\mathbf{g}_j \in \mathcal{G}(A), \mathbf{g}_j \sqsubseteq \mathbf{x} - \mathbf{x}'$ for every j (see Proposition 1). We will show that in fact there can be only one element in the conformal decomposition. If not, then we can take two distinct vectors of the conformal decomposition and use them to find another optimum solution $\hat{\mathbf{x}}$ that is closer to $\mathbf{x}$ than $\mathbf{x}'$, and the approach would be similar as in the approach of the proof of Lemma 1. Thus $\|\mathbf{x} - \mathbf{x}'\|_1 \leq g_1(A)$. To prove it for any δ, we can repeatedly apply the result where $\delta \in \{\mathbf{e}_i, -\mathbf{e}_i\}$ and use the triangle inequality, which gives the $\|\delta\|_1$ factor in the bound in Theorem 5.

Consider an instance of IP with $\mathbf{b} = \mathbf{0}$. For any $I \subseteq [n]$, the *I-scaled IP* is an IP instance with input $\bar{\mathbf{l}}^I, \bar{\mathbf{u}}^I, \bar{f}^I$ and with the constraint matrix A^I defined as follows. If $i \in I$, then $\bar{l}_i^I = \lceil l_i/2 \rceil$, $\bar{u}_i^I = \lfloor u_i/2 \rfloor$ and $\bar{f}_i^I(x_i) = f_i(2x_i)$; if $i \notin I$, then $\bar{l}_i^I = l_i$, $\bar{u}_i^I = u_i$ and $\bar{f}_i^I(x_i) = f_i(x_i)$. The i-th column of the constraint matrix $A^I \in \mathbb{Z}^{m \times n}$ is either twice the i-th column of A if $i \in I$, otherwise it is simply the i-th column of A. We use $I \triangle J$ to denote the symmetric difference of sets I, J, i.e., $I \triangle J = (I \setminus J) \cup (J \setminus I)$. Our goal is to prove the following theorem:

Theorem 6. *Let $I, J \subseteq [n]$, and let $\mathbf{x}^I$ be an optimum of the I-scaled* IP. *Let $\bar{A} \in \mathbb{Z}^{(n+m)\times 3n}$ be the matrix $\left(\begin{smallmatrix} A & 2A & 0 \\ I & 2I & -I \end{smallmatrix}\right)$. Then if the J-scaled IP is feasible, it has an optimum $\mathbf{x}^J$ satisfying $\|\mathbf{x}^I - \mathbf{x}^J\|_1 \leq 2|I \triangle J| g_1(\bar{A}) \leq 4|I \triangle J| g_1(A)$.*

We want to use Theorem 5 to prove Theorem 6. In order to do that, we will construct an auxiliary IP instance which will be able to encode the I-scaled instance for any $I \subseteq [n]$ by only a small change in its bounds.

Specifically, the constraint matrix is $\bar{A} = \left(\begin{smallmatrix} A & 2A & 0 \\ I & 2I & -I \end{smallmatrix}\right)$. The bounds $\hat{\mathbf{l}}^I, \hat{\mathbf{u}}^I$ are defined as follows: for $j \in [n]$, we set

- $\hat{l}^I_{2n+j} = l_j$ and $\hat{u}^I_{2n+j} = u_j$,
- $\hat{l}^I_{n+j} = \lceil l_j/2 \rceil$ and $\hat{u}^I_{n+j} = \lfloor u_j/2 \rfloor$,
- if $j \in I$, $\hat{l}^I_j = \hat{u}^I_j = 0$,
- otherwise $j \notin I$, then $\hat{l}^I_i = -1$ and $\hat{u}^I_i = 1$,

$\hat{f}$ is simply $\hat{f}(\mathbf{z}) = f((z_i)_{i \in [2n+1,3n]})$. In summary:

$$\min\left\{\hat{f}(\mathbf{z}^I) : \bar{A}\mathbf{z}^I = \mathbf{0},\ \hat{\mathbf{l}}^I \leq \mathbf{z} \leq \hat{\mathbf{u}}^I,\ \mathbf{z}^I \in \mathbb{Z}^{3n}\right\}. \qquad \text{(aux-}I\text{-IP)}$$

Given a solution $\mathbf{x}^I$ of the I-scaled IP, a vector $\mathbf{z}^I \in \mathbb{Z}^{3n}$ intuitively encodes $\mathbf{x}^I$ in the following way: if $i \notin I$, the variable i is already in the refined grid and we split its value into the least significant digit (z^I_i) and the rest (z^I_{i+n}); otherwise if $i \in I$, the variable i is not in the refined grid, so we just set $z^I_{i+n} = x^I_i$. Formally, if $i \in I$, set $z^I_{i+n} = x^I_i$ and $z^I_i = 0$, otherwise we need to distinguish whether x^I_i is positive or negative: if $x^I_i \geq 0$, set $z^I_{i+n} = \lfloor x^I_i/2 \rfloor$ and $z^I_i = x^I_i \bmod 2$; otherwise set $z^I_{i+n} = \lceil x^I_i/2 \rceil$ and $z^I_i = -x^I_i \bmod 2$. For each $i \in [n]$, set $z^I_{i+2n} = z^I_i + 2z^I_{n+i}$. In the other direction, given a solution $\mathbf{z}^I$ of the aux-I-IP, we construct a solution $\mathbf{x}^I$ of the I-scaled IP as follows. If $i \in I$, set $x^I_i = z^I_{n+i}$, otherwise if $i \notin I$, set $x^I_i = z^I_i + 2z^I_{n+i}$. Notice that the value of $\mathbf{z}^I$ with respect to $\hat{f}$ is equal to the value of $\mathbf{x}^I$ with respect to $\bar{f}^I$:

- $\forall i \in I : \bar{f}^I_i(x^I_i) = f_i(2x^I_i) = \hat{f}_i(z_{2n+i})$ since $z^I_{i+n} = x^I_i$, $z^I_i = 0$ and $z^I_{i+2n} = z^I_i + 2z^I_{n+i} = 2x^I_i$,
- $\forall i \notin I : \bar{f}^I_i(x^I_i) = f_i(x^I_i) = \hat{f}_i(z_{2n+i})$ since $z^I_{i+2n} = z^I_i + 2z^I_{n+i} = x^I_i$ by definition.

By the construction above we immediately get:

Lemma 2. *A solution $\mathbf{x}^I$ of the I-scaled IP is optimal if and only if the corresponding solution $\mathbf{z}^I$ of* aux-I-IP *is optimal.*

Proof. of Theorem 6 Let $\mathbf{z}^I$ be the solution of aux-I-IP corresponding to the optimum $\mathbf{x}^I$ from the statement. By Lemma 2, $\mathbf{z}^I$ is optimal for aux-I-IP. Observe that when we replace I with J, the bounds of the aux-I-IP instance only change in coordinates $I \triangle J$, and in each it changes by 2 (for a coordinate in $J \setminus I$, the lower bound changes from 0 to -1, and the upper bound from 0 to $+1$, and for a coordinate in $I \setminus J$, the bounds change from $+1$ or -1 to 0). Thus, by Theorem 5, there exists an optimum $\mathbf{z}^J$ at ℓ_1-distance at most $2|I \triangle J| g_1(\bar{A})$. Finally, the corresponding optimum $\mathbf{x}^J$ of the J-scaled IP is at the claimed distance, again, by construction. It remains to prove that $g_1(\bar{A}) \leq 2g_1(A)$, which we defer to the full version of the paper. □

Open Problems Our results primarily point in two interesting research directions. First, is it possible to shave off the $\log n$ factor in the dual algorithm (Theorem 2), at least for some classes of IP with small $\text{td}_D(A) + \|A\|_\infty$? If not, how could the $n \log \|\mathbf{u} - \mathbf{l}\|_\infty$ lower bound be strengthened to match our results? Second, the lower bound does not apply outside of the comparison oracle model; for which classes of objectives beyond linear can one design faster, possibly strongly polynomial algorithms?

Acknowledgments. Hunkenschröder acknowledges support by the Einstein Foundation Berlin. Levin is partially supported by ISF - Israel Science Foundation grant number 1467/22. Koutecký and Vu are partially supported by Charles Univ. project UNCE 24/SCI/008 and by the ERC-CZ project LL2406 of the Ministry of Education of Czech Republic. Koutecký is additionally supported by projects 22-22997S and 25-17221S of GA ČR, Vu by the project 24-10306S of GA ČR.

References

1. Achterberg, T., Bixby, R.E., Gu, Z., Rothberg, E., Weninger, D.: Presolve reductions in mixed integer programming. INFORMS J. Comput. **32**(2), 473–506 (2020)
2. Ahuja, R.K., Magnanti, T.L., Orlin, J.B.: Network Flows. Prentice Hall, Inc (1993)
3. Alumur, S.A., Kara, B.Y.: Network hub location problems: the state of the art. Eur. J. Oper. Res. **190**(1), 1–21 (2008)
4. Anderson, R., Huchette, J., Ma, W., Tjandraatmadja, C., Vielma, J.P.: Strong mixed-integer programming formulations for trained neural networks. Math. Program. **183**(1), 3–39 (2020)
5. Aschenbrenner, M., Hemmecke, R.: Finiteness theorems in stochastic integer programming. Found. Comput. Math. **7**(2), 183–227 (2007)
6. Bertsimas, D., King, A., Mazumder, R.: Best subset selection via a modern optimization lens. Ann. Stat. **44**(2), 813–852 (2016)
7. van den Briel, M., Vossen, T., Kambhampati, S.: Reviving integer programming approaches for AI planning: a branch-and-cut framework. In: Automated Planning and Scheduling, ICAPS 2005. Proceedings, pp. 310–319. AAAI (2005)
8. Chen, L., Kyng, R., Liu, Y.P., Meierhans, S., Gutenberg, M.P.: Almost-linear time algorithms for incremental graphs: Cycle detection, sccs, s-t shortest path, and minimum-cost flow. In: Proceedings of the 56th Annual ACM Symposium on Theory of Computing, STOC 2024, pp. 1165–1173. ACM (2024)

9. Chen, L., Kyng, R., Liu, Y.P., Peng, R., Gutenberg, M.P., Sachdeva, S.: Maximum flow and minimum-cost flow in almost-linear time. In: 63rd IEEE Annual Symposium on Foundations of Computer Science, FOCS 2022, pp. 612–623. IEEE (2022)
10. Chen, L., Marx, D., Ye, D., Zhang, G.: Parameterized and approximation results for scheduling with a low rank processing time matrix. In: 34th Symposium on Theoretical Aspects of Computer Science, STACS 2017. LIPIcs, vol. 66, pp. 1–14. Schloss Dagstuhl—Leibniz-Zentrum für Informatik (2017)
11. Cslovjecsek, J., Eisenbrand, F., Hunkenschröder, C., Rohwedder, L., Weismantel, R.: Block-structured integer and linear programming in strongly polynomial and near linear time. In: 32nd Annual ACM-SIAM Symposium on Discrete Algorithms, SODA 2021, pp. 1666–1681. SIAM (2021)
12. Cslovjecsek, J., Eisenbrand, F., Pilipczuk, M., Venzin, M., Weismantel, R.: Efficient sequential and parallel algorithms for multistage stochastic integer programming using proximity. In: 29th Annual European Symposium on Algorithms, ESA 2021. LIPIcs, vol. 204, pp. 1–14. Schloss Dagstuhl—Leibniz-Zentrum für Informatik (2021)
13. Cslovjecsek, J., Koutecký, M., Lassota, A., Pilipczuk, M., Polak, A.: Parameterized algorithms for block-structured integer programs with large entries. In: Proceedings of the 2024 ACM-SIAM Symposium on Discrete Algorithms, SODA 2024, pp. 740–751. SIAM (2024)
14. De Loera, J.A., Hemmecke, R., Köppe, M.: Algebraic and Geometric Ideas in the Theory of Discrete Optimization, MOS-SIAM Series on Optimization, vol. 14. SIAM (2013)
15. De Loera, J.A., Hemmecke, R., Onn, S., Weismantel, R.: N-fold integer programming. Discr. Optim. **5**(2), 231–241 (2008). in Memory of George B. Dantzig
16. Dong, S., Goranci, G., Li, L., Sachdeva, S., Ye, G.: Fast algorithms for separable linear programs. In: Proceedings of the 2024 ACM-SIAM Symposium on Discrete Algorithms, SODA 2024, pp. 3558–3604. SIAM (2024)
17. Dong, S., Lee, Y.T., Ye, G.: A nearly-linear time algorithm for linear programs with small treewidth: a multiscale representation of robust central path. In: STOC '21: 53rd Annual ACM SIGACT Symposium on Theory of Computing, pp. 1784–1797. ACM (2021)
18. Eisenbrand, F., Hunkenschröder, C., Klein, K.: Faster algorithms for integer programs with block structure. In: 45th International Colloquium on Automata, Languages, and Programming, ICALP 2018. LIPIcs, vol. 107, pp. 1–13. Schloss Dagstuhl — Leibniz-Zentrum für Informatik (2018)
19. Eisenbrand, F., Hunkenschröder, C., Klein, K., Koutecký, M., Levin, A., Onn, S.: Reducibility bounds of objective functions over the integers. Oper. Res. Lett. **51**(6), 595–598 (2023)
20. Eisenbrand, F., Hunkenschröder, C., Klein, K.M., Koutecký, M., Levin, A., Onn, S.: Sparse integer programming is fixed-parameter tractable. Math. Oper. Res. **0**(0) (2024)
21. Floudas, C.A., Lin, X.: Mixed integer linear programming in process scheduling: modeling, algorithms, and applications. Ann. Oper. Res. **139**(1), 131–162 (2005)
22. Gavenčiak, T., Koutecký, M., Knop, D.: Integer programming in parameterized complexity: five miniatures. Discret. Optim. **44**(Part), 100596 (2022)
23. Graver, J.E.: On the foundations of linear and integer linear programming I. Math. Program. **9**(1), 207–226 (1975)

24. Grötschel, M., Lovász, L., Schrijver, A.: Geometric algorithms and combinatorial optimization. Algorithms and Combinatorics, vol. 2, second edn. Springer-Verlag, Berlin (1993)
25. Hemmecke, R., Köppe, M., Weismantel, R.: Graver basis and proximity techniques for block-structured separable convex integer minimization problems. Math. Programm. **145**(1-2, Ser. A), 1–18 (2014)
26. Hemmecke, R., Onn, S., Romanchuk, L.: n-fold integer programming in cubic time. Math. Program. **137**(1–2), 325–341 (2013)
27. Hemmecke, R., Schultz, R.: Decomposition of test sets in stochastic integer programming. Math. Program. **94**, 323–341 (2003)
28. Hochbaum, D.S.: Lower and upper bounds for the allocation problem and other nonlinear optimization problems. Math. Oper. Res. **19**(2), 390–409 (1994)
29. Hochbaum, D.S., Shanthikumar, J.G.: Convex separable optimization is not much harder than linear optimization. JACM **37**(4), 843–862 (1990)
30. Hunkenschröder, C., Klein, K., Koutecký, M., Lassota, A., Levin, A.: Tight lower bounds for block-structured integer programs. In: Integer Programming and Combinatorial Optimization - 25th International Conference, IPCO 2024. Lecture Notes in Computer Science, vol. 14679, pp. 224–237. Springer (2024)
31. Jansen, K., Klein, K., Lassota, A.: The double exponential runtime is tight for 2-stage stochastic ILPs. Math. Program. **197**(2), 1145–1172 (2023)
32. Jansen, K., Klein, K., Maack, M., Rau, M.: Empowering the configuration-IP: new PTAS results for scheduling with setup times. Math. Program. **195**(1), 367–401 (2022)
33. Jansen, K., Lassota, A., Rohwedder, L.: Near-linear time algorithm for n-fold ILPs via color coding. SIAM J. Discret. Math. **34**(4), 2282–2299 (2020)
34. Karp, R.M.: Reducibility among combinatorial problems. In: Complexity of Computer Computations, The IBM Research Symposia Series, pp. 85–103. Springer (1972)
35. Klein, K.: About the complexity of two-stage stochastic IPs. Math. Program. **192**(1), 319–337 (2022)
36. Klein, K., Reuter, J.: Collapsing the tower — on the complexity of multistage stochastic IPs. In: 33rd Annual ACM-SIAM Symposium on Discrete Algorithms, SODA 2022, pp. 348–358. SIAM (2022)
37. Knop, D., Koutecký, M., Levin, A., Mnich, M., Onn, S.: Multitype integer monoid optimization and applications. CoRR **abs/1909.07326** (2019)
38. Knop, D., Koutecký, M.: Scheduling meets n-fold integer programming. J. Sched. **21**(5), 493–503 (2018)
39. Knop, D., Koutecký, M., Levin, A., Mnich, M., Onn, S.: Parameterized complexity of configuration integer programs. Oper. Res. Lett. **49**(6), 908–913 (2021)
40. Knop, D., Koutecký, M., Levin, A., Mnich, M., Onn, S.: High-multiplicity N-fold IP via configuration LP. Math. Program. **200**(1), 199–227 (2023)
41. Knop, D., Koutecký, M., Mnich, M.: Combinatorial n-fold integer programming and applications. Math. Program. **184**(1), 1–34 (2020)
42. Knop, D., Koutecký, M., Mnich, M.: Voting and bribing in single-exponential time. ACM Trans. Econ. Comput. **8**(3), 1–28 (2020)
43. Knop, D., Pilipczuk, M., Wrochna, M.: Tight complexity lower bounds for integer linear programming with few constraints. ACM Trans. Comput. Theory **12**(3), 1–19 (2020)
44. Knueven, B., Ostrowski, J., Watson, J.: On mixed-integer programming formulations for the unit commitment problem. INFORMS J. Comput. **32**(4), 857–876 (2020)

45. Koutecký, M., Levin, A., Onn, S.: A parameterized strongly polynomial algorithm for block structured integer programs. In: Proceedings of ICALP 2018. Leibniz International Proceedings Informatics, vol. 107, pp. 1–14 (2018)
46. Lau, L.C., Ravi, R., Singh, M.: Iterative Methods in Combinatorial Optimization, vol. 46. Cambridge University Press (2011)
47. Lodi, A., Martello, S., Monaci, M.: Two-dimensional packing problems: a survey. Eur. J. Oper. Res. **141**(2), 241–252 (2002)
48. Murota, K., Tamura, A.: Proximity theorems of discrete convex functions. Math. Program. **99**(3), 539–562 (2004)
49. Norton, C.H., Plotkin, S.A., Tardos, É.: Using separation algorithms in fixed dimension. J. Algorithms **13**(1), 79–98 (1992)
50. Onn, S.: Nonlinear discrete optimization. Zurich Lectures in Advanced Mathematics, European Mathematical Society (2010)
51. Reidl, F., Rossmanith, P., Villaamil, F.S., Sikdar, S.: A faster parameterized algorithm for Treedepth. In: Esparza, J., Fraigniaud, P., Husfeldt, T., Koutsoupias, E. (eds.) ICALP 2014. LNCS, vol. 8572, pp. 931–942. Springer, Heidelberg (2014). https://doi.org/10.1007/978-3-662-43948-7_77
52. Schultz, R., Stougie, L., van der Vlerk, M.H.: Two-stage stochastic integer programming: a survey. Stat. Neerl. **50**(3), 404–416 (1996)
53. Toth, P., Vigo, D. (eds.): The Vehicle Routing Problem. Society for Industrial and Applied Mathematics (2001)
54. Vossen, T., Ball, M.O., Lotem, A., Nau, D.S.: On the use of integer programming models in AI planning. In: International Joint Conference on Artificial Intelligence, IJCAI 99. Proceedings, pp. 304–309. Morgan Kaufmann (1999)

An $O(\log n)$-Approximation Algorithm for (p, q)-Flexible Graph Connectivity via Independent Rounding

Sharat Ibrahimpur[1(✉)] and László A. Végh[2]

[1] Research Institute for Discrete Mathematics, University of Bonn, 53113 Bonn, Germany
ibrahimpur@dm.uni-bonn.de

[2] Hertz Chair in the Transdisciplinary Research Area Modelling, University of Bonn, 53115 Bonn, Germany
lvegh@uni-bonn.de

Abstract. In the (p, q)-Flexible Graph Connectivity problem, the input is a graph $G = (V, E)$ whose edges have nonnegative costs and the edge set $E = \mathcal{S} \cup \mathcal{U}$ is partitioned into safe and unsafe edges. The goal in this problem is to find a minimum cost set of edges F such that the subgraph (V, F) remains p-edge-connected after removing any q unsafe edges from F. We give a new integer programming formulation for the problem by adding knapsack cover constraints to the $p(p+q)$-connected capacitated edge-connectivity formulation studied in previous work and show that the corresponding linear relaxation can be solved in polynomial time by giving an efficient separation oracle. Furthermore, we show that independent randomized rounding yields an $O(\log n)$-approximation for arbitrary values of p and q, improving the state of the art $O(q \log n)$. For both separation and rounding, a key insight is to use Karger's bound on the number of near-minimum cuts.

Keywords: Approximation Algorithms · Network Design · LP Rounding · Knapsack Cover Inequalities

1 Introduction

We consider the *Flexible Graph Connectivity* (FGC) model introduced by Adjiashvili, Hommelsheim and Mühlenthaler [2] as a means to capture nonuniformity of edges in survivable network design problems. In this model, we are given an undirected graph $G = (V, E)$ on n vertices, a partition of $E = \mathcal{S} \cup \mathcal{U}$ into *safe* and *unsafe* edges, respectively, and nonnegative edge-costs $\{c_e\}_{e \in E}$. Safe edges are assumed to be always reliable, whereas an unsafe edge may fail. As typical in survivable network design problems, we want to buy a cheap set of edges $F \subseteq E$

The first author is grateful for the support received from the Deutsche Forschungsgemeinschaft (DFG, German Research Foundation) – 537750605.

N. Megow and A. Basu (Eds.): IPCO 2025, LNCS 15620, pp. 312–325, 2025.
https://doi.org/10.1007/978-3-031-93112-3_23

such that the subgraph (V, F) has some specified edge-connectivity properties, and these properties are robust against certain failures among unsafe edges.

Formally, in an instance of the (p, q)-FGC problem, we are given an edge connectivity parameter $p \in \mathbb{Z}_{\geq 1}$ and a robustness parameter $q \in \mathbb{Z}_{\geq 0}$. A subset $F \subseteq E$ of edges is feasible if for any set F' consisting of at most q unsafe edges, the subgraph $(V, F \setminus F')$ is p-edge connected, i.e., for any pair of distinct vertices $u, v \in V$, there are p edge-disjoint (u, v)-paths in $(V, F \setminus F')$. The algorithmic goal in (p, q)-FGC is to find a feasible solution F that minimizes $c(F) := \sum_{e \in F} c_e$. We assume that E is feasible, which can be checked efficiently (see Lemma 2). Throughout the paper, let $\text{OPT} := c(F^*)$ denote the cost of an optimal solution to this instance. Note that this problem is already APX-hard [10] as the classical problem of finding the smallest 2-edge connected spanning subgraph (abbreviated as unit-cost/unweighted 2-ECSS) of an undirected graph G arises as a special case of $(1, 1)$-FGC where every edge is unsafe and has unit cost. Thus, it is natural to seek approximation algorithms for this problem that can produce a solution F whose cost is at most $\alpha \cdot \text{OPT}$ for some small $\alpha > 1$, or establish complementary hardness-of-approximation results.

In the work [2] that pioneered this model, Adjiashvili et al. gave a 2.527-approximation algorithm for $(1, 1)$-FGC. Observe that $(1, 1)$-FGC already captures prominent survivable network design problems such as the 2-ECSS problem, the tree augmentation problem (TAP), and the more general forest augmentation problem (FAP); note that $(1, 0)$-FGC is the minimum cost spanning tree problem, which can be solved exactly. In some sense, (p, q)-FGC can be viewed as an interpolation between the p-ECSS problem (when all edges are safe) and the $(p + q)$-ECSS problem (when all edges are unsafe).

The FGC model attracted considerable interest in the network design community because of its elegance in capturing the non-uniformity of edges that appear in network design applications. Boyd et al. [7] gave several approximation algorithms for different choices of p and q. Their approximation ratios were the following: 4 when $q = 1$, $q + 1$ when $p = 1$, and $O(q \log n)$ for general p and q. Subsequently, two independent groups obtained improved approximation algorithms when p or q is small. Bansal et al. [5] gave an $O(1)$-approximation when $q = 2$. Concurrently, Chekuri and Jain [9] gave $O(p)$-approximations for the parameter settings $(p, 2)$, $(p, 3)$ and $(2p, 4)$, and an $O(q)$-approximation when $p = 2$. Bansal [4] removed the dependence on p in the approximation ratio for the $q = 3$ case, thus obtaining a constant approximation ratio. Nutov [18] gave improved approximation algorithms and/or analysis for $(p, 1)$-/$(p, 2)$-/$(2p, 3)$-FGC. Recently, Bansal, Cheriyan, Khanna, and Simmons [6] gave a 7-approximation algorithm when $p = 1$, thus removing the dependence on q for the simplest (1-)connectivity case in the FGC model.

1.1 Our Contributions

The focus of this work is (p, q)-FGC when p and q are arbitrary integer parameters. As mentioned above, the current best approximation algorithm for the general case has an approximation ratio of $O(q \log n)$ and is due to Boyd et al.

[7]. Our headline result is an improved $O(\log n)$-approximation algorithm. We note that the $O(q \log n)$-approximation algorithm is deterministic, whereas our algorithm is randomized.

Theorem 1. *Let $\mathcal{I} = (G = (V, E), \mathcal{S}, \mathcal{U}, c, p, q)$ be an instance of the (p, q)-FGC problem. There is a randomized algorithm that outputs, with probability at least $1/3$, a feasible solution whose cost is at most $200 \log n$ times the optimum value. The running time of the algorithm is polynomial in the input encoding size.*

Our algorithm in Theorem 1 is based on a new linear integer programming (IP) formulation for (p, q)-FGC. We present this formulation in Sect. 2 and also give an efficient separation oracle for the corresponding linear relaxation. The algorithm is then a standard independent randomized rounding. We solve the linear relaxation to obtain an optimal fractional solution x and then independently include each edge $e \in E$ in the final solution with probability $\min(1, 100 \cdot \log n \cdot x_e)$.

We make two remarks here. First, Chekuri and Jain [9] also gave a valid IP formulation for (p, q)-FGC along with an efficient separation oracle for the associated linear program (LP). However, their approximation algorithms for $(p, 2)$-FGC and $(2, q)$-FGC are not based on this formulation. Second, our LP is based on writing knapsack cover inequalities for some careful choice of edge-capacities that may depend on the cut. We will elaborate on this shortly. Although our LP is a bit more involved, it was inspired by the work of Bansal, Cheriyan, Khanna, and Simmons [6] where they give a 7-approximation algorithm for $(1, q)$-FGC for a general q.

Technical Overview and Comparison with Prior Work. We now give an overview of our results with a brief discussion of relevant prior work. Our algorithm builds on the techniques and ideas developed in a sequence of prior works [6–9].

The following notation and terminology will be used throughout the paper. The graphs we consider in this work are connected and undirected, with no self-loops. We allow the input graph G to have parallel edges; however, any solution F can contain at most one copy of any edge e that appears in E. Any vertex-set R induces a *cut* $\delta(R) := \{e \in E : |R \cap e| = 1\}$ consisting of all edges with exactly one endpoint in R. We call a vertex-set R *nontrivial* if R is neither empty nor all of V. Note that $\delta(\cdot)$ is symmetric, i.e., $\delta(R) = \delta(V \setminus R)$, and satisfies $\delta(R) = \emptyset$ if and only if R is trivial. For an edge-set F, we define $\delta_F(R) := F \cap \delta(R)$ as the set of F-edges on the cut $\delta(R)$. For any real number $z \in \mathbb{R}$, we use z^+ to denote $\max(0, z)$.

We say that an algorithm is *efficient* if its running time is polynomial in the input encoding size. In the context of (p, q)-FGC, the runtime of an efficient algorithm has a polynomial dependence on the encoding size of the edge-costs. Note that in any nontrivial instance, n, p, and q are bounded by $|E|$.

We start with the integer programming formulation of (p, q)-FGC that admits an efficient separation oracle for the associated linear relaxation. The following cut-characterization of feasible solutions follows immediately from the problem statement.

Proposition 1. *An edge-set $F \subseteq E$ is feasible for (p, q)-FGC if and only if for every nontrivial vertex-set R, the edge-set $\delta_F(R)$ contains p safe edges or $p+q$ (safe or unsafe) edges.*

In [7], Boyd et al. observed a useful connection between (p, q)-FGC and the capacitated version of the k-ECSS problem, denoted Cap-k-ECSS. In an instance of the Cap-k-ECSS problem, we have a graph $G = (V, E)$ with nonnegative edge-costs $\{c_e\}_e$, integer edge-capacities $\{u_e\}_e$, and a global edge-connectivity parameter $k \in \mathbb{Z}_{\geq 1}$. A feasible solution here is an edge-set $F \subseteq E$ such that for every $\emptyset \neq R \subsetneq V$ we have $u(\delta_F(R)) \geq k$, i.e., the global minimum cut in the capacitated graph $(V, F, \{u_e\}_{e \in F})$ is at least k. The algorithmic goal is then to find a feasible solution with the minimum cost. For the given instance of (p, q)-FGC, one can construct an instance of the Cap-k-ECSS problem such that feasible solutions to the former problem remain feasible for the latter problem: we take $k = p(p+q)$, $u_e = p+q$ for safe edges, and $u_e = p$ for unsafe edges. Although the two problems are not equivalent in general, the covering constraints in the Cap-k-ECSS problem are valid for (p, q)-FGC and therefore the optimum value for this instance of Cap-k-ECSS is at most OPT. Furthermore, Boyd et al. observed that a feasible solution to this Cap-k-ECSS instance is not far from feasible to (p, q)-FGC. In fact, they show that applying at most q rounds of augmenting steps is sufficient to achieve (p, q)-FGC feasibility, where in each step a suitable instance of the hitting-set problem is solved to fix cuts that are still violated. Overall, this leads to an $O(q \log n)$-approximation guarantee.

By the above discussion, a natural set of valid inequalities to include in an IP formulation for (p, q)-FGC is the following:

$$(p+q) \cdot x(\delta(R) \cap \mathcal{S}) + p \cdot x(\delta(R) \cap \mathcal{U}) \geq p(p+q) \quad \forall \emptyset \neq R \subsetneq V\,. \tag{1}$$

This is also the starting point for Chekuri and Jain [9]. However, to obtain a valid IP, they additionally include the following "robustness" constraint:

$$x(\delta(R) \setminus F') \geq p \quad \forall \emptyset \neq R \subsetneq V, \forall F' \subseteq \delta(R) \cap \mathcal{U} \text{ with } |F'| \leq q\,. \tag{2}$$

Chekuri and Jain also show that the separation oracle for the associated LP can be implemented efficiently (see Lemma 7 in [9]).

Our approach, instead of (2), is to include knapsack cover inequalities arising from (1). As mentioned before, this insight was inspired by the work of [6] on $(1, q)$-FGC. We also highlight that Chakrabarty et al. [8] used knapsack cover inequalities in the context of designing approximation algorithms for Cap-k-ECSS. The natural IP formulation for Cap-k-ECSS has a cut-capacity constraint $\sum_{e \in \delta(R)} u_e x_e \geq k$ for each $\emptyset \neq R \subsetneq V$. It is not hard to see that the associated LP may have an integrality gap of $\Omega(k)$. Chakrabarty et al. observed that including knapsack cover inequalities strengthens the associated LP.

Before formulating the knapsack cover inequalities in Lemma 1 below, we provide some motivation. For this informal discussion, let us fix a nontrivial vertex-set R and focus solely on the edges that are in $\delta(R)$. We assume that q is a large integer multiple of $p \geq 2$. Consider the following two solutions F_1 and F_2: F_1 has 1 safe edge and $p+q-1-q/p$ unsafe edges, while F_2 has $p-1$ safe edges and $1+q/p$ unsafe edges from $\delta(R)$. Clearly, both F_1 and F_2 satisfy the basic cut-capacity constraint (1) for R, while both are infeasible for (p,q)-FGC. We now show how a slightly modified version of (1) can detect the infeasibility of F_1. Let $J_1 := \delta_{F_1}(R) \cap \mathcal{S}$ consist of the unique safe edge in F_1. Since J_1 already contributes a safe edge to $\delta(R)$, the residual requirement (w.r.t. (p,q)-FGC feasibility) from $F_1 \setminus J_1$ is to have $p-1$ safe edges or $p+q-1$ total edges. Thus, we could rewrite (1) by giving a capacity of $p+q-1$ to safe edges and $p-1$ to unsafe edges and requiring that a solution has total capacity at least $(p-1)(p+q-1)$ on this cut. Since $(p-1)(p+q-1-q/p) < (p-1)(p+q-1)$, F_1 violates this modified version of (1). Similarly, let $J_2 := \delta_{F_2}(R) \cap \mathcal{U}$ consist of exactly $1+q/p$ unsafe edges in F_2. Since J_2 already contributes $1+q/p$ edges to $\delta(R)$, the residual requirement from $F_2 \setminus J_2$ is to have p safe edges or $p+q-1-q/p$ total edges. Again, as $(p-1)(p+q-1-q/p) < p(p+q-1-q/p)$ holds, F_2 violates a different version of (1). Thus, adjusting the capacities of safe and unsafe edges in a cut depending on the residual requirement can strengthen the formulation. In fact, we show that these adjustments are sufficient to obtain a valid IP formulation.

Lemma 1. *An edge-set $F \subseteq E$ is a feasible solution to (p,q)-FGC if and only if for every nonempty $R \subsetneq V$ and every partition $\delta_F(R) = J \sqcup K$, the following inequality holds:*

$$(p-|J \cap \mathcal{S}|)^+ \cdot |K| + (q-|J \cap \mathcal{U}|)^+ \cdot |K \cap \mathcal{S}| \geq (p-|J \cap \mathcal{S}|)^+ \cdot (p+q-|J|)^+. \quad (3)$$

Proof. For the implication, suppose F is a feasible solution, R is a nontrivial subset of vertices, and J is a subset of edges in $\delta_F(R)$. Since the left-hand side (LHS) of (3) is always nonnegative, it suffices to focus on the cases when $|J \cap \mathcal{S}| < p$ and $|J| < p+q$ hold. By the feasibility of F, either $|\delta_F(R)| \geq p+q$ or $|\delta_F(R) \cap \mathcal{S}| \geq p$ must hold (see Proposition 1). Since $K = \delta_F(R) \setminus J$, we have that either $|K| \geq p+q-|J|$ or $|K \cap \mathcal{S}| \geq p-|J \cap \mathcal{S}|$ holds. The inequality is trivial in the former case and holds in the latter case because $(p-|J \cap \mathcal{S}|)+(q-|J \cap \mathcal{U}|)^+ \geq p+q-|J|$. Hence, (3) holds for every choice of R and J when F is feasible.

For the converse, suppose F is an infeasible solution. By Proposition 1, there is a nontrivial vertex-set R such that both $|\delta_F(R)| < p+q$ and $|\delta_F(R) \cap \mathcal{S}| < p$ hold. Taking $J = \delta_F(R)$ and $K = \emptyset$, we see that the LHS of (3) is 0, while the right-hand side (RHS) is positive. This gives a violation in (3), as desired. □

Our $O(\log n)$-approximation algorithm for (p,q)-FGC is based on independent randomized rounding of an optimal LP solution. The output of our algorithm is a random edge-set F that is feasible with constant probability. The following result, borrowed from [7], will be useful in testing the feasibility of F.

Lemma 2. ***(Proposition 5.1 in [7]).*** *For any $F \subseteq E$, we can efficiently check the feasibility of F for the given instance of (p,q)-FGC.*

In a capacitated network, we say that a cut $\delta(R)$ is an *α-approximate min-cut*, if its capacity is within α from the minimum cut value. In a seminal work, Karger [15] gave the following bound on the number of near-minimum cuts.

Theorem 2. *([15]). For any $\alpha \geq 1$, any capacitated network has at most $O(n^{2\alpha})$ α-approximate minimum cuts.*

This paper also gave a randomized polynomial-time algorithm to enumerate all near-minimum cuts with high probability. Subsequently, Nagamochi et al. [17] gave a deterministic $n^{O(\alpha)}$-time algorithm to enumerate all α-approximate minimum cuts. These results are applied several times in our proofs. First, in Lemma 2, to check the feasibility of a candidate (p, q)-FGC solution by relating it to an instance of the Cap-k-ECSS problem; see [7] for more details. Next, in our separation oracle for checking if a candidate fractional solution is feasible to the LP relaxation (see Lemma 6). Last, in the proof of the rounding result (Theorem 4), where we use a union-bound argument over all cuts to show that the output of our algorithm is feasible with constant probability.

1.2 Other Related Work

The notion of introducing non-uniformity in network design problem via safe and unsafe edges is due to Adjiashvili [1], where they initially considered flexible versions of edge-disjoint (s, t)-paths problem. Also, see the follow-up work of Adjiashvili et al. [3]. In [9], Chekuri and Jain introduced the more general (p, q)-Flex-ST problem and gave approximation algorithms when p or q is sufficiently small. Here, the input consists of source-sink pairs $\{(s_i, t_i)\}_i$, each with prescribed connectivity p_i and robustness q_i parameters satisfying $p_i \leq p$ and $q_i \leq q$, and the goal is to find a minimum-cost $F \subseteq E$ such that for any i, the subgraph (V, F) contains p_i edge-disjoint (s_i, t_i)-paths regardless of which q_i unsafe edges fail.

The FGC model has also been studied in the unweighted setting, i.e., when all edges have unit cost. Let α_k denote the current best approximation ratio for the unweighted k-ECSS problem. Boyd et al. [7] gave a $4\alpha_2/(2\alpha_2+1)$-approximation for unweighted $(1, 1)$-FGC. Using the result of [11], this implies an approximation ratio of ≈ 1.45. Recently, Hyatt-Denesik, Jabal-Ameli, and Sanità [13] tightened the analysis of [7] to obtain an approximation ratio of $10/7$. They also gave a $1 + O(1/\sqrt{q})$-approximation algorithm for unweighted $(1, q)$-FGC. Nutov [18] gave an $(\alpha_p + 2q/p)$-approximation algorithm for unweighted (p, q)-FGC.

In their work, Hyatt-Denesik et al. [13] introduced and studied the vertex connectivity version of FGC. Here, vertices are either safe or unsafe, and the goal is to find a minimum-size spanning subgraph that is connected and has the property that no unsafe vertex is a cut-vertex in this subgraph.

The study of the FGC model gave a new impetus to a decades-long open problem in survivable network design. For example, this model motivated Bansal et al. [5] to extend the analysis of the primal-dual method to handle augmentation problems when the cut-requirement function does not have desirable uncrossing properties.

2 Integer Programming Formulation for (p, q)-FGC

As mentioned in the introduction, a key contribution in this work is a new IP formulation for (p, q)-FGC based on Lemma 1. For each edge $e \in E$, we have a Boolean variable x_e that denotes the inclusion of e in a feasible solution F, i.e., $x_e = 1$ if and only if $e \in F$. Linearization of (3) yields the following valid IP. By Lemma 1, a 0–1 solution is feasible if and only if it corresponds to a feasible set F.

$$\min \quad \sum_{e \in E} c_e x_e \tag{4}$$

$$\text{s.t.:} \quad (p - |J \cap \mathcal{S}|)^+ \cdot x(K) + (q - |J \cap \mathcal{U}|)^+ \cdot x(K \cap \mathcal{S}) \geq (p - |J \cap \mathcal{S}|)^+ \cdot (p + q - |J|)^+ \quad \forall \emptyset \neq R \subsetneq V, \forall \text{ partitions } \delta(R) = J \sqcup K \tag{5}$$

$$0 \leq x_e \leq 1, \quad x_e \in \mathbb{Z} \qquad \forall e \in E \tag{6}$$

Our linear program (LP) for (p, q)-FGC is obtained by relaxing the integrality condition on the x_e's as $0 \leq x_e \leq 1$. Let LP-OPT denote the optimum value of this LP.

Lemma 3. *We have* LP-OPT $\leq$ OPT.

A few remarks are in order. Although the number of constraints in the LP is exponential in $|E|$, we will show that it admits an efficient separation oracle. Next, for any choice of a nontrivial vertex-set R and the fixed choice of $J = \emptyset$, constraint (5) reduces to the covering inequality: $p \cdot x(\delta(R)) + q \cdot x(\delta(R) \cap \mathcal{S}) \geq p(p+q)$. These constraints are much simpler to work with and will be crucial in designing an efficient separation oracle; see Lemma 6.

For an $x \in [0,1]^E$, we define the capacity function $u_x : E \to \mathbb{R}_{\geq 0}$ on the edges of G as follows: $u_x(e) := (p + q \cdot \mathbf{1}_{e \in \mathcal{S}}) x_e$ for any $e \in E$. For convenience, we refer to the graph G equipped with the capacity function u_x as the capacitated graph H_x. The following lemma gives two conditions, one necessary and the other sufficient, for x to be a feasible LP solution.

Lemma 4. *Let $x \in [0,1]^E$ and $H_x = (G, u_x)$ be the capacitated graph derived from G and x. We have the following:*

1. *If $u_x(\delta(R)) < p(p+q)$ for some nontrivial $R \subset V$, then x violates constraint* (5) *for this choice of R and $J = \emptyset$. Therefore, if the minimum cut in H_x is less than $p(p+q)$, then x is infeasible to the LP.*
2. *If $u_x(\delta(R)) \geq 2p(p+q)$ for some nontrivial $R \subset V$, then x satisfies constraint* (5) *for this choice of R and every choice of $J \subseteq \delta(R)$. Therefore, x is feasible for the LP if the minimum cut in H_x is at least $2p(p+q)$.*

Proof. For the first part, observe that when $J = \emptyset$ the LHS of (5) is exactly $u(\delta(R))$ and the RHS is $p(p+q)$. So we get a violated inequality if $u_x(\delta(R)) < p(p+q)$ holds.

For the second part, suppose R is such that $u_x(\delta(R)) \geq 2p(p+q)$ holds, and $J \subseteq \delta(R)$ is an arbitrary subset of edges in $\delta(R)$. Clearly, constraint (5) is satisfied when $|J \cap \mathcal{S}| \geq p$ or $|J| \geq p+q$, so we assume otherwise. Thus, the RHS of the constraint is $(p - |J \cap \mathcal{S}|) \cdot (p+q-|J|)$. By definition, $u_x(\delta(R)) = (p+q) \cdot x(\delta(R) \cap \mathcal{S}) + p \cdot x(\delta(R) \cap \mathcal{U})$. Since $u_x(\delta(R)) \geq 2p(p+q)$, we must have either $x(\delta(R) \cap \mathcal{S}) \geq p$ or $x(\delta(R) \cap \mathcal{U}) \geq p+q$. In the former case, we have:

$$\begin{aligned}
\text{LHS of (5)} &= (p - |J \cap \mathcal{S}|)^+ \cdot x(\delta(R) \setminus J) + (q - |J \cap \mathcal{U}|)^+ \cdot x((\delta(R) \setminus J) \cap \mathcal{S}) \\
&\geq (p+q-|J|) \cdot x((\delta(R) \setminus J) \cap \mathcal{S}) \\
&\geq (p+q-|J|) \cdot \big(x(\delta(R) \cap \mathcal{S}) - |J \cap S|\big) && (x \in [0,1]^E) \\
&\geq (p - |J \cap \mathcal{S}|) \cdot (p+q-|J|) = \text{RHS of (5)}
\end{aligned}$$

In the latter case, we have:

$$\begin{aligned}
\text{LHS} &\geq (p - |J \cap \mathcal{S}|) \cdot x(\delta(R) \setminus J) \\
&\geq (p - |J \cap \mathcal{S}|) \cdot \big(x(\delta(R)) - |J|\big) && (x \in [0,1]^E) \\
&\geq (p - |J \cap \mathcal{S}|) \cdot (p+q-|J|) = \text{RHS}
\end{aligned}$$

Note that the final equality in both chain of inequalities is because of our assumption that $|J \cap \mathcal{S}| < p$ and $|J| < p+q$. □

The next lemma is based on a well-known trick that is used to check covering inequalities en masse.

Lemma 5. *Let $x \in [0,1]^E$ and $\emptyset \neq R \subseteq V$. Let L_s and L_u be ordered lists of the safe (respectively, unsafe) edges in $\delta(R)$ in nonincreasing order of their x_e values, with ties broken arbitrarily. For nonnegative integers $a \in \{0, 1, \ldots, \min(p-1, |L_s|)\}$ and $b \in \{0, 1, \ldots, \min(p+q-1, |L_u|)\}$, let $J_{a,b}$ denote the edge-set consisting of the first a edges in L_s and the first b edges in L_u. Let $\mathcal{J}$ be the collection of all $J_{a,b}$. If x satisfies constraint (5) for every choice of $J \in \mathcal{J}$, then x satisfies the same constraint for every choice of $J \subseteq \delta(R)$.*

Proof. By the form of constraint (5), it is clear that the numerical quantities that are relevant to us are $\alpha_1 := (p - |J \cap \mathcal{S}|)^+$, $\alpha_2 := (q - |J \cap \mathcal{U}|)^+$, and $\alpha_3 := (p - |J \cap \mathcal{S}|)^+ \cdot (p+q-|J|)^+$. Since the values of $a := |J \cap \mathcal{S}|$ and $b := |J \cap \mathcal{U}|$ completely determine the values of α_is, we can check constraint (5) by grouping the possible choices of J for each $a \in \{0, 1, \ldots, \min(p-1, |\delta(R) \cap \mathcal{S}|)\}$ and $b \in \{0, 1, \ldots, \min(p+q-1, |\delta(R) \cap \mathcal{U}|)\}$; note that if $\alpha_3 = 0$, then the constraint is trivially satisfied. Fix one such choice of a and b. By grouping variables according to whether they are safe or unsafe, constraint (5) can be re-written as:

$$(\alpha_1 + \alpha_2) \cdot \sum_{e \in \delta_{\mathcal{S}}(R) \setminus J} x_e + \alpha_1 \cdot \sum_{e \in \delta_{\mathcal{U}}(R) \setminus J} x_e \overset{?}{\geq} \alpha_3.$$

Conditioned on $|J \cap \mathcal{S}| = a$ and $|J \cap \mathcal{U}| = b$, the set $J_{a,b}$ minimizes the LHS of the above inequality. So, if constraint (5) is satisfied by $J_{a,b}$, then it is satisfied by all such J. It follows that if constraint (5) is satisfied by $J_{a,b}$ for all choices of a and b, then it is satisfied for every choice of $J \subseteq \delta(R)$. □

The above two lemmas imply a natural strategy for devising a separation oracle for our LP.

Lemma 6. *Let $x \in [0,1]^E$. In time polynomial in n, p, and q, we can determine whether x satisfies constraint* (5) *for every choice of a $\emptyset \neq R \subsetneq V$ and a set of edges $J \subseteq \delta(R)$, or find a violated inequality.*

Proof. Let $H_x = (G, u_x)$ denote the capacitated graph derived from G and x. Using standard tools, we can efficiently compute a minimum cut in H_x. Let R^* denote one of the shores of this minimum cut, and $\lambda := u_x(\delta(R^*))$. If $\lambda < p(p+q)$, then we get a violated inequality for $(R = R^*, J = \emptyset)$ by using the first part of Lemma 4.

Now suppose $\lambda \geq p(p+q)$. By the second part of Lemma 4, a vertex-set R with $u_x(\delta(R)) \geq 2p(p+q)$ cannot give a violated inequality for any choice of $J \subseteq \delta(R)$. Thus, it suffices to focus only on those R with $u_x(\delta(R)) < 2p(p+q)$. Define $\mathcal{R} := \{R \in 2^V \setminus \{\emptyset, V\} : u_x(\delta(R)) < 2p(p+q)\}$. Note that $|\mathcal{R}| = O(|V|^4)$ (Theorem 2) by our assumption that $\lambda \geq p(p+q)$. In fact, using [17], we can efficiently enumerate $\mathcal{R}$ in time polynomial in n. Now fix some $R \in \mathcal{R}$. Let L_s and L_u be the ordered lists of the safe (respectively, unsafe) edges in $\delta(R)$ in nonincreasing order of their x_e values, with ties broken arbitrarily. By Lemma 5, if there is a violation in (5) for some J, then there is a violation for some $J_{a,b}$ with $a \in \{0,1,\ldots,\min(p-1,|L_s|)\}$ and $b \in \{0,1,\ldots,\min(p+q-1,|L_u|)\}$, where $J_{a,b}$ denotes the set of edges consisting of the first a edges in L_s and the first b edges in L_u. Thus, we either find a violated inequality for this choice of R, if one exists, or we have a certificate that there are no violations for any choice of J. □

In summary, x is feasible if and only if all the following conditions holds: (i) $x \in [0,1]^E$; (ii) the minimum cut in H_x is at least $p(p+q)$; and (iii) constraint (5) holds for every (R, J), where R has $u_x(\delta(R)) < 2p(p+q)$ and J consists of up to $p-1$ safe edges and up to $p+q-1$ unsafe edges with the largest x_e-values. Since the feasible region has bounded facet complexity (and is in particular in the box $[0,1]^E$), and an efficient separation oracle is available, we can use the Ellipsoid method to obtain an optimal solution in polynomial time; see [12, Chapter 6].

Theorem 3. *Suppose that we are given a feasible instance of the (p,q)-FGC problem. We can efficiently compute a vector $x \in [0,1]^E$ that is feasible for the linear relaxation of the integer program given by* (4)–(6) *and also satisfies $c^\top x := \sum_{e \in E} c_e x_e \leq \text{OPT}$.*

3 Independent Rounding Algorithm for (p, q)-FGC

Our main result in this section is an algorithm that takes as input a fractional LP solution x and with constant probability outputs a random edge-set F that is an $O(\log n)$-approximate solution to the given instance of (p, q)-FGC. Our algorithm is conceptually simple, relying on independent randomized rounding. Define $y \in [0,1]^E$ as follows: for each edge e, $y_e := \min(1, 100 \cdot \log n \cdot x_e)$. Let F be a random set of edges that is obtained by independently including each edge $e \in E$ in F with probability y_e.

Theorem 4. *With probability at least* $1/3$, *F is feasible and has cost at most* $200 \cdot \log n \cdot c^\top x$.

Running the algorithm with an optimal LP solution x obtained from Theorem 3 gives us an $O(\log n)$ approximation algorithm for (p, q)-FGC, thus proving our headline result (Theorem 1). We can efficiently check whether $F \subseteq E$ is a feasible solution to the given (p, q)-FGC instance (Lemma 2). (This involves enumerating all near-minimum cuts, as in the proof of Lemma 6). Thus, the algorithm can be run repeatedly until a feasible solution with cost $O(\log n) \cdot c^\top x$ is obtained. This gives a Las Vegas $O(\log n)$-approximation algorithm with expected running time that is polynomial in the input encoding size.

We give a proof of Theorem 4 at the end of this section. Like in [8,9], we use the well-known technique of applying a union-bound argument over all cuts in a graph. The following version of the Chernoff lower tail bound will be useful to us; for example, see [16].

Lemma 7. ***(Chernoff bound).*** *Let $\{X_j\}_j$ be a finite collection of independent random variables taking values in $\{0, 1\}$, and let Z denote their sum. For any $\mu \le \mathbb{E}Z$ and $\eta \in (0, 1)$, we have:*

$$\mathbb{P}[Z < (1 - \eta)\mu] \le \exp(-\mu\eta^2/2).$$

In particular, by taking $\eta = 2/3$, we can derive $\mathbb{P}[Z < \mu/3] \le \exp(-\mu/5)$.

Define $A := \{e \in E : y_e = 1\}$ as the set of edges (safe or unsafe) that are always included in F. By definition, for any $e \notin A$ we have $y_e = 100 \cdot \log n \cdot x_e < 1$. For any nontrivial vertex-set R, define B_R to be the indicator random variable for the *bad* event $\{|\delta_F(R) \cap \mathcal{S}| < p$ and $|\delta_F(R)| < p + q\}$. Note that $B_R = 1$ whenever the cut $\delta(R)$ certifies the infeasibility of F. The following lemma is trivial.

Lemma 8. *If $|\delta_A(R) \cap \mathcal{S}| \ge p$ or $|\delta_A(R)| \ge p + q$ holds, then $\mathbb{P}[B_R = 1] = 0$.*

Given the above lemma, it suffices to focus only on those R that satisfy both $|\delta_A(R) \cap \mathcal{S}| < p$ and $|\delta_A(R)| < p + q$. Since x is a feasible LP solution, it satisfies constraint (5) for this choice of R and all choices of $J \subseteq \delta(R)$. For convenience, we restate constraint (5) below:

$$\begin{aligned}\left((p - |J \cap \mathcal{S}|) + (q - |J \cap \mathcal{U}|)^+\right) \cdot x(K \cap \mathcal{S}) + (p - |J \cap \mathcal{S}|) \cdot x(K \cap \mathcal{U}) \\ \ge (p - |J \cap \mathcal{S}|)(p + q - |J|). \qquad (7)\end{aligned}$$

As the edges in A have been purchased fully and the LP solution x satisfies all knapsack cover inequalities, it is natural to consider the partition $J \sqcup K = \delta(R)$ where $J := A \cap \delta(R)$. As none of the A-edges appear in K, we have $y_e < 1$ for every $e \in K$. Changing the variable in (7) from x to y gives:

$$\begin{aligned}\big((p - |J \cap \mathcal{S}|) + (q - |J \cap \mathcal{U}|)^+\big) \cdot y(K \cap \mathcal{S}) + (p - |J \cap \mathcal{S}|) \cdot y(K \cap \mathcal{U}) \\ \geq 100 \cdot \log n \cdot (p - |J \cap \mathcal{S}|)(p + q - |J|). \qquad (8)\end{aligned}$$

For an edge $e \in K$, let Y_e denote the indicator random variable for the event $\{e \in F\}$. Note $\mathbb{E}Y_e = y_e = 100 \cdot \log n \cdot x_e$. Define the integer random variables $Z_s := \sum_{e \in K \cap \mathcal{S}} Y_e$ and $Z_u := \sum_{e \in K \cap \mathcal{U}} Y_e$ to be the number of safe and unsafe k-edges, respectively, that are included in F. The following lemma gives a crude upper bound on $\mathbb{P}[B_R = 1]$ for any R.

Lemma 9. *We have* $\mathbb{P}[B_R = 1] \leq n^{-10}$.

Proof. The lemma is trivial if R satisfies the hypothesis of Lemma 8, so we assume otherwise. By the above discussion, (8) implies that at least one of the following three cases occurs.

1. Suppose $|J \cap \mathcal{U}| < q$ and $y(K \cap \mathcal{S}) \geq 50 \cdot \log n \cdot (p - |J \cap \mathcal{S}|) =: \mu_1$ hold. Observe that:

$$\begin{aligned}\mathbb{P}[B_R = 1] &\leq \mathbb{P}[|\delta_F(R) \cap \mathcal{S}| < p] \\ &= \mathbb{P}[|F \cap K \cap \mathcal{S}| < p - |J \cap \mathcal{S}|] \\ &= \mathbb{P}[Z_s < p - |J \cap \mathcal{S}|] \leq \mathbb{P}[Z_s < \mu_1/3] && (\mathbb{E}Z_s = y(K \cap \mathcal{S})) \\ &\leq \exp(-10 \cdot \log n \cdot (p - |J \cap \mathcal{S}|)) && \text{(Lemma 7)} \\ &\leq \exp(-10 \cdot \log n) = n^{-10}. && (p - |J \cap \mathcal{S}| \geq 1)\end{aligned}$$

2. Suppose $|J \cap \mathcal{U}| < q$ and $y(K \cap \mathcal{U}) \geq 50 \cdot \log n \cdot (p + q - |J|) =: \mu_2$ hold. We repeat the same calculation as above for the random variable Z_u, whose expectation $y(K \cap \mathcal{U})$ is at least μ_2, to get:

$$\begin{aligned}\mathbb{P}[B_R = 1] &\leq \mathbb{P}[|\delta_F(R)| < p + q] = \mathbb{P}[|F \cap K| < p + q - |J|] \\ &\leq \mathbb{P}[|F \cap K \cap \mathcal{U}| < p + q - |J|] = \mathbb{P}[Z_u < p + q - |J|] \\ &\leq \mathbb{P}[Z_u < \mu_2/3] \leq \exp(-10 \cdot \log n \cdot (p + q - |J|)) \leq n^{-10}.\end{aligned}$$

3. Suppose $|J \cap \mathcal{U}| \geq q$ and $y(K) \geq 100 \cdot \log n \cdot (p + q - |J|) =: \mu_3$ hold. We repeat the same calculation for the random variable $Z_s + Z_u$, whose expectation $y(K)$ is at least μ_3, to obtain $\mathbb{P}[B_R = 1] \leq \mathbb{P}[Z_s + Z_u < p + q - |J|] \leq n^{-20}$.

Thus, the bad event B_R occurs with probability at most n^{-10}. □

We now derive a much stronger tail bound for cuts with large u_x capacity.

Lemma 10. *Suppose* $u_x(\delta(R)) \geq \ell p(p + q)$ *holds for some* $\ell \geq 4$. *We have* $\mathbb{P}[B_R = 1] \leq n^{-5\ell}$.

Proof. We follow the same template as in the proof of Lemma 9. We first show that $x(K \cap \mathcal{S}) \geq \ell p/4$ or $x(K \cap \mathcal{U}) \geq \ell(p+q)/4$ holds, and then use Chernoff tail bounds to establish the claim. The lemma is trivial if R satisfies the hypothesis of Lemma 8, so we assume otherwise. By the definition of u_x, we have $u_x(\delta(R)) = (p+q) \cdot x(\delta(R) \cap \mathcal{S}) + p \cdot x(\delta(R) \cap \mathcal{U})$. Since $u_x(\delta(R)) \geq \ell p(p+q)$, we must have that $x(\delta(R) \cap \mathcal{S}) \geq \ell p/2$ or $x(\delta(R) \cap \mathcal{U}) \geq \ell(p+q)/2$ holds. By our assumption on R, we have $x(J \cap \mathcal{S}) \leq |J \cap \mathcal{S}| < p$ and $x(J \cap \mathcal{U}) \leq x(J) \leq |J| < p+q$ since $x \in [0,1]^E$. As J and K partition $\delta(R)$, either $x(K \cap \mathcal{S}) \geq \ell p/4$ or $x(K \cap \mathcal{U}) \geq \ell(p+q)/4$ holds. Here, we use $\ell \geq 4$. It remains to apply Chernoff bounds in each of the two cases. We recall that $y_e = 100 \log n \cdot x_e < 1$ for each $e \in K$.

1. Suppose $y(K \cap \mathcal{S}) \geq 25 \cdot \ell \cdot \log n \cdot p =: \mu_1$ holds. We have:

$$\begin{aligned} \mathbb{P}[B_R = 1] &\leq \mathbb{P}[|\delta_F(R) \cap \mathcal{S}| < p] \leq \mathbb{P}[|F \cap K \cap \mathcal{S}| < p] \\ &\leq \mathbb{P}[Z_s < \mu_1/3] && (\mathbb{E}Z_s = y(K \cap \mathcal{S})) \\ &\leq \exp(-5 \cdot \ell \cdot \log n \cdot p) \leq n^{-5\ell} && \text{(Lemma 7)} \end{aligned}$$

2. Suppose $y(K \cap \mathcal{U}) \geq 25 \cdot \ell \cdot \log n \cdot (p+q) =: \mu_2$ holds. Again:

$$\begin{aligned} \mathbb{P}[B_R = 1] &\leq \mathbb{P}[|\delta_F(R)| < p+q] \leq \mathbb{P}[|F \cap K \cap \mathcal{U}| < p+q] \\ &\leq \mathbb{P}[Z_u < \mu_2/3] && (\mathbb{E}Z_u = y(K \cap \mathcal{U})) \\ &\leq \exp(-5 \cdot \ell \cdot \log n \cdot (p+q)) \leq n^{-5\ell}. && \text{(Lemma 7)} \end{aligned}$$

Thus, in both cases, the bad event B_R occurs with probability at most $n^{-5\ell}$. □

We are now ready to prove the main result (Theorem 4) of this section.

Proof. First, observe that the expected cost of F is at most $100 \log n \cdot c^\top x$ since each edge e is included in F with probability $y_e \leq 100 \cdot \log n \cdot x_e$. By Markov's inequality, $c(F) \geq 200 \log n \cdot c^\top x$ holds with probability at most $1/2$.

Second, we use a union-bound argument to show that F is infeasible with probability at most $1/6$. For F to be infeasible, a bad event B_R must occur for at least one nontrivial R. We partition the nontrivial cuts in H_x as $\mathcal{C}_{<4} \cup \mathcal{C}_4 \cup \mathcal{C}_5 \cup \ldots$, where $\mathcal{C}_{<4}$ is the set of all R with $u_x(\delta(R)) \in [p(p+q), 4p(p+q))$ and for $\ell \in \mathbb{Z}_{\geq 4}$, $\mathcal{C}_\ell$ consists of all R with $u_x(\delta(R)) \in [\ell p(p+q), (\ell+1)p(p+q))$. Note $\mathcal{C}_\ell = \emptyset$ when $\ell > |E|$. By Theorem 2 we have $|\mathcal{C}_{<4}| = O(n^8)$, where we use the fact that the minimum cut in H_x is at least $p(p+q)$ (see the first part of Lemma 4). In Lemma 9 we showed that $\mathbb{P}[B_R = 1] \leq n^{-10}$ for any R. Thus, $\sum_{R \in \mathcal{C}_{<4}} \mathbb{P}[B_R = 1] \leq O(n^{-2})$. Again, by Theorem 2, we have $|\mathcal{C}_\ell| \leq O(n^{2\ell+2})$ for any integer $\ell \geq 4$. In Lemma 10 we showed that $\mathbb{P}[B_R = 1] \leq n^{-5\ell}$ for any $R \in \mathcal{C}_\ell$. Thus, for any integer $\ell \geq 4$ we have $\sum_{R \in \mathcal{C}_\ell} \mathbb{P}[B_R = 1] \leq O(n^{-3\ell+2}) = O(n^{-2\ell})$. Overall, for sufficiently large n, we get:

$$\sum_R \mathbb{P}[B_R = 1] = \sum_{R \in \mathcal{C}_{<4}} \mathbb{P}[B_R = 1] + \sum_{\ell=4}^{\infty} \Big(\sum_{R \in \mathcal{C}_\ell} \mathbb{P}[B_R = 1] \Big)$$
$$\leq O(n^{-2}) + \sum_{\ell \geq 4} O(n^{-2\ell}) \leq 1/6.$$

Thus, F is an $O(\log n)$-approximate solution with constant probability. □

4 Conclusions

In this paper, we give the first linear programming relaxation for general (p, q)-FGC, and obtained an $O(\log n)$ approximation by an independent randomized rounding of this relaxation. Obtaining an $f(p, q)$-approximation algorithm for (p, q)-FGC for some function $f(\cdot, \cdot)$ remains open. It remains an open question whether the integrality gap of our LP relaxation can be bounded as $f(p, q)$, or even by a universal constant.

In the context of network design, a standard and powerful technique is iterative rounding, first introduced by Jain for survivable network design [14]. There are two main obstacles to using this approach in our linear program. First, the demand function does not satisfy skew submodularity or other properties that enable uncrossing. Recall that while we can solve the LP efficiently, it relies on checking all near-minimum cuts R and $O(m)$ edge sets J for each. Second, iterative rounding is typically applicable for covering and packing LPs with uniform coefficients for each variable; in our constraints, the coefficients of variables depend on the set J.

Acknowledgments. The authors thank Joseph Cheriyan for discussions on the topic which was the main inspiration for this work.

References

1. Adjiashvili, D.: Fault-tolerant shortest paths - beyond the uniform failure model. CoRR abs/1301.6299 (2013). https://doi.org/10.48550/arXiv.1301.6299
2. Adjiashvili, D., Hommelsheim, F., Mühlenthaler, M.: Flexible graph connectivity. In: Proceedings of the 21st Conference on Integer Programming and Combinatorial Optimization, vol. 12125, pp. 13–26 (2020). https://doi.org/10.1007/978-3-030-45771-6_2
3. Adjiashvili, D., Hommelsheim, F., Mühlenthaler, M., Schaudt, O.: Fault-tolerant edge-disjoint s-t paths — beyond uniform faults. In: Proceedings of the 18th Scandinavian Symposium and Workshops on Algorithm Theory (SWAT), vol. 227, pp. 1–19 (2022). https://doi.org/10.4230/LIPIcs.SWAT.2022.5
4. Bansal, I.: A Global Analysis of the Primal-Dual Method for Edge Augmentation Problems. In: Megow, N., Basu, A. (eds.) IPCO2025, vol. 15620, pp.xx–xx (2025). https://doi.org/10.1007/978-3-031-93112-3_5

5. Bansal, I., Cheriyan, J., Grout, L., Ibrahimpur, S.: Improved approximation algorithms by generalizing the primal-dual method beyond uncrossable functions. In: Proceedings of the 50th International Colloquium on Automata, Languages, and Programming, vol. 261, article 15, pp. 1–19 (2023). https://doi.org/10.4230/LIPIcs.ICALP.2023.15
6. Bansal, I., Cheriyan, J., Khanna, S., Simmons, M.: Improved approximation algorithms for flexible graph connectivity and capacitated network design (2024). https://arxiv.org/abs/2411.18809
7. Boyd, S.C., Cheriyan, J., Haddadan, A., Ibrahimpur, S.: Approximation algorithms for flexible graph connectivity. Math. Program. **204**, 493–516 (2024). https://doi.org/10.1007/s10107-023-01961-5
8. Chakrabarty, D., Chekuri, C., Khanna, S., Korula, N.: Approximability of capacitated network design. Algorithmica **72**(2), 493–514 (2015). https://doi.org/10.1007/s00453-013-9862-4
9. Chekuri, C., Jain, R.: Approximation algorithms for network design in non-uniform fault models. In: Proceedings of the 50th International Colloquium on Automata, Languages, and Programming, vol. 261, article 36, pp. 1–20 (2023). https://doi.org/10.4230/LIPIcs.ICALP.2023.36
10. Gabow, H.N., Goemans, M.X., Tardos, É., Williamson, D.P.: Approximating the smallest k-edge connected spanning subgraph by LP-rounding. Networks **53**(4), 345–357 (2009). https://doi.org/10.1002/net.20289
11. Garg, M., Grandoni, F., Ameli, A.J.: Improved approximation for two-edge-connectivity. In: Proceedings of the 36th Symposium on Discrete Algorithms, pp. 2368–2410 (2023). https://doi.org/10.1137/1.9781611977554.ch92
12. Grötschel, M., Lovász, L., Schrijver, A.: The ellipsoid method and its consequences in combinatorial optimization. Combinatorica **1**, 169–197 (1981). https://doi.org/10.1007/BF02579273
13. Hyatt-Denesik, D., Jabal-Ameli, A., Sanità, L.: Improved approximations for flexible network design. In: Proceedings of the 32nd European Symposium on Algorithms, vol. 308, pp. 1–14 (2024). https://doi.org/10.4230/LIPIcs.ESA.2024.74
14. Jain, K.: A factor 2 approximation algorithm for the generalized Steiner network problem. Combinatorica **21**(1), 39–60 (2001). https://doi.org/10.1007/s004930170004
15. Karger, D.R.: Global min-cuts in *RNC*, and other ramifications of a simple mincut algorithm. In: Proceedings of the 4th Symposium on Discrete Algorithms, pp. 21–30 (1993). https://doi.org/10.5555/313559.313605
16. Mitzenmacher, M., Upfal, E.: Probability and Computing: Randomized Algorithms and Probabilistic Analysis. Cambridge University Press (2005)
17. Nagamochi, H., Nishimura, K., Ibaraki, T.: Computing all small cuts in an undirected network. SIAM J. Discret. Math. **10**(3), 469–481 (1997). https://doi.org/10.1137/S0895480194271323
18. Nutov, Z.: improved approximation algorithms for covering pliable set families and flexible graph connectivity. In: Proceedings of the 22nd International Workshop on Approximation and Online Algorithms, pp. 151–166 (2025). https://doi.org/10.1007/978-3-031-81396-2_11

A Black-Box Approach for Exogenous Replenishment in Online Resource Allocation

Suho Kang, Ziyang Liu(✉), and Rajan Udwani

University of California, Berkeley, USA
{suho_kang,ziyang_liu,rudwani}@berkeley.edu

Abstract. In a typical online resource allocation problem, we start with a fixed inventory of resources and make online allocation decisions in response to resource requests that arrive sequentially over a finite horizon. We consider settings where the inventory is replenished over time according to an unknown exogenous process. We introduce black-box methods that extend any existing algorithm, originally designed without considering replenishment, into one that works with an arbitrary (adversarial or stochastic) replenishment process and asymptotically retains the competitive ratio for sufficiently large starting inventory. Our proofs rely only on properties of the general problem formulation, allowing seamless integration of exogenous replenishment into a large body of existing algorithmic results for both adversarial and stochastic arrival models.

Keywords: Online resource allocation · Exogenous replenishment · Competitive ratio

1 Introduction

Online decision making problems where a fixed inventory of resources is allocated to sequentially arriving resource requests, have been extensively studied, with applications in online advertising (Mehta et al. [13]), airline pricing (Talluri and Van Ryzin [15]), personalized recommendation (Golrezaei et al. [7]), sharing economy (Rusmevichientong et al. [14]) and many other domains. In the settings we consider, the goal is to design an online algorithm that maximizes a global objective, such as total reward, over a possibly unknown planning horizon. A particularly interesting theoretical goal is to find an algorithm that has the highest possible competitive ratio, which is the worst-case ratio of the expected total reward of an online algorithm to the expected total reward of the optimal offline solution (suitably defined for each problem).

There is a vast body of literature on such algorithms for various settings, we refer to [12] and [10] for recent surveys. Almost all of the settings in the literature assume that the inventory is fixed initially and is not replenished during the planning horizon. In contrast, we are interested in settings where resources may

N. Megow and A. Basu (Eds.): IPCO 2025, LNCS 15620, pp. 326–340, 2025.
https://doi.org/10.1007/978-3-031-93112-3_24

be replenished through an (adversarial or stochastic) exogenous process. From this perspective, classical settings without replenishment are a special case where the replenishment is identically zero. The main question we try to answer in this paper is as follows.

Is there a way to generalize existing algorithms for adversarial or stochastic arrivals while preserving their performance guarantees in the presence of exogenous replenishment?

1.1 Previous Work

The essence of this question can be traced back to Mehta et al. [13], who introduced the classical Adwords problem for search ads. In their model, a fixed set of advertisers, each with a limited budget, is determined at the beginning of the planning horizon, and opportunities for displaying ads arrive sequentially in an online manner. They proposed the Inventory Balancing (IB) algorithm, which achieves a competitive ratio of $(1 - 1/e)$ in the *large inventory regime*[1]. As a practical extension of this model, they observed that IB's competitive ratio guarantees can be generalized to settings where new advertisers join the platform in an online manner and where budgets may be replenished during the planning horizon, provided *each replenishment is sufficiently large.*

Chan and Farias [2] introduced a broader framework, the stochastic depletion problem, which captures a wide range of online allocation problems, including the Adwords problem. Their framework allows for modeling an exogenous replenishment process in which arbitrary (possibly small) amount of new inventory may arrive in each period. In this context, they showed that a simple greedy (myopic) algorithm is $\frac{1}{2}$ competitive even in adversarial replenishment scenarios.

While the IB algorithm outperforms greedy algorithms in the large inventory regime for many settings (including Adwords), its performance guarantee does not extend to scenarios where the replenishment process is arbitrary, and small amounts of inventory may arrive over time. Recently, Feng et al. [5] provided the first generalization of IB, achieving a competitive ratio of $(1 - 1/e)$ in the large inventory regime under an arbitrary exogenous replenishment process. They employed a novel primal-dual analysis to overcome analytical challenges posed by the replenishment process. However, their approach is highly tailored to the IB algorithm and a specific problem formulation, and may not easily generalize to other settings and algorithms. In particular, there are many settings in which IB is not the algorithm with the highest competitive ratio (see, for example, [4,8], and [16]). Additionally, their analysis is limited to adversarial replenishment and arrival processes.

Our goal is to develop a black-box method that treats exogenous replenishment as a general feature rather than a setting-specific challenge, and to

[1] The large budget/inventory assumption, in which the starting inventory of every resource is assumed to be much larger than the amount of resource that can be allocated to any individual request, is ubiquitous in the literature due to its practical relevance and technical simplicity. This regime is essentially equivalent to considering a fractional relaxation of the problem.

extend known algorithmic results from fixed-inventory settings to those where the replenishment process may be either adversarial or stochastic. We also consider settings where the arrival process is stochastic, i.e., where requests arrive independently according to a known, non-stationary probability distribution. To the best of our knowledge, no prior work has studied inventory replenishment in stochastic settings within the online resource allocation framework.

1.2 Our Contributions

We propose a generalization of the online resource allocation framework of Devanur et al. [3], which includes many well-studied settings, and introduce a model of exogenous replenishment within the general framework. For the problems in this framework, we propose black-box methods that extend any online algorithm designed for settings without replenishment into one that retains the same performance guarantee with exogenous replenishment when the starting inventory is sufficiently large.

Our black-box algorithms extend the idea of "batching" introduced in [5], and we consider both adversarial and stochastic settings for arrival processes and replenishment processes. In particular, when the arrival process is adversarial, we sequentially transform an instance with replenishment into an instance without replenishment and stream the transformed instance as input to the online algorithm. In the transformation, we accumulate (batch) new inventory of a resource until it hits a certain threshold, after which we introduce it as a "new" type of resource that (one can imagine) was available since the beginning but incompatible with all prior arrivals. We prove that this transformation is lossless under the large starting inventory, regardless of amount of replenishment in any period.

We then extend this idea to settings where the arrival process is stochastic. When the replenishment process is also stochastic (with a known, possibly non-stationary distribution), we replace each random replenishment with a deterministic value, treating the process as if it were fully predictable. Since this transformation can be performed before the algorithm begins, we can directly incorporate existing algorithms to effectively utilize the "deterministic" replenishment using the same batching idea. When the quantity of each replenishment is bounded above, we prove that this results in a lossless transformation.

However, the same approach cannot be applied when the arrival process is stochastic and the replenishment process is adversarial since there is no prior knowledge of future replenishment. In this setting, we show that a lossless transformation is inherently impossible. Table 1 summarizes the outcomes across all four combinations of arrival and replenishment processes.

Overview of Proof Technique: We analyze our black-box transformation in two steps: (i) we compare the performance of the transformed algorithm with the optimal offline solution of the transformed instance, and (ii) we compare the optimal offline solution of the transformed instance with that of the original instance. In Step (i), we show that the competitive ratio is lower bounded by that of the original algorithm. In Step (ii), the analysis becomes independent of the online algorithm and depends solely on the properties of the general problem formulation. This two-step approach enables a unified analysis of the batching extensions.

Table 1. Existence of Lossless Transformations in the large inventory Regime. 'Yes' indicates that it is possible to transform any existing competitive ratio result from the setting without replenishment to the setting with exogenous replenishment. 'No' indicates that such an extension is, in general, impossible.

Exogenous Replenishment	Arrivals (Requests)	
	Adversarial	Stochastic
Adversarial	Yes	No
Stochastic	Yes	Yes

1.3 Outline

In Sect. 2, we formally define a general framework for online resource allocation problems and discuss how it can be used to incorporate classical problems with exogenous replenishment. In Sect. 3, we introduce black-box algorithms applicable to problems with various arrival and replenishment processes. In Sect. 4, we analyze these extensions, relying solely on the general problem formulation. Finally, in Sect. 5, we conclude the paper by summarizing our findings and discussing open problems. *Due to space constraints, all proofs are deferred to the full version of this paper.*

2 A General Model with Stochastic Usage and Exogenous Replenishment

In this section, we formally present the online resource allocation framework of interest along with the relevant notation. The model has five main components: resources, (online) requests, actions, exogenous replenishment, and reward. In the following, we denote the set $\{1, \dots, m\}$ by $[m]$, for any natural number m.

Resources. Let $\mathcal{I} = [n]$ represent the set of resource types, each resource type $i \in \mathcal{I}$ having an initial inventory c_i. Denote the minimum inventory among all resources as $c_{\min} = \min_{i \in \mathcal{I}} c_i$.

Requests. Let $\mathcal{J} = [m]$ denote the set of requests, with the details of each request revealed upon its arrival. The sequence of requests is indexed from 1 (earliest) to m (latest). Each request belongs to a request type that encapsulates its defining characteristics, such as resource requirements and rewards (which will be formally defined later). Let Z denote the set of possible request types. When a request $j \in \mathcal{J}$ arrives, its request type $z \in Z$ is revealed to the online algorithm.

Our prior knowledge of $\mathcal{J}$ depends on the arrival process. When the arrival process is stochastic, the total number of requests m and the categorical distribution of request types are fully known in advance, prior to the arrival of the first request. Specifically, for each request $j \in \mathcal{J}$, the request type $z \in Z$ is independently drawn according to a known distribution $(p_{jz})_{z \in Z}$, where $\sum_{z \in Z} p_{jz} = 1$. The type of request is revealed to the online algorithm as soon as the request arrives. When the arrival process is adversarial, the online algorithm has no prior knowledge of the number of requests m or the distribution over request types. In this case, for simplicity, we assume that each request j is of type j, and we omit explicit notation for the request type in the adversarial setting.

Actions. Let $\mathcal{K} = [K]$ be the set of possible actions. For each request, we must select exactly one action from $\mathcal{K}$. Some actions may be infeasible if there is not enough inventory when the request arrives (discussed later). For illustration, in the online bipartite matching problem [11], each action corresponds to either matching a request with a resource or ignoring the request. In the context of online assortment optimization [7], each action involves offering an assortment (set of resources) in response to a given request.

The selection of actions is irrevocable, and each action consumes a random amount of (possibly multiple) resources over time, provided that there is enough available inventory. Specifically, if action $k \in \mathcal{K}$ is selected for request j of type z, it requires $A_{ijkz}(l)$ units of resource type $i \in \mathcal{I}$ when request l $(\geq j)$ arrives. We assume that $A_{ijkz}(l)$ is a two-point random variable with support $\{0, b_{ijkz}(l)\}$, where $b_{ijkz}(l)$ is a deterministic quantity less than or equal to 1. The online algorithm observes its realization only after choosing action k for request j of request type z if $l = j$, or upon the arrival of request l if $l > j$.

The algorithm can choose action k for request j of request type z only if the remaining inventory of each resource type $i \in \mathcal{I}$ is at least $b_{ijkz}(j)$. If $b_{ijkz}(j)$ is nonzero, we say that action k for request j of request type z uses the resource type i. In other words, action k is implementable only if the "capacity constraints" are satisfied for every resource used by the action. To ensure feasibility, we assume the existence of a trivial action $k_0 \in \mathcal{K}$, which requires no resources: $b_{ijk_0z}(l) = 0$ for all $i \in \mathcal{I}$, $j \in \mathcal{J}$, $l \in \mathcal{J}$, $j \leq l$, and $z \in Z$. Also, to ensure that our allocation remains feasible after it is selected, we assume that

$$A_{ijkz}(l) \geq A_{ijkz}(l') \quad \forall l' > l,\, z \in Z.$$

Lastly, we assume that resource consumption is independent across distinct arrivals and past decisions of the online algorithm. Specifically, we assume that $A_{ijkz}(l)$ is independent of $A_{ij'k'z'}(l)$ whenever $j \neq j'$, $k \neq k'$, or $z \neq z'$. Under these assumptions, for each request, we consider the sequence of diminishing resource consumption $\{A_{ijkz}(l)\}_{l \geq j, l \in \mathcal{J}}$, where the random variables in the sequence may be dependent due to monotonicity but are independent of past decisions and the resource consumption from other requests. Prior knowledge of actions also depends on the arrival process. If the arrival process is stochastic, the full distribution of $A_{ijkz}(l)$ is known at the beginning. If the arrival process is adversarial, we omit the index for request type z, and the distribution of $A_{ijk}(l)$ is revealed upon arrival of the request j for all $l \geq j$.

Exogenous Replenishment. Without loss of generality, we assume that all exogenous replenishment occurs upon the arrival of each request, as decisions are only made at these moments. When request $j \in \mathcal{J}$ arrives, the additional amount of exogenous replenishment for each resource type $i \in \mathcal{I}$, denoted as $\zeta_i(j)$, is provided.

When the exogenous replenishment process is stochastic, the distribution of $\zeta_i(j)$ is known at the beginning. In this case, $\zeta_i(j)$ is a non-negative two-point random variable with support $\{0, w_{ij}\}$ and its expected amount is q_{ij} Let M be an upper bound of each $\zeta_i(j)$, that is, $w_{ij} \leq M$ for every $i \in \mathcal{I}$ and $j \in \mathcal{J}$. The realization of $\zeta_i(j)$ is revealed upon the arrival of request j and is assumed to be exogenous—i.e., independent of previously chosen actions and resource consumption. Beyond that, we do not impose any further assumptions on the dependencies among $\{\zeta_i(j)\}_{i\in\mathcal{I},j\in\mathcal{J}}$. When the exogenous replenishment process is adversarial, we have no prior knowledge of $\zeta_i(j)$. Without loss of generality, we assume that $\zeta_i(j)$ is a fixed, but unknown, non-negative constant with no additional assumptions beyond its non-negativity.

Rewards. Upon the arrival of request $j \in \mathcal{J}$ of request type $z \in Z$, choosing an implementable action $k \in \mathcal{K}$ gives a random non-negative reward of R_{ijkz} for each resource type $i \in \mathcal{I}$. The realization of reward R_{ijkz} is independent of prior decisions and is revealed after we select an action k for request j of request type z. For example, in the context of the online matching problem with stochastic rewards or the assortment optimization, this assumption (for both actions and rewards) captures scenarios where the reward is received only if the chosen action is matched successfully. If the arrival process is stochastic, the distribution of R_{ijkz} is known a priori. If the arrival process is adversarial, we omit z from the notation, and the distribution of reward R_{ijkz} is revealed upon the arrival of request j.

The objective of the online algorithm is to maximize the minimum profit earned across all resource types, subject to the capacity constraints. Formally, for each $j \in \mathcal{J}$, let $z(j) \in Z$ denote its realized request type, and let $k(j) \in \mathcal{K}$ be the action chosen for request j by the online algorithm. The profit associated with resource type i is $\sum_{j\in\mathcal{J}} R_{ijk(j)z(j)}$, and the objective function value of the algorithm becomes $\min_{i\in\mathcal{I}}(\sum_{j\in\mathcal{J}} R_{ijk(j)z(j)})$.

Offline Benchmark and Competitive Ratio. Throughout this paper, we focus on worst-case (competitive ratio) analysis over a family of instances $\mathcal{H}$. We compare online algorithms with a clairvoyant algorithm, OPT, which has full knowledge of the sequence of requests, the set of feasible actions, and the replenishment process before the first request arrives. However, while OPT has prior knowledge of the distributions of resource consumption ($A_{ijkz}(l)$'s) and rewards (R_{ijkz}'s), it does not know their realizations before the arrival of the corresponding request[2].

Let ALG denote an online algorithm for a family of instances $\mathcal{H}$. For each instance $H \in \mathcal{H}$, let ALG(H) and OPT(H) denote the expected objective function value achieved by ALG and OPT, respectively. Then, the competitive ratio of ALG for $\mathcal{H}$, is defined as follows:

$$\text{Competitive Ratio of ALG for}\mathcal{H} := \min_{H \in \mathcal{H}} \frac{\text{ALG}(H)}{\text{OPT}(H)}.$$

2.1 Main Results

To describe our main result, we need some notation. Given a family of instances $\mathcal{H}$, consider a subset of instances $\mathcal{H}_0 \subset \mathcal{H}$ that consists of instances without exogenous replenishment, that is, $\zeta_i(j) = 0$ for all $i \in \mathcal{I}$ and $j \in \mathcal{J}$. We assume that for every instance in $\mathcal{H}_0$, adding or removing a resource type results in a modified instance that remains in $\mathcal{H}_0$. For example, given an instance of the online bipartite matching problem, adding or removing a specific resource type still produces an instance of the online bipartite matching problem. To the best of our knowledge, this assumption holds for all families of instances discussed in the literature. Depending on the choice of $\mathcal{H}$, $\mathcal{H}_0$ may correspond to a well-studied problem.

For any $c > 0$, let $\mathcal{H}(c) \subset \mathcal{H}$ represent the set of instances with a minimum initial inventory, $c_{\min}$, of at least c. Similarly, let $\mathcal{H}_0(c) \subset \mathcal{H}_0$ be the set of instances whose minimum initial inventory is also at least c. Given an algorithm ALG operating on instances in $\mathcal{H}_0(c)$, let ALG(H) represent the expected objective function value achieved by ALG on $H \in \mathcal{H}_0(c)$. Define $\alpha(c)$ as the parametric competitive ratio of ALG for $\mathcal{H}$, where c is the minimum initial inventory among all resource types:

$$\alpha(c) := \min_{H \in \mathcal{H}_0(c)} \frac{\text{ALG}(H)}{\text{OPT}(H)}.$$

Without loss of generality, we assume that for each request j of type z, ALG selects either a trivial action k_0 or a non-trivial action $k \in \mathcal{K}$ with a non-zero reward R_{ijkz} for at least one resource type $i \in \mathcal{I}$. In other words, we assume that ALG does not intentionally consume resources without any reward. Note that if an algorithm selects an action that consumes resources without reward, we can replace the selected action with the trivial action k_0 without decreasing the objective function value.

Recall that each stochastic replenishment $\zeta_i(j)$ for any resource type i and request j is bounded above by M. Let d be the maximum number of resource types used by a single action $k \in \mathcal{K}$ in an instance of $\mathcal{H}$. Theorems 1 and 2 state the theoretical guarantees obtained through our black-box extensions. The

[2] No interesting competitive ratio results are possible against the benchmark that knows the realization of $A_{ijkz}(l)$'s in advance. See [9] for an example.

corresponding extensions are described in Sect. 3. As noted earlier, formal proofs are provided in the full version of the paper.

Theorem 1. *Given a family of instances $\mathcal{H}$ with exogenous replenishment and an adversarial arrival process, suppose that ALG achieves a competitive ratio of at least $\alpha(c)$ for $\mathcal{H}_0(c)$. There exists an algorithm, ALG-B, that achieves a competitive ratio of at least $\alpha(\sqrt{c})\left(1 - \mathcal{O}\left(\sqrt{\frac{\log(cd)}{c}}\right)\right)$ for $\mathcal{H}(c)$.*

Theorem 2. *Given a family of instances $\mathcal{H}$ and $c > 0$ under a stochastic arrival process and a stochastic replenishment process, suppose that ALG achieves a competitive ratio of at least $\alpha(c)$ for $\mathcal{H}_0(c)$. There exists an algorithm, ALG-B, that achieves a competitive ratio of at least $\alpha(\epsilon c)(1 - \mathcal{O}(\epsilon))$, where $\epsilon = \left(\frac{M}{c}\log(cd)\right)^{\frac{1}{3}}$ for the set of instances $\mathcal{H}(c)$.*

In the large inventory regime ($c \to +\infty$), the above theorems imply that there exists an algorithm with asymptotic competitive ratio equal to $\lim_{c\to+\infty}\alpha(c)$, meaning that we can losslessly transform an algorithm from a setting without replenishment into an algorithm for the corresponding generalization of the setting with exogenous replenishment. We summarize the competitive ratio achievable by ALG-B in some specific cases in Table 2.[3] Our transformation technique (discussed in Sect. 3) is inspired by the algorithm of Feng et al. (2021) [5], who proposed a generalization of the Inventory Balancing algorithm with a batching of exogenous replenishment. However, rather than relying on an ad hoc primal-dual analysis, we provide a generic method to construct and analyze a new algorithm (ALG-B) based on an existing algorithm (ALG), which we refer to as the "batching extension".

Table 2. Lower bound on the competitive ratio of the batching algorithm in the large inventory regime (under adversarial or stochastic arrival processes).

Problems	Adversarial	Stochastic
Adwords [13],[1],[3]	$(1-\frac{1}{e})(1-\mathcal{O}(\sqrt{\frac{\log(c)}{c}}))$	$1-\mathcal{O}\left(\frac{M}{c}(\log(c))^{\frac{1}{3}}\right)$
Assortment Optimization [7],[3]	$(1-\frac{1}{e})(1-\mathcal{O}(\sqrt{\frac{\log(nc)}{c}}))$	$1-\mathcal{O}\left(\frac{M}{c}(\log(nc))^{\frac{1}{3}}\right)$
Hypergraph matching [16],[3]	$\frac{1}{d+1}(1-\mathcal{O}(\sqrt{\frac{\log(cd)}{c}}))$	$1-\mathcal{O}\left(\frac{M}{c}(\log(cd))^{\frac{1}{3}}\right)$
(Stochastic) Reusable Resources [8],[6]	$(1-\frac{1}{e})(1-\mathcal{O}(\sqrt{\frac{\log(c)}{c}}))$	$1-\mathcal{O}\left(\frac{M}{c}(\log(c))^{\frac{1}{3}}\right)$

Finally, when the arrival sequence is stochastic and the replenishment process is adversarial, we show that a lossless transformation is, in general, impossible, as stated in Theorem 3.

[3] In the case of a stochastic arrival process, we assume the replenishment process is also stochastic.

Theorem 3. *There exists a family of instances $\mathcal{H}$ and $c > 0$ under a stochastic arrival process and an adversarial replenishment process such that there exists a near-optimal algorithm ALG for $H_0(c)$, but no online algorithm can have the near-optimal competitive ratio for $\mathcal{H}$.*

3 Black-Box Algorithmic Extensions

Consider a set of instances $\mathcal{H}_0 \subset \mathcal{H}$, where no exogenous replenishment occurs in $H \in \mathcal{H}_0$. Given an algorithm ALG designed for $\mathcal{H}_0$, our goal is to construct a black-box extension, denoted by ALG-B, that can handle any instance $H \in \mathcal{H}$. The core idea behind ALG-B is simple: (i) wait until the cumulative exogenous replenishment for each resource type reaches a threshold c_B, and (ii) treat the batched replenishment as a new resource type and make allocation decisions guided by ALG. These two steps will be discussed in Sects. 3.1 and 3.2.

3.1 Batching Strategy for Exogenous Replenishment

Choosing an appropriate threshold c_B is critical for the batching strategy. A larger value of c_B allows us to leverage existing results in the large-inventory regime, even when the replenishment process is adversarial. Conversely, a smaller value of c_B reduces excessive delays caused by waiting for batching.

We set $c_B = \sqrt{c_{\min}}$ for an adversarial replenishment process, and $c_B = \epsilon c_{\min}$, where $\epsilon = \left(\frac{M}{c_{\min}} \log\left(c_{\min} d\right)\right)^{\frac{1}{3}}$ for a stochastic replenishment process. These choices balance the trade-off and yield the best performance guarantees in our framework. Once we have the threshold, we can define the update rule of the batched replenishment sequence accordingly.

Batching Adversarial Replenishment We formally define a batched replenishment sequence $\left\{\zeta_i^B(j)\right\}_{j\in\mathcal{J}}$ as follows. Initialize $\zeta_i^B(0) = 0$ and $\bar{c}_i = 0$. The variable $\bar{c}_i$ will capture the new inventory of i that has been received but has not been batched to use in the online algorithm. On arrival of a request j, we update $\bar{c}_i = \sum_{l\in[j]} \zeta_i(l) - \sum_{l\in[j-1]} \zeta_i^B(l)$. If $\bar{c}_i \geq c_B$, then we set $\zeta_i^B(j) = \bar{c}_i$, otherwise we let $\zeta_i^B(j)$ stay equal to 0. In the batched replenishment process, on arrival of request j we receive $\zeta_i^B(j)$ units of resource type i. Notice that $\zeta_i^B(j)$ is either 0 or at least c_B.

Batching Stochastic Replenishment In the stochastic setting, the probability distributions of future replenishment are known as a prior. We construct a deterministic batched replenishment process so that it can be seamlessly incorporated into existing algorithms for stochastic arrivals.

To incorporate stochastic exogenous replenishment without knowing its exact realization, we firstly consider a fluid sequence $\left\{\tilde{\zeta}_i(j)\right\}_{j\in\mathcal{J}}$ with adjusted mean, $\tilde{\zeta}_i(j) = (1-\epsilon)q_{ij}$. Then, we apply batching similar to that used in the adversarial setting, except $c_B = \epsilon c_{\min}$. Since we have full knowledge of $\tilde{\zeta}_i(j)$, we compute the batched replenishment sequence $\left\{\zeta_i^B(j)\right\}_{j\in\mathcal{J}}$ before the first request arrives.

This procedure transforms each stochastic exogenous replenishment into a deterministic value, taking either 0 or at least c_B. When c_B is large relative to the realized replenishment amounts, concentration bounds can be used to show that the resulting loss from this transformation is small.

3.2 Integrating Batched Replenishment Into Online Decisions

After defining the batched replenishment, the next step is to figure out how to make allocation decisions with the batched replenishment. We first describe how batched replenishment is handled within the online algorithm and provide a complete definition of the algorithmic extension in the adversarial setting. The stochastic setting is deferred to the full version of the paper, as batched replenishment is treated identically in both adversarial and stochastic cases.

Recall that under an adversarial arrival process, the request type of request j is set to j, allowing the request type to be omitted. Before choosing an action for request j, we generate copies of each action $k \in \mathcal{K}$ by replacing the original resources in $\mathcal{I}$ with their identical counterparts, as additional batched resource types may now be available. Formally, for each request $j \in \mathcal{J}$, we update the available set of actions $\mathcal{K}(j)$ recursively. First, initialize $\mathcal{K}(0) := \mathcal{K}$ and assume we have $\mathcal{K}(j-1)$. When request j arrives, we construct $\mathcal{K}(j)$ as follows: set $\mathcal{K}_0(j) := \mathcal{K}(j)$, and for each $i \in \mathcal{I}$, define $\mathcal{K}_i(j) := \mathcal{K}_{i-1}(j)$ and check if a new batched resource type i_j is created. If so, for each action $k \in \mathcal{K}_{i-1}(j)$, we create a duplicate action $k(i_j)$ by replacing the usage of resource type i with resource type i_j and add it to $\mathcal{K}_{i-1}(j)$ with the same reward. After checking all resource types, set $\mathcal{K}(j) := \mathcal{K}_n(j)$. Based on the current state and available actions $\mathcal{K}(j)$, we use ALG to select an action as if batched resource types and actions in $\mathcal{K}(j)$ were present from the start.

Although the identical resource i_j is introduced later, we can treat i_j and its associated actions in $\mathcal{K}(j)$ as if they were present from the beginning by controlling rewards. Assume that we already know all batched resource types and available actions $\mathcal{K}(m)$ at the start of the algorithm. If an action $k \in \mathcal{K}(m)$ is not activated for the current request j (i.e., $k \in \mathcal{K}(m) \backslash \mathcal{K}(j)$), we set its reward to 0. Conversely, if an action k is activated when request j arrives (i.e., $k \in \mathcal{K}(j)$), selecting an action k for request j yields the original reward. In other words, we obtain a nonzero reward only if the action uses resource types defined at or before the arrival of request j. Recall that the algorithm selects either a trivial action k_0 or a non-trivial action $k \in \mathcal{K}(m)$ with a non-zero reward R_{ijk} for some $i \in \mathcal{I}$ and each request $j \in \mathcal{J}$. Although we create the identical resource i_j later, by the assumptions underlying ALG, the output of the algorithm remains consistent up to request j, as if we generated the resource type i_j and its associated actions from the beginning.

It is worth noting that ALG-B remains an online algorithm. By abuse of notation, let the allocation action of ALG for request j be denoted by $\text{ALG}\left(\{S^l\}_{l=1}^j\right)$, where S^l is the state when request l arrives. When ALG-B introduces a newly batched resource type and allocates resources, the state is updated accordingly. For each arrival $l \in [j]$, denote the updated state when request l arrives by $S_B^l([j])$. Then, for request j, let the action chosen by ALG-B be represented as $\text{ALG}\left(\{S_B^l([j])\}_{l=1}^j\right)$. Given that we can assume the additional resource types and actions were always part of the system, this interpretation shows that ALG-B is a well-defined online algorithm. Indeed, the definition of our batching extension and its state update rule ensure that

$$\text{ALG}\left(\{S_B^l([j])\}_{l=1}^j\right) = \text{ALG}\left(\{S_B^l([j'])\}_{l=1}^j\right)$$

for any $j < j'$. This is because actions using the batched resource types added to the state at request j' have zero reward prior to request j'. Thus, the action chosen by ALG (or ALG-B) for request j remains unaffected, even as the states continue to update for $j' > j$.

Therefore, every time the cumulative replenishment reaches the threshold, we create an identical resource and corresponding actions. From that point onward, we can proceed as if these resources and actions had always been part of the system. The formal definition of ALG-B is presented in Algorithm 1.

Algorithm 1. Definition of ALG-B

Input: A set of resource types $\mathcal{I}$, the initial inventory c_i for each $i \in \mathcal{I}$, and a set of actions $\mathcal{K}$.
Initialize $\mathcal{K}(0) := \mathcal{K}$.
for each resource type $i \in \mathcal{I}$ **do**
  Initialize $\bar{c}_i := 0$
end for
for each request $j \in \mathcal{J}$ **do** ▷ Process each arrival online
  Initialize $K_0(j) := \mathcal{K}(j-1)$.
  for each resource type $i \in \mathcal{I}$ **do**
    $\bar{c}_i := \bar{c}_i + \zeta_i(j)$. ▷ Keep track of cumulative replenishment
    $\mathcal{K}_i(j) := \mathcal{K}_{i-1}(j)$. ▷ Keep track of available actions
    if $\bar{c}_i \geq c_B$ **then** ▷ Cumulative replenishment exceeds threshold
      $\mathcal{I} := \mathcal{I} \cup \{i_j\}$. ▷ Create an identical resource type
      $c_{i_j} := \bar{c}_i$.
      $\mathcal{K}_i(j) := \mathcal{K}_i(j) \cup \{k(i_j) : k \in \mathcal{K}_i(j)\}$. ▷ Add copies of actions
      $\bar{c}_i := 0$. ▷ Reset cumulative replenishment
    end if
  end for
  $\mathcal{K}(j) := \mathcal{K}_n(j)$. ▷ Define a set of available actions for request j
  Choose an action $\text{ALG}\left(\{S_B^l([j])\}_{l=1}^j\right)$ based on $\mathcal{I}$ and $\mathcal{K}(j)$.
end for

When the replenishment process is stochastic, the batched replenishment can be computed at the beginning, as described in Sect. 3.1. Each batched replenishment is then processed in the same way as in the adversarial setting, resulting in a transformed instance H_B without exogenous replenishment. The algorithm proceeds by following the same allocation decisions as ALG(H_B), except in cases where the required batched replenishment is not realized (which occurs with low probability). In such cases, the algorithm skips the allocation. In this way, the black-box batching extension under stochastic arrivals and replenishment closely follows the structure of Algorithm 1. The complete details can be found in the full version of this paper.

3.3 An Illustrative Example of Batched Replenishment

Example 1. To illustrate how a batching extension dynamically constructs transformed instances, consider an instance with two types of resources $\mathcal{I} = \{1, 2\}$ with starting inventory $c_1 = c_2 = 4$. Both the arrival and replenishment processes are adversarial, and the batch size is set to $c_B = \sqrt{c_{\min}} = 2$. Assume that the original set of actions $\mathcal{K}$ is the power set of $\mathcal{I}$:

$$\mathcal{K} = \{\{1, 2\}, \{1\}, \{2\}, k_0\},$$

where $\{\emptyset\}$ is replaced by a trivial action k_0.

Consider an online algorithm ALG and its batching extension, ALG-B. Upon the arrival of the first request, suppose the exogenous replenishment is $\zeta_1(1) = 1$ and $\zeta_2(1) = 2$. Since $\zeta_1(1) < c_B$, ALG-B ignores the replenishment for resource type 1. However, because $\zeta_2(1) = c_B$, a new batched resource type 2_1 is introduced with different starting inventory $c_{2_1} = 2$. Recall that the name 2_1 indicates a batched resource type identical to the original resource type 2, introduced upon the arrival of the first request. ALG-B then runs ALG on the modified instance with the extend resource set $\{1, 2, 2_1\}$ and chooses an action from the updated set of actions $\mathcal{K}(1) = \{\{1, 2\}, \{1, 2_1\}, \{1\}, \{2\}, \{2_1\}, k_0\}$. Suppose ALG-B (or ALG) selects action $\{2_1\}$. Note that the reward and resource consumption of this action are stochastically identical to those of action $\{2\}$.

Now, upon the second request, assume that we receive additional replenishment $\zeta_1(2) = 2$ and $\zeta_2(2) = 1$. The updated resource set becomes $\{1, 1_2, 2, 2_1\}$ with starting inventories $c_1 = 4$, $c_{1_2} = 3$, $c_2 = 4$, and $c_{2_1} = 2$. The updated set of actions becomes $\mathcal{K}(2) = \{\{1, 2\}, \{1_2, 2\}, \{1, 2_1\}, \{1_2, 2_1\}, \{1\}, \{1_2\}, \{2\}, \{2_1\}, k_0\}$.

We may interpret that the new actions involving resource type 1_2 ($\{1_2, 2\}$, $\{1_2, 2_1\}$, and $\{1_2\}$) were also available for the first request, but they yield no reward at that time since the resource 1_2 was not yet available. ALG-B now proceeds by running ALG on this transformed instance, given that it chose an action $\{2_1\}$ for the first request.

4 Analysis of the Algorithmic Extensions

The high-level idea of the analysis of batching extensions is to bound the ratio of ALG-B(H) to OPT(H) using the known competitive ratio $\alpha(c_B)$ of ALG (H_B) to OPT (H_B). We first bridge OPT(H) and OPT (H_B) by introducing the expected LP formulations of H and H_B as intermediate benchmarks in Sect. 4.1. The remaining steps of the proof are outlined in Sect. 4.2.

4.1 Expected LP Formulation

To isolate the impact of batching on the offline benchmark, we introduce an expected LP formulation, which provides a tractable and comparable relaxation for both instances H and H_B. We unify the expected LP formulations for instances under both adversarial and stochastic settings. Note that when the replenishment process is adversarial, we can omit the request type index, as request j is assumed to be of type j, i.e. $p_{jj} = 1$, for every $j \in \mathcal{J}$ and denote realized replenishment by $\zeta_i(j) = q_{ij}$.

Let $a_{ijkz}(l)$ denote the expectation of $A_{ijkz}(l)$ and r_{ijkz} denote the expectation value of R_{ijkz}, respectively. Given an instance H, the following formulation (LP(H)) can be used to get an upper bound of OPT(H):

$$
\begin{aligned}
(\mathrm{LP}(H)) \quad \max \quad & \lambda && (1)\\
\text{s.t.} \quad & \lambda \le \sum_{j\in\mathcal{J}}\sum_{k\in\mathcal{K}}\sum_{z\in Z} r_{ijkz}x_{jkz}, \quad \forall i \in \mathcal{I}, && (2)\\
& \sum_{j\in[l]}\sum_{k\in\mathcal{K}}\sum_{z\in Z} a_{ijkz}(l)x_{jkz} \le c_i + \sum_{j\in[l]} q_{ij}, \quad \forall i \in \mathcal{I}, l \in \mathcal{J}, && (3)\\
& \sum_{k\in\mathcal{K}} x_{jkz} \le p_{jz}, \quad \forall j \in \mathcal{J}, z \in Z && (4)\\
& x_{jkz} \ge 0, \quad \forall j \in \mathcal{J}, k \in \mathcal{K}, z \in Z. && (5)
\end{aligned}
$$

In the original problem, the variable x_{jkz} can be interpreted as a binary decision variable that takes a value 1 only if action k is chosen for the request t of request type z, and only one action can be chosen for each request. In (LP(H)), however, this condition is relaxed as shown in (4) and (5). Additionally, the usage of resources and the reward for each action are deterministic. Constraint (1) and (2) define λ as the minimum expected reward among all resource types, and constraint (3) ensures that the total amount of resource type i used does not exceed the accumulative inventory for any resource type.

Let LP(H) denote the optimal objective function value of (LP(H)). Recall that OPT(H) represents the expected reward achieved by the optimal policy for instance H. The optimal policy can be represented as $(X_{jkz}(\omega_j))_{j\in\mathcal{J},k\in\mathcal{K},z\in Z}$, where X_{jkz} has a value of 1 if action k is selected for request j of type z and has a value of 0 otherwise, based on the filtration ω_j up to the arrival of request j. Consequently, OPT(H) can be expressed as $\min_{i\in\mathcal{I}}\left(\sum_{j\in\mathcal{J}}\sum_{k\in\mathcal{K}}\sum_{z\in Z} r_{ijkz}\mathbb{E}[X_{jkz}(\omega_j)]\right)$. The following lemma establishes that the linear program (LP) provides an upper bound to the expected optimal reward.

Lemma 1. *For any instance H, LP(H) $\ge$ OPT(H).*

Next, we show that as the initial inventory c_i increases, the optimal reward OPT(H) approaches LP(H). This result leverages the attenuated independent rounding technique and the well-known concentration bounds. Recall that $c_{\min} = \min_{i \in \mathcal{I}} c_i$ is the minimum initial inventory across all resource types and d is the maximum number of resource types affected by any action $k \in \mathcal{K}$. Lemma 2 shows that the gap between LP(H) and OPT(H) depends on $c_{\min}$ and d.

Lemma 2. *For any instance H, $OPT(H) \geq (1 - \mathcal{O}(\sqrt{\frac{\log(c_{\min} d)}{c_{\min}}}))LP(H)$.*

4.2 Proof Idea of Theorems 1 and 2

Our goal is to bound the ratio of ALG-B(H) to OPT(H). By definition, the batching extension ensures that ALG-B (H) is close to ALG (H_B), and the ratio of ALG (H_B) to OPT (H_B) is at least $\alpha(c_B)$.

To relate OPT (H) and OPT (H_B), Lemmas 1 and 2 imply that these values are close whenever LP (H) is close enough to LP (H_B). Since the only difference between LP (H) and LP (H_B) is the right-hand side of constraint (2), we have LP $(H) \leq \left(1 + \mathcal{O}\left(\frac{c_B}{c_{\min}}\right)\right)$ LP (H_B). This overall idea is illustrated in Fig. 1, and the full details of the proof are deferred to the full version of this paper.

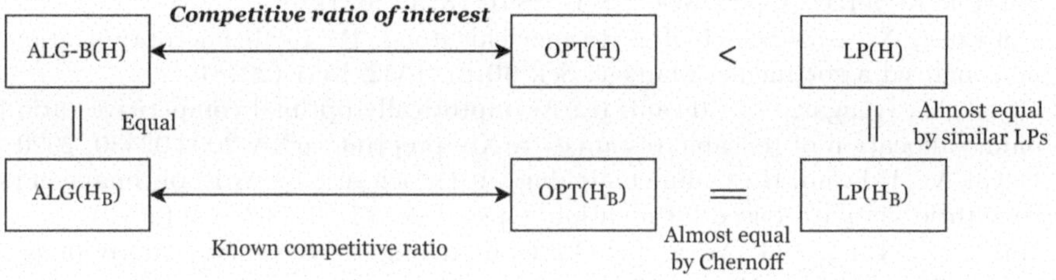

Fig. 1. Analysis of ALG-B with batching

5 Conclusion and Open Problems

We consider a general framework for incorporating exogenous replenishment in online resource allocation. Our black-box approach leverages existing algorithms and known performance guarantees for settings without replenishment to obtain algorithms with the same guarantees for settings with replenishment. The key idea is to batch new inventory of a resource until it reaches a threshold value, at which point we plug-in all the new inventory as a "new" resource in the online algorithm.

An interesting open question for future work is to examine the need for batching inventory. In practice, a major drawback of batching is that resources may remain unused simply because the cumulative new inventory has not yet reached a threshold. Can we achieve the same general theoretical guarantees using a myopic approach that readily uses new inventory as it arrives? Another challenging open problem is to remove the large inventory assumption and generalize (randomized) algorithms for the general inventory to settings with exogenous replenishment.

References

1. Alaei, S., Hajiaghayi, M., Liaghat, V.: Online prophet-inequality matching with applications to ad allocation, pp. 18–35 (2012)
2. Chan, C.W., Farias, V.F.: Stochastic depletion problems: effective myopic policies for a class of dynamic optimization problems. Math. Oper. Res. **34**(2), 333–350 (2009)
3. Devanur, N.R., Jain, K., Sivan, B., Wilkens, C.A.: Near optimal online algorithms and fast approximation algorithms for resource allocation problems. J. ACM (JACM) **66**(1), 1–41 (2019)
4. Ekbatani, F., Feng, Y., Kash, I., Niazadeh, R.: Online job assignment. Available at SSRN (2024)
5. Feng, Y., Niazadeh, R., Saberi, A.: Robustness of online inventory balancing algorithm to inventory shocks. Available at SSRN 3795056 (2021)
6. Feng, Y., Niazadeh, R., Saberi, A.: Near-optimal Bayesian online assortment of reusable resources. Oper. Res. **72**(5), 1861–1873 (2024)
7. Golrezaei, N., Nazerzadeh, H., Rusmevichientong, P.: Real-time optimization of personalized assortments. Manage. Sci. **60**(6), 1532–1551 (2014)
8. Goyal, V., Iyengar, G., Udwani, R.: Asymptotically optimal competitive ratio for online allocation of reusable resources. arXiv preprint: arXiv:2002.02430, 2020
9. Goyal, V., Udwani, R.: Online matching with stochastic rewards: optimal competitive ratio via path-based formulation. Oper. Res. **71**(2), 563–580 (2023)
10. Huang, Z., Tang, Z.G., Wajc, D.: Online matching: a brief survey. arXiv preprint: arXiv:2407.05381, 2024
11. Karp, R. M., Vazirani, U. V., Vazirani, V.V.: An optimal algorithm for on-line bipartite matching. In: Proceedings of the Twenty-second Annual ACM Symposium on Theory of computing, pp. 352–358 (1990)
12. Mehta, A.: Online matching and ad allocation. Found. Trends® Theor. Comput. Sci. **8**(4), 265–368 (2013)
13. Mehta, A., Saberi, A., Vazirani, U., Vazirani, V.: Adwords and generalized online matching. J. ACM (JACM) **54**(5), 22–es (2007)
14. Rusmevichientong, P., Sumida, M., Topaloglu, H.: Dynamic assortment optimization for reusable products with random usage durations. Manage. Sci. **66**(7), 2820–2844 (2020)
15. Talluri, K., Van Ryzin, G.: An analysis of bid-price controls for network revenue management. Manage. Sci. **44**(11-part-1), 1577–1593 (1998)
16. Tröbst, T., Udwani, R.: Almost tight bounds for online hypergraph matching. arXiv preprint: arXiv:2402.08775 (2024)

Generalized Assignment and Knapsack Problems in the Random-Order Model

Max Klimm(✉) and Martin Knaack

Technische Universität Berlin, 10623 Berlin, Germany
{klimm,knaack}@math.tu-berlin.de

Abstract. We study different online optimization problems in the random-order model. There is a finite set of bins with known capacity and a finite set of items arriving in a random order. Upon arrival of an item, its size and its value for each of the bins is revealed and it has to be decided immediately and irrevocably to which bin the item is assigned, or to not assign the item at all. In this setting, an algorithm is α-competitive if the total value of all items assigned to the bins is at least an α-fraction of the total value of an optimal assignment that knows all items beforehand. We give an algorithm that is α-competitive with $\alpha = (1 - \ln(2))/2 \approx 1/6.52$ improving upon the previous best algorithm with $\alpha \approx 1/6.99$ for the generalized assignment problem and the previous best algorithm with $\alpha \approx 1/6.65$ for the integral knapsack problem. We then study the fractional knapsack problem where we have a single bin and it is also allowed to pack items fractionally. For that case, we obtain an algorithm that is α-competitive with $\alpha = 1/e \approx 1/2.71$ improving on the previous best algorithm with $\alpha = 1/4.39$. We further show that this competitive ratio is the best-possible for this model.

Keywords: Generalized Assignment Problem · Random-Order Model · Knapsack Problem · Online Optimization

1 Introduction

In the secretary problem, there is a sequence of n applicants with unknown values that arrive in a random order. Upon arrival of an applicant, the decision maker observes the value of the applicant and has to decide immediately and irrevocably whether to hire the applicant, or not. The famous $1/e$-rule stipulates that in order to maximize the probability of hiring the applicant with the highest value, the optimal policy is to not hire any of the first n/e applicants and to hire any later applicant whose value exceeds all values observed so far (Dynkin [13]; Lindley [28]). In the worst case, it achieves a probability of hiring the best applicant of $1/e$. Surprisingly, this bound of $1/e$ also translates to a setting where the values of the applicants are drawn i.i.d. from a distribution not known to the decision maker and the objective is to maximize the expected value of the hired applicant (Correa et al. [12, Theorem 2]).

N. Megow and A. Basu (Eds.): IPCO 2025, LNCS 15620, pp. 341–354, 2025.
https://doi.org/10.1007/978-3-031-93112-3_25

In this paper, we study a generalization of the secretary problem known as the *generalized assignment problem* in the random-order model. In this problem, there is a sequence of n items that arrive in a random order and a set of m bins. Every bin i has a known capacity $C_i > 0$. Upon arrival of item j, the value $v_{i,j} \geq 0$ and size $s_{i,j} \geq 0$ of item j for bin i is revealed. After having observed these values, the decision maker has to decide immediately and irrevocably to which bin this item j is assigned to, or to not assign the item to any bin at all. Formally, we introduce a binary variable $x_{i,j} \in \{0,1\}$ for each bin i and item j which is equal to 1 if item j is assigned to bin i and 0 otherwise. Since every item can be assigned to at most one bin, we have the inequality $\sum_{i=1}^{m} x_{i,j} \leq 1$ for all items j. In addition, the assignment of the items to the bins has to obey the capacity constraint of each bin, that is, $\sum_{j=1}^{n} s_{i,j}\, x_{i,j} \leq C_i$ for all bins i. The goal of decision maker is to maximize the expected value of the items assigned to the bins, i.e., $\sum_{i=1}^{m} \sum_{j=1}^{n} v_{i,j}\, x_{i,j}$. Let $\textsc{Opt}$ denote the value of an optimal offline solution of the generalized assignment problem, and let $x_{i,j}$, $i \in [m]$, $j \in [n]$ denote the variables set by an online algorithm. Then, for $\alpha \in [0,1]$, we call an online algorithm *α-competitive* if, on all instances, we have $\sum_{i=1}^{m} \sum_{j=1}^{n} v_{i,j}\, x_{i,j} \geq (\alpha - o(1))\, \textsc{Opt}$. We then also call α the *competitive ratio* or *competitiveness* of the algorithm.

The generalized assignment problem in the random-order model has been studied before by Kesselheim et al. [24] who gave an algorithm that is 1/8.1-competitive. Naori and Raz [33] and, independently, Albers et al. [3] gave an algorithm with an improved competitive ratio of 1/6.99.

Our Contribution and Techniques. In this paper, we give an algorithm for the online generalized assignment problem in the random-order model with an improved competitiveness of $(1 - \ln(2))/2 \approx 1/6.52$. Similar to previous algorithms [3,24,33], our algorithm is based on a fractional relaxation of the generalized assignment problem. After a sampling phase of $n/2$ items that will not be assigned, after each item arrived, our algorithm solves the fractional allocation of the generalized assignment problem and assigns the item randomly according to the fractional variables corresponding to them item as long as the capacity of this bin is not exceeded. A main challenge in the analysis of this algorithm is to bound the probability that the assignment of an item to a bin is successful, i.e., there is still enough capacity in the bin to accommodate the item. Previous algorithms handle this challenge by distinguishing between large and small assignments. Kesselheim et al. [24] call an assignment of an item j to a bin i *heavy* if it uses more than half of the capacity of the bin, i.e., $s_{i,j} > C_i/2$, and *light* otherwise. Their algorithm then either only considers heavy assignments or only light assignments. If only heavy assignments are considered, the instance reduces to an edge-weighted matching problem in the random-order model for which an $1/e$-competitive algorithm is known (Kesselheim et al. [23]). When only light assignments are considered, it suffices to bound the probability that in the assignment based on the fractional relaxation, the selected bin i is at most filled up to $C_i/2$. Albers et al. [3] and Naori and Raz [33] use the same distinction between heavy and light assignments, but use a sequential approach. In a first phase of the algorithm from round 1 to round $\lfloor 0.5261n \rfloor$ no item is packed. In a second phase from round $\lfloor 0.5261n \rfloor + 1$ to round $\lfloor 0.6906n \rfloor$ only heavy assignments are considered, and in a third phase from round $\lfloor 0.6906n \rfloor + 1$ on only light assignments are considered. The heavy assignments are again handled by a bipartite matching algorithm while the small items are assigned based on the fraction relaxation.

Clearly, the distinction between heavy and light assignments degrades the performance of these algorithms since the packing of otherwise lucrative items maybe prevented based on the somewhat arbitrary distinction between heavy and light assignments. Our algorithm does not use any distinction between heavy and light assignments at all. This is problematic for the heavy assignments since it is not clear how to bound the probability that the bin is sufficiently empty to pack the item. We circumvent this issue by considering preliminary assignments where the capacity of a bin may be violated by a single item. This makes it possible to bound the probability of a successful preliminary assignment by bounding the probability that in previous rounds the sum of the weights of the items assigned to bin i is at most C_i which can be done with similar techniques as by Kesselheim et al. [24]. We show that this infeasible assignment obtains a competitive ratio of $1 - \ln(2) \approx 1/3.26$. To turn this infeasible assignment to a feasible one, we have to sacrifice an additional factor of $1/2$. With probability $1/2$, we run the original algorithm but do *not* add the last item violating the capacity constraint to each bin. With the remaining probability of $1/2$, we run this algorithm only virtually and assign *only* the last item that would violate the capacity of the bin in the original algorithm. The resulting algorithm always obtains a feasible assignment and achieves a competitive ratio of $(1-\ln(2))/2 \approx 6.52$. The technique of overpacking a bin and selecting with probability $1/2$ either all items but the last or the last is reminiscent of the two-bin algorithm used by Han et al. [17] for the online knapsack problem with adversarial order and unit density.

We further study the fractional knapsack problem in the random-order model. This problem corresponds to the generalized assignment problem with a single bin where items can also be packed fractionally, i.e., we have $x_j \in [0, 1]$ for all items j. For this problem, we give an algorithm with competitive ratio $1/e$ improving on the previous best competitive ratio of $1/4.39$ due to Giliberti and Karrenbauer [16]. Our algorithm has a sample phase of $\lfloor n/e \rfloor$ items. After that phase, in each round ℓ, the algorithm solves the fractional knapsack problem once with all items revealed so far and once with all items revealed so far except the current item. It then packs the current item fractionally corresponding to the total volume of items revealed during the sample phase that where removed from the fractional knapsack solution by including the current item. This procedure ensures that the fractional knapsack solution computed by the algorithm does not exceed the knapsack capacity. This general idea has been used before by the virtual algorithm devised by Babaioff et al. [6] for the k-secretary problem, but the latter problem is conceptually easier since all items have a unit size and items are always packed integrally. We further show that the competitive ratio of $1/e$ is best-possible for deterministic algorithms. We first note that this does not readily follow from the lower bound of $1/e$ for the standard secretary problem since items may also be packed fractionally which may help an online algorithm. For the proof, we exploit a similarity between deterministic algorithms for the fractional knapsack problem in the random-order model and randomized stopping rules together with Correa et al. [12, Theorem 2]. Due to space constraints, all proofs are deferred to the full version of this paper (Table 1).

Table 1. Results obtained in this paper.

Model	Competitive Ratios	
	Previous Results	Our Results
Generalized Assignment Problem	1/6.99 [3,33]	$\frac{1-\ln(2)}{2} \approx 1/6.52$ [Thm. 1]
Fractional Knapsack	1/4.39 [16]	$1/e \approx 1/2.71$ [Thm. 2]*

* best-possible by [Thm. 3]

Related Work. For a general introduction to offline knapsack and generalized assignment problems, see Martello and Toth [31]. Chekuri and Khanna [11] showed that the generalized assignment problem is APX-hard. For a minimization version of the problem, Shmoys and Tardos [34] gave a 1/2-approximation. Chekuri and Khanna [11] noted that this 1/2-approximation translates also to the maximization version of the problem. The generalized assignment problem contains as a special case the knapsack problem when there is a single bin only. This problem is NP-hard (Karp [20]) but admits a fully polynomial-time approximation scheme (FPTAS) as shown by Ibarra and Kim [19]. It further contains as a special case the multiple knapsack problem when the value and size of each item does not depend on the bin. Chekuri and Khanna [11] showed that this problem does not admit an FPTAS even for 2 bins, unless P = NP, and gave a polynomial-time approximation scheme (PTAS). The generalized assignment problem further contains as a special case the maximum bipartite matching problem when all bin capacities are 1, and all item sizes are 1 or 2. A further special case of the generalized assignment problem is the AdWords problem that corresponds to the case when $s_{i,j} = v_{i,j}$ for all bins i and items j. Here, the bins correspond to advertisers with a given daily budget C_i and the items correspond to queries to a search engine. The value $s_{i,j} = v_{i,j}$ is the revenue generated from assigning query j to advertiser i.

For the online bipartite matching problem where items arrive in an adversarial order, Karp et al. [21] gave an algorithm with a competitive ratio of $1 - 1/e \approx 0.63$. Interestingly, this is also the competitive ratio of the natural greedy algorithm in the random-order model. In the random-order model, an algorithm with a competitive ratio of $\alpha \approx 0.69$ was devised by Mahdian and Yan [30]. The edge-weighted variant of the maximal online bipartite matching problem was first studied by Korula and Pál [26] who gave a 1/8-competitive algorithm. This was improved by Kesselheim et al. [23] to a $1/e$-competitive algorithm. This is best-possible since this contains the secretary problem as a special case. The node-weighted variant of the maximal online bipartite matching problem corresponds to the matroid secretary problem when the underlying matroid is a transversal matroid. For general matroids, Babaioff et al. [5,6] gave a $O(1/\log k)$-competitive algorithm where k is the rank of the matroid. This bound has been improved to $O(1/\sqrt{\log k})$ by Chakraborty and Lachish [10] and to $O(1/\log\log k)$ by Lachish [27] and Feldman et al. [15]. The question whether there is a constant-competitive algorithm for general matroids remains open [7]. Algorithms with constant competitiveness are only known for special cases of matroids such as a $1/(2e)$-competitive algorithm by Korula and Pál [26], a $1/4$-competitive algorithm by Soto et al. [35], and a $1/3.71$-competitive algorithm by Bérczi et al. [9] for graphic matroids, as well as a $1/9.6$-competitive algorithm by Ma et al. [29], a 5.19-competitive algorithm by Soto et al. [35], a $1/4.75$-competitive algorithm by Huang et al. [18], and a $1/3.26$-competitive algorithm by Bérczi et al. [9] for laminar matroids. Mehta et al. [32] study the online AdWords problem in the random-order model. Under a large markets assumption that no item contributes a constant fraction to the capacity of an advertiser, they provide an algorithm with a competitiveness of $1 - 1/e \approx 0.63$.

The random-order knapsack problem was first studied by Babaioff et al. [4] who gave an α-competitive algorithm with $\alpha = 1/(10e) \approx 1/27.18$. For the same problem, Kesselheim et al. [24] gave an algorithm with a competitiveness of $\alpha \approx 1/8.06$. The competitive ratio was further improved by Albers et al. [3] to $\alpha \approx 1/6.65$. Under a large market assumption that no item contributes a constant fraction to the optimum solution, Vaze [36] gave a α-competitive algorithm with $\alpha = 1/(2e) \approx 5.44$. Abels et al. [1] considered the special case in which the capacity is 2 and all items have a size of either 1 or 2. They obtain an algorithm that matches the best-possible competitiveness of $1/e$.

The fractional knapsack problem in the random-order model was first studied by Karrenbauer and Kovalevskaya [22]. They developed a framework that turns an α-competitive algorithm for the integral knapsack problem into an $1/(\alpha^{-1} + e)$-competitive algorithm for the fractional knapsack problem. Together with the algorithm of Albers et al. [3], this yields a competitiveness of $\alpha \approx 1/9.37$. Giliberti and Karrenbauer [16] gave an algorithm with an improved competitive ratio of $\alpha \approx 1/4.39$.

The random-order online generalized assignment problem was first studied by Kesselheim et al. [24] who gave an α-competitive algorithm with $\alpha \approx 1/8.06$. Albers et al. [3] gave an algorithm with an improved competitiveness of $\alpha \approx 1/6.99$. A variant of the generalized assignment problem where capacities can be violated has been studied by Feldman et al. [14] under a large markets assumption. The generalized assignment problem can be further generalized to online packing LPs where the main difference is that packing an item may consume capacity of multiple bins. The competitiveness of online algorithms depends on various parameters of the problem such as the maximum number of bins used by packing an item and the ratio of the capacity of a bin and the capacity consumption of an item [2,8,24].

2 Preliminaries

We let $\mathbb{N}$ denote the strictly positive natural numbers and, for $n \in \mathbb{N}$, let $[n] = \{1, \ldots, n\}$. The generalized assignment problem (GAP) is given by a set $[m]$ of bins, each with a capacity $C_i > 0$ for $i \in [m]$, and a set $[n]$ of items. Assigning an item $j \in [n]$ to a bin $i \in [m]$ raises the total size of the bin by $s_{i,j} > 0$ and generates a profit or value of $v_{i,j} > 0$. The goal is to find an assignment of items to bins that maximizes the total value and fulfils the capacity constraint for every bin. The problem can be stated as the following integer linear program:

$$\max \sum_{i=1}^{m} \sum_{j=1}^{n} v_{i,j}\, x_{i,j},$$

$$\text{s.t.} \sum_{j=1}^{n} s_{i,j}\, x_{i,j} \leq C_i \qquad \text{for all } i \in [m], \tag{C1}$$

$$\sum_{i=1}^{m} x_{i,j} \leq 1 \qquad \text{for all } j \in [n], \tag{C2}$$

$$x_{i,j} \in \{0,1\} \qquad \text{for all } i \in [m], j \in [n]. \tag{C3}$$

We also use the linear programming relaxation where (C3) is replaced by

$$x_{i,j} \geq 0 \qquad \text{for all } i \in [m], j \in [n]. \tag{C4}$$

We denote an assignment by $\mathbf{x} \in \{0,1\}^{m \times n}$ with $\mathbf{x} = (x_{i,j})_{i \in [m], j \in [n]}$ and we let $v(\mathbf{x}) = \sum_{i=1}^{m} \sum_{j=1}^{n} v_{i,j}\, x_{i,j}$ denote the total value of an assignment. Let

$$\mathbf{x}^* \in \arg\max\{v(\mathbf{x}) \mid \mathbf{x} \text{ satisfies (C1)} - \text{(C3)}\}, \qquad \text{and}$$
$$\tilde{\mathbf{x}} \in \arg\max\{v(\mathbf{x}) \mid \mathbf{x} \text{ satisfies (C1), (C2), (C4)}\}$$

denote an optimal binary assignment and an optimal fractional assignment, respectively. For any assignment we let $\mathbf{x}_j$ be the jth column of $\mathbf{x}$. In addition let $\mathbf{v}_j = (v_{1,j}, \ldots, v_{m,j})^\top$ and $\mathbf{s}_j = (s_{1,j}, \ldots, s_{m,j})^\top$ be the vectors with values and sizes of item $j \in [n]$. Without loss of generality, we assume that $s_{i,j} \leq C_i$ for each bin $i \in [m]$ and each item $j \in [n]$.

In the random-order model an algorithm knows the number of items n, but the values $\mathbf{v}_j$ and sizes $\mathbf{s}_j$ of each item $j \in [n]$ are previously unknown. They are revealed to an algorithm one by one in rounds $1, \ldots, n$ and an algorithm has to decide immediately and irrevocably whether to assign the item to a bin or to skip that item. The order in which the items are revealed is given by a random permutation $\pi \colon [n] \to [n]$ that assigns to each round $\ell \in [n]$ the item $\pi(\ell)$ that is revealed. The random permutation is drawn uniformly at random from the set of all possible permutations on $[n]$, which we denote by Π. We say that an algorithm in the random-order model is α-competitive, if for the assignment $\mathbf{x}$

of the algorithm we have $\mathbb{E}[v(\mathbf{x})] \geq (\alpha - o(1))\, v(\mathbf{x}^*)$ for every instance of the problem. The expectation is taken over the random permutation and over the randomized decisions of the algorithm.

2.1 Knapsack and Fractional Knapsack

The knapsack problem arises as a special case of GAP for $m = 1$. Given are a knapsack capacity $C > 0$ and a set of items $[n]$ where each item $j \in [n]$ has a value $v_j > 0$ and a size $s_j > 0$. The goal is to find a subset of items that maximizes the total value while the total size does not exceed the capacity. We use the same notation as for GAP and denote a subset of items by a binary assignment $\mathbf{x} \in \{0,1\}^n$ with $\mathbf{x} = (x_1, \ldots, x_n)$ and an optimal knapsack solution by

$$\mathbf{x}^* \in \arg\max \left\{ v(\mathbf{x}) \;\middle|\; \mathbf{x} \in \{0,1\}^n,\ \sum_{j=1}^{n} s_j\, x_j \leq C \right\}.$$

Further, we define the density of an item $j \in [n]$ by $d_j = v_j / s_j$ and we assume without loss of generality that $0 < s_j \leq C$ for all items $j \in [n]$. We further assume that any ordering of the items by value or density yields a unique ordering. This could be achieved by arbitrarily small perturbations of the values and does not affect our results, but makes the presentation easier. For the fractional variant of the knapsack problem, we use $\mathbf{x} \in [0,1]^n$ to denote a fractional assignment. We further assume that the items are sorted by their density in non-increasing order, i.e., $d_1 > d_2 > \cdots > d_n$. Additionally, let $\rho = \max\{q \in [n] \mid \sum_{p=1}^{q} s_p \leq C\}$ denote the largest index, such that the first ρ items do not exceed the knapsack capacity. By $\tilde{\mathbf{x}}$ we denote the fractional greedy solution which is defined for all $j \in [n]$ by $\tilde{x}_j = 1$ if $j \leq \rho$, $\tilde{x}_j = (C - \sum_{q=1}^{\rho} s_q)/s_j$ if $j = \rho + 1$, and $\tilde{x}_j = 0$ otherwise. Note that the fractional greedy solution is uniquely defined under our assumptions and that it defines an optimal solution of the fractional knapsack problem [25, § 17.1]. We state some properties of fractional greedy solution for later reference.

Proposition 1. *For the fractional greedy solution* $\tilde{\mathbf{x}}$*, it holds that*

1. $\tilde{\mathbf{x}} = \arg\max\{v(\mathbf{x}) \mid \mathbf{x} \in [0,1]^n,\ \sum_{j=1}^{n} s_j\, x_j \leq C\}$,
2. $\sum_{j=1}^{n} s_j\, \tilde{x}_j = \min\{C,\ \sum_{j=1}^{n} s_j\}$.

3 The Generalized Assignment Problem

In this section, we study the generalized assignment problem in the random-order model. We improve over the currently best-known competitive ratio of 1/6.99 shown independently by Albers et al. [3] and Naori and Raz [33]. We state our main result of this section in the following theorem.

Theorem 1. *There exists an α-competitive randomized algorithm for the generalized assignment problem in the random-order model where* $\alpha = \frac{1-\ln(2)}{2} \approx \frac{1}{6.52}$.

Algorithm 1. InfeasibleGAP

Require: random permutation π, number of items n, set of bins $[m]$ with capacities $C_i, i \in [m]$
Ensure: assignment $\mathbf{x}$ satisfying GAP constraints (C2)–(C3)

$\mathbf{x} \leftarrow 0$
$Q_0 \leftarrow \emptyset$
for rounds $\ell = 1, \dots, n$ **do**
  $Q_\ell \leftarrow Q_{\ell-1} \cup \{\pi(\ell)\}$
  if $\ell > t \leftarrow \lfloor n/2 \rfloor$ **then**
    $\tilde{\mathbf{x}}(Q_\ell) \leftarrow$ optimal fractional assignment of revealed items Q_ℓ
    $i^{(\ell)} \leftarrow$ select a bin $i \in [m]$ where $\mathbb{P}[i^{(\ell)} = i] = \tilde{x}_{i,\pi(\ell)}(Q_\ell)$ ($i^{(\ell)} = 0$ if none)
    Define $\mathbf{x}^{(\ell)}$ with $x_{i,j}^{(\ell)} = \begin{cases} 1 \text{ if } i = i^{(\ell)}, j = \pi(\ell), \\ 0 \text{ otherwise} \end{cases}$ for $i \in [m], j \in [n]$.
    if $i^{(\ell)} = 0$ **or** $\sum_{k=1}^{\ell-1} s_{i^{(\ell)},\pi(k)} \, x_{i^{(\ell)},\pi(k)} \leq C_{i^{(\ell)}}$ **then**
      $\mathbf{x} \leftarrow \mathbf{x} + \mathbf{x}^{(\ell)}$
return $\mathbf{x}$

We obtain the result by analyzing a variant of an algorithm proposed by Kesselheim et al. [24]. The original algorithm consists of a sampling phase and an assignment phase. The sampling phase lasts for $t \in [n]$ rounds and during these rounds, the items are not assigned to a bin. Afterwards, in the assignment phase, the algorithm computes in each round $\ell > t$ the optimal fractional assignment $\tilde{\mathbf{x}}(Q_\ell)$ of all revealed items Q_ℓ. Then it uses the fractional assignment of the current item $\pi(\ell)$ as a probability distribution over bins to randomly determine a bin $i^{(\ell)}$. Finally, the item $\pi(\ell)$ is assigned to the bin $i^{(\ell)}$ if this does not violate the capacity constraint of the bin.

To achieve a constant competitive ratio for this algorithm, the authors of [24] restrict the options $(v_{i,j}, s_{i,j})$ in the GAP instance to those with $s_{i,j} \leq C_i/2$. This allows to bound the probability that item $\pi(\ell)$ can be assigned to the bin $i^{(\ell)}$ by the probability that the total size of items assigned to the bin in previous rounds is at most $C_{i^{(\ell)}}/2$, which particularly is independent of $s_{i^{(\ell)},\pi(\ell)}$. The instance with the remaining options can be seen as an edge-weighted matching problem in the random-order model and is covered with a suitable algorithm. In the end, each algorithm is executed with a certain probability.

To improve the competitive ratio, Albers et al. [3] and Naori and Raz [33] combined two algorithms for the different options by running them during different rounds of a single algorithm.

Our version of the algorithm, presented in Algorithm 1, does not restrict the instance. Instead, we allow the algorithm to exceed the capacity of every bin by at most one item. More precisely, we permit the assignment of item $\pi(\ell)$ to the bin $i^{(\ell)}$, if the total size of items assigned to the bin in previous rounds is at most $C_{i^{(\ell)}}$. It turns out, that this is sufficient to handle all options $(v_{i,j}, s_{i,j})$ together, but the assignment returned by Algorithm 1 may be infeasible. Therefore, we split the algorithm at the end of this chapter into two separate algorithms. The first algorithm does the same as Algorithm 1, but maintains feasibility by assigning items only if they do not violate the capacity constraint of the bin.

Then, the second algorithm imitates the first algorithm and assigns the first item to each bin that the first algorithm could not assign due to the capacity constraint.

For the analysis of Algorithm 1 let $\mathbf{x}$ denote the final assignment of the algorithm, let $t \in [n]$, and consider a fixed round $\ell > t$. As in the algorithm, we let $i^{(\ell)}$ denote the bin that the algorithm selects randomly and we let $\mathbf{x}^{(\ell)}$ denote the tentative assignment. Let $V^{(\ell)}$ denote the value that is obtained in round $\ell > t$. If the assignment of the tentative allocation $\mathbf{x}^{(\ell)}$ is successful, then $V^{(\ell)}$ equals $v(\mathbf{x}^{(\ell)})$, and if the assignment is not successful, then $V^{(\ell)}$ equals 0. To bound the expected value of $V^{(\ell)}$, we determine the random order up to round ℓ in three steps as done by Kesselheim et al. [24]: (i) select a random subset Q_ℓ of $[n]$ with $|Q_\ell| = \ell$ which are the items revealed in the first ℓ rounds; (ii) the item that is revealed in round ℓ is drawn uniformly at random from the set Q_ℓ; (iii) the order of the remaining $\ell - 1$ items is determined by drawing an item uniformly at random from the remaining items.

By steps (i) and (ii) together with the random decision of the algorithm in round ℓ, we can bound the expected value of the tentative assignment in round ℓ in terms of an optimal offline assignment. The crux is that the probability that the assignment is successful in round ℓ is bounded independently to the outcomes of steps (i) and (ii) and the random decision in round ℓ. This allows us to bound the expected value of $V^{(\ell)}$ by the bound we derive for the probability that the assignment is successful times the expected value we obtain for the tentative assignment. In the following lemma, we obtain the bound for the value of the tentative assignment.

Lemma 1. *For a round $\ell \in \{t+1, \ldots, n\}$, let Q_ℓ be a random subset of $[n]$ with $|Q_\ell| = \ell$ and let $\pi(\ell)$ be an item drawn uniformly at random from Q_ℓ. For the tentative assignment $\mathbf{x}^{(\ell)}$ we have $\mathbb{E}\left[v(\mathbf{x}^{(\ell)})\right] \geq v(\mathbf{x}^\star)/n$.*

The tentative assignment can be realized if the accumulated size of the selected bin has not exceeded the capacity in previous rounds. The next lemma shows a bound on the probability that the first $\ell - 1$ assignments exceed the capacity of a bin $i \in [m]$. As in Kesselheim et al. [24], we prove the bound for the tentative assignments in previous rounds instead of the realized assignments. The bound only depends on the order of the first $\ell - 1$ items and holds for every possible subset of items $Q_{\ell-1}$. Note that we slightly overload the notation here and use $Q_{\ell-1}$ as the set of items revealed in the first $\ell - 1$ rounds and for the event that this set of items is revealed in the first $\ell - 1$ rounds.

Lemma 2. *Consider a round $\ell \in \{t+1, \ldots, n\}$ and a bin $i \in [m]$. Let $Q_{\ell-1}$ be any subset of $[n]$ with $|Q_{\ell-1}| = \ell - 1$. We have*

$$\mathbb{P}\left[\sum_{k=t+1}^{\ell-1} s_{i,\pi(k)}\, x_{i,\pi(k)}^{(k)} > C_i \,\middle|\, Q_{\ell-1}\right] \leq \sum_{k=t+1}^{\ell-1} \frac{1}{k}.$$

With the two lemmas at hand, we proceed to bound the expected value of the assignment returned by Algorithm 1.

Lemma 3. *Let $\mathbf{x}$ be the infeasible assignment computed by Algorithm 1. Then,* $\mathbb{E}[v(\mathbf{x})] \geq (1 - \ln(2))\, v(\mathbf{x}^*)$.

Algorithm 2. FRACTIONALKNAPSACK

Require: random permutation π, number of items n, capacity C.
Ensure: fractional knapsack solution $\mathbf{x}$

$\mathbf{x} \leftarrow 0$
$Q_0 \leftarrow \emptyset$
for rounds $\ell = 1, \ldots, n$ **do**
  $Q_\ell \leftarrow Q_{\ell-1} \cup \{\pi(\ell)\}$
  $\tilde{\mathbf{x}}(Q_\ell) \leftarrow$ fractional greedy solution of revealed items Q_ℓ
  if $\ell > t \leftarrow \lfloor n/e \rfloor$ **then**
    $x_{\pi(\ell)} = \tilde{x}_{\pi(\ell)}(Q_\ell) - \dfrac{\sum_{k=t+1}^{\ell-1} s_{\pi(k)} \left(\tilde{x}_{\pi(k)}(Q_{\ell-1}) - \tilde{x}_{\pi(k)}(Q_\ell)\right)}{s_{\pi(\ell)}}$
return $\mathbf{x}$

Recall that Algorithm 1 adds an item to a bin if the capacity constraint of the bin is not violated *before* adding the item. In order to obtain a feasible assignment, we consider the variant where items are only added to a bin if the capacity constraint is not violated *after* adding the item. Like this, we obtain a feasible assignment, but we potentially loose the values contributed by the last items added to each bin in Algorithm 1. To compensate for this loss, we run the feasible variant only with probability 1/2 and with the remaining probability, we run a second variant of Algorithm 1. This algorithm tries to assign to each bin the last item that Algorithm 1 assigned to the bin. In order to achieve this, it creates an auxiliary assignment in the same way as Algorithm 1. The actual assignment is then obtained by assigning only the first item to each bin that would violate the capacity constraint in the auxiliary assignment. We obtain our final algorithm, by running both variants with a probability of 1/2. A formal description of the mentioned algorithms is deferred to the full version. For the proof of Theorem 1 we show that both algorithms together are as good as Algorithm 1. The idea is the following: if we fix a permutation π and if we fix the random decisions of the algorithms for that permutation, i.e., fixing the bin that is selected in each round, then the statement follows immediately. The rest follows from linearity of expectation.

4 The Fractional Knapsack Problem

In this section, we address the fractional variant of the knapsack problem in the random-order model. In this setting, the currently best-known competitive ratio is $\frac{1}{4.39}$ by Giliberti and Karrenbauer [16]. We state our main result of this section in the following theorem.

Theorem 2. *There exists an α-competitive algorithm for the fractional knapsack problem in the random-order model where $\alpha = \frac{1}{e} \approx \frac{1}{2.71}$.*

We obtain the competitive ratio by Algorithm 2. The algorithm is built around the ideas of the virtual algorithm by Babaioff et al. [4]. The virtual algorithm was designed for packing items subject to a cardinality constraint of size k in the random-order model, also referred to as the k-secretary problem.

In each round ℓ after the sampling phase, the algorithm packs the current item if two conditions are fulfilled: the item has to be one of the k most valuable items that have been revealed up to this round and the item that the current item removes from the set of the k most valuable items must have been revealed during the sampling phase. Note that the second condition ensures feasibility, since each of the k most valuable items after the sampling phase is removed at most once.

For the fractional knapsack problem we replace the set of the k most valuable items by the fractional greedy solution. For each round ℓ, let $\tilde{\mathbf{x}}(Q_\ell)$ denote the fractional greedy solution of the revealed items Q_ℓ. For an arbitrary round ℓ and a fixed permutation π, we know that the total size of $\tilde{\mathbf{x}}(Q_\ell)$ cannot decrease compared to the total size of $\tilde{\mathbf{x}}(Q_{\ell-1})$ by part (2) of Propostion 1, i.e., $\sum_{k=1}^{\ell} s_{\pi(k)}\,\tilde{x}_{\pi(k)}(Q_\ell) \geq \sum_{k=1}^{\ell-1} s_{\pi(k)}\,\tilde{x}_{\pi(k)}(Q_{\ell-1})$ and we obtain

$$s_{\pi(\ell)}\,\tilde{x}_{\pi(\ell)}(Q_\ell) \geq \sum_{k=1}^{\ell-1} s_{\pi(k)}\left(\tilde{x}_{\pi(k)}(Q_{\ell-1}) - \tilde{x}_{\pi(k)}(Q_\ell)\right). \tag{1}$$

Compared to the setting with a cardinality constraint, it is now possible that the revelation of item $\pi(\ell)$ in round ℓ removes multiple items from the fractional greedy solution or just a fraction of a single item. But note that it never increases the fraction by which other items are contained, i.e., $\tilde{x}_{\pi(j)}(Q_{\ell-1}) \geq \tilde{x}_{\pi(j)}(Q_\ell)$ for each item $j \in Q_{\ell-1}$.

Therefore, we have to adapt the packing conditions. In Algorithm 2, we pack the item $\pi(\ell)$ in round ℓ by a fraction $x_{\pi(\ell)}$, such that

$$s_{\pi(\ell)}\,x_{\pi(\ell)} = s_{\pi(\ell)}\,\tilde{x}_{\pi(\ell)}(Q_\ell) - \sum_{k=t+1}^{\ell-1} s_{\pi(k)}\left(\tilde{x}_{\pi(k)}(Q_{\ell-1}) - \tilde{x}_{\pi(k)}(Q_\ell)\right).$$

Thus, we take the size that the new item fills in the fractional greedy solution and for each item that was revealed after the sampling phase we subtract the size by which it is removed from the fractional greedy solution.

From the previous observations, we obtain that $x_{\pi(\ell)} \in [0, 1]$. Moreover, we get that the total size of the fractional assignment $\mathbf{x}$ is at most C, since

$$\begin{aligned}
\sum_{\ell=1}^{n} s_{\pi(\ell)}\,x_{\pi(\ell)} &= \sum_{\ell=t+1}^{n}\left(s_{\pi(\ell)}\,\tilde{x}_{\pi(\ell)}(Q_\ell) - \sum_{k=t+1}^{\ell-1} s_{\pi(k)}\left(\tilde{x}_{\pi(k)}(Q_{\ell-1}) - \tilde{x}_{\pi(k)}(Q_\ell)\right)\right)\\
&= \sum_{\ell=t+1}^{n}\left(s_{\pi(\ell)}\,\tilde{x}_{\pi(\ell)}(Q_\ell) - \sum_{k=\ell+1}^{n} s_{\pi(\ell)}\left(\tilde{x}_{\pi(\ell)}(Q_{k-1}) - \tilde{x}_{\pi(\ell)}(Q_k)\right)\right)\\
&= \sum_{\ell=t+1}^{n} s_{\pi(\ell)}\,\tilde{x}_{\pi(\ell)} \leq C,
\end{aligned}$$

where we changed the summation order of the inner sum and the inequality at the end follows from the feasibility of fractional greedy solution $\tilde{\mathbf{x}}$ of all items Q_n.

For the proof of Theorem 2, we show in the following lemma a bound on the expectation of x_j for each item $j \in [n]$.

Lemma 4. *Let* $\mathbf{x}$ *be the fractional knapsack assignment of Algorithm 2. For each item* $j \in [n]$ *we have that* $\mathbb{E}[x_j] \geq \frac{\tilde{x}_j}{n} \sum_{\ell=t+1}^{n} \frac{t}{\ell-1}$.

With the lemma, we are able to prove Theorem 2. The statement of the theorem follows from summing over all items and optimization over t.

We proceed to give an upper bound of $1/e$ on the competitiveness of any deterministic algorithm for the fractional knapsack problem in the random-order model. The proof exploits the similarity between deterministic algorithms for the fractional knapsack problem in the random-order model and randomized stopping rules together with Correa et al. [12, Theorem 2].

Theorem 3. *Let* $\varepsilon > 0$. *No deterministic algorithm for the fractional knapsack problem in the random-order model is* $(\frac{1}{e} + \varepsilon)$*-competitive.*

References

1. Abels, A., Ladewig, L., Schewior, K., Stinzendörfer, M.: Knapsack secretary through boosting. In: Chalermsook, P., Laekhanukit, B. (eds.) Proceedings of the 20th International Workshop on Approximation and Online Algorithms (WAOA), pp. 61–81 (2022). https://doi.org/10.1007/978-3-031-18367-6_4
2. Agrawal, S., Wang, Z., Ye, Y.: A dynamic near-optimal algorithm for online linear programming. Oper. Res. **62**(4), 876–890 (2014). https://doi.org/10.1287/OPRE.2014.1289
3. Albers, S., Khan, A., Ladewig, L.: Improved online algorithms for knapsack and GAP in the random order model. Algorithmica **83**(6), 1750–1785 (2021). https://doi.org/10.1007/S00453-021-00801-2
4. Babaioff, M., Immorlica, N., Kempe, D., Kleinberg, R.: A knapsack secretary problem with applications. In: Charikar, M., Jansen, K., Reingold, O., Rolim, J.D.P. (eds.) Proceedings of the 10th International Workshop on Approximation, Randomization, and Combinatorial Optimization (APPROX), pp. 16–28 (2007). https://doi.org/10.1007/978-3-540-74208-1_2
5. Babaioff, M., Immorlica, N., Kempe, D., Kleinberg, R.: Matroid secretary problems. J. ACM **65**(6), 1–26 (2018). https://doi.org/10.1145/3212512
6. Babaioff, M., Immorlica, N., Kleinberg, R.: Matroids, secretary problems, and online mechanisms. In: Bansal, N., Pruhs, K., Stein, C. (eds.) Proceedings of the 18th Annual ACM-SIAM Symposium on Discrete Algorithms (SODA), pp. 434–443 (2007)
7. Bahrani, M., Beyhaghi, H., Singla, S., Weinberg, S.M.: Formal barriers to simple algorithms for the matroid secretary problem. In: Feldman, M., Fu, H., Talgam-Cohen, I. (eds.) Proceedings of the 17th International Conference on Web and Internet Economics (WINE), pp. 280–298 (2021). https://doi.org/10.1007/978-3-030-94676-0_16
8. Buchbinder, N., Naor, J.: Online primal-dual algorithms for covering and packing. Math. Oper. Res. **34**(2), 270–286 (2009). https://doi.org/10.1287/MOOR.1080.0363
9. Bérczi, K., Livanos, V., Soto, J., Verdugo, V.: Matroid secretary via labeling schemes (2024). https://arxiv.org/abs/2411.12069

10. Chakraborty, S., Lachish, O.: Improved competitive ratio for the matroid secretary problem. In: Rabani, Y. (ed.) Proceedings of the 23rd Annual ACM-SIAM Symposium on Discrete Algorithms (SODA), pp. 1702–1712 (2012). https://doi.org/10.1137/1.9781611973099.135
11. Chekuri, C., Khanna, S.: A polynomial time approximation scheme for the multiple knapsack problem. SIAM J. Comput. **35**(3), 713–728 (2005). https://doi.org/10.1137/S0097539700382820
12. Correa, J., Dütting, P., Fischer, F.A., Schewior, K.: Prophet inequalities for independent and identically distributed random variables from an unknown distribution. Math. Oper. Res. **47**(2), 1287–1309 (2022). https://doi.org/10.1287/MOOR.2021.1167
13. Dynkin, E.B.: Optimal choice of the stopping moment of a Markov process. Dokl. Akad. Nauk SSSR **150**, 238–240 (1963)
14. Feldman, J., Korula, N., Mirrokni, V.S., Muthukrishnan, S., Pál, M.: Online ad assignment with free disposal. In: Leonardi, S. (ed.) Proceedings of the 5th International Workshop on Internet and Network Economics (WINE), pp. 374–385 (2009). https://doi.org/10.1007/978-3-642-10841-9_34
15. Feldman, M., Svensson, O., Zenklusen, R.: A simple $o(\log\log(\text{rank}))$-competitive algorithm for the matroid secretary problem. Math. Oper. Res. **43**(2), 638–650 (2018). https://doi.org/10.1287/MOOR.2017.0876
16. Giliberti, J., Karrenbauer, A.: Improved online algorithm for fractional knapsack in the random order model. In: Könemann, J., Peis, B. (eds.) Proceedings of the 19th International Workshop on Approximation and Online Algorithms (WAOA), pp. 188–205 (2021). https://doi.org/10.1007/978-3-030-92702-8_12
17. Han, X., Kawase, Y., Makino, K.: Randomized algorithms for online knapsack problems. Theor. Comput. Sci. **562**, 395–405 (2015). https://doi.org/10.1016/J.TCS.2014.10.017
18. Huang, Z., Parsaeian, Z., Zhu, Z.: Laminar matroid secretary: greedy strikes back. In: Chan, T.M., Fischer, J., Iacono, J., Herman, G. (eds.) Proceedings of the 32nd Annual European Symposium on Algorithms (ESA), pp. 1–8 (2024). https://doi.org/10.4230/LIPICS.ESA.2024.73
19. Ibarra, O.H., Kim, C.E.: Fast approximation algorithms for the knapsack and sum of subset problems. J. ACM **22**(4), 463–468 (1975). https://doi.org/10.1145/321906.321909
20. Karp, R.M.: Reducibility among combinatorial problems. In: Miller, R., Thatcher, J.W., Bohlinger, J. (eds.) Complexity of Computer Computations. Springer, Boston (1972). https://doi.org/10.1007/978-1-4684-2001-2_9
21. Karp, R.M., Vazirani, U.V., Vazirani, V.V.: An optimal algorithm for on-line bipartite matching. In: Ortiz, H. (ed.) Proceedings of the 22nd Annual ACM Symposium on Theory of Computing (STOC), pp. 352–358 (1990). https://doi.org/10.1145/100216.100262
22. Karrenbauer, A., Kovalevskaya, E.: Reading articles online. In: Wu, W., Zhang, Z. (eds.) Proceedings 14th International Conference on Combinatorial Optimization and Applications (COCOA), pp. 639–654. Springer (2020). https://doi.org/10.1007/978-3-030-64843-5_43
23. Kesselheim, T., Radke, K., Tönnis, A., Vöcking, B.: An optimal online algorithm for weighted bipartite matching and extensions to combinatorial auctions. In: Bodlaender, H.L., Italiano, G.F. (eds.) Proceedings of the 21st Annual European Symposium on Algorithms (ESA), pp. 589–600 (2013). https://doi.org/10.1007/978-3-642-40450-4_50
24. Kesselheim, T., Radke, K., Tönnis, A., Vöcking, B.: Primal beats dual on online packing LPs in the random-order model. SIAM J. Comput. **47**(5), 1939–1964 (2018). https://doi.org/10.1137/15M1033708

25. Korte, B., Vygen, J.: Combinatorial Optimization: Theory and Algorithms, Algorithms and Combinatorics, vol. 21. Springer, Berlin, Heidelberg (2018). https://doi.org/10.1007/978-3-662-56039-6_17
26. Korula, N., Pál, M.: Algorithms for secretary problems on graphs and hypergraphs. In: Albers, S., Marchetti-Spaccamela, A., Matias, Y., Nikoletseas, S.E., Thomas, W. (eds.) Proceedings of the 36th International Colloquium on Automata, Languages and Programming (ICALP), pp. 508–520 (2009). https://doi.org/10.1007/978-3-642-02930-1_42
27. Lachish, O.: $O(\log\log \text{Rank})$ competitive ratio for the matroid secretary problem. In: Proceedings of the 55th IEEE Annual Symposium on Foundations of Computer Science (FOCS), pp. 326–335 (2014). https://doi.org/10.1109/FOCS.2014.42
28. Lindley, D.V.: Dynamic programming and decision theory. Appl. Statist. **10** (1961). https://doi.org/10.2307/2985407
29. Ma, T., Tang, B., Wang, Y.: The simulated greedy algorithm for several submodular matroid secretary problems. Theory Comput. Syst. **58**(4), 681–706 (2016). https://doi.org/10.1007/S00224-015-9642-4
30. Mahdian, M., Yan, Q.: Online bipartite matching with random arrivals: an approach based on strongly factor-revealing LPs. In: Fortnow, L., Vadhan, S.P. (eds.) Proceedings of the 43rd ACM Symposium on Theory of Computing (STOC), pp. 597–606 (2011). https://doi.org/10.1145/1993636.1993716
31. Martello, S., Toth, P.: Knapsack Problems: Algorithms and Computer Implementations. John Wiley & Sons, New York (1990)
32. Mehta, A., Saberi, A., Vazirani, U.V., Vazirani, V.V.: Adwords and generalized online matching. In: Proceedings of the 46th Annual IEEE Symposium on Foundations of Computer Science (FOCS), pp. 264–273 (2005). https://doi.org/10.1109/SFCS.2005.12
33. Naori, D., Raz, D.: Online multidimensional packing problems in the random-order model. In: Lu, P., Zhang, G. (eds.) Proceedings of the 30th International Symposium on Algorithms and Computation (ISAAC), pp. 1–15 (2019). https://doi.org/10.4230/LIPICS.ISAAC.2019.10
34. Shmoys, D.B., Tardos, É.: An approximation algorithm for the generalized assignment problem. Math. Program. **62**, 461–474 (1993). https://doi.org/10.1007/BF01585178
35. Soto, J.A., Turkieltaub, A., Verdugo, V.: Strong algorithms for the ordinal matroid secretary problem. Math. Oper. Res. **46**(2), 642–673 (2021). https://doi.org/10.1287/MOOR.2020.1083
36. Vaze, R.: Online knapsack problem and budgeted truthful bipartite matching. In: Proceedings of the IEEE Conference on Computer Communications (INFOCOM), pp. 1–9 (2017). https://doi.org/10.1109/INFOCOM.2017.8057223

Totally Delta-Modular IPs with Two Non-zeros in Most Rows

Stefan Kober(✉)

Université Libre de Bruxelles, Bruxelles, Belgium
kober.stefan@gmail.com

Abstract. Integer programs (IPs) on constraint matrices with bounded subdeterminants are conjectured to be solvable in polynomial time. We give a strongly polynomial time algorithm to solve IPs where the constraint matrix has bounded subdeterminants and at most two non-zeros per row after removing a constant number of rows and columns. This result extends the work by Fiorini, Joret, Weltge & Yuditsky (J. ACM 72(1), 1–50 (2025)) by allowing for additional, unifying constraints and variables.

Keywords: Integer Programming · Parametrized Integer Programming · Subdeterminants · Structural Graph Theory · Dynamic Programming

1 Introduction

In this work, we are concerned with integer programs (IPs) of the following form:

$$\min \{c^\intercal x : Mx \leq b,\ x \in \mathbb{Z}^n\}, \tag{IP}$$

where $c \in \mathbb{Z}^n$, $M \in \mathbb{Z}^{m \times n}$, and $b \in \mathbb{Z}^m$ for $m, n \in \mathbb{N}$. In general, assuming P $\neq$ NP, one cannot hope for a polynomial-time algorithm to solve (IP). Still, there are certain subclasses of IPs that can be solved efficiently. The most prominent example for this is IPs with *totally unimodular* (TU) constraint matrices, i.e., if M is TU. Recall that a matrix $M \in \mathbb{Z}^{m \times n}$ is TU, if all its *subdeterminants* are in $\{0, \pm 1\}$, where a subdeterminant is the determinant of a square submatrix of M.

Totally unimodular matrices possess the nice geometric property that the convex hull of the feasible solutions to (IP) is equal to the space of feasible solutions of its linear relaxation $\min \{c^\intercal x : Mx \leq b,\ x \in \mathbb{R}^n\}$. Therefore, we can solve such IPs by solving a single linear program. Such (combinatorial) linear programs can be solved in strongly polynomial time due to a result by Tardos [29]. This implies polynomial solvability for a collection of combinatorial optimization problems that can be stated as IPs on TU constraint matrices,

N. Megow and A. Basu (Eds.): IPCO 2025, LNCS 15620, pp. 355–370, 2025.
https://doi.org/10.1007/978-3-031-93112-3_26

such as the maximum-flow problem, the matching problem in bipartite graphs, or the stable set problem in bipartite graphs.

Almost 30 years ago, Shevchenko [28] asked for generalizations of this result to IPs on constraint matrices with bounded subdeterminants. We call a matrix M *totally Δ-modular*, if all its subdeterminants are in $[-\Delta, \Delta] \cap \mathbb{Z}$. A fundamental open question in combinatorial optimization is the following:

Can we solve (IP) *in polynomial time, when M is totally Δ-modular?*

In a breakthrough paper, Artmann, Weismantel & Zenklusen [3] answered the question in the affirmative for $\Delta = 2$. Generalizing their ideas to $\Delta \geq 3$ has proven to be difficult, despite some progress, see e.g. Nägele, Santiago & Zenklusen [26]. Fiorini, Joret, Weltge & Yuditsky [16] fix Δ to be an arbitrary constant, and further require a structural condition on M, see Theorem 1.

Theorem 1. ***(Fiorini, Joret, Weltge & Yuditsky** [16]). Let $\Delta \in \mathbb{N}$. There is a polynomial-time algorithm to solve* (IP) *if M is totally Δ-modular and has at most two non-zeros per row (or column).*

Another example for restricting the structure of the constraint matrix are *nearly totally unimodular matrices* as studied by Aprile et al. [2]. The same idea has also proven helpful for a strongly polynomial-time algorithm for linear programming, see Dadush et al. [8], who also study the case of two non-zeros per row (or column).

We follow this second stream of research and investigate the question in the case where Δ is arbitrary but fixed, and there is a further structural condition on the constraint matrix. In the context of this work, we study the following parametrized problem, and provide a strongly polynomial-time algorithm. In particular, our result combines ideas from Fiorini et al. [16] and Aprile et al. [2].

problem 2. Fix $\Delta, k \in \mathbb{N}$. Let $m_1, n_1 \in \mathbb{N}$ and $m_2, n_2 \leq k$, and define $m := m_1 + m_2$, $n := n_1 + n_2$. Let $M = \left[\begin{smallmatrix} A & B \\ W & D \end{smallmatrix}\right] \in \mathbb{Z}^{m \times n}$ be totally Δ-modular, where $A \in \mathbb{Z}^{m_1 \times n_1}$, $B \in \mathbb{Z}^{m_1 \times n_2}$, $W \in \mathbb{Z}^{m_2 \times n_1}$ and $D \in \mathbb{Z}^{m_2 \times n_2}$. Assume A has at most two non-zeros per row. Let $c \in \mathbb{Z}^n$ and $b \in \mathbb{Z}^m$. Solve (IP).

We reduce Problem 2 to Problem 3 below, in which the main technical difficulty for this project lies.

problem 3. Fix $\Delta, k \in \mathbb{N}$. Let $A \in \{0, \pm 1\}^{m \times n}$, $W \in \mathbb{Z}^{k \times n}$, let $M = \left[\begin{smallmatrix} A \\ W \end{smallmatrix}\right]$ be totally Δ-modular and assume A has two non-zeros per row. Let $c, \ell, u \in \mathbb{Z}^n$, $b \in \mathbb{Z}^m$ and $d \in \mathbb{Z}^k$. Solve $\min \{c^\intercal x : Ax \leq b,\, Wx = d,\, \ell \leq x \leq u,\, x \in \mathbb{Z}^n\}$.

In the *(k-*dimensional) partially ordered knapsack problem, we are given an integer $k \in \mathbb{N}$, a set of elements V with an underlying partial order, a profit function $p : V \to \mathbb{Z}$, k weight-functions $W_i : V \to \mathbb{Z}$ for $i \in [k]$, and k budgets $b_i \in \mathbb{Z}$ for $i \in [k]$. The goal is to find the maximum profit subset of V that respects the budgets with respect to the weight functions and adheres to the partial order as precedence constraints, i.e., in order to choose any element, we need to choose all its predecessors with respect to the partial order. This problem is known to be NP-complete even if there is only one weight function with bounded entries, and the partial order is bipartite [19]. It can be formulated as an IP on a totally unimodular constraint matrix that has two non-zeros per row, and k extra constraints. In the following, we refer to this formulation as the respective IP. This makes it an interesting problem to consider in the context of this work. Note that the Hasse diagram of the partial order induces a directed graph G on V.

1.1 Contribution

Our main result is a strongly polynomial-time algorithm for IPs as defined in Problem 2, see Theorem 4. By combining our structural results with a proximity result by Aprile et al. [2], our proof technique implies an FPT-algorithm for Problem 3 if A is totally unimodular, see Corollary 5 and in particular for the k-dimensional partially order knapsack problem, see Corollary 6. For this problem, only few non-trivial polynomially solvable cases are known [22]. A proof for Corollary 5 has previously been given in the author's thesis [21].

Proofs for the statements are deferred to Sect. 6.

Theorem 4. *There is a strongly polynomial-time algorithm to solve Problem 2.*

Corollary 5. *There is an FPT-algorithm to solve Problem 3 with respect to the largest subdeterminant if A is totally unimodular.*

Corollary 6. *There is an FPT-algorithm for the k-dimensional partially ordered knapsack problem with parameter*

$$\Delta = \max\{\sum_{v\in V(H)} W_i(v) : i \in [k], H \textit{ connected subgraph of } G\}.$$

Note that for $k = 1$, Δ in Corollary 6 corresponds to the largest subdeterminant of the corresponding constraint matrix, see [21, Corollary 4.24]. For constant k, the subdeterminants of the respective IP are bounded by a function in k and Δ.

Our results contribute to the research on IPs with bounded subdeterminants in the following ways. We combine ideas from two previous technical papers [2,16] and push the boundary of IPs with bounded subdeterminants in terms of polynomial-time algorithms. In particular, we relate the subdeterminants of matrices that have at most two non-zeros after removing one row to a novel invariant of weighted (signed) graphs, called the *alternating weight of a tree*, see Proposition 11. We show that a bound on the alternating weight of a tree in an integer-weighted graph gives a structural restriction on the structure of non-zero weights within the graph. More specifically, given an instance of Problem 3, we obtain a weighted graph and show that graphs corresponding to matrices with bounded subdeterminants admit a structured decomposition, see Theorem 16. We exploit this decomposition in a dynamic program that allows us to solve Problem 2, see Theorem 4.

We remark that it is unclear whether a constant proximity and decomposition result similar to the one in [2] holds in our case. Therefore, in Sect. 5 we crucially need to exploit the fact that we work in a different graph class within the context of this work, and provide a substantially stronger decomposition. In particular, Aprile et al. [2] show that graphs associated with their instances admit a decomposition such that the bags of the tree-decomposition either contain a *small* number of vertices, correspond to a leaf of the tree and contain a *small* number of roots, or are *almost embeddable in a surface of bounded genus.* We can show that the latter type is not necessary in graphs occuring in the context of this work. Finally, our structural results are largely self-contained and in particular do not make use of the graph minor structure theorem for K_t-minor free graphs as required in [2, 16].

1.2 Background and Related Work

The main focus of our work lies in integer programming with bounded subdeterminants, see Fig. 1 for an overview of recent results. There has been a large number of exciting results in the last years that consider the question of polynomial solvability, e.g., the aforementioned papers [2,3,16,26], if the matrix is $\{a, b, c\}$-modular [17], if the problem corresponds to the total matching problem [15], or for restricted subdeterminants over a polynomial ring [6]. Further, there is a randomized polynomial-time algorithm for feasibility of IPs on constraint matrices that are strictly Δ-modular for $\Delta \in \{3, 4\}$ [25, 26].

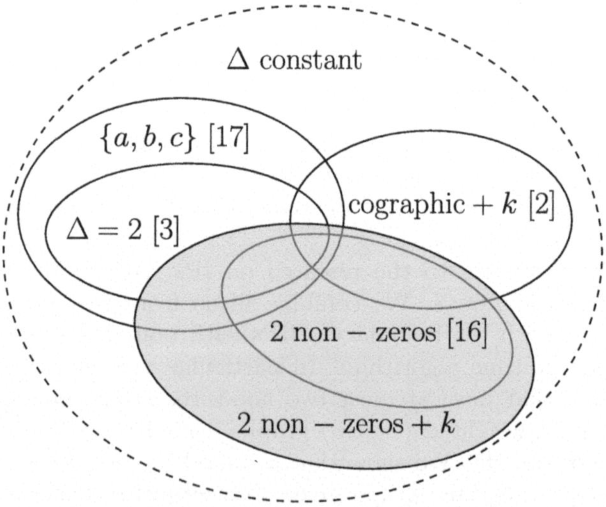

Fig. 1. Recent results on IPs with bounded subdeterminants: bimodular matrices [3], $\{a, b, c\}$-modular matrices [17], two non-zeros per row (2 non-zeros) [16], nearly totally unimodular matrices, the cographic case (cographic $+k$) [2], and this work (2 non-zeros $+k$, shaded in green).(Color figure online)

Secondly, we work on IPs over a class of constraint matrices *with a constant number of extra rows*. Such problems appear naturally within combinatorial optimization or operations research, where the extra rows can play the role of budget or exactness constraints. Since already the knapsack problem is NP-hard, which only has one budget-constraint, one typically needs bounds on the entries or subdeterminants to design polynomial-time algorithms for such problems. We specifically mention the exact matching problem [27], integer programming on few constraints [1,13], the exact matroid basis problem [12,24], integer programming over TU matrices with few extra constraints [2], or the partially ordered knapsack problem [22].

One important problem in the class of IPs with bounded subdeterminants and two non-zeros per row is the maximum stable set problem with bounded *odd cycle packing number*, i.e., there is a bound on the maximum number of disjoint odd cycles. This problem has an efficient algorithm due to Fiorini et al. [16], which we use in order to solve our base case. Two exemplary problems that are covered by Problem 2 are the budgeted stable set problem studied by Dhahan [9], and the partially ordered knapsack problem, where we further restrict to bounded subdeterminants.

1.3 Overview

We begin by setting up definitions that are used throughout the paper in Sect. 2. In Sect. 3, we reduce Problem 2 to Problem 3. We proceed to exploit the bound on the subdeterminants in order to obtain bounds on the parameters of a related graph, as well as a structural description of such graphs in Sects. 4 and 5. We conclude the proof of our main result in Sect. 6 by providing a dynamic program. In the final Sect. 7, we mention related problems and directions for further research.

Any missing proof can be found in the full version [20].

2 Preliminaries

Throughout the paper, we use the notation $[k] := \{1, \ldots, k\}$. Given an integer matrix $A \in \mathbb{Z}^{m \times n}$, we denote its largest subdeterminant by

$$\Delta(A) := \max\{|\det(B)| : B \text{ is a submatrix of } A\}.$$

We refer to Diestel [10] for more background on the notions from graph theory that we use. We proceed to give an overview of the most important tools and definitions we use in the remainder of the paper.

Given a graph $G = (V, E)$, and a subset of vertices $U \subseteq V$, we define the *neighborhood* $N(U) \subseteq V - U$ of U in G, such that for every $v \in N(U)$, there is an edge $vv' \in E$ with $v' \in U$. Given an edge $e = vv'$ of G, we denote by G/e the graph obtained by *contracting* an edge e, where v and v' are replaced by one vertex v_e that is adjacent to all vertices in $N(\{v, v'\})$. We say for graphs G and H that G contains an H-*minor*, if we can obtain a graph isomorphic to H by deleting vertices, deleting edges, and contracting edges of G. We call a graph G a *subdivision* of a graph H, if we can obtain a graph isomorphic to G by replacing some of the edges of H by paths.

Given a graph G, we define a *tree-decomposition* of G as a tuple $(T, \mathcal{B})$, such that T is a tree, and a collection of bags $\mathcal{B} = (B_t)_{t \in V(T)}$ where $B_t \subseteq V(G)$ for each $t \in V(T)$ is a subset of the vertices of G such that the following properties are satisfied:

- $\bigcup_{t \in V(T)} B_t = V(G)$,
- for each edge $vv' \in E(G)$, there is a vertex $t \in V(T)$ such that $\{v, v'\} \subseteq B_t$, and
- for each $v \in V(G)$, the set $\{t \in V(T) : v \in B_t\}$ induces a connected graph in T.

If T is a star, we also call $(T, \mathcal{B})$ a *star-decomposition.* We define the *adhesion-size* of a tree-decomposition as $\max\{|B_t \cap B_{t'}| : tt' \in E(T)\}$. Note that we allow G to not be connected, and B_t to be empty for some $t \in V(T)$.

3 Reduction

Theorem 7. ***(Cook, Gerards, Schrijver, & Tardos** [7]). There exists a function $f_7 : \mathbb{Z}^2 \to \mathbb{Z}$ with the following properties. Fix $\Delta \in \mathbb{N}$ and consider an integer program as in* (IP)*. If M is totally Δ-modular,* (IP) *is feasible, and x^* is an optimal solution of the linear relaxation, then there exists an optimal solution z^* of* (IP) *with $\|x^* - z^*\|_\infty \leq f_7(n, \Delta)$.*

Cook, Gerards, Schrijver, & Tardos [7] show that $f_7(n, \Delta) \leq n\Delta$. This was recently improved by Celaya, Kuhlmann, Paat & Weismantel [5] to $f_7(n, \Delta) < \frac{4n+2}{9}\Delta$. If the constraint matrix has a special structure, it is possible to give a bound not depending on the dimension of the problem, see e.g. Aliev, Henk & Oertel [1], or Aprile et al. [2]. In fact, it is not clear that the dimension should appear in the proximity function at all.

The following statement relates the two problems from the introduction. Clearly, any instance of Problem 3 is also an instance of Problem 2. But in fact, there is a polynomial-time reduction from Problem 2 to Problem 3. Note that we need to solve a polynomial number of instances of Problem 3 in order to solve Problem 2.

Proposition 8. *If there is a polynomial-time algorithm to solve Problem 3, then there is a polynomial-time algorithm for Problem 2.*

The proof of the statement is based on the proximity result by Cook et al. [7], see Theorem 7. This theorem allows us to guess on a constant number of variables in any totally Δ-modular instance of (IP). Fiorini et al. [16, Lemma 4] show that A has at most $\log_2(\Delta)$ columns with entries larger than 1 in absolute value. Thus, we can efficiently enumerate all possible assignments to the corresponding variables. Therefore, we can solve a polynomial number of instances of Problem 3 in order to solve an instance of Problem 2.

In addition, we recenter the problem around $\mathbf{0}_n$, i.e., given a solution $x^* \in \mathbb{R}^n$ of the linear relaxation, we redefine our variables $x' := x - \lfloor x^* \rfloor$. By Theorem 7 there is an optimal solution z^* to the integer program, such that $\|z^*\|_\infty \leq f_7(n, \Delta)$. Thus, we can assume without loss of generality that the lower and upper bounds have bounded entries, i.e., $\ell, u \in [-f_7(n, \Delta), f_7(n, \Delta)]^n \cap \mathbb{Z}^n$.

For a detailed description of the reduction, we refer to Fiorini et al. [16], see Lemma 4 and the following discussion.

4 Subdeterminants

Let $M = \left[\begin{smallmatrix} A \\ w^\intercal \end{smallmatrix}\right]$, where $A \in \{0, \pm 1\}^{m \times n}$ with at most two non-zeros per row, and $w \in \mathbb{Z}^n$. In this section, we show a relationship between the subdeterminants of M, and the structure of a corresponding graph.

Let $G = (V, E)$ be a graph, and $\sigma : E \to \{-1, +1\}$ be a signing of its edges. We call the triple $S = (V, E, \sigma)$ a *signed graph*. We call a subgraph S' of S *odd* if $\prod_{e \in E(S')} \sigma(e) = -1$, and *even* otherwise. Observe that if $\sigma(e) = -1$ for all $e \in E$, then the notion of an *odd cycle* coincides with the notion of an odd cycle for unsigned graphs. The *odd cycle packing number* of a signed graph S is denoted by $\text{ocp}(S)$, and is defined as the maximum number of vertex-disjoint odd cycles. For a signed graph S, we define an edge-vertex incidence matrix $M \in \{0, \pm 1\}^{E \times V}$ in the following way. Given an edge $e = vv'$, $M_{e,v} M_{e,v'} = -\sigma(e)$, i.e., if e is odd, then $M_{e,v} = +1$ and $M_{e,v'} = +1$, or $M_{e,v} = -1$ and $M_{e,v'} = -1$, and if e is even, then $M_{e,v} = +1$ and $M_{e,v'} = -1$, or $M_{e,v} = -1$ and $M_{e,v'} = +1$. For v'' not incident to e, we set $M_{e,v''} = 0$. Note that given a signed graph, its edge-vertex incidence matrix is not necessarily unique, but any $\{0, \pm 1\}$-matrix with two non-zeros per row is the edge-vertex incidence matrix of a signed graph, and thus has a unique associated signed graph.

The following result is well-known.

Proposition 9. (Grossman, Kulkarni, & Schochetman *[18]). Let $A \in \{0, \pm 1\}^{m \times n}$ be the edge-vertex incidence matrix of a signed graph S. Then $2^{\text{ocp}(S)} = \Delta(A)$.*

Let S be a signed graph, denote the set of subgraphs of S that are trees by $\mathcal{T}(S)$ and let $T \in \mathcal{T}(S)$. For v, v' vertices of T, we denote the edges of the unique $v - v'$-path in T by $P_T(v, v')$. Note that T induces a partition on $V(T)$ as follows. We define $X_T, Y_T \subseteq V(T)$, such that $X_T \cap Y_T = \emptyset$, $X_T \cup Y_T = V(T)$, any even edge of T is in X_T or Y_T, and any odd edge of T crosses between X_T and Y_T. Hence, $P_T(v, v')$ is even if and only if v and v' are both in X_T or in Y_T. This partition can be found greedily.

Definition 10. *Let $S = (V, E, \sigma)$ be a signed graph and $w : V \to \mathbb{Z}$ be a weight function on the vertices of S. Let $T \in \mathcal{T}(S)$. We define the* alternating weight *of a tree as*

$$\text{alt}(T, w) := \left| \sum_{v \in X_T} w(v) - \sum_{v \in Y_T} w(v) \right|.$$

We define $\text{alt}(S, w) := \max_{T \in \mathcal{T}(S)} \text{alt}(T, w)$ *to be the maximum alternating weight of a tree in* S*.*

Note that $\text{alt}(T, w)$, and $\text{alt}(S, w)$ are well-defined, since the bipartition is unique up to exchanging the sides, and we take the absolute value. The alternating weight of a tree appears naturally in the context of subdeterminants.

Proposition 11. *Let* $A \in \{0, \pm 1\}^{m \times n}$ *be the edge-vertex incidence matrix of a signed graph* $S = (V, E, \sigma)$ *and* $w : V \to \mathbb{Z}$ *be a weight function on the vertices of* S*. Then* $\text{alt}(S, w) \leq \Delta\left(\left[\begin{smallmatrix} A \\ w^\intercal \end{smallmatrix}\right]\right)$*.*

Proof. We show that any tree T of S gives rise to a subdeterminant of absolute value $\text{alt}(T, w)$ within the matrix $\left[\begin{smallmatrix} A \\ w^\intercal \end{smallmatrix}\right]$. We prove this by induction on the number of vertices of T. Given a tree T, we denote by M the square submatrix of $\left[\begin{smallmatrix} A \\ w^\intercal \end{smallmatrix}\right]$, where the rows are indexed by $E(T)$ and w, and the columns are indexed by $V(T)$.

If T has just a single vertex v_1 then $\text{alt}(T, w) = |w(v_1)| = |\det(M)|$. Assume that T has $t \geq 2$ vertices. Then T has a leaf v_1 with an edge $e = v_1 v_2$ to some vertex $v_2 \in V(T)$.

Recall for the following that for an edge $e = (v, v')$, $\sigma(e) = -M_{e,v} M_{e,v'}$. In M, we subtract the row indexed by e scaled with a factor of $w(v_1) \cdot M_{e,v_1}$ from the row indexed by w to obtain M'. To be precise, we define $M'_{w,v_1} := 0$, $M'_{w,v_2} := w(v_2) + \sigma(e) \cdot w(v_1)$, and $M'_{i,j} := M_{i,j}$ otherwise, for $i \in E(T) \cup \{w\}$, and $j \in V(T)$. Since M' is obtained from an elementary row operation on M, we observe that $\det(M') = \det(M)$. In M', the column indexed by v_1 has exactly one non-zero entry of absolute value one, so we can remove the row indexed by e and the column indexed by v_1 to obtain a matrix of size $(t-1) \times (t-1)$ while preserving the determinant of M' in absolute value.

We define $T' := T - \{v_1\}$, and $w' : V(T') \to \mathbb{Z}$, where $w'(v_2) := w(v_2) + \sigma(e) \cdot w(v_1)$, and $w'(v) := w(v)$ for $v \in V(T') - \{v_2\}$. By the induction hypothesis, it follows that

$$|\det(M)| = |\det(M')| = \left| \sum_{v \in A_{T'}} w'(v) - \sum_{v \in B_{T'}} w'(v) \right| = \left| \sum_{v \in X_T} w(v) - \sum_{v \in Y_T} w(v) \right|.$$

□

We remark that Proposition 11 also follows from the *max-circuit imbalance* of an associated matrix, see [14, Proposition 3.2] for a relation between subdeterminants and the imbalance of circuits. In contrast, we are not aware of a proof of Proposition 9 based solely on the structure of circuits of the matrix.

5 Structure

Given a graph $G = (V, E)$ and k weight-functions $w_1, \ldots, w_k : V \to \mathbb{Z}$, we define an associated set of terminals $R := \{v \in V \mid \exists i \in [k] : w_i(v) \neq 0\}$ as the set of vertices whose weight is non-zero with respect to at least one of the weight functions. For an integer $r \in \mathbb{N}$, we call G *r-decorated* if G is a tree, and at least r of its leaves are terminals.

Lemma 12. *Let $S = (V, E, \sigma)$ be a signed graph, $k, r \in \mathbb{N}$, and let $w_1, \ldots, w_k : V \to \mathbb{Z}$ be k weight-functions on the vertices. Let $G := (V, E)$, and $R := \{v \in V \mid \exists i \in [k] : w_i(v) \neq 0\}$. If G has an r-decorated tree as a subgraph, then there is some $i \in [k]$ such that $\operatorname{alt}(S, w_i) \geq \frac{r}{2k}$.*

Proof. Let T be an r-decorated tree in G with respect to R. Let $L(T)$ denote the set of leaves of T, and define the set of leaves which are terminals as $L' := L(T) \cap R$. Note that $|L'| \geq r$. By the pigeon-hole principle, there is some $i \in [k]$ such that $w_i(v) \neq 0$ for at least $\frac{r}{k}$ vertices in L'.

Fix a bipartition X_T, Y_T of T in order to compute the alternating weight of T. We denote the set of leaves with a positive contribution to $\operatorname{alt}(T, w_i)$ by $L^+ := \{v \in L \cap X_T \mid w_i(v) > 0\} \cup \{v \in L \cap Y_T \mid w_i(v) < 0\}$, and the set of leaves with a negative contribution to $\operatorname{alt}(T, w_i)$ by $L^- := \{v \in L \cap X_T \mid w_i(v) < 0\} \cup \{v \in L \cap Y_T \mid w_i(v) > 0\}$. Clearly, $T^+ := T \setminus L^-$ and $T^- := T \setminus L^+$ are trees, and $|\operatorname{alt}(T^+, w_i) - \operatorname{alt}(T^-, w_i)| \geq \frac{r}{k}$. Thus,

$$\operatorname{alt}(S, w_i) \geq \max\{|\operatorname{alt}(T^+, w_i)|, |\operatorname{alt}(T^-, w_i)|\} \geq \frac{r}{2k}.$$

□

Theorem 13. (Ding [11, Lemma 3.1]). *Let $G = (V, E)$ be a graph with some set of terminals $R \subseteq V$, and $r \in \mathbb{N}$. There exists a function $f_{13} \in O(r^2)$ with the following properties. If G has no r-decorated tree as a subgraph, then there is a star-decomposition $(T, \mathcal{B})$ of G (with t_0 the center of T), such that*

- $|B_{t_0} \cap R| \leq f_{13}(r)$,
- $|B_{t_0} \cap B_t| = 2$ *for all* $t \in V(T) \setminus \{t_0\}$, *and both vertices in* $B_{t_0} \cap B_t$ *are in* R,
- $B_t \cap B_{t'} = \emptyset$ *for all* $t, t' \in V(T)$ *leaves of* T,
- *for all* $t \in V(T) \setminus \{t_0\}$ *it holds that* $G[B_t]$ *is connected, and each* $r \in (B_t \setminus B_{t_0}) \cap R$ *is a cut vertex in* $G[B_t]$ *separating the two vertices in* $B_{t_0} \cap B_t$.

Note that this implies that the maximum degree of T is bounded from above by $f_{13}(r)/2$. We further remark that all steps in the proof of [11, Lemma 3.1] are constructive, so one can find the above decomposition in polynomial time.

In the proof of [11, Lemma 3.1], the concrete order of $f_{13}(r)$ is shown to be in $O(s(r)^2)$, where $s(r)$ denotes the minimal number such that every connected graph excluding a $K_{1,r}$-minor is a subdivision of a graph on $s(r)$ vertices.

Lemma 14. *Let $r \in \mathbb{N}$, and G be a connected graph without $K_{1,r}$-minor. There exists a function $s_{14}(r) \in O(r)$, such that G is a subdivision of a graph on at most $s_{14}(r)$ vertices.*

We proceed to modify the decomposition from Theorem 13 in order to obtain a graph decomposition that allows us to employ a dynamic programming framework in order to solve Problem 3.

Lemma 15. *Let G be a connected graph with set of terminals $R = \{v_0, \ldots, v_\alpha\}$ for some $\alpha \in \mathbb{N}$. Assume that for all $i \in [\alpha - 1]$, v_i is a cut vertex of G separating v_0 from v_α.*

Then there is an ordering of R, such that every $v_0 - v_\alpha$-path in G visits all terminals in sequence with respect to the ordering. Further, any component of $G - R$ attaches to at most two terminals. These terminals must be consecutive with respect to the ordering. See Fig. 2 for a sketch of the graph structure in Lemma 15:

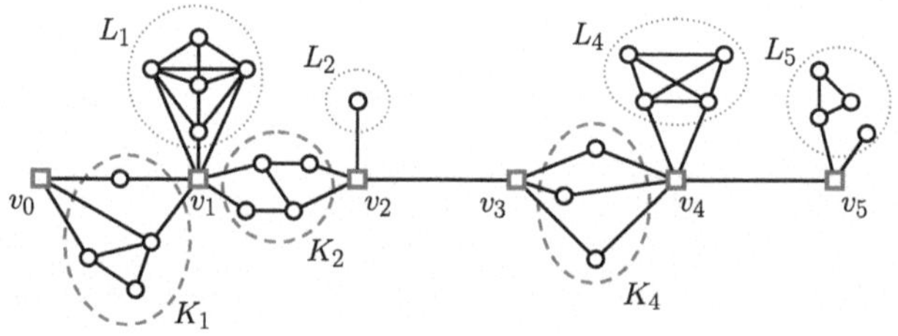

Fig. 2. A depiction of the structure of graphs in Lemma 15. The red vertices are terminals. The collections of components K_i (dashed) and L_j (dotted) are used in the proof of Theorem 16.(Color figure online)

Theorem 16. *Let $r \in \mathbb{N}$ and G be a graph with set of terminals R. There is a function $f_{16}(r) \in O(r^2)$ with the following properties. If G has no r-decorated tree as a subgraph, then there is a tree-decomposition $(T, \mathcal{B})$ of G with adhesion-size and degree bounded by $f_{16}(r)$, such that each bag B_t for $t \in V(T)$ is of at least one of the following two types:*

(i) $|B_t| \leq f_{16}(r)$
(ii) t *is a leaf of* T *with neighbor* t', *and* $B_t \cap R \subseteq B_{t'}$

Further, we can compute such a decomposition in polynomial time.

Proof. Let $f_{16} := \max\{3, f_{13}(r)/2 + 2, f_{13}(r)\}$. We first assume that G is a connected graph. Then, we can apply Theorem 13 and obtain a decomposition $(T', \mathcal{B}')$ of G. We proceed to describe how to modify such a decomposition in order to obtain a decomposition $(T, \mathcal{B})$ that fulfills the conditions of the lemma.

For $k \geq 1$, we define a *k-comb* as a tree on $2k$ vertices $c_1, \dots, c_k, d_1, \dots, d_k$, consisting of a path $c_1, \dots, c_k$, and edges $c_i d_i$ for $i \in [k]$. We first describe how we modify T' in order to obtain T. Let w_0 be the center vertex of T'. For any leaf w of T', we define $\lambda_w := |B'_w \cap R| - 1$. We replace each w by a λ_w-comb on the vertex set $\{c_1^w, \dots, c_{\lambda_w}^w, d_1^w, \dots, d_{\lambda_w}^w\}$, such that there is an edge $w_0 c_1^w$. Finally, we attach a vertex t_{-1} to the center vertex w_0, and denote the resulting tree by T with center vertex t_0.

We set $B_{t_{-1}} := B'_{w_0}$ and $B_{t_0} := B'_{w_0} \cap R$. Observe that B_{t_0} has type 16, $B_{t_{-1}}$ has type 16, and their intersection is bounded by $f_{13}(r)$.

For any leaf w of T', we denote the two terminals in $B'_w \cap B'_{w_0}$ by $v_0(w)$ and $v_{\lambda_w}(w)$. By Lemma 15 applied to B'_w, we obtain an ordering $v_0(w), v_1(w), \dots, v_{\lambda_w}(w)$ of the terminals, such that any $v_0(w) - v_{\lambda_w}(w)$-path visits the terminals in this ordering. We set $B_{c_i^w} := \{v_{i-1}(w), v_i(w), v_{\lambda_w}(w)\}$ for $i \in [\lambda_w]$. Observe that each such bag has at most 3 elements and is of type 16.

It remains to describe the bags $B_{d_i^w}$ for $i \in [\lambda_w]$. By Lemma 15, each component of $G[B'_w] \setminus R$ attaches to at most two consecutive terminals in $v_0(w), v_1(w), \dots, v_{\lambda_w}(w)$. For $i \in [\lambda_w]$, we denote by K_i the union of all components that attach to $v_{i-1}(w)$ and $v_i(w)$. For $j \in \{0, \dots, \lambda_w\}$, we denote by L_j the union of all components that attach only to $v_j(w)$. We define $B_{d_1^w} := \{v_0(w), v_1(w), v_{\lambda_w}(w)\} \cup K_1 \cup L_0 \cup L_1$ and $B_{d_i^w} := \{v_{i-1}(w), v_i(w), v_{\lambda_w}(w)\} \cup K_i \cup L_i$ for $i \in \{2, \dots, \lambda_w\}$. Observe that all d_i^w are leaves of T, and any terminal that they contain is also present in c_i^w for all $i \in [\lambda_w]$. Therefore, each such bag is of type 16 and the decomposition $(T, \mathcal{B})$ is of the desired form.

If G is not connected, we consider the connected components $G_1, \dots, G_\gamma$ of G for some $\gamma \in \mathbb{N}$. By the previous arguments, we obtain a tree-decomposition $(T_i, \mathcal{B}_i)$ for each $i \in [\gamma]$. We combine the tree-decompositions into one central decomposition in the following way.

We construct the tree T as a γ-comb on the vertex set $\{c_1, \dots, c_\gamma, d_1, \dots, d_\gamma\}$, where d_i is replaced by T_i, such that c_i attaches to the central vertex of T_i. This increases the maximum degree of the tree-decomposition by at most one. The collection of bags $\mathcal{B}$ is obtained by copying the $\mathcal{B}_i$ for $i \in [\gamma]$. Note that in this way, the bags corresponding to c_i for $i \in [\gamma]$ remain empty. □

6 Dynamic Program

In this section, we describe how to exploit the structural results from Sects. 4 and 5 into an efficient algorithm for Problem 3. We propose a *bottom-up* dynamic program on the tree-decomposition given by Theorem 16, which follows the ideas of the dynamic program presented in [2]. We define a set of *rooted instances* corresponding to subtrees of the tree-decomposition. In each of these subtrees, we compute optimal solutions for the possible assignments on the right hand sides of the extra rows as well as the possible assignments on variables in the intersection with the parent in the tree-decomposition.

In bags of type (ii), we fix all assignments on the terminals of the subgraph, and solve the remaining integer program with at most two non-zeros per row.

Bags of type (i) have bounded size. Hence, we can enumerate all possible assignments to the corresponding variables. Since the degree of the tree-decomposition is bounded, we can also combine the optimal solutions in the corresponding subtrees efficiently. Note that by Theorem 7 and Theorem 16, the number of possible assignments is polynomial in n and Δ.

In the context of the dynamic program, we work with *rooted tree-decompositions*. A rooted tree-decomposition of a graph G is a tree-decomposition $(T, \mathcal{B})$ with the additional property that some vertex $t_0 \in V(T)$ is identified as the root of T. In a rooted tree-decomposition $(T, \mathcal{B})$ with root t_0, any vertex t of T induces a *rooted subtree*, i.e., the tree induced on the vertices $t' \in V(T)$, such that the unique $t' - t_0$-path in T contains t. We denote this rooted subtree by $T(t)$, and its associated collection of vertices of G by $B(t) := \{v \in V(G) \mid \exists t' \in V(T(t)) : v \in B_{t'}\}$. Further, for any vertex of a rooted tree-decomposition $t \in V(T)$, we define its *adhesion* as $\mathrm{adh}(t) := \emptyset$ if $t = t_0$, and $\mathrm{adh}(t) := B_t \cap B_{t'}$ for t' being the unique parent of t in the rooted tree-decomposition otherwise.

We begin by recalling the setup for Problem 3. Let $\Delta, k \in \mathbb{N}$, $A \in \{0, \pm 1\}^{m \times n}$, $W \in \mathbb{Z}^{k \times n}$, such that A has two non-zeros per row. Let $M := \left[\begin{smallmatrix} A \\ W \end{smallmatrix}\right]$ be totally Δ-modular, $c, \ell, u \in \mathbb{Z}^n$, $b \in \mathbb{Z}^m$, and $d \in \mathbb{Z}^k$, which together define an instance of Problem 3. Let $S = (V, E, \sigma)$ be a signed graph with corresponding incidence matrix A, and denote $G := (V, E)$ the unsigned graph with set of terminals $R \subseteq V$, where $v \in V$ is a terminal if the column vector of W corresponding to v is non-zero. By Proposition 11 and Lemma 12, we conclude that G has no $(2k\Delta + 1)$-decorated tree as a subgraph.

Let $(T, \mathcal{B})$ be the tree-decomposition of G given by Theorem 16 with respect to the set of terminals R. We turn $(T, \mathcal{B})$ into a rooted tree-decomposition by identifying the root as c_1 (in the notation of the proof of Theorem 16), i.e., the first vertex on the path of the comb that combines the decompositions of the components.

Further, we define assignment spaces for the right hand side and the variables respectively: $\mathcal{D} := [-n\Delta f_7(n, \Delta), n\Delta f_7(n, \Delta)]^k \cap \mathbb{Z}^k$, and $\mathcal{X}_t := [-f_7(n, \Delta), f_7(n, \Delta)]^{\mathrm{adh}(t)} \cap \mathbb{Z}^{\mathrm{adh}(t)}$.

Definition 17. *Given a vertex of the tree-decomposition $t \in V(T)$, an assignment on the adhesion to the parent $\chi_t \in \mathcal{X}_t$, and a target for the extra constraints $d_t \in \mathcal{D}$, we define a rooted instance of Problem 3. We denote by A_t the restriction of A to the columns corresponding to vertices of $B(t)$ and to the rows corresponding to edges induced on $G[B(t)]$. Similarly, we denote by W_t, c_t, ℓ_t, and u_t the restriction of W, c, ℓ and u to the vertices of $B(t)$, where $W_t(i, v) = 0$ and $c_t(i, v) = 0$ for $v \in \mathrm{adh}(t)$ and $i \in [k]$. Finally, we denote by b_t the restriction of b to the edges induced on $G[B(t)]$.*

The rooted problem is defined as

$$\min \left\{ c_t^\intercal x : A_t x \leq b_t,\ W_t x = d_t,\ \ell_t \leq x \leq u_t,\ x(v) = \chi_t(v) \forall v \in \mathrm{adh}(t),\ x \in \mathbb{Z}^{B(t)} \right\}.$$

We denote an instance of the rooted problem by $I := I(t, \chi_t, d_t)$, and its optimal value by $\mathrm{OPT}(I)$. By restricting to the vertices and edges in a single bag B_t

instead of a rooted subtree $B(t)$, we define a local *instance $I^0(t, \chi_t, d_t)$ with optimal value $F_t^0(\chi_t, d_t)$. We denote any optimal solution of $I^0(t, \chi_t, d_t)$ by $x_t^0(\chi_t, d_t)$.*

Note that $I(t_0, \emptyset, d)$ is the original instance of Problem 3. Thus, in order to solve Problem 3 it is sufficient to give a recursive algorithm on how to solve rooted problems. Further, for any $t \in V(T)$, there is $(2n\Delta f_7(n, \Delta) + 1)^k \cdot (2f_7(n, \Delta) + 1)^{O((k\Delta)^2)}$ different possible assignments for rooted instances, which is polynomial for fixed $\Delta, k \in \mathbb{N}$.

Definition 18. *Let $I(t, \chi_t, d_t)$ be a rooted instance corresponding to vertex $t \in V(T)$. Let $t_1, \ldots, t_q$ denote the children of t in T (where $q = 0$ if t is a leaf). We define*

$$F_t(\chi_t, d_t) := \min \Bigg\{ F_t^0(\chi_t, d_t^0) + \sum_{i \in [q]} (F_{t_i}(\chi_{t_i}, d_{t_i})) \; : \; \chi_{t_i} \in \mathcal{X}_{t_i} \forall i \in [q],\ \chi_t(v) = \chi_{t_i}(v)$$

$$\forall i \in [q],\ v \in B_t \cap B_{t_i},\ d_t^0, d_{t_1}, \ldots, d_{t_q} \in \mathcal{D},\ d_t^0 + \sum_{i \in [q]} d_{t_i} = d_t \Bigg\}.$$

We denote by $x_t(\chi_t, d_t)$ the composition of $x_t^0(\chi_t, d_t)$ with $x_{t_i}(\chi_{t_i}, d_{t_i})$ for $i \in [q]$.

Proof (Proof of Theorem 4). By Proposition 8, it suffices to give a polynomial-time algorithm for instances of Problem 3. By Theorem 16 we obtain a tree-decomposition $(T, \mathcal{B})$. We show that $F_t(\chi_t, d_t) = \mathrm{OPT}(I(t, \chi_t, d_t))$ for all $t \in T$, all $\chi_t \in \mathcal{X}_t$, and $d_t \in \mathcal{D}$. We proceed by induction on the distance of a vertex $t \in T$ from the root t_0, starting with the largest distance.

The claimed algorithm corresponds to computing $F_{t_0}(\emptyset, d)$ as in Definition 18.

By definition, the statement is true for any $t \in V(T) - t_0$, where t is a leaf. Assume that t is not a leaf. Then, t has q children $t_1, \ldots, t_q$, such that for each $i \in [q]$, and each $\chi_{t_i} \in \mathcal{X}_{t_i}$ and $d_{t_i} \in \mathcal{D}$, it holds by induction that $F_{t_i}(\chi_{t_i}, d_{t_i}) = \mathrm{OPT}(I(t_i, \chi_{t_i}, d_{t_i}))$. Recall from Sect. 3, that $\ell, u \in [-f_7(n, \Delta), f_7(n, \Delta)]^n \cap \mathbb{Z}^n$. Consider an optimal solution $x_t^*(\chi_t, d_t) \in [-f_7(n, \Delta), f_7(n, \Delta)]^n \cap \mathbb{Z}^n$ of $I(t, \chi_t, d_t)$. For $i \in [q]$, we can decompose $x_t^*(\chi_t, d_t)$ into solutions for $I_{t_i}(\chi_{t_i}, d_{t_i})$, where χ_{t_i} and d_{t_i} are induced by the corresponding assignments of $x_t^*(\chi_t, d_t)$. By definition of the spaces $\mathcal{X}_{t_i}$ for $i \in [q]$ and $\mathcal{D}$, we can see that the corresponding optimal values have been pre-computed. Further, the dynamic program explores all valid assignments for variables in B_t. Thus, $c_t^\intercal x_t^*(\chi_t, d_t) = c_t^\intercal x_t(\chi_t, d_t)$, certifying optimality.

We conclude the proof with an analysis of the runtime. The total size of the tree-decomposition is linear in n, i.e., $|V(T)| \in O(n)$. For each vertex $t \in V(T)$, we obtain at most $(2n\Delta f_7(n, \Delta) + 1)^k \cdot (2f_7(n, \Delta) + 1)^{O((k\Delta)^2)}$ relevant rooted instances. Each local instance can be solved in polynomial time. For bags of type (i), this corresponds to enumerate all possible assignments within the proximity bound. For bags of type (ii), this corresponds to solving an integer program with bounded subdeterminants and two non-zeros per row, see Theorem 1. Solving a rooted instance corresponds to solving a local instance and combining it with a polynomial number of rooted instances, which can again be done in polynomial time.

Proof (Proof of Corollary 5). The statement follows from Theorem 4 and its proof with the following additional observations.

Since A is totally unimodular, we can solve the local problem for bags of type 16 such that the exponent of the running time does not depend on Δ, see e.g. Tardos [29]. Further, here the proximity function $f_7(n, \Delta)$ does not depend on the dimension of the problem n, but just on Δ and k, see Aprile et al. [2, Theorem 4].

7 Conclusion

We study IPs with bounded subdeterminants and give a strongly polynomial-time algorithm to solve a new structured class of IPs. We achieve our results by combining structural results with a dynamic programming approach. Within the dynamic program, we use the algorithm to solve integer programs with bounded subdeterminants and two non-zeros per row [16] in a blackbox fashion. In particular, for our main result, the proximity result by Cook et al. [7] can be used directly. This is in contrast to other recent papers, that make use of auxiliary proximity or solution decomposition results, see e.g. [2,3,12,13,16]. Still, this leaves open, whether one can prove a stronger proximity result for Problem 3 or its generalizations. Such a proximity result could open the door to more general results on the solvability of totally Δ-modular IPs.

In the work by Fiorini et al. [16], the majority of the technical difficulty lies in constraint matrices with bounded subdeterminants and two non-zeros per *row*. They also solve the case with two non-zeros per *column*. We do not extend this part of their result to additional rows and columns. Note that such integer programs can model a version of the *exact matching* problem [27] with a structural restriction. In the exact matching problem, we are given a graph with red and blue edges, and $k \in \mathbb{N}$. The goal is to decide whether there exists a perfect matching with exactly k red edges. This problem is not known to be NP-hard, but no efficient deterministic algorithm is known. Subproblems involving the exact matching problem have proven to be complex to solve deterministically in the literature on subdeterminants, see [2,25,26].

Finally, we remark on a structural question for matrices with bounded subdeterminants. In many recent papers, it has been helpful to exploit a relation between such matrices and certain graph classes, see [2,3,16,25,26]. We ask whether this connection can be made more precise, and if the question of solving integer programs with bounded subdeterminants can be solved by analyzing the right graph classes. One indication for such a relation can be found in recent work on the number of disjoint columns in matrices with bounded subdeterminants, see [4,23].

Acknowledgements. The author thanks Manuel Aprile, Samuel Fiorini, Caleb McFarland, Stefan Weltge and Yelena Yuditsky for helpful discussions. The author acknowledges funding by the ERC Consolidator Grant 615640-ForEFront and the *Fonds de la Recherche Scientifique* - FNRS through research project BD-DELTA (PDR 20222190, 2021–24).

In addition, we would like to thank the anonymous reviewers for their careful reading and their helpful comments which greatly helped to improve the paper.

References

1. Aliev, I., Henk, M., Oertel, T.: Distances to lattice points in knapsack polyhedra. Math. Program. **182**(1), 175–198 (2020)
2. Aprile, M., et al.: Integer programs with nearly totally unimodular matrices: the cographic case. ACM-SIAM Symp. Discrete Algorithms (SODA25) (2025)
3. Artmann, S., Weismantel, R., Zenklusen, R.: A strongly polynomial algorithm for bimodular integer linear programming. In: Proceedings of the 49th Annual ACM SIGACT Symposium on Theory of Computing. pp. 1206–1219 (2017)
4. Averkov, G., Schymura, M.: On the maximal number of columns of a Δ-modular integer matrix: bounds and computations. Math. Program. 1–29 (2023)
5. Celaya, M., Kuhlmann, S., Paat, J., Weismantel, R.: Proximity and flatness bounds for linear integer optimization. Math. Oper. Res. (2023)
6. Celaya, M., Kuhlmann, S., Weismantel, R.: On matrices over a polynomial ring with restricted subdeterminants. In: International Conference on Integer Programming and Combinatorial Optimization. pp. 43–56. Springer (2024)
7. Cook, W., Gerards, A.M., Schrijver, A., Tardos, É.: Sensitivity theorems in integer linear programming. Math. Program. **34**(3), 251–264 (1986)
8. Dadush, D., Koh, Z.K., Natura, B., Olver, N., Végh, L.A.: A strongly polynomial algorithm for linear programs with at most two nonzero entries per row or column. In: Proceedings of the 56th Annual ACM Symposium on Theory of Computing. pp. 1561–1572 (2024)
9. Dhahan, J.: The maximum weight stable set problem with a budget constraint. Master's thesis, Simon Fraser University (2020)
10. Diestel, R.: Graph Theory, Graduate Texts in Mathematics, vol. 173. Springer, Heidelberg; New York, fourth edn (2010)
11. Ding, G.: Graphs without large $K_{2,n}$-minors. arXiv:1702.01355 (2017)
12. Eisenbrand, F., Rohwedder, L., Węgrzycki, K.: Sensitivity, proximity and FPT algorithms for exact matroid problems. In: 2024 IEEE 65th Annual Symposium on Foundations of Computer Science (FOCS). pp. 1610–1620. IEEE (2024)
13. Eisenbrand, F., Weismantel, R.: Proximity results and faster algorithms for integer programming using the Steinitz lemma. ACM Trans. Algorithms (TALG) **16**(1), 1–14 (2019)
14. Ekbatani, F., Natura, B., Végh, L.A.: Circuit imbalance measures and linear programming. Surv. Comb. 64–114 (2022)
15. Ferrarini, L., Fiorini, S., Kober, S., Yuditsky, Y.: Total matching and subdeterminants. In: International Symposium on Combinatorial Optimization. pp. 192–204. Springer (2024)
16. Fiorini, S., Joret, G., Weltge, S., Yuditsky, Y.: Integer programs with bounded subdeterminants and two nonzeros per row. J. ACM **72**(1), 1–50 (2025)
17. Glanzer, C., Stallknecht, I., Weismantel, R.: Notes on {a, b, c}-modular matrices. Vietnam J. Math. **50**(2), 469–485 (2022)
18. Grossman, J.W., Kulkarni, D.M., Schochetman, I.E.: On the minors of an incidence matrix and its Smith normal form. Linear Algebra Appl. **218**, 213–224 (1995)
19. Johnson, D.S., Niemi, K.: On knapsacks, partitions, and a new dynamic programming technique for trees. Math. Oper. Res. **8**(1), 1–14 (1983)

20. Kober, S.: Totally Δ-modular IPs with two non-zeros in most rows (2024). arXiv:2411.15282
21. Kober, S.A.: Three perspectives on integer programming: practical and theoretical applications, and the case of bounded subdeterminants. Ph.D. thesis, Technische Universität München (2023)
22. Kolliopoulos, S.G., Steiner, G.: Partially ordered knapsack and applications to scheduling. Discret. Appl. Math. **155**(8), 889–897 (2007)
23. Lee, J., Paat, J., Stallknecht, I., Xu, L.: Polynomial upper bounds on the number of differing columns of Δ-modular integer programs. Math. Oper. Res. (2022)
24. Liu, S., Xu, C.: On the congruency-constrained matroid base. In: International Conference on Integer Programming and Combinatorial Optimization. pp. 280–293. Springer (2024)
25. Nägele, M., Nöbel, C., Santiago, R., Zenklusen, R.: Advances on strictly Δ-modular IPs. Math. Program. 1–30 (2024)
26. Nägele, M., Santiago, R., Zenklusen, R.: Congruency-constrained TU problems beyond the bimodular case. Math. Oper. Res. **49**(3), 1303–1348 (2024)
27. Papadimitriou, C.H., Yannakakis, M.: The complexity of restricted spanning tree problems. J. ACM (JACM) **29**(2), 285–309 (1982)
28. Shevchenko, V.: Qualitative topics in integer linear programming. Am. Math. Soc. (1996), original manuscript in russian
29. Tardos, É.: A strongly polynomial algorithm to solve combinatorial linear programs. Oper. Res. **34**(2), 250–256 (1986)

Sparse Approximation in Lattices and Semigroups

Stefan Kuhlmann[1(✉)], Timm Oertel[2], and Robert Weismantel[1]

[1] Department of Mathematics, ETH Zürich, Zürich, Switzerland
{stefan.kuhlmann,robert.weismantel}@ifor.math.ethz.ch
[2] Department of Data Science, FAU Erlangen-Nürnberg, Erlangen, Germany
timm.oertel@fau.de

Abstract. Given an integer or a non-negative integer solution $\boldsymbol{x}$ to a system $\boldsymbol{Ax} = \boldsymbol{b}$, where the number of non-zero components of $\boldsymbol{x}$ is at most n. This paper addresses the following question: How closely can we approximate $\boldsymbol{b}$ with $\boldsymbol{Ay}$, where $\boldsymbol{y}$ is an integer or non-negative integer solution constrained to have at most k non-zero components with $k < n$? We establish upper bounds for this question in general. In specific cases, these bounds are tight. The key finding is that the quality of the approximation increases exponentially as k goes to n.

1 Introduction

Let $\boldsymbol{A} = (\boldsymbol{a}_1, \ldots, \boldsymbol{a}_n) \in \mathbb{Z}^{m\times n}$ have full row rank. Fix a set $X \subseteq \mathbb{R}^n$. We will mostly consider the case where $X \in \{\mathbb{Z}^n, \mathbb{Z}^n_{\geq 0}\}$. This paper is concerned with studying sparse approximations of elements in $\boldsymbol{A} \cdot X := \{\boldsymbol{b} \in \mathbb{R}^m : \boldsymbol{b} = \boldsymbol{Ax} \text{ for } \boldsymbol{x} \in X\}$ using only elements from X whose support is small. More precisely, let $\|\cdot\|$ be a norm on $\mathbb{R}^m$ and let $k \in [n]$ be a given sparsity parameter, where $[n] := \{1, \ldots, n\}$ for $n \geq 1$. Our target is to determine the best possible bound in terms of n, m, k, and eventually some parameter associated with $\boldsymbol{A}$ for the problem

$$\mathrm{app}_{X,k,\|\cdot\|}(\boldsymbol{A}) := \max_{\boldsymbol{b}\in \boldsymbol{A}\cdot X} \min \{\|\boldsymbol{Ax} - \boldsymbol{b}\| : \boldsymbol{x} \in X,\ |\mathrm{supp}(\boldsymbol{x})| \leq k\},$$

where $\mathrm{supp}(\boldsymbol{x}) := \{i \in [n] : \boldsymbol{x}_i \neq 0\}$ for $\boldsymbol{x} \in \mathbb{R}^n$. This is a very broad definition and bounds vary significantly with different X and $\|\cdot\|$.

The task of determining $\mathrm{app}_{X,k,\|\cdot\|}(\boldsymbol{A})$ is related to sparse recovery problems in the context of signal processing. One typical high-level question in this area is to find conditions on $\boldsymbol{A}$ such that one can replace the constraint $|\mathrm{supp}(\boldsymbol{x})| \leq k$ by $\|\boldsymbol{x}\|_1 \leq k$ when k is small, i.e., $k \in \mathcal{O}(\log_2(m))$. A fundamental assumption for this to work is to know that a sparse solution exists, or even that it is unique. Then one can try to recover it by applying a tractable algorithm. This line of research is very active, with extensive literature addressing the cases when X contains real-valued or complex-valued vectors; see for instance [5–8,11] or the book by Foucart and Rauhut [14] for a comprehensive overview. In our

N. Megow and A. Basu (Eds.): IPCO 2025, LNCS 15620, pp. 371–384, 2025.
https://doi.org/10.1007/978-3-031-93112-3_27

work, we are concerned with the case when X contains integer-valued vectors, see [1,13,15,19] for some results on an integer version of sparse recovery. The direction that we take in this paper deviates from the classical research questions addressed in the field of sparse recovery in signal processing. In particular, we consider a fixed matrix $\boldsymbol{A}$ and have no a-priori knowledge about the existence of a sparse solution.

Our point of departure is that there exists a representation of $\boldsymbol{b} \in \boldsymbol{A} \cdot X$ in form of $\boldsymbol{b} = \boldsymbol{A}\boldsymbol{x}$ for a given $\boldsymbol{x} \in X$ with $|\mathrm{supp}(\boldsymbol{x})| \leq n$. For different values of $k \in [n]$, we want to establish deterministic worst-case bounds for $\mathrm{app}_{X,k,\|\cdot\|}(\boldsymbol{A})$. For instance, if $X = \mathbb{R}^n$ and $k \geq m$, then we have $\mathrm{app}_{X,k,\|\cdot\|}(\boldsymbol{A}) = 0$, since every $\boldsymbol{b} \in \boldsymbol{A} \cdot X = \mathbb{R}^m$ can be expressed by an invertible $m \times m$ submatrix of $\boldsymbol{A}$.

Another natural choice is $X = \mathbb{R}^n_{\geq 0}$. Here, a theorem by Carathéodory implies that if $k \geq m$, then $\mathrm{app}_{X,k,\|\cdot\|}(\boldsymbol{A}) = 0$. For $k < m$ and $X = \mathbb{R}^n$ or $X = \mathbb{R}^n_{\geq 0}$, finding an optimal sparse approximation is equivalent to identifying a k-dimensional subspace spanned by a subset of the set of columns of $\boldsymbol{A}$ that is close to a target vector $\boldsymbol{b}$. Of course, if we maximize over all feasible vectors in $\boldsymbol{A} \cdot X$, the approximation error $\mathrm{app}_{X,k,\|\cdot\|}(\boldsymbol{A})$ then diverges to infinity as $\boldsymbol{b}$ may lie arbitrarily far away from any of these k-dimensional subspaces. This is also true when $X \in \{\mathbb{Z}^n, \mathbb{Z}^n_{\geq 0}\}$. To avoid this, we assume throughout that $k \geq m$.

Let us now turn to the focus of this paper where $X \in \{\mathbb{Z}^n, \mathbb{Z}^n_{\geq 0}\}$. A first natural question is to ask for an upper bound on k that guarantees $\mathrm{app}_{X,k,\|\cdot\|}(\boldsymbol{A}) = 0$. When $X = \mathbb{Z}^n$, such an upper bound on k has been established in [2, Theorem 1]: Let

$$\delta(\boldsymbol{A}) := \min\{|\det \boldsymbol{B}| : \boldsymbol{B} \text{ is an invertible } m \times m \text{ submatrix of } \boldsymbol{A}\}. \quad (1)$$

Then the bound in [2, Theorem 1] gives essentially[1] $\mathrm{app}_{\mathbb{Z}^n,k,\|\cdot\|_\infty}(\boldsymbol{A}) = 0$ whenever $k \geq m + \log_2(\delta(\boldsymbol{A}))$. We show a result that is more general as it states that $\mathrm{app}_{\mathbb{Z}^n,k,\|\cdot\|_\infty}(\boldsymbol{A}) \leq \delta(\boldsymbol{A})/2^{k-m+1}$. In fact, for $k \geq m + \log_2(\delta(\boldsymbol{A}))$ we obtain $\mathrm{app}_{\mathbb{Z}^n,k,\|\cdot\|_\infty}(\boldsymbol{A}) = 0$ and thus rediscover the bound in [2, Theorem 1].

In the case $X = \mathbb{Z}^n_{\geq 0}$, the analysis of the quantity $\mathrm{app}_{\mathbb{Z}^n_{\geq 0},k,\|\cdot\|}(\boldsymbol{A})$ is significantly more involved. As before, let us first summarize the results on an upper bound on k that guarantees that $\mathrm{app}_{X,k,\|\cdot\|}(\boldsymbol{A}) = 0$. It has been shown in [12] that $\mathrm{app}_{X,k,\|\cdot\|}(\boldsymbol{A}) = 0$ whenever $k \geq 2m \log_2(4m\|\boldsymbol{A}\|_\infty)$. This bound has been improved to $2m \log_2(2\sqrt{m}\|\boldsymbol{A}\|_\infty)$ in [3]. Similiar bounds are available when the parameter $\|\boldsymbol{A}\|_\infty$ is replaced by $\sqrt{\det \boldsymbol{A}\boldsymbol{A}^\top}$; see [2] and [3] for further details.

To the best of our knowledge there are no studies of the approximation parameter $\mathrm{app}_{\mathbb{Z}^n_{\geq 0},k,\|\cdot\|}(\boldsymbol{A})$ for general values of k available. In the special case when $\mathrm{pos}\{\boldsymbol{a}_1, \ldots, \boldsymbol{a}_n\} = \mathbb{R}^m$ with $\mathrm{pos}\, Y$ denoting the set of all non-negative combinations of elements in $Y \subseteq \mathbb{R}^m$, it turns out that the question of determining $\mathrm{app}_{\mathbb{Z}^n_{\geq 0},k,\|\cdot\|}(\boldsymbol{A})$ can be tackled similar to the lattice setting where $X = \mathbb{Z}^n$. On the contrary, if $\mathrm{pos}\{\boldsymbol{a}_1, \ldots, \boldsymbol{a}_n\}$ is a pointed cone, we fix a basis $\boldsymbol{B}$ consisting of linearly independent column vectors of $\boldsymbol{A}$, say $\boldsymbol{a}_1, \ldots, \boldsymbol{a}_m$, and further

[1] The result in [2, Theorem 1] is more refined and incorporates the prime factorization of the involved determinants.

assume that $\boldsymbol{a}_i \in \text{pos}\{\boldsymbol{a}_1, \ldots, \boldsymbol{a}_m\}$ for each $i \in [n]$. For our analysis it will be convenient to work with the norm $\|\cdot\|_{P(\boldsymbol{B})}$ induced by the symmetric parallelepiped $P(\boldsymbol{B}) := \boldsymbol{B} \cdot [-1,1]^m$. The dependence of our approximation bound on the matrix $\boldsymbol{A}$ is given by the parameter $\mu := \max_{i\in[n]} \|\boldsymbol{a}_i\|_{P(\boldsymbol{B})}$. We will show that $\text{app}_{\mathbb{Z}^n_{\geq 0},k,\|\cdot\|}(\boldsymbol{A})$ decreases exponentially in k. The precise dependence of $\text{app}_{\mathbb{Z}^n_{\geq 0},k,\|\cdot\|}(\boldsymbol{A})$ on the parameters m, k, μ, and $\boldsymbol{A}$ is presented in Theorem 6.

2 Lattices

Let $X = \mathbb{Z}^n$ and $\boldsymbol{A} \in \mathbb{Z}^{m\times n}$ have full row rank. The set $\boldsymbol{A} \cdot X = \boldsymbol{A}\mathbb{Z}^n$ is the lattice generated by the columns of $\boldsymbol{A}$. We refer the reader to [17] for more about lattices, sublattices, and their determinants. Our aim is to find sparse approximations of lattice vectors in $\boldsymbol{A}\mathbb{Z}^n$. We measure the approximation error with respect to the ℓ_∞-norm. Observe that if $k \geq m$, we can always express $\boldsymbol{b}$ using a basis of the lattice $\boldsymbol{A}\mathbb{Z}^n$, which consists of m elements. However, it is not necessarily given that such a basis exists among the columns of $\boldsymbol{A}$. Even stronger, it could be that no proper subset of the set of columns is a generating set for the lattice $\boldsymbol{A}\mathbb{Z}^n$. In this situation, our approximation result below applies. Recall from (1) that $\delta(\boldsymbol{A})$ denotes the smallest subdeterminant in absolute value of an invertible $m \times m$ submatrix of $\boldsymbol{A}$.

Theorem 1. *Let $\boldsymbol{A} \in \mathbb{Z}^{m\times n}$ have full row rank and $m \leq k \leq n$. Then there exists a $m \times k$ submatrix $\boldsymbol{D}$ of $\boldsymbol{A}$ such that*

$$\text{app}_{\mathbb{Z}^n,k,\|\cdot\|_\infty}(\boldsymbol{A}) \leq \max_{\boldsymbol{b}\in\boldsymbol{A}\mathbb{Z}^n} \min_{\boldsymbol{x}\in\mathbb{Z}^k} \|\boldsymbol{D}\boldsymbol{x} - \boldsymbol{b}\|_\infty \leq \frac{1}{2^{k-m+1}} \cdot \delta(\boldsymbol{A}).$$

The first inequality holds by definition. To show the second inequality, we rely on the fact that proper subsets of the set of columns of $\boldsymbol{A}$ still generate some sublattice of $\boldsymbol{A}\mathbb{Z}^n$. Our point of departure is the following lemma that, in particular, holds for $m \times l$ submatrices of $\boldsymbol{A}$ with full row rank.

Lemma 1. *Let $\boldsymbol{D} \in \mathbb{Z}^{m\times l}$ have full row rank and $\Lambda = \boldsymbol{D}\mathbb{Z}^l$. Then*

$$\min_{\boldsymbol{x}\in\mathbb{Z}^l} \|\boldsymbol{D}\boldsymbol{x} - \boldsymbol{d}\|_\infty \leq \frac{1}{2} \cdot \det \Lambda$$

for all $\boldsymbol{d} \in \mathbb{R}^m$.

Proof. There exists a unimodular matrix $\boldsymbol{U} \in \mathbb{Z}^{l\times l}$ such that $\boldsymbol{D}\boldsymbol{U} = [\boldsymbol{B}, \boldsymbol{0}]$ is in Hermite normal form; cf. [22, Chapter 4]. In particular, the matrix $\boldsymbol{B}$ is a lower triangular matrix and $\det \Lambda = B_{1,1} \cdot \ldots \cdot B_{m,m}$, where $B_{i,i}$ refers to the i-th diagonal entry of $\boldsymbol{B}$ for $i \in [m]$. Define $\boldsymbol{z} = (z_1, \ldots, z_m)^\top \in \mathbb{Z}^m$ recursively such that

$$z_i := \left\lceil \frac{d_i - \sum_{j=1}^{i-1} B_{i,j} z_j}{B_{i,i}} \right\rfloor,$$

where $\lceil x \rfloor$ denotes the integer closest to $x \in \mathbb{R}$. Then we have

$$|(\boldsymbol{DUz} - \boldsymbol{d})_i| = |(\boldsymbol{Bz} - \boldsymbol{d})_i| = \left| B_{i,i} \left(\frac{\sum_{j=1}^{i-1} B_{i,j} z_j - d_i}{B_{i,i}} + z_i \right) \right| \leq \tfrac{1}{2} B_{i,i}$$

for all $i \in [m]$. The last term above is upper bounded by $1/2 \cdot \det \Lambda$. So the integer vector $\boldsymbol{Uz}$ satisfies the claimed bound. □

Note, one may also interpret $\min_{\boldsymbol{x} \in \mathbb{Z}^l} \|\boldsymbol{Dx} - \boldsymbol{d}\|_\infty$ in Lemma 1 as the inhomogeneous minimum or covering radius of the unit cube $[-1, 1]^m$ with respect to the lattice Λ. Upper bounds on the inhomogeneous minima for general convex bodies typically rely on the so-called successive minima. In the special case of the unit cube, our bound surpasses these general results. Furthermore, the bound in Lemma 1 is tight in general, which follows from the special case when $\Lambda = \mathbb{Z}^n$. We refer the interested reader to [17, Chapter 2, Sect. 13] for details concerning inhomogeneous minima, successive minima, and their relation.

Proof (of Theorem 1). As mentioned, the first inequality holds by definition. We prove the second inequality by recursively constructing submatrices $\boldsymbol{D}$ of $\boldsymbol{A}$ to which Lemma 1 applies. Let $\boldsymbol{B}$ be an $m \times m$ submatrix of $\boldsymbol{A}$ such that $|\det \boldsymbol{B}| = \delta(\boldsymbol{A})$. We set $\boldsymbol{D}_0 := \boldsymbol{B}$ and $\Lambda_0 := \boldsymbol{D}_0 \mathbb{Z}^m$. For $i \in [n-m]$, choose a column of $\boldsymbol{A}$, say $\boldsymbol{v}_i$, such that $\boldsymbol{v}_i \notin \Lambda_{i-1}$. If there is no such column, we simply select an arbitrary column. Then we define $\boldsymbol{D}_i := [\boldsymbol{D}_{i-1}, \boldsymbol{v}_i] \in \mathbb{Z}^{m+i}$ and $\Lambda_i := \boldsymbol{D}_i \mathbb{Z}^{m+i}$. By Lemma 1, we have

$$\min_{\boldsymbol{x} \in \mathbb{Z}^{m+i}} \|\boldsymbol{D}_i \boldsymbol{x} - \boldsymbol{b}\|_\infty \leq \frac{1}{2} \cdot \det \Lambda_i. \tag{2}$$

for all $i \in \{0, \dots, n-m\}$ and $\boldsymbol{b} \in \boldsymbol{A}\mathbb{Z}^n$. To finish the proof, we analyze $\det \Lambda_i$ for $i \in \{0, \dots, n-m\}$. If $i = 0$, we get $\det \Lambda_0 = |\det \boldsymbol{B}| = \delta(\boldsymbol{A})$ and the claim follows. Fix some $i \in [n-m]$ and recall that $\boldsymbol{v}_i$ is the column that we added in the i-th step. If $\boldsymbol{v}_i \in \Lambda_{i-1}$, our construction implies that all columns of $\boldsymbol{A}$ that are not columns of $\boldsymbol{D}_{i-1}$ are contained in Λ_{i-1}. It follows that $\det \boldsymbol{A}\mathbb{Z}^n = \det \Lambda_{i-1} = \det \Lambda_i$ and $\min_{\boldsymbol{x} \in \mathbb{Z}^{m+i}} \|\boldsymbol{D}_i \boldsymbol{x} - \boldsymbol{b}\|_\infty = 0$ for all $\boldsymbol{b} \in \boldsymbol{A}\mathbb{Z}^n$, which implies the claim. So suppose that $\boldsymbol{v}_i \notin \Lambda_{i-1}$. This is equivalent to $\Lambda_{i-1} \subsetneq \Lambda_i$. Since Λ_{i-1} is a proper sublattice of Λ_i, we obtain that $\det \Lambda_i \neq \det \Lambda_{i-1}$ and $\det \Lambda_i$ divides $\det \Lambda_{i-1}$. Thus, we have $\det \Lambda_i \leq 1/2 \cdot \det \Lambda_{i-1}$. By recursively applying this inequality, we recover

$$\det \Lambda_i \leq \frac{1}{2^i} \cdot \det \Lambda_0 = \frac{1}{2^i} \cdot \delta(\boldsymbol{A}).$$

Plugging this into (2) and choosing $i = k - m$ proves the claim. □

From Theorem 1 we conclude that the approximation factor drops exponentially in k. One can refine our argument from above by choosing the $\boldsymbol{v}_i$ that minimizes $\det \Lambda_i$ when constructing $\boldsymbol{D}$. This relates to incorporating the prime factorization of $\delta(\boldsymbol{A})$ similar to the techniques in the proof of [2, Theorem 1].

We omit this here for the sake of brevity and since we are mainly interested in proving the exponential behavior in k. The bound in Theorem 1 also implies bounds on the exact representation of vectors in lattices in terms of some generating set $\boldsymbol{A}$: For $k \geq m + \log_2(\delta(\boldsymbol{A}))$ we obtain $\text{app}_{\mathbb{Z}^n,k,\|\cdot\|_\infty}(\boldsymbol{A}) \leq 1/2$. Since $\text{app}_{\mathbb{Z}^n,k,\|\cdot\|_\infty}(\boldsymbol{A})$ is an integer, we conclude that $\text{app}_{\mathbb{Z}^n,k,\|\cdot\|_\infty}(\boldsymbol{A}) = 0$. This is essentially the order of magnitude presented in [2, Theorem 1], which is not a coincidence as our techniques extend the approach used in [2].

3 Semigroups

Let $X = \mathbb{Z}^n_{\geq 0}$ and $\boldsymbol{A} = (\boldsymbol{a}_1, \ldots, \boldsymbol{a}_n) \in \mathbb{Z}^{m \times n}$ have full row rank. Then $\boldsymbol{A} \cdot X$ is the semigroup generated by the columns of $\boldsymbol{A}$. If $\text{pos}\{\boldsymbol{a}_1 \ldots, \boldsymbol{a}_n\} = \mathbb{R}^m$, each subset $I \subseteq [n]$ with $\text{pos}\{\boldsymbol{a}_i : i \in I\} = \mathbb{R}^m$ generates non-negatively the lattice spanned by $\boldsymbol{a}_i$ for $i \in I$. Therefore, one can apply the lattice technique presented in the previous section. So we suppose that $\text{pos}\{\boldsymbol{a}_1, \ldots, \boldsymbol{a}_n\}$ generates a pointed cone. We begin our analysis with the case when $m = 1$ and write $\boldsymbol{a}^\top = (a_1, \ldots, a_n)$ for the matrix $\boldsymbol{A}$. The assumption that $a_1, \ldots, a_n$ generate a pointed cone implies without loss of generality that $0 < a_1 \leq a_2 \leq \ldots \leq a_n$. Our goal is to determine the approximation error with respect to the absolute value, that is, $\text{app}_{\boldsymbol{a}^\top\mathbb{Z}^n,k,|\cdot|}(\boldsymbol{a})$ for a fixed $k \geq 1$. We abbreviate this parameter by $\text{app}_k(\boldsymbol{a})$. To bound $\text{app}_k(\boldsymbol{a})$, we apply the following strategy: We fix the smallest integer a_1 and bound the slightly refined parameter

$$\text{app}_{a_1,k}(\boldsymbol{a}) := \max_{b \in \boldsymbol{a}^\top\mathbb{Z}^n_{\geq 0}} \min\left\{\left|\boldsymbol{a}^\top \boldsymbol{x} - b\right| : \boldsymbol{x} \in \mathbb{Z}^n_{\geq 0},\ |\text{supp}(\boldsymbol{x}) \backslash \{1\}| \leq k-1\right\},$$

which corresponds to the best approximation that uses $\boldsymbol{a}_1$. By definition, we have $\text{app}_k(\boldsymbol{a}) \leq \text{app}_{\boldsymbol{a}_1,k}(\boldsymbol{a})$. Using the pigeonhole principle one can derive an upper bound on $\text{app}_{n-1}(\boldsymbol{a})$, which even holds for real valued $\boldsymbol{a}$.

Theorem 2 *Let* $\boldsymbol{a} \in \mathbb{R}^n$ *with* $0 < a_1 \leq \ldots \leq a_n$. *Then we have*

$$\text{app}_{n-1}(\boldsymbol{a}) \leq \text{app}_{a_1,n-1}(\boldsymbol{a}) \leq \frac{1}{2^{n-1}} \cdot a_1.$$

Next we apply Theorem 2 repeatedly to obtain a bound for general values of k. This bound shows that the approximation factor decreases exponentially as k goes to n.

Theorem 3 *Let* $\boldsymbol{a} \in \mathbb{R}^n$ *with* $0 < a_1 \leq \ldots \leq a_n$ *and* $1 \leq k \leq n$. *Then we have*

$$\text{app}_k(\boldsymbol{a}) \leq \text{app}_{a_1,k}(\boldsymbol{a}) \leq \left(\frac{1}{2^{k-1}} - \frac{1}{2^{n-1}}\right) \cdot a_1.$$

Theorems 2 and 3 under the assumption that $\boldsymbol{a} \in \mathbb{Z}^n$ follow from their extensions to general values of m that we present below.

Let us first comment on the question whether the bounds in Theorems 2 and 3 are tight. It can be shown that the bound in Theorem 2 is tight, even when $\boldsymbol{a}$

is integral. For $k = 1$, Theorem 3 can be strengthened to $\text{app}_{a_1,1}(\boldsymbol{a}) \leq 1/2 \cdot a_1$, based on the observation that we can write every integer as an integer multiple of a_1 plus or minus a remainder of size at most $1/2 \cdot a_1$. This construction is tight. These observations for $k \in \{1, n-1\}$ suggest that $\text{app}_k(\boldsymbol{a}) \leq 1/2^k \cdot a_1$ might be the correct bound for all $1 \leq k \leq n$. This is, however, not true, even when $k = 2$. In fact, Theorem 4 below presents a tight bound for $k = 2$. To prepare for the presentation of this bound, let us define the sequence $t_0 := 1$ and $t_i := t_{i-1} \cdot (t_{i-1} + 1)$ for $i \geq 1$. This sequence $\{t_i\}_{i \in \mathbb{N}}$ is well-known and appears in several papers related to combinatorial optimization problems; see, for instance, [16,18,23]. Moreover, the sequence $\{t_i + 1\}_{i \in \mathbb{N}}$ is known as *Sylvester's sequence.* It gives among others the optimal solution to approximating 1 by egyptian fractions from below; see [9,20,21,24] for some appearances of Sylvester's sequence in the literature. For $m \in \mathbb{N}$, we define

$$\varphi(m) := \sum_{i=1}^{m} \frac{1}{t_i}.$$

Theorem 4 *Let $\boldsymbol{a} \in \mathbb{R}^n$ with $0 < a_1 \leq \ldots \leq a_n$. Then we have*

$$\text{app}_2(\boldsymbol{a}) \leq \text{app}_{a_1,2}(\boldsymbol{a}) \leq \frac{\varphi(n-2)}{2\varphi(n-2)+1} \cdot a_1.$$

Note that for $n = 3$ we get an upper bound of $1/4 \cdot a_1$, for $n = 4$ we get $2/7 \cdot a_1$ and $29/100 \cdot a_1$ for $n = 5$. In the limit, we obtain $\lim_{n \to \infty} \varphi(n) \approx 0.691$ which gives us a bound of approximately $0.2901 \cdot a_1$. This improves upon the upper bound of $1/2 \cdot a_1$ for $k = 2$ and $n \to \infty$ given in Theorem 3. We prove Theorem 4 in Sect. 5.

Let us now turn to the extensions of Theorem 2 and 3 to general values of $m \geq 2$. Our strategy to extend these results is motivated by the analysis when $m = 1$: We fix an invertible $m \times m$ submatrix $\boldsymbol{B}$ of $\boldsymbol{A}$. Without loss of generality let $\boldsymbol{B} = (\boldsymbol{a}_1, \ldots, \boldsymbol{a}_m)$. In addition, we assume that $\boldsymbol{a}_i \in \text{pos}\{\boldsymbol{a}_1, \ldots, \boldsymbol{a}_m\}$ for each $i \in [n]$. This assumption resembles the case $m = 1$, where we fixed $\boldsymbol{B} = a_1$ and also assumed that $a_i \in \text{pos}\{a_1\}$. We measure our approximation error in terms of the symmetric parallelepiped spanned by $\boldsymbol{a}_1, \ldots, \boldsymbol{a}_m$, which is defined by $P(\boldsymbol{B}) := \boldsymbol{B} \cdot [-1,1]^m$. Our errors will be taken with respect to the norm $\|\cdot\|_{P(\boldsymbol{B})}$, the norm with unit ball $P(\boldsymbol{B})$. This also is in correspondence with what we did when $m = 1$ because, if $m = 1$, we have $\boldsymbol{B} = a_1$ and the rescaled norm induced by $1/a_1 \cdot P(a_1) = [-1,1]$ equals simply the absolute value. As before, let us abbreviate $\text{app}_{\boldsymbol{A}\mathbb{Z}^n_{\geq 0}, k, \|\cdot\|_{P(\boldsymbol{B})}}(\boldsymbol{A})$ by $\text{app}_k(\boldsymbol{A})$ for a fixed value of $k \geq 1$. We are now prepared to investigate the refined parameter

$$\text{app}_{\boldsymbol{B},k}(\boldsymbol{A}) := \max_{\boldsymbol{b} \in \boldsymbol{A}\mathbb{Z}^n_{\geq 0}} \min\left\{\|\boldsymbol{A}\boldsymbol{x} - \boldsymbol{b}\|_{P(\boldsymbol{B})} : \boldsymbol{x} \in \mathbb{Z}^n_{\geq 0},\ |\text{supp}(\boldsymbol{x}) \backslash [m]| \leq k - m\right\}.$$

Theorem 5 *Let $\boldsymbol{A} = (\boldsymbol{a}_1, \ldots, \boldsymbol{a}_n) \in \mathbb{Z}^{m \times n}$ have full row rank, $m \leq n-1$, $\boldsymbol{B} = (\boldsymbol{a}_1, \ldots, \boldsymbol{a}_m)$ be invertible, $\mu = \max_{i \in [n]} \|\boldsymbol{a}_i\|_{P(\boldsymbol{B})}$, and $\boldsymbol{a}_i \in \text{pos}\{\boldsymbol{a}_1, \ldots, \boldsymbol{a}_m\}$ for all $i \in [n]$. Then we have*

$$\text{app}_{n-1}(\boldsymbol{A}) \leq \text{app}_{\boldsymbol{B},n-1}(\boldsymbol{A}) \leq \frac{1}{2^{\frac{n-m}{m}}} \cdot \mu^{\frac{m-1}{m}} \cdot |\det \boldsymbol{B}|^{\frac{m-1}{m}}.$$

Theorem 5 implies Theorem 2 when $m = 1$ and $\boldsymbol{a} \in \mathbb{Z}^n$. As for $m = 1$, Theorem 5 applied repeatedly allows us to observe that the approximation factor decreases exponentially as k goes to n.

Theorem 6 *Let* $\boldsymbol{A} = (\boldsymbol{a}_1, \ldots, \boldsymbol{a}_n) \in \mathbb{Z}^{m\times n}$ *have full row rank. Let* $\boldsymbol{B} = (\boldsymbol{a}_1, \ldots, \boldsymbol{a}_m)$ *be invertible,* $\mu = \max_{i\in[n]} \|\boldsymbol{a}_i\|_{P(\boldsymbol{B})}$*, and* $\boldsymbol{a}_i \in \mathrm{pos}\{\boldsymbol{a}_1, \ldots, \boldsymbol{a}_m\}$ *for all* $i \in [n]$*. Then we have*

$$\mathrm{app}_k(\boldsymbol{A}) \leq \mathrm{app}_{\boldsymbol{B},k}(\boldsymbol{A}) \leq \frac{1}{2^{\frac{1}{m}} - 1} \cdot \left(\frac{1}{2^{\frac{k-m}{m}}} - \frac{1}{2^{\frac{n-m}{m}}} \right) \cdot \mu^{\frac{m-1}{m}} \cdot |\det \boldsymbol{B}|^{\frac{m-1}{m}}$$

for all $m \leq k \leq n$*.*

By evaluating the inequality in Theorem 6 at $m = 1$ and rescaling the norm, we recover the bound given in Theorem 3 provided that $\boldsymbol{a} \in \mathbb{Z}^n$.

A few comments on the bound in Theorem 6 are in order. Notice that $\mathrm{app}_k(\boldsymbol{A}) \leq \mathrm{app}_{\boldsymbol{B},k}(\boldsymbol{A}) \leq 1/2$ is a trivial upper bound that holds for all $k \geq m$. For small values of k, the bound in Theorem 6 is worse than $1/2$. In particular, one can calculate that the bound in Theorem 6 improves upon $1/2$ if we have $k \geq 2m + m \log_2(cm\mu^{(m-1)/m} |\det \boldsymbol{B}|^{(m-1)/m})$ for some constant $c > 0$. These calculations utilize the fact that $1/(2^{1/m} - 1) = \mathcal{O}(m)$ for $m \geq 1$. However, recall that a sparsity result in [12, Theorem 1 (ii)] implies that we have an exact sparse representation using only $2m \log_2(4m\|\boldsymbol{A}\|_\infty)$ vectors. Adapting the argument in [12] to the parameters μ and $|\det \boldsymbol{B}|$ one can derive that there is always an exact sparse representation using $2m \log_2(4m\mu\, |\det \boldsymbol{B}|^{1/m})$ vectors. Hence, our bounds are relevant whenever we have

$$2\, m + m \log_2(cm\mu^{\frac{m-1}{m}}\, |\det \boldsymbol{B}|^{\frac{m-1}{m}}) \leq k \leq 2\, m \log_2 \left(4\, m\mu\, |\det \boldsymbol{B}|^{\frac{1}{m}}\right).$$

This is the case when μ is relatively large compared to m and $|\det \boldsymbol{B}|$. Similar to the case $X = \mathbb{Z}^n$, the results presented above imply bounds on the size of an exact representation when $X = \mathbb{Z}^n_{\geq 0}$. This follows from Theorem 5.

Corollary 1 *Let* $\boldsymbol{A} = (\boldsymbol{a}_1, \ldots, \boldsymbol{a}_n) \in \mathbb{Z}^{m\times n}$ *have full row rank. Let* $\boldsymbol{B} = (\boldsymbol{a}_1, \ldots, \boldsymbol{a}_m)$ *be invertible,* $\mu = \max_{i\in[n]} \|\boldsymbol{a}_i\|_{P(\boldsymbol{B})}$*, and* $\boldsymbol{a}_i \in \mathrm{pos}\{\boldsymbol{a}_1, \ldots, \boldsymbol{a}_n\}$ *for all* $i \in [n]$*. Then we have* $\mathrm{app}_{n-1}(\boldsymbol{A}) = 0$ *for*

$$n \geq m + m \log_2(\mu^{(m-1)/m}\, |\det \boldsymbol{B}|^{(2\, m-1)/m}).$$

When $m = 1$, we recover a tight bound on the sparse representation of integers in semigroups; cf. [2, Theorem 3]. For $m \geq 2$, the bound in Corollary 1 outperforms the bound $2m \log_2(4m\mu\, |\det \boldsymbol{B}|^{1/m})$ in [12] whenever μ is sufficiently large.

It is important to point out that our bounds for $m \geq 2$ depend on the factor μ whereas the bounds for $m = 1$ are independent of μ. This is not a coincidence. In fact, for $m \geq 2$, it is possible to show that a dependence on μ (or some related parameter measuring the size of $\boldsymbol{A}$) is necessary to obtain approximation factors that decrease in k.

Finally, let us mention that our upper bounds are based on the assumption that $\boldsymbol{a}_i \in \text{pos}\{\boldsymbol{a}_1, \ldots, \boldsymbol{a}_m\}$ for all $i \in [n]$. A closer examination of the techniques in Sect. 4 reveals that this assumption is not necessary. However, eliminating it requires us to introduce an additional factor of $2^{(m-1)/m}$ in Theorem 5 and Theorem 6. We state this here without a proof.

Remark 1 Let $\boldsymbol{A} = (\boldsymbol{a}_1, \ldots, \boldsymbol{a}_n) \in \mathbb{Z}^{m \times n}$ with $m \leq k \leq n$ have full row rank, $\boldsymbol{B} = (\boldsymbol{a}_1, \ldots, \boldsymbol{a}_m)$ be invertible and $\mu = \max_{i \in [n]} \|\boldsymbol{a}_i\|_{P(\boldsymbol{B})}$. Then we have

$$\text{app}_k(\boldsymbol{A}) \leq \text{app}_{\boldsymbol{B},k}(\boldsymbol{A}) \leq \frac{2^{\frac{m-1}{m}}}{2^{\frac{1}{m}} - 1} \cdot \left(\frac{1}{2^{\frac{k-m}{m}}} - \frac{1}{2^{\frac{n-m}{m}}} \right) \cdot \mu^{\frac{m-1}{m}} \cdot |\det \boldsymbol{B}|^{\frac{m-1}{m}} .$$

The proof techniques will be presented in the next section. In fact, they extend to arbitrary real matrices. This requires further technicalities that we refrain from discussing in the paper.

4 Proofs of Theorem 5, Theorem 6, and Corollary 1

Let us start to describe the major steps required to prove Theorem 5. The first part follows a similar approach to the techniques in [12]: We fix some $\boldsymbol{b} \in \boldsymbol{A} \cdot \mathbb{Z}^n_{\geq 0}$. This vector has a representation of the form $\boldsymbol{b} = \lambda_1 \boldsymbol{a}_1 + \ldots + \lambda_n \boldsymbol{a}_n$ with $\lambda_i \in \mathbb{N}$ for all $i \in [n]$. We tile $\text{pos}\{\boldsymbol{a}_1, \ldots, \boldsymbol{a}_m\}$ with sufficiently small copies of the parallelepiped spanned by $\boldsymbol{a}_1, \ldots, \boldsymbol{a}_m$. Then we count how many subsums of $\boldsymbol{b} = \lambda_1 \boldsymbol{a}_1 + \ldots + \lambda_n \boldsymbol{a}_n$ are contained in the copies of the sufficiently small parallelepipeds. It turns out that the number of subsums grows exponentially in n and the number of parallelepipeds that we need to consider only grows exponentially in m. Once this has been established one could derive a first bound that is, however, weaker than the one stated in Theorem 5. (The key distinction is that one would replace the expression $\mu^{(m-1)/m} \cdot |\det \boldsymbol{B}|^{(m-1)/m}$ by the expression $\mu \cdot |\det \boldsymbol{B}|$.) To get Theorem 5, a second ingredient must be incorporated, one that, to the best of our knowledge, has not previously been utilized: the idea is to show that we only need to consider so-called incomparable subsums.

To make this notion precise, let $(P_{m,s}, \preceq)$ be the partially order set given by $P_{m,s} := \mathbb{Z}^m \cap [0, s]^m$ and $\boldsymbol{y} \preceq \boldsymbol{z}$ if and only if $y_1 \leq z_1, \ldots, y_m \leq z_m$ for $m, s \geq 1$. An antichain of $(P_{m,s}, \preceq)$ is a subset of $P_{m,s}$ whose elements are pairwise incomparable. An upper bound on the size of a largest antichain in $(P_{m,s}, \preceq)$ can be established easily.

Lemma 2 *Let $m, s \in \mathbb{N}_{\geq 1}$. The largest antichain of $(P_{m,s}, \preceq)$ is bounded above by $(s+1)^{m-1}$. It is in the order of $\Theta(s^{m-1})$.*

Proof We determine the upper bound first. By Dilworth's theorem [10], the largest antichain equals the minimal size of a chain cover. Thus, we can construct a chain cover to get an upper bound on the size of the largest antichain. Define the chains $C_{z_1,\ldots,z_{m-1}} := \{(z_1, \ldots, z_{m-1}, 0)^\top, \ldots, (z_1, \ldots, z_{m-1}, s)^\top\}$ for fixed values $z_1, \ldots, z_{m-1} \in \{0, \ldots, s\}$. Those chains cover $P_{m,s}$. In total, there are $(s+1)^{m-1}$ chains of the type $C_{z_1,\ldots,z_{m-1}}$. This gives the desired upper bound.

For the asymptotic bound, note that the upper bound implies that the size of a largest antichain is in $\mathcal{O}(s^{m-1})$. Hence, it suffices to construct an antichain of size $\Omega(s^{m-1})$. Define the subset of $P_{m,s}$ given by $S := \{\boldsymbol{z} \in P_{m,s} : \sum_{i=1}^{m} z_i = s\}$. The elements in S are incomparable by construction. From Ehrhart theory, cf. [4, Chapter 3], it follows that $|S| = \Omega(s^{m-1})$ since the elements in S lie on an $(n-1)$-dimensional affine hyperplane. □

Let $\text{par}\{\boldsymbol{a}_1, \ldots, \boldsymbol{a}_m\}$ denote the half-open parallelepiped spanned by the basis vectors $\boldsymbol{a}_1, \ldots, \boldsymbol{a}_m$. Taking the parameter μ into consideration one has that

$$\boldsymbol{a}_i \in \mu \cdot \text{par}\{\boldsymbol{a}_1, \ldots, \boldsymbol{a}_m\} \text{ for all } i \in [n].$$

Proof (of Theorem 5). Let $\boldsymbol{b} := \lambda_1\boldsymbol{a}_1 + \ldots + \lambda_n\boldsymbol{a}_n$ with $\lambda_i \in \mathbb{N}$ for all $i \in [n]$. Since we fix $\boldsymbol{a}_1, \ldots, \boldsymbol{a}_m$ in our non-negative integral combination, we assume without loss of generality $\lambda_i = 0$ for $i \in [m]$. So we have $\boldsymbol{b} = \lambda_{m+1}\boldsymbol{a}_{m+1} + \ldots + \lambda_n\boldsymbol{a}_n$. Let us abbreviate $\lambda := \lambda_{m+1} + \ldots + \lambda_n$. We have that $\boldsymbol{b} \in \mu \cdot \lambda \cdot \text{par}\{\boldsymbol{a}_1, \ldots, \boldsymbol{a}_m\}$ as $\boldsymbol{a}_i \in \mu \cdot \text{par}\{\boldsymbol{a}_1, \ldots, \boldsymbol{a}_m\}$ for all $i \in [n]$. We tile $\mu \cdot \lambda \cdot \text{par}\{\boldsymbol{a}_1, \ldots, \boldsymbol{a}_m\}$ using $\mu^m \cdot \lambda^m$ copies of the parallelepiped $\text{par}\{\boldsymbol{a}_1, \ldots, \boldsymbol{a}_m\}$. We refine this tiling by further dividing each $\text{par}\{\boldsymbol{a}_1, \ldots, \boldsymbol{a}_m\}$ into l^m parellelepipeds of size and shape $1/l \cdot \text{par}\{\boldsymbol{a}_1, \ldots, \boldsymbol{a}_m\}$. In total, we have $\mu^m \cdot \lambda^m \cdot l^m$ parallelepipeds tiling $\mu \cdot \lambda \cdot \text{par}\{\boldsymbol{a}_1, \ldots, \boldsymbol{a}_m\}$. Let $\tilde{\boldsymbol{a}}_i := \lambda_i\boldsymbol{a}_i$. Suppose that two subsums of $\tilde{\boldsymbol{a}}_{m+1} + \ldots + \tilde{\boldsymbol{a}}_n$ are in the same tile, up to adding multiples of $\boldsymbol{a}_1, \ldots, \boldsymbol{a}_m$ to one of the subsums. In other words, suppose that we have two different sets $I, J \subseteq \{m+1, \ldots, n\}$ such that $\mu_1\boldsymbol{a}_1 + \ldots + \mu_m\boldsymbol{a}_m + \sum_{i\in I}\tilde{\boldsymbol{a}}_i$ and $\sum_{j\in J}\tilde{\boldsymbol{a}}_j$ are in the same tile for $\mu_1, \ldots, \mu_m \in \mathbb{N}$. This implies $\|\mu_1\boldsymbol{a}_1 + \ldots + \mu_m\boldsymbol{a}_m + \sum_{i\in I}\tilde{\boldsymbol{a}}_i - \sum_{j\in J}\tilde{\boldsymbol{a}}_j\|_{P(B)} \leq 1/l$. Since I, J are different sets, we assume without loss of generality that there exists $j^* \in J$ with $j^* \notin I$. If this is not the case, we have $J \subsetneq I$ which implies $\sum_{j\in J}\tilde{\boldsymbol{a}}_j < \sum_{i\in I}\tilde{\boldsymbol{a}}_i$. So we have $\mu_i = 0$ for all $i \in [m]$ and we can interchange the role of I and J. Using this, we construct a new approximate vector

$$\bar{\boldsymbol{b}} := \boldsymbol{b} + \mu_1\boldsymbol{a}_1 + \ldots + \mu_m\boldsymbol{a}_m + \sum_{i\in I}\tilde{\boldsymbol{a}}_i - \sum_{j\in J}\tilde{\boldsymbol{a}}_j.$$

This is an element of the semigroup that is generated by $n-1$ elements since we do not use the generator $\tilde{\boldsymbol{a}}_{j^*}$. We get

$$\|\bar{\boldsymbol{b}} - \boldsymbol{b}\|_{P(B)} = \|\mu_1\boldsymbol{a}_1 + \ldots + \mu_m\boldsymbol{a}_m + \sum_{i\in I}\tilde{\boldsymbol{a}}_i - \sum_{j\in J}\tilde{\boldsymbol{a}}_j\|_{P(B)} \leq \frac{1}{l}. \tag{3}$$

It remains to find a bound on l. To accomplish this task, we investigate how fine we want to tile $\text{par}\{\boldsymbol{a}_1, \ldots, \boldsymbol{a}_m\}$ while simultaneously ensuring that two subsums of $\tilde{\boldsymbol{a}}_{m+1} + \ldots + \tilde{\boldsymbol{a}}_n$ are in the same tile, up to adding multiples of $\boldsymbol{a}_1, \ldots, \boldsymbol{a}_m$ to one of the subsums. We have l^m tiles in $\text{par}\{\boldsymbol{a}_1, \ldots, \boldsymbol{a}_m\}$. Fix one of those tiles and call it T. We define $T(\mu_1, \ldots, \mu_m) := T + \sum_{i=1}^{n}\mu_i\boldsymbol{a}_i$ for $\mu_1, \ldots, \mu_m \in \{0, \ldots, \mu \cdot \lambda - 1\}$. These are the translates of T modulo $\boldsymbol{a}_1, \ldots, \boldsymbol{a}_m$ that are contained in $\mu \cdot \lambda \cdot \text{par}\{\boldsymbol{a}_1, \ldots, \boldsymbol{a}_m\}$. There are $\mu^m \cdot \lambda^m$ of those translates. We claim that if there are more than $\mu^{m-1} \cdot \lambda^{m-1}$ subsums in different $T(\mu_1, \ldots, \mu_m)$, we can add multiples of $\boldsymbol{a}_1, \ldots, \boldsymbol{a}_m$ to one of the subsums and end up in the same translated tile. To see this, we define a partial order on the set $\mathcal{T}$ of the translates of $T(\mu_1, \ldots, \mu_m)$ for $\mu_1, \ldots, \mu_m \in \{0, \ldots, \mu \cdot \lambda - 1\}$. Let $T(\mu_1, \ldots, \mu_m) \preceq T(\mu'_1, \ldots, \mu'_m)$ if and only if $\mu_1 \leq \mu'_1, \ldots, \mu_m \leq \mu'_m$. This partial order implies that if two subsums are contained in translated tiles that obey the ordering above, we are done since $T(\mu_1, \ldots, \mu_m) + \sum_{i=1}^{m}(\mu'_i - \mu_i) = T(\mu'_1, \ldots, \mu'_m)$ and $\mu'_i - \mu_i \geq 0$ for all $i \in [m]$. Hence, we want to find the largest subset of $\mathcal{T}$ whose elements are pairwise incomparable with respect to the partial order. This is equivalent to determining the largest antichain in $(\mathcal{T}, \preceq)$. Observe that $(\mathcal{T}, \preceq)$ is isomorphic to $(P_{m,\mu\cdot\lambda-1}, \preceq)$. So we can apply Lemma 2 to obtain the upper bound $\mu^{m-1} \cdot \lambda^{m-1}$ on the size of the largest antichain in $(\mathcal{T}, \preceq)$. This holds for all l^m tiles. Hence, by the pigeonhole principle we must ensure that

$$2^{n-m} \geq \mu^{m-1} \cdot \lambda^{m-1} \cdot l^m$$

to guarantee that two subsums are, up to adding multiples of $\boldsymbol{a}_1, \ldots, \boldsymbol{a}_m$, in the same tile. To get the best possible such bound, we choose l to be as large as possible such that the inequality holds. Since we have ignored for simplicity a couple of improving steps in the analysis above (such as the empty subsum ruling out more tiles), we do not round l to the nearest integer, but simply select $l := 2^{n-m} \cdot (\mu \cdot \lambda)^{-(m-1)}$. After plugging this into (3) it remains to show that $\lambda \leq |\det \boldsymbol{B}|$. To verify the latter inequality, we use again the pigeonhole principle.

For the purpose of deriving a contradiction assume that $\lambda \geq |\det \boldsymbol{B}| + 1$. We analyze the sum

$$\underbrace{\boldsymbol{a}_{m+1} + \ldots + \boldsymbol{a}_{m+1}}_{\lambda_{m+1} \text{ times}} + \ldots + \underbrace{\boldsymbol{a}_n + \ldots + \boldsymbol{a}_n}_{\lambda_n \text{ times}}.$$

Let $\boldsymbol{y}_i$ denote the subsum of the first i terms from above for $i \in [\sum_{i=m+1}^{n} \lambda_i]$. There are at least $|\det \boldsymbol{B}|+1$ such subsums. Moreover, there exist $|\det \boldsymbol{B}|$ different elements of the form $\boldsymbol{y} + \boldsymbol{B}\mathbb{Z}^m$ for $\boldsymbol{y} \in \mathbb{Z}^m$. By the pigeonhole principle, we can find two different subsums $\boldsymbol{y}_i$ and $\boldsymbol{y}_j$ with $i < j$ such that $\boldsymbol{y}_j - \boldsymbol{y}_i \in \boldsymbol{B}\mathbb{Z}^m$ and $\boldsymbol{y}_j - \boldsymbol{y}_i \in \text{pos}\{\boldsymbol{a}_1, \ldots, \boldsymbol{a}_m\}$ as $i < j$. So we have $\boldsymbol{y}_j - \boldsymbol{y}_i = \sum_{i=1}^{m} \mu_i \boldsymbol{a}_i$ for non-negative integral values $\mu_1, \ldots, \mu_m$. Instead of approximating $\boldsymbol{b}$, we approximate the new integer vector $\boldsymbol{b} - (\boldsymbol{y}_j - \boldsymbol{y}_i) \in \text{pos}\{\boldsymbol{a}_1, \ldots, \boldsymbol{a}_m\}$. Once we have an approximation of this new vector, we add the vector $\sum_{i=1}^{m} \mu_i \boldsymbol{a}_i$ back and obtain an approximation of $\boldsymbol{b}$ with the same error. Since $\boldsymbol{b} - (\boldsymbol{y}_j - \boldsymbol{y}_i)$ has a strictly smaller coordinate sum than $\boldsymbol{b}$, one can repeat these steps until the coordinate sum is bounded by $|\det \boldsymbol{B}|$. □

The lower bound in Lemma 2 implies that we cannot asymptotically improve upon the bound $\mu^{m-1} \cdot \lambda^{m-1}$ when estimating the size of a largest antichain.

Proof (of Theorem 6). The claim follows immediately when $k = n$. So suppose that $k \leq n - 1$. Let $\boldsymbol{b} = \boldsymbol{A}\boldsymbol{x}$ for $\boldsymbol{x} \in \mathbb{Z}^n_{\geq 0}$ be given. We set $\boldsymbol{x}^0 := \boldsymbol{x}$ and $I_0 := \text{supp}(\boldsymbol{x}) \setminus [m]$. For $j \geq 1$, we define I_j and $\boldsymbol{x}^j$ recursively as follows:

1. Let $\boldsymbol{A}_{I_{j-1}}$ be the matrix with columns in $[m] \cup I_{j-1}$ and $\boldsymbol{b}^{j-1} = \boldsymbol{A}\boldsymbol{x}^{j-1}$.

2. Apply Theorem 5 to $\boldsymbol{b}^{j-1}$ and $\boldsymbol{A}_{I_{j-1}}$. We obtain an approximation $\boldsymbol{b}^j = \boldsymbol{A}\boldsymbol{x}^j$ for some $\boldsymbol{x}^j \in \mathbb{Z}^n_{\geq 0}$ with $\mathrm{supp}(\boldsymbol{x}^j) \subseteq [m] \cup I_{j-1}$ and $\left|\mathrm{supp}(\boldsymbol{x}^j)\backslash[m]\right| < |I_{j-1}|$.
3. If $\left|\mathrm{supp}(\boldsymbol{x}^j)\backslash[m]\right| \leq k-m$, we are done. Otherwise, define $I_j := \mathrm{supp}(\boldsymbol{x}^j)\backslash[m]$ and return to the first step.

Since we reduce $|I_j|$ by at least one in each step, this procedure terminates with some $\boldsymbol{x}^l$ such that $\left|\mathrm{supp}(\boldsymbol{x}^l)\backslash[m]\right| \leq k-m$ and $1 \leq l \leq n-k$. We observe

$$\mathrm{app}_k(\boldsymbol{A}) \leq \mathrm{app}_{\boldsymbol{B},k}(\boldsymbol{A}) \leq \sum_{j=0}^{l-1} \mathrm{app}_{\boldsymbol{B},|I_j|+m-1}(\boldsymbol{A}_{I_j}). \tag{4}$$

It remains to bound $\mathrm{app}_{\boldsymbol{B},|I_j|+m-1}(\boldsymbol{A}_{I_j})$ using Theorem 5. To this end let $\tau := \mu^{(m-1)/m} \left|\det \boldsymbol{B}\right|^{(m-1)/m}$. Inequality (4) and Theorem 5 give

$$\mathrm{app}_k(\boldsymbol{A}) \leq \tau \cdot \sum_{j=0}^{l-1} \left(\frac{1}{2^{|I_j|}}\right)^{\frac{1}{m}} \leq \tau \cdot \sum_{j=0}^{n-k-1} \left(\frac{1}{2^{n-m-j}}\right)^{\frac{1}{m}}$$

since $l \leq n-k$. With the geometric sum formula one obtains

$$\sum_{j=0}^{n-k-1} \left(\frac{1}{2^{n-m-j}}\right)^{\frac{1}{m}} = \frac{1}{2^{\frac{n-m}{m}}} \cdot \sum_{j=0}^{n-k-1} 2^{\frac{j}{m}} = \frac{1}{2^{\frac{1}{m}}-1} \cdot \left(\frac{1}{2^{\frac{k-m}{m}}} - \frac{1}{2^{\frac{n-m}{m}}}\right).$$

□

Proof (of Corollary 1). The polytope $1/\left|\det \boldsymbol{B}\right| \cdot P(\boldsymbol{B})$ does not contain non-zero integer vectors in its interior since each non-zero integer vector can be expressed in terms of $\boldsymbol{B}$ and non-zero coefficients which are at least $1/\left|\det \boldsymbol{B}\right|$. Hence, if $\mu^{(m-1)/m} \cdot \left|\det \boldsymbol{B}\right|^{(2m-1)/m} < 2^{(n-m)/m}$, Theorem 5 implies $\mathrm{app}_{\boldsymbol{B},n-1}(\boldsymbol{A}) < 1/\left|\det \boldsymbol{B}\right|$. This means that we can find an approximation of some $\boldsymbol{b} \in \boldsymbol{A}\mathbb{Z}^n_{\geq 0}$ that is contained in $\boldsymbol{b} + 1/\left|\det \boldsymbol{B}\right| \cdot \mathrm{int}(P(\boldsymbol{B}))$. Since $1/\left|\det \boldsymbol{B}\right| \cdot \mathrm{int}(P(\boldsymbol{B}))$ does not contain non-zero integer vectors, the approximation error has to be zero. □

5 Proof of Theorem 4

Let $n \geq 3$. We abbreviate the bound in Theorem 4 by

$$\epsilon := \frac{\varphi(n-2)}{2\varphi(n-2)+1} \cdot a_1.$$

We first argue that without loss of generality we can suppose that $b := a_2 + \ldots + a_n$: If $b = \lambda_1 a_1 + \ldots + \lambda_n a_n$ for $\lambda_i \in \mathbb{N}$, we can assume that $\lambda_i > 0$ for all $i \geq 2$ since, otherwise, we pass to a setting with fewer integers. Similarly, if $\lambda_i > 1$ for an index $i \geq 2$, replace the integer a_i by $\tilde{a}_i := \lambda_i a_i$. Since we fix a_1 in our non-negative combination, we also assume that $\lambda_1 = 0$.

Let $a_i = z_i a_1 + r_i$ with $r_i \in (-a_1/2, a_1/2]$ and $z_i \in \mathbb{N}$ for $i \in \{2, \ldots, n\}$. Our target is to approximate $a_2 + \ldots + a_n$ using a_1 and at most one additional integer a_i. To do so, we work with the residues r_i and approximate the sum of residues $r_2 + \ldots + r_n$ using only one r_i. Without loss of generality we can assume that $r_2 + \ldots + r_n > 0$. Define $I := \{i \in \{2, \ldots, n\} : r_i \geq 0\}$ and, hence, $r_2 + \ldots + r_n \leq \sum_{i \in I} r_i$. We define $\tilde{s}_i := \lfloor \sum_{j \in I} a_j / a_i \rfloor$ for all $i \in I$. Consider the following integer knapsack problem:

$$\max \sum_{i \in I} r_i x_i \quad \text{s.t.} \quad \sum_{i \in I} \frac{1}{\tilde{s}_i + 1} x_i < 1, \;\; x_i \in \mathbb{Z}_{\geq 0} \text{ for all } i \in I. \tag{5}$$

The knapsack constraint is chosen such that the all-ones vector is a feasible solution to the problem. This is true since

$$\sum_{i \in I} \frac{1}{\tilde{s}_i + 1} < \sum_{i \in I} \frac{a_i}{\sum_{j \in I} a_j} = 1.$$

Let $\boldsymbol{z}^*$ be an optimal solution for (5) and $\tilde{s}_{i^*} r_{i^*} = \max_{i \in I} \tilde{s}_i r_i$. From a result of Kohli and Krishnamurti [18, Theorem 2], one obtains

$$\sum_{i \in I} r_i z_i^* \leq (1 + \varphi(|I| - 1))\, \tilde{s}_{i^*} r_{i^*}.$$

Since the all-ones vector is a feasible integral solution to (5), we conclude that

$$\sum_{i \in I} r_i - \tilde{s}_{i^*} r_{i^*} \leq \sum_{i \in I} r_i z_i^* - \tilde{s}_{i^*} r_{i^*} \leq \varphi(|I| - 1) \tilde{s}_{i^*} r_{i^*}. \tag{6}$$

We next derive an upper bound on $\tilde{s}_{i^*} r_{i^*}$.

Let $s_{i^*} := \lfloor \sum_{j=2}^{n} a_j / a_{i^*} \rfloor$ and denote by $\bar{b}$ the least positive value in $r_2 + \ldots + r_n + a_1 \mathbb{Z} = b + a_1 \mathbb{Z}$. Suppose that there exists an integer $k \in \{0, \ldots, s_{i^*}\}$ such that $k r_{i^*} \in [\bar{b} - \epsilon, \bar{b} + \epsilon]$. This implies that there exists an integer λ such that $|\lambda a_1 + k a_{i^*} - b| = |\bar{b} - k r_{i^*}| \leq \epsilon$. The integer λ is non-negative because $k a_{i^*} \leq a_2 + \ldots + a_n = b$ by definition of s_{i^*}. Hence, the theorem is correct under this assumption.

From now on we assume that $k r_{i^*} \notin [\bar{b} - \epsilon, \bar{b} + \epsilon]$ for all $k \in \{0, \ldots, s_{i^*}\}$. We claim that we can strengthen this assumption to $k r_{i^*} < \bar{b} - \epsilon$ for all $k \in \{0, \ldots, s_{i^*}\}$. Suppose this is not true. Let $k^* \in \{0, \ldots, s_{i^*}\}$ be the smallest index such that $k^* r_{i^*} > \bar{b} + \epsilon$. We obtain

$$r_{i^*} = k^* r_{i^*} - (k^* - 1) r_{i^*} > \bar{b} + \epsilon - (\bar{b} - \epsilon) = 2\epsilon \geq \frac{1}{2} a_1,$$

where we use $\epsilon \geq 1/4 a_1$ and $(k^* - 1) r_{i^*} < \bar{b} - \epsilon$ due to the minimality of k^* and $(k^* - 1) r_{i^*} \notin [\bar{b} - \epsilon, \bar{b} + \epsilon]$. This contradicts that $r_{i^*} \leq 1/2 a_1$. Thus, we can assume that $k r_{i^*} < \bar{b} - \epsilon$ for all $k \in \{0, \ldots, s_{i^*}\}$.

Since $\tilde{s}_{i^*} = \lfloor \sum_{j \in I} a_j / a_i^* \rfloor \leq \lfloor \sum_{j=2}^{n} a_j / a_i^* \rfloor = s_{i^*}$ we obtain the relation

$$\tilde{s}_{i^*} r_{i^*} < \bar{b} - \epsilon. \tag{7}$$

By definition, we have $\bar{b} < a_1$. This bound can be strengthened to $\bar{b} < a_1 - \epsilon$ since, if $\bar{b} \geq a_1 - \epsilon$, we can approximate b with an error of size at most ϵ by using $\lceil b/a_1 \rceil a_1$. Inserting (7) and $\bar{b} < a_1 - \epsilon$ into (6) results in

$$\sum_{i \in I} r_i - \tilde{s}_{i^*} r_{i^*} \leq \varphi(|I|-1)\tilde{s}_{i^*} r_{i^*} < \varphi(|I|-1)\left(\bar{b} - \epsilon\right) < \varphi(|I|-1)(a_1 - 2\epsilon).$$

Note that $\varphi(|I|-1) \leq \varphi(n-2)$ as $|I| \leq n-1$. So we obtain

$$\sum_{i \in I} r_i - \tilde{s}_{i^*} r_{i^*} < \varphi(n-2)(a_1 - 2\epsilon) = \varphi(n-2)\left(1 - \frac{2\varphi(n-2)}{2\varphi(n-2)+1}\right) a_1 = \epsilon.$$

Since $\bar{b} \leq \sum_{i \in I} r_i$, we deduce

$$\bar{b} - \tilde{s}_{i^*} r_{i^*} \leq \sum_{i \in I} r_i - \tilde{s}_{i^*} r_{i^*} < \epsilon.$$

However, from (7), we get $\tilde{s}_{i^*} r_{i^*} < \bar{b} - \epsilon$, which implies $\epsilon < \bar{b} - \tilde{s}_{i^*} r_{i^*}$. This contradicts the inequality above and proves the upper bound on $\mathrm{app}_{a_1,2}(\boldsymbol{a})$. □

Acknowledgements. The authors are grateful to the reviewers for their valuable suggestions and comments.

References

1. Alasmari, A., Aliev, I.: On unique recovery of finite-valued integer signals and admissible lattices of sparse hypercubes. Optim. Lett. **17**(3), 739–751 (2023)
2. Aliev, I., Averkov, G., de Loera, J.A., Oertel, T.: Sparse representation of vectors in lattices and semigroups. Math. Program. Seri. B **192**, 05 (2021)
3. Aliev, I., De Loera, J.A., Oertel, T., O'Neill, C.: Sparse solutions of linear diophantine equations. SIAM J. Appl. Algebra Geom. **1**, 239–253 (2017)
4. Beck, M., Robins, S.: Computing the continuous discretely: Integer-point enumeration in polyhedra, 2nd edn. Undergraduate Texts in Mathematics. Springer, New York, NY (2015)
5. Candès, E.J., Romberg, J., Tao, T.: Robust uncertainty principles: exact signal reconstruction from highly incomplete frequency information. IEEE Trans. Inf. Theor. **52**(2), 489–509 (2006)
6. Candès, E.J., Romberg, J., Tao, T.: Stable signal recovery from incomplete and inaccurate measurements. Commun. Pure Appl. Math. **59**(8), 1207–1223 (2006)
7. Candès, E.J., Tao, T.: Decoding by linear programming. IEEE Trans. Inf. Theor. **51**(12), 4203–4215 (2005)
8. Candès, E.J., Tao, T.: The Dantzig selector: statistical estimation when p is much larger than n. Ann. Stat. **35**(6), 2313–2351 (2007)
9. Curtiss, D.R.: On Kellogg's diophantine problem. Am. Math. Mon. **29**(10), 380–387 (1922)
10. Dilworth, R.P.: A decomposition theorem for partially ordered sets. Ann. Math. **51**(1), 161–166 (1950)

11. Donoho, D.L.: Compressed sensing. IEEE Trans. Inf. Theor. **52**(4), 1289–1306 (2006)
12. Eisenbrand, F., Shmonin, G.: Carathéodory bounds for integer cones. Oper. Res. Lett. **34**, 564–548 (2006)
13. Flinth, A., Kutyniok, G.: Promp: a sparse recovery approach to lattice-valued signals. Appl. Comput. Harmon. Anal. **45**(3), 668–708 (2018)
14. Foucart, S., Rauhut, H.: A Mathematical Introduction to Compressive Sensing. Springer, Applied and Numerical Harmonic Analysis (2013)
15. Fukshansky, L., Needell, D., Sudakov, B.: An algebraic perspective on integer sparse recovery. Appl. Math. Comput. **340**, 31–42 (2019)
16. Galambos, G., Woeginger, G.J.: On-line bin packing – a restricted survey. Math. Methods Oper. Res. **42**, 25–42 (1995)
17. Gruber, P.M., Lekkerkerker, C.G.: Geometry of numbers, North-holland mathematical library. North-Holland Publishing Co., vol. 37 (1987)
18. Kohli, R., Krishnamurti, R.: A total-value greedy heuristic for the integer knapsack problem. Oper. Res. Lett. **12**(2), 65–71 (1992)
19. Konyagin, S.V.: On the recovery of an integer vector from linear measurements. Math. Notes **104**(5), 859–865 (2018)
20. Nathanson, M.B.: Underapproximation by egyptian fractions. J. Number Theor. **242**, 208–234 (2023)
21. OEIS Foundation Inc. The On-Line Encyclopedia of Integer Sequences (2024). https://oeis.org/A000058
22. Schrijver, A.: Theory of Linear and Integer Programming. Wiley (1986)
23. Seiden, S.S., Woeginger, G.J.: The two-dimensional cutting stock problem revisited. Math. Program. Seri. A **102**, 519–530 (2005)
24. Soundararajan, K.: Approximating 1 from below using n egyptian fractions. arxiv.org/abs/math/0502247

Unsplittable Multicommodity Flows in Outerplanar Graphs

David Alemán-Espinosa(✉) and Nikhil Kumar

Department of Combinatorics and Optimization, University of Waterloo, Waterloo, Canada
{dalemanespinosa,nikhil.kumar2}@uwaterloo.ca

Abstract. We consider the problem of multicommodity flows in outerplanar graphs. Okamura and Seymour [7] showed that the cut-condition is sufficient for routing demands in outerplanar graphs. We consider the unsplittable version of the problem and prove that if the cut-condition is satisfied, then we can route each demand along a single path by exceeding the capacity of an edge by no more than $\frac{18}{5} \cdot d_{\max}$, where $d_{\max}$ is the value of the maximum demand.

Keywords: Multicommodity Flows · Unsplittable Flows · Planar Graphs

1 Introduction

Given a graph G with edge capacities and multiple source-sink pairs, each with an associated demand, the multicommodity flow problem consists of routing all demands simultaneously without violating edge capacities. The graph obtained by including an edge (s_i, t_i) for a demand with source-sink s_i, t_i is called the demand graph and will be denoted by H. The problem was first formulated in the context of VLSI routing in the 70's and since then it has been the subject of a long and impressive line of work.

A multicommodity flow is called unsplittable if all the flow between a source-sink pair is routed along a single path. In general, existence of a feasible (fractional) flow does not imply the existence of an unsplittable flow, even in very simple settings. Consider a four-cycle $v_1v_2v_3v_4v_1$ with uniform edge capacity $d \geq 0$, where the demand edges are (v_1, v_3) and (v_2, v_4), both of value d. A feasible flow is given by evenly splitting the demand across the two possible paths for each demand edge. However, any unsplittable flow would violate the capacity of at least one edge by d. The above leads to a natural question: given a feasible flow, does there exist an unsplittable flow which satisfies all the demands and violates the edge capacities (in an additive sense) by at most a small factor times the value of the maximum demand $d_{\max}$? This problem was first considered by Kleinberg [6] in the setting where all the demands are incident at a single source. He also explored connections to some well-studied combinatorial

N. Megow and A. Basu (Eds.): IPCO 2025, LNCS 15620, pp. 385–399, 2025.
https://doi.org/10.1007/978-3-031-93112-3_28

optimization problems such as the generalized assignment problem [12]. Dinitz, Garg and Goemans [3] proved that in the single-source setting, any feasible fractional flow can be converted into an unsplittable flow that violates the edge capacities by no more than $d_{\max}$.

The problem is significantly less understood when the demand graph H is arbitrary. Schrijver, Seymour and Winkler [9] proved that if G is a cycle, then any feasible multicommodity flow can be converted into an unsplittable one that violates the edge capacities by at most $\frac{3}{2} \cdot d_{\max}$. Before our work, cycles were the only known nontrivial class of graphs for which an unsplittable flow was guaranteed to exist, incurring at most an additive $\mathcal{O}(d_{\max})$ violation of edge capacities, whenever a feasible flow existed. Our main contribution is extending this result to the class of *outerplanar graphs*.

In particular, we show that any feasible multicommodity flow in an outerplanar graph can be converted into an unsplittable one that violates the edge capacities by no more than $\frac{18}{5} \cdot d_{\max}$. This improves a recent result of Shapley and Shmoys [11], who proved that if G is an outerplanar graph, then one can convert any feasible multicommodity flow into an unsplittable one that violates the edge capacities by at most $\mathcal{O}(\log k) \cdot d_{\max}$, where k denotes the number of faces in the graph.

To prove our result, we crucially rely on the fact that there is a nice characterization for the existence of a feasible flow in outerplanar graphs. A necessary condition for the existence of a feasible flow is that the total demand across a cut should not exceed the capacity of the cut. This condition is called the *cut-condition* and is known to be sufficient in many cases, such as: (a) when all demand edges in H are incident on a single vertex (the single source setting) [4], (b) when G is planar and all demand edges in H are incident on one face (the Okamura-Seymour setting) [7], and (c) when $G+H$ is planar [10]. In outerplanar graphs, all the vertices are incident on one face. Hence, the result of Okamura-Seymour [7] implies that the cut-condition is sufficient for a feasible routing if G is outerplanar and H is arbitary. In general, the cut-condition is not sufficient for the existence of a feasible flow, even for very small instances. It is worthwhile to point out that all the results mentioned above for unsplittable flows are for instances where the cut-condition is also sufficient for the existence of a feasible multicommodity flow, although none of the proofs use this fact explicitly. Our result can be stated equivalently as follows: if G is outerplanar and the cut-condition is satisfied, then there exists an unsplittable flow satisfying all the demands that does not violate the capacity of any edge by more than $\frac{18}{5} \cdot d_{\max}$.

2 Definitions and Preliminaries

For an integer $n \geq 1$, we use $[n]$ to denote $\{1, \ldots, n\}$. For any universe U, function $c : U \mapsto \mathbb{R}$, set $S \subseteq U$ and an element $e \in S$, we use $S - e$ and $S + e$ to denote $S \setminus \{e\}$ and $S \cup \{e\}$, respectively. We use $c(S)$ to denote $\sum_{g \in S} c(g)$.

2.1 Multicommodity Flows

An instance $\mathcal{I}(G, c, D, d)$ of *multicommodity flow* is given by an undirected graph $G = (V, E)$ with edge capacities $c : E \mapsto \mathbb{R}_{\geq 0}$, and a graph $H = (V, D)$ with demands $d : D \mapsto \mathbb{R}_{\geq 0}$. We note that H could have parallel edges but no loops. We assume that G has no parallel edges or loops. We refer to G and E as the supply graph and supply edges, respectively. We refer to H and D as the demand graph and demand edges, respectively. For a demand edge $g \in D$, we denote by $\mathcal{P}_g$ the set of all (simple) paths between endpoints of g in G. For a $g \in D$, a demand flow x_g is an assignment of non-negative real numbers to paths in $\mathcal{P}_g$, i.e. $x_g : \mathcal{P}_g \mapsto \mathbb{R}_{\geq 0}$. A collection of demand flows $x = \{x_g \mid g \in D\}$ constitutes a flow for the instance. For $D' \subseteq D$, we use $\sum_{g \in D'} x_g$ to denote the collection of demand flows in D'.

For an edge $e \in E$ and demand $g \in D$, we use $x_g(e)$ to denote the total flow going through an edge e in x_g, i.e. $x_g(e) = \sum_{p: e \in p} x_g(p)$. For an edge $e \in E$ and flow x, we use $x(e)$ to denote the total flow going through an edge e in x, i.e. $x(e) = \sum_{g \in D} x_g(e)$. We say that a flow x is *feasible* if it does not violate the edge capacity constraints, i.e. $x(e) \leq c(e)$ for all $e \in E$, and satisfies all the demands, i.e. $\sum_{p \in \mathcal{P}_g} x_g(p) = d(g)$ for all $g \in D$.

We say that a flow x is *unsplittable* if at most one of the paths in $\mathcal{P}_g$ is assigned a non-zero value in x_g. We use $d_{\max} := \max_{g \in D} d(g)$ to denote the maximum demand value. We say that that an unsplittable flow x is *α-feasible* if $x(e) \leq c(e) + \alpha \cdot d_{\max}$ for all $e \in E$ and $\sum_{p \in \mathcal{P}_g} x_g(p) = d(g)$ for all $g \in D$.

2.2 The Cut Condition

For $S \subseteq V$, a cut $(S, V \setminus S)$ is a bi-partition of the vertex set. We refer to the cut $(S, V \setminus S)$ by S for convenience. Given a set of edges $X \subseteq E$, we will use $c(X)$ to denote the sum of capacities of edges in X. Given a set of edges $Y \subseteq D$, we will use $d(Y)$ to denote the sum of demands edges in Y. The set of supply edges going across the cut S is denoted by $\delta_G(S)$. Similarly, $\delta_H(S)$ denotes the set of demand edges going from S to $V \setminus S$. One necessary condition for the existence of a feasible flow is as follows: $c(\delta_G(S)) \geq d(\delta_H(S))$ for every $S \subseteq V$. In other words, across every cut, the total capacity of supply edges should be at least the total value of demand edges. This condition is also known as the *cut-condition*. In general, the cut-condition is not sufficient for a feasible routing, see [7] for an example. The following classic results identify settings where the cut-condition is also sufficient for routing demands in planar graphs. We will be invoking these to prove our main result.

Theorem 1 (Okamura-Seymour [7]). *Let $\mathcal{I}(G, c, D, d)$ be an instance of multicommodity flow such that G is planar and all the edges of H are incident on a fixed face*[1]*, then the cut-condition is necessary and sufficient for the existence of a feasible flow.*

[1] i.e. there exists a face f such that for each $(u, v) \in D$, both u and v lie on f.

Theorem 2 (Seymour [10]). *Let $\mathcal{I}(G, c, D, d)$ be an instance of multicommodity flow such that $G + H$ is planar, then the cut-condition is necessary and sufficient for the existence of a feasible flow.*

We say that $g \in D$ is *good* if there exists an edge $e \in E$ with exactly the same end-points as g. A demand is *bad* if it is not good. Given a bad demand $g \in D$ with end-points s, t and a vertex $u \in V$, the process of forming a new instance by replacing g with two new demands g_1, g_2, with end-points s, u and u, t is called *pinning*. Note that g_1, g_2 above have the same demand value as g. Given an instance where the cut-condition is satisfied, we say that a given pinning operation is *feasible* if the resulting instance also satisfies the cut-condition.

2.3 Planar Graphs

We assume a fixed planar embedding of G. Without loss of generality, one can assume that G is 2-vertex connected. If there is a cut-vertex v and ab is a demand separated by the removal of v, then replacing ab by av, vb maintains the cut-condition. By doing this for every cut-vertex and each demand separated by them, we get separate smaller instances for each 2-vertex connected component. Hence, every vertex is a part of a cycle corresponding to some face. We abuse notation and use f to also denote the edges and vertices associated with the cycle of face f. Given a set $S \subseteq V$, we denote the subgraph induced by vertices in S as $G[S]$. We call a subset $A \subseteq V$ *central* if both $G[A]$ and $G[V - A]$ are connected. A cut $(A, V \setminus A)$ is called central if A is central. The following is well-known.

Lemma 1 ([8]). *(G, H) satisfies the cut-condition if and only if $c(\delta_G(A)) \geq d(\delta_H(A))$ for all central sets $A \subseteq V$.*

We use $\mathcal{F}$ to denote the set of all faces of G. The **dual** of a planar graph, denoted by $G^D = (V^D, E^D)$, is defined as follows: The vertex set V^D of G^D is the face set $\mathcal{F}$ of G. There is a one-to-one correspondence between E and E^D, where every edge $e \in E$ induces precisely one edge e^D of E^D incident with the two faces of G that contain e. The **weak dual** of G, denoted by G^*, is defined as the induced graph $G^D[V^D - f_\infty]$, where f_∞ denotes the unbounded face of G. The following are well-known facts; see for instance Sect. 14.3 in [5].

Theorem 3. *Given a planar graph $G = (V, E)$ and $F \subseteq E$, F is the set of edges of a central cut in G if and only if $\{e^D : e \in F\}$ is the set of edges of a simple circuit[2] in G^D.*

Fact 4 *Let G be a 2-vertex connected outerplanar graph. Then G^* is a tree.*

---

[2] A simple circuit is a closed walk with at least one edge, where the only repeated vertices are the first and the last.

2.4 The Ring Loading Problem

When G is a cycle, we refer to the unsplittable multicommodity flow instance as a *ring-loading* instance. Such instances were first studied by Cosares and Saniee [1]. For such instances, we assume that G has vertex set $V := [n]$ and edge set $E = \{\{i, i+1\}\}_{i \in [n]}$, where we define $n+1 := 1$. We use $\mathcal{RL}([n], c, D, d)$ to denote a ring-loading instance where c, D, d are defined as above.

Theorem 1 implies that the cut-condition is sufficient for the existence of a feasible flow if G is a cycle and H is arbitrary. In their seminal paper, Schrijver, Seymour and Winkler [9] showed that if the cut-condition is satisfied for a ring-loading instance, then there exists a $\frac{3}{2}$-feasible unsplittable flow. This was later improved to $\frac{19}{14}$ by Skutella [13], and ultimately to $\frac{13}{10}$ by Däubel [2]. Skutella [13] also showed that it is not possible to obtain a α- feasible unsplittable flow for $\alpha \geq \frac{11}{10}$, improving upon the previous best lower bound of $\frac{101}{100}$ provided in [9]. Let α_{RL} denote the minimum value of α that can be achieved by a polynomial time algorithm for the ring-loading problem. We know that $\alpha_{\text{RL}} \in [\frac{11}{10}, \frac{13}{10}]$, and it is an interesting open question to pin down its precise value.

3 Our Contributions

As mentioned above, there exists a feasible flow for a multicommodity instance on outerplanar graphs if and only if the cut-condition is satisfied. We study the unsplittable version of the problem, and show that if the cut-condition is satisfied for a multicommodity flow instance on an outerplanar graph, then there exists a polynomial time computable $(2 \cdot \alpha_{\text{RL}} + 1)$-feasible unsplittable flow. By the result of Däubel [2], $\alpha_{\text{RL}} \leq \frac{13}{10}$, and hence we obtain a $\frac{18}{5}$-feasible unsplittable flow. We use the results on unsplittable flow for the ring-loading problem as a blackbox, and any improvement in the upper bound for α_{RL} would also imply an improved bound for outerplanar graphs.

Theorem 5. *Let $\mathcal{I}(G, c, D, d)$ be an instance of the multicommodity flow problem such that G is outerplanar and the cut-condition is satisfied. Then there exists a polynomial time computable $(2 \cdot \alpha_{\text{RL}} + 1)$-feasible unsplittable flow for the instance.*

We also prove the following theorem, which is useful in proving Theorem 5. We defer its proof to the full version of the paper.

Theorem 6. *Let $\mathcal{I}(G, c, D, d)$ be an instance of multicommodity flow problem such that G is outerplanar, all the demand edges are* good, *and the cut-condition is satisfied. Then there exists a polynomial time computable 1-feasible unsplittable flow for the instance.*

3.1 Overview of Techniques

We give an overview of the proof of Theorem 5, and defer the details of proof of Theorem 6 to the full version. Demand pinning is a well known technique that has been used to prove similar results. Recall that given a bad demand g with end-points s, t and an arbitrary vertex u, a feasible pinning involves creating a new instance that satisfies the cut-condition by replacing g by g_1, g_2 with end-points s, u and u, t. In many cases, it is possible to show that such a feasible pinning always exists and this immediately implies that the cut-condition is sufficient for feasibly routing all the demands. For example, when G is an outerplanar graph such that $G + H$ is Eulerian, all the demand values are one and there are no good demand edges[3], then there always exists a feasible pinning. Unfortunately, the value of demands could be arbitrary in our setting and such a pinning is not guaranteed to exist. Fortunately, this does not rule out a feasible pinning if we are also allowed to increase the edge capacities, and we crucially exploit this fact to design our algorithm.

We deviate from the traditional pinning paradigm described above, and instead of pinning one demand in an iteration, we pin a set of demands simultaneously. As far as we know, this is the first time such an operation has been used in this setting. This allows us to combine the operation of pinning and increasing edge capacities in a careful manner, and ensure that the capacity of an edge is not increased by too much during the whole process. In particular, we show that there exists a careful sequence of pinning and edge capacity increases such that (i) the resulting intermediate instances satisfy the cut-condition (ii) the capacity of no edge is increased by more than $2 \cdot \alpha_{\mathrm{RL}} \cdot d_{\max}$ during the whole process. Furthermore, each demand h in the final instance is *good*, i.e., there is an edge of G with the same end-points as h. This immediately implies that the resulting instance satisfies the conditions of Theorem 6, and that there exists a $(2 \cdot \alpha_{\mathrm{RL}} + 1)$-feasible unsplittable flow. All of the missing proofs from this extended abstract can be found in the full version of the paper.

4 Unsplittable Flows in Outerplanar Graph

We assume that $G = (V, E)$ is a simple outerplanar graph for the remainder of this section. As noted in Sect. 2.3, we work with a fixed planar embedding of G and assume that G is 2-vertex connected. Since G is 2-vertex connected, every face of G is a cycle. We assume that the vertex set of G is $V = [n] := \{1, \ldots, n\}$. In order to simplify the notation, we assume that $n + 1$ and 0 denote vertices 1 and n, respectively. We denote the set of faces of G by $\mathcal{F}$, and use f_∞ to denote the unbounded face of G. We assume that $f_\infty = 12 \ldots n1$, i.e. the nodes of the cycle f_∞ are ordered in a clockwise manner and the edges of f_∞ are $\{i, i + 1\}$ for $i \in [n]$.

We view the nodes of every face $f = i_1 i_2 \ldots i_\ell i_1 \in \mathcal{F}$ as being ordered clockwise around f as well. In particular, we assume that $1 \leq i_1 < i_2 < \ldots < i_j \leq n$, and $1 \leq i_{j+1} < i_{j+2} < \ldots < i_\ell < i_1$ for some $j \in [\ell]$. If it is clear from

[3] if there is a good demand edge, we can always create a smaller instance by deleting that demand edge and reducing the capacity of the corresponding supply edge by one unit.

the context which face is being considered, we will sometimes use $i_{\ell+1}$ and i_0 to denote i_1 and i_ℓ, respectively.

For any $i, j \in [n]$, we use $[i, j]$ to denote $\{i, i+1, \ldots, j\}$ if $i \leq j$, and $\{i, i+1, \ldots, n, 1, \ldots j\}$ if $j < i$. Recall that a set $S \subseteq V$ is called central if both $G[S]$ and $G[V \setminus S]$ are connected. Since G is outerplanar, all central sets are of the form $[i, j]$ for $i, j \in [n]$. Observe that $[i, j] = [n]$ if $i = j + 1$, so we may assume that $i \neq j + 1$ when considering central cuts.

Let G^* be the weak dual of G. By Fact 4, G^* is a tree. For $f_1, f_2 \in \mathcal{F} - f_\infty$, we use $G^*_{f_1,f_2}$ to denote the unique path between f_1 and f_2 in G^*. We overload notation, and use $G^*_{f_1,f_2}$ as a set of edges, or a set of faces; the meaning will be clear from the context. When considering any $e \in E \setminus f_\infty$, we also use e to denote the edge corresponding to e in G^*; it will be clear from the context which edge and graph (G or G^*) we are considering. The following simple yet crucial lemma will prove to be extremely useful.

Lemma 2. *Let $i, j \in [n]$ with $i \neq j + 1$, and $f = i_1 i_2 \ldots i_\ell i_1 \in \mathcal{F} - f_\infty$. Let f_1 and f_2 denote the unique faces in $\mathcal{F} - f_\infty$ that contain $\{i - 1, i\}$ and $\{j, j + 1\}$, respectively. Then,*

(a) $\delta_G([i, j]) = G^*_{f_1,f_2} + \{i - 1, i\} + \{j, j + 1\}$.
(b) $\delta_G([i, j]) \subseteq G^*_{f_1,f} \cup G^*_{f,f_2} + \{i - 1, i\} + \{j, j + 1\}$.
(c) $\delta_G([i, j]) = G^*_{f_1,f} \cup G^*_{f,f_2} + \{i-1, i\} + \{j, j+1\}$ *if both* $[i, j]$ *and* $[j+1, i-1] = [n] \setminus [i, j]$ *contain a node of* f.

As mentioned earlier, our plan is to first transform the given multicommodity flow instance into a more suitable instance consisting of only good demands, which allows us to invoke Theorem 6 in order to prove our main result.

Theorem 7. *Given an outerplanar graph $G = (V, E)$ and a multicommodity flow instance $\mathcal{I}(G, c, D, d)$ such that the cut-condition is satisfied, there exists a polynomial time algorithm that computes an instance $\mathcal{I}(G, c', D', d')$ and a partition $\{\mathcal{W}_g\}_{g \in D}$ of D' such that the following hold:*

(a) (Feasibility) *The cut-condition is satisfied for* $\mathcal{I}(G, c', D', d')$,
(b) (Capacity) $c'(e) \leq c(e) + \alpha_{\mathrm{RL}} \cdot d_{\max}$ *for* $e \in f_\infty$ *and* $c'(e) \leq c(e) + 2 \cdot \alpha_{\mathrm{RL}} \cdot d_{\max}$ *for* $e \in E \setminus f_\infty$,
(c) (Path property) *For each $g \in D$, all the edges in $\mathcal{W}_g$ are* good *and form a path between the endpoints of g. Furthermore, $d'(h) = d(g)$ for each $h \in \mathcal{W}_g$.*

We devote the remainder of this extended abstract to describing the proof of Theorem 7. Before that, we see how Theorems 6 and 7 together imply Theorem 5.

Proof. **[Theorem** 5**]** Given an instance $\mathcal{I}(G, c, D, d)$ satisfying the conditions of the theorem, we construct a feasible instance $\mathcal{I}(G, c', D', d')$ using Theorem 7. The path property (c) in Theorem 7 implies that every demand in D' is good. It then follows from Theorem 6 that $\mathcal{I}(G, c', D', d')$ admits a 1-feasible unsplittable flow z. For a demand $h \in D'$, let $z(h)$ be the edge set of the path on which z routes h. We construct an unsplittable flow y for instance $\mathcal{I}(G, c, D, d)$ as follows: for $g \in D$, let p_g be a path between the end-points of g using the edges in

$\cup_{h \in W_g} z(h)$. The path property (c) ensures that such a path p_g always exists. We set $y_g(p_g) = d(g)$ and $y_g(p) = 0$ for all $p \in \mathcal{P}_g \setminus \{p_g\}$. For each $g \in D$ and $h \in W_g$, the path property (c) states that $d'(h) = d(g)$. Hence, the following holds for all $e \in E$:

$$y(e) \leq z(e) \leq c'(e)+d_{\max} \leq c(e)+2\cdot\alpha_{\text{RL}}\cdot d_{\max}+d_{\max} = c(e)+(2\cdot\alpha_{\text{RL}}+1)\cdot d_{\max}.$$

The third inequality follows from item (b) in Theorem 7. We conclude that y is a $(2\cdot\alpha_{\text{RL}}+1)$-feasible unsplittable flow for $\mathcal{I}(G, c, D, d)$, proving the theorem. □

We now describe our main algorithm which will lead to a proof of Theorem 7.

4.1 Description of the Algorithm

Overview: In this section, we give a detailed description of the algorithm. The input of the algorithm is an instance $\mathcal{I}(G, c, D, d)$ of multicommodity flow satisfying the cut-condition. We also assume that we are given an algorithm $\mathcal{A}_{RL}$ which given a feasible ring-loading instance outputs an α-feasible unsplittable flow in polynomial time.

The algorithm works in iterations. At the end of every iteration, we use a *pinning* operation to create a new (feasible) instance. Recall that a demand h is *bad* if no edge of G has the same end-points as h. In each iteration, we pin a carefully chosen set of bad demands and increase the capacity of carefully chosen edges to create a new feasible instance. The algorithm terminates when there are no bad demands left.

To keep track of the bad demands and edge capacity increases, we use edge-labels. The label of an edge changes during the course of the algorithm and takes value in $\{0, 1, 2\}$. Initially, all edges on the unbounded face f_∞ are assigned a label of 1, while the rest of the edges are assigned a label of 0. The label of an edge cannot decrease during the course of the algorithm. The algorithm terminates when all the edges have a label of 2. The label of an edge keeps track of the increase in capacity of an edge as follows: whenever the label of an edge e increases by one, we increase the capacity of e by $\alpha \cdot d_{\max}$. Since the label of any edge can only increase by at most 2, we do not increase the capacity of any edge by more than $2 \cdot \alpha \cdot d_{\max}$ during the course of the algorithm.

We now list some useful properties of edge labels, which are also summarized in Lemma 3. The edges with label 1 form a simple cycle C and the label of an edge $e \in E$ is 0 if and only if both of its end-points are on C and $e \notin C$. All the edges which do not have label 1 or 0 have a label of 2. The algorithm terminates when no edge has label 1. This implies that there are no edges of label 0 and all the edges have label 2 at the termination. It is easy to see that these invariants hold at the beginning of the algorithm.

The cycle bounded by label 1 edges is used to keep track of the bad demands. More precisely, both the end points of a bad demand are incident on the cycle formed by label 1 edges. At the termination of the algorithm, there are no label 1 edges and hence the final instance comprises only of *good* demands.

Description of an Iteration: At the start of the algorithm we define the following:

$$c^{(1)} := c\ ;\ D^{(1)} := D\ ;\ \mathcal{W}_g^{(1)} := \{g\} \text{ for } g \in D\ ;\ \text{LABEL}^{(1)}(e) := \begin{cases} 1, & \text{if } e \in f_\infty \\ 0, & \text{otherwise.} \end{cases}$$

Let $x^{(1)} = \sum_{g \in D} x_g^{(1)}$ be a feasible flow of $\mathcal{I}(G, c, D, d)$.[4] We now give a detailed description of an iteration of the algorithm. Suppose we are at the start of iteration k of the algorithm. Let $D^{(k)}$ be the current set of demands and $\{\mathcal{W}_g^{(k)}\}_{g \in D}$ be the partition of $D^{(k)}$. We note that the demand edges in $\mathcal{W}_g^{(k)}$ form a walk between the end points of g for $g \in D$. Let $x^{(k)}$ be the feasible flow for the current instance $\mathcal{I}(G, c^{(k)}, D^{(k)}, d^{(k)})$, and $\text{LABEL}^{(k)}$ be the current labels of the edges. The current demand values $d^{(k)}$ are always given by $d^{(k)}(h) = d(g)$, if $h \in \mathcal{W}_g^{(k)}$.

If there are no label 1 edges, then the algorithm returns the instance $\mathcal{I}(G, c' = c^{(k)}, D' = D^{(k)}, d' = d^{(k)})$ and the set $\{\mathcal{W}_g = \mathcal{W}_g^{(k)}\}_{g \in D}$, and terminates.

Otherwise, let $C^{(k)}$ be the (simple) cycle formed by the edges $e \in E$ with $\text{LABEL}^{(k)}(e) = 1$ and $G[C^{(k)}]$ be the induced graph on the vertices of $C^{(k)}$. Observe that $G[C^{(1)}] = G$. We overload notation and use $C^{(k)}$ to denote both the edge and vertex set of the cycle $C^{(k)}$.

Since $G[C^{(k)}]$ is 2-vertex connected and outerplanar, its weak dual $G^*[C^{(k)}]$ is a tree. Also, the edges of $G^*[C^{(k)}]$ correspond precisely to the edges $e \in E$ with $\text{LABEL}^{(k)}(e) = 0$. Let f be a leaf vertex of $G^*[C^{(k)}]$. Note that f corresponds to a bounded face of G.

If $G[C^{(k)}] = C^{(k)}$, then $f = C^{(k)}$ and $\text{LABEL}^{(k)}(e) = 1$ for each $e \in f$. Otherwise, $\text{LABEL}^{(k)}(e') = 0$ for precisely one edge $e' \in f$ and $\text{LABEL}^{(k)}(e) = 1$ for each $e \in f - e'$. Thus, suppose that f is given by the vertex sequence $f = i_1 i_2 \dots i_\ell i_1$ such that $\ell \geq 3$ and the label of all edges of f distinct from $\{i_\ell, i_1\}$ is 1. The label of edge $\{i_\ell, i_1\}$ could be 0 or 1.

At the end of the current iteration, we increase the label of all edges of f by 1. Hence the only edge of f which could have a label of 1 after this iteration is $\{i_\ell, i_1\}$. This implies that in iteration $(k+1)$, either there are no edges of label 1, or the cycle formed by label 1 edges is $(C^{(k)} \setminus E(f)) \cup \{\{i_\ell, i_1\}\}$. As mentioned in the last section, we want to maintain the invariant that both the end-points of a bad demand are incident on the cycle formed by the label 1 edges. To this end, we will pin a set of demands so that no bad demands are incident on the vertices $\{i_2, i_3, \dots, i_{\ell-1}\}$ after the end of the current iteration.

Let $D_1^{(k)} \subseteq D^{(k)}$ be the set of bad demands containing one end-point in $\{i_2, i_3, \dots, i_{\ell-1}\}$ and one end-point in $C^{(k)} \setminus V(f)$.[5] Let $D_2^{(k)} \subseteq D^{(k)}$ be the set of bad demands both whose end-points are incident on $\{i_1, i_2, \dots, i_\ell\}$. Define:

[4] By Theorem 1, the cut-condition is sufficient for the existence of a feasible flow on outerplanar graphs.

[5] By outerplanarity, any demand with an endpoint in $\{i_2, i_3, \dots, i_{\ell-1}\}$ and an endpoint in $C^{(k)} \setminus V(f)$ is bad.

$$x^* := \sum_{h \in D_1^{(k)} \cup D_2^{(k)}} x_h^{(k)}. \tag{1}$$

We would like to find an appropriate pinning of the demands in $D_1^{(k)} \cup D_2^{(k)}$, and to this end, we create a ring-loading instance as follows. For $i \in [n]$, let f_i be the unique face in $\mathcal{F} - f_\infty$ containing the edge $\{i, i+1\}$. We define the following ring-loading instance,

$$\mathcal{RL}([n], c_{\text{rl}}, D_1^{(k)} \cup D_2^{(k)}, d^{(k)}) \text{ where } c_{\text{rl}}(\{i, i+1\}) := x^*(G^*_{f,f_i}) + x^*(\{i, i+1\}). \tag{2}$$

See Fig. 1 for an example. In Lemma 4 we show that the cut-condition is satisfied for this ring-loading instance, and hence we can use $\mathcal{A}_{RL}$ to obtain an α-feasible unsplitable flow $z = \sum_{h \in D_1^{(k)} \cup D_2^{(k)}} z_h$. For $h \in D_1^{(k)} \cup D_2^{(k)}$, we overload notation and use z_h to also denote the (unique) path along which h is routed in z. The manner in which we pin a demand h depends on whether the path z_h uses the vertices i_1, i_ℓ or not. More precisely, we replace each demand $h \in D_1^{(k)} \cup D_2^{(k)}$ by a set of demands $D_h^{(k)}$ as described below.

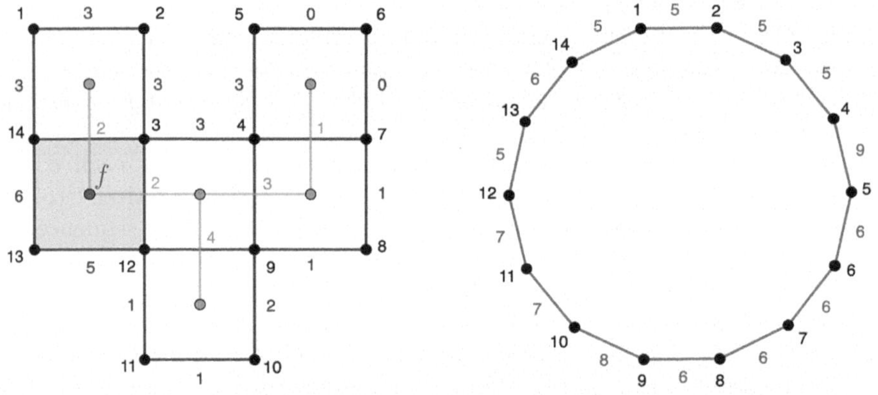

Fig. 1. The figure on the left depicts an outerplanar graph G where each edge $e \in E$ is labelled by $x^*(e)$, where x^* is defined as in (1). The shaded region represents the face f being considered during the iteration, and the yellow tree corresponds to the weak dual G^* of G. The figure on the right depicts the resulting capacities c_{rl} assigned to the ring loading instance $\mathcal{RL}([n], c_{\text{rl}}, D_1^{(k)} \cup D_2^{(k)}, d^{(k)})$.

Case 1: Suppose that $h \in D_1^{(k)}$ and $V(h) = \{s, t\}$; without loss of generality, we may assume that $s = i_p$ for some $1 < p < \ell$, and $t \in C^{(k)} \setminus V(f)$. If $i_1 \in z_h$, then

$$D_h^{(k)} := \{\{i_j, i_{j+1}\} : 1 \le j < p\} \cup \{\{i_1, t\}\}. \tag{3}$$

Otherwise, if $i_\ell \in z_h$, then

$$D_h^{(k)} := \{\{i_j, i_{j+1}\} : p \le j < \ell\} \cup \{\{i_\ell, t\}\}. \tag{4}$$

Case 2: Now suppose that $h \in D_2^{(k)}$ and $V(h) = \{s, t\}$; without loss of generality, we may assume that $s = i_p$ and $t = i_q$ for $1 \le p < q \le \ell$. If both $i_1, i_\ell \in z_h$, then:

$$D_h^{(k)} := \{\{i_j, i_{j+1}\} : 1 \le j < p\} \cup \{\{i_\ell, i_1\}\} \cup \{\{i_j, i_{j+1}\} : q \le j < \ell\}. \tag{5}$$

Otherwise; if either $i_1 \notin z_h$ or $i_\ell \notin z_h$,[6] then:

$$D_h^{(k)} := \{\{i_j, i_{j+1}\} : p \le j < q\}. \tag{6}$$

In item (c) of Lemma 3 we show that there is at most one bad demand edge in $\mathcal{W}_g^{(k)}$. If $h \in \mathcal{W}_g^{(k)}$ for $h \in D_1^{(k)} \cup D_2^{(k)}$, then we update $\mathcal{W}_g^{(k)}$ by removing h and adding $D_h^{(k)}$.

$$\begin{aligned} \mathcal{W}_g^{(k+1)} &\leftarrow (\mathcal{W}_g^{(k)} - h) \cup D_h^{(k)} &&, \text{ if there exists } h \in \mathcal{W}_g^{(k)} \text{ with } h \in D_1^{(k)} \cup D_2^{(k)}; \\ \mathcal{W}_g^{(k+1)} &\leftarrow \mathcal{W}_g^{(k)} &&, \text{ otherwise.} \end{aligned} \tag{7}$$

$$\begin{aligned} \mathcal{W}_g^{(k+1)} &\leftarrow (\mathcal{W}_g^{(k)} - h) \cup D_h^{(k)}, && \text{if there exists } h \in \mathcal{W}_g^{(k)} \text{ with } h \in D_1^{(k)} \cup D_2^{(k)}; \\ \mathcal{W}_g^{(k+1)} &\leftarrow \mathcal{W}_g^{(k)}, && \text{otherwise.} \end{aligned}$$

We update $D^{(k)}$ as follows:

$$D^{(k+1)} \leftarrow \cup_{g \in D} \mathcal{W}_g^{(k+1)} \tag{8}$$

We also update $d^{(k+1)}$ accordingly. Let $\mathbb{1}_f \in \{0,1\}^E$ be the characteristic vector of the edge set of f. In Lemma 4, we show that the cut-condition is satisfied for the following instance:

$$\mathcal{I}\left(G, x^* + \alpha \cdot d_{\max} \cdot \mathbb{1}_f, D_f, d^{(k+1)}\right) \quad \text{where} \quad D_f := \cup_{h \in D_1^{(k)} \cup D_2^{(k)}} D_h^{(k)}. \tag{9}$$

We compute a feasible flow $y = \sum_{r \in D_f} y_r$ for this instance and set,

$$x^{(k+1)} \leftarrow x^{(k)} - x^* + y \tag{10}$$

As already mentioned, we increase the label of all the edges of f by 1, and also increase their capacity by $\alpha \cdot d_{\max}$. More precisely,

$$c^{(k+1)} \leftarrow c^{(k)} + \alpha \cdot d_{\max} \cdot \mathbb{1}_f \quad \text{and} \quad \text{LABEL}^{(k+1)} \leftarrow \text{LABEL}^{(k)} + \mathbb{1}_f. \tag{11}$$

This completes the description of the algorithm. See Fig. 2 for an example that illustrates how the demand edges could evolve on each iteration of the algorithm.

[6] It may be the case that $|\{i_1, i_\ell\} \cap V(z_h)| = 1$, if either $p = 1$ or $q = \ell$.

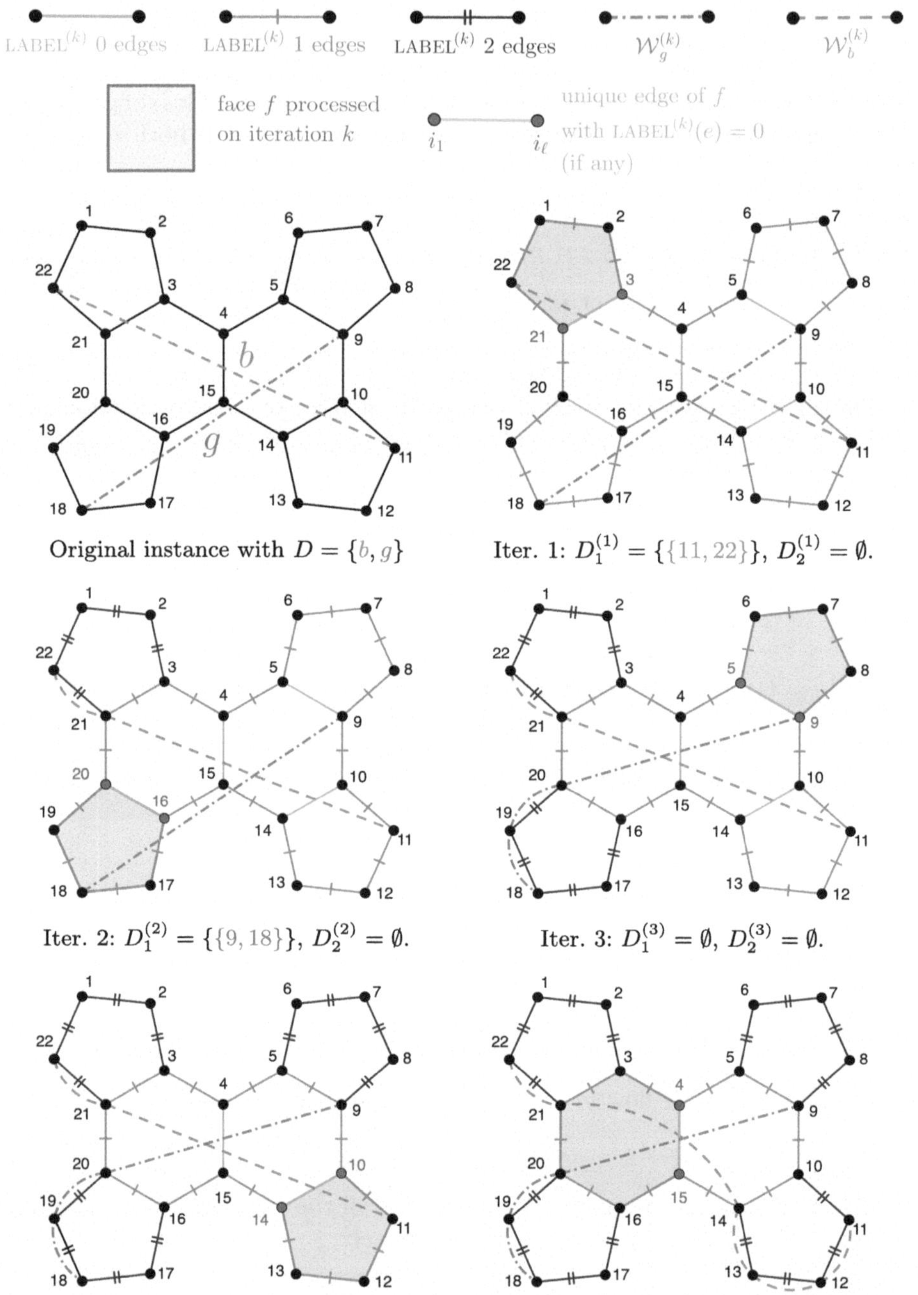

Original instance with $D = \{b, g\}$

Iter. 1: $D_1^{(1)} = \{\{11, 22\}\}$, $D_2^{(1)} = \emptyset$.

Iter. 2: $D_1^{(2)} = \{\{9, 18\}\}$, $D_2^{(2)} = \emptyset$.

Iter. 3: $D_1^{(3)} = \emptyset$, $D_2^{(3)} = \emptyset$.

Iter. 4: $D_1^{(4)} = \{\{11, 21\}\}$, $D_2^{(4)} = \emptyset$.

Iter. 5: $D_1^{(5)} = \{\{14, 21\}, \{9, 20\}\}$, $D_2^{(5)} = \emptyset$.

Fig. 2. The algorithm terminates at the start of iteration 7. The demand edges of the resulting instance returned by the algorithm are $D^{(7)} = \mathcal{W}_b^{(7)} \cup \mathcal{W}_g^{(7)}$, where $\mathcal{W}_b^{(7)} = \{\{11, 12\}, \{12, 13\}, \{13, 14\}, \{14, 15\}, \{15, 16\}, \{16, 20\}, \{20, 21\}, \{21, 22\}\}$, $\mathcal{W}_g^{(7)} = \{\{9, 5\}, \{5, 4\}, \{4, 3\}, \{3, 21\}, \{21, 20\}, \{20, 19\}, \{19, 18\}\}$.

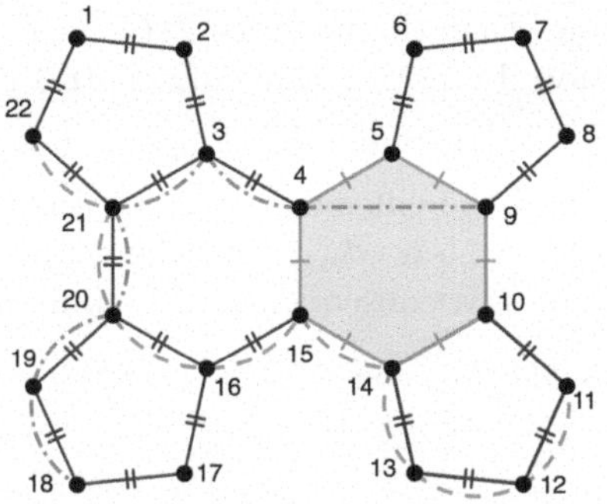

Iter. 6: $D_1^{(6)} = \emptyset$, $D_2^{(6)} = \{\{4, 9\}\}$.

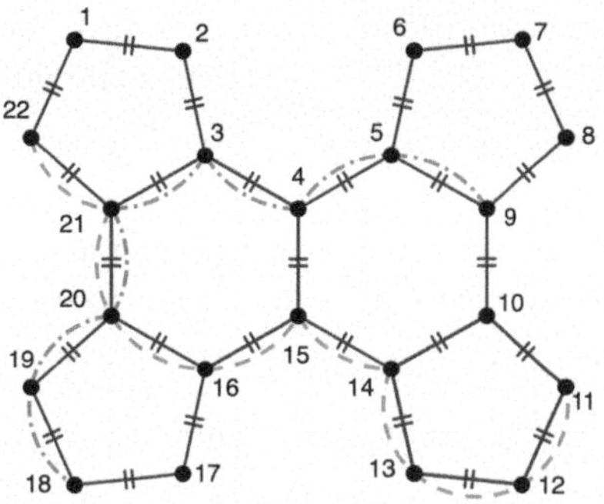

Iter. 7: $\text{LABEL}^{(7)}(e) = 2$ for each $e \in E$.

Fig. 2. (continued)

4.2 Proof of Correctness

We first note a few key lemmas which are useful in proving Theorem 7. For any iteration $k \geq 1$ and $j \in \{0, 1, 2\}$, let $E_j(k) := \{e \in E : \text{LABEL}^{(k)}(e) = j\}$. Recall that $\text{LABEL}^{(k)}(e)$ denotes the label of e at the start of iteration k; where we initially define $\text{LABEL}^{(1)}(e) = 1$ if $e \in f_\infty$, and $\text{LABEL}^{(1)}(e) = 0$ if $e \in E \setminus E(f_\infty)$.

Lemma 3. *The following holds for each iteration $k \geq 1$ and $g \in D$.*

(a) $E_1(k)$ is either the edge-set of a simple cycle $C^{(k)}$ in G, or the empty set.
(b) Consider the induced graph $G[C^{(k)}]$ on the vertex-set of $C^{(k)}$. Then $E_0(k)$ are precisely the edges of $G[C^{(k)}]$ that do not lie on the cycle $C^{(k)}$, and $E_2(k)$ are precisely the edges of G that are not contained in $G[C^{(k)}]$.
(c) The demand edges in $\mathcal{W}_g^{(k)}$ form a walk between the end-points of g. Furthermore, there is at most one bad demand edge $h \in \mathcal{W}_g^{(k)}$. If such h exists, then both of its end-points lie on the cycle $C^{(k)}$.

Lemma 4. *Let x be a feasible flow for $\mathcal{I}(G, c^{(k)}, D^{(k)}, d^{(k)})$, then the following hold:*

(a) The ring-loading instance $\mathcal{RL}\left([n], c_{rl}, D_1^{(k)} \cup D_2^{(k)}, d^{(k)}\right)$ defined at eq. (2) satisfies the cut-condition.
(b) The instance $\mathcal{I}\left(G, x^* + \alpha \cdot d_{\max} \cdot \mathbb{1}_f, D_f, d^{(k+1)}\right)$ defined at eq. (9) satisfies the cut-condition.

We sketch the proof of the critical Lemma 4, which makes a repeated use of Lemma 2. The ring-loading instance $\mathcal{RL}([n], c_{\text{rl}}, D_1^{(k)} \cup D_2^{(k)}, d^{(k)})$ satisfies the cut-condition (i.e., it is feasible) because (1) $\mathcal{I}(G, x^*, D_1^{(k)} \cup D_2^{(k)}, d^{(k)})$ satisfies the cut-condition, and (2) the capacity of any central cut is at least as large in the ring with respect to c_{rl} as it is in G with respect to x^*. Because of this and by definition of D_f, the instance $\mathcal{RL}\left([n], c_{\text{rl}} + \alpha \cdot d_{\max}, D_f, d^{(k+1)}\right)$ satisfies the cut-condition. From this and (1), it follows that the instance $\mathcal{I}(G, x^* + \alpha \cdot d_{\max} \cdot \mathbb{1}_f, D_f, d^{(k+1)})$ satisfies the cut-condition, by noting that any central cut induced by some $[i, j]$ for which $d^{(k+1)}\left(\delta_{D_f}([i, j])\right) > d^{(k)}\left(\delta_{D_1^{(k)} \cup D_2^{(k)}}([i, j])\right)$, satisfies the crucial property that the capacity of the cut induced by $[i, j]$ is the same in both the ring with capacities given by $c_{\mathrm{rl}} + \alpha \cdot d_{\max}$ and the graph G with capacities $x^* + \alpha \cdot d_{\max} \cdot \mathbb{1}_f$. The following is a consequence of Lemma 4.

Lemma 5. *Let* $x^{(k)} = \sum_{h \in D^{(k)}} x_h$ *be a feasible flow for* $\mathcal{I}\left(G, c^{(k)}, D^{(k)}, d^{(k)}\right)$, *then* $x^{(k+1)}$ *is a feasible flow for* $\mathcal{I}\left(G, c^{(k+1)}, D^{(k+1)}, d^{(k+1)}\right)$.

We are now ready to give a proof of Theorem 7.

Proof. **[Theorem 7]** By Lemma 5, we have that the cut-condition is satisfied for each $\mathcal{I}(G, c^{(k)}, D^{(k)}, d^{(k)})$, and hence the cut-condition is also satisfied for the final output $\mathcal{I}(G, c', D', d')$. Note that the capacity of an edge e is increased by $\alpha \cdot d_{\max}$ only when its label is also increased by 1. Initially, the label of all the edges on the unbounded face f_∞ is 1 while the label of the remaining edges is 0. At the end of the algorithm, the label of every edge is 2. This implies that the capacity of edges on f_∞ increases by $\alpha \cdot d_{\max}$ and the capacity of the remaining edges increases by $2 \cdot \alpha \cdot d_{\max}$. By item (c) of Lemma 3, it follows that the edges in W_g form a walk between the end points of g. Furthermore, if there are bad edges in W_g, then both its end-points are incident on the cycle formed by label 1 edges. At the end of the algorithm, there are no label 1 edges and hence there are no bad edges in W_g. Whenever we insert a new demand edge in $\mathcal{W}_g^{(k)}$, we set its value to $d(g)$. Hence all the demand edges in W_g have value $d(g)$, and this completes the proof of the theorem. □

5 Conclusion

We proved that if G is an outerplanar graph and the cut-condition is satisfied, then there exists a $\mathcal{O}(1)$-feasible unsplittable flow. It is an interesting open problem to generalize this result to other settings where the cut-condition is also sufficient for a feasible routing of all the demands, for example when all the source-sink pairs are incident on the unbounded face of a planar graph. It would also be interesting to see if a similar result holds for a more general class of graphs, where the cut-condition is not sufficient for a feasible routing, such as series-parallel graphs.

Acknowledgments. This work was partially supported by the NSERC Discovery Grant Program, grant number RGPIN-2024-04532.

References

1. Cosares, S., Saniee, I.: An optimization problem related to balancing loads on sonet rings. Telecommun. Syst. **3**, 165–181 (1994)
2. Däubel, K.: An improved upper bound for the ring loading problem. In: Proceedings of the 17th International Workshop on Approximation and Online Algorithms, WAOA2019, pp. 89–105 (2019)
3. Dinitz, Y., Garg, N., Goemans, M.X.: On the single-source unsplittable flow problem. Combinatorica **19**(1), 17–41 (1999)
4. Ford, L.R., Fulkerson, D.R.: Maximal flow through a network. Can. J. Math.**8**, 399–404 (1956)
5. Godsil, C., Royle, G.F.: Algebraic graph theory. Springer, New York, USA (2001)
6. Kleinberg, J.M.: Single-source unsplittable flow. In: Proceedings of 37th Conference on Foundations of Computer Science, pp. 68–77. IEEE (1996)
7. Okamura, H., Seymour, P.D.: Multicommodity flows in planar graphs. J. Comb. Theor. Seri. B **31**(1), 75–81 (1981)
8. Schrijver, A.: Combinatorial optimization: polyhedra and efficiency. Springer Sci. Bus. Media **24** (2003)
9. Schrijver, A., Seymour, P.D., Winkler, P.: The ring loading problem. SIAM J. Discrete Math. **11**(1), 1–14 (1998)
10. Seymour, P.D.: On odd cuts and plane multicommodity flows. Proc. Lond. Math. Soc. **3**(1), 178–192 (1981)
11. Shapley, R., Shmoys, D.B.: Small additive error for unsplittable multicommodity flow in outerplanar graphs. In: Proceedings of the 22nd International Workshop on Approximation and Online Algorithms, WAOA2024, pp. 167–182 (2024)
12. Shmoys, D.B., Éva, T.: An approximation algorithm for the generalized assignment problem. Math. Program. **62**(1), 461–474 (1993)
13. Skutella, M.: A note on the ring loading problem. SIAM J. Discrete Math. **30**(1), 327–342 (2016)

A Simple and Fast Algorithm for Fair Cuts

Jason Li(✉) and Owen Li(✉)

Carnegie Mellon University, Pittsburgh 15213, PA, USA
jmli@cs.cmu.edu, tianwei2@andrew.cmu.edu

Abstract. We present a simple and faster algorithm for computing *fair cuts* on undirected graphs, a concept introduced in recent work of Li et al. (SODA 2023). Informally, for any parameter $\epsilon > 0$, a $(1+\epsilon)$-fair (s,t)-cut is an (s,t)-cut such that there exists an (s,t)-flow that uses $1/(1+\epsilon)$ fraction of the capacity of *every* edge in the cut. Our algorithm computes a $(1+\epsilon)$-fair cut in $\tilde{O}(m/\epsilon)$ time, improving on the $\tilde{O}(m/\epsilon^3)$ time algorithm of Li et al. and matching the $\tilde{O}(m/\epsilon)$ time algorithm of Sherman (STOC 2017) for standard $(1+\epsilon)$-approximate min-cut.

Our main idea is to run Sherman's approximate max-flow/min-cut algorithm iteratively on a (directed) *residual* graph. While Sherman's algorithm is originally stated for undirected graphs, we show that it provides guarantees for directed graphs that are good enough for our purposes.

1 Introduction

The (s,t)-min-cut and (s,t)-max-flow problems are among the most basic and well-studied problems in combinatorial optimization. A long line of research on fast algorithms [4–6,9] culminated in the recent breakthrough $m^{1+o(1)}$ time algorithm of Chen et al. [3]. A separate line of research has focused on applying max-flow to solve other cut-based problems in combinatorial optimization, most notably Steiner min-cut [8], Gomory-Hu tree [1], and expander decomposition [10]. Using the algorithm of Chen et al. as a black box, all of these problems are now solvable in $m^{1+o(1)}$ time, which is optimal up to the factor $m^{o(1)}$.

On the other hand, the algorithm of Chen et al. (and subsequent improvements [12]) have a few downsides. First, the algorithms do not quite achieve "near"-linear time, which colloquially means $\tilde{O}(m)$ time where $\tilde{O}(\cdot)$ suppresses polylogarithmic factors. In fact, a near-linear time algorithm appears out of reach with the current techniques, which exploit recursion at the cost of $m^{o(1)}$ factors. Also, the algorithms are inherently sequential, leaving open the question of parallel max-flow in $m^{1+o(1)}$ work and sublinear time. These downsides carry over to any algorithm that requires max-flow as a black box, and hence to the cut-based problems mentioned above.

To address these issues, Li et al. [7] introduced the concept of *fair cuts*, a robust form of *approximate* min-cut. They present an algorithm for $(1+\epsilon)$-fair

N. Megow and A. Basu (Eds.): IPCO 2025, LNCS 15620, pp. 400–411, 2025.
https://doi.org/10.1007/978-3-031-93112-3_29

cut in $\tilde{O}(m/\epsilon^3)$ time that can be parallelized,[1] and then show how to solve $(1+\epsilon)$-approximate Steiner min-cut and Gomory-Hu tree using fair cut as a black box, leading to $\tilde{O}(m/\epsilon^{O(1)})$ time algorithms for both problems that can be parallelized. They also establish the first $\tilde{O}(m)$ time algorithm for expander decomposition that can also be parallelized.

The fair cut problem should be viewed as a generalization of $(1+\epsilon)$-approximate min-cut, which can be solved in $\tilde{O}(m/\epsilon)$ time by a recent breakthrough of Sherman [11] but is not robust enough for the above applications.[2] Nevertheless, there was a gap between the $\tilde{O}(m/\epsilon)$ time algorithm for $(1+\epsilon)$-approximate min-cut (and max-flow) and the $\tilde{O}(m/\epsilon^3)$ time algorithm for $(1+\epsilon)$-fair cut.

In this paper, we close the gap between the two problems by solving $(1+\epsilon)$-fair cut in $\tilde{O}(m/\epsilon)$ time. Conceptually, we present evidence that fair cut is no harder than approximate min-cut despite being more robust and powerful.

Theorem 1. *There is an $\tilde{O}(m/\epsilon)$ time randomized algorithm that, with high probability,*[3] *solves $(1+\epsilon)$-fair cut on an undirected graph with integral and polynomial capacities.*

Our algorithm is iterative, sending flow on each iteration and updating the *residual* graph, which is directed. Our main idea is observing that Sherman's approximate max-flow/min-cut algorithm (for undirected graphs) actually performs well on certain *directed* graphs, such as residual graphs of originally undirected graphs.

1.1 Preliminaries

We work with both undirected and directed graphs in this paper. For an undirected graph $G = (V, E)$, let $\overleftrightarrow{G} = (V, \overleftrightarrow{E})$ be the directed version of G with each edge replaced by bidirectional arcs of the same capacity. Given a vertex set $S \subseteq V$, let ∂S in an undirected graph be the set of edges with exactly one endpoint in S, and let $\vec{\partial} S$ in a directed graph be the set of arcs whose tail is in S and whose head is not in S. We may also use $\vec{\partial} S$ for an undirected graph G, in which case we are referring to the bidirected $\overleftrightarrow{G}$.

Throughout the paper, we use $c_G(\cdot)$ to denote edge and arc capacities. For an arc/edge set F, let $c_G(F)$ denote the total capacity of arcs/edges in F. We assume that all capacities are integers and polynomially bounded; in general, we

[1] The parallelization requires $m^{1+o(1)}$ work and $m^{o(1)}$ time, but the $m^{o(1)}$ factors can be improved to polylogarithmic by recent work [2]. For simplicity, we do not discuss parallelization in this paper.

[2] In more technical terms, the concept of *uncrossing two cuts* breaks down for arbitrary approximate min-cuts. Fair cuts are designed to satisfy an approximate version of uncrossing, which suffices for the applications.

[3] We adopt the convention that *with high probability* means with probability $1 - 1/n^{O(1)}$ for arbitrarily large polynomial in n.

would incur extra $\log W$ terms where W is the maximum integral capacity, but we stick with polynomially bounded for simplicity.

We represent a *flow* as a nonnegative vector $f \in \mathbb{R}^{\vec{E}}$ for an undirected graph and $f \in \mathbb{R}^E$ for a directed graph. The *congestion* of a flow is $\max_{(u,v)\in\vec{E}} f(u,v)/c_G(u,v)$, where $\vec{E}$ is replaced by E for a directed graph. Sometimes we abuse notation and say the flow *has congestion* κ if the congestion of the flow is *at most* κ. A flow is *feasible* if its congestion is at most 1. A *demand* is a vector $d \in \mathbb{R}^V$ with $\sum_v d_v = 0$. A flow *satisfies* or *routes* demand $d \in \mathbb{R}^V$ if for each vertex $v \in V$, $\sum_{(v,w)} f(v,w) - \sum_{(u,v)} f(u,v) = d_v$, i.e., the net flow out of v is exactly d_v. A flow is an (s,t)-flow of *value* τ if it satisfies demand $\tau(\mathbb{1}_s - \mathbb{1}_t)$. Here, $\mathbb{1}_v$ is the vector with entry 1 at v and entry 0 elsewhere. We also use $\mathbb{1}$ as the all-ones vector.

Given an undirected/directed graph $G = (V,E)$ and a flow f, the *residual* graph G' of G for flow f is the directed graph with arc capacities $c_{G'}(u,v) = c_G(u,v) - f(u,v) + f(v,u)$ for each (u,v) where either $(u,v) \in E$ or $(v,u) \in E$. Here, $c_G(u,v), f(u,v)$ are zero if $(u,v) \notin E$ and likewise for (v,u).

A *cut* is a bipartition $(S, V \setminus S)$ of the vertex set where $S, V \setminus S \neq \emptyset$. It is an (s,t)-cut if $s \in S$ and $t \notin S$. For a directed graph G, the *value* of the cut $(S, V \setminus S)$ is $c_G(\vec{\partial} S)$. We require the following fact about the *submodularity* of the directed cut function: for any directed graph G and two sets $A, B \subseteq V$, $c_G(\vec{\partial} A) + c_G(\vec{\partial} B) \geq c_G(\vec{\partial}(A \cup B)) + c_G(\vec{\partial}(A \cap B))$.

We now define the object of study in this paper, a *fair cut.*

Definition 1. *Let s,t be two vertices in V. For any parameter $\alpha \geq 1$, we say that a cut $(S, V \setminus S)$ is an α-*fair (s,t)-cut *if there exists a feasible (s,t)-flow f such that $f(u,v) \geq \frac{1}{\alpha} \cdot c(u,v)$ for every arc $(u,v) \in \vec{\partial} S$.*

Definition 2. *For any* undirected *graph G, an α-congestion approximator of G is a matrix $R \in \mathbb{R}^{[r]\times V}$ (where dimension r is unspecified) such that for any demand d whose optimal flow has congestion $OPT(d)$, it holds that $\|Rd\|_\infty \leq OPT(d) \leq \alpha \|Rd\|_\infty$.*

We defer the matrix notation from Sherman's approximate max-flow/min-cut algorithm to its relevant Sect. 3.

2 Fair Cut Algorithm

In this section, we present our fair cut algorithm, establishing Theorem 1. It will be more convenient to prove the following version, where ϵ is replaced by $O(\epsilon \log n)$.

Theorem 2. *There is an $\tilde{O}(m/\epsilon)$ time randomized algorithm that, with high probability, solves $(1 + O(\epsilon \log n))$-fair cut on an undirected graph with integral and polynomial capacities.*

We will use the following approximate max-flow/min-cut primitive for *residual* graphs, which we present in Sect. 3.

Theorem 3. *Given an undirected graph G, a residual graph G' of G, two vertices s, t, and a parameter $\tau > 0$, there is a randomized algorithm that runs in time $\tilde{O}(m/\epsilon)$ and computes, with high probability,*

1. *Either an (s,t)-cut of value less than τ, or*
2. *A feasible flow f in G' routing a demand d such that the residual demand $\tau(\mathbb{1}_s - \mathbb{1}_t) - d$ can be routed in G with congestion ϵ.*

Equipped with this flow/cut primitive, the fair cut algorithm is quite simple. We iteratively maintain a cut $(S_i, V \setminus S_i)$ and a flow f_i that both gradually improve over time. On each iteration, we temporarily remove the edges in ∂S_i that are nearly saturated in the right direction, and then call the flow primitive on the residual graph (minus the removed edges) with a careful choice of τ. Depending on whether the flow primitive returns a flow or cut, we either update the current flow or the current cut, leaving the other unchanged.

We present the formal algorithm below.

1. Let f_1 be the empty flow and $(S_1, V \setminus S_1)$ be an arbitrary (s,t)-cut.
2. For iteration $i = 1, 2, \ldots, L = \Theta(\log n)$:
   (a) Let $\boldsymbol{U}_i \subseteq \overleftrightarrow{E}$ be all arcs $(u,v) \in \boldsymbol{\partial} S_i$ satisfying $f_i(u,v) \leq (1-4\epsilon)c_G(u,v)$, i.e., the "unsaturated" arcs in $\boldsymbol{\partial} S_i$.
   (b) Let $U_i \subseteq E$ be $\boldsymbol{U}_i$ with all arcs undirected (and parallel edges removed).
   (c) Let $G_i \subseteq G$ be the undirected graph $G \setminus (\partial S_i \setminus U_i)$, i.e., remove all edges in ∂S_i that are "saturated" in the right direction.
   (d) Let G'_i be the residual graph of G_i for the restricted flow $f_i|_{G_i}$, defined as the flow f_i with flow on arcs outside $\overleftrightarrow{G}_i$ removed.
   (e) Call Theorem 3 on graph G_i, its residual graph G'_i, vertices s, t, and parameter $\tau = 0.5c_{G'_i}(\boldsymbol{\partial} S_i)$.
   (f) If Theorem 3 returns a flow h:
      - Set $f_{i+1} = f_i + h$ and $S_{i+1} = S_i$, i.e., add the new flow but keep the current cut.
   (g) If Theorem 3 returns a cut $(X_i, V \setminus X_i)$:
      - Set $f_{i+1} = f_i$ and S_{i+1} as either $S_i \cup X_i$ or $S_i \cap X_i$, i.e., update the cut but keep the current flow. Of the two choices, pick the S_{i+1} minimizing $c_{G'_i}(\boldsymbol{\partial} S_{i+1})$.
3. Output the (s,t)-cut $(S_{L+1}, V \setminus S_{L+1})$.

It is clear that the algorithm makes $O(\log n)$ calls to Theorem 3 and runs in $\tilde{O}(m)$ time outside these calls, for an overall running time of $\tilde{O}(m/\epsilon)$. For the rest of this section, we prove its correctness.

Our measure of progress is the quantity $c_{G'_i}(\boldsymbol{\partial} S_i)$, i.e., the total residual capacity of all "unsaturated" arcs in $\boldsymbol{\partial} S_i$, which we show drops by a constant factor on each iteration.

Lemma 1. *If Theorem 3 returns a flow h, then $c_{G'_{i+1}}(\boldsymbol{\partial} S_{i+1}) \leq 0.75c_{G'_i}(\boldsymbol{\partial} S_i)$.*

Proof. Let d be the demand routed by flow h. By Theorem 3, there is a feasible flow r in G_i routing the residual demand $\tau(\mathbb{1}_s - \mathbb{1}_t) - d$ with congestion ϵ. Then, the flow $h + r$ in G_i' routes demand $\tau(\mathbb{1}_s - \mathbb{1}_t)$ (with arbitrary congestion). This flow must pass through the cut $\boldsymbol{\partial} S_i$ in G_i', so $h + r$ sends a net flow of τ across $\boldsymbol{\partial} S_i$. Since the flow r has congestion ϵ in G_i, removing it from $h + r$ affects the net flow across $\boldsymbol{\partial} S_i$ by at most $\epsilon c_{G_i}(\partial S_i)$, so the flow h sends at least $\tau - \epsilon c_{G_i}(\partial S_i)$ across $\boldsymbol{\partial} S_i$. Each arc $(u, v) \in \boldsymbol{\partial} S_i$ in G_i' satisfies $(u, v) \in U_i$, so $f_i(u, v) \leq (1 - 4\epsilon)c_G(u, v)$ and $c_{G_i'}(u, v) \geq 4\epsilon c_G(u, v) = 4\epsilon c_{G_i}(u, v)$. Summing over all arcs $(u, v) \in \boldsymbol{\partial} S_i$ gives $c_{G_i'}(\boldsymbol{\partial} S_i) \geq 4\epsilon c_{G_i}(\partial S)$, so the flow h sends at least $\tau - 0.25c_{G_i'}(\boldsymbol{\partial} S_i) = 0.25c_{G_i'}(\boldsymbol{\partial} S_i)$ flow across $\boldsymbol{\partial} S_i$.

Let H be the residual graph of G_i for the restricted flow $f_{i+1}|_{G_i}$. By definition of residual graph, the quantity $c_{G_i}(\partial S_i) - c_H(\boldsymbol{\partial} S_i)$ is exactly the net flow that $f_{i+1}|_{G_i}$ sends across $\boldsymbol{\partial} S_i$. Since G_i' is the residual graph of G_i for the restricted flow $f_i|_{G_i}$, the quantity $c_{G_i}(\partial S_i) - c_{G_i'}(\boldsymbol{\partial} S_i)$ is exactly the net flow that $f_i|_{G_i}$ sends across $\boldsymbol{\partial} S_i$. Since $f_{i+1}|_{G_i} = f_i|_{G_i} + h$, the difference of quantities $(c_{G_i}(\partial S_i) - c_H(\boldsymbol{\partial} S_i)) - (c_{G_i}(\partial S_i) - c_{G_i'}(\boldsymbol{\partial} S_i))$ is exactly the net flow that h sends across $\boldsymbol{\partial} S_i$, which is at least $0.25c_{G_i'}(\boldsymbol{\partial} S_i)$. In other words, $c_H(\boldsymbol{\partial} S_i) \leq c_{G_i'}(\boldsymbol{\partial} S_i) - 0.25c_{G_i'}(\boldsymbol{\partial} S_i) = 0.75c_{G_i'}(\boldsymbol{\partial} S_i)$. Any previously "saturated" arc $(u, v) \in \boldsymbol{\partial} S_i \setminus \boldsymbol{U}_i$ is still "saturated" in flow f_{i+1} (i.e., $f_{i+1}(u, v) > (1 - 4\epsilon)c_G(u, v)$) since the arc is absent from G_i' and hence carries no flow in h. Since $S_{i+1} = S_i$, we have $\partial S_i \setminus U_i \subseteq \partial S_{i+1} \setminus U_{i+1}$, which means that $G_i \supseteq G_{i+1}$. In particular, the arcs in $\boldsymbol{\partial} S_i$ present in G_{i+1}' are also present in H, and they have the same capacity since both G_{i+1}' and H are residual graphs for a restriction of f_{i+1}. We conclude that $c_{G_{i+1}'}(\boldsymbol{\partial} S_i) \leq c_H(\boldsymbol{\partial} S_i)$, and

$$c_{G_{i+1}'}(\boldsymbol{\partial} S_{i+1}) = c_{G_{i+1}'}(\boldsymbol{\partial} S_i) \leq c_H(\boldsymbol{\partial} S_i) \leq 0.75c_{G_i'}(\boldsymbol{\partial} S_i),$$

as promised.

Lemma 2. *If Theorem 3 returns a cut* $(X_i, V \setminus X_i)$, *then* $c_{G_{i+1}'}(\boldsymbol{\partial} S_{i+1}) \leq 0.75c_{G_i'}(\boldsymbol{\partial} S_i)$.

Proof. By Theorem 3, the cut $(X_i, V \setminus X_i)$ satisfies $c_{G_i'}(\boldsymbol{\partial} X_i) \leq \tau = 0.5c_{G_i'}(\boldsymbol{\partial} S_i)$. By submodularity of the cut function $c_{G_i'}(\boldsymbol{\partial} S)$,

$$c_{G_i'}(\boldsymbol{\partial} S_i) + c_{G_i'}(\boldsymbol{\partial} X_i) \geq c_{G_i'}(\boldsymbol{\partial}(S_i \cup X_i)) + c_{G_i'}(\boldsymbol{\partial}(S_i \cap X_i)),$$

and by the choice of S_{i+1},

$$c_{G_i'}(\boldsymbol{\partial} S_{i+1}) \leq \frac{1}{2}(c_{G_i'}(\boldsymbol{\partial}(S_i \cup X_i)) + c_{G_i'}(\boldsymbol{\partial}(S_i \cap X_i))) \leq \frac{1}{2}(c_{G_i'}(\boldsymbol{\partial} S_i) + c_{G_i'}(\boldsymbol{\partial} X_i)) \leq 0.75c_{G_i'}(\boldsymbol{\partial} S_i).$$

We now claim that $c_{G_{i+1}'}(\boldsymbol{\partial} S_{i+1}) \leq c_{G_i'}(\boldsymbol{\partial} S_{i+1})$. Note that arcs present in both G_i' and G_{i+1}' must have the same capacity since $f_{i+1} = f_i$, so it suffices to show that the arcs in $\boldsymbol{\partial} S_{i+1}$ present in G_{i+1}' are a subset of those present in G_i'. Any arc $(u, v) \in \boldsymbol{\partial} S_{i+1}$ present in G_{i+1}' satisfies $(u, v) \in \boldsymbol{U}_{i+1}$, so $f_i(u, v) = f_{i+1}(u, v) \leq (1 - 4\epsilon)c_G(u, v)$. If $(u, v) \in \boldsymbol{\partial} S_i$ as well, then $(u, v) \in \boldsymbol{U}_i$ and the arc belongs to G_i'. Otherwise, if $(u, v) \notin \boldsymbol{\partial} S_i$, then there are two cases. If

$S_{i+1} = S_i \cap X_i$, then since $u \in S_{i+1} \subseteq S_i$, we must have $v \in S_i$ as well. So the edge (u, v) is not in ∂S_i, which means the arc (u, v) belongs to G'_i, establishing the claim. If $S_{i+1} = S_i \cup X_i$, then since $v \in V \setminus S_{i+1} \subseteq V \setminus S_i$, we must have $u \in V \setminus S_i$ as well. So the edge (u, v) is not in ∂S_i, and the same argument follows.

Putting everything together, we conclude that $c_{G'_{i+1}}(\boldsymbol{\partial} S_{i+1}) \leq c_{G'_i}(\boldsymbol{\partial} S_{i+1}) \leq 0.75 c_{G'_i}(\boldsymbol{\partial} S_i)$.

Finally, we prove the correctness of the algorithm, establishing Theorem 2.

Lemma 3. *The output* $(S_{L+1}, V \setminus S_{L+1})$ *is a* $(1 + O(\epsilon \log n))$*-fair cut with high probability.*

Proof. By Lemmas 1 and 2, we have $c_{G'_{i+1}}(\boldsymbol{\partial} S_{i+1}) \leq 0.75 c_{G'_i}(\boldsymbol{\partial} S_i)$ for each iteration i. Since capacities are polynomially bounded, we start with $c_{G'_1}(\boldsymbol{\partial} S_1) \leq n^{O(1)}$, so for large enough $L = \Theta(\log n)$ we have $c_{G'_{L+1}}(\boldsymbol{\partial} S_{L+1}) < 4\epsilon$ with high probability. Any arc $(u, v) \in \boldsymbol{U}_{L+1}$ belongs to $\boldsymbol{\partial} S_{L+1}$ and satisfies $c_{G'_{L+1}}(u, v) \geq 4\epsilon c_G(u, v) \geq 4\epsilon$, so no such arcs exist. In other words, $\boldsymbol{U}_{L+1} = \emptyset$, and it follows that $f_{L+1}(u, v) \geq (1 - 4\epsilon) c_G(u, v)$ for all arcs $(u, v) \in \boldsymbol{\partial} S_{L+1}$. To establish fairness, it remains to augment f_{L+1} to an (s, t)-flow.

By construction, $f_{L+1} = h_1 + h_2 + \cdots + h_L$, and there exist flows $r_1, \ldots, r_L$ in G of congestion ϵ such that each $h_i + r_i$ is an (s, t)-flow. In particular, the flow $f' = f_{L+1} + r_1 + \cdots + r_L$ is an (s, t)-flow. Since $r_1 + \cdots + r_L$ has congestion $O(\epsilon \log n)$, we have $|f'(u, v) - f_{L+1}(u, v)| \leq O(\epsilon \log n) \cdot c_G(u, v)$ for all arcs (u, v). In particular, for each arc $(u, v) \in \boldsymbol{\partial} S_{L+1}$,

$$\begin{aligned} f'(u, v) &\geq f_{L+1}(u, v) - O(\epsilon \log n) \cdot c_G(u, v) \\ &\geq (1 - 4\epsilon) c_G(u, v) - O(\epsilon \log n) \cdot c_G(u, v) \\ &\geq \frac{1}{1 + O(\epsilon \log n)} c_G(u, v), \end{aligned}$$

so the (s, t)-flow f' certifies that $(S_{L+1}, V \setminus S_{L+1})$ is a $(1 + O(\epsilon \log n))$-fair cut.

3 Approximate Max-Flow on Residual Graphs

In this section, we show how to apply Sherman's approximate max-flow/min-cut algorithm on directed *residual graphs* of an underlying undirected graph. The flow may not satisfy the input demand, but the leftover demand will be routable with low congestion on the *undirected* graph. Our main goal is to prove Theorem 3, restated below.

Theorem 3. *Given an undirected graph* G*, a residual graph* G' *of* G*, two vertices* s, t*, and a parameter* $\tau > 0$*, there is a randomized algorithm that runs in time* $\tilde{O}(m/\epsilon)$ *and computes, with high probability,*

1. *Either an* (s, t)*-cut of value less than* τ*, or*

2. *A feasible flow f in G' routing a demand d such that the residual demand $\tau(\mathbb{1}_s - \mathbb{1}_t) - d$ can be routed in G with congestion ϵ.*

We first introduce some preliminaries from Sherman [11]. For a matrix $A \in \mathbb{R}^{n\times m}$, define the matrix norm $\|A\|_{\infty\to\infty} = \max_{\|v\|_\infty=1} \|Av\|_\infty$, and define $\mathrm{nnz}(A)$ as the number of nonzero entries in A. Define $\Delta_k^m \subseteq \mathbb{R}^{m\times k}$ as the set of matrices $X \in \mathbb{R}^{m\times k}$ with $X \geq \mathbf{0}$ and $\sum_j X_{ij} = 1$ for all $i \in [m]$. We now present a key subroutine from [11]:

Theorem 4 (Corollary 1.8 of [11]). *There is an algorithm that, given $B \in \mathbb{R}^{n\times k}$, and $A \in \mathbb{R}^{n\times m}$ with $\|A\|_{\infty\to\infty} \leq 1$, takes $\tilde{O}(k\mathrm{nnz}(A)\epsilon^{-1})$ time and outputs either,*

(1) $X \in \Delta_k^m$ such that $AX \leq B + \epsilon R$ where $R \in \Delta_k^n$.
(2) $Y \in \mathbb{R}^{k\times n}, Y \geq \mathbf{0}$ such that $\mathrm{tr}(Y(AX - B)) > 0$ for all $X \in \Delta_k^m$.

This result can be used to solve approximate multi-commodity flow, as indicated by Lemma 4.2 of [11]. Sherman only provided a sketch proof in the original paper. For specificity and completeness, we state and prove a two-commodity flow version of Sherman's Lemma 4.2 for some residual network G' of G, where the matrix R is a congestion approximator for G. In particular, it may be instructive to compare Theorem 3 to Lemma 4.2 of [11].

For a given directed graph $G = (V, \boldsymbol{E})$, we represent a flow f by its vector of congestions on edges, which is a nonnegative vector in $\mathbb{R}^{\boldsymbol{E}}$ where for each arc $(u, v) \in \boldsymbol{E}$, the flow f sends $f_{(u,v)}c_G(u, v)$ flow. Note that $\|f\|_\infty$ is exactly the congestion of the flow. We define $C_G \in \mathbb{R}^{\boldsymbol{E}\times\boldsymbol{E}}$ as the diagonal matrix whose entry (u, v) is the capacity of arc $(u, v) \in \boldsymbol{E}$. We define $D_G \in \mathbb{R}^{\boldsymbol{E}\times V}$ as the matrix whose row $(u, v) \in \boldsymbol{E}$ has vector $(\mathbb{1}_u - \mathbb{1}_v)^\top$, also called the discrete divergence operator which maps any vector $C_G f$ to the demand satisfied by the flow f. If $G = (V, E)$ is an undirected graph, we treat it as a directed graph with bi-directed edge set $\overleftrightarrow{E}$.

Given a directed graph $G = (V, E)$, we say $G' = (V, E')$ is a subgraph of G (denoted as $G' \leq G$) iff for all $(u, v) \in V \times V$, it holds that $c_{G'}(u, v) \leq c_G(u, v)$. In particular, if G' is some residual network for undirected G, then it holds that $G' \leq 2G$.

Lemma 4. *Given an undirected graph G, a subgraph $G' \leq \overleftrightarrow{G}$, and an α-congestion approximator R of G, it holds that $\|RD_{G'}C_{G'}\|_{\infty\to\infty} \leq 2$.*

Proof. Fix any vector $f \in \mathbb{R}^{\overleftrightarrow{E}}$ with $\|f\|_\infty = 1$, and let $d := D_{\overleftrightarrow{G}}C_{\overleftrightarrow{G}}f$. Notice that each pair of anti-parallel arcs e^+, e^- of $\overleftrightarrow{G}$ has the same capacity, and adding a constant to f_{e^+}, f_{e^-} will not change the demand routed by the flow; we perform such operations on each pair of anti-parallel arcs to obtain some f' such that $\min(f'_{e^+}, f'_{e^-}) = 0$ for all pairs of anti-parallel arcs. Now it is evident that $\|f'\|_\infty \leq 2$, and that f' corresponds to a flow f_G of G with demand d and congestion ≤ 2, so

$$\|RD_GC_Gf\|_\infty = \|Rd\|_\infty \leq OPT(d) \leq \|f_G\|_\infty = 2$$

and it follows that

$$\|RD_GC_Gf\|_\infty = \|Rd\|_\infty \leq 2$$

so

$$\|RD_GC_G\|_{\infty\to\infty} \leq 2.$$

Let $F := \{f \in \mathbb{R}^{\vec{E}} : \|f\|_\infty \leq 1\}$ be the set of all directed flows of congestion 1. Since $c_{G'}(u,v) \leq c_G(u,v)$ for all arcs $(u,v) \in \vec{E}$, we have the set inclusion

$$\{D_{G'}C_{G'}f : f \in F\} \subseteq \{D_GC_Gf : f \in F\}$$

so

$$\max_{f\in F} \|RD_{G'}C_{G'}f\|_\infty \leq \max_{f\in F} \|RD_GC_Gf\|_\infty \leq 2$$

and it follows that $\|RD_{G'}C_{G'}\|_{\infty\to\infty} \leq 2$.

Note that Sherman's Stochastic Matrix Algorithm, as per Corollary 1.8 of [11], may output a dual, and when leveraging Sherman's algorithm to solve approximate (s,t)-max-flow, we may need to explicitly compute a corresponding (s,t)-cut that is integral. The following lemma constructs such a cut in a directed graph, given an infeasibility criterion from Sherman's algorithm:

Lemma 5. *Given a directed graph G and a "potential" vector $\phi \in \mathbb{R}^n$ on vertices, we define a corresponding flow f_ϕ via the following:*

$$(f_\phi)_{uv} = \begin{cases} 1, & \text{if } \phi_u > \phi_v \\ 0, & \text{otherwise} \end{cases}$$

We also suppose that

$$\phi^\top(d - D_GC_Gf_\phi) > 0$$

for some demand d. Then if we sort the vertices by decreasing potential ϕ_v, there must be some prefix $S \subset V$, such that $\sum_{v\in S} d_v > c_G(\boldsymbol{\partial} S)$, certifying the infeasibility of such a demand. Furthermore, the cut $(S, V \setminus S)$ can be computed in $O(m + n\log n)$ time.

Moreover, if $d = \tau(\mathbb{1}_s - \mathbb{1}_v)$ for two vertices $s,t \in V$ and parameter $\tau > 0$, then the cut $(S, V \setminus S)$ is an (s,t)-cut with $c_G(\boldsymbol{\partial} S) < \tau$.

Proof. We begin with some notation. Let $V_{>x} := \{v \in V(G) : \phi_v > x\}$ denote the set of vertices of G whose potential is strict greater than x, and let $\Delta(S) := \sum_{v\in S} d_v$ denote the sum of demands in the set S of vertices.

Let M be some positive real number such that $|\phi_v| < M$ for all $v \in V(G)$. We seek to prove

$$\int_{-M}^{M} \Delta(V_{>x})dx = \phi^\top d > \phi^\top(B_GC_Gf_\phi) = \int_{-M}^{M} c_G(\boldsymbol{\partial} V_{>x})dx$$

because then there must be some $-M \leq x \leq M$ s.t. $\Delta(V_{>x}) > c_G(\boldsymbol{\partial} V_{>x})$, and setting $S = V_{>x}$ achieves the desired $\Delta(S) > c_G(\boldsymbol{\partial} S)$.

1. For the first equality, it holds that

$$\int_{-M}^{M} \Delta(V_{>x})dx = \int_{-M}^{M} \sum_{v \in V} d_v \cdot \mathbb{1}[\phi_v > x]dx = \sum_{v \in V} \int_{-M}^{M} d_v \cdot \mathbb{1}[\phi_v > x]dx = \sum_{v \in V} d_v(\phi_v + M)$$

and since d is a demand, it holds that $\sum_v d_v = 0$, so the above equals

$$\sum_{v \in V} d_v \phi_v = \phi^\top d$$

2. For the second equality, we start from the definition of f_ϕ and have

$$\begin{aligned}
\phi^\top(B_G C_G f_\phi) &= \sum_v \phi_v \cdot \Big(-\sum_{(u,v)\in E} c_G(u,v)\mathbb{1}[\phi_u > \phi_v] + \sum_{(v,w)\in E} c_G(v,w)\mathbb{1}[\phi_v > \phi_w]\Big) \\
&= \sum_{(u,v)\in E} c_G(u,v)\mathbb{1}(\phi_u > \phi_v)(\phi_u - \phi_v) \\
&= \sum_{(u,v)\in E} c_G(u,v)\max(0, \phi_u - \phi_v) \\
&= \int_{-M}^{M} \sum_{(u,v)\in E} c_G(u,v)\mathbb{1}((u,v) \in \partial(V_{>x}))dx \\
&= \int_{-M}^{M} c_G(\partial V_{>x})dx
\end{aligned}$$

3. The inequality follows from the assumption $\phi^\top(d - B_G C_G f_\phi) > 0$.

Furthermore, if $d = \tau(\mathbb{1}_s - \mathbb{1}_t)$, then notice that

$$\Delta(S) > c_G(\partial S) \geq 0$$

and the only way for $\Delta(S) > 0$ is if $s \in S$ and $t \notin S$, in which case $\Delta(S) = \tau$, so $(S, V \setminus S)$ is an (s,t)-cut with $c_G(\partial S) < \tau$.

To identify the set $S = V_{>x}$, it suffices to iterate through the cuts corresponding to all $O(n)$ prefixes $S = V_{>x}$, keeping track of $\Delta(S)$ and $c_G(\partial S)$. We may start with $|S| = 1$, which includes the vertex v with the highest potential ϕ_v, identify edges in the cut, and compute the value of $c_G(\partial S)$; then we keep adding vertices to S, one at a time, in the order of decreasing potentials, and for each added vertex, we re-compute the set of edges in the new cut, then compute the associated capacity. Each update after adding some vertex v requires iterating through $\deg(v)$ many edges, and since $\sum_v \deg(v) = 2m$, the overall time complexity to find the minimum threshold cut is $O(m)$ (after the initial sorting by ϕ_v in $O(n \log n)$).

Using Lemmas 4 and 5, we finally prove Theorem 3:

Proof. To convert a single commodity flow into a right stochastic matrix problem, we consider the "empty demand" as another type of commodity. Specifically, we fix $k = 2$ and $X = [f, f_\emptyset]$, where f is the vector of a flow of congestion 1, with its entries indicating congestion on edges, and $f_\emptyset$ is the vector indicating remaining congestion, defined as $f_\emptyset := \mathbb{1} - f$; we also define $A := D_{G'}C_{G'}$, and $B := [d, d_\emptyset]$, where $d_\emptyset := D_G C_G \mathbb{1} - d$ is the empty demand vector. Then $AX = B$ encodes a solution for the single commodity problem.

Next, we cite Theorem 4.4 from Sherman [11] to compute an α-congestion approximator R of G with $\alpha = (\log n)^{O(1)}$ in $\tilde{O}(m)$ time with high probability, where R is specifically column sparse, i.e. each column contains $O(\log n)$ many nonzero entries. Observe that $(RD'_G C'_G)_{r,e} \neq 0$ iff the directed edge e is in the cut represented by row r of R, and since R is column sparse, there can only be $O(\log n)$ many cuts in R that contain e; overall we obtain

$$nnz(RD'_G C'_G) \leq O(m \log n)$$

Since G' is a residual graph of G, we have $G' \leq 2G$, and applying Lemma 4 on subgraph $G'/2$ (and then scaling up by factor 2) gives $\|RD_{G'}C_{G'}\|_{\infty\to\infty} \leq 4$. Now define $R' := R/4$ and $A_2 := \begin{bmatrix} R'A \\ -R'A \end{bmatrix}$ and $B_2 := \begin{bmatrix} R'B \\ -R'B \end{bmatrix}$, so that

$$\|A_2\|_{\infty\to\infty} = \|R'A\|_{\infty\to\infty} = \|RA/4\|_{\infty\to\infty} = \|RD_{G'}C_{G'}/4\|_{\infty\to\infty} \leq 1$$

and we invoke Theorem 4 to obtain, in time $\tilde{O}(k\,\mathrm{nnz}(A_2)\epsilon^{-1}) \leq \tilde{O}(k\,\mathrm{nnz}(RD_{G'}C_{G'})\epsilon^{-1}) \leq \tilde{O}(m\epsilon^{-1})$, either

1. some X and some $S_1, S_2 \in \Delta_2^n$ such that

$$\begin{cases} R'AX - R'B \leq \frac{\epsilon}{\alpha} S_1 \\ R'(-A)X - R'(-B) \leq \frac{\epsilon}{\alpha} S_2 \end{cases}$$

2. some $Y = [y_1, y_2] > 0$ such that

$$\mathrm{tr}(Y(A_2 X - B_2)) \geq 0$$

for all feasible flows $X = [f, f_\emptyset]$

In the first case, combining the two inequalities gives

$$-\frac{\epsilon}{\alpha} S_2 \leq R'(AX - B) \leq \frac{\epsilon}{\alpha} S_1$$

thus each row $r_i \in \mathbb{R}^2$ of $R'(AX-B)$ satisfies $\|r_i\|_\infty \leq \frac{\epsilon}{\alpha}$, so that $\|r_i\|_1 \leq 2\frac{\epsilon}{\alpha}$, and

$$\begin{aligned} \|R'(Af - d)\|_{\infty\to\infty} &= \max_{\|v\|_\infty = 1} \|R'(Af - d)v\|_\infty \\ &= \max_{\|v\|_\infty = 1} \max_i |(R'(Af - d)v)_i| \\ &= \max_i \max_{\|v\|_\infty = 1} |(R'(Af - d)v)_i| \\ &= \max_i \|(R'(Af - d))_i\|_1 \\ &\leq \max_i \|(R'(AX - B))_i\|_1 \\ &\leq \frac{2\epsilon}{\alpha} \end{aligned}$$

and it follows from definition of R that if the algorithm returns some flow $X = [f, f_\emptyset]$, then the residual demand $Af - d$ can be routed with congestion $\leq O(\epsilon)$ with respect to the undirected graph G. We may retroactively set ϵ a constant factor smaller so that the congestion is $\leq \epsilon$.

In the second case, it equivalently holds for all feasible flows $X = [f, f_\emptyset]$ that

$$y_1^\top (\begin{bmatrix} A \\ -A \end{bmatrix} f - \begin{bmatrix} d \\ -d \end{bmatrix}) + y_2^\top (\begin{bmatrix} A \\ -A \end{bmatrix} f_\emptyset - \begin{bmatrix} d_\emptyset \\ -d_\emptyset \end{bmatrix}) > 0$$

By construction of duplicated rows, we have $y_i = \begin{bmatrix} w_i \\ z_i \end{bmatrix}$, and either one of

(i) $w_1^\top (Af - d) + w_2^\top (Af_\emptyset - d_\emptyset) > 0$
(ii) $z_1^\top (Af - d) + z_2^\top (Af_\emptyset - d_\emptyset) < 0$

is true. We further notice that

$$A(f + f_\emptyset) = d + d_\emptyset \iff Af - d = -(Af_\emptyset - d_\emptyset)$$

so we substitute and either one of

(i) $(w_2 - w_1)^\top (d - Af) > 0$
(ii) $(z_1 - z_2)^\top (d - Af) > 0$

is true; either of which, according to Lemma 5, implies the existence of some (s, t)-cut $(S, V \setminus S)$ with $c_G(\partial S) < \tau$ that can be computed in $O(m + n \log n)$ time.

Acknowledgement. The authors have no competing interests to declare that are relevant to the content of this article.

References

1. Abboud, A., Li, J., Panigrahi, D., Saranurak, T.: All-pairs max-flow is no harder than single-pair max-flow: gomory-hu trees in almost-linear time. In: 2023 IEEE 64th Annual Symposium on Foundations of Computer Science (FOCS). pp. 2204–2212. IEEE (2023)
2. Agarwal, A., et al.: Parallel approximate maximum flows in near-linear work and polylogarithmic depth. In: Proceedings of the 2024 Annual ACM-SIAM Symposium on Discrete Algorithms (SODA). pp. 3997–4061. SIAM (2024)
3. Chen, L., Kyng, R., Liu, Y.P., Peng, R., Probst Gutenberg, M., Sachdeva, S.: Maximum flow and minimum-cost flow in almost-linear time (2022)
4. Dinitz, Y.: Algorithm for solution of a problem of maximum flow in networks with power estimation. Soviet Math. Doklady **11**, 1227–1280 (1970)
5. Ford, L.R., Fulkerson, D.R.: Maximal flow through a network. Can. J. Math. **8**(3), 399–404 (1956)
6. Goldberg, A.V., Rao, S.: Beyond the flow decomposition barrier. J. ACM **45**(5), 783–797 (1998). https://doi.org/10.1145/290179.290181, http://doi.acm.org/10.1145/290179.290181

7. Li, J., Nanongkai, D., Panigrahi, D., Saranurak, T.: Near-linear time approximations for cut problems via fair cuts. In: Proceedings of the 2023 Annual ACM-SIAM Symposium on Discrete Algorithms (SODA). pp. 240–275. SIAM (2023)
8. Li, J., Panigrahi, D.: Deterministic min-cut in poly-logarithmic max-flows. In: 61st IEEE Annual Symposium on Foundations of Computer Science, FOCS 2020. IEEE Computer Society (2020)
9. Liu, Y.P., Sidford, A.: Faster energy maximization for faster maximum flow. In: Makarychev, K., Makarychev, Y., Tulsiani, M., Kamath, G., Chuzhoy, J. (eds.) Proccedings of the 52nd Annual ACM SIGACT Symposium on Theory of Computing, STOC 2020, Chicago, IL, USA, June 22-26, 2020. pp. 803–814. ACM (2020)
10. Saranurak, T., Wang, D.: Expander decomposition and pruning: Faster, stronger, and simpler, to appear in SODA'19 (2019)
11. Sherman, J.: Area-convexity, l_∞ regularization, and undirected multicommodity flow. In: Proceedings of the 49th Annual ACM SIGACT Symposium on Theory of Computing. pp. 452–460 (2017)
12. Van Den Brand, J., et al.: A deterministic almost-linear time algorithm for minimum-cost flow. In: 2023 IEEE 64th Annual Symposium on Foundations of Computer Science (FOCS). pp. 503–514. IEEE (2023)

The Augmented Factorization Bound for Maximum-Entropy Sampling

Yongchun Li(✉)

Department of Industrial and Systems Engineering,
The University of Tennessee, Knoxville, USA
ycli@utk.edu

Abstract. The maximum-entropy sampling problem (MESP) aims to select the most informative principal submatrix of a prespecified size from a given covariance matrix. This paper proposes an augmented factorization bound for MESP based on the concave relaxation. By leveraging majorization and Schur-concavity theory, we demonstrate that this new bound dominates the classic factorization bound of [27] and a recent upper bound proposed by [24]. Furthermore, we provide theoretical guarantees that quantify how much the proposed augmented factorization bound improves the two existing ones and establish sufficient conditions for when the improvement is strictly attained. These results allow us to refine the celebrated approximation bounds for the two approximation algorithms of MESP. Motivated by the strength of the augmented factorization bound, we develop a variable fixing logic for MESP from a primal perspective. Finally, our numerical experiments demonstrate that the augmented factorization bound achieves smaller integrality gaps and fixes more variables than the tightest bounds in the MESP literature on most benchmark instances, with the improvement being particularly significant when the condition number of the covariance matrix is small.

1 Introduction

The maximum-entropy sampling problem (MESP) arises in spatial statistics and information theory, which was introduced by the celebrated work of [32]. MESP aims to select the most informative subset of s variables from a total of n variables to maximize the collected information, measured by entropy [20]. It has been widely applied to designing environmental monitoring networks [8,28]. When dealing with variables that follow Gaussian or more general multivariate elliptical distributions, MESP is recast as a combinatorial optimization problem:

$$z^* := \max_{S} \left\{ \log\det\left(\boldsymbol{C}_{S,S}\right) : S \subseteq [n], |S| = s \right\}, \tag{MESP}$$

where log det denotes the natural logarithm of the determinant function, $\boldsymbol{C}$ is a sample covariance matrix obtained from the observations of n random variables, $s \leq n$ is a user-specified parameter, and for a subset $S \subseteq [n]$, $\boldsymbol{C}_{S,S}$ denotes a principal submatrix of $\boldsymbol{C}$ indexed by S. We assume that the matrix $\boldsymbol{C}$ is positive

N. Megow and A. Basu (Eds.): IPCO 2025, LNCS 15620, pp. 412–426, 2025.
https://doi.org/10.1007/978-3-031-93112-3_30

definite, a condition commonly used in the MESP literature and well satisfied by the benchmark instances (see [1,2,20,31] and references therein).

MESP is computationally challenging and known to be NP-hard, as [20] demonstrated. Indeed, MESP has no constant-factor polynomial-time approximation algorithm [12]. The primary solution method for solving MESP to optimality is branch-and-bound (see the excellent survey by [14, chapter 2] and many references they contain). In practice, branch-and-bound can quickly find a (near-)optimal solution after only a few iterations; however, proving its optimality can be substantially time-consuming. A tight upper bound of MESP is often desired to accelerate this process. Inspired by the Schur-concavity of the objective function in MESP, as discussed in Sect. 3, this paper develops a tighter upper bound than directly factorizing $\boldsymbol{C}$, a technique commonly used in the literature on MESP (see, e.g., [10,25,27]).Our method starts with subtracting a scaled identity matrix from $\boldsymbol{C}$ before factorization.

Remark 1. For any t, $0 \leq t \leq \lambda_{\min}(\boldsymbol{C})$, the matrix $\boldsymbol{C}-t\boldsymbol{I}$ is positive semidefinite, and we denote by $\boldsymbol{A}(t) \in \mathbb{R}^{n\times n}$ the Cholesky factor of $\boldsymbol{C} - t\boldsymbol{I}$, i.e.,

$$\boldsymbol{C} - t\boldsymbol{I} = [\boldsymbol{A}(t)]^{\top} \boldsymbol{A}(t). \tag{1}$$

Note that the rank of the matrix $\boldsymbol{C}-t\boldsymbol{I}$ varies with the value of t. Specifically, for $0 \leq t < \lambda_{\min}(\boldsymbol{C})$, the matrix is full-rank; however, at $t = \lambda_{\min}(\boldsymbol{C})$, it becomes singular. The Choleskey factor of a rank-r matrix contains $n-r$ rows of all zeros. These zero rows can be removed to yield a Choleskey factor of size $r \times n$. Thus, the Cholesky factor is not unique in this context. Fortunately, the bound derived from the matrix factorization is independent of the particular Choleskey factor employed, based on a result of [10]. For consistency, we therefore compute a sized-$n \times n$ Cholesky factor for any t.

1.1 Related Work

The upper bounds of MESP have been derived in various ways. An eigenvalue-based upper bound was first introduced by [20]. Following this, a variety of eigenvalue-based bounding methods have subsequently been developed and investigated by [6,7,19,21]. Another seminal approach to developing upper bounds for MESP is based on the continuous relaxations of its equivalent concave integer programs. The classic work of [3,5] first developed a concave relaxation for MESP. Since then, researchers have actively developed different concave integer programs to achieve strong concave relaxations and improve existing methods (see [1,2,24,27]). These bounding methods were further combined and refined by [9–11]. While no bounding technique wins in all test instances, the "Linx" bound, as introduced by [2], and the Fact bound of [27] seem to provide the tightest upper bounds for MESP from a computational perspective. Later, [10] applied the known mixing technique to combine them, which can further decrease the bound, especially for values of s close to $n/2$. We refer to the mixing bound as "Mix-LF" in Sect. 4.

Our Aug-Fact bound can be viewed as an augmentation of Fact of [27] by employing a general t, $0 \leq t \leq \lambda_{\min}(\boldsymbol{C})$, in (1), leading to a notable reduction in integrality gaps on most test instances. Nevertheless, [27] focused on $t = 0$ in (1) to directly factorize the matrix $\boldsymbol{C}$. Fact has been widely recognized for its high effectiveness and computational efficiency. Hence, Fact and its properties have been extensively studied by [10,15,25] and applied to different variants of MESP, including generalized MESP ([30]), the D-optimal data fusion (DDF) problem ([24]), and the D-optimal design problem ([29]). In particular, [24] decomposed the matrix $\boldsymbol{C}$ into the form of (1) in which they set $t = \lambda_{\min}(\boldsymbol{C})$, establishing the equivalence between MESP and DDF. DDF results from an application to sensor placement in power systems [22]. DDF aims to select a subset of rank-one positive semidefinite matrices to add to the initial Fisher information matrix, so as to maximize the D-optimality of the updated matrix. The continuous relaxation of DDF is concave and naturally provides a practical upper bound- DDF-R for MESP. DDF-R has been widely used to provide an upper bound in branch-and-bound-based methods for finding an optimal solution to DDF (see, e.g., [17,24]). However, DDF-R may not serve as a strong upper bound, as demonstrated in the numerical results of [24]. They also demonstrated that DDF-R with $t = \lambda_{\min}(\boldsymbol{C})$ is not comparable with Fact.

1.2 Contributions and Outline

The main contributions and an outline of the rest of the paper are summarized below.

In Sect. 2, we convert MESP into a concave integer program for any t using Remark 1 and the concave envelope technique, which leads to a new upper bound based on the concave relaxation- Aug-Fact.

Section 3 highlights the advantages of the proposed factorization method in Remark 1 for improving the upper bound, through analyzing how the parameter t affects the performance of Aug-Fact. Specifically,

(i) By leveraging the theory of majorization and Schur-concave functions, we establish that the Aug-Fact bound decreases as t increases in Subsect. 3.1. We also demonstrate that Aug-Fact is invariant under scaling;
(ii) We demonstrate that for any strictly positive t, Aug-Fact is tighter than the two existing upper bounds, Fact and DDF-R, and provide theoretical guarantees for their difference in optimal values in Subsects. 3.2 and 3.3, respectively. Besides, Aug-Fact allows us to enhance the best-known and first-known approximation bounds of the sampling and local search algorithms for MESP, respectively; and
(iii) From a primal perspective, Subsect. 3.4 develops a variable fixing logic based on a feasible solution of Aug-Fact at $t = \lambda_{\min}(\boldsymbol{C})$.

The numerical experiments in Sect. 4 demonstrate the superior performance of Aug-Fact across various test instances, compared to most promising bounds in the literature on MESP.

The proofs of all the results presented in this paper are provided in the full-length version [23].

Notations: We let $\mathcal{S}^n, \mathcal{S}^n_+$ denote the set of all the $n \times n$ symmetric real matrices and the set of all the $n \times n$ symmetric positive semidefinite matrices, respectively. We let $\mathbb{R}^n$ denote the set of all the n-dimensional vectors and let $\mathbb{R}^n_+$ denote the set of all the n-dimensional nonnegative vectors. We let $\mathbf{1}$ denote the all-ones vector and let $\boldsymbol{I}$ denote the identity matrix, with their dimensions being clear from the context. Given a positive integer n and a positive integer $s \leq n$, we let $[n] := \{1, 2, \cdots, n\}$, let $[s, n] := \{s, s+1, \cdots, n\}$, and let $\mathbb{I}_s$ be a vector with the first s elements as 1 and the rest as 0. For a vector $\boldsymbol{y} \in \mathbb{R}^n$, we let $y_i^{\downarrow}$ denote the i-th largest element of $\boldsymbol{y}$ for each $i \in [n]$, let $\mathrm{Diag}(\boldsymbol{y})$ denote a diagonal matrix whose diagonal entries consist of the vector $\boldsymbol{y}$, let $\sqrt{\boldsymbol{y}}$ denote a vector where each entry is the square root of that in $\boldsymbol{y}$. For a symmetric matrix $\boldsymbol{A}$, we let $\lambda_{\min}(\boldsymbol{X})$ and $\lambda_{\max}(\boldsymbol{X})$ denote the smallest and largest eigenvalues of $\boldsymbol{X}$, respectively, and let $\boldsymbol{\lambda}(\boldsymbol{X})$ denote the eigenvalue vector, with eigenvalues sorted in nonincreasing order, that is, $\lambda_i(\boldsymbol{X})$ is the i-th largest eigenvalue of $\boldsymbol{X}$ for each $i \in [n]$. Additional notation will be introduced later as needed.

2 Reformulating MESP as a Concave Integer Program

This section develops an equivalent concave integer program for MESP, based on the proposed factorization method of $\boldsymbol{C}$ in Remark 1 and the concave envelope technique.

2.1 A Naive Reformulation of MESP

This subsection presents an equivalent reformulation of MESP, but it is not a concave integer program. We begin by introducing the following matrix and function.

Definition 1. *For any t, $0 \leq t \leq \lambda_{\min}(\boldsymbol{C})$ and a vector $\boldsymbol{x} \in [0,1]^n$, we define $\mathcal{M}_t(\boldsymbol{x}) \in \mathcal{S}^n_+ := \sum_{i \in [n]} x_i \boldsymbol{a}_i(t) \left[\boldsymbol{a}_i(t)\right]^\top$, where $\boldsymbol{a}_i(t) \in \mathbb{R}^n$ is the i-th column of $\boldsymbol{A}(t)$ for each $i \in [n]$.*

Definition 2. *For a matrix $\boldsymbol{X} \in \mathcal{S}^n_+$ with the eigenvalues $\boldsymbol{\lambda}(\boldsymbol{X}) \in \mathbb{R}^n_+$, an integer $s \in [n]$, and a constant $t \geq 0$, we define a function $\Phi_s(\boldsymbol{X}; t) := \sum_{i \in [s]} \log\left(\lambda_i(\boldsymbol{X}) + t\right)$.*

Next, we rewrite MESP as the problem (2) using the function Φ_s. When $t = 0$, (2) reduces to a known reformulation derived by [25] (see also [10]). We extend this formulation to any t within the range $[0, \lambda_{\min}(\boldsymbol{C})]$. The most striking result derived from this extension is that as t increases, the Lagrangian dual bound of (2) becomes tighter, as demonstrated in Sect. 3.

Proposition 1. *For any t with $0 \leq t \leq \lambda_{\min}(\boldsymbol{C})$, MESP can be reduced to*

$$z^* = \max_{\boldsymbol{x} \in \{0,1\}^n} \left\{ \Phi_s\left(\mathcal{M}_t(\boldsymbol{x}); t\right) : \sum_{i \in [n]} x_i = s \right\}. \tag{2}$$

The objective function of (2) is not concave. This motivates us to concavify the function $\varPhi_s$ in the following subsection.

2.2 The Concave Envelope

This subsection presents the concave envelope of $\varPhi_s$, denoted by $\widehat{\varPhi}_s$, which allows us to reformulate (2) as a concave integer program. The *concave envelope* of a function is the pointwise infimum of all its concave over-estimators. First, let us introduce a function.

Definition 3. *For a vector $\boldsymbol{y} \in \mathbb{R}^n_+$ and an integer $s \in [n]$, suppose that k, $0 \leq k \leq s-1$, is an integer such that $y^{\downarrow}_k > \frac{1}{s-k}\sum_{i\in[k+1,n]} y^{\downarrow}_i \geq y^{\downarrow}_{k+1}$, with the convention $y^{\downarrow}_0 = \infty$. Then, we define $\psi_s(\boldsymbol{y}) = \sum_{i\in[k]} \log\left(y^{\downarrow}_i\right) + (s-k)\log\left(\frac{\sum_{i\in[k+1,d]} y^{\downarrow}_i}{s-k}\right)$.*

Note that the integer k above is unique, a technical result from Nikolov [27, lemma 14]. They also established that ψ_s is a concave function with its supergradient defined below.

Remark 2 ([27]). Let $\boldsymbol{y} \in \mathbb{R}^n_+$ be $y_1 \geq \cdots \geq y_n \geq 0$, and k follows from Definition 3. Then, $\boldsymbol{g} \in \mathbb{R}^n_+$ is a supergradient of the function ψ_s at $\boldsymbol{y}$, where

$$g_i = \frac{1}{y_i}, \quad \forall i \in [k], \quad g_i = \frac{s-k}{\sum_{i\in[k+1,n]} y_i}, \quad \forall i \in [k+1,n].$$

For $t = 0$, previous works have established that the concave envelope $\widehat{\varPhi}_s(\boldsymbol{X};0)$ equals $\psi_s(\boldsymbol{\lambda}(\boldsymbol{X}))$ for a matrix $\boldsymbol{X} \in \mathcal{S}^n_+$ (see [25,27]). They followed the approach of [18] that computed the *bi-conjugate* of a function to derive its concave envelope. However, applying their proof directly to a general t can be intricate, specifically due to the complexity of solving the two underlying nonlinear optimization problems. It is surprising that by using the eigen-decomposition technique and perturbing the eigenvalue vector, we can smoothly extend the established concave envelope result at $t=0$ to explicitly describe $\widehat{\varPhi}_s$ for any t, $0 \leq t \leq \lambda_{\min}(\boldsymbol{C})$.

Proposition 2. *For any t with $0 \leq t \leq \lambda_{\min}(\boldsymbol{C})$ and a matrix $\boldsymbol{X} \in \mathcal{S}^n_+$, the concave envelope of the function $\varPhi_s(\boldsymbol{X};t)$ is $\widehat{\varPhi}_s(\boldsymbol{X};t) = \psi_s(\boldsymbol{\lambda}(\boldsymbol{X}) + t\mathbb{I}_s)$.*

As a side product, Proposition 2 leads to an equivalent concave integer program for MESP.

Corollary 1. *For any t with $0 \leq t \leq \lambda_{\min}(\boldsymbol{C})$, MESP is equivalent to*

$$z^* = \max_{\boldsymbol{x}\in\{0,1\}^n} \left\{\widehat{\varPhi}_s\left(\mathcal{M}_t(\boldsymbol{x});t\right) : \sum_{i\in[n]} x_i = s\right\}. \qquad \text{(MESP-I)}$$

The continuous relaxation of MESP-I offers a practical upper bound- Aug-Fact. It is worth noting that Aug-Fact meets the Lagrangian dual bound of (2). This is because the concave envelope $\widehat{\Phi}_s$ is precisely the bi-conjugate of the objective function Φ_s of (2). By duality and concave conjugate theory, the dual of (2) and the continuous relaxation of MESP-I form a primal-dual pair.

3 The Augmented Factorization Bound and its Properties

Relaxing the binary variables in MESP-I leads to an upper bound:

$$z^* \leq \widehat{z}(t) := \max_{\boldsymbol{x} \in [0,1]^n} \left\{ \widehat{\Phi}_s \left(\mathcal{M}_t(\boldsymbol{x}); t \right) : \sum_{i \in [n]} x_i = s \right\}. \qquad \text{(Aug-Fact)}$$

For $t = 0$, Aug-Fact reduces to the known factorization bound (Fact), proposed by [27]:

$$z^* \leq \widehat{z}(0) := \max_{\boldsymbol{x} \in [0,1]^n} \left\{ \widehat{\Phi}_s \left(\mathcal{M}_0(\boldsymbol{x}); 0 \right) : \sum_{i \in [n]} x_i = s \right\}. \qquad \text{(Fact)}$$

In this section, we establish that Aug-Fact decreases monotonically with t, $0 \leq t \leq \lambda_{\min}(\boldsymbol{C})$ and is invariant under scaling. We also demonstrate that Aug-Fact is tighter than the two existing upper bounds- Fact and DDF-R for any $t > 0$ and quantify how much smaller Aug-Fact is compared to them at $t = \lambda_{\min}(\boldsymbol{C})$. Motivated by the strength of Aug-Fact, we propose a variable fixing logic from a primal perspective.

3.1 The Monotonicity of Aug-Fact and its Dominance over Fact

We begin by introducing *Schur-concave* and *strictly Schur-concave* functions.

Definition 4 ([13,33]). *A function $f : \mathbb{R}^n \to \mathbb{R}$ is Schur-concave if for all $\boldsymbol{\nu}, \boldsymbol{\mu} \in \mathbb{R}^n$ such that $\boldsymbol{\mu}$ majorizes $\boldsymbol{\nu}$ (denoted $\boldsymbol{\mu} \succ \boldsymbol{\nu}$), i.e.,*

$$\sum_{i \in [\ell]} \mu_i^{\downarrow} \geq \sum_{i \in [\ell]} \nu_{[i]}^{\downarrow}, \forall \ell \in [n-1], \quad \sum_{i \in [n]} \mu_i = \sum_{i \in [n]} \nu_i,$$

one has that $f(\boldsymbol{\mu}) \leq f(\boldsymbol{\nu})$. The function f is strictly Schur-concave if $f(\boldsymbol{\mu}) < f(\boldsymbol{\nu})$ holds for any $\boldsymbol{\nu}, \boldsymbol{\mu} \in \mathbb{R}^n$, such that $\boldsymbol{\mu} \succ \boldsymbol{\nu}$ but $\boldsymbol{\nu}$ is not a permutation of $\boldsymbol{\mu}$.

Note that every concave and symmetric function is Schur-concave (see [26]). We observe that the concave function ψ_s is permutation-invariant with the arguments. Therefore,

Observation 1 *The function ψ_s in Definition 3 is Schur-concave.*

For a matrix $\boldsymbol{X} \in \mathcal{S}_+^n$, by Proposition 2, the objective function of Aug-Fact is based on the perturbed eigenvalue vector $\boldsymbol{\lambda}(\boldsymbol{X}) + t\mathbb{I}_s$. The following lemma presents several technical results about the perturbed eigenvalue vector across different t.

Lemma 1. *Given a vector $\boldsymbol{x} \in [0,1]^n$ satisfying $\sum_{i\in[n]} x_i = s$, for all t_1, t_2 such that $0 \le t_1 \le t_2 \le \lambda_{\min}(\boldsymbol{C})$, suppose that $\boldsymbol{\nu}^{t_1}$ and $\boldsymbol{\mu}^{t_2}$ are the eigenvalues of $\mathcal{M}_{t_1}(\boldsymbol{x})$ and $\mathcal{M}_{t_2}(\boldsymbol{x})$, respectively, sorted in nonincreasing order. Then, the following hold:*

(i) $\boldsymbol{\mu}^{t_2} + t_2\mathbb{I}_s \succ \boldsymbol{\nu}^{t_1} + t_1\mathbb{I}_s$;
(ii) $\sum_{i\in[\ell]} \mu_i^{t_2} + (t_2 - t_1)\sum_{i\in[\ell]} x_i^{\downarrow} \ge \sum_{i\in[\ell]} \nu_i^{t_1}$ *for each* $\ell \in [s]$; *and*
(iii) $\mu_i^{t_2} + t_2 - t_1 \ge \nu_i^{t_1}$ *for each* $i \in [s]$.

Theorem 1. *Aug-Fact is monotonically decreasing with t, $0 \le t \le \lambda_{\min}(\boldsymbol{C})$. That is, for all t_1, t_2 such that $0 \le t_1 \le t_2 \le \lambda_{\min}(\boldsymbol{C})$, the inequality $\widehat{z}(t_1) \ge \widehat{z}(t_2)$ holds.*

By leveraging the Schur-concavity of the function ψ_s, a property not previously explored in the literature, Theorem 1 demonstrates the monotonicity of Aug-Fact over t.

As Fact is a special case of Aug-Fact at $t = 0$, a notable side product of Theorem 1 is that Aug-Fact dominates Fact whenever $t > 0$, as summarized below. In addition, Aug-Fact has a similar computational efficiency with Fact.

Theorem 2. *For any $0 < t \le \lambda_{\min}(\boldsymbol{C})$, Aug-Fact dominates Fact, i.e., $\widehat{z}(0) \ge \widehat{z}(t)$.*

We close this subsection by showing that the performance of the Aug-Fact bound is invariant under multiplication of $\boldsymbol{C}$ by a scale factor γ, a property that is not found in other concave-optimization bounds.

Proposition 3. *For any $0 \le t \le \lambda_{\min}(\boldsymbol{C})$, the difference between Aug-Fact and MESP, i.e., $\widehat{z}(t) - z^*$, is invariant under scaling of $\boldsymbol{C}$.*

3.2 Theoretical Guarantees for the Improvement of Aug-Fact over Fact

By leveraging Lemma 1 and the concavity of the function ψ_s, this subsection establishes a lower bound for the difference $\widehat{z}(0) - \widehat{z}(\lambda_{\min}(\boldsymbol{C}))$ and propose a sufficient condition where Aug-Fact strictly improves Fact at $t = \lambda_{\min}(\boldsymbol{C})$.

Theorem 3. *Suppose that $\boldsymbol{x}^*$ is an optimal solution of Aug-Fact at $t = \lambda_{\min}(\boldsymbol{C})$ and $\boldsymbol{\beta}^*$ is the vector of eigenvalues of $\mathcal{M}_0(\boldsymbol{x}^*)$ in nonincreasing order.*

(i) Let $(x^*)_0^{\downarrow} = 0$ by default. We have that

$$\widehat{z}(0) - \widehat{z}(\lambda_{\min}(\boldsymbol{C})) \ge \Delta^{lb} \ge 0,$$

where $\Delta^{lb} := \lambda_{\min}(\boldsymbol{C})\left(k - \sum_{i\in[k]}(x^*)_i^{\downarrow}\right)\left(\frac{s-k}{\sum_{i\in[k+1,n]}\beta_i^*} - \frac{1}{\beta_k^*}\right)$; *and*

(ii) Aug-Fact with $t = \lambda_{\min}(\boldsymbol{C})$ *strictly dominates Fact if* $k \geq 1$ *and* $(x^*)^{\downarrow}_k < 1$, *where* $0 \leq k \leq s-1$ *is an integer, such that* $\beta^*_k > \frac{1}{s-k}\sum_{i\in[k+1,n]}\beta^*_i \geq \beta^*_{k+1}$.

Theorem 3 provides important insights into how the condition number of $\boldsymbol{C}$ affects the performance of Aug-Fact at $t = \lambda_{\min}(\boldsymbol{C})$. As seen in Theorem 3, the lower bound Δ^{lb} is an increasing function of $\lambda_{\min}(\boldsymbol{C})$. Thus, a larger $\lambda_{\min}(\boldsymbol{C})$ is desired to guarantee a greater improvement. Besides, the lower bound is determined by the difference between the reciprocals of the eigenvalues of $\mathcal{M}_0(\boldsymbol{x}^*)$, specifically $(s-k)/(\sum_{i\in[k+1,n]}\beta^*_i) - 1/\beta^*_k$. The difference generally decreases as we scale up all the eigenvalues $\boldsymbol{\beta}^*$. It is, therefore, likely that a negative relationship between Δ^{lb} and $\lambda_{\max}(\boldsymbol{C})$ exists, given that the eigenvalues of $\mathcal{M}_0(\boldsymbol{x}^*)$ are bounded by $\lambda_{\max}(\boldsymbol{C})$ according to Remark 1. Then, a possible implication is that Aug-Fact is more effective at improving Fact at $t = \lambda_{\min}(\boldsymbol{C})$ when the condition number of $\boldsymbol{C}$, denoted $\lambda_{\max}(\boldsymbol{C})/\lambda_{\min}(\boldsymbol{C})$, is smaller. Our numerical results further support this hypothesis.

A tighter concave relaxation is often beneficial to enhance the theoretical guarantees of approximation algorithms. By leveraging Fact, [25] derived the best-known and first-known approximation bounds when applying the randomized sampling and local search algorithms to MESP, respectively. Their randomized sampling algorithm samples a size-s subset $S \subset [n]$ with a pre-specified probability that is based on an optimal solution of Fact. In addition, the local search algorithm swaps an element from the chosen set S with an element in the complement $[n] \setminus S$ and check if such a movement increases the objective value or not at each iteration- if YES, then accept the swap, and otherwise continue. We show that Aug-Fact with $t = \lambda_{\min}(\boldsymbol{C})$ allows us to enhance those approximation bounds of [25] by Δ^{lb}. Analogously, a strict improvement occurs when the condition in Part (ii) of Theorem 3 is satisfied.

Corollary 2. *The randomized sampling [25, algorithm 2] returns a* $(s\log(s/n) + \log(\binom{n}{s}) - \Delta^{lb})$*-approximation bound for MESP. The local search algorithm returns a* $(s\min\{\log(s), \log(n-s-n/s+2)\} - \Delta^{lb})$*-approximation bound for MESP.*

3.3 Theoretical Guarantees for the Improvement of Aug-Fact over DDF-R

This subsection generalizes the existing upper bound- DDF-R for MESP and demonstrates that Aug-Fact produces a tighter upper bound than DDF-R.

By setting $t = \lambda_{\min}(\boldsymbol{C})$ in (1), [24] transformed MESP into the form of the D-optimality data fusion (DDF) problem. We begin with a slight generalization of Li et al. [24, theorem 1] to any t, $0 < t \leq \lambda_{\min}(\boldsymbol{C})$.

Corollary 3. *For any* t *with* $0 < t \leq \lambda_{\min}(\boldsymbol{C})$, *MESP reduces to*

$$z^* = \max_{\boldsymbol{x}\in\{0,1\}^n}\left\{\log\det\left(\mathcal{M}_t(\boldsymbol{x}) + t\boldsymbol{I}\right) : \sum_{i\in[n]} x_i = s\right\} - (n-s)\log(t). \qquad (3)$$

Corollary 3 immediately provides a concave integer program for MESP, and it falls into the DDF framework. A concave relaxation bound "DDF-R" can be naturally obtained from relaxing the binary variables $\boldsymbol{x}$ of (3) to be continuous.

$$z^* \leq \widehat{z}^D(t) := \max_{\boldsymbol{x} \in [0,1]^n} \left\{ \log\det\left(\mathcal{M}_t(\boldsymbol{x}) + t\boldsymbol{I}\right) : \sum_{i \in [n]} x_i = s \right\} - (n-s)\log(t). \tag{DDF-R}$$

Note that for $t = 0$, DDF-R approaches negative infinity due to the rank deficiency of the objective matrix. Therefore, the condition $0 < t \leq \lambda_{\min}(\boldsymbol{C})$ must be satisfied. Our proposed Aug-Fact bound outperforms DDF-R, as shown below.

Theorem 4. *For any t with $0 < t \leq \lambda_{\min}(\boldsymbol{C})$, the following hold:*

(i) Aug-Fact dominates DDF-R;
(ii) Aug-Fact strictly dominates DDF-R if s is strictly less than the rank of $\boldsymbol{C} - t\boldsymbol{I}$ and Aug-Fact is not an exact concave relaxation of MESP, i.e., $\widehat{z}(t) > z^*$; and
(iii) Aug-Fact meets DDF-R if the integer s is no less than the rank of $\boldsymbol{C} - t\boldsymbol{I}$.

It is worth mentioning that both conditions in Part (ii) of Theorem 4 can be readily satisfied. That is, Aug-Fact strictly dominates DDF-R in most cases. As Aug-Fact is stronger than Fact for any t with $0 < t \leq \lambda_{\min}(\boldsymbol{C})$, Part (iii) of Theorem 4 results in a sufficient condition under which DDF-R dominates Fact.

Corollary 4. *Suppose that the integer s is no less than the rank of $\boldsymbol{C} - t\boldsymbol{I}$. Then, DDF-R dominates Fact for any t, $0 < t \leq \lambda_{\min}(\boldsymbol{C})$.*

Analogous to Aug-Fact, we show that DDF-R decreases monotonically as t increases by leveraging the theory of Schur-concavity. This indicates that setting $t = \lambda_{\min}(\boldsymbol{C})$ yields the best DDF-R bound, which is exactly the one proposed by [24].

Proposition 4. *DDF-R is monotonically decreasing with t, $0 < t \leq \lambda_{\min}(\boldsymbol{C})$. That is, for all t_1, t_2 such that $0 < t_1 \leq t_2 \leq \lambda_{\min}(\boldsymbol{C})$, the inequality $\widehat{z}^D(t_1) \geq \widehat{z}^D(t_2)$ holds.*

Next, we derive a theoretical bound for the difference in optimal values between Aug-Fact and DDF-R with $t = \lambda_{\min}(\boldsymbol{C})$.

Theorem 5. *Suppose that $\boldsymbol{x}^*$ is an optimal solution of Aug-Fact $t = \lambda_{\min}(\boldsymbol{C})$ and the vector $\boldsymbol{\lambda}^* \in \mathbb{R}_+^n$ contains the eigenvalues of $\mathcal{M}_{\lambda_{\min}(\boldsymbol{C})}(\boldsymbol{x}^*)$ in nonincreasing order. Then, we have that*

$$\widehat{z}^D(\lambda_{\min}(\boldsymbol{C})) - \widehat{z}(\lambda_{\min}(\boldsymbol{C})) \geq \Theta^{lb} \geq 0,$$

where $\Theta^{lb} := \left(\frac{1}{\lambda_{s+1}^* + \lambda_{\min}(\boldsymbol{C})} - \frac{1}{\lambda_s^* + \lambda_{\min}(\boldsymbol{C})} \right) \sum_{i \in [s+1,n]} \lambda_i^*$.

We close this subsection by discussing how the lower bound Θ^{lb} in Theorem 5 varies with the condition number of $\boldsymbol{C}$. Following Theorem 3, we note that Aug-Fact with $t = \lambda_{\min}(\boldsymbol{C})$ may be more effective at improving Fact when the condition number of $\boldsymbol{C}$ is small. Conversely, Theorem 5 suggests that the improvement of Aug-Fact over DDF-R becomes notable given a large condition number, as detailed below. Our numerical studies also demonstrate that Aug-Fact is significantly tighter than DDF-R when its improvement over Fact is minor, and vice versa.

As seen in Theorem 5, the lower bound Θ^{lb} decreases as $\lambda_{\min}(\boldsymbol{C})$ increases. Besides, the bound Θ^{lb} generally increases as we scale up all the eigenvalues $\boldsymbol{\lambda}^*$ of $\mathcal{M}_t(\boldsymbol{x}^*)$. Note that these eigenvalues are bounded by $\lambda_{\max}(\boldsymbol{C})$ based on Remark 1. It is possible, therefore, that there exists a positive relationship between Θ^{lb} and $\lambda_{\max}(\boldsymbol{C})/\lambda_{\min}(\boldsymbol{C})$. Thus, a large condition number of $\boldsymbol{C}$ is desirable for achieving a notable improvement of Aug-Fact over DDF-R at $t = \lambda_{\min}(\boldsymbol{C})$, as shown in Sect. 4.

3.4 A Primal Certificate for Variable Fixing Using Aug-Fact

Variable fixing has been extensively studied for MESP in the context of various concave relaxation bounds (see, e.g., [1,2,4,10] and references therein). It is often used to accelerate the computation of exact solution methods [24]. However, previous research has focused on deriving dual certificates, which requires computing (near-)optimal dual solutions of those concave relaxations. By contrast, this subsection introduces a primal certificate for variable fixing using the property of concave functions, independent of the dual problem of Aug-Fact.

To begin, we need an expression for the supergradient of the objective function of Aug-Fact. Note that $\widehat{\Phi}_s(\mathcal{M}_t(\boldsymbol{x}); t)$ is a spectral function that only depends on the eigenvalues of $\mathcal{M}_t(\boldsymbol{x})$. Based on the spectral property and the supergradient of ψ_s in Remark 2, [25] derived the supergradient of $\widehat{\Phi}_s(\mathcal{M}_t(\boldsymbol{x}); t)$ over $\boldsymbol{x}$ at $t = 0$ (see also [10]). Their result can directly extend to any t, $0 \le t \le \lambda_{\min}(\boldsymbol{C})$.

Remark 3. For any feasible solution $\boldsymbol{x}$ of Aug-Fact and any t, $0 \le t \le \lambda_{\min}(\boldsymbol{C})$, suppose that $\mathcal{M}_t(\boldsymbol{x}) = \boldsymbol{Q}\,\mathrm{Diag}(\boldsymbol{\lambda})\boldsymbol{Q}^\top$ is the eigen-decomposition of $\mathcal{M}_t(\boldsymbol{x})$, where $\lambda_1 \ge \cdots \ge \lambda_n \ge 0$ are the eigenvalues of $\mathcal{M}_t(\boldsymbol{x})$. Then, a supergradient of $\widehat{\Phi}_s(\mathcal{M}_t(\boldsymbol{x}); t)$ at $\boldsymbol{x}$ can be defined as

$$\frac{\partial \widehat{\Phi}_s\left(\mathcal{M}_t(\boldsymbol{x}); t\right)}{\partial x_i} = [\boldsymbol{a}_i(t)]^\top \boldsymbol{Q}\,\mathrm{Diag}(\boldsymbol{g})\boldsymbol{Q}^\top \boldsymbol{a}_i(t), \quad \forall i \in [n],$$

where $\boldsymbol{g}$ is a supergradient of the function ψ_s at $\boldsymbol{\lambda}$, as defined in Remark 2.

Theorem 6. *For any feasible solution $\tilde{\boldsymbol{x}}$ of Aug-Fact at $t = \lambda_{\min}(\boldsymbol{C})$, let $\tilde{\boldsymbol{g}}$ be a supergradient of the function $\widehat{\Phi}_s$ at $\tilde{\boldsymbol{x}}$, as defined in Remark 3. Then, any optimal solution $\boldsymbol{x}^*$ of MESP must satisfy*

$$x_i^* = 1 \;\; \text{if} \;\; \tilde{g}_i - \tilde{g}_{s+1}^{\downarrow} > UB - LB, \quad \forall i \in [n], \;\; \text{and}$$
$$x_i^* = 0 \;\; \text{if} \;\; \tilde{g}_s^{\downarrow} - \tilde{g}_i > UB - LB, \quad \forall i \in [n],$$

where $UB = \widehat{\Phi}_s\left(\mathcal{M}_{\lambda_{\min}(\boldsymbol{C})}(\tilde{\boldsymbol{x}}); \lambda_{\min}(\boldsymbol{C})\right) - \tilde{\boldsymbol{g}}^\top \tilde{\boldsymbol{x}} + \sum_{i \in [s]} \tilde{g}_i^{\downarrow}$ *and LB is is the objective value of a feasible solution of MESP.*

We make the following remarks about Theorem 6.

(i) The main advantage of our dual-free variable fixing in Theorem 6 is its ease of implementation- it can be easily integrated into any first-order algorithm for solving Aug-Fact. At each iteration, to fix variables, it suffices to sort the elements of the supergradient at the current solution;
(ii) Our variable fixing conditions in Theorem 6 align well with the cardinality constraint in MESP. Given that UB – LB ≥ 0 and the supergradient vector has at most s entries larger than its $s+1$ largest entry, we can fix up to s variables to 1. Likewise, we can fix at most $n-s$ variables to 0; and
(iii) Our theoretical analysis of Theorem 6 can be directly generalized to other upper bounds based on concave relaxations.

4 Numerical Experiments

In this section, we numerically compare Aug-Fact with the existing upper bounds of MESP and verify its dominance over Fact and DDF-R with varying-scale instances. As both Aug-Fact and DDF-R are decreasing with t, we set $t = \lambda_{\min}(\boldsymbol{C})$ for them throughout this section. Besides, we use the Frank-Wolfe algorithm to compute the upper bounds. To obtain a high-quality lower bound of MESP, we employ the local search algorithm proposed by [25]. All the experiments are conducted in Python 3.6 with calls to Gurobi 9.5.2 and MOSEK 10.0.29 on a PC with 10-core CPU, 16-core GPU, and 16GB of memory.

To evaluate Aug-Fact, we first consider three benchmark covariance matrices with $n = 63, 90, 124$ that have been repeatedly used in the literature on MESP [2,16]. Their corresponding condition numbers are 48.42, 200.45, and 78340.48, respectively. Figures 1 to 3 display the gaps between several upper bounds and a lower bound generated by the local search algorithm. We note that the gap values for Fact, Linx, and Mix-LF are taken from the computational results of [10]. For each benchmark instance, gap values are given for $s \in [2, n-1]$. Their computational time is negligible (i.e., less than one minute), so we do not report and compare them.

Figures 1(a) and 2(a) show that Aug-Fact gives the best upper bound for MESP on the $n = 63$ and $n = 90$ data sets. Surprisingly, our Aug-Fact reduces the integrality gaps effectively for the most difficult instances, with intermediate values of s, where Fact and Linx are nearly identical. Consistent with our analysis of Theorem 3, Aug-Fact is only a bit better than Fact for $n = 124$, as displayed in Fig. 3(a). This is because the $n = 124$ covariance matrix has a huge condition number. We present the comparison between Aug-Fact and DDF-R in different figures, since DDF-R often results in much worse integrality gaps. We see from Figs. 1(b), 2(b) and 3(b) that Aug-Fact is much tighter than DDF-R for all the test cases. As s approaches $n-1$, the gaps produced by Aug-Fact and DDF-R become nearly identical. It is interesting to observe that the difference of gaps between Aug-Fact and DDF-R is increasing with the condition number of $\boldsymbol{C}$. Especially for $n = 124$, we observe a significant reduction in the gaps in Fig. 3(b). These comparison results parallel our theoretical findings in Subsect. 3.3.

We verify the enhanced capacity of Aug-Fact to fix variables for MESP in Figs. 1(c) 2(c) and 3(c), when compared to Aug-Fact and DDF-R. Note that we employ the primal conditions in Theorem 6 to check whether to fix a variable. For all other bounds, their fixed variables are sourced from Chen et al. [10, section 3], using the dual certificates. For $n = 64$ and $n = 90$, we see that Aug-Fact fixes many variables for large values of s, while Fact fails to fix any variables at all. For $n = 124$, Aug-Fact still leads to more variables fixed than Fact, even at points in which they have very similar gaps. In addition, we observe a slightly different comparison result between Aug-Fact and Linx for $n = 90$ and $n = 124$. Specifically, for small values of s, Aug-Fact has a smaller integrality gap and a stronger fixing power than the Linx bound, whereas the reverse holds when s is large.

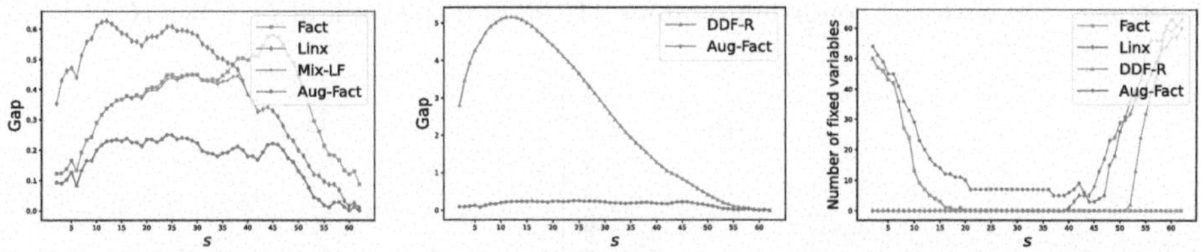

(a) Gaps (b) Gaps (c) Number of variables fixed

Fig. 1. $n = 63$ with the condition number 48.42

Finally, we present Figs. 4 and 5 using synthetic data to illustrate cases where DDF-R becomes a strong bound. Based on our theoretical results in Theorem 4 and Corollary 4, DDF-R meets Aug-Fact and dominates Fact when the integer s is at least the rank of $\boldsymbol{C} - t\boldsymbol{I}$. To enforce a fixed rank r of $\boldsymbol{C} - t\boldsymbol{I}$, we construct the matrix $\boldsymbol{C}$ as $\boldsymbol{C} = \rho \boldsymbol{I} + \boldsymbol{U}\boldsymbol{U}^\top$, where $\rho > 0$ is a constant and $\boldsymbol{U} \in \mathbb{R}^{n \times r}$ consists of i.i.d. standard normal entries. In Figs. 4 and 5, we set $n = 100$, $r = 30$ with

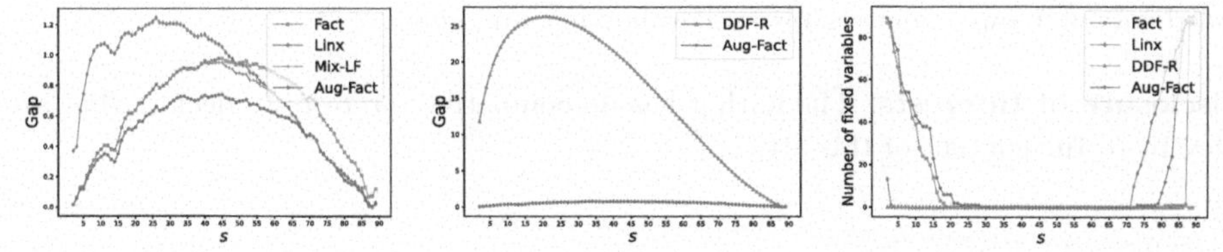

(a) Gaps (b) Gaps (c) Number of variables fixed

Fig. 2. $n = 90$ with with the condition number 200.45

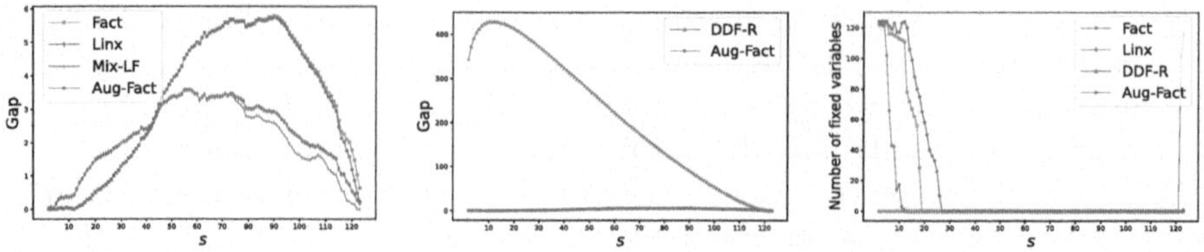

(a) Gaps (b) Gaps (c) Number of variables fixed

Fig. 3. $n = 124$ with the condition number 78340.48

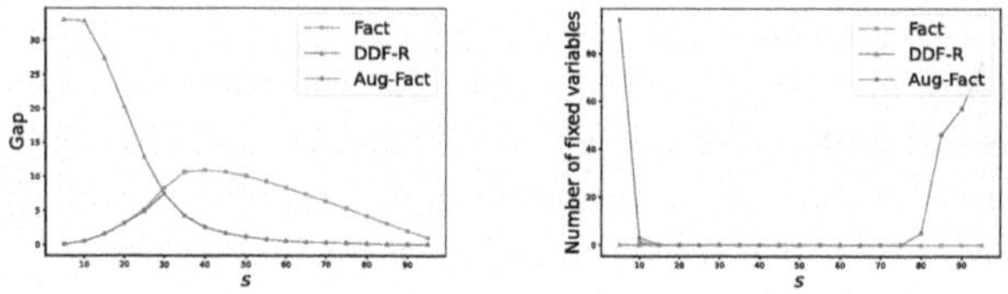

(a) Gaps (b) Number of variables fixed

Fig. 4. $n = 100$ with the condition number 241.96 and the rank of $\boldsymbol{C} - \lambda_{\min}(\boldsymbol{C})\boldsymbol{I}$ is 30

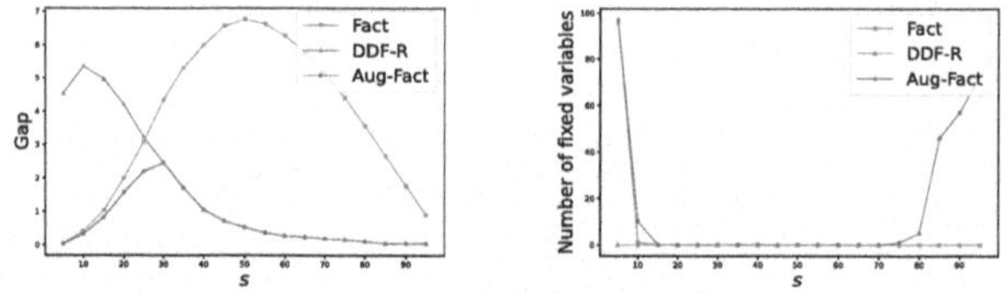

(a) Gaps (b) Number of variables fixed

Fig. 5. $n = 100$ with the condition number 25.10 and the rank of $\boldsymbol{C} - \lambda_{\min}(\boldsymbol{C})\boldsymbol{I}$ is 30

$\rho = 1$ and $\rho = 10$, respectively, which leads to different condition numbers for $\boldsymbol{C}$. As the conditional number decreases in Fig. 5(a), DDF-R can even dominate Fact in some cases where $s < r$. In addition, the comparison between Figs. 4(a) and 5(a) reinforces our analysis of how the condition number of $\boldsymbol{C}$ affects the improvement of Aug-Fact over Fact and DDF-R.

Acknowledgements. The author thanks Gabriel Ponte for his valuable suggestions, and the anonymous reviewers for their valuable comments.

Disclosure of Interests. The author has no competing interests to declare that are relevant to the content of this article

References

1. Anstreicher, K.M.: Maximum-entropy sampling and the boolean quadric polytope. J. Global Optim. **72**, 603–618 (2018)
2. Anstreicher, K.M.: Efficient solution of maximum-entropy sampling problems. Oper. Res. **68**(6), 1826–1835 (2020)
3. Anstreicher, K.M., Fampa, M., Lee, J., Williams, J.: Continuous relaxations for Constrained Maximum-Entropy Sampling. In: Cunningham, W.H., McCormick, S.T., Queyranne, M. (eds.) IPCO 1996. LNCS, vol. 1084, pp. 234–248. Springer, Heidelberg (1996). https://doi.org/10.1007/3-540-61310-2_18
4. Anstreicher, K.M., Fampa, M., Lee, J., Williams, J.: Maximum-entropy remote sampling. Discret. Appl. Math. **108**(3), 211–226 (2001)
5. Anstreicher, K.M., Fampa, M., Lee, J., Williams, J., et al.: Using continuous nonlinear relaxations to solve constrained maximum-entropy sampling problems. Math. Program. **85**(2), 221–240 (1999)
6. Anstreicher, K.M., Lee, J.: A masked spectral bound for maximum-entropy sampling. In: mODa 7—Advances in Model-Oriented Design and Analysis: Proceedings of the 7 th International Workshop on Model-Oriented Design and Analysis held in Heeze, The Netherlands, June 14–18, 2004. pp. 1–12. Springer (2004)
7. Burer, S., Lee, J.: Solving maximum-entropy sampling problems using factored masks. Math. Program. **109**(2), 263–281 (2007)
8. Caselton, W.F., Zidek, J.V.: Optimal monitoring network designs. Stat. Probab. Lett. **2**(4), 223–227 (1984)
9. Chen, Z., Fampa, M., Lambert, A., Lee, J.: Mixing convex-optimization bounds for maximum-entropy sampling. Math. Program. **188**(2), 539–568 (2021). https://doi.org/10.1007/s10107-020-01588-w
10. Chen, Z., Fampa, M., Lee, J.: On computing with some convex relaxations for the maximum-entropy sampling problem. INFORMS J. Comput. **35**(2), 368–385 (2023)
11. Chen, Z., Fampa, M., Lee, J.: Masking anstreicher's linx bound for improved entropy bounds. Oper. Res. **72**(2), 591–603 (2024)
12. Civril, A., Magdon-Ismail, M.: Exponential inapproximability of selecting a maximum volume sub-matrix. Algorithmica **65**, 159–176 (2013)
13. Constantine, G.M.: Schur convex functions on the spectra of graphs. Discret. Math. **45**(2–3), 181–188 (1983)
14. Fampa, M., Lee, J.: Maximum-entropy sampling: algorithms and application. Springer Nature (2022)
15. Fampa, M., Lee, J.: An outer-approximation algorithm for maximum-entropy sampling. Discret. Appl. Math. **347**, 271–284 (2024)
16. Guttorp, P., Le, N.D., Sampson, P.D., Zidek, J.V.: Using entropy in the redesign of an environmental monitoring network. Multivar. Environ. Stat. **6**, 175–202 (1993)
17. Hendrych, D., Besançon, M., Pokutta, S.: Solving the optimal experiment design problem with mixed-integer convex methods. arXiv preprint arXiv:2312.11200 (2023)
18. Hiriart-Urruty, J.B., Lemaréchal, C.: Convex analysis and minimization algorithms ii: Advanced theory and bundle methods. Grundlehren der mathematischen Wissenschaften (1993)
19. Hoffman, A., Lee, J., Williams, J.: New upper bounds for maximum-entropy sampling. In: mODa 6—Advances in Model-Oriented Design and Analysis: Proceedings of the 6th International Workshop on Model-Oriented Design and Analysis held in Puchberg/Schneeberg, June 25–29, 2001. pp. 143–153. Springer (2001)

20. Ko, C.W., Lee, J., Queyranne, M.: An exact algorithm for maximum entropy sampling. Oper. Res. **43**(4), 684–691 (1995)
21. Lee, J., Williams, J.: A linear integer programming bound for maximum-entropy sampling. Math. Program. **94**, 247–256 (2003)
22. Li, Q., Negi, R., Ilić, M.D.: Phasor measurement units placement for power system state estimation: a greedy approach. In: 2011 IEEE Power and Energy Society General Meeting, pp. 1–8. IEEE (2011)
23. Li, Y.: The augmented factorization bound for maximum-entropy sampling. arXiv preprint arXiv:2410.10078 (2024)
24. Li, Y., Fampa, M., Lee, J., Qiu, F., Xie, W., Yao, R.: D-optimal data fusion: exact and approximation algorithms. INFORMS J. Comput. **36**(1), 97–120 (2024)
25. Li, Y., Xie, W.: Best principal submatrix selection for the maximum entropy sampling problem: scalable algorithms and performance guarantees. Oper. Res. **72**(2), 493–513 (2024)
26. Marshall, A.W., Olkin, I., Arnold, B.C.: Inequalities: theory of majorization and its applications (1979)
27. Nikolov, A.: Randomized rounding for the largest simplex problem. In: Proceedings of the Forty-Seventh Annual ACM Symposium on Theory of Computing, pp. 861–870 (2015)
28. Ozkul, S., Harmancioglu, N.B., Singh, V.P.: Entropy-based assessment of water quality monitoring networks. J. Hydrol. Eng. **5**(1), 90–100 (2000)
29. Ponte, G., Fampa, M., Lee, J.: Branch-and-bound for integer d-optimality with fast local search and variable-bound tightening. arXiv preprint arXiv:2309.00117 (2023)
30. Ponte, G., Fampa, M., Lee, J.: Convex relaxation for the generalized maximum-entropy sampling problem. arXiv preprint arXiv:2404.01390 (2024)
31. Sebastiani, P., Wynn, H.P.: Maximum entropy sampling and optimal bayesian experimental design. J. R. Stat. Soc. Series B (Stat. Methodol.) **62**(1), 145–157 (2000)
32. Shewry, M.C., Wynn, H.P.: Maximum entropy sampling. J. Appl. Stat. **14**(2), 165–170 (1987)
33. law St, C.: An effective characterization of schur-convex functions with applications. J. Convex Anal. **14**(1), 103–108 (2007)

Integer and Unsplittable Multiflows in Series-Parallel Digraphs

Mohammed Majthoub Almoghrabi[1(✉)], Martin Skutella[1], and Philipp Warode[2]

[1] Technische Universität Berlin, Berlin, Germany
{majthoub,skutella}@math.tu-berlin.de
[2] Humboldt-Universität zu Berlin, Berlin, Germany
philipp.warode@hu-berlin.de
http://coga.tu-berlin.de , https://hu.berlin/warode

Abstract. An unsplittable multiflow routes the demand of each commodity along a single path from its source to its sink node. As our main result, we prove that in series-parallel digraphs, any given multiflow can be expressed as a convex combination of unsplittable multiflows, where the total flow on any arc deviates from the given flow by less than the maximum demand of any commodity. This result confirms a 25-year-old conjecture by Goemans for single-source unsplittable flows, as well as a stronger recent conjecture by Morell and Skutella, for series-parallel digraphs—even for general multiflow instances where commodities have distinct source and sink nodes. Previously, no non-trivial class of digraphs was known for which either conjecture holds. En route to proving this result, we also establish strong integrality results for multiflows on series-parallel digraphs, showing that their computation can be reduced to a simple single-commodity network flow problem.

1 Introduction

Network flow theory focuses on the optimal routing of commodities through networks and is fundamental to a wide array of applications in fields such as logistics, telecommunications, and computer networking; see, e.g., [1,19]. In many scenarios, splitting a commodity might degrade the quality of service, increase complexity, or is simply infeasible. For example, in optical networks, dividing data streams across multiple paths requires specialized equipment. Similarly, in freight logistics, splitting loads across multiple routes is often impractical.

This motivates *unsplittable* multiflows. Here, for a given digraph with k source-sink pairs, the given demand d_i of every commodity $i \in [k] := \{1, \dots, k\}$ must be sent along a single path P_i from its source s_i to its sink t_i. To differentiate unsplittable flows from traditional flows, where the demand of each commodity may be divided and sent along multiple s_i-t_i-paths, we refer to the latter as *fractional* flows.

Unsplittable flow problems represent a compelling extension of disjoint path problems and have received considerable attention in the literature. Even in the

N. Megow and A. Basu (Eds.): IPCO 2025, LNCS 15620, pp. 427–441, 2025.
https://doi.org/10.1007/978-3-031-93112-3_31

apparently straightforward case of a (series-parallel) digraph consisting of one common source and one common sink connected by parallel arcs, determining the existence of a *feasible* unsplittable multiflow (i.e., one that adheres to given arc capacities) is NP-complete. In fact, several classical problems in combinatorial optimization, such as, e.g., Bin Packing, Partition, or parallel machine scheduling with makespan objective occur as special cases; for more details we refer to Kleinberg's PhD thesis [4].

Integer multiflows. When every commodity $i \in [k]$ has unit demand $d_i := 1$, unsplittable multiflows precisely correspond to integer multiflows, which have been extensively studied in the literature. In fact, in unit-capacity networks, feasible integer multiflows model arc-disjoint paths, and therefore cannot be computed efficiently, unless P=NP. For example, Vygen [18] shows that the arc-disjoint paths problems is NP-complete even in acyclic and planar digraphs; see [14, Chapter 70.13b+c] for further related complexity results. On the other hand, single-commodity network flows (i.e., $k = 1$) are well-known to feature strong integrality properties, and there are efficient combinatorial algorithms to compute integer single-commodity flows; see, e.g., [1,19]. In particular, the network flow polytope is integral, that is, any single-commodity flow $(x_e)_{e \in E}$ on a digraph $G = (V, E)$ with integer excess at every node, is a convex combination of integer flows $(y_e)_{e \in E}$ that maintain the same excesses, such that

$$\lfloor x_e \rfloor \le y_e \le \lceil x_e \rceil \qquad \text{for every arc } e \in E. \tag{1}$$

Arguably, the most important tool for proving the existence of integer multiflows in special graph classes is the Nagamochi-Ibaraki Theorem [10] (see also [14, Theorem 70.8]), which relies on the sufficiency of the *cut condition.* This (necessary) condition requires that the capacity of every cut exceeds the total demand of commodities that must cross it; see (4) below. The concept of cut-sufficiency is of interest in its own right and is an active research topic; see, e.g., the recent work of Poremba and Shepherd [12] and references therein. For a fixed digraph, the Nagamochi-Ibaraki Theorem asserts that if, for any integer arc capacities and demands, the cut condition guarantees the existence of a feasible multiflow, it also guarantees the existence of a feasible *integer* multiflow. For most classes of digraphs in which the existence of a feasible multiflow implies the existence of a feasible integer multiflow, the cut condition is indeed sufficient. However, the multiflow instance in Fig. 1 illustrates that the cut condition is generally not sufficient for the class of series-parallel digraphs studied in this paper.

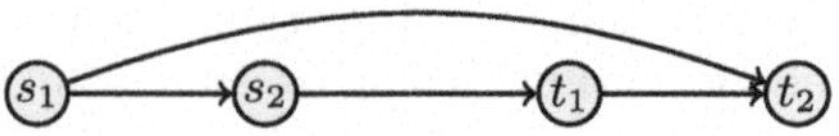

Fig. 1. An infeasible multiflow instance with two unit-demand commodities on a series-parallel digraph with unit arc capacities satisfying the cut condition

Single-source unsplittable flows. Even though our results hold for general multiflow instances in series-parallel digraphs, where each commodity is routed from a distinct source to a distinct sink, our main result is primarily inspired by prior research on single-source unsplittable flows, which we review next. For a comprehensive overview of results on general unsplittable multiflows, we refer to Kolliopoulos' survey [5].

Single-source unsplittable flows, where all commodities share a common source node s, have first been studied by Kleinberg [4]. Dinitz, Garg, and Goemans [3] prove that a given fractional flow $(x_e)_{e \in E}$ can always be turned into an unsplittable flow $(y_e)_{e \in E}$ (given by s-t_i-paths P_i with $y_e := \sum_{i:e \in P_i} d_i$) such that

$$y_e \leq x_e + d_{\max} \qquad \text{for every arc } e \in E, \tag{2}$$

where $d_{\max} := \max_{i \in [k]} d_i$. A famous conjecture of Goemans says that flow $(x_e)_{e \in E}$ can even be expressed as a convex combination of unsplittable flows $(y_e)_{e \in E}$ that satisfy (2).[1] Skutella [15], based on a flow augmentation method also used by Kolliopoulos and Stein [6], proves that Goemans' Conjecture is valid when the demands of the commodities are multiples of one another. For acyclic digraphs, Morell and Skutella [9] show that any fractional flow $(x_e)_{e \in E}$ can be turned into an unsplittable flow $(y_e)_{e \in E}$ that satisfies the lower bound

$$y_e \geq x_e - d_{\max} \qquad \text{for every arc } e \in E. \tag{3}$$

Furthermore, they conjecture the existence of an unsplittable flow that satisfies both the upper bounds (2) and the lower bounds (3). Only recently, for the special case of acyclic and planar digraphs, this conjecture has been proven by Traub, Vargas Koch, and Zenklusen [16]. Morell and Skutella also propose the following strengthening to Goemans' conjecture.

Conjecture 1 (Morell and Skutella [9]*).* For the single-source unsplittable flow problem on acyclic digraphs, any fractional flow $(x_e)_{e \in E}$ can be expressed as a convex combination of unsplittable flows $(y_e)_{e \in E}$ that satisfy both (2) and (3).

Using techniques of Martens, Salazar, and Skutella [8], they show that their Conjecture 1 is valid when the demands of the commodities are multiples of one another. Moreover, for the special case of acyclic and planar digraphs, and under the slightly relaxed lower and upper bounds $x_e - 2d_{\max} \leq y_e \leq x_e + 2d_{\max}$ for $e \in E$, Conjecture 1 is proved by Traub, Vargas Koch, and Zenklusen [16], based on a reduction to a well-structured discrepancy problem.

Contribution and outline After discussing preliminaries and notation for multiflows on series-parallel digraphs[2] in Sect. 2, in Sect. 3 we present the following observations on (integer) multiflows in series-parallel digraphs.

[1] Goemans' original conjecture asserts that, for any arc costs, an unsplittable flow can be found that satisfies (2) and whose cost does not exceed that of $(x_e)_{e \in E}$. This is equivalent to the stated existence of a convex combination; see, e.g., [8].

[2] We emphasize that we only consider series-parallel *directed* graphs in this paper and, thus, do not contribute to the line of research on multiflows in (series-parallel) undirected graphs; see, e.g., [2].

Theorem 1. *Consider a multiflow instance on a series-parallel digraph.*

(a) For integer demands, the total arc flows $(x_e)_{e \in E}$ *of any multiflow are a convex combination of total arc flows* $(y_e)_{e \in E}$ *of integer multiflows satisfying (1).*
(b) For (integer) arc capacities, a feasible (integer) multiflow, if one exists, can be efficiently found through a single-commodity flow computation.

Remark 1. Notice that Theorem 1(a) is only concerned with the total arc flows of multiflows instead of the multiflows itself. In fact, as stated in the theorem, every *total flow* of a multiflow can be expressed as a convex combination of *total flows* of integer multiflows satisfying (1). However, this is in general not true for the multiflow vector, i.e., the vector containing all commodity flows. In Fig. 2 we give an example of a multiflow that cannot be expressed as a convex combination of integer multiflows, while its total flow can.

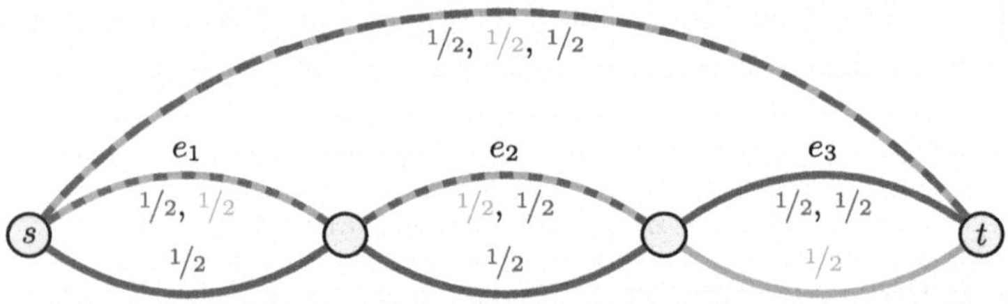

Fig. 2. An instance of a multiflow that cannot be expressed as the convex combination of integer multiflows that satisfy (1). In this instance, three commodities (red, yellow, blue) route a demand of 1 from s to t each. The fractional flow is given by the numbers above. In particular, the fractional multiflow routes a flow of $1/2$ of exactly two commodities over each of the arcs e_1, e_2, e_3 resulting in a total flow of $x_e = 1$ on these arcs. By (1), every integer flow must therefore also send a total flow of $y_e = 1$ over these arcs. In order to achieve a flow $y_e \geq 1$ on e_1, e_2, e_3 simultaneously, at least two commodities need to be routed through the lower part of the graph. But then at least one of the arcs e_1, e_2, e_3 has a total flow of at least 2, which violates the constraint (1)(Color figure online).

In light of the fact that multiflow instances on series-parallel digraphs generally do not satisfy the assumptions of the Nagamochi-Ibaraki Theorem (i.e., the cut condition is not sufficient; see Fig. 1), the strong integrality property in Theorem 1(a) might seem surprising. On the other hand, the proof of Theorem 1 relies on the simple observation that, by carefully subdividing commodities, multiflow instances on series-parallel digraphs can be efficiently reduced to a certain subclass of instances which can be solved by single-commodity flow techniques. The integrality property in Theorem 1(a) is then inherited from the integrality of the network flow polytope. To the best of our knowledge, and somewhat surprisingly, these observations have not appeared in the literature before. The general idea of subdividing commodities, however, was used before, for example, in the proof of the Okamura-Seymour Theorem [11].

Finally, in Sect. 4 we obtain our main result, which generalizes the strong integrality result in Theorem 1(a) towards unsplittable multiflows.

Theorem 2. *The total arc flows* $(x_e)_{e \in E}$ *of a fractional multiflow in a series-parallel digraph can be expressed as a convex combination of total arc flows* $(y_e)_{e \in E}$ *of unsplittable multiflows that satisfy*

$$x_e - d_{\max} < y_e < x_e + d_{\max} \qquad \textit{for every arc } e \in E.$$

This result implies, in particular, that Conjecture 1 holds for series-parallel digraphs, even for general multiflow instances where commodities have individual source and sink nodes. Even for the weaker conjecture of Goemans, Theorem 2 provides the first proof for a non-trivial class of digraphs.

Theorem 2 is obtained as follows. We first show that, without altering the total arc flows, any multiflow can be transformed into an *almost unsplittable* multiflow, where no more than two commodities are fractionally routed through any component of the series-parallel digraph. Such an almost unsplittable multiflow can be readily expressed as a convex combination of unsplittable multiflows, with total arc flows deviating by no more than $2d_{\max}$ (Theorem 4). Note that for the specific case of single-source multiflows, Traub, Vargas Koch, and Zenklusen [16] prove the same result for the broader class of acyclic and planar digraphs. Reducing the bound from $2d_{\max}$ to $d_{\max}$, however, requires more elaborate arguments and refined techniques. It is worthwhile to mention that, while Theorem 2 is formulated as an existence result, these unsplittable multiflows can be efficiently computed.

Due to space constraints, some more detailed explanations, examples, and proofs are omitted in this extended abstract; we refer to the full paper [7].

2 Preliminaries and Notation

Multiflows. We consider a directed graph $G = (V, E)$ together with k commodities given by source-sink pairs $(s_i, t_i) \in V \times V$, $i \in [k] := \{1, \dots, k\}$. For given demands $d_i \in \mathbb{R}_{\geq 0}$, $i \in [k]$, a multiflow $\mathbf{X}$ sends d_i units of flow from s_i to t_i for every commodity $i \in [k]$. That is, $\mathbf{X}$ specifies for every commodity $i \in [k]$ and every arc $e \in E$ a flow value $X_{e,i} \geq 0$, such that the following flow conservation constraints hold:

$$\sum_{e \in \delta^+(v)} X_{e,i} - \sum_{e \in \delta^-(v)} X_{e,i} = \begin{cases} d_i & \text{if } v = s_i, \\ -d_i & \text{if } v = t_i, \\ 0 & \text{otherwise,} \end{cases} \qquad \text{for every } v \in V, i \in [k].$$

Here, $\delta^+(v)$ and $\delta^-(v)$ denote the sets of arcs $e \in E$ leaving and entering node v, respectively. In what follows we always assume that $d_i > 0$, $s_i \neq t_i$, and that there exists an s_i-t_i-path in G for every $i \in [k]$. A multiflow $\mathbf{X}$ is *unsplittable* if the demand d_i is routed along a single s_i-t_i-path, for every $i \in [k]$. More precisely, $X_{e,i} \in \{0, d_i\}$ for every $i \in [k]$, $e \in E$, and the arcs $e \in E$ with $X_{e,i} = d_i$ form an s_i-t_i-path.

For a multiflow $\mathbf{X} = (X_{e,i})_{e \in E, i \in [k]}$, the *total flow* $\mathbf{x} = (x_e)_{e \in E}$ is given by the sum of the flows of all commodities, that is, $x_e := \sum_{i \in [k]} X_{e,i}$. A multiflow $\mathbf{X}$ satisfies given *arc capacities* $c_e \in \mathbb{R}_{\geq 0}$, $e \in E$, if $x_e \leq c_e$ for every arc $e \in E$. In this case we say that multiflow $\mathbf{X}$ is *feasible*. A multiflow instance is called *feasible* if there is a feasible multiflow.

Series-parallel digraphs. A digraph $G = (V, E)$ is *series-parallel*[3] if it has a designated start node $s \in V$ and a designated end node $t \in V$ that satisfy one of the following three (recursive) conditions:

- G consists of a single arc from s to t, that is, $V = \{s, t\}$ and $E = \{(s, t)\}$.
- G is a *series composition*, that is, there are two series-parallel digraphs $G_1 = (V_1, E_1)$ with designated $s_1, t_1 \in V_1$ and $G_2 = (V_2, E_2)$ with designated $s_2, t_2 \in V_2$ such that $V = V_1 \cup V_2$, $E = E_1 \cup E_2$, $s = s_1$, $t_1 = s_2$, $t = t_2$, and $V_1 \cap V_2 = \{t_1\}$.
- G is a *parallel composition*, that is, there are two series-parallel digraphs $G_1 = (V_1, E_1)$ with designated $s_1, t_1 \in V_1$ and $G_2 = (V_2, E_2)$ with designated s_2, t_2 such that $V = V_1 \cup V_2$, $E = E_1 \cup E_2$, $s = s_1 = s_2$, $t = t_1 = t_2$, and $V_1 \cap V_2 = \{s, t\}$.

We consider *decompositions* of series-parallel digraphs represented by *sp-trees.* Given a series-parallel digraph $G = (V, E)$, the corresponding *sp*-tree $T = (V_T, E_T)$ is a rooted binary tree where every node $\omega \in V_T$ has one of the types *p-node*, *s-node*, or *q-node*, and a label $(u_\omega, v_\omega) \in V \times V$. Then, node ω represents a series-parallel subdigraph $G_\omega = (V_\omega, E_\omega)$ of G with start node u_ω and end node v_ω. The root of the *sp*-tree has label (s, t) and represents the entire series-parallel digraph G. The leafs of the *sp*-tree are of type q and represent the arcs of G. More precisely, a leaf node ω corresponds to the subdigraph G_ω that consists of a single arc from u_ω to v_ω, that is, $V_\omega = \{u_\omega, v_\omega\}$ and $E_\omega = \{(u_\omega, v_\omega)\}$. All non-leaf nodes of T are either of type s or p. If node $\omega \in V_T$ is of type p, then G_ω is obtained by parallel composition of two series-parallel subdigraphs that are represented by the two children ω_1 and ω_2 of ω. If node ω is of type s, then G_ω is a series composition of two series-parallel digraphs represented by the two children ω_1 and ω_2 of ω. For a node ω of the *sp*-tree, let V_ω° denote the set of *inner nodes* of subdigraph G_ω, that is, $V_\omega^\circ := V_\omega \setminus \{u_\omega, v_\omega\}$.

Multiflows in series-parallel digraphs. For a given multiflow $\mathbf{X}$ in G, we are interested in the amount of flow of every commodity that is routed through every subcomponent of the series-parallel graph. To this end, we use the decomposition of the series-parallel graph given by the *sp*-tree T. Formally, we define for every $\omega \in V_T$ with associated subdigraph G_ω the *demand share* $z_{\omega,i}(\mathbf{X}) \in [0, 1]$

[3] Series–parallel digraphs come in two versions in the literature, the 'edge version' and the 'vertex version' [17]; here we refer to the more common 'edge version'.

of commodity $i \in [k]$ that is routed through the subdigraph G_ω by

$$z_{\omega,i}(\mathbf{X}) := \begin{cases} 1 & \text{if } V_\omega^\circ \cap \{s_i, t_i\} \neq \emptyset, \\ \frac{1}{d_i} \sum_{e \in \delta_\omega^+(u_\omega)} X_{e,i} & \text{otherwise.} \end{cases}$$

Here, $\delta_\omega^+(\cdot)$ denotes the set of outgoing arcs from a particular node within G_ω. Moreover, let $\mathbf{z}^\omega(\mathbf{X}) := (z_{\omega,i}(\mathbf{X}))_{i \in [k]}$.

Definition 1. *For $\omega \in V_T$ and $i \in [k]$, we say that commodity i is* routed unsplittably *through the subdigraph G_ω if $z_{\omega,i}(\mathbf{X}) \in \{0,1\}$; otherwise, commodity i is* routed fractionally *through G_ω. Let $I_\omega(\mathbf{X}) := \{i \in [k] : 0 < z_{\omega,i}(\mathbf{X}) < 1\}$ denote the subset of commodities that are routed fractionally through G_ω.*

Notice that a multiflow $\mathbf{X}$ is an unsplittable flow if every commodity is routed unsplittably through every subdigraph G_ω or, equivalently, if $I_\omega(\mathbf{X}) = \emptyset$ for every node ω of the *sp*-tree. We outline three fundamental properties of the relationship between the demand shares in the subdigraph of a non-leaf node and the subdigraphs corresponding to its child nodes.

Observation 1. *Let $\omega \in V_T$ be a non-leaf node with child nodes ω_1, ω_2 and let $i \in [k]$ be some commodity.*

(a) If ω is a p-node, then $z_{\omega,i}(\mathbf{X}) = z_{\omega_1,i}(\mathbf{X}) + z_{\omega_2,i}(\mathbf{X})$.
(b) If ω is an s-node and $s_i, t_i \notin V^\circ(\omega)$, then $z_{\omega,i}(\mathbf{X}) = z_{\omega_1,i}(\mathbf{X}) = z_{\omega_2,i}(\mathbf{X})$.
(c) If ω is an s-node and $\{s_i, t_i\} \cap V^\circ(\omega) \neq \emptyset$, then it holds that $z_{\omega,i}(\mathbf{X}) = 1$ and $z_{\omega_1,i}(\mathbf{X}), z_{\omega_2,i}(\mathbf{X}) \in \{0,1\}$.

3 The Cut Condition and Integer Multiflows

The following *cut condition* is a necessary criterion for the existence of a feasible (fractional) multiflow (see, e.g., [14, Chapter 70.7]):

$$c(\delta^+(X)) \geq \sum_{i \in [k] : s_i \in X \subseteq V \setminus \{t_i\}} d_i \qquad \text{for all } X \subseteq V. \tag{4}$$

However, in series-parallel digraphs, the cut condition generally does not guarantee the existence of a feasible multiflow. The instance in Fig. 1 is a counterexample, in which the cut condition (4) is satisfied, yet no feasible multiflow exists. Note that the only s_1-t_1 path in the given network passes through node s_2, meaning that all flow of commodity 1 must be routed through s_2. We thus obtain an equivalent multiflow instance by subdividing commodity 1 into two subcommodities 1_1 and 1_2 with $s_{1_1} = s_1$, $t_{1_1} = s_2 = s_{1_2}$, and $t_{1_2} = t_1$, where both subcommodities inherit the unit demand value of the original commodity, i.e., $d_{1_1} = d_{1_2} = d_1 = 1$. The resulting multiflow instance violates the cut condition since $c(\delta^+(\{s_2\})) = 1$ but commodities 1_2 and 2 both need to send one unit of flow across cut $\delta^+(\{s_2\})$. In the context of multiflows in undirected graphs, the trick of subdividing commodities was used, for example, in [2,11].

Aligned Multiflow Instances on Series-Parallel Digraphs. We generalize the simple idea of subdividing commodities outlined above. If, for some commodity $i \in [k]$, there is a node $v \in V \setminus \{s_i, t_i\}$ that is on every s_i-t_i-path in G, commodity i can be subdivided into two commodities i_1 and i_2 with $s_{i_1} = s_i$, $t_{i_1} = v = s_{i_2}$, $t_{i_2} = t_i$, and $d_{i_1} = d_{i_2} = d_i$. There is a straightforward one-to-one correspondence between multiflows for the original instance and multiflows for the subdivided instance. In particular, unsplittable multiflows for the original instance are mapped to unsplittable multiflows for the subdivided instance, and vice versa. If we iteratively subdivide commodities as far as possible, we arrive at an equivalent multiflow instance on G that features a particular structure described in the following definition.

Definition 2. *The commodities of a multiflow instance on a series-parallel digraph G with sp-tree T are* aligned *with the series-parallel decomposition if, for every commodity $i \in [k]$, there is a component G_ω, $\omega \in V_T$, with $s_i = u_\omega$ and $t_i = v_\omega$. In this case, we also say that the multiflow instance is* aligned.

Note that the choice of $\omega \in V_T$ for commodity $i \in [k]$ in Definition 2 is not necessarily unique. If, for example, ω is a p-node, then its children have the same start and end node, and are thus alternative choices. Whenever we refer to *the node ω corresponding to commodity i* in an aligned multiflow instance, we mean the unique node $\omega \in V_T$ with $s_i = u_{\omega_i}$ and $t_i = v_{\omega_i}$ for which all s_i-t_i-paths in G lie in G_{ω_i} or, equivalently, which has minimum distance to the root. Note that commodity i must then be routed entirely within G_ω.

Any multiflow instance on a series-parallel digraph can be efficiently reduced to an aligned instance by repeatedly subdividing commodities.

Observation 2. *For a non-aligned multiflow instance on a series-parallel digraph, there is an $i \in [k]$ and $v \in V \setminus \{s_i, t_i\}$ such that v is on every s_i-t_i-path.*

Solving Aligned Instances via Single-Commodity Flows. A relaxation of a given multiflow instance can be obtained by allowing flow to be sent from any source s_i to any sink t_j where j is not necessarily equal to i. This relaxation yields a single-commodity b-transshipment instance where

$$b_v := \sum_{i \in [k]: s_i = v} d_i - \sum_{i \in [k]: t_i = v} d_i.$$

Then, for any multiflow $\mathbf{X}$, its total flow $\mathbf{x}$ is a b-transshipment. In particular, for a feasible multiflow instance, the corresponding b-transshipment instance is feasible as well. The opposite direction, however, is not true. Here is a simple counterexample on a unit-capacity series-parallel digraph:

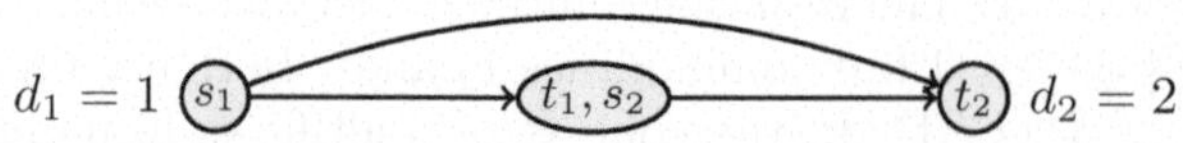

Note that any multiflow must send the entire demand $d_2 = 2$ across the unit-capacity arc (s_2, t_2), and is thus infeasible. In the relaxation, however, the supply of value 2 at s_2 together with the unit demand at the same node $t_1 = s_2$ results in a total supply of 1 that can be feasibly sent across arc (s_2, t_2), while s_1 sends its unit supply via arc (s_1, t_2) to t_2. What makes the b-transshipment instance feasible is thus the canceling of supply and demand at node $t_1 = s_2$. But if we split this node into two nodes that are connected by an infinite capacity arc, we arrive at an equivalent (infeasible) multiflow instance on a series-parallel digraph, whose corresponding b-transshipment instance is also infeasible:

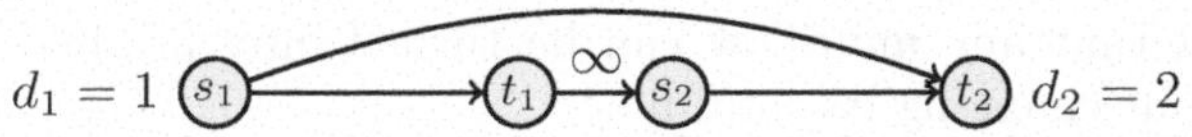

The simple idea of splitting a node v that is both a source and a sink works in general. Given an aligned multiflow instance on a series-parallel digraph with such a node v, we obtain an equivalent aligned multiflow instance on a modified series-parallel digraph as follows: split node v into two nodes v_1 and v_2 joined by an infinite capacity arc (v_1, v_2); the head node of every arc $e \in \delta^-(v)$ is set to v_1, and the tail node of every arc $e \in \delta^+(v)$ is set to v_2; moreover, for $i \in [k]$, if $t_i = v$, we set $t_i := v_1$, and if $s_i = v$, we set $s_i := v_2$. Note that the resulting multiflow instance is indeed equivalent to the original one, the modified digraph is still series-parallel, and the new multiflow instance is still aligned.

Lemma 1. *For an aligned multiflow instance on a series-parallel digraph with $s_i \neq t_j$ for all $i \neq j$, every b-transshipment $\mathbf{y} = (y_e)_{e \in E}$ can be efficiently turned into a multiflow $\mathbf{X}$ whose total flow $\mathbf{x}$ equals $\mathbf{y}$.*

Lemma 1 implies that aligned multiflow instances with $s_i \neq t_j$ for all $i \neq j$ inherit the sufficiency of the cut condition as well as strong integrality properties from single-commodity flows. Moreover, as we argue above, any multiflow instance on a series-parallel digraph can be reduced to such an aligned instance in a straightforward manner. This concludes the proof of Theorem 1.

4 Almost Unsplittable and Unsplittable Multiflows

In this section, as an intermediate step towards our main result, Theorem 2, we first prove a slightly weaker version, allowing unsplittable multiflows to deviate from the given fractional multiflow by at most $2d_{\max}$ on each arc.[4]

[4] Traub, Vargas Koch, and Zenklusen obtained the same bound for the special case of a single source node $s = s_i$, $i \in [k]$, on the more general class of planar and acyclic digraphs; see [16, Theorem 1.8].

Almost Unsplittable Multiflows. We say that a multiflow $\mathbf{X}$ in G is *almost unsplittable* if at most two commodities are routed fractionally through every subdigraph G_ω, $\omega \in V_T$, and the two components in a parallel composition may share at most one fractionally routed commodity.

Definition 3. *A multiflow* $\mathbf{X}$ *in* G *is* almost unsplittable *if it satisfies:*

(i) for every node $\omega \in V_T$ *of the sp-tree* T*, we have* $|I_\omega(\mathbf{X})| \leq 2$*;*
(ii) $|I_{\omega_1}(\mathbf{X}) \cap I_{\omega_2}(\mathbf{X})| \leq 1$ *for every p-node with two children* $\omega_1, \omega_2 \in V_T$*. If* $i \in I_{\omega_1}(\mathbf{X}) \cap I_{\omega_2}(\mathbf{X})$*, we say that commodity* i *is* split between ω_1 and ω_2.

It turns out that any multiflow can be turned into an almost unsplittable multiflow without changing its total flow.

Theorem 3. *Any multiflow in* G *can efficiently be turned into an almost unsplittable multiflow that satisfies the same demands and has the same total flow.*

The proof of Theorem 3 relies on a simple swapping operation that exchanges the flow of two commodities between the parallel components of G corresponding to the two children of a p-node. Details and the full proof can be found along with all other missing proofs and further illustrating examples in the full version of the paper [7].

Theorem 4. *For any multiflow* $\mathbf{X}$ *in* G*, its total arc flows* $(x_e)_{e \in E}$ *can be expressed as a convex combination of total arc flows* $(y_e)_{e \in E}$ *of unsplittable multiflows satisfying*

$$x_e - 2d_{\max} < y_e < x_e + 2d_{\max} \qquad \text{for every arc } e \in E. \tag{5}$$

Moreover, the unsplittable multiflows and the coefficients of the convex combination can be computed efficiently.

We discuss the proof of Theorem 4 in detail since the construction of the convex decomposition also forms the basis of the proof of our main result Theorem 2. In view of Theorem 3, it suffices to prove Theorem 4 for an almost unsplittable multiflow $\mathbf{X}$ with the same total flow $\mathbf{x}$. Although the statement of Theorem 4 is only about the total flows, we proceed by showing the stronger statement that an almost unsplittable multiflow $\mathbf{X}$ itself (i.e., the full matrix $\mathbf{X}$ and not only the total flow vector $\mathbf{x}$) can be expressed as a convex combination of unsplittable multiflows. Thus, we may assume that, by Definition 3, for each node $\omega \in V_T$ of the sp-tree T, multiflow $\mathbf{X}$ routes (at most) two commodities fractionally through the corresponding component. To simplify notation and avoid case distinctions, we assume that $|I_\omega| = 2$ for every ω, and let $p^\omega, q^\omega \in [k]$ be the two distinct commodities routed fractionally through ω, i.e., $I_\omega = \{p^\omega, q^\omega\}$. The special case $|I_\omega| < 2$ can be handled by introducing dummy commodities and is fully covered in the full version [7].

The proof of Theorem 4 uses a recursive approach. We start in the leaves $\omega \in V_T$ of the *sp*-tree representing single arcs e of the graph G, and construct unsplittable flows that either do or do not route the commodities in I_ω over the arc e. Then, these unsplittable flows are recursively combined in *s*- or *p*-nodes of the *sp*-tree. While the general approach is straightforward, in particular the process of combining given convex decompositions of child nodes in *s*- and *p*-nodes is quite involved and requires careful definitions in order to ensure that the resulting convex combinations are compatible between parent and child nodes. Due to the nature of this recursive approach, we construct a convex decomposition for each node $\omega \in V_T$ of the *sp*-tree T, with the decomposition at the root node ω_0 serving as the convex decomposition asserted in Theorem 4. Formally, for each $\omega \in V_T$, we define unsplittable multiflows $\mathbf{X}_\ell^\omega$, $\ell = 1, \ldots, L_\omega$, within the subdigraph G_ω, and coefficients $\rho_\ell^\omega \in [0,1]$, $\ell = 1, \ldots, L_\omega$, such that the fractional multiflow restricted to G_ω is the convex combination of the flows $\mathbf{X}_\ell^\omega$ with coefficients ρ_ℓ^ω and the corresponding total flows $\mathbf{x}_\ell^\omega$ satisfy (5).

Unsplittable Routings. In order to construct suitable multiflows $\mathbf{X}_\ell^\omega$ for some node $\omega \in V_T$, we leverage the properties of almost unsplittable multiflows. Since in an almost unsplittable multiflow all but the two commodities $p^\omega, q^\omega \in I_\omega$ must be routed through the subdigraph G_ω, we are only left with four possible routing options. We represent these options with the subsets

$$J_1^\omega := \emptyset, \qquad J_2^\omega := \{p^\omega\}, \qquad J_3^\omega := \{q^\omega\}, \qquad \text{and} \quad J_4^\omega := \{p^\omega, q^\omega\}, \tag{6}$$

where routing option $j \in \{1,2,3,4\}$ means that exactly the commodities in $\{i \in [k] \mid z_{i,\omega} = 1\} \cup J_j^\omega$ are routed through the subdigraph G_ω. If node ω is a leaf of the *sp*-tree, i.e., ω represents an arc e of G, every routing option corresponds immediately to an unsplittable flow on this arc defined as

$$X_{\ell,e,i}^\omega := d_i \quad \text{if } z_{\omega,i} = 1 \text{ or } i \in J_\ell^\omega, \quad \text{and} \quad X_{\ell,e,i}^\omega := 0 \text{ otherwise.}$$

In view of the desired bound (5), note that the difference of the total flow induced by a flow $\mathbf{X}_\ell^\omega$ and the given total flow x_e on any arc is bounded from above by

$$\Big|\sum_{i \in J_j^\omega} d_i - \sum_{i \in I_\omega} z_{\omega,i} d_i\Big| < d_{p^\omega} + d_{q^\omega} \le 2d_{\max}, \qquad \text{for } j = 1,2,3,4. \tag{7}$$

If $\omega \in V_T$ is an *s*- or a *p*-node, there is no one-to-one correspondence between routing options and unsplittable multiflows. Rather, we will group the unsplittable multiflows in the convex decomposition into four groups $K_1^\omega, \ldots, K_4^\omega$, where all multiflows in group K_j^ω route commodities in $\{i \in [k] \mid z_{i,\omega} = 1\} \cup J_j^\omega$ through the subgraph G_ω. In order to obtain a convex decomposition of the multiflow $\mathbf{X}^\omega$ in the component, we need to define proper coefficients. To this end, we distribute a total weight of 1 over the different routing options. Formally, we need to find numbers $\mu_j^\omega \in [0,1]$, $j = 1, \ldots, 4$, for every ω such that

$$\mu_2^\omega + \mu_4^\omega = z_{\omega,p^\omega}, \quad \mu_3^\omega + \mu_4^\omega = z_{\omega,q^\omega}, \quad \text{and} \quad \mu_1^\omega + \mu_2^\omega + \mu_3^\omega + \mu_4^\omega = 1. \tag{8}$$

If the unsplittable multiflows in group K_j^ω are weighted with a total weight of μ_j^ω, this ensures that the unsplittable multiflows in fact form a convex decomposition of the multiflow. The following lemma gives an explicit solution to the linear system (8). For a concise representation, we denote for any number $z \in \mathbb{R}$ by $[z]^+$ the *positive part* of z, that is, $[z]^+ := z$ if $z > 0$, and $[z]^+ := 0$ if $z \leq 0$.

Lemma 2. *For every node ω of the sp-tree, the numbers*

$$\mu_1^\omega := \left[(1 - z_{\omega,q^\omega}) - z_{\omega,p^\omega}\right]^+, \qquad \mu_2^\omega := \min\{z_{\omega,p^\omega}, 1 - z_{\omega,q^\omega}\},$$
$$\mu_3^\omega := 1 - \max\{z_{\omega,p^\omega}, 1 - z_{\omega,q^\omega}\}, \qquad \mu_4^\omega := \left[z_{\omega,p^\omega} - (1 - z_{\omega,q^\omega})\right]^+ \tag{9}$$

are a solution to the linear system (8). Moreover, $\mu_j^\omega \in [0,1], j = 1,\ldots,4$.

Note that, by definition, at most three of the four coefficients defined in (9) are non-zero. Thus, there are at most three non-empty groups of unsplittable multiflows for each node $\omega \in V_T$. Moreover, if ω represents a single arc $e \in E$, we can immediately use the numbers μ_j^ω as coefficients ρ_ℓ for the convex decomposition.

Recursive Refinement of Convex Decompositions. To complete the recursive construction, we discuss how to refine the convex decompositions of two child nodes ω_1, ω_2 to a valid convex decomposition in an *s*- or *p*-node ω. We outline the main ideas and definitions here, while details can be found in the full version [7].

First, let $\omega \in V_T$ be an *s*-node, i.e., let ω be a representation of a serial component. In this case, the routings within the child components dependent on each other in the sense that, if an unsplittable multiflow routes commodity $i \in I_{\omega_1}$ through G_{ω_1}, it needs to be routed through G_{ω_2} as well to not contradict flow conservation. This implies, that flows from the group $K_j^{\omega_1}$ for some $j = 1,\ldots,4$ must be combined with flows from the group $K_j^{\omega_2}$ for the same j. Observation 1 (b) immediately yields $I_\omega = I_{\omega_1} = I_{\omega_2}$. Hence, the sets J_j^ω and the numbers μ_j^ω are the same for parent and child nodes. This allows for a straightforward combination of the respective flows as well as a straightforward refinement of the convex decomposition.

If ω is a *p*-node with children ω_1 and ω_2, the situation is more complex. By definition of almost unsplittable flows, there is (at most) one commodity in $I_{\omega_1} \cap I_{\omega_2}$ that is split between ω_1 and ω_2. For simplicity, we assume that $|I_{\omega_1} \cap I_{\omega_2}| = 1$, i.e., we assume that a split commodity exists. We refer to this commodity by r^ω. We also assume that $r^\omega \notin \{p^\omega, q^\omega\}$. The remaining (simpler) cases, where these assumptions do not hold, are discussed in the full version [7]. Finally, we may assume without loss of generality that $I_{\omega_1} = \{p^\omega, r^\omega\}$ and $I_{\omega_2} = \{r^\omega, q^\omega\}$. That is, $p^{\omega_1} = p^\omega$, $q^{\omega_1} = r^\omega = p^{\omega_2}$, and $q^{\omega_2} = q^\omega$. With the additional choice of whether to route r^ω through either ω_1 or ω_2, we obtain for every set J_j^ω in the parent component two additional possibilities for the child nodes. This yields in total eight possible routings in a parallel component. These eight cases are shown in Table 1.

Table 1 shows which groups of flows of the child nodes must be combined and which weights need to be used. For example, the first row states that we need to

Table 1. Routing options in a parallel component. Every routing option J_j^ω of the parent node can be achieved by two combinations of routing options $J_{j_1}^{\omega_1}$ and $J_{j_2}^{\omega_2}$ in the child nodes. Each of the eight routings has a weight λ_i that must be consistent with the weights $\mu_{j_i}^{\omega_i}$ of the respective routings in the child nodes.

parent ω routing	μ	group	child ω_1 routing	μ	group	child ω_2 routing	μ	group	
$J_1^\omega = \emptyset$	μ_1^ω	K_1^ω	$J_3^{\omega_1} = \{r^\omega\}$	$\mu_3^{\omega_1}$	$K_3^{\omega_1}$	$J_1^{\omega_2} = \emptyset$	$\mu_1^{\omega_2}$	$K_1^{\omega_2}$	λ_1
			$J_1^{\omega_1} = \emptyset$	$\mu_1^{\omega_1}$	$K_1^{\omega_1}$	$J_2^{\omega_2} = \{r^\omega\}$	$\mu_2^{\omega_2}$	$K_2^{\omega_2}$	λ_2
$J_2^\omega = \{p^\omega\}$	μ_2^ω	K_2^ω	$J_4^{\omega_1} = \{p^\omega, r^\omega\}$	$\mu_4^{\omega_1}$	$K_4^{\omega_1}$	$J_1^{\omega_2} = \emptyset$	$\mu_1^{\omega_2}$	$K_1^{\omega_2}$	λ_3
			$J_2^{\omega_1} = \{p^\omega\}$	$\mu_2^{\omega_1}$	$K_2^{\omega_1}$	$J_2^{\omega_2} = \{r^\omega\}$	$\mu_2^{\omega_2}$	$K_2^{\omega_2}$	λ_4
$J_3^\omega = \{q^\omega\}$	μ_3^ω	K_3^ω	$J_3^{\omega_1} = \{r^\omega\}$	$\mu_3^{\omega_1}$	$K_3^{\omega_1}$	$J_3^{\omega_2} = \{q^\omega\}$	$\mu_3^{\omega_2}$	$K_3^{\omega_2}$	λ_5
			$J_1^{\omega_1} = \emptyset$	$\mu_1^{\omega_1}$	$K_1^{\omega_1}$	$J_4^{\omega_2} = \{r^\omega, q^\omega\}$	$\mu_4^{\omega_2}$	$K_4^{\omega_2}$	λ_6
$J_4^\omega = \{p^\omega, q^\omega\}$	μ_4^ω	K_4^ω	$J_4^{\omega_1} = \{p^\omega, r^\omega\}$	$\mu_4^{\omega_1}$	$K_4^{\omega_1}$	$J_3^{\omega_2} = \{q^\omega\}$	$\mu_3^{\omega_2}$	$K_3^{\omega_2}$	λ_7
			$J_2^{\omega_1} = \{p^\omega\}$	$\mu_2^{\omega_1}$	$K_2^{\omega_1}$	$J_4^{\omega_2} = \{r^\omega, q^\omega\}$	$\mu_4^{\omega_2}$	$K_4^{\omega_2}$	λ_8

combine flows from group $K_3^{\omega_1}$ of the first child and flows from the group $K_1^{\omega_2}$ of the second child, assign a total weight of λ_1 to these flows and then assign the resulting flows in group K_1^ω of the parent node. The total weight λ_i needs to satisfy certain conditions so that the new convex decomposition of the parent node is consistent with the convex decompositions of the child nodes. For a detailed description of this process and illustrating examples, we again refer to the full version [7].

Overall, we are now able to recursively combine the unsplittable multiflows and find suitable coefficients in every s- and p-node. The proof of Theorem 4 can be turned into an efficient algorithm that computes a family of unsplittable flows together with convex coefficients. The algorithm goes through the sp-tree from top to bottom (i.e., from the root to the leafs), successively refining a common convex combination of the flows routed through the components corresponding to the considered tree nodes. Moreover, by regularly applying Carathéodory's Theorem (see, e.g., [13, Chapter 7.7]) to intermediate convex combinations, one can keep the number of unsplittable routings bounded by $O(k \cdot m)$, where m is the number of arcs of digraph G. A similar argument is used in the work of Martens et al. [8].

Finally, we argue that the bounds from (7) can be strengthened by showing that for every component ω and $j \in \{1, \ldots, 4\}$,

$$\Big|\sum_{i \in J_j^\omega} d_i - \sum_{i \in I_\omega} z_{\omega,i} d_i\Big| < d_{\max}, \qquad \text{for } j = 1, 2, 3, 4, \tag{10}$$

holds true, whenever $\mu_j^\omega > 0$. This implies that the difference between the flow through a component for a given unsplittable routing J_j^ω and the fractional flow is at most $d_{\max}$ whenever the corresponding coefficient in the convex combination is non-zero.

Lemma 3. *For* $\omega \in V_T$ *and* $j \in \{1, 2, 3, 4\}$ *with* $\mu_j^\omega > 0$ *inequality (10) holds.*

Finally, this allows us to proof our main result.

Proof (of Theorem 2). If we use Lemma 3 instead of (7) in the proof of Theorem 4, the result follows immediately.

□

Acknowledgments. We thank the anonymous referees for their insightful comments that helped to improve the presentation of the paper.

Disclosure of Interests. The authors have no competing interests to declare that are relevant to the content of this article

References

1. Ahuja, R.K., Magnanti, T.L., Orlin, J.B.: Network Flows. Algorithms, and Applications. Prentice Hall, Englewood Cliffs, NJ, Theory (1993)
2. Chekuri, C., Shepherd, F.B., Weibel, C.: Flow-cut gaps for integer and fractional multiflows. J. Comb. Theor. Ser. B **103**(2), 248–273 (2013). https://doi.org/10.1016/J.JCTB.2012.11.002
3. Dinitz, Y., Garg, N., Goemans, M.X.: On the single source unsplittable flow problem. Combinatorica **19**, 17–41 (1999). https://doi.org/10.1007/s004930050043
4. Kleinberg, J.M.: Approximation algorithms for disjoint paths problems. Ph.D. thesis, M.I.T. (1996). http://hdl.handle.net/1721.1/11013
5. Kolliopoulos, S.G.: Edge-disjoint paths and unsplittable flow. In: Gonzalez, T.F. (ed.) Handbook of Approximation Algorithms and Metaheuristics. Chapman and Hall/CRC (2007). https://doi.org/10.1201/9781420010749.CH57
6. Kolliopoulos, S.G., Stein, C.: Approximation algorithms for single-source unsplittable flow. SIAM J. Comput. **31**, 919–946 (2002). https://doi.org/10.1137/S0097539799355314
7. Majthoub Almoghrabi, M., Skutella, M., Warode, P.: Integer and unsplittable multiflows in series-parallel digraphs (2024). https://doi.org/10.48550/arXiv.2412.05182
8. Martens, M., Salazar, F., Skutella, M.: Convex Combinations of Single Source Unsplittable Flows. In: Arge, L., Hoffmann, M., Welzl, E. (eds.) ESA 2007. LNCS, vol. 4698, pp. 395–406. Springer, Heidelberg (2007). https://doi.org/10.1007/978-3-540-75520-3_36
9. Morell, S., Skutella, M.: Single source unsplittable flows with arc-wise lower and upper bounds. Math. Program. **192**, 477–496 (2022). https://doi.org/10.1007/s10107-021-01704-4
10. Nagamochi, H., Ibaraki, T.: On max-flow min-cut and integral flow properties for multicommodity flows in directed networks. Inf. Process. Lett. **31**(6), 279–285 (1989). https://doi.org/10.1016/0020-0190(89)90087-2
11. Okamura, H., Seymour, P.D.: Multicommodity flows in planar graphs. J. Comb. Theory Ser. B **31**(1), 75–81 (1981). https://doi.org/10.1016/S0095-8956(81)80012-3

12. Poremba, J., Shepherd, F.B.: Cut-sufficient directed 2-commodity multiflow topologies. In: International Conference on Integer Programming and Combinatorial Optimization, pp. 408–422. Springer (2023). https://doi.org/10.1007/978-3-031-32726-1_29
13. Schrijver, A.: Theory of Linear and Integer Programming. John Wiley & Sons, Chichester (1986)
14. Schrijver, A.: Combinatorial Optimization: Polyhedra and Efficiency. Springer (2003)
15. Skutella, M.: Approximating the single source unsplittable min-cost flow problem. Math. Program. **91**(3), 493–514 (2001). https://doi.org/10.1007/s101070100260
16. Traub, V., Vargas Koch, L., Zenklusen, R.: Single-source unsplittable flows in planar graphs. In: Proceedings of the 2024 Annual ACM-SIAM Symposium on Discrete Algorithms, pp. 639–668. SIAM (2024). https://doi.org/10.1137/1.9781611977912.24
17. Valdes, J., Tarjan, R.E., Lawler, E.L.: The recognition of series parallel digraphs. SIAM J. Comput. **11**, 298–313 (1982). https://doi.org/10.1137/0211023
18. Vygen, J.: NP-completeness of some edge-disjoint paths problems. Discret. Appl. Math. **61**(1), 83–90 (1995). https://doi.org/10.1016/0166-218X(93)E0177-Z
19. Williamson, D.P.: Network Flow Algorithms. Cambridge University Press (2019)

Space-Efficient Algorithm for Integer Programming with Few Constraints

Lars Rohwedder[1] and Karol Węgrzycki[2(✉)]

[1] University of Southern Denmark, Odense, Denmark
rohwedder@sdu.dk
[2] Max Planck Institute for Informatics, Saarbrücken, Germany
kwegrzyc@mpi-inf.mpg.de

Abstract. Integer linear programs $\min\{c^T x : Ax = b, x \in \mathbb{Z}^n_{\geq 0}\}$, where $A \in \mathbb{Z}^{m \times n}$, $b \in \mathbb{Z}^m$, and $c \in \mathbb{Z}^n$, can be solved in pseudopolynomial time for any fixed number of constraints $m = O(1)$. More precisely, in time $(m\Delta)^{O(m)}\text{poly}(I)$, where Δ is the maximum absolute value of an entry in A and I the input size.

Known algorithms rely heavily on dynamic programming, which leads to a space complexity of similar order of magnitude as the running time. In this paper, we present a polynomial space algorithm that solves integer linear programs in $(m\Delta)^{O(m(\log m + \log\log \Delta))}\text{poly}(I)$ time, that is, in almost the same time as previous dynamic programming algorithms.

Keywords: Integer Programming · Polynomial Space · Pseudopolynomial Time · Parameterized Algorithms

1 Introduction

We consider integer linear programs (ILP) in equality form

$$\min\left\{c^\mathsf{T} x : Ax = b, x \in \mathbb{Z}^n_{\geq 0}\right\} \tag{1}$$

where $A \in \mathbb{Z}^{m \times n}$, $b \in \mathbb{Z}^m$, and $c \in \mathbb{Z}^n$. Integer linear programs are a widely used model to describe problems in combinatorial optimization and the algorithms available for them are used as general purpose optimization suites. Unlike its continuous counterpart, where $x \in \mathbb{R}^n_{\geq 0}$, integer linear programming is NP-complete even in the case of $m = 1$. Hence, theoretical research is mainly focused on identifying structural assumptions, under which ILPs can still be solved efficiently. Here, two dominant directions can be distinguished.

Based on the geometry of numbers, Lenstra [17] showed that ILPs of the form $\min\{c^\mathsf{T} x : Ax \leq b, x \in \mathbb{Z}^n\}$ can be solved in polynomial time for constant n, in fact, that they are fixed-parameter tractable (FPT) in parameter n. Note that this form is equivalent to (1) by a standard transformation, but this would lead to an increase in parameter n. Subsequent improvements by Kannan [14] and most recently Reis and Rothvoss [22] lead to the currently best running time of $(\log n)^{O(n)}\,\text{poly}(I)$, where I is the input size.

N. Megow and A. Basu (Eds.): IPCO 2025, LNCS 15620, pp. 442–451, 2025.
https://doi.org/10.1007/978-3-031-93112-3_32

The second direction is mainly based on dynamic programming approaches and goes back to Papadimitriou [20]. He considered (1) in the case when the absolute values of the entries of A and b are bounded by Δ and designed a pseudopolynomial time algorithm when m is fixed. This generalizes the classical pseudopolynomial time algorithm for Knapsack problems [1]. Later, his algorithm was improved by Eisenbrand and Weismantel [6]. The current best algorithm is due to Jansen and Rohwedder [11] with a running time of $O(\sqrt{m}\Delta)^{2m}\,\mathrm{poly}(I)$ and is optimal up to the constant in the exponent assuming Exponential Time Hypothesis [16]. In the latter two results the bound of Δ is only required on A, but not b. Similar approaches were also applied successfully to constraint matrices A, where both dimensions n and m can be large but satisfy certain structures of their non-zero entries, see e.g. [3].

Space efficiency of integer linear programs. NP-completeness tells us that any algorithm that solves general ILPs needs to have a super-polynomial running time at least in some parameter unless P=NP. But not only running time, also space usage can be a limiting factor. It is well known that for practical ILP solvers based on Branch-and-Bound it is often a bottleneck to store the list of active nodes, see e.g. [24, Section 7.5.1]. This motivates the theoretical question of whether the aforementioned results can also be achieved in polynomial space or, more generally, what tradeoffs between running time and space one can achieve.

Frank and Tardos [8] showed that Lenstra's algorithm and the improvement by Kannan with running time $n^{O(n)}\,\mathrm{poly}(I)$ can be implemented in polynomial space. For the improvement by Reis and Rothvoss this is not known (see discussion in ??)

Regarding the algorithms based on dynamic programming, the space usage is also pseudopolynomial, which is inherent in the method. Alternative, space-efficient methods are known for Subset Sum, a very restricted variant of ILP, where the task is to find a solution to $\{x \in \{0,1\}^n : a^{\mathsf{T}}x = b\}$ with $a \in \mathbb{Z}^n_{\geq 0}$ and $b \in \mathbb{Z}_{\geq 0}$. Note that here $m = 1$ and there is no objective function. For the Subset Sum problem, Lokshtanov and Nederlof [18] introduced an algebraization technique and designed an $\tilde{O}(n^3 b \log b)$ time and $\tilde{O}(n^2)$ space algorithm (see also [7,13]). This algorithm was later improved by Bringmann [2], and the current best low-space algorithm for Subset Sum is due to Jin et al. [12], which runs in $\tilde{O}(nb)$ time and $O(\log n \log\log n + \log b)$ space. Despite the intensive research on polynomial space algorithms for Subset Sum, little is known for more general problems. The seminal result of Lokshtanov and Nederlof [18] is the only one that considers the more general Knapsack problem, that is, $\max\{c^{\mathsf{T}}x : a^{\mathsf{T}}x \leq b, x \in \{0,1\}^n\}$ with $c, a \in \mathbb{Z}^n_{\geq 0}$, and $b \in \mathbb{Z}_{\geq 0}$. Nevertheless, their method suffers from a pseudopolynomial factor in both b and $\|c\|_\infty$ in the running time, see [18, Theorem 4.5], whereas the classical dynamic programming algorithm only requires pseudopolynomial time in one of the two parameters. A $\mathrm{poly}(n, b)$ time and polynomial space algorithm is widely open for the Knapsack problem and Unbounded Knapsack (the variant with $x \in \mathbb{Z}^n_{\geq 0}$) [19].

Savitch's theorem [23] can be used to avoid the high space usage of dynamic programming, but often comes at a significant cost for the running time. In the case of (1), this method leads to an algorithm in time $(m\Delta)^{O(m^2 \log(m\Delta))}$ poly(I) and polynomial space, see Sect. 2 for details.

Our contribution. In this paper, we show that integer programming (1) can be solved in polynomial space and a running time that nearly matches the best known running time of the dynamic programming approach by [6,11].

Theorem 1. *Let $A \in \mathbb{Z}^{m\times n}$, $b \in \mathbb{Z}^m$, and $c \in \mathbb{Z}^n$ and suppose that the absolute value of each entry of A is bounded by $\Delta > 1$. Then, in time*

$$(m\Delta)^{O(m(\log m+\log\log \Delta))} \cdot \mathrm{poly}(I)$$

and polynomial space, one can compute a solution to (1) or conclude that such a solution does not exist. Here, I is the encoding size of $A, b,$ and c.

Compared to the running time in [6,11], this is an additional factor of $O(\log m + \log\log \Delta)$ in the exponent of $m\Delta$, whereas all previously known techniques lead to an additional factor of at least $O(m \log \Delta)$, see Sect. 2.

2 Overview of Previous and New Techniques

Let us briefly introduce some notation and the model of computation. Throughout this paper, we assume all logarithms are base 2. We write $\tilde{O}(f) = O(f \cdot \mathrm{polylog}(f))$. Further, we make use of the shorthand notation $[n] = \{1, \ldots, n\}$. For a vector $x \in \mathbb{Z}^n$, let $\|x\|_p$ denote the ℓ_p norm of vector x. In the special case of $p = 0$, recall that $\|x\|_0 = |\operatorname{supp}(x)|$, where $\operatorname{supp}(x) = \{i \in [n] : x_i \neq 0\}$ are the non-zero coefficients.

For two vectors $x, y \in \mathbb{Z}^n_{\geq 0}$, we say that x is lexicographically smaller than y ($x \prec_{\mathsf{lex}} y$) if there exists an index $i \in [n]$ such that (i) $x_j = y_j$ for every $j < i$, and (ii) $x_i < y_i$.

As in related works, see [12], we work with a random-access model with word length $\Theta(\log n + \log W)$, where W is the largest absolute value of an input integer (see [9] for a formal definition). We measure space complexity in terms of the total number of words in the working memory. We say that an algorithm runs in polynomial space if the space complexity is bounded by a polynomial in the input size, that is, in $\mathrm{poly}(I) := \mathrm{poly}(n, m, \log(\Delta), \log(\|b\|_\infty), \log(\|c\|_\infty))$. Similarly, the running time of an algorithm is measured in the number of arithmetic operations on words.

Before we explain the ideas behind Theorem 1 we briefly review previous techniques used to achieve polynomial space dependency.

The standard tool to reduce space usage are the ideas in Savitch's theorem, which transforms algorithms with non-deterministic space complexity to slightly higher deterministic space complexity. Here, the key technique is to decide (s,t)-reachability in a directed graph with N vertices in $N^{O(\log N)}$ time and $O(\log N)$ space. This is straightforward by recursively guessing the middle vertex of the path from s to t. The space complexity comes from the $O(\log N)$ vertices stored in the stack. This can easily be adapted to finding the shortest path in a weighted graph in the same time and space. One can already apply this to ILP. Namely, the algorithm of Eisenbrand and Weismantel [6] constructs a directed acyclic graph with $N = (m\Delta)^{O(m)} \cdot \|b\|_\infty$ vertices and then computes the shortest path between two of the vertices. With standard polynomial time preprocessing described in [11], one can reduce $\|b\|_\infty$ to $(m\Delta)^{O(m)}$. This shortest path query is the main source of the high space usage of [6] and using Savitch's theorem, the ILP can therefore be solved in $N^{O(\log N)} = (m\Delta)^{O(m^2 \log(m\Delta))}$ time and $O(\log N) = O(m \log(m\Delta))$ space.

A more intricate method is the algebraization technique [18], which relies on the following coefficient test lemma.

Theorem 2 (cf. Lemma 3.2 in [12]). *Consider a polynomial $f(x) = \sum_{i=0}^{d} \alpha_i x^i \in \mathbb{F}[x]$ of degree at most d and assume there is an algorithm running in time T and space S for evaluating $f(a)$ for a given $a \in \mathbb{F}$. Then, there is an algorithm running in $\tilde{O}(dT)$ time and $\tilde{O}(S)$ space that for a given $t \in \{0, \ldots, d\}$ decides if $\alpha_t = 0$.*

The idea to solve the Subset Sum problem, is to consider the polynomial $p(x) = \prod_{i=1}^{n}(1 + x^{a_i})$ and use Lemma 2 to test whether the coefficient α_t is nonzero, where $a_1, \ldots, a_n \in \mathbb{Z}_{\geq 0}$ are the input integers, $t \in \mathbb{Z}_{\geq 0}$ is the target, and α_t is a coefficient in front of monomial x^t of polynomial p. Note that the polynomial $p(x)$ can be evaluated in $\text{poly}(n, \|a\|_\infty)$ time, which, combined with Lemma 2 yields the result of [18].

This technique can be applied to check feasibility of (1) by aggregating the m constraints of A into a single one $a^\intercal x = b'$ where $\|a\|_\infty = (m\Delta)^{O(m^2)}$, see e.g. [15], and reducing the non-negative integer variables to binary ones, which can be done by applying proximity techniques [6]. We omit the details here, since the method has two inherent disadvantages for our application: it has a higher running time with an additional factor of $O(m)$ in the exponent compared to [6, 11] and to use Lemma 2 to solve the optimization variant in (1) one would need to additionally encode the coefficient of the objective function c in a, which would yield a pseudopolynomial running time in $\|c\|_\infty$ that is not necessary in [6,11].

Another straightforward approach is based on the notable property of (1) that it always has sparse optimal solutions if it is bounded and feasible.[1]

Theorem 3 (cf. Corollary 5 in [5]). *If (1) is bounded and feasible, then the lexicographically minimal optimal solution $x^* \in \mathbb{Z}^n_{\geq 0}$ satisfies*

$$\|x^*\|_0 \leq \Gamma,$$

where $\Gamma := 2(m+1)(\log(m+1) + \log\Delta + 2)$.[2]

---

[1] We say that (1) is *bounded* if the optimal solution is finite.

[2] Eisenbrand and Shmonin [5] stated only that there exists a solution x^* with the mentioned support bound. To derive the statement, one can replace the objective by $c'_i = c_i + \epsilon^i$ for each $i \in [n]$, where $\epsilon > 0$ is sufficiently small. This simulates a lexicographical tie-breaking rule.

This property is sometimes used in the following way: one can guess the support of an optimal solution from the $O(n^\Gamma)$ potential variable sets and then apply Lenstra's algorithm for integer linear programming from the $O(n^\Gamma)$ potential variable sets and then apply Lenstra's algorithm for integer linear programming in small dimension. This idea was used for example in [10]. Recall that Lenstra's algorithm and the improvement by Kannan can be implemented in polynomial space.

Theorem 4 (cf. Theorem 5.2 in [8]). *There is an algorithm that solves*

$$\max\{c^\mathsf{T} x : Ax \leq b, x \in \mathbb{Z}^n\}$$

in time $n^{O(n)}$ poly(I) *and polynomial space.*

Note that (1) can be translated to the form in Theorem 4 without increasing the number of variables, hence the same result holds there. Together with the bound on Γ, this yields a running time of $n^\Gamma \cdot \Gamma^{O(\Gamma)} \cdot \text{poly}(I) \leq (m\Delta)^{O(m^2 \log \Delta)} \cdot \text{poly}(I)$ and polynomial space, using that without loss of generality all columns of A are different and, in particular, $n \leq (2\Delta + 1)^m$.

For our result we introduce two new ideas. In Sect. 3 we introduce the following branching algorithm that can be seen as an application of Savitch's theorem to the support of the solution. Consider the lexicographically minimal solution x^* and let $x^* = x^*_\ell + x^*_r$ be such that x^*_ℓ and x^*_r have half the support each. We then guess a vector $b_\ell \in \mathbb{Z}^m$ such that $Ax^*_\ell = b_\ell$ and for each such guess, recurse into subproblems with the right-hand sides b_ℓ and $b - b_\ell$ and then retrieve both x^*_ℓ and x^*_r. Note, that in each recursive call, the support halves, so the total depth of the recursion is $O(\log \Gamma) = O(\log m + \log\log \Delta)$. This idea already yields a $(m\Delta)^{O(m^2 \log\log \Delta)}$ poly(I) time and polynomial space algorithm. However, the dependence on the number of constraints m is much higher than what is promised in Theorem 1. The bottleneck of the earlier mentioned idea of guessing the support and applying Lenstra's algorithm is that one enumerates $O(n^\Gamma)$ support sets. In Sect. 4 we show that the number of relevant support sets is much lower and that they can be enumerated using the branching algorithm as a subroutine on small instances where it runs more efficiently. This is then combined with running Lenstra's algorithm on each support set.

3 Branching Algorithm

As a subroutine in Theorem 1, we first introduce an algorithm based on the idea of branching on the support, which has the following properties.

Lemma 1. *Let* $A \in \mathbb{Z}^{m\times n}$, $b \in \mathbb{Z}^m$, $c \in \mathbb{Z}^n$, *and* $\Delta \geq \|A\|_\infty$. *Assume that (1) is bounded and feasible and let* x^* *be the lexicographically minimal optimal solution.*

For any $\sigma \in \mathbb{N}$ *we can in time* $(\sigma\Delta)^{O(m \log \Gamma)} \cdot \text{poly}(n)$ *and polynomial space, compute* x^* *under the promise that* $\|x^*\|_1 \leq \sigma$. *If this promise is not satisfied, there is no guarantee of correctness.*

Note that the running time matches Theorem 1 except for the dependence on σ. There are known bounds on σ, see e.g., [21], and after the preprocessing in [11] one can guarantee a bound of $\sigma = (m\Delta)^{O(m)}$. This would therefore give a polynomial space algorithm with running time $(m\Delta)^{O(m^2 \log \Gamma)} \operatorname{poly}(I)$, but this is larger than the running time claimed in Theorem 1. We will apply it more efficiently in Sect. 4.

Now, we present the algorithm behind Lemma 1. For the pseudocode, see Algorithm 1. This recursive algorithm has an additional parameter s, which is a bound on the support. On the initial call we use the bound Γ from Theorem 3. The base case consists of the scenario when $s \leq 1$. If $s = 0$, we return the all-zero vector if and only if b is an all-zero vector; otherwise, we return $\perp$ to indicate that no solution exist. If $s = 1$ let e_i, $i \in [n]$, be the indicator vector, i.e., a vector with 1 at position i and 0 otherwise. We consider the set $\{z \cdot e_i : z \cdot Ae_i = b, \text{ z} \in \mathbb{Z}, i \in [n]\}$ and return the last element that minimizes the dot product with c^T (or $\perp$ if no such element exists). Observe that for each i, such an integer z can be determined in $O(m)$ time; hence, in total, the base case can be decided in $O(mn)$ time.

When $s > 1$, we intuitively split the support of the solution into two (almost) equal-size parts and guess the target b_ℓ for the first part in the original solution. More precisely, we guess a vector $b_\ell \in \{-\sigma\Delta, \ldots, \sigma\Delta\}^m$, recursively compute x_ℓ and x_r, which are lexicographically minimal optimal solutions to $Ax_\ell = b_\ell$ and $Ax_r = b - b_\ell$ with half the support of the previous call (rounded up for one recursion and down for the other), and if both of them exist, return $x_r + x_\ell$, breaking ties according to lexicographical order. If all of the guesses for b_ℓ fail in at least one recursion, we return $\perp$. This concludes the description of the algorithm.

function branch(A, b, c, σ, s):
if $s \leq 1$ **then** // Base case
  return lexicographically minimal optimal solution to $\min\{c^\mathsf{T}x : x \in \mathbb{Z}^n_{\geq 0}, Ax = b, \|x\|_0 = s\}$ or $\perp$ if none exists
$z \leftarrow \perp$
foreach $b_\ell \in \{-\sigma\Delta, \ldots, \sigma\Delta\}^m$ **do**
  $x_\ell \leftarrow$ branch$(A, b_\ell, c, \sigma, \lfloor s/2 \rfloor)$
  $x_r \leftarrow$ branch$(A, b - b_\ell, c, \sigma, \lceil s/2 \rceil)$
  if $x_\ell + x_r$ *is* better *than* z **then**
    $z \leftarrow x_\ell + x_r$
return z

Algorithm 1: Pseudocode of the algorithm behind Lemma 1. Input values A, b, c, σ are defined as in Lemma 1. Parameter $s \in \mathbb{N}$ denotes a bound on the support. The condition that $x_\ell + x_r$ is *better* than z means: $x_\ell \neq \perp$ and $x_r \neq \perp$ and ($z = \perp$ or $c^\mathsf{T}(x_\ell + x_r) < c^\mathsf{T}z$ or ($c^\mathsf{T}(x_\ell + x_r) = c^\mathsf{T}z$ and $x_\ell + x_r \prec_{\text{lex}} z$)).

Proof (Lemma 1). For the running time, observe that the integer s decreases by a constant factor in each recursive call and starts with value Γ. Thus, the depth of the recursion is $O(\log \Gamma)$. The number of possible b_ℓ vectors in each node of the recursion tree is $(2\sigma\Delta + 1)^m$, so the total size of the branching tree is $(\sigma\Delta)^{O(m \log \Gamma)}$. The base case takes $O(mn)$ time; hence, the total running time is as claimed. Similarly, the space complexity is $\mathrm{poly}(n, \log \Gamma)$ as the algorithm needs to store the stack of the recursion of height $O(\log \Gamma)$; the base case requires $\mathrm{poly}(n)$ space.

Next, we argue about the correctness. We prove by induction over s that branch (A, b, c, σ, s) returns the lexicographically minimal optimal solution $x^* \in \mathbb{Z}^n_{\geq 0}$ to (1), provided that $\|x^*\|_1 \leq \sigma$ and $\|x^*\|_0 \leq s$.

In the base case, $s \leq 1$, the algorithm straightforwardly checks over all possible non-zero coefficients of x^* and returns the lexicographically minimal one. Note that if b is the all-zero vector then assuming the integer program is bounded there cannot be a non-zero solution with negative objective value. Hence, the all-zero solution is optimal and lexicographically minimal.

Now assume that $s > 1$. Let $x^* \in \mathbb{Z}^n_{\geq 0}$ be the lexicographically minimal solution to (1) and assume that $\|x^*\|_1 \leq \sigma$ and $\|x^*\|_0 \leq s$. Let $x^*_\ell, x^*_r \in \mathbb{Z}^n_{\geq 0}$ be arbitrary vectors such that $x^*_\ell + x^*_r = x^*$, $\|x_\ell\|_0 \leq \lfloor \|x^*\|_0/2 \rfloor$, and $\|x^*_r\|_0 \leq \lceil \|x^*\|_0/2 \rceil$. For example, x^*_ℓ could be the first half of the non-zero coefficients of x^*, and x^*_r could be the remaining non-zero coefficients. Finally, let $b^*_\ell = Ax^*_\ell$. Observe that $\|x^*_\ell\|_1 \leq \|x^*\|_1 \leq \sigma$, $\|x^*_r\|_1 \leq \|x^*\|_1 \leq \sigma$, and $b^*_\ell \in \{-\sigma\Delta, \ldots, \sigma\Delta\}^m$. Furthermore, x^*_ℓ and x^*_r must be lexicographically minimal optimal solutions for right-hand sides b^*_ℓ and $b - b^*_\ell$ since otherwise this would contradict x^* being the lexicographically minimal optimal solution for b.

Hence, if in Line 5 we guess $b_\ell = b^*_\ell$, the induction hypothesis guarantees that $\mathrm{branch}(A, b_\ell, c, \sigma, \lfloor s/2 \rfloor)$ and $\mathrm{branch}(A, b - b_\ell, c, \sigma, \lceil s/2 \rceil)$ return x^*_ℓ and x^*_r. In particular, $x^* = x^*_\ell + x^*_r$ must be returned by the algorithm, since no lexicographically smaller optimal solution exists that could dominate x^*.

4 Candidate Support Sets and Main Theorem

A naive enumeration of all support sets of cardinality at most Γ would require time $O(n^\Gamma)$, where Γ is a bound from Theorem 3. In this section we design a more efficient enumeration and prove Theorem 1.

Definition 1 (Candidate support set). *For a matrix $A \in \mathbb{Z}^{m \times n}$, we say that a set $\mathcal{C} \subseteq \{0,1\}^n$ is a* candidate support set *if:*

(i) $\|x\|_1 \leq \Gamma$ for every $x \in \mathcal{C}$, and
(ii) for every $b \in \mathbb{Z}^m$ and $c \in \mathbb{Z}^n$ such that (1) is bounded and feasible there exists an optimal solution x^* to (1) with $\mathrm{supp}(x^*) \in \mathcal{C}$.

Lemma 2. *For every $A \in \mathbb{Z}^{m \times n}$ with $\|A\|_\infty \leq \Delta$, there exists a candidate support set of size $(m\Delta)^{O(m)}$. Moreover, this set can be enumerated in time $(m\Delta)^{O(m \log \Gamma)} \cdot \mathrm{poly}(I)$ and polynomial space.*

We note that, with Lemma 2, the proof of Theorem 1 is a straightforward application of known results.

Proof (Theorem 1 *assuming Lemma* 2*).* We enumerate a candidate support set $\mathcal{C}$ of A using Lemma 2. Then for every $c \in \mathcal{C}$, we consider the instance truncated to variables in $\text{supp}(c)$ and solve it using the exact algorithm for integer programming by Lenstra and Kannan, i.e., Theorem 4.

The correctness follows from Property (ii) of $\mathcal{C}$. For the running time, note that by Lemma 2 we have $|\mathcal{C}| \leq (m\Delta)^{O(m)}$ and it requires overhead of $(m\Delta)^{O(m \log \Gamma)}$ poly(I) to enumerate $\mathcal{C}$. Recall that the algorithm of Lenstra and Kannan [14,17] solves (1) in $\Gamma^{O(\Gamma)}$ poly(I) time and polynomial space for at most Γ variables. In total, the running time of the algorithm is

$$\begin{aligned}(m\Delta)^{O(m \log \Gamma)} \text{poly}(I) + |\mathcal{C}| \cdot \Gamma^{O(\Gamma)} \cdot \text{poly}(I) &\leq (m\Delta)^{O(m \log \Gamma)} \cdot \Gamma^{O(\Gamma)} \cdot \text{poly}(I) \\ &\leq (m\Delta)^{O(m(\log m + \log\log \Delta))} \cdot \text{poly}(I).\end{aligned}$$

It remains to prove Lemma 2. This relies on the following property that Deng, Mao, and Zhong [4] noticed for Unbounded Knapsack and which easily generalizes to ILP.

Lemma 3. *Let* $x^* \in \mathbb{Z}^n_{\geq 0}$ *be the lexicographically minimum optimal solution to (1). Then, there exists* $x' \in \{0,1\}^n$ *and* $b' \in \mathbb{Z}^m$ *such that (i)* x' *is the lexicographically minimal optimal solution to* $\min\{c^\intercal x' : Ax' = b'\}$ *and (ii)* $\text{supp}(x') = \text{supp}(x^*)$.

Proof. Let x' be the indicator vector of $\text{supp}(x^*)$ and let $b' = Ax'$. Clearly, Property (ii) holds. Hence, it remains to prove (i). For the sake of contradiction, assume that y is a solution to the integer program with right-hand side b' and $c^\intercal y < c^\intercal x'$ Consider $z := y + x^* - x'$. Then z is a solution to $\{Ax = b', x \in \mathbb{Z}^n_{\geq 0}\}$. But $c^\intercal z < c^\intercal x^*$, which contradicts the optimality of x^*. Similarly, if there is a solution y with $c^\intercal y = c^\intercal x'$ and $y \prec_{\text{lex}} x'$. Then $z = y + x^* - x'$ is an optimal solution to $\min\{c^\intercal x, Ax = b', x \in \mathbb{Z}^n_{\geq 0}\}$ that is lexicographically smaller than x^*.

Proof (Lemma 2*).* First, we iterate over every $b' \in \mathbb{Z}^m$ such that $\|b'\|_\infty \leq \Gamma \cdot \Delta$. Note that the number of such b' is $(2\Gamma\Delta + 1)^m$. Next, for each such b', we invoke Lemma 1 with $\sigma = \Gamma$, which returns the lexicographically minimal optimal solution x^* to

$$\min\{c^\intercal x : Ax = b', x \in \mathbb{Z}^n_{\geq 0}\}$$

if $\|x^*\|_1 \leq \sigma$. Finally, for each of these guesses, we output x if $x \in \{0,1\}^n$. This concludes the description of the algorithm. The algorithm runs in time

$$(2\Gamma\Delta + 1)^m \cdot (\Gamma\Delta)^{O(m \log \Gamma)} \cdot \text{poly}(I) \leq (m\Delta)^{O(m \log \Gamma)} \cdot \text{poly}(I)$$

and uses polynomial space. Property (i) of Definition 1 holds because every x^* considered is lexicographically minimal and therefore by Theorem 3 has support at most Γ. For Property (ii) consider some $b \in \mathbb{Z}^m, c \in \mathbb{Z}^n$ for which (1) is bounded and feasible. Let x^* be the lexicographically minimal optimal solution.

By Theorem 3 we have $\|x^*\|_0 \leq \Gamma$. Furthermore, by Lemma 3 there is some $x' \in \{0,1\}^n$ with $\mathrm{supp}(x') = \mathrm{supp}(x^*)$, which is the lexicographically minimal optimal solution for $b' = Ax'$. Because $\|b'\|_\infty \leq \Gamma\Delta$ and $\|x'\|_1 \leq \sigma$ the algorithm above will consider b' and correctly output x'.

5 Conclusion

In this paper, we show that an integer linear program can be solved in time

$$O(m\Delta)^{O(m(\log m + \log\log \Delta)} \mathrm{poly}(I)$$

and polynomial space. It remains open whether the additional factor of $O(\log m + \log\log \Delta)$ in the exponent compared to the algorithm without space limitation can be avoided. We note that our algorithm has two bottlenecks that lead to this running time: (i) the depth of $O(\log m + \log\log \Delta)$ for our branching and (ii) because we use the algorithm of Lenstra and Kannan that runs in time $N^{O(N)}\mathrm{poly}(I)$ on an instance with $N = O(m \log(m\Delta))$. Thus, even if the improvement to $(\log N)^{O(N)}\mathrm{poly}(I)$ by Reis and Rothvoss [22] could be implemented in polynomial space, this would not immediately lead to an improvement. Already for the special case of Unbounded Knapsack it is unclear how to avoid the exponent of $\log\log \Delta$ and achieve a pseudopolynomial running time.

Finally, we want to mention the case of binary integer linear programs, i.e., where we substitude the domain constraint for $x \in \{0,1\}^n$ in (1). Here, the best known algorithm runs in time $(m\Delta)^{O(m^2)}\mathrm{poly}(I)$ [6] and naive application of Savitch's theorem yields polynomial space at significant higher running time of $(m\Delta)^{O(m^3 \log(m\Delta))}\mathrm{poly}(I)$. It is unclear how to come close to an exponent of $O(m^2)$ in polynomial space and the ideas in this paper do not seem to be applicable, since the solutions need not be sparse.

References

1. Bellman, R.: Dynamic programming. Science **153**(3731), 34–37 (1966)
2. Bringmann, K.: A near-linear pseudopolynomial time algorithm for subset sum. In: Proceedings of SODA, pp. 1073–1084. SIAM (2017)
3. Cslovjecsek, J., Eisenbrand, F., Hunkenschröder, C., Rohwedder, L., Weismantel, R.: Block-structured integer and linear programming in strongly polynomial and near linear time. In: Proceedings of SODA, pp. 1666–1681 (2021)
4. Deng, M., Mao, X., Zhong, Z.: On problems related to unbounded subsetsum: a unified combinatorial approach. In: Proceedings SODA, pp. 2980–2990 (2023)
5. Eisenbrand, F., Shmonin, G.: Carathéodory bounds for integer cones. Oper. Res. Lett. **34**(5), 564–568 (2006)
6. Eisenbrand, F., Weismantel, R.: Proximity results and faster algorithms for integer programming using the Steinitz lemma. ACM Trans. Algorithms (TALG) **16**(1), 1–14 (2019)
7. Elberfeld, M., Jakoby, A., Tantau, T.: Logspace versions of the theorems of Bodlaender and Courcelle. In: Proceedings of FOCS, pp. 143–152. IEEE (2010)

8. Frank, A., Tardos, É.: An application of simultaneous diophantine approximation in combinatorial optimization. Combinatorica **7**, 49–65 (1987)
9. Hagerup, T.: Sorting and searching on the word ram. In: Proceedings of STACS, pp. 366–398. Springer, Cham (1998)
10. Jansen, K.: An EPTAS for scheduling jobs on uniform processors: using an MILP relaxation with a constant number of integral variables. SIAM J. Discret. Math. **24**(2), 457–485 (2010)
11. Jansen, K., Rohwedder, L.: On integer programming, discrepancy, and convolution. Math. Oper. Res. **48**(3), 1481–1495 (2023)
12. Jin, C., Vyas, N., Williams, R.: Fast low-space algorithms for subset sum. In: Proceedings of SODA, pp. 1757–1776. SIAM (2021)
13. Kane, D.M.: Unary subset-sum is in logspace. arXiv preprint (2010). arXiv:1012.1336
14. Kannan, R.: Improved algorithms for integer programming and related lattice problems. In: Proceedings of STOC, pp. 193–206 (1983)
15. Kannan, R.: Polynomial-time aggregation of integer programming problems. J. ACM (JACM) **30**(1), 133–145 (1983)
16. Knop, D., Pilipczuk, M., Wrochna, M.: Tight complexity lower bounds for integer linear programming with few constraints. ACM Trans. Comput. Theory **12**(3), 19:1–19:19 (2020)
17. Lenstra, H.W., Jr.: Integer programming with a fixed number of variables. Math. Oper. Res. **8**(4), 538–548 (1983)
18. Lokshtanov, D., Nederlof, J.: Saving space by algebraization. In: Proceedings of STOC, pp. 321–330. ACM (2010)
19. Nederlof, J.: Personal communication
20. Papadimitriou, C.H.: On the complexity of integer programming. J. ACM (JACM) **28**(4), 765–768 (1981)
21. Papadimitriou, C.H., Yannakakis, M.: The complexity of restricted spanning tree problems. J. ACM (JACM) **29**(2), 285–309 (1982)
22. Reis, V., Rothvoss, T.: The subspace flatness conjecture and faster integer programming. In: Proceedings of FOCS, pp. 974–988. IEEE (2023)
23. Savitch, W.J.: Relationships between nondeterministic and deterministic tape complexities. J. Comput. Syst. Sci. **4**(2), 177–192 (1970)
24. Wolsey, L.A.: Integer programming. John Wiley & Sons (2020)

Implied Integrality in Mixed-Integer Optimization

Rolf van der Hulst(✉) and Matthias Walter

University of Twente, Enschede, The Netherlands
{r.p.vanderhulst,m.walter}@utwente.nl

Abstract. Implied-integer detection is a well-known presolving technique that is used by many Mixed-Integer Linear Programming solvers. Informally, a variable is said to be *implied integer* if its integrality is enforced implicitly by integrality of other variables and the constraints of a problem. In this paper we formalize the definition of implied integrality by taking a polyhedral perspective. Our main result characterizes implied integrality as occurring when a subset of integer variables is fixed to integer values and the polyhedron on the remaining variables is integral. While integral polyhedra are well-understood theoretically, existing detection methods infer implied integrality only for one variable at a time. We introduce new detection methods based on the detection of integral polyhedra, extending existing techniques to multiple variables. Additionally, we discuss the computational complexity of recognizing implied integers. We conduct experiments using a new detection method that uses totally unimodular submatrices to identify implied integrality. For the MIPLIB 2017 collection dataset our results indicate that, on average, 18.8% of the variables are classified as implied integer after presolving, compared to just 3.3% identified by state-of-the-art techniques. We are able to reduce the average percentage of variables whose integrality needs to be enforced after presolving from 70.2% to 59.0%.

Keywords: Mixed-Integer Linear Programming · Implied Integers · Integral Polyhedra · Total Unimodularity

1 Introduction

In order to solve practical Mixed-Integer Linear Programming (MILP) problems, many solvers apply a presolve procedure with the aim of reducing the model's size, to strengthen the linear programming (LP) relaxation and to infer additional information that can be exploited further during execution of a branch-and-bound algorithm [21]. This work is motivated by the technique of *implied integers*, which falls under the latter category of presolving methods.

In previous works, most notably by Achterberg, Bixby, Gu, Rothberg and Weninger [2], a continuous variable was considered to be *implied integer* if an optimal solution exists in which that variable attains an integer value. This

N. Megow and A. Basu (Eds.): IPCO 2025, LNCS 15620, pp. 452–465, 2025.
https://doi.org/10.1007/978-3-031-93112-3_33

definition lacks an indication of which integer variables cause this implication. Before providing a more rigorous definition based on a polyhedral point of view we introduce useful notation.

We consider an MILP problem given by a polyhedron $P = \{x \in \mathbb{R}^N \mid Ax \leq b\}$, where $A \in \mathbb{R}^{M \times N}$ and $b \in \mathbb{R}^M$ for index sets M and N, and a set $I \subseteq N$ of variables x_I that are required to be integral. For any $S \subseteq N$, we use $\mathbb{MI}^S := \{x \in \mathbb{R}^N \mid x_i \in \mathbb{Z}\ \forall i \in S\}$ to denote the corresponding mixed-integer set. For $J \subseteq M$ and $S \subseteq N$, we use $A_{J,S}$ to denote the $|J| \times |S|$ submatrix of A induced by I and J. If J or S is $\star$ then this shall indicate all rows or columns, respectively. Now, let us define implied integrality.

Definition 1 (implied integrality). *Let $P \subseteq \mathbb{R}^N$ be a rational polyhedron and let $S, T \subseteq N$. We say that x_T is implied integer by x_S for P if* $\operatorname{conv}(P \cap \mathbb{MI}^S) = \operatorname{conv}(P \cap \mathbb{MI}^{S \cup T})$ *holds. If $S = \emptyset$, we also say that x_T is implied integer and that P is T-integral. Furthermore, we say that S are the implying variables and T are the implied variables.*

Definition 1 differs from the previous [2]. It is more general since the implied set T is not necessarily restricted to continuous variables, and since T can be of arbitrary size, rather than just a single-variable set. However, we do not consider only optimal solutions with respect to some given objective, but all extreme solutions. We deal with this limitation in an extended version of the paper.

Note that Definition 1 for $S = \emptyset$ and $T = N$ is equivalent to the usual definition of integral polyhedra, such as in [26, Chapter 16]. We briefly review a few basic properties of integral polyhedra; for more detailed definitions and explanations, we refer to [26, Chapter 16]. The *recession cone* $\operatorname{rec}(X)$ of a set $X \subseteq \mathbb{R}^N$ is the set of directions $y \in \mathbb{R}^N$ such that $x + y \in X$ for all $x \in X$. A polyhedron P is *rational* if there exist rational $A \in \mathbb{Q}^{m \times n}$ and $b \in \mathbb{Q}^m$ such that $P = \{x \in \mathbb{R}^n \mid Ax \leq b\}$ holds. A rational polyhedron $P \subseteq \mathbb{R}^N$ is said to be *integral* if $P = \operatorname{conv}(P \cap \mathbb{Z}^N)$ holds. Note that we require rationality of P, as otherwise $\operatorname{conv}(P \cap \mathbb{Z}^N)$ may not be a polyhedron. A fundamental theorem by Meyer [23] shows that if P is a rational polyhedron then $\operatorname{conv}(P \cap \mathbb{MI}^I)$ is a rational polyhedron with the same recession cone unless it is empty. From a practical point of view, rationality is not a limitation and hence we focus on rational polyhedra. There are many known classes of integral polyhedra, such as those generated by totally unimodular matrices [17] or k-regular matrices [3].

Implied integrality can also be interpreted as a method to reduce the number of necessary integrality constraints, which has been explored in prior research. Bader, Hildebrand, Weismantel and Zenklusen [4] introduce the notion of an affine TU decomposition, which decomposes a constraint matrix $A \in \mathbb{Z}^{m \times n}$ as $A = \tilde{A} + UW$, such that $\left[\tilde{A}^\intercal\ W^\intercal\right]$ is totally unimodular and $U \in \mathbb{Z}^{m \times k}$ holds. If an integer program admits such an affine TU decomposition then it can be reformulated as a mixed-integer program with only k integrality constraints. Hupp [20] and the authors of [4] also describe specific affine TU decompositions for several families of integer programs. In [24], Paat, Schlöter and Weismantel show lower bounds on the number of necessary integrality constraints for integer programs after reformulation as a mixed-integer program for integer programs with bounded subdeterminants.

Implied integers are used in state-of-the-art MILP solvers such as SCIP [9], HiGHS [18] and Gurobi [2]. In contrast to the theoretical results for integral polyhedra, the detection of implied integers for general mixed-integer programming solvers received little attention in the literature. To the best of our knowledge, it has only been investigated in [2], and received a quick mention in Achterberg's thesis [1, Section 14.5.1]. In [2] a primal and a dual detection method are presented, and the computational impact of enabling implied integer detection for the Gurobi solver is highlighted. We illustrate the primal and dual detection method from [2] in Example 1 and Example 2.

Example 1 (Primal detection). We consider a mixed-integer program that has the constraint $ax + by + z = c$ for $a, b, c \in \mathbb{Z}$ with integer variables $x, y \in \mathbb{Z}$. After fixing $x = \bar{x}$ and $y = \bar{y}$ for arbitrary integer values $\bar{x}, \bar{y}$, we can infer that $z = c - a\bar{x} - b\bar{y}$ must be integral. Since this holds for any choice of integral $\bar{x}, \bar{y}$, we can freely choose to omit or enforce the constraint $z \in \mathbb{Z}$.

Example 2 (Dual detection). Consider the following mixed-integer program:

$$\begin{aligned} \max \quad & 10x + 10y + z \\ \text{s.t.} \quad & 3x + 2y + z \leq 4,\ x + 3y - z \leq 3,\ z \geq 0,\ x, y \in \mathbb{Z}_{\geq 0} \end{aligned} \tag{1}$$

Assume that x and y are fixed to integral values $\bar{x}$ and $\bar{y}$, respectively. Then the remaining linear program that maximizes z only has integral bounds for the single variable z, since $4 - 3\bar{x} - 2\bar{y}$ and $3 - 3\bar{y} - \bar{x}$ and 0 are integral. This yields existence of an optimal solution with integral z, and we may enforce integrality.

By using $z \in \mathbb{Z}$, stronger cutting planes can be derived. The initial solution of the LP relaxation of (1) is $(x^\star, y^\star, z^\star) = (\frac{6}{7}, \frac{5}{7}, 0)$ with objective $\frac{110}{7} \approx 15.71$. The Gomory-Mixed-Integer cut [15] from the Simplex tableau for variable y is given by $21x + 49y \leq 43 + 33z$. Adding this cut to (1) and optimizing the LP relaxation yields $(x_1^\star, y_1^\star, z_1^\star) = (\frac{1}{2}, 1, \frac{1}{2})$, which has objective value 15.5. Taking integrality of z into account yields the stronger cut $21x + 49y \leq 43 + 19z$, which cuts off $(x_1^\star, y_1^\star, z_1^\star)$. Resolving the augmented LP relaxation in (1) yields a better objective value of approximately 14.94.

The main benefit of implied integers is that integrality of a variable can be used during the solution process without having to explicitly enforce its integrality. Implied integer variables that are continuous variables in the original model may be used in branching and can use integrality to derive stronger coefficients in cutting planes (see Example 2). Moreover, new classes of cutting planes that require rows with integer variables only may become usable. Furthermore, propagation techniques and conflict analysis can infer stronger bounds by using the detected integrality. For implied integer variables that are integer variables in the original model it is not necessary to enforce their integrality during branch-and-bound, which can reduce the size of branch-and-bound tree. This can be particularly useful in primal heuristics that successively solve LPs and fix variables, such as diving methods or local-neighborhood search methods, to limit the number of branching candidates that are to be fixed or explored. One result in [2] is that enabling implied integers reduces Gurobi's total solve time by 13% and the total node count by 18% on average over all models in their testset that take longer than 10 s to solve. Compared to the other presolving methods tested in [2], this is a significant reduction and places implied integer detection as one of the more important presolving methods for mixed-integer optimization.

Contribution and Outline. As the focus in [2] is clearly on computational impact, a formal treatment of the topic is missing. In this paper, we close this gap by studying implied integrality from a polyhedral perspective and apply it in various ways:

- We formally define implied integers as the equivalence of two mixed-integer hulls with two different sets of integer variables in Sec. 2.
  As our main tool, we apply the concept of a *fiber* [7], which arises from the LP relaxation by fixing a subset of the variables to integers. We characterize implied integrality for a polyhedron in terms of (partial) integrality of all fibers.
- In Section 3 we present conditions under which integrality of all fibers can be established by well-known concepts for establishing perfect formulations, such as (total) unimodularity. We show that this generalizes the dual detection method.
- In Section 4 we show that partial integrality of the fibers may arise from fixed variables in the fibers, which generalizes the primal detection method.
- Our new methods yield questions related to the computational complexity of recognizing implied integers, some of which we answer in Section 5.
- In Section 6 we complement our theoretical results with a computational study by detecting implied integers using totally unimodular submatrices. For the MIPLIB 2017 collection dataset our results indicate that, on average, 18.8% of the variables are classified as implied integer after presolving, compared to just 3.3% identified by the primal and the dual detection method. Moreover, we can significantly reduce the average percentage of variables whose integrality needs to be enforced after presolving from 70.2% to 59.0%.

2 A Polyhedral View on Implied Integrality

The primal and dual detection methods in Examples 1 and 2 use integrality of variables x_S to conclude that another variable x_t can be assumed to be integral. In Definition 1 we generalized this idea to multiple variables x_T and phrased it in terms of polyhedra (see Fig. 1a).

As noted in the introduction, Definition 1 contains the definition of integral polyhedra with $S = \emptyset$ and $T = N$. For the latter there exist several equivalent statements (see [10, Theorem 4.1] and [26, Section 16.3]). Note that for $S = \emptyset$ we typically have a description of $\operatorname{conv}(P \cap \mathbb{MI}^S) = P$ in terms of linear inequalities. We will add two more equivalent statements, both of which reduce implied integrality to this case using the concept of a fiber.

Definition 2 (fiber; see [7]). *Let $P \subseteq \mathbb{R}^N$ be a polyhedron and let $S \subseteq N$. The* S-integral fibers *are $Q_S(\bar{x}) := \{x \in P \mid x_S = \bar{x}\}$, parameterized by $\bar{x} \in \mathbb{Z}^S$.*

The fibers for the example depicted in Fig. 1a are shown in Fig. 1b. Note that a fiber $Q_S(\bar{x})$ is non-empty if and only if $\bar{x}$ lies in the projection of P onto the x_S-variables. Hence, to deal with the odd case of $S = \emptyset$, it is convenient to define $\mathbb{Z}^\emptyset$ as a set consisting of a single element $\bar{x}^\emptyset$ that satisfies $Q_\emptyset(\bar{x}^\emptyset) = P$ in this case. Moreover, if the constraints defining P are $Ax_S + Bx_{N \setminus S} \leq c$ then $Q_S(\bar{x}) = \{x \in \mathbb{R}^N \mid x_S = \bar{x},\ Bx_{N \setminus S} \leq c - A\bar{x}\}$ is an inequality description of the fiber.

We now turn to the characterization of implied integrality.

Theorem 1. *Let $P \subseteq \mathbb{R}^N$ be a rational polyhedron and let $S, T \subseteq N$. Then the following are equivalent:*

(i) x_T is implied integer by x_S for P, i.e., $\operatorname{conv}(P \cap \mathbb{MI}^S) = \operatorname{conv}(P \cap \mathbb{MI}^{S \cup T})$.
(ii) Every minimal face of $\operatorname{conv}(P \cap \mathbb{MI}^S)$ *contains a point $x \in \mathbb{MI}^{S \cup T}$.*
(iii) $\max\{c^\intercal x \mid x \in \operatorname{conv}(P \cap \mathbb{MI}^S)\}$ *has some optimal solution $x^\star \in \mathbb{MI}^{S \cup T}$ for each c for which the maximum is finite.*
(iv) For each $\bar{x} \in \mathbb{Z}^S$ and every minimal face F of $Q_S(\bar{x})$, $F \subseteq$ $\operatorname{conv}(P \cap \mathbb{MI}^{S \cup T})$.
(v) $Q_S(\bar{x}) \subseteq$ $\operatorname{conv}(P \cap \mathbb{MI}^{S \cup T})$ *holds for each $\bar{x} \in \mathbb{Z}^S$.*

The following corollary follows from Theorem 1 since $x \in Q_S(\bar{x}) \cap \mathbb{MI}^T$ (for $\bar{x} \in \mathbb{Z}^S$) implies $x \in P \cap \mathbb{MI}^{S \cup T}$, and is useful for establishing implied integrality.

Corollary 1. *Let $S, T \subseteq N$ and let $P \subseteq \mathbb{R}^N$ be a rational polyhedron such that $Q_S(\bar{x})$ is T-integral for each $\bar{x} \in \mathbb{Z}^S$. Then x_T is implied integer by x_S for P.*

Note that T-integrality of all fibers is, in general, not a necessary condition, as shown in Fig. 1b. However, in case of binary x_S-variables, T-integrality of all fibers turns out to also be necessary:

Corollary 2. *Let $S, T \subseteq N$ and let $P \subseteq [0,1]^S \times \mathbb{R}^{N \setminus S}$ be a rational polyhedron. Then the following are equivalent:*

(a) x_T is implied integer by x_S for P.
(b) For each $\bar{x} \in \{0,1\}^S$, every minimal face of $Q_S(\bar{x})$ contains a point in $\mathbb{MI}^T$.
(c) For each $\bar{x} \in \{0,1\}^S$, $Q_S(\bar{x})$ is T-integral.

In our definition of implied integrality in Definition 1 we consider precisely the two sets S and T. One downside of this characterization is that we cannot additionally require integrality of x_j for any $j \in N \setminus (S \cup T)$ and add j to S since the derived implied integrality of x_T may no longer be valid (see Fig. 1c). This motivates the definition of a stronger implied integrality concept.

Definition 3 (total implied integrality). *Let $P \subseteq \mathbb{R}^N$ be a rational polyhedron. For $S, T \subseteq N$, we say that x_T is totally implied integer by x_S for P if x_T is implied integer by $x_{S'}$ for every $S' \supseteq S$ (for P).*

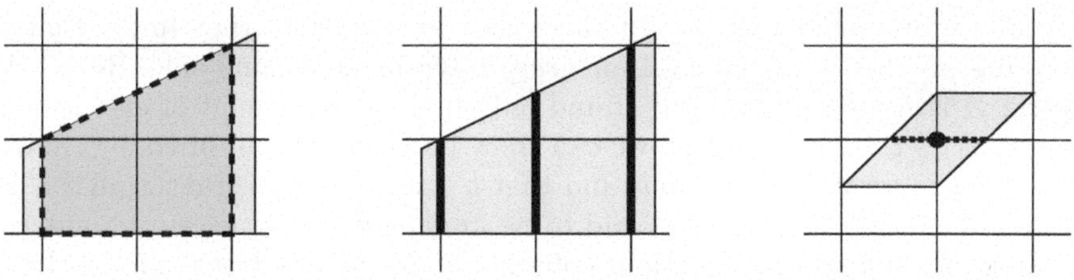

(a) $P_I = P_1 \subsetneq P$ (b) Three $\{1\}$-fibers of (a); one has a fractional vertex. (c) $P_I \subsetneq P_2 \subsetneq P = P_1$

Fig. 1. Examples of polyhedron $P \subseteq \mathbb{R}^2$ (thin line) together with $P_1 := \text{conv}(P \cap (\mathbb{Z} \times \mathbb{R}))$, $P_2 := \text{conv}(P \cap (\mathbb{R} \times \mathbb{Z}))$ (dotted line) and $P_I := \text{conv}(P \cap \mathbb{Z}^2)$ (thick dashed line in (a), dot in (c)). In (a), x_2 is implied integer by x_1. The corresponding fibers are depicted in (b). The second fiber has a fractional vertex, which is a convex combination of points from other fibers. In (c), x_1 is implied integer by $x_\emptyset$ due to $P_1 = P$, but not totally implied integer due to $P_I \neq P_2$.

The main use case of total implied integrality is the combination of several implied integrality relations. Suppose for example, for disjoint sets $S, T, U \subseteq N$, that x_U is totally implied integer by x_T, and x_T is totally implied integer by x_S. It follows that $x_{T \cup U}$ is implied integer by x_S, i.e., that $\text{conv}(P \cap \mathbb{MI}^S) = \text{conv}(P \cap \mathbb{MI}^{S \cup T}) = \text{conv}(P \cap \mathbb{MI}^{S \cup T \cup U})$ holds.

In Theorem 2 we show that, to establish total implied integrality, it suffices to consider sets obtained from S by adding elements not in T whose variables are not binary. Subsequent corollaries highlight its consequences.

Theorem 2. *Let $S, T, U \subseteq N$ and consider a rational polyhedron $P \subseteq \mathbb{R}^{N \setminus U} \times [0,1]^U$. If x_T is implied integer by x_S then x_T is implied integer by every $x_{S'}$ with $S \subseteq S' \subseteq S \cup T \cup U$.*

Corollary 3. *Let $S, T, U \subseteq N$ and consider a rational polyhedron $P \subseteq \mathbb{R}^{N \setminus U} \times [0,1]^U$. Then x_T is totally implied integer by x_S for P if and only if x_T is implied integer by every $x_{S'}$ (for P) that satisfies $S \subseteq S' \subseteq S \cup (N \setminus (T \cup U))$.*

Corollary 4. *Let $S, T \subseteq N$ and consider a rational polyhedron $P \subseteq \mathbb{R}^{S \cup T} \times [0,1]^{N \setminus (S \cup T)}$. Then x_T is totally implied integer by x_S for P if and only if x_T is implied integer by x_S (for P).*

3 Implied Integrality for Block-Structured Problems

In Corollary 1, we observed that, in order to derive that x_T is implied integer by x_S, it suffices to show T-integrality of every fiber $Q_S(\bar{x})$. One can show that if such a fiber is the Cartesian product of two polyhedra then we can treat integrality separately for both, and use known results on integral polyhedra.

For $P = \{x \in \mathbb{R}^N \mid Ax \leq b\}$ an inequality description is given by $Q_S(\bar{x}) = \{A_{\star,N \setminus S} x_{N \setminus S} \leq b - A_{\star,S} \bar{x}\}$. Note that $A_{\star,N \setminus S}$ is identical for each fiber, and that the fibers' descriptions thus only differ in the right-hand sides.

Totally unimodular matrices fit this structure well since they imply that the underlying polyhedron is integral for *every* integral right hand side. Recall that a matrix A is *totally unimodular* if and only if the determinant of every square submatrix is -1, 0 or $+1$. Thus, we can show integrality of all fibers by showing that $A_{\star,N\setminus S}$ is totally unimodular and that $b - A_{\star,S}\bar{x}$ is integral for all $\bar{x} \in \mathbb{Z}^S$. A matrix $A \in \mathbb{Z}^{M\times N}$ of rank r is said to be *unimodular* if for every submatrix B consisting of r linearly independent columns of A, the greatest common divisor of the subdeterminants of B of order r equals 1. Hoffman and Kruskal showed that (totally) unimodular matrices are closely related to integral polyhedra [17]:

Proposition 1 ([17])**.** *For $A \in \mathbb{Z}^{m\times n}$ the following hold:*

(i) The polyhedron $\{x \in \mathbb{R}^n \mid Ax = b,\ \ell \le x \le u\}$ is integral for all $b \in \mathbb{Z}^m$ and all $\ell, u \in (\mathbb{Z} \cup \{\pm\infty\})^n$ if and only if A is unimodular.
(ii) The polyhedron $\{x \in \mathbb{R}^n \mid Ax \le b,\ \ell \le x \le u\}$ is integral for all $b \in \mathbb{Z}^m$ and all $\ell, u \in (\mathbb{Z} \cup \{\pm\infty\})^n$ if and only if A is totally unimodular.

Our main result for showing how total unimodularity yields (total) implied integrality reads as follows.

Theorem 3. *Consider a rational polyhedron of the form*

$$P = \{(x, y, z) \in \mathbb{R}^S \times \mathbb{R}^T \times \mathbb{R}^U \mid Ax + By \le c, \ell \le y \le u,\ Dx + Ez \le f\}, \quad (2)$$

where $A \in \mathbb{Z}^{M\times S}$, $c \in \mathbb{Z}^M$ and $\ell_j, u_j \in \mathbb{Z} \cup \{\pm\infty\}$ for all $j \in T$. If B is totally unimodular then y is totally implied integer by x for P.

As a special case of Theorem 3 we obtain the dual detection method from [2] by observing that, after scaling, it treats a single-column totally unimodular submatrix.

Corollary 5 (dual detection). *Consider a rational polyhedron $P = \{x \in \mathbb{R}^N \mid Ax \le b\}$ with $A \in \mathbb{Q}^{M\times N}$ and $b \in \mathbb{Q}^M$. For some $k \in N$, let $M_k := \{j \in M \mid A_{j,k} \neq 0\}$ be the rows that contain a nonzero of column k. Let $S := \{k' \in N \setminus \{k\} \mid$ there exists $j \in M_k$ with $A_{j,k'} \neq 0\}$ be the other columns having nonzeros in M_k. If $\frac{A_{j,s}}{A_{j,k}} \in \mathbb{Z}$ and $\frac{b_j}{A_{j,k}} \in \mathbb{Z}$ holds for all $j \in M_k$ and all $s \in S$ then x_k is totally implied integer by x_S for P.*

Theorem 3 is strictly more general than Corollary 5 because we can apply it to rows of the MILP that involve multiple continuous variables. In contrast to this, Corollary 5 requires a column that may only appear in rows where all other variables are integral.

It might seem unlikely that the conditions of Theorem 3 hold, as it imposes strong integrality requirements and requires a totally unimodular submatrix. However, many MILP formulations of common problems actually satisfy these requirements. For example, in the canonical formulation of fixed-charge network flow problems [16] that have integer capacities, the flow variables are implied integer by the decision variables that activate an arc. Other examples include the canonical formulation of the uncapacitated facility location problem and the natural dates formulation of the single machine scheduling problem [5].

Note that in the proof of Theorem 3, we only use total unimodularity to show that $Q_y(\bar{x})$ is an integral polyhedron. Using almost exactly the same proof, we can derive a similar result for unimodular matrices.

Theorem 4. *Let P be a rational polyhedron*

$$P = \{(x, y, z) \in \mathbb{R}^S \times \mathbb{R}^T \times \mathbb{R}^U \mid Ax + By = c,\ \ell \leq y \leq u,\ Dx + Ez \leq f\},$$

where $A \in \mathbb{Z}^{M \times S}$, $c \in \mathbb{Z}^M$ and $\ell_j, u_j \in \mathbb{Z} \cup \{\pm\infty\}$ for all $j \in T$. If B is unimodular then y is totally implied integer by x for P.

4 Implied Fixed Integrality

The primal detection method from [2], as illustrated in Example 1, works on any equation of the MILP with suitable coefficients. Importantly, it does not require a diagonal block structure as in (2), which is the key idea that we used in the previous section to detect implied integrality. Moreover, observe that in Example 1, z has a *unique* solution that is integral after x and y are fixed. This motivates us to define *implied fixed integers.*

Definition 4 (implied fixed integrality). *Let $P \subseteq \mathbb{R}^N$ be a rational polyhedron and let $S, T \subseteq N$. We say that x_T is implied fixed integer by x_S for P if for each $\bar{x} \in \mathbb{Z}^S$, the set $\operatorname{proj}_{x_T}(Q_S(\bar{x}))$ is either empty or consists of a unique vector that is integer.*

First, we show that implied fixed integrality is a form of implied integrality, in fact even of total implied integrality.

Lemma 1. *Let $P \subseteq \mathbb{R}^N$ be a rational polyhedron and let $S, T \subseteq N$. If x_T is implied fixed integer by x_S for P then, for any $S' \subseteq N$ with $S' \supseteq S$, x_T is implied fixed integer by $x_{S'}$.*

Corollary 6. *Let $P \subseteq \mathbb{R}^N$ be a rational polyhedron and let $S, T \subseteq N$. If x_T is implied fixed integer by x_S for P then x_T is totally implied integer by x_S for P.*

As uniqueness often arises as a result of solving a full-rank system of linear equations, Theorem 4 stands out. In particular, if, for $By = c - A\bar{x}$, B is unimodular and has full column rank then the solution y is not just integral, but also unique for the given $\bar{x}$.

Theorem 5. *Consider, for $A \in \mathbb{Z}^{M \times S}$ and $c \in \mathbb{Z}^M$, the rational polyhedron*

$$P = \{(x, y, z) \in \mathbb{R}^S \times \mathbb{R}^T \times \mathbb{R}^U \mid Ax + By = c,\ Dx + Ey + Fz \leq g\}.$$

If B is unimodular and has full column rank then y is implied fixed integer by x. for P.

Theorem 5 generalizes the primal detection method from [2] since the 1-by-1 matrix with entry ± 1 is unimodular and has full column rank.

Corollary 7 (primal detection). *Let $S \subseteq N$ and $k \in N \setminus S$. Consider the polyhedron $P = \{x \in \mathbb{R}^N \mid a_S x_S + a_k x_k = b,\ Gx \leq d\}$ with $a_S \in \mathbb{Q}^S$, $a_k \in \mathbb{Q} \setminus \{0\}, b \in \mathbb{Q}$, $G \in \mathbb{Q}^{M \times N}$ and $d \in \mathbb{Q}^M$. If $\frac{a_S}{a_k} \in \mathbb{Z}^S$ and $\frac{b}{a_k} \in \mathbb{Z}$, then x_k is implied fixed integer by x_S for P.*

Although Corollary 7 focuses on only one implied variable, it succeeds quite often in practice. Nevertheless, an example where it fails is easy to construct: if $3x + 2y = 3$ and $5x + 3y = 4$ are part of an MILP model, the associated 2-by-2 matrix is unimodular and has full column rank, so x and y are implied fixed integers, which is not detected by Corollary 7.

In Theorem 5, we used given equations and observed that implied fixings for y have a direct linear relationship with the x-variables. The following example shows that implied fixed integers may also arise from inequalities, and that the unique fixings for y may depend nonlinearly on x.

Example 3 Consider $P := \{x \in \mathbb{R}^n \mid Ax \leq b\}$ and suppose that Fortet's inequalities [12] for linearizing $x_3 = x_1 x_2$ for binary variables x_1, x_2, x_3 are valid for P, e.g., $Ax \leq b$ contains a subsystem of the form

$$x_3 \leq x_1, \qquad x_3 \leq x_2, \qquad x_3 \geq x_1 + x_2 - 1, \qquad 0 \leq x_i \leq 1\ \forall i \in \{1,2,3\}.$$

Then x_3 is implied fixed integer by (x_1, x_2).

5 Complexity of Recognizing Implied Integers

We now turn to questions related to the complexity of recognizing implied integers. Since these generalize integral polyhedra, many of our results (implicitly) rely on the work of Papadimitriou and Yannakakis [25]. They considered the following decision problem, and showed coNP-completeness.

Integrality Recognition: For a rational matrix $A \in \mathbb{R}^{M \times N}$ and rational vector $b \in \mathbb{R}^M$, determine if $\{x \in \mathbb{R}^N \mid Ax \leq b\}$ is integral.

Ding, Feng and Zang [11] also show coNP-hardness by reducing a specific binary integer program to a graph problem. The NO-certificate for showing containment in coNP is given by a non-integral (minimal) face F of the polyhedron P. Such a face is an affine subspace, and Schrijver [26, Corollary 5.3b] shows that there always exists an equation system for this subspace and an objective vector defining F that both have polynomial encoding length. A similar result holds in the mixed-integer setting.

The first question we consider is the recognition of implied integrality, phrased as a decision problem.

Implied Integer Recognition: For a rational matrix $A \in \mathbb{R}^{M \times N}$, a rational vector $b \in \mathbb{R}^M$ and subsets $S, T \subseteq N$, determine whether

$$\operatorname{conv}(\{x \in \mathbb{MI}^S \mid Ax \leq b\}) = \operatorname{conv}(\{x \in \mathbb{MI}^{S \cup T} \mid Ax \leq b\}).$$

Note that the recognition of integral polyhedra is the special case of $S = \emptyset$ and $T = N$, which immediately yields coNP-hardness of Implied Integer Recognition. Completeness does not immediately follow, because we do not necessarily have a polynomial-size certificate due to lack of an accessible description of $\operatorname{conv}(P \cap \mathbb{MI}^S)$. Note that if we restrict our problem to binary variables then it reduces to one over the fibers using Corollary 2, which yields a certificate. This motivates the dedicated treatment of this case:

Binary Implied Integer Recognition: For a rational matrix $A \in \mathbb{Q}^{M \times N}$, a rational vector $b \in \mathbb{Q}^M$ and subsets $S, T \subseteq N$, determine if

$$\operatorname{conv}(\{x \in \mathbb{MI}^S \mid \mathbb{0} \leq x_S \leq \mathbb{1},\ Ax \leq b\}) = \operatorname{conv}(\{x \in \mathbb{MI}^{S \cup T} \mid \mathbb{0} \leq x_S \leq \mathbb{1},\ Ax \leq b\}).$$

Theorem 6. Binary Implied Integer Recognition *is* coNP*-complete.*

While we were able to treat the binary case, our certificate does not work for the general integer case. However, we establish containment in Π_2^p and conjecture hardness. For more details on the complexity class Π_2^p, we refer to [13].

Theorem 7. Implied Integer Recognition *is in* Π_2^p.

Conjecture 1. Implied Integer Recognition *is* Π_2^p*-complete.*

6 Implied Integrality in Practice

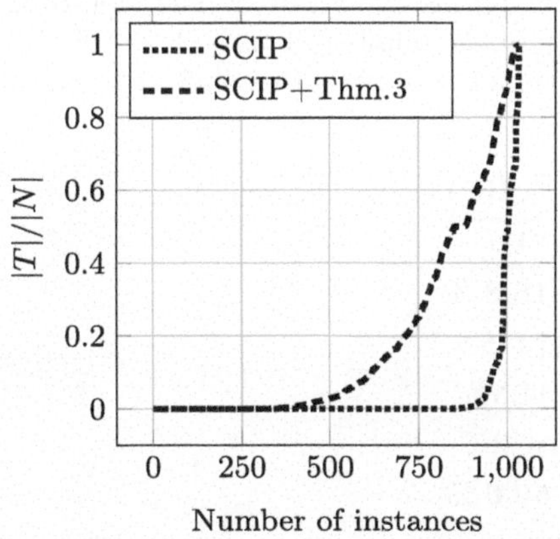

(a) Distribution of ratio of implied integers.

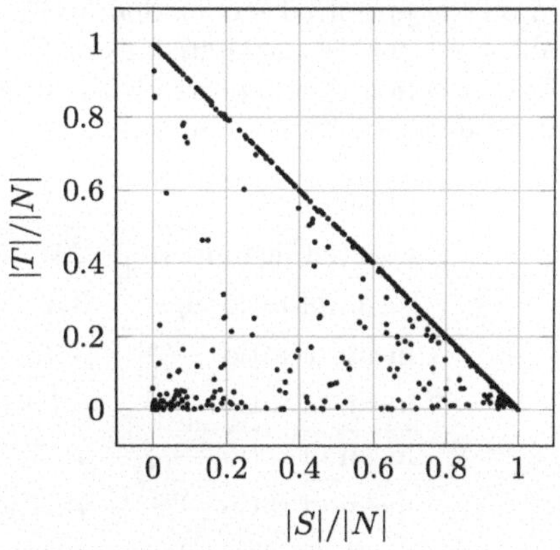

(b) The ratios $\frac{|S|}{|N|}$ and $\frac{|T|}{|N|}$ per instance, represented by a dot.

Fig. 2. Results on implied integers for the MIPLIB 2017 collection dataset [14].

Although the presented concepts are of theoretical interest, they are also practically relevant, which is highlighted in this section. In light of Theorem 3 it is tempting to search for totally unimodular matrices, which can be recognized in polynomial time [27], however, with significant computational cost [28]. Hence, we restrict our detection to the special case of network matrices and their transposes (see [26, Chapter 19]). To this end we use adjusted versions of the algorithms developed in [8] and [19] to greedily grow network submatrices one column or one row at a time. This way, we detect submatrices with the structure of Theorem 3.

As a baseline, we use the default SCIP 9.2.1 implementation [9], which detects implied integers using Corollaries 5 and 7. This is compared to a version that, in addition, uses a presolver plugin that detects implied integers using the structure of Theorem 3, using the greedy methods of [8,19] to grow (transposed) network matrices. The plugin and its source code will be part of the upcoming SCIP 10 release. These computations were done on an Intel Gold 5217 CPU with 64 GB of memory.

In Table 1 we report about results on the MIPLIB 2017 collection set [14], where 5 models were excluded because they could not be presolved using 64 GB of memory, and 25 further models were excluded because they were solved during presolve with both methods. We observe that using Theorem 3 one finds nearly six times as many implied integer variables at the cost of only a small detection time. Figure 2a shows the distribution of the implied integers over all the instances. Furthermore, the relationship between the sizes of S and T is highlighted in Fig. 2b.

Table 1. Implied integer detection on 1035 instances of the MIPLIB 2017 collection set [14]. For a set $X \subseteq N$, X_{int} (X_{con}) indicates the set of variables in X that (do not) have integrality constraints in the original model. The numbers and ratios reported are based on the presolved model, and averaged over all instances tested. The number of variables reported is the shifted geometric mean of the number of variables with shift 10. The detection time reported is the shifted geometric mean with a shift of 1 ms. All other means are arithmetic means.

Method	SCIP 9.2.1	SCIP 9.2.1 + Theorem 3
affected instances	203	**712**
mean of ratio $\frac{\lvert T\rvert}{\lvert N\rvert}$	3.3%	**18.8 %**
mean of ratio $\frac{\lvert T_{\text{con}}\rvert}{\lvert N\rvert}$	3.2%	7.5%
mean of ratio $\frac{\lvert T_{\text{int}}\rvert}{\lvert N\rvert}$	0.1%	11.3%
mean of ratio $\frac{\lvert N_{\text{con}}\setminus T\rvert}{\lvert N\rvert}$	26.5%	22.2%
mean of ratio $\frac{\lvert N_{\text{int}}\setminus T\rvert}{\lvert N\rvert}$	70.2%	**59.0 %**
number of variables	6937	6944
mean detection time	-	**13 ms**

Table 2 shows results of preliminary performance experiments. In this comparison, we ignore the detected implied integrality for originally integer variables, because preliminary tests showed a negative performance impact for using implied integrality for them. The performance results are inconclusive: both methods use on average the same amount of time and nodes. Although we are 5% faster on instances that take more than 1000 seconds to solve by detecting more implied integer variables, this is achieved only on a small sample of 32 instances of the 240 instances tested, and cannot be considered conclusive. The inconclusive results can be explained by the sensitivity of SCIP to the distribution of (implied) integers. In fact, we already had to adapt many of its algorithms and parameters to account for the changed variable type distribution in order to obtain these neutral performance results.

Table 2. Performance comparison on the MIPLIB 2017 benchmark set [14]. The subsets indicated by [x, tilim] contain all instances solved by at least one of the two runs, that took at least x seconds to solve. The time and nodes reported are shifted geometric means with 1 s and 100 nodes, respectively. These experiments were run on a AMD Ryzen 2600 CPU with a time limit of 3600 s and a memory limit of 12 GB.

Subset	instances	SCIP 9.2.1			SCIP 9.2.1 + Thm. 3			relative	
		solved	time	nodes	solved	time	nodes	time	nodes
all	240	118	730.6	3450	118	733.3	3425	1.00	0.99
[0,tilim]	120	118	147.6	2292	118	148.7	2307	1.01	1.01
[1,tilim]	118	116	159.5	2388	116	160.7	2405	1.01	1.01
[10,tilim]	105	103	246.0	3459	103	248.4	3525	1.01	1.01
[100,tilim]	73	71	617.5	9460	71	617.4	9559	1.00	1.01
[1000,tilim]	32	30	1768.8	33695	30	1678.7	33664	0.95	1.00

7 Conclusion and Discussion

In line with the structure of the paper we start with a discussion of our theoretical insights. One of our main contributions is the reduction of implied integrality to partial integrality of fibers via Corollaries 1 and 2. This property, however, does not seem to be well-understood, even in contexts where integrality characterizations exist. We are only aware of one line of research in that direction, namely on strengths of Dantzig-Wolfe reformulations for stable-set polytopes [22].

Besides showing (partial) integrality, our insights relate to many other topics in MILP since important techniques like cutting planes, branch-and-bound and domain propagation typically only use the integrality of a subset of the integral variables. Implied integrality could be used in these contexts to strengthen results that only use partial integrality to larger sets of integer variables. For example, Balas, Ceria and Cornuejols' lift-and-project procedure [6] provides a way to strengthen $P \subseteq [0,1]^N$ to $\mathrm{conv}(P \cap \mathbb{MI}^S)$ for some $S \subseteq N$ by adding valid inequalities in a procedure that iterates over the variables S. If x_T is implied integer by x_S then it suffices to apply the lift-and-project procedure only on the x_S-variables to obtain $\mathrm{conv}(P \cap \mathbb{MI}^{S \cup T})$, rather than on the $x_{S \cup T}$-variables.

Theorem 3 can be seen as a special case of an affine TU decomposition [4] in which one of the matrices is an identity matrix augmented with zero columns. A natural question is whether more general decompositions can be detected as well. Several other questions remain open. For example, given that T is (totally) implied integer by S, it is interesting to ask for the size of the smallest subset $S' \subseteq S$ such that T remains (totally) implied integer by S'. Naturally, the resolution of Conjecture 1 as well as extensions of implied integrality to mixed-integer nonlinear optimization constitute further research directions.

On the practical side we believe that proper usage of the additional implied integrality information in SCIP will eventually yield a performance boost. This is justified by the impressive result in [2] where a speed-up of 13% using the basic version of implied integrality is reported and by the extremely small detection times that we observe in Table 1. Finally, we see a lot of unexplored potential in exploiting implied integrality in the context of decomposition approaches. In Benders' decomposition, implied integrality may provide a major advantage since continuous variables in a subproblem are much easier to deal with than integer ones. For Dantzig-Wolfe reformulations, however, constraints that induce integrality properties should not be reformulated as they do not contribute to strengthening of the relaxation. Hence, we hope that presence of implied integrality can be exploited to lead to better decompositions in the future.

Acknowledgements. Both authors acknowledge funding support from the Dutch Research Council (NWO) on grant number OCENW.M20.151. We thank three anonymous reviewers for their valuable suggestions and constructive feedback that led to an improved manuscript.

References

1. Achterberg, T.: Constraint Integer Programming. Ph.D. thesis, TU Berlin (2008). https://doi.org/10.14279/depositonce-1634
2. Achterberg, T., Bixby, R.E., Gu, Z., Rothberg, E., Weninger, D.: Presolve reductions in mixed integer programming. INFORMS J. Comput. **32**(2), 473–506 (2020). https://doi.org/10.1287/ijoc.2018.0857
3. Appa, G., Kotnyek, B.: Rational and integral k-regular matrices. Discret. Math. **275**(1–3), 1–15 (2004). https://doi.org/10.1016/S0012-365X(03)00095-5
4. Bader, J., Hildebrand, R., Weismantel, R., Zenklusen, R.: Mixed integer reformulations of integer programs and the affine TU-dimension of a matrix. Math. Program. **169**(2), 565–584 (2017). https://doi.org/10.1007/s10107-017-1147-2
5. Balas, E.: On the facial structure of scheduling polyhedra. Math. Program. Study **24**, 179–218 (1985)
6. Balas, E., Ceria, S., Cornuéjols, G.: A lift-and-project cutting plane algorithm for mixed 0–1 programs. Math. Program. **58**(1–3), 295–324 (1993). https://doi.org/10.1007/BF01581273

7. Billera, L.J., Sturmfels, B.: Fiber polytopes. Ann. Math. **135**(3), 527–549 (1992). https://doi.org/10.2307/2946575
8. Bixby, R.E., Wagner, D.K.: An almost linear-time algorithm for graph realization. Math. Oper. Res. **13**(1) (1988). https://doi.org/10.1287/moor.13.1.99
9. Bolusani, S., et al.: The SCIP Optimization Suite 9.0 (2024)
10. Integer Programming. GTM, vol. 271. Springer, Cham (2014). https://doi.org/10.1007/978-3-319-11008-0
11. Ding, G., Feng, L., Zang, W.: The complexity of recognizing linear systems with certain integrality properties. Math. Program. **114**(2), 321–334 (2008). https://doi.org/10.1007/s10107-007-0103-y
12. Fortet, R.: L'algebre de Boole et ses applications en recherche operationnelle. Trabajos de Estadistica **11**(2), 111–118 (1960). https://doi.org/10.1007/BF03006558
13. Garey, M.R., Johnson, D.S.: Computers and Intractability A guide to the Theory of NP-completeness. W.H. Freeman and Company, New York (1979)
14. Gleixner, A., et al.: MIPLIB 2017: data-driven compilation of the 6th mixed-integer programming library. Math. Program. Comput. **13**(3), 443–490 (2021). https://doi.org/10.1007/s12532-020-00194-3
15. Gomory, R.E.: An algorithm for the mixed-integer problem. Tech. Rep. RM-2597, RAND Corporation (1960)
16. Hirsch, W.M., Dantzig, G.B.: The fixed charge problem. Naval Res. Logistics Q. **15**(3), 413–424 (1968). https://doi.org/10.1002/nav.3800150306
17. Hoffman, A.J., Kruskal, J.B.: Integral boundary points of convex polyhedra. In: Linear Inequalities and Related Systems, vol. 38, pp. 223–246. Princeton University Press (1956). https://doi.org/10.1515/9781400881987-014
18. Huangfu, Q., Hall, J.: Parallelizing the dual revised simplex method. Math. Program. Comput. **10**(1), 119–142 (2017). https://doi.org/10.1007/s12532-017-0130-5
19. van der Hulst, R., Walter, M.: A row-wise algorithm for graph realization. arXiv:2408.12869 (2024). http://arxiv.org/abs/2408.12869
20. Hupp, L.M.: Integer and Mixed-Integer Reformulations of Stochastic, Resource-Constrained, and Quadratic Matching Problems. Ph.D. thesis, Friedrich-Alexander-Universität Erlangen-Nürnberg (2017)
21. Land, A.H., Doig, A.G.: An automatic method of solving discrete programming problems. Econometrica **28**(3), 497 (1960). https://doi.org/10.2307/1910129
22. Lübbecke, M.E., Witt, J.T.: The strength of Dantzig-Wolfe reformulations for the stable set and related problems. Discret. Optim. **30**, 168–187 (2018). https://doi.org/10.1016/j.disopt.2018.07.001
23. Meyer, R.R.: On the existence of optimal solutions to integer and mixed-integer programming problems. Math. Program. **7**, 223–235 (1974). https://doi.org/10.1007/BF01585518
24. Paat, J., Schlöter, M., Weismantel, R.: The integrality number of an integer program. Math. Program. 1–21 (2021). https://doi.org/10.1007/s10107-021-01651-0
25. Papadimitriou, C.H., Yannakakis, M.: On recognizing integer polyhedra. Combinatorica **10**(1), 107–109 (1990). https://doi.org/10.1007/BF02122701
26. Schrijver, A.: Theory of Linear and Integer Programming. John Wiley & Sons Inc, New York (1986)
27. Truemper, K.: A decomposition theory for matroids. V. Testing of matrix total unimodularity. J. Comb. Theory, Series B **49**(2), 241–281 (1990). https://doi.org/10.1016/0095-8956(90)90030-4
28. Walter, M., Truemper, K.: Implementation of a unimodularity test. Math. Program. Comput. **5**(1), 57–73 (2013). https://doi.org/10.1007/s12532-012-0048-x

Author Index

N. Megow and A. Basu (Eds.): IPCO 2025, LNCS 15620, pp. 467–468, 2025.
https://doi.org/10.1007/978-3-031-93112-3

The manufacturer's authorised representative in the EU is Springer Nature Customer Service Centre GmbH, Europaplatz 3, 69115 Heidelberg, Germany. If you have any concerns regarding our products, please contact ProductSafety@springernature.com

Printed and bound by CPI Group (UK) Ltd, Croydon, CR0 4YY
30/09/2026
02225585-0014